Recent Trends in Engineering and Science for Resource Optimization and Sustainable Development

Recent Trends in Engineering and Science for Resource Optimization and Sustainable Development

Editors

Prof. (Dr.) Dorota Jelonek
Prof. (Dr.) Narendra Kumar
Prof. (Dr.) Mamta Chahar
Prof. (Dr.) Rusudan Kinkladze
Prof. (Dr.) Lilla Knop

CRC Press
Taylor & Francis Group
Boca Raton London New York

CRC Press is an imprint of the
Taylor & Francis Group, an **informa** business

First edition published 2024
by CRC Press
4 Park Square, Milton Park, Abingdon, Oxon, OX14 4RN

and by CRC Press
2385 NW Executive Center Drive, Suite 320, Boca Raton FL 33431

British Library Cataloguing-in-Publication Data
A catalogue record for this book is available from the British Library

ISBN: 978-1-032-98029-4 (hbk)
ISBN: 978-1-032-98030-0 (pbk)
ISBN: 978-1-003-59672-1 (ebk)

DOI: 10.1201/9781003596721

Typeset in Times LT Std
by Aditiinfosystems

Recent Trends in Engineering and Science for Resource Optimization and
Sustainable Development – Prof. (Dr.) Dorota Jelonek et al. (eds)
© 2024 Taylor & Francis Group, London, ISBN 978-1-032-98030-0

Contents

Recent Trends in Engineering and Science for Resource Optimization and
Sustainable Development – Prof. (Dr.) Dorota Jelonek et al. (eds)
© 2024 Taylor & Francis Group, London, ISBN 978-1-032-98030-0

List of Figures

Recent Trends in Engineering and Science for Resource Optimization and
Sustainable Development – Prof. (Dr.) Dorota Jelonek et al. (eds)
© 2024 Taylor & Francis Group, London, ISBN 978-1-032-98030-0

List of Tables

Recent Trends in Engineering and Science for Resource Optimization and
Sustainable Development – Prof. (Dr.) Dorota Jelonek et al. (eds)
© 2024 Taylor & Francis Group, London, ISBN 978-1-032-98030-0

Message from Vice-chancellor, BIT Mesra, Ranchi

I am extremely happy that BIT Jaipur, an External Center of BIT Mesra is or- ganizing an International Conference on Contemporary Trends in Multidisciplinary Research and Innovation (ICCTMRI-2023) during July 27-28, 2023.

Any conference offers a platform shared by peers, researchers and scholars to showcase important and recent contributions made by them. Fruitful and objective interaction between experts and students from various disciplines enables furthering of knowledge and opening new opportunities of research in a given subject domain. I am told that this event, the first in a chain of conferences has already drawn very encouraging response from the scientific community. I extend my earnest appreciation to all the organizers and stakeholders for their useful contributions to the conference.

Like all other units of BIT Mesra, the BIT External Centre at Jaipur is dedicated to constantly innovate and adapt as per the demand of the rapidly changing land- scape of technology. The Institute is keen to provide the students and scholars the best of Engineering, Computer Applications and Management education and skills they need to succeed in their professional career. We look forward to continuing our efforts collectively in pursuit of excellence in education, research and innovation in Engineering and integrate the same with the ovemll scope and mandate of engineering for national and Societal growth.

Wishing you all a fulfilling and productive experience at the International Conference on Contemporary Trends in Multidisciplinary Research and Innovation (ICCTMRI- 2023).

(Indranil Manna)
Patron: ICCTMRI - 23

Recent Trends in Engineering and Science for Resource Optimization and
Sustainable Development – Prof. (Dr.) Dorota Jelonek et al. (eds)
© *2024 Taylor & Francis Group, London, ISBN 978-1-032-98030-0*

Director, BIT Mesra, Off-Campus Jaipur

It is a privilege to welcome you all to the two-day INTERNATIONAL CONFERENCE ON CONTEMPORARY TRENDS IN MULTIDISCIPLINARY RESEARCH& INNOVATION (ICCTMRI- 2023) being conducted by Birla Institute of Technology Mesra, Ranchi Off-Campus Jaipur, Rajasthan, India.

ICCTMRI– 2023 provides a unique opportunity to interact with researchers, academicians, scientists, and specialists in the various research and development fields of Biotechnology and Bioengineering, Management, Animation, Science and Technology across the globe.

ICCTMRI– 2023 offers a platform for global experts to gather and interact intensively on the topics of Animation, Biotechnology, Computer Science, Electronics Engineering, Electrical Engineering, Environmental Engineering, Management Practices, and Multimedia and Sciences.

I am assured of the success of ICCTMRI– 2023 due to the perseverance and efforts of the conveners and associated departments who started planning a long time back to make this conference a huge success. I hope the ICCTMRI – 2023 will be a grand success.

Dr. Peeyush Tewari

Chairman: ICCTMRI - 23

Acknowledgment

Dear Attendees,

It is with immense pleasure that I extend a warm welcome to all of you to the recently concluded Conference, International Conference on Contemporary Trends in Multidisciplinary Research and Innovation (ICCTMRI-2023) during July 27-28, 2023, organized by BIT Mesra, Jaipur campus, Jaipur (India), with special sessions on Animation, Biotechnology, Computer Science, Electronics & Electrical Engineering, Management Practices and Sciences.

We would like to extend our deepest gratitude to each and every one of you for your participation in our international conference. We understand the time and effort required to attend such events, especially amidst your busy schedules, and we are immensely grateful for your commitment and contribution.

A special thanks go to the Shanti Educational Research Foundation, Jaipur, India for the successful completion of this conference. A huge gratitude to all the contributors/authors who displayed their exceptional patience and commitment to research and development. Your contributions have significantly enriched the knowledge base and sparked important discussions that will undoubtedly lead to progress in research domains.

We would also like to specifically acknowledge our keynote speakers: *Prof. Ramesh Narayanan*, IIT Delhi, *Dr. Raj Kiran*, NTU, Singapore, *Dr. Ramkrishna Rane*, IPR, Gandhinagar, Gujarat, India, *Dr Andrew M Lynn,* JNU New Delhi, *Prof Gajendra Pal Singh Raghava*, IIIT, New Delhi, *Prof. Soumya Banerjee* from INRIA, Paris, France, *Dr. Nishant Kumar Verma*, IIM, Banglore and *Mr. Sunil Dutta*, Jellyfish Pictures, London UK. Your insightful presentations were a highlight of the conference, providing invaluable perspectives and stimulating thought-provoking dialogue.

Our heartfelt thanks go out to the organizing committee. Your tireless efforts, dedication, and meticulous planning have made this conference a resounding success. *Prof. M. Krishan Mohan, Dr. Madhavi Sinha, Dr. Roopali Sharma, Dr. P S Rathore, Dr. Vibhuti Pandya, Dr. Anand K. Srivastava, Dr. Ritu Pareek, Mrs. Jyoti Sharma, Dr. Ravindra Kumar, Mrs. Nidhi Singh, Mr. Gautam Goswami, Mr. Santosh Sharma* and *Mr. Gaurav Choudhary* without your efforts, this event would not have been possible.

A very special thanks to reviewers and experts panel headed by Prof. Dr. Dorota Jelonek, (Mentor, Shanti Educational Research Foundation, Jaipur, India) Czestochowa University of Technology, Czestochowa, Poland for the intelligent efforts to make available everything at publication process and a team of gems **Dr Aleksandra Ptak, Dr Alok Aggarwal, Dr Anna Dunay, Dr Andrii Galkin, Dr Damian Dziembek, Dr Dejan Mircetic, Dr Deepa Parasar, Dr Ike Umejesi, Dr Ilona Paweloszek, Dr Jelena Franjkovic, Dr Leszek Ziora, Dr Liviu Rosca, Dr Mahesh Joshi, Dr Paula Bajdor, Dr Mihaela Lupeanu, Dr Nataliia Pavlikha, Dr Naveen K Sharma, Dr Olena Pozniakova, Dr Raya Karlibaeva, Dr Shivani Dixit, Dr Tatsiana Verezubova, Dr Liliana Gurran, Dr. Aysun Kahraman, Dr Iwona Chomiak-Orsa, and Dr Rusudan** Kinkladze

Last, but certainly not least, we want to thank all attendees and volunteers. Your active participation, insightful questions, and shared experiences truly enriched the event. We hope that you found the conference informative and worthwhile, and we look forward to seeing you at our next event.

Once again, thank you for your valuable contribution to this successful conference.

Best Regards,
Dr. Anju Sharma, Conference Chair (ICCTMRI 23)
Faculty of Computer Science & Engineering
BIT Mesra, Jaipur Campus, Jaipur

About the Editors

Prof. Dorota Jelonek: is a full professor of Economics at the Faculty of Management of the Czestochowa University of Technology in Poland. Her scientific and research interests focus on solving problems related to the implementation of management information systems in enterprises and improving management information processes. She is the author of 6 books and the editor of 10. Additionally, she has authored or coauthored 250 articles in Polish and foreign journals and book chapters. She previously held the positions of Associate Dean for Science and Dean of the Faculty. Since 2019, Prof. Dorota Jelonek has been the President of the Scientific Society for Economic Informatics and is a member of many societies, including the Polish Association for Innovation Management and the Informing Science Institute

Dr. Narendra Kumar is a doctor of Computer Science & Engineering and Mathematics. He is working in NIMS Institute of Engineering and Technology, NIMS University Rajasthan, Jaipur. He has completed his M.Phil. with the gold medal and Ph.D. from Dr. Bhimrao Ambedkar University, Agra. He has an academic experience for more than 28 years. He worked as Dean, Joint Director and Director in various universities. He has published more than two dozen books in the domain of mathematics, statistics and computer science and engineering, more than 80 research papers in national/ international journals. He has guided more than a dozen students for research degrees. He has many patents in his credit and member of Board of Studies in many universities. His key areas of research work are Data science, Big data and brand management, Mathematical modeling, and Theory of relativity.

Prof. (Dr.) Mamta Chahar is an Associate Professor at the NIET, NIMS University, Jaipur, India. She has completed her Ph.D. from the Indian Institute of Technology Delhi (IITD). She did her Postdoctoral research at the University of Florida, USA and Touro University USA from 2009-2012. She has published 20 research papers in reputed International Journals and 6 Book chapters with international publishers. Her area of interest lies in Supramolecular Chemistry, Heterocyclic Chemistry, Peptide Chemistry, Medicinal Chemistry, Nano Chemistry and Green Chemistry.

Dr. Rusudan Kinkladze is **an** academic doctor of Economics. She graduated from Tbilisi State University in 1986. In 2008, she defended her doctoral dissertation. In 1994-2006 she was a lecturer at the Department of Statistics of Tbilisi State University, in 2006-2008 she was an assistant professor at the Department of Economic Statistics of the Faculty of Economics and Business at Ivane Javakhishvili Tbilisi State University, and from 2009 until now she was an associate professor of the Department of Tourism and Marketing of the Faculty of Business Technologies of the Technical University of Georgia. She is the author of 127 scientific papers, 12 Books and supporting manuals. Her field of research is tourism statistics, gender statistics, social statistics, and business statistics. She is a participant in various international conferences and seminars, a UN expert on census statistics.

Prof (Dr) Lilla Knop, is an associate professor at the Faculty of Organization and Management, Silesian University of Technology, Poland and Chairperson of the Discipline Council for Management and Quality Studies. Her research interests cover the areas of strategic and innovation management, inter-organizational networks, clusters and partnership management, and business ecosystems. She has realized two scientific projects for Polish National Science Center and she was an expert in several international projects. She has been a member of the Executive Board of the Scientific Society of Organization and Management (SSOaM) since 2017, and since 2001 he has been active in the authorities of the SSoM Katowice Branch. She has been author, co-author and consultant on several dozen strategies and restructuring plans of industrial and public organizations. She is the author of more than 150 scientific publications on management, innovation and strategy.

*Recent Trends in Engineering and Science for Resource Optimization and
Sustainable Development – Prof. (Dr.) Dorota Jelonek et al. (eds)*
© 2024 Taylor & Francis Group, London, ISBN 978-1-032-98030-0

1

Smart Grid Systems with Secure Blockchain-based Communication

S. Ninisha Nels[1]

Department of Computer Science and Engineering,
Noorul Islam Centre for Higher Education,
Kumaracoil, India

Narendra Mohan[2]

Assistant Professor, Computer Engineering &
Applications, GLA University,
Mathura, India

P. John Augustine[3]

Professor, Department of Information Technology,
Sri Eshwar College of Engineering,
Kondampatti, India

M. Sundar Rajan[4]

Associate Professor, Faculty of Electrical and
Computer Engineering, Arbaminch Institute of Technology,
Arbaminch University, Ethiopia

Balakrishna Kothapalli

School of Engineering, SR University,
Warangal, Telangana

Shaik Rehana Banu[5]

Post Doctoral Fellowship, Business Management,
Lincoln University College, Malaysia

Abdelmonaim Fakhry Kamel Mohammed[6]

Ain Shams University, Egypt

Abstract

An urgent global concern is the transformation of the energy landscape towards a smarter, more efficient, and sustainable grid system. By boosting security, transparency, and efficiency, the use of blockchain technology in Smart Grid technology has the potential to completely alter the energy industry. Because of their capacity to isolate problematic segments and maintain the functionality of crucial loads, transmission and distribution protection systems are seen as essential

Corresponding authors: [1]sninishanels01@gmail.com, [2]narendra.mohan@gla.ac.in, [3]pjohnaugustine@gmail.com, [4]sundar.rajan@amu.edu.et,
[5]drshaikrehanabanu@gmail.com, [6]abdelmonaimfakhry601@gmail.com

DOI: 10.1201/9781003596721-1

components of contemporary smart grid ecosystems. In order to proactively adjust their behaviour in response to extreme power system fluctuations, current protection solutions increasingly use cognitive techniques. However, the reliability, veracity, and confidentiality of the information and control signals conveyed via the communication networks with underlying relay are crucial for the long-term success and survival of these information-driven solutions. In this study, we provide a brand-new distributed network architecture built on the blockchain that improves the security of data transfer between smart grid protection relays and serves as a scalable platform for adaptive distribution system protection. The findings offer a potential strategy for safe communication amongst smart grid protection systems. Finally, a blockchain benchmarking framework is used to simulate the performance of the suggested approach. In order to give a blueprint for a future where energy is more sustainably produced, reliable, and secure, the study focus on the creative synergy between blockchain-based communication and smart grid systems.

▬▬▬▬ **Keywords**

Blockchain, Secure smart grid, Distributed network

1. INTRODUCTION

The conventional electrical grid is undergoing a significant shift in an era characterized by quick technology breakthroughs, giving rise to what is known as the "Smart Grid." The Smart Grid is a cutting-edge method for organizing, distributing, and optimizing energy resources with the goal of improving the reliability, sustainability, and efficiency of our energy infrastructure. Secure and effective communication methods are essential for this shift, and blockchain technology is one newly developing solution that has enormous potential.

The adoption of blockchain technology by Smart Grid systems is a ground-breaking innovation that has the potential to completely transform how we generate, distribute, and use electricity. Blockchain is a distributed ledger system that is decentralized and offers a transparent, secure, and impenetrable method of recording and validating transactions.

Due to the rising energy demand in populous residential and industrial sectors, the requirement for power is growing daily [1]. In addition to providing an efficient way to produce and consume energy, renewable energy also of-

fers a way to meet energy needs while reducing environmental pollution and energy shortages with the advent of modern and technical breakthroughs. In order to produce new, less expensive energy sources from natural resources, renewable energy sources must be used. The unpredictable nature of renewable natural resources like solar and wind energy presents another difficulty. An energy grid's main responsibility is to reliably and efficiently supply electricity to users. The upkeep of these decentralised energy sources and the provision of energy with improved grid security in outlying areas can both be accomplished with the use of smart grids (SG). For a smart grid with centralised infrastructure for energy transactions between prosumers and consumers, expensive and complex communication infrastructure is required. Decentralised technologies are preferred for addressing these issues [2]. At various levels of sophistication, blockchain technology combined with decentralisation methods can offer a solution for the smart grid. Figure 1.1 provides a brief overview of the history of blockchain development and usage.

In order to shed light on the possible advantages, uses, and increased security it provides to the contemporary energy landscape, this study investigates the convergence between

Fig. 1.1 Blockchain evaluation

Smart Grid technologies with blockchain-based communication. In the pages that follow, we'll examine the principles of blockchain and smart grid technology; go over the benefits and drawbacks of combining the two, and look at several practical applications where this ground-breaking strategy is already having an impact.

It is possible to increase grid resilience, enable effective peer-to-peer energy trading, and create an immutable ledger for tracking and optimizing energy consumption by combining Smart Grid technology with secure blockchain-based communication.

The combination of smart grids and blockchain technology promises a promising path towards a more efficient, dependable, and secure energy future in a time when energy sustainability, grid resilience, and security are of the utmost significance. In order to give a thorough introduction of this fascinating topic, this paper will explore the opportunities for change as well as the difficulties that will be encountered while developing secure blockchain-based communication for smart grid systems.

2. LITERATURE REVIEW

2.1 Blockchain

Blockchain is a distributed database made up of chains of interconnected blocks that serves as a ledger. The connected chain's individual blocks each include data as well as a segment that connects them to the blocks before and after them [3]. Since all blocks are linked together in a chain and cannot be changed or removed, this technology's key benefit is that it preserves all differences in the blocks. Because contracts and money may be transferred on the blockchain without the interference of a third party like a bank, it becomes immutable and more secure [4].

2.2 Smart Grid

The conventional power grid has been modernised as the Smart Grid (SG). As opposed to a one-way transaction flow, a smart grid allows for the two-way exchange of electrical data. Smart grids gather real-time data in accordance with supply and demand while generating and distributing electricity with efficient electricity generation, consumption, and maintenance. To deliver trustworthy, affordable, and efficient electrical services to consumers and prosumers, smart grid technology is currently being developed. Intelligent sensors and a quick dispersed communication network can quickly collect a variety of data to balance the supply and demand for power [5].

2.3 Smart Grids Applications for Blockchain

Some observers claim that the development of blockchain technology and its capabilities will speed up the move to energy trading in smart grids. Decentralised strategies aid in the expansion of the smart grid [6]. There is now a lot of research being done on electric grids with distributed renewable energy sources and electric smart cars [7]. Smart grid issues can be resolved with blockchain technology, which has become more widely accepted [8]. Figure 1.2 depicts the smart grid architecture that incorporates blockchain technology.

3. METHODOLOGY

For data collection and relay setting propagation, the methodology of this work introduces an Adaptive Protection

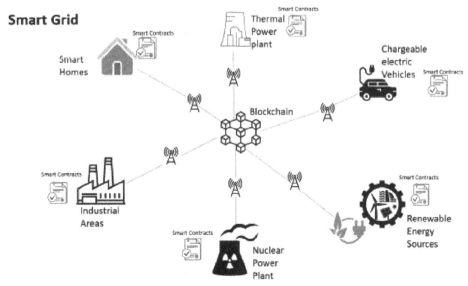

Fig. 1.2   Smart grid using Blockchain

Platform (APP) in conjunction with a network architecture that takes inspiration from the blockchain. The so-called APP management system (APPMS), which interfaces with Intelligent Electronic Devices (IEDs) through a communication hub, is responsible for managing the application programming interface. IEDs include devices like DERs and protection relays installed on distribution circuits [9]. The communication hub allows APPMS to receive the most recent circuit measurements and statuses, make calculations, and send the correct commands and settings to the IEDs in the area under its supervision.

The distribution system operator (DSO) will be in charge of overseeing the APPMS when it is implemented in the distribution feeder substation [10]. The APPMS is composed of four primary modules: setting calculation, short circuit analysis, protection coordination study, and adaptive circuit model management. The advanced protection analyser (APA), which is made up of the three latter components, continuously adjusts the settings of protective relays. Remember that the APPMS only updates the protection system; fault localization or detection is not something that it truly provides.

4. RESULTS AND DISCUSSION

The adoption of blockchain technology by Smart Grid systems is a ground-breaking innovation that has the potential to completely transform how we generate, distribute, and use electricity. Blockchain is a distributed ledger system that is decentralized and offers a transparent, secure, and impenetrable method of recording and validating transactions.

In our results, the Leader, three transaction issuers, and two side chains function concurrently. As each relay sent transactions at a higher rate for the duration of the performance measures, we monitored their performance for a total of 10 minutes. The possible transaction throughput and delay for the two side chains, averaged over the previous 10 minutes, are shown in Fig. 1.3 as each relay's request rate increases.

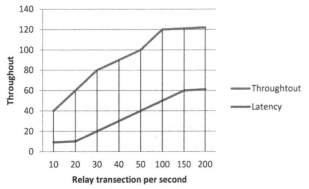

Fig. 1.3 Comparison of efficiency and rising relay transaction rates

As a result, the throughput is saturated and the delay for more demanding data transmission increases. The intrinsic security feature that the blockchain design adds to the security system, however, offsets this expense. By limiting the number of transactions created by each relay to 20 per second, we can then examine how chain size influences system performance side. The average throughput and delay are plotted against the number of active relays in Fig. 1.4.

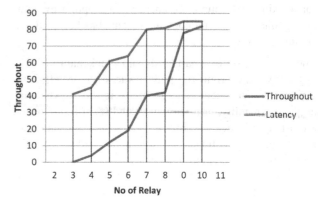

Fig. 1.4 Performance versus an increase in relays

Evidently, the performance scalability of a single-tier blockchain is constrained by the growing quantity of local network resources required to broadcast transactions and blocks to more relays. Our suggested solution's modular architecture enables further division of crowded relay systems to increase communication efficiency all around because throughput and latency are considerably higher with fewer nodes engaged.

5. CONCLUSION

The field of electronics engineering is significantly and extensively shaped by 5G. Rather than being a slight improvement, this fifth-generation technology represents a huge shift in the design, production, and usage of electronic devices. It is now feasible, thanks to 5G, a revolutionary change in communication technologies. From autonomous cars and smart cities to remote healthcare and augmented reality, it has contributed to the development of state-of-the-art electronic applications. Problems with infrastructure development, security, and privacy are just a few of the specific challenges that come with 5G's broad adoption. Overcoming these challenges and making the most of 5G's potential is the responsibility of electronics engineers. As a whole, 5G is expected to have a significant impact on Electronics Engineering. More dependable Internet of Things (IoT) and novel consumer experiences are just two examples of the new possibilities that 5G networks have for us in terms of speed, latency, dependability, and data rates. Electronics engineers will be able to create better tools that can handle larger data sets with more efficiency and speed thanks to these developments. It is thrilling to think about how 5G technology could change the face of electronics engineering as it advances. As 5G evolves and expands its global reach, electronics engineering is poised to lead the way in innovation. The importance of comprehending and getting ready for the revolutionary effects of 5G on electronics engineering is highlighted by this study, which highlights the need for ongoing research and cooperation in order to completely embrace this technological transformation.

References

1. "Communication, Control and Security Challenges for the Smart Grid | Guide books." [Online]. Available: https://dl.acm.org/doi/book/10.5555/3153696. [Accessed: 26-Jan-2020].
2. "Decentralized Controls and Communications for Autonomous Distribution Networks in Smart Grid - IEEE Journals & Magazine." [Online]. Available: https://ieeexplore.ieee.org/abstract/document/6395792. [Accessed: 26-Jan-2020].
3. Kaur, G., & Gandhi, C. (2020). Scalability in blockchain: Challenges and solutions. In Handbook of Research on Blockchain Technology (pp. 373–406). Academic Press.
4. Wohrer, M., & Zdun, U. (2018, March). Smart contracts: security patterns in the ethereum ecosystem and solidity. In 2018 International Workshop on Blockchain Oriented Software Engineering (IWBOSE) (pp. 2–8). IEEE.
5. Shaukat, N., Ali, S. M., Mehmood, C. A., Khan, B., Jawad, M., Farid, U., ... & Majid, M. (2018). A survey on consumers empowerment, communication technologies, and renewable generation penetration within Smart Grid. Renewable and Sustainable Energy Reviews, 81, 1453–1475.
6. G. Kim, J. Park, and J. Ryou, "3-A Study on Utilization of Blockchain for Electricity Trading in Microgrid," in 2018 IEEE International Conference on Big Data and Smart Computing (BigComp), Shanghai, 2018, pp. 743–746, doi: 10.1109/BigComp.2018.00141.
7. R. Vettriselvan, C. Vijai, J. D. Patel, S. Kumar.R, P. Sharma and N. Kumar, "Blockchain Embraces Supply Chain Optimization by Enhancing Transparency and Traceability from Production to Delivery," *2024 International Conference on Trends in Quantum Computing and Emerging Business Technologies*, Pune, India, 2024, pp. 1-6, doi: 10.1109/TQCEBT59414.2024.10545308.

Note: All the figures in this chapter were made by the author.

Recent Trends in Engineering and Science for Resource Optimization and
Sustainable Development – Prof. (Dr.) Dorota Jelonek et al. (eds)
© 2024 Taylor & Francis Group, London, ISBN 978-1-032-98030-0

2

Using Machine Learning to Enable Edge Computing for IoT Applications

Vinay Kumar Nassa[1]

Department of Information Communication
Technology (ICT), Tecnia Institute of Advanced Studies(Delhi),
Affiliated with Guru Gobind Singh Indraprastha University

Narendra Mohan[2]

Assistant Professor, Computer Engineering & Applications,
GLA UNIVERSITY, Mathura, India

Farrukh Arslan[3]

Assistant Professor, University of Engineering and
Technology, Lahore

M. Sai Kumar

Department of EEE, School of Engineering,
SR University, Warangal, India

Amit Dutt[4]

Lovely Professional University,
Phagwara

Thirupathi Durgam[5]

Assistant Professor, Department of ECE,
St. Martin's Engineering College Secunderabad, India

Aleksandra Ptak[6]

Czestochowa University of Technology, Poland

▰▰▰ Abstract

An enormous quantity of data is being generated at the edge of networks as a result of the Internet of Things' (IoT) exponential expansion, creating problems with efficiency, latency, and data processing. As a response to these issues, edge computing has arisen, bringing data processing closer to the data source. By enabling intelligent data analytics and edge decision-making in this environment, machine learning plays a key role. This study explores the advantages, difficulties, and practical ramifications of the convergence of machine learning and edge computing for IoT applications. For IoT devices used in complicated scenarios, machine learning is a potential method for deriving trustworthy information from unprocessed sensor data. Machine learning is suitable for the edge computing environment because to its layered structure.

Corresponding authors: [1]vn.nassa@gmail.com, [2]narendra.mohan@gla.ac.in, [3]Farrukh_arslan@hotmail.com, [4]amit.dutt@lpu.co.in,
[5]thirupathidurgamece@smec.ac.in, [6]aleksandra.ptak@pcz.pl

DOI: 10.1201/9781003596721-2

Rigidity in IoT edge computing is resolved through flexible edge computing architecture. The suggested model integrates machine learning with flexible edge computing architecture and many agents. We also provide a novel offloading technique to boost the performance of IoT machine learning applications due to the restricted processing power of the present edge nodes. A flexible and cutting-edge paradigm for IoT systems, FEC architecture is notable for its capacity for focusing on users and environment adaption. Testing machine learning operations in the edge computing environment using the FEC architecture is part of performance evaluation. The evaluation's results show that our method outperforms other machine learning solutions for IoT optimisation.

▃▃▃▃▃ Keyword

Machine learning, IoT, Flexible edge computing

1. INTRODUCTION

A data-rich era has been ushered in by the growth of the Internet of Things (IoT), in which gadgets and sensors continuously produce information, frequently at the very edge of networks. Both opportunities and difficulties are presented by this enormous and expanding data flood. Smart cities and industrial automation are just two areas where IoT applications promise to have a transformative impact, yet the traditional cloud-based architecture of data processing can also cause latency, bandwidth issues, and privacy problems. Edge computing has evolved as a key response to these problems, bringing data processing and analytics into the edge devices themselves, closer to the data source.

In parallel, machine learning has become a game-changer by making it possible to extract patterns, forecasts, and insights from enormous datasets. In the world of IoT applications, the fusion of machine learning with edge computing has the potential to unleash a new level of effectiveness, autonomy, and intelligence. In order to overcome latency and bandwidth issues while promoting real-time decision-making, this study investigates how machine learning approaches might be combined with edge computing for IoT applications. As a result of its potential to offer services based on instantly accessible contextual data, the Internet of Things (IoT) has emerged as a crucial study area. IoT devices generate a large amount of data because they have several endpoints [1]. Processing must take place close to the end device where the data is created in order to get insights from the massive volume of IoT data in real time.

A big problem is how to transfer enormous amounts of data from end devices to the main base station. Edge computing can help with the solution to this issue. The two IoT network tiers that can be used to connect IoT devices and cloud services are the edge layer and the cloud layer. Internet access and cloud servers are both part of the cloud layer. Data is generated and sent from the edge layer to the cloud layer. The cloud layer process must be used to process data. As contrast to cloud computing, edge computing is the practise of performing computation at the edge layer [2]. By just requiring the intermediate data or outcomes to be transmitted in an edge computing environment, the quantity of data that must be transmitted from the end devices to the cloud service is decreased. When the size of the intermediate data is less than the size of the input data, edge computing performs best. Because the filters in machine learning network layers limit the quantity of retrieved characteristics, edge computing is helpful for machine learning applications.

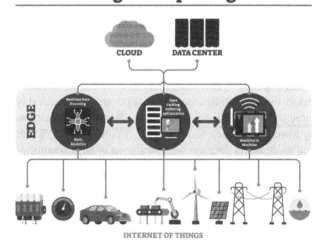

Fig. 2.1 Edge computing IOT

1.1 Objective of Study

1. The use of machine learning to enable edge computing for IoT applications is the goal of this project.
2. This study aims to ascertain whether machine learning for the Internet of Things has been shown useful in a variety of IoT applications.

2. LITERATURE REVIEW

The proliferation of inexpensive IoT device has shown fast growth recently [3]. Healthcare solutions powered by IoT edge computing are a relatively new industry, but they developments in machine learning and edge computing, and the demand for improved efficiency and effectiveness in healthcare delivery are the main reasons behind this trend [4].

The implementation of wireless sensor network (WSN) for remote patient to monitor the people with chronic illnesses like cardiovascular and diabetes were one of the earliest study subjects in this field [5]. In addition to proving that WSNs may be used to gather and transmit patient data, these tests also highlighted the need for more sophisticated analysis capabilities and data processing at the edge [6].

The computer industry is seeing a lot of interest in the concept of edge computing. Many traditional apps utilise distributed cloud computing at the edge to finish tasks [7]. Due to resource limitations, functionality [8], efficiency for transmission [9], and various other edge network-based factors, the system is more complex than cloud computing [10]. Some studies introduced a fresh paradigm for cooperative network edge optimisation in this area of research [11].

3. METHODOLOGY

The concept behind edge computing is to move processing power from central servers in the cloud to edge nodes close to the user end. In comparison to current cloud computing, edge computing offers two significant enhancements. Edge computing suggests a machine learning-based food recognition application that makes use of a service architecture based on edge computing. This application considerably increases the efficiency of machine learning applications by lowering reaction time and energy consumption. Scalar and multimedia data are both produced in large quantities by IoT devices, which then send them to a cloud server.

3.1 Flexible Edge Computing

Edge computing with flexibility An important area for research is the Design of The FEC, which optimises the job allocation between the edge and cloud in IoT systems. To solve the issues brought on by the stiff aspects of the current EC outlined in the preceding section, we advise using the Flexible EC (FEC) design. Figure 2.2 shows a five-layer IoT design.

3.2 FEC Based on Several Agents and Machine Learning

The agent framework and a few particular protocols implement the FEC architecture's capacity to adapt to the environment between the agents. Each agent receives instructions on how to create a business based on contractual privacy. This allows for the dynamic reconfiguration of a virtual organisation in response to environmental changes. Regardless of the EC design, task assignment of components based on conditions is achievable since a multi agent system often adopts a flat structure. Through contact with the user, the user agent created by abstracting a user gains knowledge of the user's preferences and service requirements. The user orientation capability is provided by this user agent, which serves as a middleman among a user and the system and is based on the "human in the loop" principle. Each agent is individually tailored using machine learning layers.

4. RESULTS AND DISCUSSION

A paradigm change in data processing and decision-making has been achieved with the integration of machine learning with edge computing for IoT applications. These technologies have converged as a result of the IoT era's requirement for real-time analytics and quick reactions. The tremendous synergy between machine learning and edge computing has been examined in this study, along with the benefits and difficulties it offers IoT implementations.

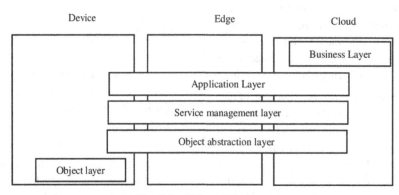

Fig. 2.2 FCl architecture

We first go over the experimental conditions before talking about the outcome of the performance assessment. We use two environments in the research, one for simulating machine learning tasks and the other for data collection from those tasks. We define 10 distinct FEC networks using Caffe as the FEC framework. We build the simulator with Python 2.7 and networkx and use a smaller proportion of the intermediate data produced by running FEC activities. We choose to include 1000 machine learning problems in the simulations. We predetermined a range of 20 to 90 edge servers for the network. Each task's input data size ranges from 100 kB to 1 MB. All FEC networks have a layer number between 5 and 10. Each edge server's bandwidth is evenly split between 10 Mb/s and 1 GB/s. The suggested approach uses several agents for edge computing while performing machine learning in FEC architecture. The edge network receives a random sequence of 1000 jobs, and this algorithm distributes tasks to edge servers. The results are contrasted with edge computing using flexible architecture and machine learning.

The suggested model combines flexible architecture and machine learning to offer improved performance when using several agents. Figure 2.3 and 2.4 demonstrate that the proposed approach, when compared to existing methodologies, produces good throughput in many scenarios.

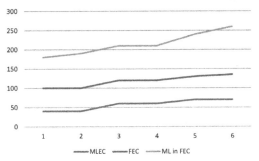

Fig. 2.3 Flexible machine learning in edge computing

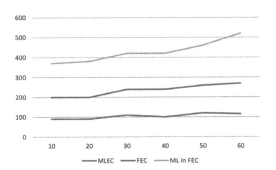

Fig. 2.4 Flexible machine learning in edge computing for IoT

It is clear from a survey of pertinent case studies and examples that edge-deployed machine learning models have the potential to increase the effectiveness, independence, and intelligence of IoT systems. They lighten the load on centralized servers and cloud platforms by enabling local data analysis, decision-making, and predictive capabilities, which leads to decreased latency and faster reaction times.

But it's necessary to be aware of the difficulties, such as resource limitations, security issues, and the demand for effective model implementation. The choice of machine learning algorithms and models that can function well in edge devices should be carefully considered in order to strike a balance between local processing and cloud interaction.

5. CONCLUSION

This article proposes a flexible framework called FEC architecture for IoT edge computing that uses machine learning. When flexible edge computing and machine learning are combined, as well as job allocation between the edge layer and cloud layer, system performance is significantly boosted. The suggested model performs better than the alternative strategy. The creation of several applications and the analysis of the contributions provided by this technology through verification exercises are among our upcoming tasks. To conclude, machine learning is a dynamic and exciting discipline that enables edge computing for IoT applications. It might open up fresh avenues for the implementation of autonomous, effective, and intelligent IoT systems. The combination of machine learning and edge computing is poised to transform how we gather, process, and act upon data as technology improves and the IoT landscape continues to change, bringing in a new era of effectiveness and innovation in IoT applications.

References

1. Maarala, A. I., Su, X., & Riekki, J. (2016). Semantic reasoning for context-aware Internet of Things applications. IEEE Internet of Things Journal, 4(2), 461–473.
2. Sittón-Candanedo, I., Alonso, R. S., Corchado, J. M., Rodríguez-González, S., & Casado-Vara, R. (2019). A review of edge computing reference architectures and a new global edge proposal. Future Generation Computer Systems, 99, 278–294.
3. Huang, J.; Kong, L.; Chen, G.; Wu, M.Y.; Liu, X.; Zeng, P. Towards secure industrial iot: Blockchain system with credit-based consensus mechanism. IEEE Trans. Ind. Inform. 2019, 15, 3680–3689. [CrossRef]
4. Wang, C.; Tan, X.; Yao, C.; Gu, F.; Shi, F.; Cao, H. Trusted Blockchain-Driven IoT Security Consensus Mechanism. Sustainability 2022, 14, 5200. [CrossRef].
5. Pu, C. A Novel Blockchain-Based Trust Management Scheme for Vehicular Networks. In Proceedings of the 2021 Wireless Telecommunications Symposium (WTS), Virtual, 21–23 April 2021.
6. Zhang, H.; Liu, J.; Zhao, H.; Wang, P.; Kato, N. Blockchain-Based Trust Management for Internet of Vehicles. IEEE Trans. Emerg. Top. Comput. 2021, 9, 1397–1409.

Note: All the figures in this chapter were made by the author.

*Recent Trends in Engineering and Science for Resource Optimization and
Sustainable Development – Prof. (Dr.) Dorota Jelonek et al. (eds)*
© 2024 Taylor & Francis Group, London, ISBN 978-1-032-98030-0

3

Evaluating the Application of Blockchain for Electronics Supply Chain Traceability

Gitanjali Singh[1]

Assistant Professor, GLA University, Mathura, India

K. Sangeeta

Department of Computer Science and Engineering,
Institute of Aeronautical Engineering,
Hyderabad, Telangana

Yogesh Kumar Sharma[2]

Assistant Professor, Mittal School of Business,
Lovely Professional University, Jalandhar, Punjab

Balakrishna Kothapalli

School of Engineering, SR University, Warangal, Telangana

Mahesh Manohar Bhanushali[3]

Assistant Professor, Management, University of Mumbai,
VPM's Dr. V. N. Bedekar Institute of Management Studies,
Thane, (Mumbai University)

Gadi Sanjeev[4]

Assistant Professor, Department of ECE,
St. Martins Engineering College Secunderabad, Telangana

Dejan Mircetic[5]

University of Novi Sad,
Faculty of Technical Sciences, Serbia

███████ **Abstract**

A effective global network, the electronics supply chain is prone to problems including fake parts, inefficiencies, and lack of transparency. This study looks into how blockchain technology might improve supply chain traceability for electronics. It examines the benefits and difficulties of using blockchain for traceability, highlighting its potential to reduce counterfeiting, boost productivity, and promote transparency in the sector.In order to ensure transparency in the supply chain traceability, access to consumer demand must be effectively communicated and product and service supplies must be regularly monitored. Specifications and reference architectures serve as the foundation for integrating business processes, and they should also allow for the full integration of product data. The necessity and promise of supply chain integration

Corresponding authors: [1]gitanjali.singh@gla.ac.in, [2]yogeshsharmame355@gmail.com, [3]maheshbhanu87@gmail.com, [4]sanjeev.gadis@gmail.com, [5]dejanmircetic@gmail.com

DOI: 10.1201/9781003596721-3

are examined in this essay. One can assume that cloud integration will give interoperable electronics supply chains traceability a commercially viable business model. We describe how the networks and traceability of electronic supply chains can undergo drastic change as a result of supply chain integration using blockchain technology.

Keywords

Blockchain, Supply chain traceability, Electronic supply chain

1. INTRODUCTION

A huge, globally connected supply chain and quick innovation are hallmarks of the electronics sector. While interconnection promotes advancement, it also creates substantial problems, notably with regard to traceability and component validity. There has never been a more pressing need for a supply chain that is open, safe, and effective.

The unchangeable ledger and decentralized architecture of blockchain technology are well known for their potential to revolutionize the electronics supply chain. Blockchain technology has the potential to significantly improve traceability, lower the danger of counterfeit parts, and increase supply chain efficiency by supplying a tamper-proof record of transactions and provenance. The current fad is to use online resources much like we would in the real world. As a result of Bitcoin's success, users can now utilise blockchain technologies for a variety of purposes, including the Internet of Things (IoT), supply chains, voting, healthcare, performance report production, and storage [1]. Supply chain management and logistics are both heavily reliant on blockchain technology. This particular technology attracted attention by showcasing some universal qualities that enable it to build a reliable, distributed, and open ledger of transactions. There is now a significant possibility for this distributive transparency as supply chain management begins to recognise all the potential of this new technology. This technology's introduction comes at a perfect time since consumers demand supply chain transparency. Consumers, for instance, frequently seek assurances that the goods they buy and consume must be of high quality, that users can verify their authenticity, and that, in the event that the goods fall short of expectations, the source of the deviation and the accountable party will be made clear [2]. The multi-tiered and increasingly global nature of supply chains exacerbates the responsibility concerns. By offering a visible and permanent record at every point of examination, blockchain technology has the potential to fundamentally alter the way transactions are conducted [3, 4]. Traceability, accountability, and transparency in the flow of goods, logistics, and commodities will be made possible through the adoption and use of public, private, and hybrid blockchain. The application of technology in logistics can lower infra-

structure costs for the supply chain and enhance operational procedures. The use of blockchain in the electronics supply chain is thoroughly investigated in this study. It explores the fundamentals of blockchain, as well as its benefits and distinctive characteristics, which make it a potent tool for boosting traceability. This study intends to offer useful insights into how blockchain can improve the traceability of electronic components by examining the difficulties and potential use cases within the electronics industry. The incorporation of blockchain technology presents a possible strategy to address these concerns as the electronics supply chain expands and faces obstacles linked to fake parts and inefficiencies. It emphasizes the value of security and transparency in the electronics sector as well as the contribution of blockchain to accomplishing these objectives.

1.1 Research Objective

1. This study's objective is to assess the use of blockchain technology for electronics supply chain traceability.
2. Examining the capabilities and requirements of supply chain integration is the goal of this study.

2. LITERATURE REVIEW

2.1 Electronic or Digital Supply Chain

The process of delivering goods using digital media is known as the digital supply chain. Digital media is utilised to pinpoint the source of the supply, and some kind of electronic route or mechanism is used to complete the entire process. A consumable product is maintained physically by a "supply chain" mechanism, which is very similar to the physical medium. That implies that nothing is happening electronically; everything is happening physically. In contrast, before a consumer may make a purchase using digital media over an electronic channel, a number of processing processes must be completed. One of the main benefits of a digital supply chain is that it eliminates any conventional contact, such as human quality control, editing, or complying, and handles all background duties that may be carried out automatically with the use of electronic media files, such as metadata [5].

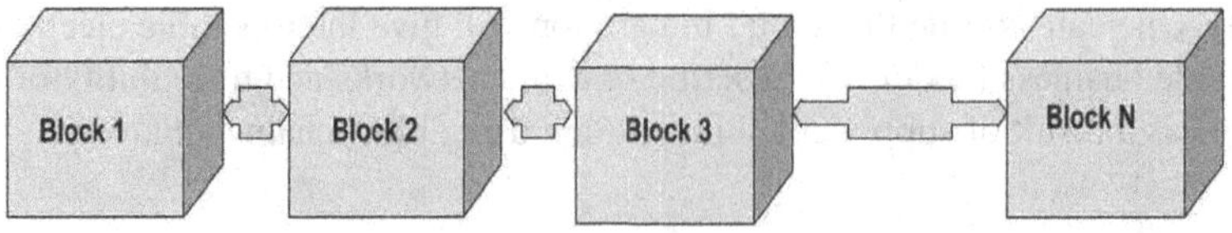

Fig. 3.1 Pictorial representation of a blockchain

2.2 Block Chain

The principle of the traditional ledger is replaced by the concept of a blockchain, which is effectively an immutable ledger [6]. Information is divided into discrete blocks, which are then cryptographically linked to form an almost impenetrable chain. Blockchain eliminates the necessity for an intermediate organisation [7]. Every transaction that has ever been performed on the blockchain is kept and shared across all nodes [8]. In the blockchain, all anonymous nodes increase the security of transaction confirmation by other nodes [9]. The first application to use Blockchain technology was "Bitcoin" [1]. Participants in a decentralised market place can transact to purchase and sell goods using Bitcoin [10]. Figure 3.1 show a visual representation of a blockchain with N connected blocks.

3. METHODOLOGY

Because it can capture the dynamics of the phenomenon and offer a multidimensional view of the situation in a particular context, the survey approach used in this study is suitable for studying business networks, and in particular business-to-business (B2B) connections within electronics supply integration.

3.1 Data Collection

Using a focus group made up of CEOs, business managers, and IT specialists, information was obtained from a well-known Finnish business consortium of 35 companies in the fields of commerce, banking, logistics, and ICT. The focus group participants all actively participated in international business networks, and the consortium functioned in 37 nations. Over the course of three seminars that lasted four hours each, data was gathered between 2015 and 2017. During the workshops, information was gathered via a web-based platform, preserving the anonymity of idea generating and ranking.

3.2 Data Analysis

During the focus group sessions, data were gathered utilising an internet-based platform that blends the respondents' anonymity with interaction involvement and organised data collection procedures. Participants in focus groups utilised this tool to generate ideas. The focus group ranked the ideas in the second phase using a Likert scale of 1 to

7. Investigations on the Delphi method indicates that using this type of group communication technique can help a group of people tackle a challenging issue together. This approach offers suggestions for designing future-focused research.

3.3 Research Design

This strategy makes recommendations for creating future-oriented research designs. Information on 87 suggestions for how blockchain technology could assist with B2B transaction integration, 47 opinions on system performance and current readiness, and 43 suggestions for prioritising requirements were gathered during the three focus group meetings on the implementation of the electronics supply chain.

4. RESULTS AND DISCUSSION

Tabscott introduced the EBE framework's first horizontal layer for blockchain design ideas. Four vertical EBE framework activities could be used to categorise functions, according to expert interviews on blockchain technology and literature review: Transaction data, processing ledgers or smart contracts, peer-to-peer network storage of blocks, expert mining management of blocks, and transaction data are the four main categories. We were able to show the current gap between what is believed to be the significance of supply chain integration and supply chain and blockchain readiness, displayed in Fig. 3.2 by the 16 most significant functionalities, by combining the results of supply chain and blockchain features into one scale and analysing the data.

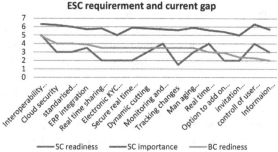

Fig. 3.2 ESC integration, present supply chains, and blockchain preparedness as perceived priorities

The overall impact of blockchain functions was examined using the quality function deployment (QFD) method, and

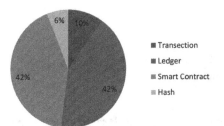

Fig. 3.3 From QFD analysis, blockchain functionality is produced

the results are significant, as shown in Fig. 3.3. Compared to hash (6%) and transactions (10%), blockchain process functions (42%) were regarded as having substantial backing for ledger and smart contract integration. This can be clarified through the requirement for a consistent data model for DSC integration, whereas blockchain integration allows for data integration but requires a data model to handle supply chain system integration as a whole.

In the last round of the study, we gathered suggestions for using blockchain. In this 2017 session, 33 participants from various organisations came up with 87 good suggestions for how blockchain could be integrated with ESC. These were connected to blockchain functionalities, and each notion's relevant BC functionalities were found using QFD.

4.1 Discussion

The current research topic of great significance and potential is the use of blockchain in the electronics supply chain to increase traceability. Key ideas and issues concerning the use of blockchain in this situation are explored in this conversation.

4.2 Tracking of Provenance

Electronic component provenance can be tracked from their source through every step of the supply chain thanks to the decentralized and tamper-proof ledger technology provided by blockchain. The possibility of counterfeit goods entering the supply chain is decreased because to this transparency, which also aids in confirming the authenticity of components.

4.3 Anti-Counterfeiting Measures

Blockchain has the potential to be an effective solution for lowering the prevalence of fake electronic parts. It is very difficult for counterfeit goods to enter the supply chain since we provide a secure, immutable record of component transactions.

4.4 Transparency and Efficiency

Supply chain processes can be streamlined thanks to blockchain's openness and effectiveness. It can lessen paperwork, get rid of middlemen, and give real-time visibility on component movement.

4.5 Challenges

Despite the significant potential of blockchain in the electronics supply chain, there are a number of issues that need to be resolved:

4.6 Integration with Existing Systems

Integrating blockchain into current supply chain systems can be difficult and calls for cooperation among businesses in the same sector.

4.7 Data Integrity and Correctness

Maintaining data integrity and correctness in the blockchain ledger is crucial.

4.8 Privacy and Security

It is crucial to safeguard private supply chain information on the blockchain.

4.9 Regulatory Compliance

It's essential to follow industry norms and regulations while putting blockchain technology into practice.

5. CONCLUSION

In conclusion, our survey was successful in eliciting fresh insights from knowledgeable corporate managers working in a global trade environment for speeding digital supply chain integration. The fact that the participating enterprises primarily represented the Finnish supply chain is a limitation of the study, though. Cloud solutions that can hasten and ease ESC integration are intriguing new research directions.

References

1. Mansfield-Devine, S. (2017). Beyond Bitcoin: using blockchain technology to provide assurance in the commercial world. Computer Fraud & Security, 2017(5), 14–18.
2. Kucharska, W., & Dąbrowski, J. (2016, September). Tacit knowledge sharing and personal branding: How to derive innovation from project teams. In Proceedings of the 11th European Conference on Innovafion and Entrepreneurship (pp. 435–443).
3. Richey Jr, R. G., & Davis-Sramek, B. (2020). Supply chain management and logistics: An editorial approach for a new era. Journal of Business Logistics, 41(2), 90–93.
4. Iansiti, M., and Lakhani, K.R., 2017. The truth about blockchain. Harvard Business Review, 95(1), pp.118–127.
5. K. Korpela, K. Mikkonen, J. Hallikas, and M. Pynnönen, "Digital business ecosystem transformation—towards cloud integration," in 2016 49th Hawaii International Conference on System Sciences (HICSS), 2016, pp. 3959–3968.
6. Khan, M.A., and Salah, K., 2018. IoT security: Review, blockchain solutions, and open challenges. Future Generation Computer Systems, 82, pp.395–411.

Note: All the figures in this chapter were made by the author.

The Development and Implementation of an Intelligent Transportation System with 5G Capability

Prithu Sarkar[1]

Assistant Professor Grade II,
Amity School of Communication,
Amity University, Kolkata

Vijay Anand Kandaswamy[2]

Associate Professor,
Department of Electronics Instrumentation Engineering,
R.M.K. Engineering College, Thiruvallur, India

Amit Kumar Jain

Assistant Professor, Department of Electrical & Electronics Engineering,
Poornima University, IS-2027-2031, Jaipur,
Rajasthan, India

Logeshwari Dhavamani[3]

Professor, Dept. of Information Technology,
St. Joseph's Institute of Technology, Chennai

Gabbeta Ramesh[4]

Assistant Professor,
ST. Martins Engineering College Secunderabad, Telangana

P. S. Ranjit[5]

Professor, Department of Mechanical Engineering,
Aditya Engineering College Surampalem, Kakinada, India

Ike Umejesi[6]

University of Fort Hare, South Africa

━━━━ **Abstract**

Technology and connection developments are causing a fundamental change in how transportation networks are evolving. The creation and deployment of an Intelligent Transportation System (ITS) outfitted with 5G capabilities are the subjects

Corresponding Author: [1]prithusarkar90@gmail.com, [2]kva.eie@rmkec.ac.in, [3]logeshgd@gmail.com, [4]grrams786@gmail.com,
[5]psranjit1234@gmail.com, [6]ikeumejesi@gmail.com

DOI: 10.1201/9781003596721-4

of this study. It examines the benefits and difficulties of this creative integration, placing special emphasis on how it may change urban mobility, improve safety, and lessen environmental effects.A growing number of gadgets are getting smarter and Internet-connected as technology develops quickly. These developments have created new difficulties for the field of intelligent transport systems (ITS), including the need to install new hardware, configure it remotely, and transfer large amounts of data quickly. Recent developments in the telecommunications industry, particularly in vehicle ad hoc networks (VANETS), have raised interest in ITS. Because of their network adaptability and programmability through the use of a logical and centralised control unit, software-defined networks (SDNs) can aid ITS. Therefore, this study proposes a novel concept for improving capabilities by exploiting the recently suggested 5G-based SDN architecture for ITS.

▬▬▬ **Keywords**

SDN, Intelligent transport systems, 5-G

1. INTRODUCTION

Urban transportation systems are experiencing unprecedented difficulties, including the need for increased safety and efficiency as well as issues with congestion and the environment. The integration of an Intelligent Transportation System (ITS) driven by 5G technology has the possibility of revolutionizing urban mobility in this era of connectedness.

A comprehensive framework called an ITS uses technology to optimize transportation systems, boost traffic control, and improve commuter experiences all around. An ITS's capabilities are improved with the development of 5G connectivity, which offers extremely quick data transmission, low latency, and broad device support. Real-time communication, data exchange, and the implementation of intelligent solutions that can lessen traffic congestion, improve safety, and lessen the environmental impact of transportation are all made possible.

The intelligent transportation system successfully combines data transmission technology, data processing technology, computer technology, independent collaborative technology, and information fusion technology for control over the infrastructure of traditional transportation systems. Intelligent transport systems use information technology to try and address issues with conventional transport offer traffic users a range of services and supervisors. When the Intermodal Surface Transportation Efficiency Act (ISTEA) was passed in 1991, the US government began a federal effort to conduct research on the creation and evaluation of intelligent transportation systems (ITS) and supported deployment [1].

Connected vehicles and very accurate real-time location services are the cornerstones of ITS development. Vehicle-to-human (V2H), Vehicle-to-Infrastructure (V2I), and Vehicle-to-Vehicle (V2V) wireless connectivity that is safe and interoperable is what the automotive network aims to provide [2].

Although ITS is capable of offering a variety of services, the requirements for user satisfaction are insufficient from a quality-of-service (QoS) perspective [2]. Furthermore, the ability to handle a sizable request is essential. Today, the term "software-defined network" (SDN) is used to describe the way wired and wireless devices communicate from a software perspective.

This study launches a thorough investigation into the creation and application of a 5G-capable intelligent transportation system. It explores the ITS guiding principles, the revolutionary potential of 5G technology, and the special benefits that result from their combination. This study intends to offer useful insights into how this amalgamation can alter transportation systems by examining the difficulties and potential use cases in the context of urban mobility.

The integration of an ITS with 5G capacity presents a viable strategy to handle these difficulties as cities all over the world struggle with the complexity of urban mobility. It emphasizes the significance of utilizing cutting-edge technologies and connections to build smarter, more effective, and environmentally friendly transportation networks.

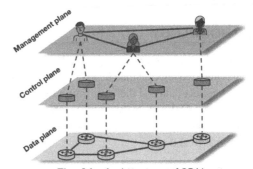

Fig. 4.1 Architecture of SDN

1.1 Research Objective

1. The creation and deployment of a 5G-capable intelligent transportation system is the primary objective of this project.

2. This work aims to give a fresh idea for improving ITS capabilities using the recently suggested ITS, SDN architecture based on 5G.

2. LITERATURE REVIEW

Ge et al.'s [3] description of the specifications for ITSs utilising software-defined networking and 5G mobile communication technology, introduction of pilotless aircraft Cloud computing, SDN, and 5G mobile connection are required for vehicle networks. They suggested combining the aforementioned technologies to create an entirely novel framework for 5G software defined vehicle networks. Wanag et al. [4] provide a decision-based tree method for selecting IoV using the vertical handoff technique. A novel choosing themselves decision tree-based VHO technique for IoVs for WiMAX, 3G, and WAVE cellular networks was created in this study. The detrimental effects of service and movement adjustments can be avoided by using a decision-making process based on feedback from services and movements on vehicles. Based on user preferences, the decision tree makes decisions. An SDN-enabled connectivity-aware geographic-aware routing system that is a performance-enhanced protocol for optimised data packet delivery was presented by Venkatraman et al. [5]. We were able to better comprehend network topology and implement the structure of SDN to transmitting data in an urban setting that includes car networks thanks to their study.

Service and connection-oriented management architecture for SD-IoV was developed by Chen et al. [19]. According to them, building IoV still presents a number of difficulties, including the need for flexible and effective connections, a QoS guarantee, and several concurrent supports. They used SD-IoV to address these problems.

3. METHODOLOGY

The suggested procedure comprises five key steps: three main stages and two intermediate layers. Various embedded IoT devices collect data and then transmit it through SDN-based (Software-Defined Networking) infrastructure to the computational layer. From there, the processed data is made available to end users.

3.1 Data Collection Unit

Multiple smart city services, such as smart transportation and healthcare, are considered for data aggregation. These services use sensors to gather data efficiently. The first intermediate layer (IL1) handles the transmission of this collected data to the upper processing layer.

3.2 Data Processing Unit

Data received from IL1 is sent to the processing level for normalization and extraction of relevant information. For example, detecting high traffic levels can assist residents in finding quicker routes. However, processing big data in real time is a complex task, as it demands significant processing power. To address this, the proposed data processing layer leverages the Hadoop ecosystem, GrapheX, and Spark, along with an optimized map-reduce methodology to ensure efficient data analysis.

3.3 Data Application

The second intermediate layer (IL2) provides an interface between the Hadoop Distributed File System (HDFS) and the application layer. This interface allows data to be transmitted from the processing stage to the application level. Though IL1 and IL2 share a common principle in handling traffic on SDN, there are notable differences at the traffic management layer. Additionally, at the application level, further modules such as Named Data Networking (NDN) and decision/event management are integrated to support decision-making processes.

4. RESULTS AND DISCUSSION

The C-coded simulation that was used to evaluate the viability and effectiveness of the suggested 5G-enabled SDN for ITS is included in this section. Each connection between the automobiles and the SDN controller is granted 181 seconds during the simulation, which is long enough to predict how the suggested plan would work. The suggested scheme is then contrasted with the current AODV scheme.

The results of our experiment shown that even if alternate routes appeared to have less traffic, the shortest route's frequency exceeds the maximum density, resulting in traffic congestion. On the other hand, our proposed strategy maintains a suitable density while equally distributing all traffic on all practical routes. As daily traffic flow is analysed on an hourly basis, different densities exist during normal and rush hours.

We assessed the suggested approach on various node counts with respect to of both time and processing in order to test the system's efficacy. As seen in Fig. 4.2, processing of the data advances more swiftly as the number of nodes rises.

This indicates that the Hadoop server analyses more data faster since it uses parallel processing. Furthermore, we were able to effectively compare our approach to the contestant's plan.

Due to the energy harvesting technology, which prolongs device lifetime, the suggested scheme's average processing time is shorter than the competitor's scheme's, as shown in Fig. 4.3. Consequently, it produces more data than the competing method.

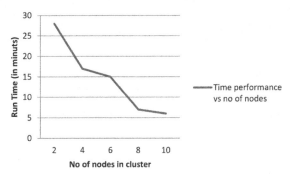

Fig. 4.2 Nodes' error rate as a function of iterations

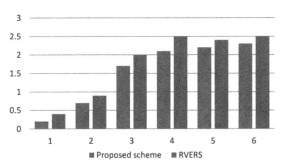

Fig. 4.3 Comparison of processing times with the current system

4.1 Discussion

The creation and deployment of a 5G-capable Intelligent Transportation System is a subject with enormous promise. This talk examines important issues and factors related to the 5G integration into an ITS.

4.2 Enhanced Connectivity and Communication

Real-time communication between vehicles, infrastructure, and traffic control systems is made possible by the rollout of 5G connection. This makes it easier for data to be shared, including updates on the weather, road dangers, and traffic conditions, enabling dynamic traffic management.

4.3 Traffic Management In Real-Time

Traffic management systems can react instantly to events and traffic jams thanks to 5G's low latency. For instance, intelligent traffic signals can modify signal timings to improve traffic flow and ease congestion.

4.4 Vehicle-to-Everything (V2X) Advanced Communication

V2X communication is made possible by 5G, allowing vehicles to talk to each other, infrastructure, and pedestrians.

This creates opportunities for enhanced traffic flow, pedestrian safety, and collision avoidance.

4.5 Challenges

Although 5G-enabled ITS has significant promise, there are a number of issues that need to be resolved:

4.6 Infrastructure Development

The installation of 5G infrastructure, such as base stations and tiny cells, is a major endeavor that demands a substantial investment.

4.7 Data Security and Privacy

Ensuring the security and privacy of data carried via 5G networks is essential, particularly in systems that handle traffic and personal data.

4.8 Regulatory Compliance

It's crucial to abide by telecommunications and transportation laws while putting 5G ITS into practice.

4.9 Interoperability

Ensuring that various ITS parts can communicate and co-operate in an effective manner is a difficult task.

5. Conclusion

The rise of ITS create a link between the physical and digital worlds, ushering in the era of intelligent transportation. Our goal is to create a system that makes it simple to find a route and other services. This article presented a 5G and SDN-based architecture based on these modern criteria. The proposed system's primary goal is to maintain constant connectivity between the SDN controller and the cars. Additionally, the role of 5G improves data rate capacitics. When timely data delivery is more crucial, such an improvement is helpful. The findings demonstrated that, under various circumstances, the proposed system design beat the current AIDV routing protocol in terms of average processing time, vehicle data processing per minute, and flow density.Finally, a critical step toward more intelligent, secure, and effective urban mobility is the development and deployment of an Intelligent Transportation System with 5G capability. Despite these difficulties, the potential of 5G-enabled ITS is too great to pass up. In order to fully realize the transformational potential of this integration in the field of transportation systems, the discussion emphasizes the significance of continual research, collaboration, and adaptation. It represents a significant change toward the transportation systems of the future rather than merely a development of technology.

References

1. T. T. Dandala, V. Krishnamurthy and R. Alwan, "Internet of Vehicles (IoV) for traffic management," In proc. of International Conference on Computer, Communication and Signal Processing (ICCCSP), Chennai, India, 2017, pp. 1–4.

2. J. Chen et al., "Service-Oriented Dynamic Connection Management for Software-Defined Internet of Vehicles," In IEEE Transactions on Intelligent Transportation Systems, vol. 18, no. 10, pp. 2826–2837, Oct. 2017.

3. X. Ge, Z. Li and S. Li, "5G Software Defined Vehicular Networks," IEEE Communications Magazine, vol. 55, no.7, pp. 87–93, 2017.

4. S. Wang, C. Fan, C. H. Hsu, Q. Sun and F. Yang, "A Vertical Handoff Method via Self-Selection Decision Tree for Internet of Vehicles," in proc. Of IEEE Systems Journal, vol. 10, no. 3, pp. 1183–1192, Sept. 2016.

5. D. K. N. Venkatramana, S. B. Srikantaiah and J. Moodabidri, "SCGRP: SDN-enabled connectivity-aware geographical routing protocol of VANETs for urban environment," in IET Networks, vol. 6, no. 5, pp. 102–111, 9 2017.

6. J. Chen et al., "Service-Oriented Dynamic Connection Management for Software-Defined Internet of Vehicles," in IEEE Transactions on Intelligent Transportation Systems, vol. 18, no. 10, pp. 2826–2837, Oct. 2017.

7. Soni, N. Kumar, Y. K. Sharma, V. Kumar and A. Aggarwal, "Generalization of Fourier Transformation of Scaling Function using Riesz basis on L2 (K)," 2022 10th International Conference on Reliability, Infocom Technologies and Optimization (Trends and Future Directions) (ICRITO), Noida, India, 2022, pp. 1–5, doi: 10.1109/ICRITO56286.2022.9965138.

Note: All the figures in this chapter were made by the author.

*Recent Trends in Engineering and Science for Resource Optimization and
Sustainable Development – Prof. (Dr.) Dorota Jelonek et al. (eds)
© 2024 Taylor & Francis Group, London, ISBN 978-1-032-98030-0*

5

Predictive Maintenance Using AI and IoT in Aerospace Engineering

Kakarlamudi V. S. Sudhakar[1]

Associate Professor, MCA Department,
R.G. Kedia College, Osmania University
Chadarghath, India

N. M. Deepika

Department of Information Technology,
Institute of Aeronautical Engineering,
Hyderabad, India

Prerana N. Khairnar[2]

Assistant Professor, Sir Visvesvaraya Institute of Technology,
Chincholi, Nashik, India

P. Satish kumar

Associate Professor, Department of
Mechanical engineering, School of Engineering,
SR University, Warangal, India

Amit Kumar Bindal[3]

Professor, CSE Department, M M Engg. College,
M M (Deemed to be University) Mullana, Ambala, India

P. S. Ranjit[4]

Professor, Department of Mechanical Engineering,
Aditya Engineering College Surampalem,
Kakinada, India

Jelena Franjkovic[5]

Osijek, J. J. Strossmayer University of Osijek, Croatia

Abstract

Industrial predictive maintenance has been used more frequently recently as a result of the expansion of data that is now available from sensors built into industrial equipment. Unparalleled accuracy, dependability, and safety are requirements for the aircraft sector. Predictive maintenance has become a game-changing technology in this environment, driven by artificial intelligence (AI) and the internet of things (IoT). In this study, the use of AI and IoT in predictive maintenance for aircraft engineering is investigated. It explores the benefits, difficulties, and consequences of this breakthrough, placing

Corresponding authors: [1]sudhakarkvs20@gmail.com, [2]autadeprerana@gmail.com, [3]amitbindal@mmumullana.org, [4]psranjit1234@gmail.com,
[5]jelena.franjkovic@efos.hr

DOI: 10.1201/9781003596721-5

special emphasis on how it could increase aircraft safety, save operating expenses, and lengthen the lifespan of essential aerospace components. Predictive maintenance has become a crucial technique for maximising maintenance in the aviation sector scheduling, decreasing downtime for aeroplanes, and finding unanticipated defects. Despite this, the field of aircraft manufacturing lacks a comprehensive review of applications and methods for predictive maintenance. The existing research on the different kinds of data, prospects, projects and applications for predictive maintenance in Arespace industry is thoroughly reviewed and up to date in this article.

■■■■■■■■ **Keywords**

Artificial intelligence, IoT, Predictive maintenance, ML

1. INTRODUCTION

Since produced goods decay over time and use and eventually fail if they are not maintained, they are all intrinsically unreliable [1]. Industrial equipment needs routine maintenance to increase its operational lifetime and lessen the money lost when it isn't working. This is crucial for aeroplanes because of the high demand from consumers and airlines for flight-ready aircraft as well as the significant revenue loss brought on by out-of-service aircraft.

Maintenance is a key component of an organization's competitiveness because actions done at this level immediately influence variables including the cost, deadlines, and quality of the goods produced or services provided [2]. In order to rapidly fix equipment breakdowns and/or prevent them from happening, proactive and reactive maintenance approaches make up the majority of the varied maintenance tactics utilised across various sectors. In a reactive maintenance strategy known as corrective maintenance (CM), work is unscheduled and carried out right away when an asset breaks down. The easiest technique to apply for technicians and the oldest way that makes the best use of component lifetime, run-to-failure methodology is not widely used in any industry because it is the most expensive [3]. The proactive approach of preventative maintenance (PM) involves scheduling and carrying out maintenance tasks at predetermined intervals in order to lower the likelihood of failure in the future.

1.1 Research Objective

1. This study's goal is to investigate Predictive Maintenance in Aerospace Engineering Using AI and IoT.
2. This study's goal is to emphasise, identify and talk about the challenges and options for additional research in aerospace engineering.

2. LITERATURE REVIEW

There are several definitions for PdM, but they all centre on analysing information regarding mechanical conditions,

operating effectiveness, and other comparable indicators of a mechanical device's state in order to optimise the time between repairs [4]. When a system is put under PdM, maintenance tasks are only carried out when a certain circumstance arises. PdM use sensor network for collection of data that may be evaluated to assess the health and degeneration of a system. With up to 25,000 sensors on an Airbus A380, aircraft are better equipped than ever to record massive amounts of sensor data while in flight across almost all of its components [5]. The growing amount of data has led to a rise in the use of data-driven PdM, which develops and trains PdM algorithms utilising information rather than domain knowledge. To identify links and produce predictions of observed characteristics, statistical models can be used to analyse the data gathered by an aircraft.

The three kinds of main application cases for PdM in the aircraft sector are prognostics, real-time diagnostics, and real-time flight aid [6]. Real-time flight assistance can direct the pilot, and real-time diagnostics make it possible to document problems encountered in flight for fast resolution upon landing. By studying the system's operating and environmental parameters, prognostics are responsible for predicting system degradation and determining the system's end-of-life (EOL) or remaining usable lifetime (RUL) [7]. These data can be used to design the optimal maintenance plans for replacement and repair [8], extending the lifespan of aircraft components [9]. Terabytes of readily available data that could be used to save costs [10], wait times, and manpower [11] are essentially wasted if PdM isn't taken advantage of.

3. METHODOLOGY

For the goal of identifying research gaps, trends, and other issues, a study compiles all of the sources that are currently available on the subject under investigation. Therefore, a methodology is needed to organise the research with the intention of doing an extensive study in order to conduct an effective survey. Figure 5.1 shows a diagram illustrating the process.

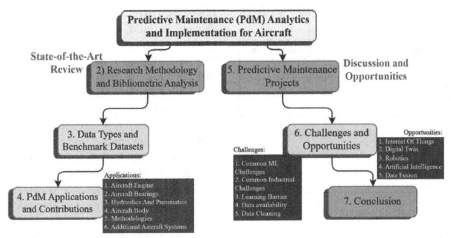

Fig. 5.1   Methodology architecture

3.1 Bibliometric Research

The papers that were identified as falling under the purview of the described search criteria were examined using a variety of criteria in order to address the first research question. The authors, journals, and country of origin can all serve as important markers for figuring out the main participants and hotbeds of this research.

3.2 Database

The first stage of a bibliometric investigation is selecting the database or databases. This study considers both of them as databases and offers thorough information on database mergers because Scopus and WoS are the two most widely used citation databases and are frequently used in bibliometric research.

3.3 Search Parameter

This section describes how inclusion and exclusion criteria work during a search. English was chosen as the publication language for both databases. No exclusions with regard to the topic, affiliation, or journal were imposed. Articles and conference papers were considered for the document categories; however editorial content and book chapters were not.

4. RESULTS AND DISCUSSION

The publications used in this study were selected using three key search criteria: being aerospace-focused where papers are available, being published after 2016 to be considered state-of-the-art, and being highly cited relative to their publication date.

Transferable industrial systems have been employed in the absence of aircraft analogues. Using search terms that have become more common as the area has developed, as shown in Figs. 5.2 and 5.3, the publications were located in the reputable research databases IEEE Xplore and Elsevier.

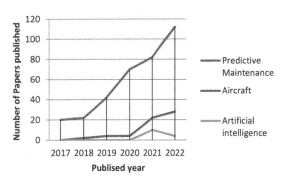

Fig. 5.2   Journal papers for PdM available from IEEE Xplore, published throughout the last five years

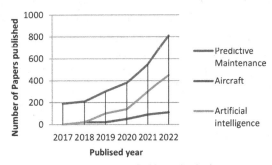

Fig. 5.3   Journal papers for PdM available from science direct (Elsevier) are publications from the previous five years

These statistics show the expansion of PdM research and, to a lesser extent, the emphasis on aeroplanes and artificial intelligence. Table 5.1 lists the top 7 journals according to a review of the journals where the papers we found were published.

Table 5.1   The top seven journals in this review for journal papers

Journal	No of papers
Chinese journal of Aeronautics	5
Reliability Engineering	4
IEEE transection instruments	4
IEEE sensor journal	3
Aerospace	3
IEEE transection on industrial electronics	3
IEEE transection on reliability	3

With 10 papers barely making the top 10, four of these came from IEEE magazines. Therefore, IEEE can be regarded as a dependable source for articles in this area in the future.

A major advance in assuring the security, dependability, and effectiveness of aircraft operations is predictive maintenance in aerospace engineering, powered by AI and IoT. A change from traditional maintenance procedures to a proactive and data-driven strategy has been made possible by the adoption of advanced sensors, data analytics, and machine learning algorithms. Predictive maintenance has many benefits for aerospace engineering. Now, aircraft operators can foresee component failures, improve maintenance plans, and prevent expensive unplanned downtime. By addressing possible problems early on, this not only lowers operational costs but also improves safety. The lifespan of vital components may be greatly increased by the use of AI and IoT in aircraft predictive maintenance. Operators can make educated judgments about repairs and replacements by continuously monitoring the state of aircraft systems and parts. This helps to ensure that parts are changed as needed rather than on predetermined schedules.

The implementation of predictive maintenance in aeronautical engineering is not without difficulties, despite its promise. It is crucial to ensure the security of data communicated through IoT devices and to address privacy issues. Additionally, careful planning and adherence to strict industry standards are required when incorporating these cutting-edge technologies into the current aircraft systems.

5. CONCLUSION

This study offers a cutting-edge survey to identify the distinctive strategies being utilised to address PdM concerns and map the current state of the field. When it comes to maximising the RUL of aircraft components, PdM can be more optimised than other maintenance approaches. Developing novel PdM approaches is now easier than ever thanks to the increasing quantity of benchmark datasets that

can be used to test prognostic methods. Given the increased demand for PdM, further development there are numerous cutting-edge techniques and prospective applications. The improvements made possible by new technology like robotics, IoT and AI will further these processes' automation and optimisation. If it is used more regularly, it might drastically reduce maintenance costs for both aircraft manufacturers and operators. Using AI and IoT for predictive maintenance will revolutionize the aerospace sector, to sum up. It provides a technique to operate aircraft in a safer, more effective, and more affordable manner. The use of predictive maintenance is evidence of the ability of cutting-edge technologies to improve safety, efficiency, and the overall flying experience as the aviation industry continues to develop. Continuous research, development, and industry collaboration are required to fully realize the promise of the AI and IoT combo in the field of aerospace maintenance.

References

1. Ben-Daya M, Kumar U, Murthy P. Introduction to Maintenance Engineering: Modelling, Optimization and Management. 1st ed. John Wiley & Sons; 2016.
2. Gilchrist, A. Industry 4.0: The Industrial Internet of Things; Springer: Berlin/Heidelberg, Germany, 2016; ISBN 978-1-4842-2047-4.
3. Selcuk, S. (2017). Predictive maintenance, its implementation and latest trends. Proceedings of the Institution of Mechanical Engineers, Part B: Journal of Engineering Manufacture, 231(9), 1670–1679.
4. Lee, J., Ni, J., Singh, J., Jiang, B., Azamfar, M., & Feng, J. (2020). Intelligent maintenance systems and predictive manufacturing. Journal of Manufacturing Science and Engineering, 142(11).
5. Shah D. Millions of data points flying in tight formation.2014. Accessed 7th July 2022. https://www.aerospacemanufacturinganddesign.com/article/millions-of-data-pointsflying- part2-121914/
6. Sadok M. Predictive maintenance in aerospace—innovative use cases. June 7, 2020. Accessed 1st July 2022. https://www.strategytransformation.com/predictive-maintenance-in-aerospace/
7. Zhao Z, Liang B, Wang X, Lu W. Remaining useful life prediction of aircraft engine based on degradation pattern learning. Reliab Eng SystSaf. 2017;164:74–83. http://www.sciencedirect.com/science/article/ pii/S0951832017302454
8. K. M. Sahu, Soni, N. Kumar and A. Aggarwal, "σ-Convergence of Fourier series & its Conjugate series," 2022 5th International Conference on Multimedia, Signal Processing and Communication Technologies (IMPACT), Aligarh, India, 2022, pp. 1–6, doi: 10.1109/IMPACT55510.2022.10029267.

Note: All the figures and table in this chapter were made by the author.

Recent Trends in Engineering and Science for Resource Optimization and Sustainable Development – Prof. (Dr.) Dorota Jelonek et al. (eds)
© 2024 Taylor & Francis Group, London, ISBN 978-1-032-98030-0

Optimising Battery Maintenance in Electric Vehicles Using Machine Learning

Nilesh Kolhalkar[1]

Assistant Professor, Department of Mechanical Engineering,
MKSSS's Cummins College of Engineering for Women,
Pune, India

Gunjan Chhabra[2]

Associate professor, Department of computer science and
engineering, Graphic Era Hill University,
Dehradun, India

Navdeep Kumar Chopra[3]

Assistant Professor, Deptt of CSE,
Seth Jai Parkash Mukand Lal Institute of Engineering and Technology (JMIT),
Radaur, Yamunanagar

Balakrishna Kothapalli

School of Engineering, SR University,
Warangal, Telangana

P. Veeramanikandan[4]

Associate Professor, Department of EEE,
Jaya Engineering College, Chennai

P. S. Ranjit[5]

Professor, Department of Mechanical Engineering,
Aditya Engineering College Surampalem,
Kakinada, India

Leszek Ziora[6]

Czestochowa University of Technology, Poland

■■■■■■ ■ **Abstract**

A crucial step toward lowering carbon emissions and achieving sustainable transportation is the switch to electric cars (EVs). The condition of the batteries in electric vehicles is crucial to their performance. This study looks into how to best maintain batteries in electric cars using machine learning approaches. It highlights how utilizing machine learning for improved battery management has the potential to increase battery lifespan, enhance performance, and lessen the

Corresponding author: [1]nrkolhalkar@rediffmail.com, [2]chhgunjan@gmail.com, [3]navdeepkumar17@gmail.com, [4]veerarmd@gmail.com,
[5]psranjit1234@gmail.com, [6]leszek.ziora@pcz.pl

DOI: 10.1201/9781003596721-6

environmental impact of electric mobility as it investigates the benefits and difficulties of doing so. Two of the primary obstacles to the widespread adoption of electric vehicles (EVs) as the future transportation infrastructure are the deterioration of rechargeable batteries and the accompanying cost of replacement. By using so-called battery-aware charging techniques, which are more effective at controlling charge cycles than discharge cycles, battery ageing can be significantly slowed down. These strategies aim to shorten standby time, or the time between when the actual charging procedure is complete and when the EV is removed from the charging station, because the battery's average level of charge is one of the key variables influencing battery ageing. In this study, we assess a sophisticated machine learning (ML)-based forecasting method's efficacy.

▬▬▬▬ Keywords

Machine learning, Electrical vehicle, Optimized battery

1. INTRODUCTION

Electric vehicles are developing as a particularly viable type of transport infrastructure for the future due to the detrimental environmental effects of transport based on petroleum and recent developments in renewable energy generating technology [1]. Because the electricity used by EVs can come from a number of sources, including numerous renewable ones, EVs are thought to be environmentally benign. Due to its geographic diversity, using renewable energy sources for transportation has the potential to drastically reduce fuel consumption and emissions while simultaneously raising the level of energy security. The bulk of both the private and public transportation sectors already use EVs, and the market for them is growing swiftly, but a serious issue with batteries is still keeping EVs from being widely used. Advanced battery optimisation techniques are urgently needed to meet customer requests for longer driving range and quicker charging times while also keeping battery replacement costs under control [2]. As a result, increasing the battery's lifespan becomes a primary objective while designing EVs.

For EV battery charging, standardised charging techniques based on established in advance current and voltage profiles are often needed [3]. CC-CV optimisations have several restrictions, but this protocol still provides a lot of degrees of freedom that can be leveraged to slow down ageing, especially the charge initiation time and the charge current [4].

1.1 Study Objective

1. Our research's goal is to provide a trustworthy prediction of how long an EV will take to charge at a residential charging station.
2. The purpose of this work is to describe how we assess the machine learning techniques used for effective forecasting.

2. LITERATURE REVIEW

2.1 CC-CV EV Battery Charging

To consider with the effect of same absolute current rate, charging a battery might reduce its lifespan more than discharging it [5]. Choosing the right charge protocol is essential in order to preserve the battery's performance as much as possible and prevent detrimental consequences like overheating and overcharging, which in addition to offering apparent concerns can also increase the battery's SOH and hasten battery ageing.

2.2 EV Battery Capacity Deterioration with Age

Temperature, DOD at each cycle, discharge and charge current and average SOC are the four main variables that have an impact on the capacity degradation of lithium-ion batteries of rechargeable type [6]. Any rise in one of these numbers accelerates ageing. Only the average SOC and charging current can be altered to optimise the process of charging because temperature is difficult to control and the discharge currents and DOD rely on how much power is consumed and how long the discharge phase lasts.

2.3 Applications of Machine Learning in EVs

Many further EV- and battery-related optimization-related applications use machine learning. Driving range estimate [2] is among the most important, and ML models that make use of linear regression and self-organizing maps [7] have been utilised to handle it. The energy that electric vehicles (EVs) request in a demand-side management system has also been optimised using deep neural network-based machine learning [4].

3. METHODOLOGY

Our study's objective is to offer a reliable estimation of how long it will take an EV to charge at a home charging station. We are particularly interested in determining whether a machine learning (ML) solution enhances prediction accuracy [3] when compared to the fundamental prediction strategies utilised in prior research [5], which we use as comparable baselines [4]. The next step is to determine whether the EV battery's aging/QoS has improved noticeably as a result of the greater precision. In order to accomplish this, we add the Aging-Optimal battery-aware CC-CV protocols [6] and As Soon As Possible (ASAP) protocols on top of our ML-based prediction and baselines from the original papers to gather extra data. Our predictor has no impact on the underlying charging method, despite testing using these two protocols of state-of-the-art enabling to evaluate its actual influence on battery ageing and QoS, as was anticipated.

4. RESULTS AND DISCUSSION

Additionally essential to improving the effectiveness and performance of electric car batteries is machine learning. EVs can provide better range, acceleration, and a more enjoyable driving experience by tuning charging algorithms and responding to driving habits. Real-time modifications are made possible by machine learning, guaranteeing that the battery works to its full potential. The use of machine learning into electric car battery management is not without difficulties, despite its promise. It is crucial to ensure data security and privacy, particularly when vehicles transfer data to cloud-based services. Continuous research and development is also needed to build robust algorithms that can adjust to various driving scenarios and battery chemistries.

4.1 Experimental Environment

The methodology outlined in part 3 was applied to this dataset. In total, 3.5 million charging events from more than 26,000 home charge-points in the UK for the entire year 2018 were gathered and saved in this dataset. We ran our studies on 5 charging stations with the identifiers BM06550, BM20268, BM34644, BM19674, and BM14114 (see Table 6.1) because there were so many charging events available.

Table 6.1 Instances of charges at each station under examination in terms of number

Station ID	Cycles
BM06550	330
BM20268	200
BM34644	191
BM19674	163
BM14114	145

4.2 Forecasting Error

In the initial studies, the pure prediction error of the suggested model and the baselines were compared. The aim statistic for this has been the mean square error (MSE). Figure 6.1 displays the findings of this experiment by displaying the MSE for each of the five charge locations under consideration as determined by all predicting techniques. As can be observed, among the five stations, the forecasting utilising LightGBM has the lowest error.

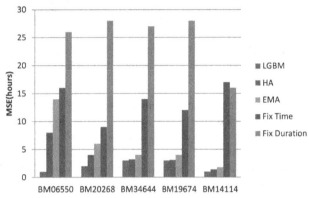

Fig. 6.1   Models' mean square error

4.3 Results of Ageing and QoS

The two ageing-conscious charging techniques outlined in Section 4.1 were updated to include our suggested plug-in duration estimator after we verified the performance of LightGBM forecasting. SOH and QoS measures have a normalised range of 0 to 1, with 1 signifying the ideal outcome. By calculating the QoS and SOH metrics' product, you may statistically assess each predictor's actual effectiveness by comparing how near they are to the ideal point (1, 1). Figures 6.2 and 6.3 display the outcomes.

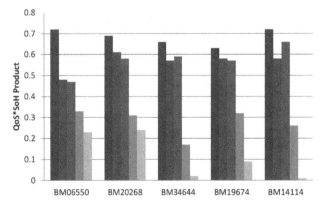

Fig. 6.2   Utilising several plug-in duration predictions and the ageing optimal charging technique, the S-SOH product

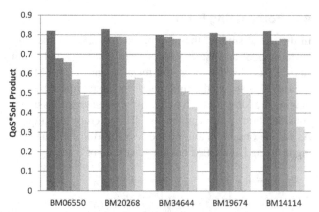

Fig. 6.3 Utilising the ASAP charging protocol and various plug-in duration predictions, the QoS-SOH product

The figures unmistakably demonstrate that all baselines are outperformed by the ML-based solution. LightGBM helps improve the QoS-SOH product by 21% and 9%, respectively, on average, using the Ageing Optimal and ASAP protocols. The improvement is typically greatest at BM06550, at 52% and 23%, respectively. This is due to the fact that the station is always the one with the lowest prediction MSE, which is brought on by the availability of more training data.

5. CONCLUSION

In an effort to increase the effectiveness of ageing sensitive charging techniques, which depend on this knowledge to provide a just-in-time charge, for evaluating the plug-in lifespan of EVs, A brand-new machine learning (ML)-based forecasting strategy has been put forth by us. Furthermore, the discovery that the charging station with the most records consistently make the most progress points to the possibility of even better outcomes with more data. A important step forward in maintaining the long-term viability and performance of EVs is the machine learning optimization of battery care. The electric mobility landscape gains a number of significant advantages from the use of machine learning techniques in battery management. The ability to increase battery lifespan is one of the main benefits of machine learning in EV battery management. Machine learning algorithms can recognize usage patterns and pro-actively advise charging and discharging procedures that reduce stress on the battery cells by continuously monitoring and analyzing battery performance. In addition to saving customers money on battery replacements, this lessens the environmental impact involved with the manufacturing and disposal of batteries.

References

1. Zhang, Q.; Ou, X.; Yan, X.; Zhang, X. Electric vehicle market penetration and impacts on energy consumption and CO_2 emission in the future: Beijing case. Energies 2017, 10, 228. [CrossRef]

2. Baek, D.; Chen, Y.; Bocca, A.; Bottaccioli, L.; Cataldo, S.D.; Gatteschi, V.; Pagliari, D.J.; Patti, E.; Urgese, G.; Chang, N.; et al. Battery-Aware Operation Range Estimation for Terrestrial and Aerial Electric Vehicles. IEEE Trans. Veh. Technol. 2019, 68, 5471–5482. [CrossRef].

3. Berckmans, G.; Messagie, M.; Smekens, J.; Omar, N.; Vanhaverbeke, L.; Van Mierlo, J. Cost projection of state of the art lithium-ion batteries for electric vehicles up to 2030. Energies 2017, 10, 1314. [CrossRef].

4. Bocca, A.; Chen, Y.; Macii, A.; Macii, E.; Poncino, M. Aging and Cost Optimal Residential Charging for Plug-In EVs. IEEE Design Test 2017, 35, 16–24. [CrossRef]

5. Bashash, S.; Moura, S.J.; Forman, J.C.; Fathy, H.K. Plug-in hybrid electric vehicle charge pattern optimization for energy cost and battery longevity. J. Power Sources 2011, 196, 541–549. [CrossRef].

6. Bocca, A.; Sassone, A.; Macii, A.; Macii, E.; Poncino, M. An aging-aware battery charge scheme for mobile devices exploiting plug-in time patterns. In Proceedings of the 2015 33rd IEEE International Conference on Computer Design (ICCD), New York, NY, USA, 18–21 October 2015; pp. 407–410.

7. Sun, S.; Zhang, J.; Bi, J.; Wang, Y. A Machine Learning Method for Predicting Driving Range of Battery Electric Vehicles. J. Adv. Transp. 2019, 2019, 4109148. [CrossRef]

8. López, K.L.; Gagné, C.; Gardner, M. Demand-Side Management Using Deep Learning for Smart Charging of Electric Vehicles. IEEE Trans. Smart Grid 2019, 10, 2683–2691. [CrossRef]

Note: All the figures and table in this chapter were made by the author.

*Recent Trends in Engineering and Science for Resource Optimization and
Sustainable Development – Prof. (Dr.) Dorota Jelonek et al. (eds)
© 2024 Taylor & Francis Group, London, ISBN 978-1-032-98030-0*

7

Assessing Blockchain's Potential for Electronics' Secure Identity Management

M. Dhinakaran

Associate Professor Department of Electronics and Communication Engineering,
Government College of Engineering, Karuppur, Salem, Tamil Nadu

Rahul Budhraja

Sr. Software Engineer, Computer Science,
Vaish College of Engineering, Maharshi Dayanand University,
Rohtak, India

B. Varasree

Department of Information Technology,
Institute of Aeronautical Engineering, Hyderabad, India

Shiv Kant Tiwari[1]

Assistant Professor, Institute of Business Management,
GLA University, Mathura, India

Amit Kumar Bindal[2]

Professor, CSE Department, M M Engg. College,
M M (Deemed to be University) Mullana, Ambala, India

Sunita Arvind Kumar[3]

Assistant Professor, Department of ECE,
St. Martin's Engineering College Secunderabad, India

Liviu Rosca[4]

Lucian Blaga University of Sibiu

▬▬▬ Abstract

A secure Identity Management (IdM) System has been developed using a variety of techniques in the past. The blockchain literature's first comprehensive map of IdM is presented here. Because of its potential to revolutionize secure identity management in the electronics industry, blockchain technology has attracted a lot of attention. The effectiveness of blockchain for boosting the security and effectiveness of electronic device identity management is evaluated in this article. We look at the main characteristics of blockchain, its possible advantages, and the difficulties it can provide in the context of electronics. This paper offers important insights into the viability of using blockchain for secure electronic identity

Corresponding authors: [1]shivkant.tiwari@gla.ac.in, [2]amitbindal@mmumullana.org, [3]sunita.a.rathod@gmail.com, [4]liviu.i.rosca@gmail.com

DOI: 10.1201/9781003596721-7

management through a thorough analysis.This article offers a comprehensive examination of IdM with a focus on how the development of blockchain has addressed the issues with IdM that have been noticed over time. A thorough review of the currently available literature has been conducted. Five databases were searched using the primary and secondary „search strings," and analysis was completed after screening. The following can be accomplished by researchers using this paper: Learn about IdM research trends that use blockchain, understand IdM challenges, and assess whether blockchain may help with those challenges. 4) Be informed of blockchain-based IdM initiatives. 3) Examine and comprehend how different IdM frameworks will solve security, integrity, and privacy challenges.

▰▰▰▰ Keywords

Block chain, Identity management, and Digital identity

1. Introduction

An individual creates a digital identity in cyberspace, which is their online persona. A digital identity acknowledges the identity holder using digital identifiers like an email address, a domain name, or certain URLs, much as how a passport's information identifies its owner for a specific purpose. A system that can recognise anticipated users and authorise their name, address, and personality is required [1]. This is because we are utilising digital technology more and more.Securing the identity management of electronic devices has become a top priority in the quickly developing electronics sector. Traditional centralized systems for controlling and tracking the identities of electronic devices are prone to data breaches and counterfeiting, among other security risks. Blockchain technology, which is renowned for its transparency, immutability, and decentralization, presents a possible answer to these problems. This study examines how blockchain technology might improve secure identity management in the electronics industry.

The service providers (SP) utilise IdM systems to verify user identities and authorise access to services. In environments for identity and access management (IAM), there are numerous users and service providers. Thanks to IAM systems, which give each user an account and a set of capabil-

ities, users can explore SP, confirm account ownership, and then access services in accordance with their capabilities. Without a solid access control system, there is little question that the security and privacy of user-provided personal data will suffer [2]. There have been numerous widely reported security lapses and data leaks, as well as allegations that unencrypted data has been compromised, hacked, and taken.

1.1 Objective of Study

1. This study uses blockchain technology to analyse current IdM research trends.
2. Examining and understanding how alternative IdM frameworks will solve security, integrity, and privacy issues is the primary goal of this study.

2. Literature Review

2.1 Identity Control

In the literature, the study of identity and identity management has always been an active area of inquiry [3]. Identity management has become a significant priority in this Internet era due to the abundance of digital identities and the users of such identities.

2.2 Models for Identity Management

We may categorise identity management models into three groups after examining the literature: isolated, federated, and centralised IdM [4]. To access the isolated service they have requested, each user in Isolated IdM is given a unique identification number by the identity supplier.

2.3 Issues with Identity Management

IdM systems in the past were centralised and managed by a single party. A ''Trusted third party'' provided assurance for the authentication and permission. In the past, a number of IdM models have been suggested. Any IdM system's primary job is to securely link the "identifier" and the "at-

Fig. 7.1   Potential of blockchain in identity management

tributes" together [5]. Poorly built IdM systems, according to some study can exacerbate current security issues and open doors for the theft of users' personal information.

2.4 Review of the Blockchain

Bitcoin was the first open-source application of blockchain technology, according to multiple research white papers that marketed blockchain as a trust-less system [6]. Blockchain reduces the need for middlemen by establishing trust, security, and data integrity through cryptography [7], peer reviews [8], and decentralised transactions [9]. A blockchain, or distributed ledger, creates a quick, secure, and reliable system by being impermeable, transparent, and redefining trust [10].

3. METHODOLOGY

A process called systematic literature mapping offers a broad perspective of any subject. Additionally, it aids in identifying current knowledge gaps and helps to create new study opportunities. The methodology outlined in is used to provide a transparent and scientific investigation.

3.1 Data Collection

A suitable set of databases must be chosen in order to execute an effective search. The studies published in IEEE xplore, ACM Digital Library, Springer Link, Wiley, and Science Direct between January 2010 and June 2021 were gathered.

3.2 Questions for Research

RQ1: How have writings on identity management and blockchain changed over time?

RQ2: What steps have been taken to use blockchain technology for identity management?

RQ3: Which blockchain consensus algorithm was used with multiple identity management frameworks?

3.3 Data Analysis

Table 7.1 displays the number of papers located in each database for an effective mapping. In the earliest stages of searches, practically all users utilise the keyword "blockchain."

Table 7.1

Database	No of paper
ACM	275
Science Direct	115
IEEE Xplore	274
Wily	219
Springer	401

3.4 Selection criteria

1284 papers were first counted, which a significant number was. The search for further pertinent papers was then conducted using an inclusion and exclusion criterion.

4. RESULTS AND DISCUSSION

This section examines research on IdM using blockchain technology and includes analysis based on RQs, trend patterns, and conversation. Through the use of Fig. 7.2, RQ1 has been interpreted. Despite the fact that research on blockchains started in 2010, the field has gained momentum since 2017. About 67% of the work on identity management and blockchain technology was finished after 2017.

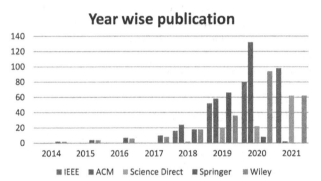

Fig. 7.2   Utilising blockchain, IdM is counted over time

The RQ2 goes over the steps used when using blockchain technology to handle IDs. According to analysis, permissioned blockchain is used more frequently than the other accessible blockchain kinds.

Blockchain cannot protect end user privacy, according to the General Data Protection Regulation (GDPR), which went into effect in 2017. More people are concerned about safeguarding their data's privacy. As a result, permissioned blockchains rather than public blockchains are becoming increasingly popular. The examination of the various types of blockchains employed is shown in Fig. 7.3.

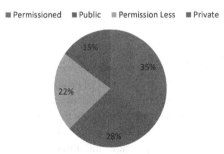

Fig. 7.3   Utilised blockchain types

Proof of Concept is another consensus mechanism that has gained popularity. Figure 7.4 makes it obvious that

POW consensus algorithm is utilised more frequently than other consensus processes. The majority of the consensus mechanisms have been found to work with the current blockchain systems.

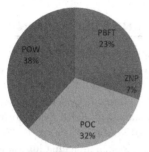

Fig. 7.4   Consensus mechanism types

The basics of blockchain technology are covered in this part, along with information on its decentralized nature, consensus procedures, and cryptographic security. We emphasize how these characteristics might help the electronics sector maintain identities securely.

4.1 Benefits of Blockchain for Electronics Identity Management

In this article, we talk about how using blockchain technology in the electronics industry may improve security, transparency, and traceability. These advantages can lessen the risk of illegal access and counterfeit electronic gadgets.

4.2 Challenges and Considerations

It is critical to recognize the difficulties and factors to be taken into account while deploying blockchain for electronic device identity management. Scalability problems, energy use problems, and regulatory compliance problems are all discussed along with potential solutions.

4.3 Use Cases

We look at actual use cases and pilot initiatives that show how blockchain may be used for electronic identity management. These illustrations highlight how useful blockchain technology may be for guaranteeing secure electronic device identities.

5. Conclusion

The electronics industry has a lot of promise for blockchain technology to transform secure identity management. The security and effectiveness of tracking the identities of electronic devices can be improved by its intrinsic properties of decentralization, transparency, and immutability. Although there are issues and concerns that must be taken into account, blockchain offers a wide range of advantages. The potential of blockchain for secure identity management in electronics has been thoroughly evaluated in this paper. Adopting blockchain technology could provide a solid and dependable solution to reduce security concerns, fight counterfeiting, and provide customers and companies more confidence in the authenticity of electronic devices as the electronics sector continues to expand and change. However, cooperation, inventiveness, and careful analysis of the particular issues offered by this will be necessary for successful execution. The usage of blockchain technology for IdM has been investigated using a complete literature mapping. Even if blockchain technology is receiving a lot of attention, it is still advisable to assess its suitability for that use case. There are solutions to every research question. The study unequivocally demonstrates that traditional IdM systems' shortcomings may be overcome by using blockchain technology in IdM. Blockchain was shown to be an effective solution for IdM problems that have developed over time. Problems with privacy and interoperability still need to be resolved, though.

References

1. Buchmann, N., Rathgeb, C., Baier, H., Busch, C., Margraf, M., 2017. Enhancing Breeder Document Long-Term Security using Blockchain Technology. https://doi.org/ 10.1109/ COMPSAC.2017.119.

2. Qiu, J., Tian, Z., Du, C., Zuo, Q., Su, S., Fang, B., 2020. A survey on access control in the age of internet of things. IEEE Internet Things J. 7, 4682–4696.

3. Jensen, J., Jaatun, M.G., 2012. Federated identity management-we built it; why won't they come? IEEE Secur. Priv. 11, 34–41.

4. Ahn, G.J., Ko, M., 2007. User-centric privacy management for federated identity management. In: Proceedings of the 3rd International Conference on Collaborative Computing: Networking, Applications and Worksharing. https:// doi. org/10.1109/COLCOM.2007.4553829.

5. Dunphy, P., Petitcolas, F.A.P., 2018. A First Look at Identity Management Schemes on the Blockchain. https://doi. org/10.1109/MSP.2018.3111247.

6. Benchoufi, M., 2017. Blockchain technology for improving clinical research quality 1–5. https://doi.org/10.1186/ s13063-017-2035-z.

7. R. Vettriselvan, C. Vijai, J. D. Patel, S. Kumar.R, P. Sharma and N. Kumar, Blockchain Embraces Supply Chain Optimization by Enhancing Transparency and Traceability from Production to Delivery," *2024 International Conference on Trends in Quantum Computing and Emerging Business Technologies*, Pune, India, 2024, pp. 1-6, doi: 10.1109/TQCEBT59414.2024.10545308.

Note: All the figures and table in this chapter were made by the author.

*Recent Trends in Engineering and Science for Resource Optimization and
Sustainable Development – Prof. (Dr.) Dorota Jelonek et al. (eds)*
© 2024 Taylor & Francis Group, London, ISBN 978-1-032-98030-0

Investigating the Use of AI in Power System Fault Identification and Diagnosis

Prateek Nigam[1]

Associate Professor, Department of Electrical and
Electronics Engineering, Rabindranath Tagore University,
Bhopal, Madhya Pradesh, India

Shaik Saddam Hussain

Department of Information Technology,
Institute of Aeronautical Engineering, Hyderabad, Telangana

Amit Baban Kasar

Assistant Professor, Engineering Sciences, International Institute of Information Technology,
Pune, Savitribai Phule Pune University, Pune, Maharashtra

Balakrishna Kothapalli

School of Engineering, SR University,
Warangal, Telangana

P. Veeramanikandan[2]

Associate Professor, Department of EEE,
Jaya Engineering College, Chennai

K. Rajeshwar[3]

Assistant Professor, Department of ECE,
St. Martin's Engineering College
Secunderabad, Telangana

Abdelmonaim Fakhry Kamel Mohammed[4]

Ain shams university – Egypt

Abstract

A new era of fault identification and diagnosis that is both effective and dependable has arrived thanks to the integration of artificial intelligence (AI) into power systems. The use of AI approaches for power system fault identification and diagnosis is examined in this research. This paper illuminates the revolutionary potential of AI in enhancing the stability and resilience of power systems through a thorough investigation of the advantages, difficulties, and real-world implementations. Unexpected breakdowns of electrical power systems occur frequently for a variety of random reasons. It is necessary to stop unanticipated problems in power systems from spreading to other parts of the protective system. The roles of the

Corresponding authors: [1]prateek.nigam@aisectuniversity.ac.in, [2]veerarmd@gmail.com, [3]kukatlarajeshwar@gmail.com,
[4]abdelmonaimfakhry601@gmail.com

DOI: 10.1201/9781003596721-8

protective systems are to identify, categorise, and then determine the when the voltage and/or current line magnitudes are defective. Finally, the protective relay must communicate with the circuit breaker in order to isolate the defective line. Powerful uses of NN employed as an intelligent technique of detection include their capacity for learning, generalisation, and parallel processing. One of the subjects that have received the most attention in the scientific community recently is fault detection. As a result, it has received a lot of attention and study recently. This scientific publication summarises some of this research, the conclusions reached, and the techniques employed.

▬▬▬ Keywords

Artificial network, ANN, Power system, Fault diagnosis

1. INTRODUCTION

Power systems are essential to modern life because they provide an ongoing supply of electricity for numerous uses. These systems are, however, prone to errors and disturbances that can cause power outages and financial losses. For power systems to remain reliable, fault identification and diagnosis must be done promptly and accurately. Machine learning and deep learning are two examples of artificial intelligence (AI) technologies that have shown promise in improving fault identification and diagnosis procedures. In-depth analysis of AI's use in this significant area is presented in this study.

The biggest danger to the continuous supply of electricity is electrical system breakdowns. Electric power system malfunctions are an inherent issue. Consequently, a well-coordinated protection system must be installed to quickly identify and isolate faults in order to minimise the harm and disturbance to the electrical system minimized. Devices that can detect faults, respond rapidly, and disconnect the problematic part are typically used to clear them. Therefore, even if they happen seldom and in random places, numerous kinds of electrical system defects is a reality of life. Active and passive faults are the two primary categories into which faults can be roughly categorised [1]. Control centres for electrical power systems house a significant number of alarms that are the result of various issues. The defects must be precisely identified and isolated in order to safeguard these systems. Overhead wires are where the majority of short-circuit failures are likely to occur [2]. To obtain the necessary information regarding the faults, the operators at the control centres must deal with a lot of data. Artificial neural networks have been developed over time with both biological and control applications in mind, and control system theory has been incorporated into theories of the nervous system and the brain. The neural network is a representation of a network with a finite number of levels, each layer having a distinct sort of connection between solitary pieces that are comparable to neurons. The number

of neurons in each layer is chosen to be adequate for delivering the necessary problem-solving capability. In order to shorten the time needed to solve a problem, the number of layers should be kept to a minimum [1, 3].

1.1 Objective of Study

1. Examining the application of AI to power system fault detection and diagnosis is the goal of this project.
2. This study's goal is to compare various artificial technologies in order to look into fault detection.

2. LITERATURE REVIEW

Several defect detecting techniques have been developed. In a typical power system, mosthas bar nodes' states (voltages and currents) are tracked and any changes over time are evaluated [4]. However, faults are frequently difficult to spot in their early stages because the data being collected is complex. These issues may first go unnoticed due to the power system's operating data complexity. The purpose of the FD system is to notify the operator as soon as any potential issues are found. To monitor the condition of crucial components of power grids, such as switchgear and transformers, the FD system uses a network of ANNs set up in a hierarchical structure. A simple connection between neurons called a synapse has the ability to either excite or inhibit the receptive neuron. The network picks up new skills by absorbing data from its environment [5].

In his study from 2004, Robert Salat introduced a novel strategy for fault detection that was more precise. He concentrated on an HVPS, or high voltage power system. To assign labels to things, he applied a Support Vector Machine (SVM) application, an AI strategy that uses a computer algorithm that learns by doing [6]. Long-distance electrical transmission uses this HVPS. The SVM employed in this study ensured that errors never exceeded 2 km, and the majority of them were less than 100 m in lines longer than 200 km. Research has become reliant on artificial inelegance (AI) according to Ernest Vhquez. He employed an approach known as expert system (E.S.). Programmes that adapt, learn, invent, and compile the collective expertise of a profession are referred to as E.S. The author recommends the E.S. structure with discerption in this study, and this approach is utilised to identify the locations where failures are most likely to occur [7]. A study on problem diagnosis using the "Backpropagation Neural Network Expert System (BPES)" was reported in a paper by Deyin Ma in 2013 [8]. From his perspective, identifying and evaluating faults is a difficult but essential issue [9].

3. METHODOLOGY

Between 2005 and 2021, a total of over 25 studies were compiled and analyzed for the purpose of this review. These studies explore a wide range of issues within the power system domain, focusing on different components such as transmission lines, transformers, generators, and distribution systems. Each component was analyzed in terms of common faults and the AI-based diagnostic techniques applied to address these issues. The research gathered spans multiple diagnostic approaches, including Artificial Neural Networks (ANN), Expert Systems (ES), Support Vector Machines (SVM), and other machine learning models. Furthermore, the studies highlighted various challenges and shortcomings in detecting and diagnosing faults within these power system components. The goal of the review is to provide a comprehensive understanding of the advancements in AI-driven fault identification and diagnosis in power systems.

3.1 Data Collection Process

To conduct a thorough review, we performed a systematic search for relevant literature across various databases. After an initial scope search, the following electronic databases were selected for their reliability and relevance to the subject matter: Business Source Complete, ScienceDirect, ABI/Inform, and Web of Science. These databases were chosen for several reasons: Comprehensive Coverage: These databases host a vast collection of peer-reviewed papers, journal articles, and conference proceedings related to AI and power systems. Reliability and Credibility: The databases are well-established in the academic community, ensuring that the sources we reviewed were from credible, authoritative publications. Ease of Access: They provide quick and efficient access to papers that discuss recent advancements in AI applications for technical fields like power systems. Relevance to AI Research: Since AI is a highly technological and evolving field, electronic databases provide up-to-date knowledge and findings that are critical for this study . The scope of the search was carefully constructed to

ensure comprehensive coverage of the relevant literature. Search terms included combinations of keywords such as "AI in power systems," "fault diagnosis using ANN," "SVM for fault detection," and "Expert Systems in energy grids." These keywords were selected to encompass the wide range of AI techniques applied in power system fault diagnosis. To ensure that the review remained current, only studies published between 2005 and 2021 were included. This timeline was selected because AI techniques have advanced significantly in the past two decades, with a notable increase in power system applications over this period. The search parameters were also refined to focus on studies that employed AI-based diagnostic tools in the context of power system fault identification, predictive maintenance, and real-time monitoring. The studies were meticulously organized by both the year of publication and the diagnostic technique employed. This categorization allowed us to track the evolution of AI methods over time and compare their effectiveness in different contexts. It also enabled us to identify patterns and trends in AI adoption across different power system components. Each study was further examined for the specific faults diagnosed (e.g., transformer failures, line outages, phase imbalances), and the AI methods used for detection were recorded. Additionally, the common limitations and gaps in the research, such as computational constraints, data availability, and model accuracy, were noted.

3.2 Data Analysis Process

Once the relevant studies were gathered, a detailed analysis was performed to evaluate the effectiveness of the AI techniques used in power system fault diagnosis. The analysis covered the following areas: Search Terms and Parameters: A thorough exploration of the keywords used for the search, the inclusion and exclusion criteria applied to filter out irrelevant studies, and the rationale behind selecting certain studies for in-depth review. For instance, we used search terms like "machine learning in power systems," "neural networks for fault detection," and "SVM applications in grid diagnostics." These terms were chosen based on their relevance to the problem of fault identification and diagnosis using AI. Comparative Analysis of AI Techniques: The studies were grouped based on the AI techniques used—ANN, SVM, ES, and others. For each category, the performance of the AI method in diagnosing specific types of faults (e.g., short circuits, transformer overheating, etc.) was analyzed. We also compared the methods based on key performance indicators such as accuracy, processing speed, and robustness under different operating conditions. This allowed us to assess which AI techniques performed better in certain fault scenarios. Trends in AI Application: The time distribution of studies

(2005–2021) provided insights into the growing reliance on AI in the power sector. Earlier studies (2005–2010) focused on simpler AI models like rule-based systems and expert systems, while more recent works (2015–2021) explored advanced techniques like deep learning, reinforcement learning, and hybrid AI models. This progression reflects the increasing complexity of power systems and the need for more sophisticated tools to diagnose faults effectively. Additionally, issues related to the interpretability of AI models and their ability to function in real time were noted. Several studies pointed out that, while deep learning models can offer high accuracy, they are computationally expensive and require significant hardware resources for real-time deployment.

4. Results and Discussion

Numerous methodologies and techniques, including data mining, spectral density, graphical, MATLAB, and mathematical morphology, require a significant amount of research. This makes it possible to broaden the focus of study on artificial intelligence methods.

This study will outline some earlier studies on fault detection and methods for reducing and changing it. What this research has discovered, as demonstrated in Figs. 8.1, 8.2, and 8.3. Despite the topic's relevance and the volume of published studies and research on it, there aren't many references to it. We were forced in this approach by the lack. This study is the first of a series on the subject. The most recent paper reviewedon this topic belongs to February 2021, as shown in Figs. 8.1 and 8.2.

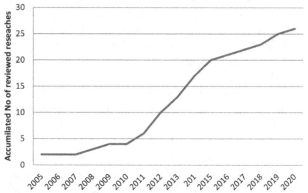

Fig. 8.1 Cumulative number of studies that have been examined each year

Figure 8.3 depicts various technologies of artificial intelligence used in different years. From Fig. 8.3 it is clear that other technologies of AI is used several time to investigate the Power System Fault Identification and Diagnosis rather than the ANN, SVM and ES.

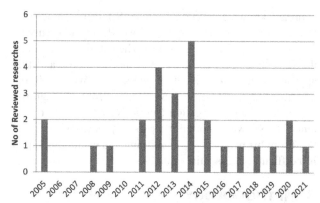

Fig. 8.2 Year-wise number of researches reviewed

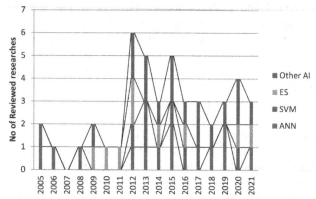

Fig. 8.3 Number of research and the method utilised

This section examines the many AI methods, such as machine learning, neural networks, and expert systems, that are employed for power system issue detection. It demonstrates their capacity for promptly identifying flaws and detecting anomalies in massive volumes of data.

4.1 Benefits of AI Integration

The advantages of using AI for fault detection and diagnostics in power systems are investigated. AI offers higher predictive ability, increased accuracy, quicker defect identification, and decreased downtime. These benefits help make electricity systems more reliable and stable overall. The difficulties and factors to be considered while using AI in power systems are highlighted. These include the need for knowledge in AI technology, algorithm complexity, and data availability and quality. It explored the Mitigation strategies.

4.2 Real-World Implementations

This section offers case studies and projects that demonstrate how AI has been successfully used to identify and diagnose power system faults in the real world. These instances demonstrate how AI can be used to enhance the robustness and operational efficacy of power systems.

The use of AI in power system failure identification and diagnosis has been examined in this research. A possible solution to the problems caused by faults and interruptions is provided by AI technology as the demand for dependable and effective power systems continues to rise. Collaboration between power system professionals and AI experts is crucial to properly utilizing AI's promise in this area. The continuing development and application of AI technologies will maintain the uninterrupted flow of electricity in our increasingly digitized and electrified world by enabling the identification and diagnosis of power system faults.

5. CONCLUSION

An important step in improving the dependability and resilience of power systems is the integration of artificial intelligence into fault identification and diagnostics in power systems. Large datasets can be processed quickly and effectively using AI technology, and they can also accurately and quickly diagnose problems. Despite these obstacles and concerns, AI offers a wide range of advantages. After many researches and numerous studies related to defect detection, treatment, categorization, and discovery, it became evident that there are numerous extensive and vital areas warranting further research and development. Artificial intelligence (AI) has emerged as a predominant focus due to its substantial advantages in terms of time and effort savings, as well as its capacity to consistently deliver superior results through self-learning and accelerated task completion. It is noteworthy that several artificial intelligence techniques have only received limited scrutiny in a few studies. As technology continues to progress, more flaws and defects will inevitably manifest themselves.

Hence, to stay in pace with technological advancements and reduce the time, effort, and resources required to rectify these issues, it is imperative for the scientific community and researchers to continue advancing the field of fault detection.

References

1. Rojas, R. (2013). Neural networks: a systematic introduction. Springer Science & Business Media.
2. Kovalev, G. F., & Lebedeva, L. M. (2019). Reliability of power systems (Vol. 1, p. 157). Springer International Publishing.
3. Mammadli, S. (2017). Financial time series prediction using artificial neural network based on Levenberg-Marquardt algorithm. Procedia computer science, 120, 602–607.
4. Dubey, D. (2022). Investigating the Use of Artificial Intelligence for Fault Detection in Electronic Circuits. Mathematical Statistician and Engineering Applications, 71(1), 313–324.
5. M. Ebid, "35 Years of (AI) in Geotechnical Engineering: State of the Art," Geotech. Geol. Eng., vol. 39, no. 2, pp. 637–690, 2021, doi: 10.1007/s10706-020-01536-7.
6. D. Chakraborty, H. Elhegazy, H. Elzarka, and L. Gutierrez, "A novel construction cost prediction model using hybrid natural and light gradient boosting," Adv. Eng. Informatics, vol. 46, no. April, p. 101201, 2020, doi: 10.1016/j.aei.2020.101201.
7. H. Elhegazy, A. M. Ebid, I. M. Mahdi, S. Y. Aboul Haggag, and I. A. Rashid, "Selecting optimum structural system for R.C. multi-story buildings considering direct cost," Structures, vol. 24, no. October 2019, pp. 296–303, 2020, doi: 10.1016/j.istruc.2020.01.039.

Note: All the figures in this chapter were made by the author.

Recent Trends in Engineering and Science for Resource Optimization and
Sustainable Development – Prof. (Dr.) Dorota Jelonek et al. (eds)
© 2024 Taylor & Francis Group, London, ISBN 978-1-032-98030-0

9

Using IoT and Machine Learning Together for Manufacturing Predictive Maintenance

Ansar Isak Sheikh[1]

Dept. of Computer Engineering, St. Vincent Pallotti College of
Engineering & Technology, Nagpur. India

T. Ch. Anil Kumar[2]

Vignan's Foundation for Science Technology and
Research, Vadlamudi, Guntur, India

Anshula Gupta[3]

Department of Computer Science,
KIET groups of Institutions, Ghaziabad

Satish G. Jangali[4]

School of Mechanical Engineering,
KLE Technological University, Hubballi, India

Amandeep Nagpal[5]

Lovely Professional University, Phagwara

P. S. Ranjit[6]

Professor, Department of Mechanical Engineering,
Aditya Engineering College Surampalem, Kakinada, India

Aleksandra Ptak[7]

Czestochowa University of Technology

▰▰▰ Abstract

For efficient production processes, manufacturing industries rely largely on machinery and equipment. Unplanned downtime brought on by equipment malfunctions might lead to considerable losses. By predicting when equipment needs maintenance, predictive maintenance has become a pro-active strategy to reduce these interruptions. When combined, the machine learning and Internet of Things present an unmatched opportunity to improve predictive maintenance. This article explores how IoT and machine learning might be combined in manufacturing to enhance maintenance procedures. A based-on data predictive maintenance solution for manufacturing production lines was created in this study. The technology uses real-time data from IoT sensors to try to spot warning signs of potential issues before they arise. As a result, it aids in problem solving by forewarning operators so that corrective action can be taken during a production stop. The current study assessed the effectiveness of the system using IoT data from the manufacturing system. The evaluation's results suggested that the predictive maintenance is considered as the successful method in identification of early warning indications of imminent faults and it has the ability to prevent several production halts.

[1]asheikh@stvincentngp.edu.in, [2]tcak_mech@vignan.ac.in, [3]anshula.gupta@kiet.edu, [4]jangalisatish@gmail.com, [5]amandeep.nagpal@lpu.co.in, [6]psranjit1234@gmail.com, [7]aleksandra.ptak@pcz.pl

DOI: 10.1201/9781003596721-9

■■■■■■ **Keywords**

IoT, Manufacturing industry, Predictive maintenance, Machine learning

1. INTRODUCTION

As a result of the present digital transformation, the usage of machine learning, big data, and the applications of AI are having an impact on our daily lives and enhancing the effectiveness of corporate processes across all industries. The fourth industrial revolution is centred on the artificial intelligence and Internet of Things, which are undergoing previously unheard-of technological and scientific advancements [1]. Due to the rapid improvements in data protocols, electronic devices, and technology, the Internet of Things has changed industrial operations by integrating systems and connecting people, devices, and systems. Manufacturing productivity is boosted because to connectivity and data interchange across production systems made possible by the Internet of Things, which is at the centre of the industry 4.0 revolution. Industry to cut back on maintenance expenses and make sure that operational management is sustainable [2]. The goal of preventative maintenance is to foresee future errors so that maintenance can be carried out before they occur. By extending the usable lives of the production, PdM has the ability to encourage sustainable production practises.

Predictive maintenance, however, can benefit from the usage of data-driven AI systems that incorporate information from IoT devices. This study aims to develop a predictive maintenance system that uses machine learning to accurately forecast potential production line issues in advance. A dataset was used to compare several solutions and identify the best effective model to handle this issue.

Fig. 9.1 IoT in predictive maintenance

In the context of manufacturing predictive maintenance, this article investigates the synergistic potential of combining the machine learning and IoT. The combination of real-time data from IoT sensors and the predictive powers of machine learning has the potential to completely transform manufacturing maintenance procedures. This study looks at the advantages, difficulties, and uses in real life of this potent combination.

1.1 Objective of Study

1. The purpose of this research is to develop a predictive maintenance system that, by employing machine learning techniques, accurately predicts future production line problems before such events occur.
2. To investigate how IoT and machine learning may be used to produce predictive maintenance.

2. LITERATURE REVIEW

The studies on the predictive maintenance of industrial machinery that have been published in the past by various researchers using various methodologies are summarised here.

For an implant-related maintenance task in semiconductor manufacture, Gian Antonio Susto suggested a method that has been shown to outperform traditional PvM techniques and a single SVM classifier alternative [3].

Predictive modelling uses the Industrial Internet of Things and data type components to glean valuable insights from machine data [4]. The researchers have looked into the usage of autoregressive integrated moving average prediction trauma plating equipment to forecast quality faults, downtime, and maintenance. In the industrial Internet of Things, machine learning has been shown to be a crucial part of quality control and management. It enhances performance and the production process. A real-world industrial group was used as an example in a distinct study by Marina Paolanti, where the concept was used experimentally and produced precise estimates [5]. Access to data that has been obtained through different sensors, machine PLCs, and communication protocols is made possible via the Data Analysis Tool. The suggested PdM approach uses Azure Machine Learning Studio to train a Random Forest technique to enable the implementation of dynamic decision rules for maintenance management.

In his study, K. Liulys demonstrates the importance of performing routine equipment checks to prevent breakdowns. Preventive maintenance, however, incurs many unnecessary and repetitive checks [6]. It results in higher prices that are frequently unaffordable [7]. A cutting-edge idea called predictive maintenance can be applied to this type of equipment to lower the cost of downtime and the number of inspections [8].

3. METHODOLOGY

In this study, the infrastructure was improved before the creation of the predictive maintenance system by setting

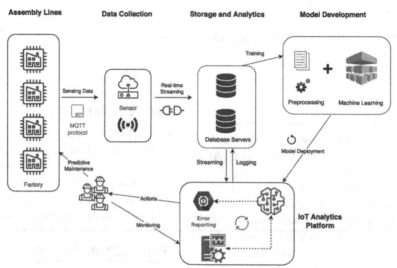

Fig. 9.2 Design of the predictive maintenance system

up the Internet of Things platform, tying equipment together with sensors and gateways, making the data constantly streaming from the production lines accessible, and combining system components.

3.1 System Architecture

The components of the suggested predictive maintenance system are shown in architectural detail in Fig. 9.2. Sensor data is gathered by the private cloud system, which is then maintained in a database to allow distributed systems to record data. "The Message Queuing Telemetry Transport (MQTT) protocol, a common messaging protocol developed by OASIS for the Internet of Things (IoT) [6], transforms data from the Internet of Things (IoT) [05] into a unified data format".

3.2 Data Collection

The information was acquired from baby nappy production lines in a real factory. The variables show the values of time-evolving device-generated data. The data contains signals from a number of Internet of Things (IoT) sensors, which are regularly integrated into a range of manufacturing line equipment and are frequently used to measure physical parameters such as movement, speed, temperature, electrical current, hoovering, and air pressure. Every 3 to 6 seconds, data was collected by IoT sensors controlling the production system. The dataset had 8,289,615 rows and 103 characteristics in total.

4. RESULTS AND DISCUSSION

The transformational potential of combining IoT and machine learning for manufacturing predictive maintenance has been highlighted in this article. This integrated

approach offers a clear route ahead as industrial companies continue to seek for increased efficiency and less downtime. For manufacturing companies to remain competitive in a market that is continually changing, it is essential that they investigate and implement these technologies.

95 sensor-generated variables about the manufacturing line and 8 data points concerning the type of failure were among the 103 elements in the data collected for the research project. The two failure categories were affected by data point information including timestamps, error types, uptime, downtime, and work shifts during stops. Additionally, based on the features' potential for prediction using the Random Forest approach, the study's Gini impurity measure was employed to estimate the variable importance. The difference in the sum of squared errors (SSE) is used to determine the variable's gini-based relevance if a variable is omitted. Top 10 significant features were determined across all variables, as shown in Fig. 9.3.

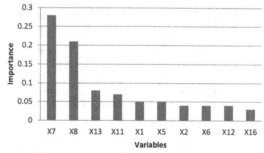

Fig. 9.3 Top ten variable importance

The R^2 (R-squared), Mean Absolute Error (MAE), Mean Absolute Percentage Error (MAPE), and Root Mean Squared Error (RMSE) metrics were used in the assessments to measure how well the algorithms predicted the outcomes. Table 9.1 contains this information.

Table 9.1 Precision of ML algorithms

Algorithms	MAE	MAPE	RSME	R^2
XGBoost	83.18	6.61	167.37	0.868
RF	50.79	4.72	156.14	0.871
MLP regressor	471.25	30.31	643.79	0.557
Gradient boosting	396.43	25.67	525.89	0.686
AdaBoost	749.85	48.36	910.58	0.285
SVR	670.34	44.66	887.05	0.297

Following assessments and comparisons, Random Forest produced the best results in the testing dataset, with an R2 value of 0.871. Even though the R2 score of 0.868 for the XGBoost approach was quite close to that of Random Forest, it fell short. The Neural Network model MLP Regressor with performed worse than the other three ensemble learning procedures and it was found to be inadequate for this prediction process. The Adaboost and SVR algorithms, with respective R2 values of 0.285 and 0.297, showed the poorest performance.

4.1 IoT in Manufacturing

In this section, the function of IoT in manufacturing is discussed in general terms, with a particular emphasis on how sensor-equipped devices can gather real-time data regarding machinery and procedures. It underscores how crucial IoT is for tracking machine health and performance.

4.2 Predictive Maintenance

The implementation of machine learning techniques to assess the enormous volumes of data produced by Internet of Things sensors is covered in the paper's section on predictive maintenance. Machine learning can suggest maintenance schedules, forecast equipment breakdowns, and improve maintenance tactics.

4.3 Machine Learning and IoT Integration

IoT and machine learning have many advantages when used together, including greater safety, less downtime, cost savings, and increased equipment reliability. Predicting maintenance requirements improves overall operational efficiency.

4.4 Challenges

Issues including data integration, security, and the necessity for qualified staff are all addressed. In order to address these issues and guarantee that IoT and machine learning are integrated seamlessly, the article investigates various techniques.

5. CONCLUSION

A revolutionary advancement in the field of industrial predictive maintenance is the combination of the Internet of Things and machine learning. The possibility exists to rethink maintenance methods and increase their effectiveness and efficiency by utilizing real-time data from IoT sensors and machine learning to forecast maintenance requirements. In this study, we developed a machine learning-based predictive maintenance solution for manufacturing environments. Using actual manufacturing IoT systems, we assessed the system's efficacy data. The expected useful time before failure is represented by the model's output. The evaluation findings show how precisely our suggested predictive maintenance system detects early warning signals of machinery breakdown using real-time sensor data, as well as how much it may help avoid potential production stops by suggesting preventive measures. According to our research, ensemble models for boosting and bagging generally perform well. We intend to use the technology for future work on various manufacturing lines in various environments.

References

1. Lee, J., Davari, H., Singh, J., & Pandhare, V. (2018). Industrial artificial intelligence for industry 4.0-based manufacturing systems. Manufacturing Letters, 18, 20–23.
2. Stock, T., & Seliger, G. (2016). Opportunities of sustainable manufacturing in industry 4.0. Procedia Cirp, 40, 536–541.
3. Gian Antonio Susto et al., "Machine Learning for Predictive Maintenance: A Multiple Classifier Approach," IEEE Transactions on Industrial Informatics, vol. 11, no. 3, pp. 812–820, 2015. [CrossRef] [Google Scholar] [Publisher Link]
4. Ameeth Kanawaday, and Aditya Sane, "Machine Learning for Predictive Maintenance of Industrial Machines Using IoT Sensor Data," in 2017 8th IEEE International Conference on Software Engineering and Service Science, pp. 87–90, 2017.
5. Marina Paolanti et al., "Machine Learning Approach for Predictive Maintenance in Industry 4.0," IEEE/ASME International Conference on Mechatronic and Embedded Systems and Applications, pp. 1–6, 2018.
6. Kaorlis Liulys, "Machine Learning Application in Predictive Maintenance," 2019 Open Conference of Electrical, Electronic and Information Sciences (es- Tream), pp. 1–4, 2019.
7. Pawełoszek, I., Kumar, N., & Solanki, U. (2022). Artificial intelligence, digital technologies and the future of law. Futurity Economics & Law, 2(2), 24–33. https://doi.org/10.57125/FEL.2022.06.25.03
8. Manpreet Singh Bhatia, Alok Aggarwal, Narendra Kumar. (2020). Speech-to-text conversion using gru and one hot vector encodings. Pal Arch's Journal of Archaeology of Egypt / Egyptology, 17(9), 8513–8524. Retrieved from https://archives.palarch.nl/index.php/jae/article/view/5796
9. Vinay Singh, Alok Aggarwal, Narendra Kumar, A. K. Saini. (2020). A Novel Approach for Pre-Validation, Auto Resiliency & Alert Notification for SVN To Git Migration Using Iot Devices. Pal Arch's Journal of Archaeology of Egypt / Egyptology, 17(9), 7131–7145. Retrieved from https://archives.palarch.nl/index.php/jae/article/view/5394

Note: All the figures and table in this chapter were made by the author.

*Recent Trends in Engineering and Science for Resource Optimization and
Sustainable Development – Prof. (Dr.) Dorota Jelonek et al. (eds)
© 2024 Taylor & Francis Group, London, ISBN 978-1-032-98030-0*

10

Evaluating Quantum Computing's Potential for Financial Engineering Optimisation

Deepali Virmani[1]

Vivekananda Institute of Professional Studies-
Technical Campus, School of Engineering and Technology

R. Anuradha

Department of Information Technology,
Institute of Aeronautical Engineering, Hyderabad, Telangana

Ahmad Y. A. Bani Ahmad[2]

Department of Accounting and Finance Science,
Faculty of Business, Middle East University, Amman 11831, Jordan

P. Satish kumar

Associate Professor, Department of Mechanical engineering,
School of Engineering, SR University, Warangal, Telangana

Ananda Ravuri[3]

Senior Software Engineer,
Intel Corporation, Hillsboro, Oregon 97124 USA

G. Poshamallu[4]

Assistant Professor, Department of ECE,
St. Martin's Engineering College Secunderabad, Telangana

Dejan Mircetic[5]

University of Novi Sad, Faculty of Technical Sciences, Serbia

Abstract

In terms of computing capacity, quantum computers are expected to surpass conventional computers throughout this decade and have a seismic impact on a number of commercial sectors, particularly finance. In actuality, it is anticipated that the financial industry would reap both immediate and long-term benefits from quantum computing. The current state of quantum computing for financial applications is thoroughly evaluated in this survey research, with a focus on stochastic modelling, optimisation, and machine learning. It demonstrates how these methods could aid in more rapidly and efficiently resolving financial issues including derivative pricing, risk modelling and fraud detection if they were updated to operate on a quantum computer. The possibility of quantum computing as a ground-breaking tool for streamlining financial engineering procedures is examined in this research. Growing interest in using quantum computing to solve these problems is being driven by the complexity of financial systems and the computational rigor of financial engineering problems. The

Corresponding author: [1]deepali.virmani@vips.edu, [2]aahmad@meu.edu.jo, [3]Ananda.ravuri@intel.com, [4]gaddi.gaddi.poshhamallu421@gmail.com, [5]dejanmircetic@gmail.com

DOI: 10.1201/9781003596721-10

effectiveness of quantum computing in advancing optimization techniques in the field of financial engineering is assessed in this paper.

Keywords

Machine learning, Quantum computing, Finance, Financial engineering

1. INTRODUCTION

A multidisciplinary area called financial engineering uses mathematical and computational methods to provide creative approaches to managing financial risks, improving portfolio performance, and enhancing trading tactics. But the complexity of financial products and the enormous volumes of data they require are pushing the boundaries of traditional computers. With its built-in parallelism and quantum characteristics, quantum computing has the potential to revolutionize how these problems are solved.

In contrast to modern classical computers, quantum computation uses a fundamentally different method of information processing and storage. The reason is because the information doesn't follow classical mechanics laws, but rather quantum physics principles. When quantum systems are properly isolated from their surroundings, quantum-mechanical effects usually only appear at incredibly tiny scales. However, given these conditions, building a quantum computer is challenging. The financial industry is anticipated to be the first to greatly benefit from quantum computing since many financial use cases may be presented as problems that may be solved by quantum algorithms suitable for future quantum computers, according to a McKinsey & Co. paper [1].

Unlike any classical computing device now in use, quantum computation has a much wider range of potential hardware platforms or physical realisations. Superconductors, trapped ions, neutral atoms, photons, and other physical systems have all been proposed, but no clear winner has yet emerged. Many businesses are vying to be the first to create a quantum computer that can execute algorithms useful in a manufacturing setting. The benefit of quantum computing could be a potential decrease in processing times and memory requirements. which could enable computations with scalability and precision that were previously impractical. Theoretically, a conventional computer is capable of doing any operation that a quantum computer is capable of performing.

The purpose of this work is to investigate how quantum computing can be used to solve optimization issues in financial engineering. We explore the special qualities of quantum computing, its potential benefits, and the particular difficulties it might help resolve in this field. This paper intends to shed light on the revolutionary possibilities that quantum technology offers for financial optimization by analyzing the synergy between quantum computing and financial engineering.

2. LITERATURE REVIEW

2.1 Quantum Optimisation

Numerous financial issues stem from optimisation issues, particularly combinatorial optimisation issues. The portfolio optimisation challenge, for instance, entails determining the best asset mix to invest in order to maximise projected returns while assuming manageable market risks [2]. To solve these issues, numerous traditional optimizers have been developed. However, the majority of NP-hard issues in financial optimisation. Conventional optimizers lose efficiency as the number of components to be optimised grows because they take longer to optimise and have a higher risk of hitting local minima. Thankfully, it is believed that quantum computing will provide a novel approach to combinatorial optimisation problems, enabling quicker and more efficient solution creation. A technique called adiabatic quantum computation is at the core of quantum optimisation [3].

2.2 Stochastic Quantum Modelling

Unpredictable events frequently occur in the financial markets, which are normally challenging to categorise by established models [4]. Stochastic modelling is investigated to assist in investing decisions in order to precisely evalu-

Fig. 10.1 Quantum computing application

ate these uncertainties; normally, the goal is to maximise return and minimise risk. To assess how uncertainty affects financial items, one statistical analysis-based method is [5] Monte Carlo. To simulate the effect of variability on a financial instrument, such as a stock, portfolio, or option, the stochastic technique is frequently used in finance. As a result, portfolio analysis, financial planning on a personal level, risks analysis, and derivatives pricing may all be done using Monte Carlo methods [6].

2.3 ML

Data-driven financial services have become a trend in the development of the sector due to the continued rise of financial historical data and the rising client desire for personalised consumption and financial management [7]. Machine learning is an effective method for extracting attributes from data and using a collection of generic architectures to address a range of issues [8]. It has been demonstrated that applying machine learning approaches rather than traditional ones to handle financial problems can produce better results [9, 10].

3. METHODOLOGY

We outline the specific applications of quantum computing in finance in this section.

3.1 Quantum Adiabatic Computing

The quantum adiabatic theorem is the foundation of the quantum computation model known as adiabatic quantum computing (AQC) [11]. This theorem asserts that if evolution from the initial nth eigenstate of the time-dependent Hamiltonian evolves slowly enough, an instantaneous nth eigenstate of a time-dependent Hamiltonian, H (t), will persist throughout time for all t [12]. The ground state of the final Hamiltonian Hf, whose fundamental state is attained, follows an adiabatic trajectory from an easy-to-prepare starting Hamiltonian Hi, which may be used to carry out a unitary operation.

3.2 Integrating Monte Carlo

Problems that are challenging to analyse or that require numerical approaches that don't scale well for multidimensional problems can be approximated using Monte Carlo techniques. For inference, integration, and optimisation, traditional Monte Carlo techniques have been employed. This subsection's main topic is Monte Carlo integration (MCI), which is essential to finance for risk assessment and pricing [12]. Considerthe number whose the function of this process at different times is what we wish to compute: $g(X_0, ..., X_T)$. Numerous sample paths are created, and g sample averages are calculated, in order to estimate the expectation. The assessment independent of the task di-

mension, error decays as $O\left(\dfrac{1}{\sqrt{N_s}}\right)$.

$$\mathbb{P}_l \,|0\rangle = \sum_{\omega \varepsilon \Omega} \sqrt{\mathrm{P}(\omega)} |(\omega)$$

3.3 Regression

Regression is the process of applying the training data set to fit a numerical function. This technique explains how value changes when features change, making it a crucial tool for economic forecasting. Finding a continuous function that substantially matches a set of N data points (x_i, y_i) is typically accomplished using least-squares fitting.

4. RESULTS AND DISCUSSION

Quantum computing offers a potentially effective method for solving some of the most complex problems in financial engineering, according to an assessment of its potential. The ongoing advancement of quantum hardware, software, and algorithms is necessary for quantum computing to have a revolutionary impact on financial engineering. It is anticipated that as quantum technology develops, among other financial engineering jobs, portfolio optimization, risk assessment, fraud detection, and asset pricing would all be considerably enhanced. Quantum computing has the potential to change financial decision-making and give companies and investors a competitive edge due to its speed and efficiency in exploring large solution areas.

In this method, the necessary eigenvalues for the rotations are first estimated using QPE. Despite the decrease in the number of controlled rotations, the loss of coherence means that this method no longer achieves the asymptotic speed-up offered by HHL.The QPE output distribution has been modified to increase efficiency and make it simpler to determine the eigenvalues, the authors created a novel method for scaling the matrix by a factor γ. The Quantinuum H1 device's experimental data are shown in Fig. 10.2 to demonstrate how the scaling technique has affected the outcomes.

Probability distributions for the eigenvalue estimates from the QCL-QPE run with values of 50 and 100 are shown on

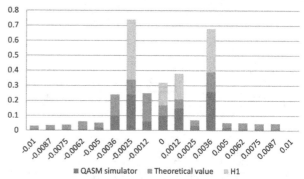

Fig. 10.2 Scaling technique graph

the left and right, respectively. The orange bar displays the assumed eigenvalues, which were normally determined via a numerical solution, while the dark orange bar display the outcomes from the Qiskit QASM simulator. The light orange bars with 1,000 shots on the left and 2,000 shots on the right represent the experiment's results on the H1 machine.

For a few compact portfolios, these procedures were run on the Quantinuum H1 device (Table 10.1). The state from the HHL and the linear issue solution developed traditionally underwent a controlled-SWAP test. Table 10.1 also shows the quantity of H1 two-qubit ZZMax gates employed.

Table 10.1 Comparison of the eigenvalue inversion's number of rotations

	NISQ-HHL	HHL	Relative change
Inner Product in simulation	.79	.60	41.70%
Inner Product on H1	.50	.45	10.49%
Rotation	125	140	-15.30%
Eigenvalue inversion	7	8	-14.06%

4.1 Discussion

A interesting convergence between cutting-edge technology and the intricate requirements of the finance industry may be seen in quantum computing's potential for financial engineering optimization. We highlight important findings from the assessment of the use of quantum computing in this field in the discussion section.

Quantum computers have shown a remarkable ability to solve challenging optimization issues in a way that classical computers are unable to match. This is especially true in the field of financial engineering, where complex mathematical models and enormous datasets are required for activities like portfolio optimization, risk management, and option pricing.

Quantum algorithms are being researched for their potential to improve financial optimization. Examples of these quantum algorithms include quantum annealing and quantum-inspired algorithms like the Variational Quantum Eigensolver (VQE).

4.2 Challenges

Despite the enormous potential of quantum computing, it is important to recognize its current restrictions. With a finite amount of qubits and a propensity for mistakes, quantum hardware is still in its infancy. Additionally, because quantum computations are so sensitive to their surroundings, the accuracy of the outputs may be impacted.

4.3 Hybrid Approaches

Many professionals are looking into hybrid systems that use classical and quantum computing to meet these constraints. This strategy makes use of the advantages of both

paradigms and is seen to be a practical way to implement quantum computing in the financial sector.

5. CONCLUSION

Financial services have a history of developing new technology and implementing them initially. The same characteristics also apply to quantum computers. It is anticipated that the financial sector will be the first to experience its global benefits because there are so many financial use cases that lend themselves to quantum computing and their potential to be solved successfully even in the face of estimates, both in the medium and long years as well as in the short term. We provide a thorough review of quantum computing for finance in this work, with a particular emphasis on quantum algorithms that might be able to solve difficult financial computations. The sector's contemporary environment has been thoroughly described.However, as quantum computing develops, the financial sector must be patient. Investment in research, development, and teamwork between quantum experts and financial engineers is necessary for quantum preparedness. In the interim, hybrid computing approaches that blend classical and quantum resources present a useful method to take use of quantum computing's advantages while minimizing its drawbacks.

In conclusion, quantum computing possesses the potential to significantly improve financial performance, risk management, and investment strategies, even though its integration into financial engineering optimization is yet in the future. Financial institutions must follow developments in quantum technology and get ready for a time when quantum computing will be essential to streamlining financial operations and improving decision-making skills.

References

1. Alexandre Ménard, Ivan Ostojic, Mark Patel, and Daniel Volz. A game plan for quantum computing, 2020
2. Zanjirdar M (2020) Overview of portfolio optimization models. Adv Math Finance Appl 5(4):419–435.
3. Albash T, Lidar DA (2018) Adiabatic quantum computation. Rev Mod Phys 90(1):015002.
4. HermanDet al (2022) Asurvey of quantum computing for finance. arXiv preprint arXiv:2201.02773.
5. Brandimarte, P. (2014). Handbook in Monte Carlo simulation: applications in financial engineering, risk management, and economics. John Wiley & Sons.
6. Orus R, Mugel S, Lizaso E (2019) Quantum computing for finance: overview and prospects. Rev Phys 4:100028
7. Zetsche DA, Buckley RP, Arner DW, Barberis JN (2017) From fintech to techfin: the regulatory challenges of data-driven finance. NYUJL Bus 14:393
8. Jordan MI, MitchellTM(2015) Machine learning: trends, perspectives, and prospects. Science 349(6245):255–260.

Note: All the figures and table in this chapter were made by the author.

Recent Trends in Engineering and Science for Resource Optimization and Sustainable Development – Prof. (Dr.) Dorota Jelonek et al. (eds)
© 2024 Taylor & Francis Group, London, ISBN 978-1-032-98030-0

11

Blockchain and AI Integration for Efficient and Secure Supply Chain Management

Manu Kumar Misra[1]
Staff Software Engineer, Walmart Global Tech, India

Pallabi Mukherjee[2]
Associate Professor, IPS Academy,
Institute of Business Management and Research,
Indore, Madhya Pradesh

Sanju Mahawar[3]
IPS Academy, Institute of Business Management and
Research Indore, Madhya Pradesh

K. Anusha[4]
Koneru Lakshmaiah Education Foundation,
Guntur, Andhra Pradesh, India

P. vamshi Krishna[5]
ST. Martins Engineering College Secunderabad,
Telangana

Kali Charan Modak[6]
IPS Academy, Institute of Business Management and
Research, Indore, Madhya Pradesh

Ike Umejesi[7]
University of Fort Hare, South Africa

Abstract

Integration of artificial intelligence and blockchain has drawn more attention recently because it can improve the security, effectiveness, and efficiency of programmes in commercial settings marked by instability, unpredictability, complexity, and ambiguity. An innovative method for improving the effectiveness and security of supply chain management has evolved through the merging of blockchain technology with artificial intelligence (AI). In order to transform supply chain operations, this study examines the synergistic potential of fusing blockchain's immutable and transparent ledger with AI's data analysis and prediction capabilities. The paper explores the advantages, difficulties, and real-world applications of an integrated strategy, illuminating its capacity to address the complexity and vulnerabilities present in contemporary supply

[1]manu706@gmail.com, [2]pallabimukherjee@ipsacademy.org, [3]sanju746@gmail.com, [4]kanagalanusha@gmail.com, [5]vamshi.pearala@gmail.com, [6]kali.modak@gmail.com, [7]ikeumejesi@gmail.com

DOI: 10.1201/9781003596721-11

chains. It is well known that artificial intelligence and blockchain technology both considerably benefit the supply chain. Among other things, these technologies improve product traceability, quicker and more affordable product delivery, and information and process resilience. In the context of supply chains, this paper undertakes a cutting-edge examination of blockchain and artificial intelligence. By conceptually and experimentally merging blockchain with AI, the study done for this work has found pertinent research efforts that have enriched and accumulated intellectual richness in the supply chain discipline.

▬▬▬▬▬▬ Keywords

AI, Blockchain, Supply chain, Security

1. Introduction

Supply chain management, which links producers, suppliers, logistics companies, and customers, is essential to worldwide trade. However, the size and complexity of modern supply chains have created a number of problems, such as problems with transparency, traceability, fake goods, and supply chain interruptions. The integration of blockchain technology with artificial intelligence has drawn a lot of interest as a potential solution to these problems since it can offer both efficiency and security. Blockchain is a perfect instrument for tracking the movement of commodities and confirming their validity since, thanks to distributed ledger technology, it provides an indelible and transparent record of transactions. On the other side, artificial intelligence excels at data analysis, predictive modeling, and automation, enabling better decision-making and proactive management of supply chain activities. In the contemporary corporate climate, traceability is becoming into a demand and competitive edge across numerous supply chain industries. Stakeholders cannot precisely assess and confirm an item's genuine value in the absence of supply chain transparency. Maintaining supply chain traceability is more difficult due to the expense of working with intermediaries as well as their transparency and dependability, which can result in competitive issues with reputation and strategy [1].

By automating every stage of the process, the use of artificial intelligence technology in the blockchain system has the ability to completely transform the supply chain. It is possible to extract helpful details from past purchase data and other sources using a hybrid AI and blockchain technique. This makes it possible to identify data features and carry out predictive analysis tasks like demand and sales forecasting in the future [2].In the context of supply chain management, the goal of this article is to investigate the convergence of blockchain and AI. It will look into how the security, effectiveness, and overall performance of supply chain activities can be improved by this integration. This study aims to offer insights into the possibility for addressing the complex issues supply chains face in the current

day by analyzing the complementary strengths of both technologies.

To the best of our knowledge, despite its importance and relevance at the time this research is being done, there isn't a comprehensive literature review of studies on the integration of AI and blockchain for supply chain. By filling this gap through a critical analysis of such works, our research advances the area of operations management. It includes literature review, methodology and result and discussion.

1.1 Research Objective

1. The project's objective is to find fresh research on the application of blockchain and AI in supply chains.

2. This study aims to uncover current blockchain and artificial intelligence application cases in the supply chain.

3. To sketch out prospective research avenues combining blockchain and AI for upcoming investigations.

2. Literature Review

Running supply chains smoothly depends on the secure, effective, and efficient sharing of information. Every step of the supply chain must be efficient and transparent, and stakeholders must have faith in one another. In order for supply chains to be sustainable, they must also improve their resilience and traceability while becoming more flexible and responsive. The success of the supply chain will once again depend heavily on innovation and technology [3]. Blockchain has emerged as a key tool in this context, capable of enhancing the adaptability and agility of supply chain operations [4]. In the first place, the term "blockchain" is credited to Satoshi Nakamoto. Author described the idea for the first peer-to-peer electronic payment system in a paper that was released in 2008 [5]. This system would be driven by the so-called "bit-coin" form of digital currency. Blocks that are connected to one another are used to record transactions. A chain of related transactions is referred to as a "blockchain" in this context. Blockchain reduces legal and transaction costs by allowing parties that do not know one another to transact securely without the requirement for a third party that is centrally trusted [6]. The term "distributed ledger" was created because records can be held in several locations and shared with numerous parties.

Some of the aforementioned issues may be resolved using artificial intelligence (AI). In fact, it is predicted that combining blockchain with AI would offer a variety of important benefits, including more reliable deliveries [7]. Without the aid of a centralised authority or third-party middlemen, parties can use such integration to share substantial volumes of data for analysis [8], learning [9], and decision-making [10].

3. METHODOLOGY

3.1 Data Collection

In this study, we looked into how blockchain and AI technologies affected SCM performance. The exact same method was applied. Only secondary data from peer-reviewed academic papers from the SCOPUS database was used.

With choices like "Document search," "Author search," "Affiliation search," and "Advanced search" for a variety of parameters like "Article Title, Abstract, Keywords," "Source Title," and "Year of Publication," among others, Scopus has a robust search engine. Between 2016 and 2021 with 285 document result, research was done by looking up findings from publications in various industries. The details that were gathered were used to create the source type, topic, and kind of document.

3.2 Bibliometric Analysis

In this study, bibliometric analysis of works on BT supply chain integration is done with a focus on traceability, logistics, and sustainability. The use of bibliometric analysis in this research is justified by the following reasons: Comparatively speaking, bibliometric analysis is more scalable and reliable than other methods (like content analysis) [11]. Bibliometric techniques can offer important and comprehensive information by providing an extensive and complete evaluation of the many relationships (such as citations, keywords, and co-citations) related with the publications under consideration. By utilising bibliometric approaches, researchers may quickly and easily visualise key study areas.

3.3 Data Analysis

Only journal papers that were complete, published, and subjected to peer review were chosen from among the 285 records after a review, leading to 77 research publications. The 75 research articles were carefully examined to see if they met the criteria and provided a clear discussion of the integration of AI and blockchain for supply chain after a set of inclusion and exclusion criteria were developed. The 45 research articles that made up the final pool after this stage were then thematically evaluated.

4. RESULTS AND DISCUSSION

Bibliometric analysis serves as an invaluable research methodology for uncovering worldwide research trends and forecasting forthcoming research directions. In the context of this study, bibliometric investigation is employed to examine the attributes and evolutions in research within the domain of supply chain studies involving the integration of artificial intelligence and blockchain. To examine the results, statistical data is generated.

As depicted in Fig. 11.1, additional studies on blockchain and AI for supply chains have been conducted throughout time. The 285 documents were all published after 2017, which means that, at least according to the Scopus database, no papers combining the three keywords existed before this year. This is significant information.

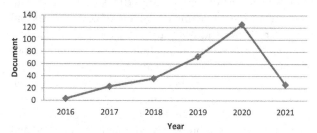

Fig. 11.1   Annual output in science

The annual document volume is shown by information source in Fig. 11.2. The top five publications for supply chain publications on blockchain and artificial intelligence can be shown.

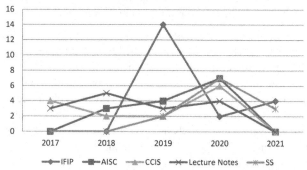

Fig. 11.2   Documents per source, per year

Figure 11.3 shows the comparison of document counts for 12 territories/nations which has numerous publications. Using the nation of the matching author, the nations of or-

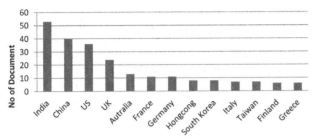

Fig. 11.3 Documents by country

igin were identified. The United States comes in third with 36 publications, and then comes China with 40, and India at the top of the list with 53 publications.

4.1 Discussion

A successful approach to address the complexities and vulnerabilities in the contemporary supply chain landscape is the combination of AI and blockchain in supply chain management. The discussion of this integration must take into consideration the following issues:

4.2 Improved Traceability and Transparency

The immutable ledger provided by blockchain technology guarantees that every transaction in the supply chain is traceable and transparent. By enhancing traceability, this transparency enables stakeholders to monitor the movement of items and confirm their legitimacy along the whole supply chain. By using this data to find trends and abnormalities, AI's data analysis capabilities can also be used to support real-time monitoring and decision-making.

4.3 Prevention of Inferior Products

The existence of counterfeit or inferior products is one of the main difficulties in supply chain management.

This problem can be addressed with the use of blockchain, which can verify the authenticity of products, and AI, which can recognize counterfeit patterns and guarantee that consumers receive genuine goods.

4.4 Supply Chain Optimization

By using AI's predictive skills, one may estimate demand, improve inventory control, and improve logistical processes. AI can offer in-the-moment insights that help with resource allocation and proactive decision-making when combined with blockchain data.

4.5 Trust and Data Security

The immutability and decentralized nature of blockchain technology ensure data security, which is essential for protecting sensitive supply chain data. The supply chain activities' overall security can be further improved by the AI

algorithms that examine this data's potential security flaws and vulnerabilities.

Supply chain management has a bright future thanks to the combination of AI and blockchain, which will help it meet the demands of the digital era. Organizations should make investments in the creation of strong blockchain infrastructure and AI capabilities, encouraging cooperation and interoperability across many stakeholders, in order to fully exploit the potential of this integration. This strategy benefits both businesses and consumers by increasing efficiency while also enhancing supply chain security and trust.

5. CONCLUSION

The use of blockchain technology and artificial intelligence in supply chain management has enormous potential for streamlining processes, boosting security, and ultimately raising supply chains' overall effectiveness. It offers the potential to deal with the industry's long-standing problems with traceability, preventing counterfeit goods, and operating efficiency. It is anticipated that this connection will become more seamless and available to a wider spectrum of supply chain stakeholders as blockchain and AI technologies continue to advance. Increased transparency, lower operating costs, better risk management, and greater consumer and partner trust are all advantages for organizations that adopt this strategy. Our findings show that conceptual instead of empirical works make up the vast majority of the evaluated works. To put it another way, there aren't sufficient research studies that concentrate on integrating blockchain-AI applications in practical settings. As a result, discussions on the subject lose some of their complete authority. Both blockchain technology and its integration with other cutting-edge technologies like artificial intelligence (AI) are still in their infancy. In order to see any benefits and long-term consequences in practise, many problems must be fixed. As a result, empirical studies that concentrate on the real-world use of blockchain technology powered by AI for supply chains and its effect on long-term performance present an intriguing and attractive future study topic.

References

1. Saberi, S., Kouhizadeh, M., Sarkis, J., & Shen, L. (2019). Blockchain technology and its relationships to sustainable supply chain management. International Journal of Production Research, 57(7), 2117–2135.
2. Zhang, Z., Song, X., Liu, L.,Yin, J.,Wang,Y.,&Lan, D. (2021b). RecentAdvances in blockchain and artificial intelligence integration: Feasibility analysis, research issues, applications, challenges, and future work. Security and Communication Networks, 2021, 9991535.

3. Baucherel, K. (2018). Blockchain from hype to help. IT-NOW, 60(4), 4–7.

4. Cole, R., Stevenson, M., & Aitken, J. (2019). Blockchain technology: Implications for operations and supply chain management. Supply Chain Management: International Journal, 24(4), 469–483.

5. Nakamoto, S. (2008). Bitcoin: A peer-to-peer electronic cash system. Retrieved from https://bitcoin.org/en/ bitcoin-paper.

6. Pilkington, M. (2016). Blockchain technology: Principles and applications. In F.X.Olleros&M. Zhegu (Eds.), Research handbook on digital transformations (pp. 225–253). Edward Elgar.

7. Odekanle, E. L., Fakinle, B. S., Falowo, O. A., &Odejobi, O. J. (2022). Challenges and benefits of combining AI with blockchain for sustainable environment. In K. Kaushik, A. Tayal, S. Dahiya,&A. O. Salau (Eds.), Sustainable and advanced applications of blockchain in smart computational technologies (pp. 43–62). Chapman and Hall/CRC.

Note: All the figures in this chapter were made by the author.

*Recent Trends in Engineering and Science for Resource Optimization and
Sustainable Development – Prof. (Dr.) Dorota Jelonek et al. (eds)
© 2024 Taylor & Francis Group, London, ISBN 978-1-032-98030-0*

12

Optimising the Allocation of Resources in Cloud Computing Using Machine Learning

Zameer Ahmed Adhoni[1]

Assistant Professor,
Department of Computer Science and Engineering,
School of Technology Gitam University,
Bangalore, India

Syeda Imrana Fatima[2]

Research Scholar,
Department of Computer Science and Engineering,
School of Technology Gitam University,
Hyderabad, India

Rajdeep Singh[3]

Consultant, Madhya Pradesh State Policy &
Planning Commission, India

S. Shalini[4]

Assistant Professor, Department of Mathematics,
J.N.N Institute of Engineering, India

Ankur Jain[5]

Computer Science & Engineering,
IFTM University, Moradabad, Uttar Pradesh, IN

Shaik Rehana Banu[6]

Post Doctoral Fellowship, Business Management,
Lincoln University College, Malaysia

Damian Dziembek[7]

Faculty of Management,
Czestochowa University of Technology, Poland

Abstract

In the digital age, optimizing resource allocation in cloud computing is of the utmost importance. In order to improve resource allocation in cloud systems, this study investigates the integration of machine learning approaches. With its

[1]zadhoni@gitam.edu, [2]syeda.imrana89@gmail.com, [3]rajdeep.singh3@mp.gov.in, [4]shalini.s.pandian@gmail.com, [5]ankur1101@gmail.com,
[6]drshaikrehanabanu@gmail.com, [7]damian.dziembek@pcz.pl

DOI: 10.1201/9781003596721-12

capacity for adaptation and learning from data, machine learning presents exciting possibilities for addressing the dynamic and intricate nature of cloud resource management. In order to increase the effectiveness and cost-effectiveness of resource allocation in cloud computing, this paper explores the advantages, difficulties, and practical applications of employing machine learning. The challenge of resource distribution in auctions is difficult for cloud computing. But because it is NP-hard, the resource allocation problem cannot be addressed in polynomial time. In this investigation, we define, formulate, and assess the multi-dimensional cloud resource allocation problem. Additionally, we provide two methods that employ both logistic and linear regressions to forecast resource allocation. The prediction model may ensure that the resource utilisation and allocation accuracy in the practical solution are startlingly close to those of the ideal allocation solution by learning a small-scale training set. The outcomes of the experiments demonstrate that the suggested approach has a positive impact on resource distribution in cloud computing.

■■■■■■■ **Keywords**

Resource allocation, Machine learning, Cloud computing, Accuracy

1. INTRODUCTION

The way people and business access and manage computational resources are changed by cloud computing. The ever-changing workloads, variety of apps, and unpredictable demands make it difficult to ensure appropriate resource allocation in cloud systems. Traditional static allocation techniques frequently lack in terms of optimization and adaptability. As a dynamic and data-driven technology, machine learning presents a fresh approach to handling the complex issues of resource allocation in cloud computing. Machine learning models can adaptively allocate resources based on demand, performance needs, and cost-effectiveness by utilizing historical data and real-time insights. This study examines the potential benefits, difficulties, and real-world applications of machine learning's incorporation into cloud resource allocation. A well-liked research topic in cloud computing is auction-based resource allocation, which has the potential to significantly improve social welfare and the usage of resource providers [1]. According to some research, a single-dimensional resource allocation issue may be NP-hard and impracticable to solve in polynomial time. Multiple resource categories must be taken into account while solving multi-dimensional resource allocation problems. A precise approach, such as PTAS or heuristic algorithms, or an efficient approach, such integer programming, can be used to address the resource allocation problem [2]. The ideal technique will struggle because of its poor processing efficiency and poor allocation accuracy because resource allocation is NP-hard and requires tiny data sets. After identifying the best allocation for user demands in a limited training set, the main idea is to match the optimum allocation using machine learning techniques. To ensure resource utilization, allocation accuracy and social welfare outcomes that come very near to allocating the optimal solutions, the successful users could be chosen using the final predict model.

The study assesses how machine learning might help improve cloud computing performance overall, save operational costs, and better allocate resources. Utilizing machine learning techniques allows for the optimization of resource allocation decisions, ensuring that computing resources are efficiently allocated while upholding service level agreements. This study addresses the changing requirements of contemporary computing environments by providing insightful information about the transformative potential of machine learning for cloud resource management.

1.1 Objective of Study

1. This study's primary goal is to use machine learning to optimise the allocation of resources in cloud computing.

2. Objective of this study is to determine in small training sets the optimal allocation strategy for a user's needs, and then to fit the optimal allocation using machine learning methods.

2. LITERATURE REVIEW

The Resource Allocation problem with an auction, according to Angelelli et al., is an optimisation problem where m customers offer their resource requirements together with the corresponding value [3]. The objective is to raise societal welfare and the income of resource providers without using up all of the available resources. The NP-hard knapsack problem is analogous to resource allocation. There are numerous approaches for finding the precise optimal answer to resource allocation issues. Integer programming was employed by some studies to resolve the resource allocation issue [4]. The similar issue was addressed by Mashayekhy et al. using a dynamic programming approach [5]. Some investigations transformed a combinatorial resource auction into a winner-decision issue to ascertain the precise response [6]. The best solution has occasionally been discovered in study using monotone branch-and-bound search. The amount of time needed for calculations grows exponentially as the number of users, resource kinds, and user requirements increase, despite the fact that the problem is modest and these tactics can be used to find the best solution. As a result, a better algorithm is needed for practical application.

There are numerous approaches to resolving the approximation of resource allocation, and they can be broadly categorised as heuristic algorithms and Polynomial Time Approximation Scheme (PTAS) algorithms. In order to address the issue of managing the resources of diverse physical devices, Liu et al. proposed an n-approximation technique [7]. A PTAS algorithm was proposed by Mashayekhy et al. to address multitask scheduling issue [8].

3. METHODOLOGY

Regression and classification methods based on machine learning are available from us for the multi-dimensional cloud resource distribution in auctions [9]. The fundamental principles involve choosing a subset of all user requirements [10], determining the optimal payment price and allocation solutions [11], fitting the best allocation solution using logistic and linear regression [12], and then utilising the learnt model to predict all user requirements. To match the optimum allocation method, we provide two separate machine learning algorithms in this study. Both the logistic prediction algorithm of resource allocation based on regression and the linear regression-based resource allocation prediction method use regression to estimate resource allocation.

4. RESULTS AND DISCUSSION

In the DAS-2 data set, we chose 6,000 user request records as the total sample set. We then separated these records into 7 training sets and 3 cross-validation sets. Each collection has 600 entries for user requests.

4.1 Result with resource allocation prediction

According to Fig. 12.1, when it comes to addressing societal welfare, the G-VMPAC-II-ALLOC algorithm performs worse than the Linear-ALLOC and Logistic-ALLOC algorithms. It is demonstrated that a machine learning algorithm may be able to match a prospective model for the optimum allocation technique. In terms of social welfare, the Linear-ALLOC method's results are quite close to the optimal allocation strategy.

The prediction accuracy is shown in Fig. 12.2. As can be observed, the greedy method's forecast accuracy for

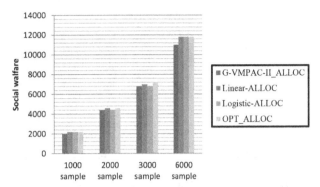

Fig. 12.1 Comparison of social welfare

G-VMPAC-II-ALLOC is lower than 91%, meaning that more than 11% of users should have their resources allocated but aren't. The accuracy of the machine learning-based prediction allocation algorithms is quite high, averaging over 96%. The Linear-ALLOC algorithm's prediction accuracy for these is greater than 99%. The accuracy rate can demonstrate the algorithm's fairness, which is a crucial factor in resource allocation.

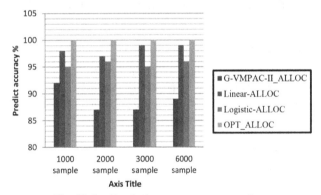

Fig. 12.2 Prediction accuracy comparison

The comparison of resource usages is shown in Fig. 12.3. The ideal solution uses up all of the CPU and storage capabilities within a certain resource capacity. Unlike the greedy-based allocation approach G-VMPAC-II-ALLOC, the machine learning-based strategy uses the resources in a way that is extremely close to the perfect solution.

To solve for the ideal payment price using Linear-ALLOC, however, necessitates the usage of VCG, which additionally requires some processing time. This is because the Linear-ALLOC cost function has the optimal payment price feature, whereas the Logistic-ALLOC method just has one dimension.

4.2 Discussion

A promising route to effective, flexible, and economical resource management is the integration of machine learning into resource allocation in cloud computing. This innova-

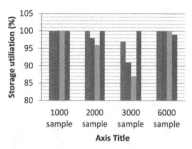 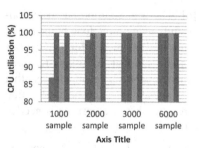

Fig. 12.3   Comparison of resource usage

tive strategy recognizes the dynamic and intricate nature of contemporary cloud settings, where workloads and requirements are always changing.

The following are the main conclusions from this investigation:

4.3 Enhanced Efficiency

Machine learning makes it possible for cloud systems to adapt in real-time, allocating resources in a way that is both effective and sensitive to changes in demand. The overall system efficiency is increased as a result of this adaptability.

4.4 Cost Savings

Machine learning can help to optimize resource allocation, which will result in lower operational expenses. Cloud providers can increase cost-effectiveness by precisely matching resources to workload demands.

4.5 Service Quality

The incorporation of machine learning enables the fine-tuning of decisions on the distribution of resources in order to preserve or enhance service quality. This is essential for ensuring that services and applications perform as expected.

The necessity for high-quality training data, model interpretability, and the dynamic nature of machine learning algorithms are some of the hurdles that must be overcome despite the enormous promise of machine learning in cloud resource allocation.

5. CONCLUSION

In order to address regression or classification machine learning issues, the multidimensional cloud computing resource allocation challenge can be reduced to one dimension. The algorithm may very well produce a workable solution that is somewhat close to the ideal answer in terms of utilisation of resources and social benefit distribution accuracy by learning from the training data. It has been demonstrated that a flawless resource allocation

model may exist, offering a cutting-edge method for carrying out this task in multi-dimensional cloud computing. The machine learning-based resource allocation method's compliance with the auction mechanism's strategy proof was not examined in this work. One of the main goals of our present study is this.It is anticipated that resource allocation in cloud computing will become more sophisticated and effective as machine learning techniques develop. For even better resource management, the combination of more complex models, reinforcement learning, and predictive analytics shows promise.

References

1. Barman, S., &Ligett, K. (2015). Finding any nontrivial coarse correlated equilibrium is hard. ACM SIGecom Exchanges, 14(1), 76–79.
2. Zhang, J., Xie, N., Zhang, X., Yue, K., Li, W., & Kumar, D. (2018). Machine learning based resource allocation of cloud computing in auction. Comput. Mater. Continua, 56(1), 123–135.
3. Angelelli, E., Bianchessi, N., & Filippi, C. (2014). Optimal interval scheduling with a resource constraint. Computers & operations research, 51, 268–281.
4. Zhang, X., Huang, Z., Wu, C., Li, Z., & Lau, F. C. (2016). Online auctions in IaaS clouds: Welfare and profit maximization with server costs. IEEE/ACM Transactions On Networking, 25(2), 1034–1047.
5. Mashayekhy, L.; Fisher, N.; Grosu, D. (2016): Truthful mechanisms for competitive reward-based scheduling. IEEE Transactions on Computers, vol. 65, no. 7, pp. 2299–2312.
6. Wu, Q.; Hao, J. K. (2016): A clique-based exact method for optimal winner determination in combinatorial auctions. Information Sciences, vol. 334, pp. 103–121.
7. Liu, X.; Li, W.; Zhang, X. (2017): Strategy-proof mechanism for provisioning and allocation virtual machines in heterogeneous clouds. IEEE Transactions on Parallel and Distributed Systems, pp. 1.
8. Mashayekhy, L.; Nejad, M. M.; Grosu, D. (2015): A ptas mechanism for provisioning and allocation of heterogeneous cloud resources. IEEE Transactions on Parallel and Distributed Systems, vol. 26, no. 9, pp. 2386-2399.

Note: All the figures in this chapter were made by the author.

Recent Trends in Engineering and Science for Resource Optimization and
Sustainable Development – Prof. (Dr.) Dorota Jelonek et al. (eds)
© 2024 Taylor & Francis Group, London, ISBN 978-1-032-98030-0

13

Using 5G for Virtual Reality Applications in Entertainment and Gaming

K. K. Ramachandran[1]
Director/Professor: Management/Commerce/
International Business, DR G R D College of Science, India

K. K. Lakshmi[2]
Head of the Department of English,
Dwaraka Doss Goverdhan Doss Vaishnav College,
Chennai, India

M Raja[3]
Department of Multimedia, VIT School of Design,
Vellore Institute of Technology, Vellore, India

J. Venkatarangan[4]
Department of CSE, St. Martin's Engineering College,
Secunderabad, India

Subhrajit Chanda
Assistant Professor, Jindal Global Law School,
OP Jindal Global University, India

Meerjumla Govind Raj[5]
Department of ECE,
St. Martin's Engineering College Secunderabad, India

Leszek Ziora[6]
Czestochowa University of Technology, Poland

■■■■■■■ **Abstract**

In recent years, virtual reality has started to use the fast speeds of wireless networks and data streaming technologies. However, due to limitations like bandwidth and latency, we are currently unable to create high-fidelity telepresence and collaborative virtual reality applications. Fortunately, these problems are known to engineers and academics, who are working to create 5G networks to ease the transition to the next generation of virtual interfaces. A significant advancement in immersive experiences will result from the combination of 5G technology and virtual reality (VR) applications in entertainment and gaming. In order to improve the performance and usability of VR in the entertainment and gaming industries, this research examines the transformational potential of using 5G networks. It explores the advantages,

Corresponding author: [1]dr.k.k.ramachandran@gmail.com, [2]sabarishnanda@gmail.com, [3]raja.m@vit.ac.in, [4]venkatarangan1986@gmail.com, [5]mgovindrajmtech@gmail.com, [6]leszek.ziora@pcz.pl

DOI: 10.1201/9781003596721-13

difficulties, and practical applications of this synergy and offers insights into how 5G is changing how entertainment and gaming are experienced in the future. This paper covers current virtual reality communications technologies and discusses current initiatives to construct a reliable, all-encompassing 5G network to support meeting the demands of virtual applications. On the basis of this new network infrastructure, we address application requirements in industries like entertainment and gaming and offer guidance and future growth paths.

■■■■■■■ **Keywords**

5G, Virtual reality, Latency, Entertainment, Gaming

1. INTRODUCTION

The gaming and entertainment industries are entering a new age of innovation and revolution thanks to the combination of 5G technology and virtual reality (VR) applications. This fusion of two cutting-edge technologies is revolutionizing how we interact with digital material by bringing never-before-seen levels of accessibility, interactivity, and immersion. The combination of 5G and VR is changing the limits of entertainment and gaming in the current digital era, where seamless, high-quality experiences are crucial.

Due to its astounding data transfer speeds, low latency, and increased network capacity, 5G, the fifth generation of mobile communication technology, has emerged as a game-changer. These qualities have created the conditions for immersive VR experiences, which were previously constrained by limitations of prior network generations.

With the advent of virtual reality, which immerses people in digital settings, completely new levels of interactive experiences are now possible. However, high-speed, low-latency connectivity is necessary for VR to reach its full potential, which is where 5G technology comes into play. With their incredible data transfer rates and low latency, 5G networks have created exciting new opportunities for advancing VR applications in the entertainment and gaming industries.

The First, Second, Third, and Fourth Generations of mobile network technologies all helped shape the Fifth-Generation mobile network technology, which aims to provide more connectivity than ever before. It was constructed with a greater capacity to deliver next-generation user experiences, offer new services, and enable new technologies. Its air interface deployment tactics are more effective and unified. 5G has the potential to expand the mobile ecosystem into new domains due to its fast speeds, extremely low latency, and higher reliability [1]. 5G will have an impact on remote healthcare, safer mobility, digitalized logistics, multimedia, entertainment, and gaming.

Virtual reality-based technology is becoming a more important type of entertainment. The term "virtual reality" (VR) refers to a computer simulation system that was developed by combining a number of related technologies, such as computer graphics and simulation, stereoscopic displays, sensor networks, voice input and output, and spatial positioning [2]. The ability of this technology to generate a virtual environment, give the user the appearance that everything in the environment is real, and allow interaction with items in the environment by simulating the user's senses of hearing, vision, and touch is its most remarkable feature.

Gaming and entertainment, particularly mobile and location-based games, is one of the first and most obvious sectors to gain from better network access. Pokemon Go has shown how VR can bring people together in a very interactive way, and 5G will improve the network-dependent mechanisms that support player localization, interactions with the outside world, as well as the sharing and cooperative manipulation of VR information. Low latency haptic devices and controllers that may be used outside will enable a flood of tools, sporting goods, (safe) virtual weapons, and magical skills. The only thing stopping them is human imagination. The development and sale of these kinds of virtual goods will probably be managed by micro transaction business methods, much how virtual purchases in MMORPGs and other games are currently managed [3].

The sections that follow examine the evolution of virtual reality (VR) and some of its most exciting applications in entertainment and gaming, provide an overview of major recent developments in the area, and then speculate on possible applications for 5G technology.

2. LITERATURE REVIEW

A new international wireless standard is set by 5G. Greater effectiveness and performance provide new user experiences and linkages to new sectors. Additionally, sub-6 GHz technology (for broad coverage and bandwidth) and mmWave technology (for quicker speeds over shorter distances) are used in 5G. Wider bandwidths will be supported by 5G due to an increased usage of spectrum resources [4].

Both mm-waves and lower bands can be used for 5G operations. Low latency, multi-Gbps throughput, and enormous capacity are brought by this. Many game creators view 5G as a pillar technology that will support advancements and the industry's expansion in the future.

Games that once required a keyboard, mouse, and fingers may now be played using the head, hands, feet, or even the complete body thanks to virtual reality technology [5]. Users can see the surrounding game sceneries in 360 degrees and take actions like walking back and squatting after opening the game interface and donning the virtual reality device. Another notable application of virtual reality technology in the gaming business is gaming wearable technology. The term "wearable technology" describes a broad range of devices that can be used as body jewellery [6].

These gadgets could be built into clothing or be wireless with microprocessors and Internet connections. Wearable technology combines virtual reality and augmented reality to produce an experience similar to a halo deck in the market for gaming accessories. The 5G network and cloud computing could be useful for wearable gaming devices [7]. With everything from glasses and vests to headphones and gloves, the gaming industry is about to pass a new sensory threshold. The present 360-degree virtual reality panorama video requires a data rate of 20 to 40 Mbps and a latency of 50 ms. Thanks to the huge boost in data speed (> 100 Mbps) and drop in latency (1 ms) that 5G technology can provide, users may experience more enjoyable virtual reality in the future [8].People will receive a similar physical thrill from lifelike haptic feedback in immersive gaming experiences as they would from actual actions and sensations.

3. METHODOLOGY

This study objective will be addressed using the following methods. In order to evaluate the current implications of 5G for virtual reality in entertainment and games, the study will employ a descriptive research design. This research approach is suitable because it enables the gathering, analysis, and interpretation of the current impacts that 5G has on virtual reality-based entertainment [9] and gaming [10]. For this study, both primary and secondary data will be gathered. Survey questions will be sent out to gaming and entertainment in many parts of the world in order to gather primary data. The poll will be made to include questions regarding the current state of virtual reality applications in entertainment and gaming, as well as how 5G technologies is affecting them [11, 12]. Interviews or focus groups will be held if more primary data needs to be collected. Through a study of prior studies and publications pertinent

to the research issue, as well as information from pertinent databases, secondary data will be gathered. Depending on the kind of data that was gathered, either quantitative or qualitative methodologies will be used to analyse the data. In order to ascertain the current state of virtual reality in entertainment and gaming and how it has been impacted by 5G technology, quantitative data will be analysed using descriptive and inferential statistics. To find any patterns or themes in the responses provided by the respondents, qualitative data will be subjected to content analysis.

4. RESULTS AND DISCUSSION

The majority of respondents (58%) chose VR entertainment and gaming as the most pertinent 5G application for consumers (see Fig. 13.1). The demand for high-quality watching experiences is seen in both linear (live streaming, for instance) and nonlinear (on demand) games. VR entertainment and gaming apps will become widely used for mobile gaming and entertainment in the mid- to long-term.

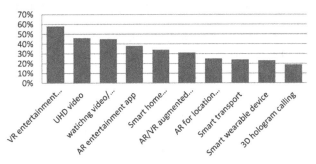

Fig. 13.1 The apps that customers are most enthusiastic about

Numerous useful applications for M&E are provided by both AR and VR (see Fig. 13.2). Across age groups and nations, interest varies. The younger millennials (ages 25–36) and members of Generation Z (Gen Z) are extremely interested in VR gaming applications. According to 62 percent and 59 percent of respondents, respectively, these applications are the most valuable.

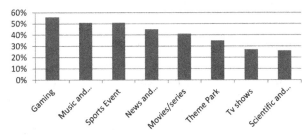

Fig. 13.2 The most valuable VR applications supported by 5G

The way we play video games is evolving, ushering in a new era of gaming and entertainment consumption. However, there are regional differences in gamers' excitement for immersive gaming media. In Canada, 32% of respondents say that VR gaming and entertainment has the most value for them, whereas in India, 69% say the same (see Fig. 13.3).

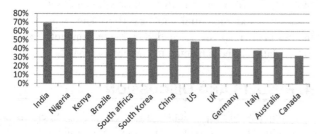

Fig. 13.3 Consumer enthusiasm for VR

4.1 Discussion

Enhanced User Experience

The smooth, high-quality VR experiences made possible by 5G's lightning-fast data transmission and minimal latency reduce motion sickness and improve immersion. Virtual environments are becoming more responsive and lifelike for gamers and other entertainment fans.

Enhanced Accessibility

5G networks offer broader coverage, opening up high-quality VR to a wider audience. The creation, distribution, and consumption of material in the gaming and entertainment sectors may change as a result of this accessibility.

Despite the potential, there are still issues to be resolved, such as infrastructure development, privacy issues, and content optimization. To fully utilize 5G-enabled VR, these problems must be resolved.

Accessibility and Ubiquity

As 5G networks quickly broaden their coverage, more people will have access to high-quality VR content. Through its global reach, this accessibility democratizes access to immersive experiences. It is now more easier to work together remotely, learn remotely, and take part in live events remotely.

Content Creation and Distribution

5G has advantages for content suppliers as well as end users. As it becomes more practical and affordable to stream high-quality VR video, new opportunities for sharing and making money from VR experiences present themselves. This might result in a plethora of cutting-edge content, ranging from interactive educational modules to live-streamed virtual concerts.

Practical Applications

The study looks at actual use cases and initiatives that successfully combined 5G and VR for gaming and entertainment.

This study provides helpful insights into the changing dynamics of immersive experiences as the confluence of 5G and VR continues to change the entertainment and gaming landscapes. In boosting user engagement, accessibility, and content delivery, 5G technology plays a crucial role. This highlights the importance of this technology and paves the way for a time when VR applications will redefine what is considered entertainment and gaming.

5. CONCLUSION

Future 5G applications could be used for a wide range of purposes, benefiting corporations, consumers, and governments alike. The gaming experience for players and entertainers will be significantly improved with 5G, including e-sports and cloud gaming. With current network technology, the gaming and entertainment business has not yet reached its full potential. The existing technology has drawbacks in the gaming and entertainment industries, including significant visual graphics lag and high latency (4G), among others. In addition to enhancing gaming visuals and experiences, 5G solves these problems. Virtual and augmented reality is being gradually incorporated into everyday life in the modern world. These cutting-edge technologies will significantly influence daily activities with the widespread adoption of 5G. Therefore, 5G technology has an impact on a country's technological development in addition to the gaming and entertainment industries. To fully utilize 5G-enabled VR in this dynamic environment, teamwork and creativity are key. By bringing these technologies together, content creators will be given more power, users will be delighted, and the gaming and entertainment industries will be pushed to their limits. The convergence of 5G with virtual reality holds interesting potential for immersive experiences in the future.

References

1. Vinayak Pujari,Mr Kajima Tambeand Rajendra Patil "Research Paper on Future of 5G Wireless System" (ISSN 2231–2137): SPECIAL ISSUE: APRIL, 2021.
2. Liang, W. (2019). Scene art design based on human-computer interaction and multimedia information system: an interactive perspective. Multimedia Tools and Applications, 78(4), 4767–4785.
3. Marder, B., Gattig, D., Collins, E., Pitt, L., Kietzmann, J., & Erz, A. (2019). The Avatar's new clothes: Understanding why players purchase non-functional items in free-to-play games. Computers in Human Behavior, 91, 72–83.

4. Marepalli Sharath Kumar "Revolution of 5g Wireless Technology-Future Direction" IOSR Journal of Computer Engineering (IOSR-JCE) e-ISSN: 2278-0661, p-ISSN: 2278–8727, Volume 21, Issue 4, Ser. III (Jul - Aug 2019)

5. Chen, H., & Fuchs, H. (2017, June). Supporting free walking in a large virtual environment: imperceptible redirected walking with an immersive distractor. In Proceedings of the Computer Graphics International Conference (pp. 1–6).

6. Indrakumari, R., Poongodi, T., Suresh, P., & Balamurugan, B. (2020). The growing role of Internet of Things in healthcare wearables. In Emergence of Pharmaceutical Industry Growth with Industrial IoT Approach (pp. 163–194). Academic Press.

7. Sun, H., Zhang, Z., Hu, R. Q., & Qian, Y. (2018). Wearable communications in 5G: Challenges and enabling technologies. ieee vehicular technology magazine, 13(3), 100–109.

8. Minopoulos, G., & Psannis, K. E. (2022). Opportunities and challenges of tangible xr applications for 5g networks and beyond. IEEE Consumer Electronics Magazine.

Note: All the figures in this chapter were made by the author.

Recent Trends in Engineering and Science for Resource Optimization and Sustainable Development – Prof. (Dr.) Dorota Jelonek et al. (eds)
© 2024 Taylor & Francis Group, London, ISBN 978-1-032-98030-0

14

Using IoT and Machine Learning Together for Agricultural Predictive Maintenance

G. Bhupal Raj

Dean, School of Argiculture, SR University,
Warangal, India

Deepak Kholiya

Professor, School of Agriculture, Graphic Era Hill University,
Dehradun, Uttarakhand, India

Vinston Raja R[1]

Assistant Professor, Department of Computational Intelligence,
Faculty of Engineering and Technology, School of Computing,
SRM Institute of Science and Technology, Kattankulathur, Chennai

Dler Salih Hasan[2]

Lecturer, Computer Science and Information tech,
College of Science / University of Salahaddin-Erbil, Erbil, Iraq

Navdeep Singh[3]

Lovely Professional University, Phagwara

Atish Mane[4]

Department of Mechanical Engineering,
Bharati Vidyapeeth's college of Engineering Lavale, Pune, India

Liviu Rosca[5]

Lucian Blaga University of Sibiu

▆▆▆▆▆ Abstract

The machine learning and Internet of Things (IoT) are being integrated to improve maintenance procedures, which is undergoing a huge transition in the agricultural sector. In order to improve agricultural predictive maintenance, this article investigates the synergistic potential of fusing IoT's real-time data collection capabilities with machine learning's predictive analytics. The report provides insights into how this integrated method is revolutionizing agriculture by raising operating efficiency and lowering downtime. It also investigates the benefits, difficulties, and real-world implementations of this approach. The development of new applications, including smart hydroponic agriculture, was made possible by the success of Internet of Things solutions. Frequent manual maintenance has historically been used to remedy this issue; however, this method is thought to be unsuccessful and could eventually harm the crops. This research' major goal was to suggest a machine learning method for automatically detecting sensor fault drifts. The solution's usability with regard to of response

Corresponding authors: [1]vinstonr@srmist.edu.in, [2]dler.hasan@su.edu.krd, [3]navdeep.dhaliwal@lpu.co.in, [5]mane.atish@bharatividyapeeth.edu, [6]liviu.i.rosca@gmail.com

DOI: 10.1201/9781003596721-14

time in a cloud computing environment was also investigated. RNN is used in the detection method proposed in this study to forecast sensor drifts from time-series data streams. Later known as Predictive Sliding Detection Window (PSDW), the detection system integrated forecasting and classification algorithms.

▬▬▬▬ Keywords

ML, IoT, Predictive maintenance, Agriculture

1. INTRODUCTION

The effortless movement of goods and services from producers to final customers is made possible by supply chain management, which serves as the foundation of contemporary business. However, there is a pressing need for cutting-edge computational solutions to optimize these sophisticated networks because of the interconnection and complexity of global supply chains. The issues posed by supply chain management could be solved by quantum computing, a ground-breaking technology at the crossroads of quantum mechanics and computer science.

The transformational potential of quantum computing is explored in this study within the framework of supply chain management optimization. The necessity for cutting-edge computational tools grows more and more important as supply networks get more complex. With its capacity to analyze huge datasets and resolve challenging optimization issues, quantum computing presents a promising way to transform supply chain operations. This study examines how quantum computing might improve supply chain efficiency, address the complex problems associated with contemporary supply chains, and provide significant value to both enterprises and consumers.

In order to balance the convergence of the physical and digital worlds, the IoT considered as the communication paradigm, expands the internet by using sensors and actuators. One of the most revolutionary inventions in contemporary history, the use of IoT technology has been growing quickly in recent years. This technical development made it possible to create brand-new "smart" applications that profit from the interaction between sensors and actuators.

Predictive maintenance is a notable application that was made possible with the aid of IoT breakthroughs in this respect. The primary concept behind this type of maintenance method is that they evaluate the assets' condition using the data collected from the sensing devices. By preventing failures and other performance difficulties, predictive maintenance procedures are crucial to maximising the lifespan of the assets [1].

Smart hydroponic agriculture is one application that makes use of IoT and predictive maintenance technologies. In this case, the farming environment is regulated using sensors and actuators from the Internet of Things to maintain the health of the crops. For example, hydroponic sensors may monitor and control the physical and chemical elements of the hydroponic environment [2]. Predictive maintenance is utilised in these situations to successfully maintain the performance of the installed IoT devices and assist in the detection of any anomalies that may have an impact on the health of the crops [3].

1.1 Objective of Study

1. To analyse the machine learning methods may be used to increase the effectiveness of failure detection in hydroponic sensor data streams.
2. This study's objective is to look at this issue using a proof-of-concept that has been put into practise.

2. LITERATURE REVIEW

In general, the interoperability of IoT devices installed on an adaptable and scalable infrastructure, like cloud computing services, is essential for the success of smart agriculture. [4] IoT devices including sensors, actuators, and embedded systems are typically used in smart agricultural setups to monitor crop status and environmental conditions. This implies that any flaws that might develop in those IoT devices could put the entire system at danger of failing and have an impact on the crops' quality. [5]

In conventional hydroponic agriculture, reactive or preventive maintenance methods are used on the sensors. These techniques are thought to be inefficient and unsustainable when applied to wider areas when the sensors have a high rate of failure recurrence. As a result, there has been an increase in interest in researching efficient and trustworthy methods for identifying sensor failures from data streams [6]. It is increasingly vital to study new techniques that give sensor nodes prediction mechanisms that boost maintenance effectiveness by automatically recognising sensor drifts. Modern maintenance strategies employ data to forecast how well deployed sensors will work. Additionally, there is an increasing need to research different machine learning algorithm configurations that can be used to address varied maintenance scenarios and produce higher forecast results [7].

3. Methodology

3.1 Data Design

By creating a solution that is appropriate for smart agriculture contexts, this will be achieved. This will be accomplished by defining hydroponic farming's characteristics and failure detection systems [8]. The data streams can be managed in two distinct stages [9], like data categorization and forecasting and data generation [10], a data pipeline will be created. A humidity sensor and low-cost temperature will perform the data generation stage by taking readings at predetermined intervals [11]. A humidity sensor and single temperature connected to AWS cloud services produced the majority of the data.

3.2 Data Analysis

To ascertain whether the results are statistically significant, the observed quantitative data will be statistically evaluated using one-way analysis of variance (ANOVA), standard deviation (SD) and confidence interval (CI).

3.3 Data Measure

Various accuracy metrics will be used to evaluate the performance of the prediction models. The overall Root-Mean-Squared Error (RMSE) for every interval size will be calculated to evaluate the performance of forecasting model.

4. Results and Discussion

Modern farming operations require effective machinery and equipment since agriculture is critical to maintaining the world's food supply. On the other hand, significant downtime and lost production might result from equipment malfunctions and maintenance problems. The agricultural sector is increasingly using the IoT and machine learning for predictive maintenance to address these issues. An interesting new step in the development of supply chain practices is the incorporation of quantum computing into supply chain management optimization. The demand for cutting-edge technologies and methods has never been higher as supply chains grow and become more complicated.

The RMSE values for each RNN model are shown in Table 14.1 along with the interval size. For 95% of the observations, the statistical computations of the mean, SD, and CI are included.

Table 14.2 summarises the F1-score performance of each of the classification models (CNN-LSTM, GRU and LSTM). The test sample's statistical mean, standard deviation, and 95% confidence range are also displayed in the table.

Table 14.1 Models for forecasting are assessed using the RMSE and the size of the prediction interval

Forecasting Model	Interval Sixe	RMSE (Avg.)	RMSE (SD)	RMSE (CI)
LSTM	4	0.6002	.0023	.0008
	6	.6247	.0025	.0010
	8	.6130	.0064	.0036
	10	.6436	.0297	.0136
	12	.6502	.0284	.0109
CNN-LSTM	4	.6006	.0025	.0010
	6	.6246	.0038	.0017
	8	.6007	.0070	.0043
	10	.6208	.0095	.0053
	12	.6644	.0455	.0216
GRU	4	.7509	.2769	.1040
	6	.6929	.0868	.0470
	8	.6202	.0400	.0259
	10	.6146	.0520	.0256
	12	.6285	.0520	.0270

Table 14.2 F1-score-based categorization model evaluation

Model Classification	Score F1 (Avg.)	Score F1 (SD)	Score F1 (95% CI)
LSTM	.7979	.1699	.0996
CNN-LSTM	.8028	.1770	.1040
GRU	.8278	.1409	.0922

Table 14.3 Accuracy-based evaluation of categorization models

Model Classification	Accuracy (Avg.)	Accuracy (SD)	Accuracy (95% CI)
LSTM	.7970	.1686	.0988
CNN-LSTM	.8256	.1820	.1070
GRU	.8355	.1595	.0925

Figure 14.1 shows how the categorization models performed in terms of their F1-score and accuracy. A higher score indicates better prediction performance.

The system performance with (and without) the deployed solution on the AWS EC2 instance is summarised in Table 14.4. The metrics of interest during the load test were the total number of HTTP requests, the total number of users making requests, and the mean, minimum, and maximum response times.

This work aims to explore how machine learning techniques may be used to enhance the efficacy of failure

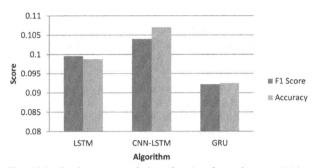

Fig. 14.1 Performance of classification for a three-point interval as measured by the F1-score and precision

Table 14.4 Comparison of EC2 instance performance in terms of reaction time

	PSDW (With-out)	PSDW (With)
Total number of user	900	900
Total number of request	204437	3180
Avg. Response Time	1247	76215
Min Response Time	58	1183
Max Response Time	4343	317346

detection in hydroponic sensor data streams. This study explores a number of significant issues and provides a thorough explanation.

4.1 Sensor Data with Hydroponics

Hydroponics is a cutting-edge and effective way to grow plants without soil. Environmental factors like temperature, humidity, and nutrition levels must be precisely controlled. In order to successfully cultivate hydroponically, real-time data from sensors that monitor these conditions is crucial.

Hydroponic systems are sensitive to changes in the environment, which presents difficulties in the failure detection process. Crop failure or subpar growth can be caused by equipment problems, sensor mistakes, or unfavorable changes in the growing environment. identifying and fixing these errors as soon as possible for crop maintenance and high yields

4.2 Agriculture and Machine Learning

Machine learning has shown that it is capable of analyzing huge datasets, finding patterns, and making predictions. It has been successfully used in agriculture to increase yields, forecast disease outbreaks, and enhance crop management. Hydroponic failure detection using machine learning has the potential to increase productivity and decrease crop losses.

4.3 Sensor Data Analysis

The sensor data produced by hydroponic systems can be processed and analyzed using machine learning methods. These algorithms have the ability to spot anomalies, anticipate equipment failures, and offer suggestions for improving environmental conditions.

4.4 Benefits of Machine Learning

Hydroponic growers should expect better crop health, lower maintenance costs, and higher yields by using machine learning for failure detection. Early warnings from machine learning models can be used to take preventative measures.

4.5 Challenges and Factors

Implementing machine learning in hydroponics has several difficulties in terms of model selection, interpretability, and data quality. For reliable projections, thorough data collection and the availability of past data are necessary.

4.6 Applications in the Real World

The study can look at case studies and applications in the actual world where machine learning techniques have been successfully incorporated into hydroponic systems. Future solutions can use these examples as models and inspiration.

5. CONCLUSION

The study was able to provide an automated drift detection technique that might be utilised to improve maintenance effectiveness. The proposed method consisted of a two-layered neural network with forecasting and classification algorithms. The findings of the developed models revealed a low level of fault categorization dependability and a moderate degree of accuracy. To assess the suggested method's applicability in a practical setting, it was then put into practise in a cloud computing environment. The load test findings show a sizable reaction time trade-off. The results suggested that additional study is necessary to improve accuracy and response time.

References

1. T. Zonta, C. A. da Costa, R. da Rosa Righi, M. J. de Lima, E. S. da Trindade, and G. P. Li, 'Predictive maintenance in the Industry 4.0: A systematic literature review', Comput. Ind. Eng., vol. 150, p. 106889, Dec. 2020, doi: 10.1016/j.cie.2020.106889.
2. V. Palande, A. Zaheer, and K. George, 'Fully Automated Hydroponic System for Indoor Plant Growth', in Procedia Computer Science, Jan. 2018, vol. 129, pp. 482–488, doi: 10.1016/j.procs.2018.03.028.

3. D. Karimanzira and T. Rauschenbach, 'Enhancing aquaponics management with IoT-based Predictive Analytics for efficient information utilization', Inf. Process. Agric., vol. 6, no. 3, pp. 375–385, Sep. 2019, doi: 10.1016/j.inpa.2018.12.003.

4. E. Navarro, N. Costa, and A. Pereira, 'A systematic review of iot solutions for smart farming', Sensors (Switzerland), vol. 20, no. 15. MDPI AG, pp. 1–29, Aug. 01, 2020, doi: 10.3390/s20154231.

5. O. Elijah, T. A. Rahman, I. Orikumhi, C. Y. Leow and M. N. Hindia, "An Overview of Internet of Things (IoT) and Data Analytics in Agriculture: Benefits and Challenges," in IEEE Internet of Things Journal, vol. 5, no. 5, pp. 3758–3773, Oct. 2018, doi: 10.1109/JIOT.2018.2844296.

6. A Gaddam, T. Wilkin, M. Angelova, and J. Gaddam, 'Detecting Sensor Faults, Anomalies and Outliers in the Internet of Things: A Survey on the Challenges and Solutions', Electronics, vol. 9, no. 3, p. 511, Mar. 2020, doi: 10.3390/electronics9030511.

7. P. P. Ray, 'Internet of things for smart agriculture: Technologies, practices and future direction', J. Ambient Intell. Smart Environ., vol. 9, no. 4, pp. 395–420, Jan. 2017, doi: 10.3233/AIS-170440.

Note: All the tables and figure in this chapter were made by the author.

*Recent Trends in Engineering and Science for Resource Optimization and
Sustainable Development – Prof. (Dr.) Dorota Jelonek et al. (eds)
© 2024 Taylor & Francis Group, London, ISBN 978-1-032-98030-0*

15

Electronic Voting Systems Using a Blockchain-Based Encrypted Identity Management

Siddhanth Prafulla Pathak[1]

IT Trainee, Client Experience (Capital Markets)
FIS Global, Pune, India

Akkaraju Sailesh Chandra[2]

Associate Professor, Faculty of Management and Commerce,
PES University Bengaluru Karnataka, India

G. Lohitha

Department of Information Technology,
Institute of Aeronautical Engineering, Hyderabad, Telangana

Mini Jain[3]

Institute of Business Management,
GLA University, Mathura, Uttar Pradesh

Christabell Joseph[4]

Associate Professor, School of Law,
Christ (Deemed to be University)

M. Ravichand[5]

Professor of English, Velagapudi Ramakrishna Siddhartha Engineering College;
Siddhartha Academy of Higher Education - A Deemed to be University;
Vijayawada, Andhra

Abdelmonaim Fakhry Kamel mohammed[6]

Ain shams university – Egypt

▄▄▄▄ Abstract

The use of electronic voting technologies has grown in popularity as a way to make elections more secure and accessible. The implementation of blockchain-based encrypted identity management in electronic voting is explored in this study, which also offers a solid option to improve the reliability and credibility of voting systems. This study explores the possibilities for anonymous and transparent electronic voting while preserving voter privacy and anonymity by incorporating blockchain technology. It has always been challenging to create an electronic voting system that properly satisfies the requirements of administrators. This problem is now being resolved by blockchain technologies, which provide a distributed database with irreversible, encrypted identity management and secure transactions. A fascinating advancement in the realms of data

Corresponding author: [1]sidpathak07@gmail.com, [2]saileshchandra@pes.edu, [3]minijain06@gmail.com, [4]Christabell.joseph@christuniversity.in, [5]ravichandenglish@gmail.com, [6]abdelmonaimfakhry601@gmail.com

DOI: 10.1201/9781003596721-15

innovation, dependability, and transparency is distributed ledger technology. Distributed ledger technology is commonly used in public blockchain. Virtually limitless potential for earning from sharing economies are provided by blockchain technology. This project aims to determine whether blockchain technology can be used to create electronic voting devices are used as a service.

▬▬▬▬ Keywords

Encryption, Electronic voting, Security, Blockchain, Distributed ledger

1. INTRODUCTION

Any modern society's foundation is its democratic system, and the accuracy and security of voting systems play a critical role in preserving this system's integrity. Electronic voting technologies have replaced traditional paper-based voting methods because they are more convenient and accessible. Electronic voting has, however, run into issues with privacy, security, and trust. Blockchain technology has emerged as a viable solution to these problems because of its transparency, immutability, and security. Electronic voting systems can guarantee the authenticity of voters, secure their privacy, and give a tamper-proof record of the electoral process by employing blockchain's encrypted identity management.

Voting is permitted in corporate, association, and even organisational events as well as in political elections. A method is built into voting systems to select one option from a range of options based on the votes cast. The resources and needs of the implementer are taken into account while deciding whether to implement a voting system. A few important considerations must be made while choosing a voting method [1]; otherwise, the outcome of the vote may be tainted. Electronic voting systems have been the subject of research for decades, with the main objectives being data protection and election cost reduction by guaranteeing safety measures and compliance with standards for election integrity. In a democracy, fair democratic assessments are required for an election to be transparent and secure. Thus, election security presents a challenge to global security. In order to apparently elect the candidates, pen-and-paper voting processes were devised or implemented, endangering the validity of the vote and socio-political evaluations [2]. Therefore, it is essential to replace the current voting system with an electronic one, which will reduce fraud and keep the voting process secure through rendering it verifiable and traceable [3].

The blockchain can be divided into four groups, as shown in Fig. 15.1 public, private, hybrid, and consortium. A consortium is one in which centralised oversight is upheld by a number of enterprises, similar to a private permissioned

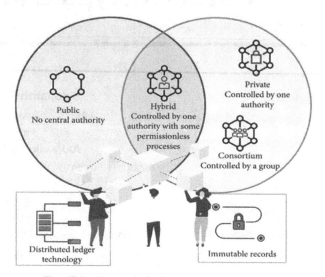

Fig. 15.1 Kinds of blockchain for secure voting

blockchain. A blockchain that combines both public and private blockchains is called a hybrid. But because voting is only available to those who are at least 18 years old, we employ permissioned blockchain in a blockchain-based electronic voting system. Therefore, to guarantee security and immutability, we restrict access to those who satisfy certain requirements.

2. LITERATURE REVIEW

In recent decades, a large number of researchers have been developing blockchain technologies. Blockchain technology was made possible by Satoshi Nakamoto's 2008 introduction of the first cryptocurrency, Bitcoin. The PoW consensus mechanism is the foundation of the Bitcoin blockchain technology, which has a public distributed ledger, irreversible transactions, and protected and limited encryption [4]. Data cannot be modified or removed from the blockchain structure; it can only be added to it as new blocks are created. Each block contains a hash of the preceding block, which is produced by adding the value of the nonce in accordance with predefined rules. This links blocks in a way that makes hashing more complex and ensures immutability [5]. A few sectors that use blockchain include healthcare, cryptocurrencies, e-voting platforms, and supply chain management systems [6]. Blockchain technology revolutionised the medical and healthcare industries by allowing for the maintenance and improvement of health-related records while also enhancing their security and accuracy. Blockchain-based e-health systems are expected to give precise medical treatments, brain research, and much more, according to a number of experts [7]. To improve the overall security of the voting process and to handle cryptographic challenges with conflicting security features, researches worked on various cryptographic primitives and consensus techniques as well, like one-time ring signature and homomorphism encryption. But these kinds of systems require extensive communication overhead and computational power [8].

3. METHODOLOGY

This paper's major research focus is on electronic voting systems that use encrypted identity management on a blockchain. In order to achieve this, we will make an effort to determine how it is used in the existing voting system and further investigate the options for its usage in safe [9], transparent [10], and trustworthy data [11]. We'll also look into blockchain's worth and effects on network, computer, and file security. Primary and secondary research will be combined in our study process. We will first use library sources, including books, journals, papers, and any other pertinent materials, to compile the current state of knowledge regarding the security implications of electronic voting. We will make use of this data to better understand how blockchain-based identity management affects security [12]. In order to further assess the potential of voting for encrypted identity management, we will then conduct surveys and interviews with industry professionals and those working in the field of blockchain technology. Our surveys and interviews will delve into the usefulness of blockchain-based encrypted identity management as well as the level of security provided by the technology. In order to assess the usefulness and impact of blockchain for safe electronic voting, we will use the data we have obtained. We will be able to better comprehend electronic voting systems using a blockchain-based encrypted identity management by analysing the current status of blockchain and its potential consequences for the secure, transparent, and reliable voting.

4. RESULTS AND DISCUSSION

Figure 15.2 and Table 15.1 below indicate the percentage of responses to the survey's closed-ended questions. The questions serve as a comparison between electronic voting technologies and the current manual paper-based approach.

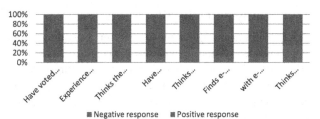

Fig. 15.2 % of responses comparing electronic voting with traditional paper ballots

Table 15.1 Percent of responses comparing electronic voting with traditional paper ballots

	Negative response	Positive response
Have voted before	10	90
Experienced challenges with manual voting	68	32
Thinks the IEC have addressed these issues	54	46
Have knowledge of e-voting	13	87
Thinks every country adopt an e- voting	27	73
Finds e- voting appealing	12	88
with e-voting results could be out sooner	12	88
Thinks voting could lead to free and fair elections	35	65

The user screen for the electronic voting system is shown in Fig. 15.3. In order to access the results, users must first log in after registering to vote.

Fig. 15.3 Login page of voting

An important step in enhancing the security and integrity of the election process is the integration of blockchain-based encrypted identity management into electronic voting systems. During the examination of this convergence, a few crucial aspects emerge, emphasizing the significant ramifications and advantages of such a combination:

4.1 Security and Integrity

Blockchain technology is recognized for its immutability and transparency security qualities. It makes sure that voter information is kept unchanged and that the entire process is transparent when used with electronic voting. Sensitive voter data is protected by encrypted identity management, allaying worries about privacy and data breaches.

Maintaining trust in the electoral process is one of the most important difficulties facing contemporary democracies. Confidence is increased by blockchain's capacity to produce tamper-resistant records.

Electronic voting technologies promote accessibility for voters, particularly for those who are unable to go to actual polling places. When blockchain technology and encrypted identity management are used together, this convenience is guaranteed to not compromise security. This is especially important in situations like remote voting for residents living abroad and absentee voting.

4.2 Privacy Protection

In every democratic system, maintaining voter privacy is crucial. Voters can cast ballots without disclosing their personal information thanks to encrypted identity management. This answer worries about voter data acquisition without authorization, vote buying, and coercion.

4.3 Transparency in Political Processes

The blockchain promotes transparency in all political processes, including choosing candidates, funding campaigns, and tabulating votes. It contributes to a more transparent election process by offering an immutable log of campaign donations and candidate qualifications.

4.4 Challenges

Integrating encrypted identity management based on blockchain technology into electronic voting is not without its difficulties. Critical factors include scalability, accessibility, and the requirement for voter education. Voter education is essential to ensure that voters understand how to utilize the system efficiently, and blockchain networks must be strong enough to handle enormous numbers of transactions.

4.5 Real-World Applications

The conversation should also cover instances where identity management based on blockchain has been successfully incorporated into electronic voting systems. Examples from nations or areas where similar systems have been implemented can shed light on the applicability and efficacy of this technology.

5. Conclusion

In order to guarantee a safe and economical election while maintaining voter anonymity, this study suggests a cutting-edge electronic voting system using smart contracts based on smart contracts. The implementation of blockchain as an electronic voting system has significant technological and legal challenges, which are examined in this essay. By comparing our findings to previous studies, we showed that blockchain technology offers democratic nations a new opportunity to transition from manual elections to a more time and money-effective voting technique while simultaneously enhancing transparency and security. We created an electronic voting method in this study that can boost voter turnout while also providing security, transparency, and user privacy.

References

1. Khan, K.M.; Arshad, J.; Khan, M.M. Secure digital voting system based on blockchain technology. Int. J. Electron. Gov. Res. 2018, 14, 53–62. [CrossRef]
2. R. Tas¸ and ¨ O. ¨ O. Tanrı¨over, "A systematic review of challenges and opportunities of blockchain for E-voting," Symmetry, vol. 12, no. 8, p. 1328, 2020.
3. N. Kshetri and J. Voas, "Blockchain-enabled e-voting," Ieee Software, vol. 35, no. 4, pp. 95–99, 2018.
4. M. Crosby, P. Pattanayak, S. Verma, and V. Kalyanaraman, "Blockchain technology: beyond bitcoin," Applied Innovation, vol. 2, no. 6–10, p. 71, 2016.
5. F. Hofmann, S. Wurster, E. Ron, and M. B¨ohmecke-Schwafert, ",e immutability concept of blockchains and benefits of early standardization," in Proceedings of the 2017 ITU Kaleidoscope: Challenges for a Data-Driven Society (ITU K), pp. 1–8, IEEE, Nanjing, China, 2017, November.
6. U. Bodkhe, S. Tanwar, K. Parekh et al., "Blockchain for industry 4.0: a comprehensive review," IEEE Access, vol. 8, Article ID 79764, 2020.
7. H. Zhao, P. Bai, Y. Peng, and R. Xu, "Efficient key management scheme for health blockchain," CAAI Transactions on Intelligence Technology, vol. 3, no. 2, pp. 114–118, 2018.
8. T. Dimitriou, "Efficient, coercion-free and universally verifiable blockchain-based voting," Computer Networks, vol. 174, p. 107234, 2020.

Note: All the figures and table in this chapter were made by the author.

Recent Trends in Engineering and Science for Resource Optimization and
Sustainable Development – Prof. (Dr.) Dorota Jelonek et al. (eds)
© 2024 Taylor & Francis Group, London, ISBN 978-1-032-98030-0

16

Investigating Quantum Computing's Potential for Supply Chain Management Optimisation

K. K. Ramachandran[1]

Director/ Professor:
Management/Commerce/International Business,
DR G R D College of Science, India

Nornajihah Nadia Hasbullah[2]

Faculty of Business and Management,
Universiti Teknologi Mara (UiTM), Cawangan Melaka,
Kampus Bandaraya Melaka, Melaka 75350, Malaysia

Nitasha Rathore[3]

Assistant Professor, Department of CSE,
Bharatiya Vidyapeeth College of Engineering

S. S Prasada Rao[4]

Dean, Centre for Teaching and Learning,
S.P Mandali's Prin. L. N Welingkar Institute of
Management Development & Research
(Weschool), Munbai

Navdeep Singh[5]

Lovely Professional University, Phagwara

P. S. Ranjit[6]

Professor, Department of Mechanical Engineering,
Aditya Engineering College Surampalem,
Kakinada, India

Tomasz Lis[7]

Faculty of Management,
Czestochowa University of Technology, Poland

▬▬▬ Abstract

Quantum computing, which has the ability to solve the most challenging logistics and supply chain management problems, is the most fascinating advancement in computation for the ensuing ten years. The current state of quantum computing is reviewed in this study, along with recommendations for further research. The transformational potential of quantum

[1]dr.k.k.ramachandran@gmail.com, [2]najihahnadia@uitm.edu.my, [3]nitasha.rathore@bharatividyapeeth.edu, [4]Profsspr@gmail.com, [5]navdeep.dhaliwal@lpu.co.in, [6]psranjit1234@gmail.com, [7]tomasz.lis@pcz.pl

DOI: 10.1201/9781003596721-16

computing is explored in this study within the framework of supply chain management optimization. The necessity for cutting-edge computational tools grows more and more important as supply networks get more complex. With its capacity to analyze huge datasets and resolve challenging optimization issues, quantum computing presents a promising way to transform supply chain operations. This study examines how quantum computing might improve supply chain efficiency, address the complex problems associated with contemporary supply chains, and provide significant value to both enterprises and consumers. First, fundamental ideas related to quantum computing and computers are introduced. The dominant quantum technologies are then described. The third section includes instances of contemporary supply chain management and logistical advancements as well as a study of the quantum sector. Fourth, suggestions for further investigation are given. This review intends to inform and motivate the application of quantum computing to supply chain and logistics optimisation, machine learning, and artificial intelligence.

▬▬▬▬ **Keywords**

Supply chain management, Quantum computing, and Logistics

1. INTRODUCTION

The effortless movement of goods and services from producers to final customers is made possible by supply chain management, which serves as the foundation of contemporary business. However, there is a pressing need for cutting-edge computational solutions to optimize these sophisticated networks because of the interconnection and complexity of global supply chains. The issues posed by supply chain management could be solved by quantum computing, a ground-breaking technology at the crossroads of quantum mechanics and computer science.

Recent assessments suggest that quantum computing holds promise for numerous industries, including banking and quantum chemical modelling [1]. In this paper, we investigate the potential benefits of using quantum computing to solve significant operations management issues. Early in the 1980s, Beniof and Feynmann developed the idea of quantum computing (QC), harnessing quantum phenomena to create a new kind of computer hardware called the quantum computer. New categories of algorithms, quantum algorithms, were created to be used with quantum computers.

A key issue in the realm of operations management is supply chain management. The study of the flow of products and services through a process is known as supply chain management. Supply chains have undergone a significant shift since the beginning of time, when inventory decisions were manually recorded with ink and paper in order to become more automated and efficient. This transformation has been made possible by the emergence of new technologies including digitization, software development, and increasing processing capacity. Investigating the possible applications of quantum computing for supply chain management is therefore logical. Because it is widely accept-

ed that the enormous decision spaces of inventory control problems in particular provide significant computational hurdles, recent research has looked into the use of deep learning algorithms for supply chain management [2].

In this study, we focus on a classical inventory control problem to conduct a fundamental examination into the potential impact of quantum computing on supply chain management. We go over the unique aspects of this inventory control issue that make quantum computing a potent solution. We also talk about hardware requirements and error-correction limits for various quantum techniques. We do IBM survey studies to make sure the validity.

1.1 Research Objective

1. The goal of this project is to look into potential uses for quantum computing in supply chain management.

2. This study's objective is to describe the hardware needs and error-correction capabilities of several quantum approaches.

2. LITERATURE REVIEW

Applications of QC pertinent to this review can be categorised into three primary domains, according to Gil Dario et al: Artificial intelligence, optimisation (such as in the areas of travel and transportation, supply chain management, network infrastructure, traffic control, and workload scheduling), and logistics, as well as machine learning (such as sampling, flexible vendor and customer interactions, decision support, training, and security) [3]. Selected projects mentioned in the evaluation serve to demonstrate the developments and applications made.

Sigma-i uses D-Wave's quantum annealer to address complicated, business-relevant problems, in this case the scheduling problem under the extraordinary circumstances of a pandemic. The molecular conformation problem, jobshop scheduling challenges, manufacturing cell construction issues, and vehicle routing issues are a few recent COVID-19 topics. To get around the combinatorial complexity of tackling large mixed-integer programming problems, Ajagekar et al. present hybrid models and methods [4]. These methods successfully combine the complementing characteristics of deterministic algorithms with QC techniques.

Sharma, 2020 analyses how to use a dataset on breast cancer to achieve classifications using classical machine learning (ML) and quantum circuits [5]. The first experimental implementation of a quantum artificial life algorithm in a QC is presented by U. Alvarez-Rodriguez, M. Sanz, et al. [6]. The quantum biomimetic protocol involves incorporating particular quantum properties from living systems, including as self-replication, mutation, inter-individual interaction, and death into IBM's QC.

Using a constrained size of qubits on IBM's quantum gate computers, Harwood et al. (2021) solve the inventory and vehicle routing problems with time windows using QC and discuss their advantages and disadvantages [7]. The authors argue that real-world business problems require tens of thousands of logical qubits in hardware [8].

3. METHODOLOGY

Investigating the Potential of Quantum Computing for Supply Chain Management Optimisation is the study's research goal. This study objective will be addressed using the following methods. The study will evaluate the existing effects of quantum computing on SDM using a descriptive research design. The collection, analysis, and interpretation of the current effects of quantum computing on supply chain management are all made possible by the research design, which is suitable [9]. For this study, both primary and secondary data will be gathered. Survey questions will be sent out to logistics in various parts of the world in order to gather primary data. The poll will be created to include inquiries regarding the current state of supply chain management and how quantum computing technology is affecting it. Interviews or focus groups will be held if more primary data needs to be collected [10]. Through a study of prior studies and publications pertinent to the research issue, as well as information from pertinent databases, secondary data will be gathered. Depending on the kind of data that was gathered, either quantitative or qualitative methodologies will be used to analyse the data [11]. Descriptive and inferential statistics will be used to analyse quantitative data in order to assess the situation of supply chain management at the moment and how quantum computing technology has affected it.To find any patterns or themes in the responses provided by the respondents, qualitative data will be subjected to content analysis.

4. RESULTS AND DISCUSSION

Several new and seasoned businesses in the computer industry are strongly tied to applied research and development in quantum computing for supply chains. Table 16.1 contains the most recent quantum gate-based computers, and Table 16.2 lists the most effective quantum annealers as a starting point for mapping the quantum computing market.

Table 16.1 Quantum gate computers overview

Manufacture	Name	Qubits
USTC	Jiuhang	77
Google	Bristlecone	73
IBM	Hummingbird	66
Google	Sycamore	54
IBM	IBM Q-53	54
Intel	Tangle Lack	50

Table 16.2 Quantum annealers overview

Manufacturer	Name	Qubits
D-wave	D-wave 2X	Above 1000
D-wave	D- wave 2000Q	Above 2000
D-wave	D- wave advantage	Above 5000

D-Wave systems have more qubits because of their unique architecture and requirements. The qubits of the quantum annealer are not entirely coupled, in contrast to a quantum gate-model computer. Unlike the qubit counts in a gate-based model quantum computer, where the number primarily represents the logical programmable qubits, D-Wave does refer to the physical qubits of their annealers.

The correlation between mistake rates and qubit counts is seen in Fig. 16.1. The arrow represents Google's goals for the study and how they want to use it to develop error-corrected quantum computers in the short term.

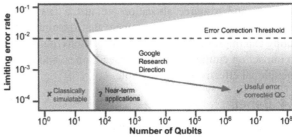

Fig. 16.1 Relationship between error and number of qubits

Figure 16.2 depicts the future goal and comparison between largest companies to reach max qubits by the end of 2023.

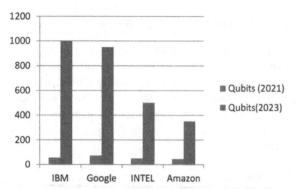

Fig. 16.2 Comparison between Companies

The effects and difficulties of utilizing this cutting-edge technology to raise supply chain operations' efficiency and effectiveness are shown by the study of quantum computing's potential for supply chain management optimization. The importance of this convergence is highlighted by a few crucial points:

4.1 Supply Chain Complexity

Today's supply networks are complicated, involving several nodes, international logistics, and a myriad of other factors. It is a difficult challenge to manage inventory, predict demand, and optimize routes and resources amid this complexity. The difficulties posed by modern supply chains can be well matched by quantum computing's capacity to manage big datasets and resolve challenging optimization issue

4.2 Computational Power of Quantum Computing

Quantum computers have the potential to carry out some computations tenfold quicker than conventional computers. Grover's and quantum annealing are examples of quantum algorithms that make it possible to explore several solutions and find the ideal arrangements. This ability can be used to make quick decisions in supply chain optimization.

4.3 Opportunities for Optimization

Demand forecasting, route optimization, inventory management, and resource allocation are just a few of the aspects of supply chain management that quantum computing can improve. Finding the best solutions quickly has an impact on cutting costs, cutting lead times, and raising operational effectiveness as a whole.

4.4 Benefits

Using quantum computing to optimize supply chains may result in increased cost effectiveness, quicker and more ac-

curate decision-making, and improved customer service, among other advantages. These benefits could give firms a competitive edge.

4.5 Challenges

Because quantum computing is still in its early stages, there are still issues to be resolved. These include issues with hardware limits, the requirement for quantum-ready algorithms, and a lack of experts in the field. Security issues about the possibility of defeating conventional encryption techniques must also be taken into account.

4.6 Applications

The study might go into actual efforts and applications where supply chain management has used quantum computing for testing and implementation. These case studies can provide useful guidance on how to successfully integrate quantum computing.

Innovative solutions are increasingly needed to manage the complexities that are inherent in supply chains and to improve operations. With its unmatched processing capacity, quantum computing offers a chance to completely rethink how supply chains are run. To realize its full potential, it also necessitates tight cooperation between specialists in supply chain management and quantum computing.

The discussion in this article highlights the revolutionary potential of quantum computing in supply chain management optimization and encourages additional investigation into the realistic application of this technology in supply chain settings.

5. Conclusion

Research on QC and its (potential) applications is growing as QC enables a leap in calculation speed to tackle even the most complex supply chain management problems. Companies like IBM, D-Wave, Google, Microsoft, and Amazon are racing to develop QC and make it available as a service in order to pave the way for other, highly specialised quantum computing services devoted to tackling certain issues because the technology is advancing at a fast acceleration rate. Furthermore, thanks to frameworks like Qiskit, Ocean, Cirq, Q#, Amazon Bracket, and others, QC is now easily accessible to anyone without a deep understanding of quantum mechanics. The quantum era is predicted by a number of studies, and according to IBM's roadmap, practical quantum hardware will be available by the end of 2023.

References

1. B. Bauer, S. Bravyi, M. Motta, and G. K.-L. Chan, "Quantum algorithms for quantum chemistry and quantum

materials science," Chemical Reviews, vol. 120, no. 22, pp. 12685–12717, 2020.

2. J. Gijsbrechts, R. N. Boute, J. A. Van Mieghem, and D. Zhang, "Can deep reinforcement learning improve inventory management? performance on dual sourcing, lost sales and multi-echelon problems," Manufacturing & Service Operations Management, 2021.

3. Gil Dario et al., "Quantum computing is coming to your business," 2018. Accessed: Sep. 01, 2021.[Online]. Available: https://www.ibm.com/thought-leadership/institute-businessvalue/report/quantumstrategy.

4. Ajagekar, T. Humble, and F. You, "Quantum computing based hybrid solution strategies for large-scale discrete-continuous optimization problems," Computers & Chemical Engineering, vol. 132, p. 106630, Jan. 2020, doi: 10.1016/J. COMPCHEMENG.2019.106630.

5. S. Sharma, "QEML (Quantum Enhanced Machine Learning): Using Quantum Computing to Enhance ML Classifiers and Feature Spaces," Feb. 2020, Accessed: Aug. 27, 2021. [Online]. Available: https://arxiv.org/abs/2002.10453v3.

6. U. Alvarez-Rodriguez, M. Sanz, L. Lamata, and E. Solano, "Quantum Artificial Life in an IBM Quantum Computer," Scientific Reports 2018 8:1, vol. 8, no. 1, pp. 1–9, Oct. 2018, doi: 10.1038/s41598-018-33125-3.

7. S. Harwood, C. Gambella, D. Trenev, A. Simonetto, D. Bernal Neira, and D. Greenberg, "Formulating and Solving Routing Problems on Quantum Computers," IEEE Transactions on Quantum Engineering, vol. 2, pp. 1–17, Jan. 2021, doi: 10.1109/TQE.2021.3049230.

Note: All the figures and tables in this chapter were made by the author.

Recent Trends in Engineering and Science for Resource Optimization and
Sustainable Development – Prof. (Dr.) Dorota Jelonek et al. (eds)
© 2024 Taylor & Francis Group, London, ISBN 978-1-032-98030-0

17

Blockchain and AI Integration for Transparent and Secured Financial Record Keeping

Devika Agrawal[1]

Assistant Professor, Institute of Business Management,
GLA University, Mathura, India

Vishal M Tidake[2]

Associate Professor, Department of MBA,
Sanjivani College of Engineering, Savitribai Phule
Pune University, India

Shaik Rehana Banu[3]

Post Doctoral Fellowship, Business Management,
Lincoln University College Malaysia

Ahmad Y. A. Bani Ahmad[4]

Department of Accounting and Finance Science,
Faculty of Business, Middle East University,
Amman 11831, Jordan

H Pal Thethi[5]

Lovely Professional University, Phagwara, India

Sanjay Singh Chauhan[6]

Uttaranchal Institute of Management,
Uttaranchal University, Dehradun, India

Dejan Mircetic[7]

University of Novi Sad, Faculty of Technical Sciences, Serbia

▬▬▬ Abstract

Artificial intelligence (AI) and blockchain technology integration has made it possible to retain financial records in new ways that favor security and transparency. In the context of maintaining financial records, this article examines the potential for synergy between blockchain and AI. It covers how AI improves data analysis, pattern identification, and anomaly detection while blockchain's immutable ledger maintains data integrity. Together, these technologies present an all-encompassing answer for open, safe, and effective financial record keeping. This article reviews the research on the potential impacts of blockchain technology on auditing that is supported by AI in general and on accounting in specific. The goal is to look into how professionals might use blockchain technology to increase the trust and transparency of accounting practises. Use

[1]devika.agrawal@gla.ac.in, [2]tidkevishal@gmail.com, [3]drshaikrehanabanu@gmail.com, [4]aahmad@meu.edu.jo, [5]h.pal@gmail.com, [6]chauhan.sanjay363@gmail.com, [7]dejanmircetic@gmail.com

DOI: 10.1201/9781003596721-17

the append-only, immutability, sharing, agreed-upon data properties and verifiability of blockchain data for improving the process of decision-making. The multi-party validation of blockchain protocols provides accurate data in real-time to the AI algorithms used by auditors, increasing efficiency and confidence. The literature on how the application of blockchain technology takes altered keeping of records in accounting has four key themes.

Keywords

Blockchain, AI, Record keeping, Accounting, Auditing

1. Introduction

Businesses that have made their systems digital have been able to reinvent their operations and adopt new technical solutions to expedite business processes [1] because they can now more easily access powerful computing resources and huge databases. Today's most valued businesses are platform-based and Internet-driven. The academic community, social media, corporate community, and government have all expressed interest in digital technologies including cloud computing, Internet of things, artificial intelligence and Blockchain. Blockchain considered as fifth pillar of digital technological revolution, is widely acknowledged as the primary technology enabling the next-generation Internet [2]. The finance, insurance, education, banking, government sectors and health care have started using it since various researches laid the foundation for what would become the technology in 2008. By 2020, 10% of the world's GDP will be monitored and recorded on a blockchain. The security of financial records using block chains is shown in Fig. 17.1.

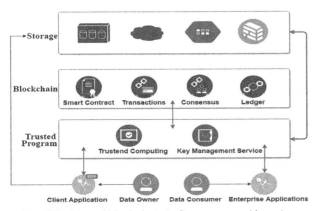

Fig. 17.1 Use of blockchain in finance record keeping

1.1 Objective of Study

This study examines how immutable, append-only, shareable, validated, and agreed-upon data based on blockchains might improve the transparency and trust in accounting practises as well as the decision-making of its practitioners. Understanding the process of blockchain that impact

accounting, in regard to AI-assisted auditing specifically, as well as uncovering trends that show the blockchain impact on implementation and record-keeping in accounting are the main goals of this investigation.

2. Literature Review

In this study, a new accounting tool called blockchain technology is being utilised to improve record-keeping that is secure, shareable, verified, and agreed upon.

2.1 Blockchain

Depending on the source, blockchain is either referred to be a type of finance technology or a type of distributed ledger technology [3]. Some people see blockchain as Sequential databases or massive spreadsheets that capture transactional data, are encrypted, and operate under a consensus process outperform traditional financial ledgers. Blockchain has a wide range of definitions that represent how diverse academic fields view it from various angles, suggesting that a singular description is elusive [4].

According to several studies, the key characteristics of blockchains are synchronised networks, consensus, robust authentication, immutability, decentralisation, transparency, and decentralisation. In other words, blockchain technology makes it possible to transfer everything of value, including money and other financial resources as well as intellectual property, health data, votes, and ideas [5].

2.2 AI-enabled Auditing Gains Credibility Thanks to Blockchain

Adopting blockchain technology for accounting record-keeping has as one of its primary goals the creation of trust and a network of confidence, with or without the assistance of a reliable third party. Before hashing and adding the block to the already-existing chain, blockchain accumulates verifiable information about a transaction's value, which it was paid to, and by whom. Blockchain is effective in modern Internet use because participants may witness completely encrypted synchronised transactions thanks to its immutability, traceability, and visibility [6].

2.3 Blockchain Integration with AI

Decision making, fraud detection, audit worksheets, tax compliance and general ledgers can all be examined by auditors using AI technology [7]. Blockchain enables an auditing approach that is more accurate and nimbler, automates assurance, and strengthens audits while delivering trust and confidence in the data, models, and analytics used in AI-based operations [8].

3. METHODOLOGY

The review process utilised in this study was the same as that in Bakker, which involved manually selecting a list of pertinent papers from trustworthy academic journals, then doing a thorough search of the relevant literature using topic-related keywords [9]. Furthermore, given the early stage of development of blockchain technology, this research also delved into relevant industry resources, such as industry magazines, online articles, and news related to the topic [1, 2]. This approach ensured a comprehensive examination of blockchain-related literature, with a specific focus on its potential applications in the establishment of novel accounting record-keeping methods and audit procedures. First, we found publications in the disciplines of accounting, finance [12], information management [3], and innovation that were ranked 5 and 4.

A closer look at each item's title, keywords, and abstract was conducted once the 433 search results were imported into Mendeley to see if they should be kept in the dataset. After eliminating duplicate entries to provide a final dataset with 180 entries, we evaluated the applicability of each input to our study.

4. RESULTS AND DISCUSSION

Due to the fact that researchers based on blockchain is still in its infancy, similar to the literature on industrial strategy, energy and business, the majority of the studies in the dataset (N = 180) are conceptual in nature. Because blockchain started to receive widespread attention in 2017, we conducted searches for "blockchain" in each of the journals from 2017 to 2020. 12 A journals were found in our initial search, and they generated 25 papers from Fig. 17.2.

Figure 17.3 depicts the number of papers that were published between 2017 and 2020 that addressed four issues, including how blockchain would impact AI-based accounting and auditing. These themes include triple-entry accounting employing blockchain technology, real-time accounting, the event-based method of continuous auditing and accounting of the accounting process. The effectiveness of auditing can be improved using data of blockchain that is auditable and traceable.

The four discussed aspects show how blockchain and AI technology can be utilised to alter accounting and auditing in order to increase accountability and boost public confidence in the accounting profession. Accountants may be able to make better judgements with the use of artificial intelligence and blockchain technology, which can both provide immutable, append-only, shareable, verifiable, and agreed-upon data.

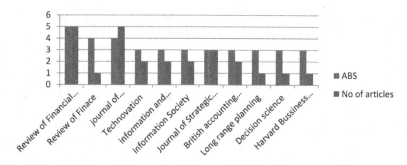

Fig. 17.2 List of reputable academic journals that has been manually screened

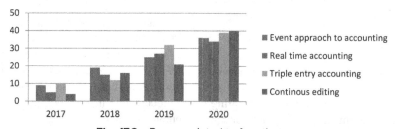

Fig. 17.3 Papers related to four themes

4.1 Discussion

An important step forward in the fields of finance and accounting is the combination of blockchain technology and artificial intelligence (AI) for transparent and secure financial record keeping. In this discussion, we examine the ramifications, significance, advantages, and issues associated with this integration.

- The improvement of transparency is one of the main benefits of merging blockchain with AI in financial record keeping. The distributed ledger technology used by blockchain ensures that all transactions are recorded in an unalterable and secure manner. In financial operations, where trust and accountability are crucial, openness is especially crucial. Stakeholders can have more faith in the accuracy of financial data, including regulators, auditors, and investors.

- Data cannot be changed or erased after it has been recorded thanks to the fundamental property of blockchain technology, immutability. This offers a high level of security and safeguards against fraud and financial record tampering. By identifying anomalies or unauthorized access and notifying the appropriate parties of suspected breaches, AI can further improve security.

- The automation of numerous financial processes, from data entry and reconciliation to complicated analytics, is made possible by the process' incorporation of AI. This lowers the possibility of human error, streamlines procedures, and improves the effectiveness of maintaining financial records.

- Blockchain and AI work together to offer real-time insights into financial data. Decision-makers now have access to the most recent information, enabling them to react to market developments and financial opportunities more quickly.

4.2 Applications and Challenges

Several financial industries, including supply chain financing, anti-money laundering, fraud detection, and regulatory compliance, are already using the integration of blockchain with AI. The way that financial organizations work and interact with data is being revolutionized by these technologies.

Although blockchain and AI integration has potential, there are obstacles that need to be overcome:

- The blockchain network may have scalability problems as the number of financial transactions rises. Exploring options like sharding and off-chain processing is necessary.

- Particularly when working with delicate financial information, ensuring data privacy and compliance with data protection standards is essential.

- It can be difficult and expensive to integrate blockchain and AI systems with legacy financial infrastructure.

- Financial operations, artificial intelligence, and blockchain each have their own rules and guidelines. Careful preparation is necessary to integrate these technologies while maintaining compliance with these rules.

- To guard against cyber-attacks and any weaknesses in both the blockchain and AI components, the integrated system's security must be strong.

An industry-changing change is taking place in the banking sector as a result of the combination of blockchain and AI for transparent and secure financial record keeping. It provides improved security, efficiency, transparency, and real-time information. To fully achieve the promise of this integration, it is necessary to overcome issues with scalability, privacy, interoperability, regulatory compliance, and cybersecurity. We may anticipate a more reliable, effective, and accountable financial ecosystem that benefits all stakeholders as these problems are overcome.

5. Conclusion

This technology makes it possible to provide shared, verified, and accepted auditable data. With the aid of AI approaches, auditing can improve audit effectiveness using traceable and auditable blockchain data. This study illustrates how, because blockchain offers fresh frameworks for managing collaboration, blockchain-enabled accounting may lessen disparities in data and involve all stakeholders. Agency and stakeholder theories are used to analyse the findings. Organizations will face difficulties when contemplating the potential risks of integrating blockchain technology into their accounting processes. To explore additional uses of blockchain-driven accounting, further research is necessary. Finally, this study raises certain queries that potential future research might attempt to address, expanding the field of blockchain literature through empirical study.

References

1. Gomber, P., Kauffman, R.J., Parker, C., Weber, B.W., 2018. On the fintech revolution: Interpreting the forces of innovation, disruption, and transformation in financial services. J. Manag. Inf. Syst. 35 (1), 220–265.
2. Thakkar, P., 2019. BOSS Magazine | How Blockchain is Redefining the Rules of Supply Chain. Boss Magazine. Available at. https://thebossmagazine.com/blockchainsupply- chain/ (Accessed: 16 January 2020).

3. Christie, L., 2018. Distributed Ledger Technology. POST-brief Houses of Parliament. Available at. https://researchbriefings.parliament.uk/ResearchBriefing/ Summary/POST-PB-0028 (Accessed: 26 November 2019).

4. Xu, M., Chen, X., Kou, G., 2019. A systematic review of blockchain. Fin. Innov. 5 (1).

5. Tapscott, D., Tapscott, A., 2016. The Impact of the Blockchain Goes Beyond Financial Services. Harvard Business Review. Available at. https://hbr.org/2016/05/theimpact-of-the-blockchain-goes-beyond-financial-services (Accessed: 25 October 2019).

6. Hughes, A., Park, A., Kietzmann, J., Archer-Brown, C., 2019. Beyond Bitcoin: What blockchain and distributed ledger technologies mean for firms. Bus. Horiz. 62 (3), 273–281.

7. Munoko, I., Brown-Liburd, H.L., Vasarhelyi, M., 2020. The ethical implications of using artificial intelligence in auditing. J. Bus. Ethics 167 (2), 209–234.

8. Cuomo, J., 2020. How blockchain adds trust to AI and IoT. IBM. Available at. https://www.ibm.com/blogs/blockchain/2020/08/how-blockchain-adds-trust-to-aiand-   iot/ (Accessed: 17 July 2021).

9. de Bakker, F.G., Rasche, A., Ponte, S., 2019. Multi-stakeholder initiatives on sustainability: A cross-disciplinary review and research agenda for business ethics. Bus. Ethics Q. 29 (3), 343–383.

Note: All the figures in this chapter were made by the author.

Recent Trends in Engineering and Science for Resource Optimization and
Sustainable Development – Prof. (Dr.) Dorota Jelonek et al. (eds)
© 2024 Taylor & Francis Group, London, ISBN 978-1-032-98030-0

18

Employing IOT and Machine Learning Together to Improve Constructive's Predictive Maintenance

S. K. UmaMaheswaran[1]

Professor, Department of Mathematics, Sri Sairam Engineering College,
Chennai, Tamil Nadu

V. S. Padmini[2]

Assistant Professor,
Department of computer science and engineering,
G.N.D.E.C, Bidar, India

Smrity Prasad

Dayananda Sagar University, Kudlu Gate,
Bengaluru, India

Ike Umejesi[3]

University of Fort Hare, South Africa

Sritam Swapnadarshi Sahu

Department of Civil Engineering,
School of Engineering, SR University, Warangal, India

H. Pal Thethi[4]

Lovely Professional University, Phagwara, India

P. S. Ranjit[5]

Professor, Department of Mechanical Engineering,
Aditya Engineering College Surampalem,
Kakinada, India

▬▬▬ Abstract

A vital component of guaranteeing the effectiveness and economy of construction operations is predictive maintenance. This study investigates how machine learning and the Internet of Things (IoT) could change predictive maintenance in the construction sector. This method enables precise equipment failure and maintenance forecasting by utilizing real-time data from sensors and machine learning algorithms. This paper outlines the potential of utilizing the Internet of Things (IoT) and Machine Learning (ML) to improve a predictive maintenance program for construction sites. The combination of an IoT system, consisting of connected sensors, and algorithms developed through ML can lead to significant improvements in

Corresponding authors: [1]umamaheswaran.maths@sairam.edu.in, [2]pvspvs30@gmail.com, [3]ikeumejesi@gmail.com, [4]h.pal@gmail.com, [5]psranjit1234@gmail.com

DOI: 10.1201/9781003596721-18

the maintenance of construction sites and better outcomes for construction professionals and developers. The use of an IoT and ML framework provides visibility to the various operational factors of a construction project and then applies various models to predict and provide alerts on potential issues. This combination of technologies enables construction professionals to mitigate risks, increase productivity, and lower operational costs associated with maintenance. Furthermore, it provides unprecedented insights into identifying potential sources of problems that would otherwise be difficult to discern. The paper also considers the potential of the technology's ability to autonomously make decisions and even possibly self-heal. The paper concludes with a discussion of the probable future applications of the combined technologies.

Keywords

IoT, Machine learning, Predictive maintenance

1. INTRODUCTION

Predictive maintenance (PdM) is an emerging field of predictive analytics that allows companies to recognize and predict product failures before they happen. This helps organizations make the most of their investments and increase operational efficiency. To obtain maximum benefit from PdM, organizations need to leverage IoT and ML to make accurate predictions with reliable data. The integration of IoT and ML is quickly becoming an essential component of PdM systems.

IoT and ML can help to derive more accurate predictions from large data sets. By combining IoT and ML, companies can gain access to data from multiple sources in an intelligently connected manner, making it easier to perform PdM activities Cheng et al, (2020). ML can then take the data and generate a better understanding of how certain parameters are related to one another and predict possible outcomes. ML can also interpret massive amounts of data quickly, while IoT technology can capture device data and transmit it securely over a network. Combining IoT and ML helps organizations to overcome the challenges posed by traditional PdM, such as high costs and inaccurate predictions Marques et al, (2020). By using IoT and ML together, construction companies can greatly improve their PdM strategies by understanding how certain components of the product interact within a larger system. IoT can be used to always monitor various components of a construction project, and ML can follow changes in that data over time to make more precise predictions about the performance of the system Siraskar et al, (2023). This helps maximize efficiency and reduce maintenance costs. Adding IoT and ML to an existing PdM system has the potential to drastically improve its accuracy and decrease turnaround times. In addition to increased accuracy, IoT-ML integration can also reduce operational costs and improve Constructive' production resources Susto et al, (2014). This would allow the company to maximize their efficiency and reduce the time and money spent on higher maintenance activities. IoT and ML are becoming increas-

ingly important components of PdM. By combining the two technologies it is possible to gather massive amounts of data from multiple sources and quickly derive accurate conclusions to help organizations understand their products better. With the use of IoT and ML, PdM activities can become more efficient and accurate.

2. LITERATURE REVIEW

Abbassi et al, (2022) Engineering infrastructure is vulnerable to catastrophic catastrophes because it frequently employs dangerous materials, sophisticated systems, and human operators. Asset management practises that have been in place for a while are necessary to continuously increase system safety for the buildings and their activities. Numerous incidents may have happened as a result of inefficient maintenance planning procedures, according to the history of hazardous events in various sectors, such as process facilities. Therefore, it is crucial to adopt ideal programmes and useful methods in maintenance planning engineering assets in order to guarantee a sufficient level of system safety and availability. Operators, and regulators have developed workable norms and recommendations for utilising quantitative methodologies in Operation and Maintenance (O&M) planning thanks to the lessons learned from prior mishaps. The current effort intends to outline previous initiatives and pinpoint the shortcomings, requirements, and difficulties in maintenance planning in engineering institutions.

Garcia Valle, (2022) Industrial machinery will ultimately fail, whether it is immediately as a result of production flaws, or over time as a result of dirt buildup, degradation of internal components, or merely wear and demand. Factory operators are forced to cope with what seems like never-ending maintenance and repair cycles by machinery, systems, and industrial lines, especially when undiagnosed problems in gear result in catastrophic failure. If we examine the behaviour of the machine before the problem takes place, maintenance and repair programmes may be arranged more effectively. With it, there can be significant financial savings in an industrial setting because fewer production stops and unneeded maintenance mechanic labour hours are required. PdM is becoming more and more well-known in production lines and is also utilised in the aerospace industry to predict when breakdowns may occur in aircraft parts. With this, it can carry out important maintenance tasks before breakdowns happen and increase the machinery's lifespan.

3. RESEARCH METHODOLOGY

This research project will have both a qualitative and quantitative component. For the qualitative component, a series of interviews, focus groups and a survey will be conducted with construction workers and industry professionals in order to understand the current Predictive Maintenance process, its strengths, and weaknesses, as well as their opinions and preferences on how it can be improved. The quantitative component of the project will further investigate the feasibility and potential of using ML and IoT technologies for PdM in the construction industry. This will be done through the analysis of historical data and simulation of different scenarios where IoT and ML can be combined. After validating the results from these simulations and analysing the data, further optimisations and recommendations can be made.

3.1 Data Collection

To capture real-time data, install IoT sensors and devices on important construction gear. Information like temperature, vibration, pressure, usage, and other pertinent metrics should be recorded by these sensors. To send the information gathered by IoT sensors to a central data repository, establish a reliable data transmission system.

3.2 Data Transmission

Wireless communication protocols like Wi-Fi, cellular networks, or specialized IoT networks can be used to do this. Put the data you've gathered in a scalable and safe database or storage architecture. Solutions that are cloud-based are frequently chosen because of their accessibility and scalability.

3.3 Data Storage

The data gathered in a scalable and safe database or storage architecture. Solutions that are cloud-based are frequently chosen because of their accessibility and scalability.

3.4 Pre-processing and feature extraction

To manage missing numbers, outliers, and maintain data consistency, clean and preprocess the raw data. For accu-

rate machine learning model training, this stage is essential. Determine and extract from the data the pertinent elements that can be applied to predictive maintenance. Features could include sensor data patterns, historical maintenance records, and equipment runtime.

4. RESULTS AND DISCUSSION

As can be seen in Fig. 18.1, after they are created, the classes are not matched since there are substantially less data in the safe class than in the safe worn class. This is typical behaviour because observations are stopped when V B reaches critical values.

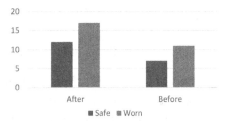

Fig. 18.1 Classes distribution

The categorization task's outcomes are displayed in Table 18.1. The Two-Class Boosted Decision Tree is the most effective algorithm in this situation. It displays the best evaluation metrics, with a nearly 96% accuracy rate.

Table 18.1 Outcomes of classification algorithms

Algorithm	SP	WP	ACC
LogR	0.689	0.657	0.652
DF	0.772	0.786	0.619
BLR	0.601	0.501	0.711
NN	0.553	0.432	0.582

The NN regression technique has the greatest performance for the regression on V B in Table 18.2. This model has the lowest error values and the highest R2, which indicates that it may account for a significant portion of the variance in V B.

Table 18.2 Outcomes of classification algorithms

Algorithm	RMSE	R^2
LogR	0.127	0.178
DF	0.117	0.169
BLR	0.213	0.156
NN	0.012	0.134

In order to guarantee the efficient operation of machinery and equipment in the construction industry, predictive maintenance is a vital component. Unexpected breakdown downtime not only necessitates expensive repairs but can also cause project delays, which reduces overall construction efficiency and profitability. In this context, utilizing IoT (Internet of Things) and machine learning technology together has proven to be a potent way to improve the predictive maintenance strategies used by construction organizations. With the help of this synergy, construction companies may proactively handle maintenance needs, cut downtime, and allocate resources more effectively. Real-time data collecting, analysis, and predictive capabilities are also provided.

It is a progressive and revolutionary strategy to use machine learning and the Internet of Things (IoT) to improve preventive maintenance in the construction sector. In the context of construction predictive maintenance, this debate focuses on the consequences, advantages, difficulties, and considerations of merging IoT with machine learning.

4.1 Integration of IoT and Machine Learning

In order to avoid expensive equipment malfunctions, reduce downtime, and maximize resource allocation, predictive maintenance is essential. This process gains intelligence with the use of IoT sensors and machine learning algorithms, enabling more accurate, data-driven decision-making.

4.2 Early Warnings and Real-Time Data

Construction gear and equipment fitted with IoT sensors continuously gather data on a range of characteristics, including temperature, pressure, vibration, and usage patterns.

Allocating Resources Most Effectively

IoT and machine learning help construction organizations allocate resources more effectively by properly forecasting maintenance needs. This entails planning maintenance as needed rather than relying on predetermined schedules. This not only lowers operating expenses but also lengthens the useful life of the equipment.

4.3 Challenges

Despite the substantial advantages of IoT and machine learning integration, there remain issues to be resolved:

4.4 Data Management

Handling the enormous amount of data that IoT devices produce can be challenging. For precise forecasts, proper data cleansing, preprocessing, and storage are necessary.

4.5 Model Training

Excellent labeled data are necessary for training machine learning models. It can take a while to compile historical data and ensure its accuracy.

4.6 Security and Privacy Issues

They are raised by the use of IoT devices in the building industry. Data security and network security are of utmost importance.

4.7 Interoperability

Since there are many different manufacturers of construction equipment, incorporating IoT solutions may need to adhere to interoperability standards to guarantee seamless data gathering and analysis.

Implementing machine learning and IoT infrastructure can need a substantial initial investment. Companies must compare these expenses to the anticipated return on investment.

4.8 Applications in Real Life

There are several useful uses for the IoT and machine learning integration in construction predictive maintenance. These consist of:

- Estimating the maintenance requirements for large equipment and machinery.
- Scheduling repairs to reduce downtime throughout important building periods.
- Identifying trends in equipment use to improve resource allocation.
- Real-time evaluation of the condition and effectiveness of construction trucks.
- Increasing safety by identifying potential problems that can cause accidents.

A paradigm shift in the industry will occur with the use of IoT and machine learning for increased predictive maintenance in the construction sector. It enhances safety and lowers costs while also increasing operational effectiveness and equipment reliability. The benefits of this integration make it an appealing path for the construction sector to explore and utilize for its long-term success, despite the obstacles to be addressed, including data management, security, and interoperability. The approach, advantages, drawbacks, and practical applications of IoT and machine learning in construction predictive maintenance are all covered in this study. According to the findings, this integrated approach has the potential to boost resource allocation, decrease downtime, and considerably increase operational efficiency in the construction industry. By examining these elements, this research hopes to advance the use of predictive maintenance techniques in the construction sector.

5. CONCLUSION

The integration of IoT and ML enables PdM of constructions sites to be more efficient and effective than ever. By using a combination of data from sensors, ML algorithms, and predictive analytics, it is possible for companies to leverage the power of the IoT to enable more precise and informed decisions on maintenance tasks. By using ML to analyse all the data points being collected by various sensors, companies can more accurately anticipate potential performance issues and address them before they impact operations. This enables them to maximize the effectiveness of any preventive maintenance activities being carried out, as well as to minimize unplanned downtime. This greatly benefits companies by enabling them to identify potential problems before they occur, helping to reduce costs and increase productivity. Additionally, by using PdM and the IoT together, companies can better identify and respond to trends in machine behaviour, helping them to adjust to unanticipated changes more readily in the market. All told, the development of IoT and ML can greatly improve PdM in constructions sites, helping to reduce costs and increase productivity while achieving greater levels of overall success.

References

1. Marques, S., Oliveira, J., and de Melo, O. (2020). Predictive Maintenance: A Review of Machine Learning Techniques for Its Diagnosis and Forecasting. IEEE Latin America Transactions, 18(6), 1429–1442.

2. Cheng, J. C., Chen, W., Chen, K., & Wang, Q. (2020). Data-driven predictive maintenance planning framework for MEP components based on BIM and IoT using machine learning algorithms. Automation in Construction, 112, 103087.

3. Susto, G. A., Wan, J., Pampuri, S., Zanon, M., Johnston, A. B., O'Hara, P. G., & McLoone, S. (2014, August). An adaptive machine learning decision system for flexible predictive maintenance. In 2014 IEEE International Conference on Automation Science and Engineering (CASE) (pp. 806–811). IEEE.

4. Siraskar, R., Kumar, S., Patil, S., Bongale, A., & Kotecha, K. (2023). Reinforcement learning for predictive maintenance: a systematic technical review. Artificial Intelligence Review, 1–63.

5. Abbassi, R., Arzaghi, E., Yazdi, M., Aryai, V., Garaniya, V., & Rahnamayiezekavat, P. (2022). Risk-based and predictive maintenance planning of engineering infrastructure: existing quantitative techniques and future directions. Process Safety and Environmental Protection.

6. Garcia Valle, C. (2022). An approximation of predictive maintenance for a plane propeller through sensorization, vibration analysis and Machine Learning (Bachelor's thesis, Universitat Politècnica de Catalunya).

Note: All the tables and figure in this chapter were made by the author.

*Recent Trends in Engineering and Science for Resource Optimization and
Sustainable Development – Prof. (Dr.) Dorota Jelonek et al. (eds)
© 2024 Taylor & Francis Group, London, ISBN 978-1-032-98030-0*

19

Trusted Payment Methods Based on Blockchain For Electronic Government Services

Chintala Lakshmana Rao[1]

Associate Professor, School of Law,
GITAM (Deemed to be University), Visakhapatnam, India

Shamik Palit

Program Director and Associate Professor,
Department of Computing Science and Software Engineering,
University of Stirling RAK Campus, UAE

Varalaksmi Dandu[2]

Assistant Professor,
JNTUA School of Management Studies, JNTUA

Gargi Verma[3]

B.Tech CS Student,
D.Y. Patil International University Pune, India

Sorabh Lakhanpal[4]

Lovely Professional University, Phagwara, India

Ahmad Y. A. Bani Ahmad[5]

Department of Accounting and Finance Science,
Faculty of Business, Middle East University, Amman, Jordan

▬▬▬ Abstract

Secure, effective, and reliable payment systems are essential as electronic government services take center stage in
contemporary governance. Blockchain technology offers a strong response to these objectives because of its reputation
for cryptographic security and transparency. This study examines the usage of blockchain-based payment systems in the
context of electronic government services with an eye on how they might improve user security and trust. This paper will
propose a new trusted payment solution for e-government services, based on distributed ledger technology. The goal of
this proposal is to improve the transparency, security, and trustworthiness of payments made to government agencies.
The proposed solution will be based on the principles of blockchain and will include a consensus mechanism to ensure
speed and accuracy of transactions. The proposed payment network will be consumer-friendly and enable both direct
payments and the satisfaction of financial obligations. The paper will also discuss the challenges that must be overcome

Corresponding authors: [1]lchintal@gitam.edu, [2]Varalakshmi.jntua@gmail.com, [3]gargi2004india@gmail.com, [4]sourabh.lakh@gmail.com,
[5]aahmad@meu.edu.jo

DOI: 10.1201/9781003596721-19

in order to maximize the potential of blockchain technology for e-government services, such as scalability, privacy, and regulatory compliance. Finally, the paper will outline how the proposed solution can be implemented, and potential options for incentivizing the adoption of the new system. This paper will be of great value to those interested in the potential of blockchain and distributed ledger technology for revolutionizing the government services sector.

Keywords

Blockchain, Payment method, Electronic government services

1. INTRODUCTION

The offering of electronic government services has evolved into a crucial element of contemporary governance in today's increasingly digital environment. Businesses and citizens alike want quick, easy, and secure ways to communicate with and pay for various government services. The selection of payment mechanisms is crucial in guaranteeing user trust, efficiency, and security as electronic government services continue to develop. Governments may offer reliable and effective payment mechanisms that not only improve user experience but also reduce risks related to fraud and data breaches by utilizing the possibilities of blockchain. In the context of electronic government services, this study investigates the usage of secure payment systems based on blockchain technology. It explores the possible benefits, difficulties, and effects of using blockchain for payments with the goal of illuminating how this ground-breaking strategy might promote trust and security for users interacting with government services online.

The development of blockchain technology has enabled many industries to evaluate and implement digital solutions, allowing them to become more efficient, secure, and transparent. One of the most valuable elements of blockchain technology is its ability to provide trust less and se-

cure payment methods for electronic government services Sujatha et al, (2020). The growth of digital government services has enabled government agencies to become more efficient, reduce costs, and provide services to citizens faster and easier than ever before. However, with the increasing use of digital services, the risk of fraud and cybercrime has also become increasingly apparent.

Traditional payment methods do not offer the necessary security to ensure secure and reliable payments, leaving citizens vulnerable to fraud or other malicious intentions. This is why the use of blockchain technology and trusted payment methods has become increasingly necessary for electronic government services Shuaib et al, (2020). Blockchain technology offers an immutable and secure ledger that can be used to easily maintain accuracy, security and trust between government organizations and citizens. This ensures that citizens' money is not exposed to fraudulent activities or any other malicious activities.

The blockchain technology also utilizes existing infrastructure to create an ecosystem where security protocols are in place to protect against cybercrime and other fraudulent activities. This makes it extremely complicated for hackers or other malicious actors to gain access to these payment methods without being detected Kuperberg et al, (2019). This ensures that the government can protect its citizens, while also providing a secure platform for citizens to engage in electronic transactions with confidence. Beyond security, blockchain technology also provides governments with options for creating payment systems that are decentralized and permissionless. This allows citizens to easily pay for services or to engage in peer-to-peer payments without having to be reliant on any centralized authorities. This makes it easier for citizens to send and receive payments with confidence in knowing that their money is safe and secure.

In essence, blockchain technology provides an efficient and secure method for governments to offer trusted payment options for electronic services Alketbi et al, (2018). By providing a secure and immutable ledger that can be used to store and transfer sensitive data, blockchain technology ensures that citizens are able to transact with confidence

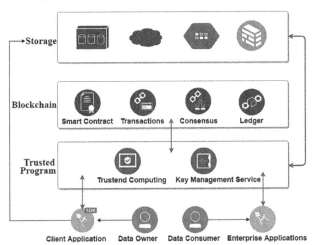
Fig. 19.1 Blockchain for trusted transactions

and without the threat of cybercrime or other malicious activity.

1.1 Objectives of the Study

1. This study aims to add to the continuing discussion about modernizing public administration and promoting more trust between governments and their citizens by examining the incorporation of blockchain into electronic government payments. It also emphasizes how crucial it is to balance cutting-edge technological advancements with strong security measures in the pursuit of more effective and user-friendly electronic government services.

2. This ensures that all digital transactions are handled securely and efficiently, while also ensuring that citizens are protected from any potential risks.

2. LITERATURE REVIEW

A research paper by Tam et al. (2020) presents a two-tier blockchain system to provide trust guarantees in the transaction of EGS. This creates a remote third-party to validate the authenticity of the transaction through consensus and recording on the blockchain which eliminates the possibility of data corruption and the risk of malicious attacks (Kumar., 2023). With this system, stakeholders in the delivery of EGS can collaborate more effectively because of the secure platform the blockchain provides (Pawełoszek., 2000).

In addition, Zheng et al. (2020) establishes a blockchain-based supplier trust system for exchanging EGS data. This algorithm enables trust and consensus of the information exchanged between different stakeholders, making the transaction secure. The system also helps reduce transaction costs by automatically generating invoices and allowing for swift and streamlined payment processing (Pawełoszek et al., 2022).

In conclusion, blockchain-based payment methods present a secure and trusted system for EGS payment processing. With its robust security protocols and distributed ledger model it provides a platform through which transactions can be transparently and privately conducted. Furthermore, its ability to facilitate collaborations between stakeholders makes it a viable option for delivering EGS (Korczak& Pawełoszek.,2022).

3. RESEARCH METHODOLOGY

The objective of this research is to analyze the use of blockchain-based secure payment systems in electronic government services and how they can provide trust and safety for users of government services. The research design for this project will be a qualitative research method, which aims to explore and explain the topic at hand.

3.1 Data Collection

The data collection process will involve interviews with experts in the field of electronic government services and blockchain technology. The sources of data for this project will include literature from academic journals, government websites, industry articles and reports, and expert interviews. Various data collection techniques will be used for this project. These techniques will include interviews, direct observation, document analysis, and online surveys.

3.2 Data Analysis

The data collected from the interviews, direct observation, document analysis, and online surveys will be analyzed using qualitative content analysis. This technique involves coding the data to reduce it to salient themes and patterns that can be evaluated and analyzed.

This methodology offers a methodical and thorough way for looking into the viability and appeal of integrating blockchain-based payment systems into online government services. It allows a comprehensive understanding of user preferences, issues, and possible benefits by combining both quantitative and qualitative data collection approaches. The goal of the research is to aid in the implementation of dependable payment systems based on blockchain in the public sector.

4. RESULTS AND DISCUSSION

The development of 5G in 2019 has led to improvements in Internet infrastructure, the maturation of the global logistics network, and a high growth trend in e-commerce. According to the current state of advancement, it is anticipated that global e-commerce, which will be impacted by the emerging crown pandemic, will continue to rise by 10% to 20% on a B2C transaction scale over the next few years. (see Fig. 19.2).

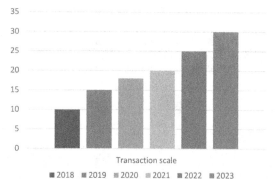

Fig. 19.2 Forecast of the volume of B2C e-commerce transactions globally from 2018 to 2023

The percentage of B2B transactions is decreasing year over year while the percentage of B2C transactions is increasing, based to the "2019 China's e-commerce market data monitoring report" and the "2018-2023 China's e-commerce transaction scale."

Table 19.1 Scale of China's B2B and B2C e-commerce exports from 2018 to 2023

Year	B2B Transaction scale	B2C Transaction scale
2018	55	60
2019	43	50
2020	34	39
2021	13	17
2022	25	28
2023	27	30

Amazon continues to hold the top spot among e-commerce platforms, with 73% of consumers choosing to conduct their transactions through the Amazon platform, 43% using Global Express (Alibaba's e-commerce network), and 14% using eBay.

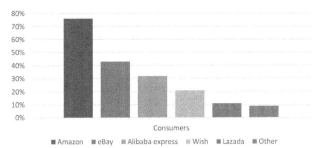

Fig. 19.3 Market share of e-commerce platforms

A important step forward in the digital transformation of government operations is the use of reliable payment systems based on blockchain for electronic government services. In the context of electronic government services, this discussion examines the main ideas and consequences of applying blockchain technology to safe and open financial transactions.

- Government payment systems' usage of blockchain dramatically improves security and trust. Because transactions are immutable and visible thanks to blockchain's decentralized ledger, there is a lower chance of fraud and unauthorized changes. Both citizens and government organizations can be confident that their payments are safely documented and untouchable.
- Blockchain is a potent weapon in the battle against fraud and corruption in government transactions thanks to its built-in security features. Blockchain can automatically enforce laws and do away with the need

for middlemen by using smart contracts and cryptographic verification, which lowers the possibility of corrupt behaviours.

- Good governance is built on the principle of transparency. All payment transactions are made available to authorized parties thanks to blockchain's transparent ledger. This openness encourages accountability within government organizations and gives citizens a clear understanding of how their tax dollars is being spent.
- Government payments might significantly reduce costs if blockchain technology is used. The government can lower transaction costs and administrative costs by doing away with intermediaries and automating procedures, thus saving money for the taxpayers.

5. Conclusion

In conclusion, blockchain-based trusted payment methods offer great opportunities for electronic government services. The blockchain technology is secure, efficient, and cost-effective, making it an attractive technology for governments looking to increase data security and improve public service delivery. Governments can capitalize on this technology to reduce costs, increase efficiency, and promote transparency. Blockchain-based trusted payment methods have the potential to revolutionize the way governments provide electronic services and can help increase trust between the government and citizens. A crucial step towards a more secure, open, and effective digital government ecosystem is the use of trustworthy payment systems based on blockchain technology for electronic government services. Enhancing security, reducing fraud and corruption, increasing transparency and accountability, streamlining operations, and maybe saving money are just a few of the advantages that this invention brings. Additionally, it encourages inclusivity and accessibility, opening up access to government services to a wider range of the populace. To ensure public confidence in these payment systems, it is crucial to solve privacy issues. To fully achieve the potential of blockchain-based payment systems for electronic government services, public education and adoption will be essential. Governments must adapt as technology develops in order to take advantage of blockchain's advantages while protecting their citizens' security and privacy. Governments may deliver more reliable and efficient services to their constituents while also advancing the development of electronic government services by putting these reputable payment mechanisms into place.

References

1. Sujatha, R., Navaneethan, C., Kaluri, R., & Prasanna, S. (2020). Optimized digital transformation in government

services with blockchain. In Blockchain Technology and Applications (pp. 79–100). Auerbach Publications.

2. Shuaib, M., Daud, S. M., Alam, S., & Khan, W. Z. (2020). Blockchain-based framework for secure and reliable land registry system. TELKOMNIKA (Telecommunication Computing Electronics and Control), 18(5), 2560–2571.

3. Kuperberg, M., Kemper, S., & Durak, C. (2019). Blockchain usage for government-issued electronic IDs: A survey. In Advanced Information Systems Engineering Workshops: CAiSE 2019 International Workshops, Rome, Italy, June 3–7, 2019, Proceedings 31 (pp. 155–167). Springer International Publishing.

4. Alketbi, A., Nasir, Q., & Talib, M. A. (2018, February). Blockchain for government services—Use cases, security benefits and challenges. In 2018 15th Learning and Technology Conference (L&T) (pp. 112–119). IEEE.

5. Tam, D. V., Tuyen, P. M. H., Cu, T. H., & Ho, H. T. (2020). Towards a secure and distributed platform for efficient electronic government services in Vietnam. Government Information Quarterly, 37(2), 279–287.

6. Zheng, Y., Sun, X., Kuang, B., & Lu, S. (2020). A trusted payment method for the e-government service data exchange based on blockchain technology. IEEE Transactions on industrial informatics, 16(3), 1744–1751.

7. Pawełoszek, I., Kumar, N., & Solanki, U. (2022). Artificial intelligence, digital technologies and the future of law. Futurity Economics & Law, 2(2), 24–33. https://doi.org/10.57125/FEL.2022.06.25.03

Note: All the figures and table in this chapter were made by the author.

Recent Trends in Engineering and Science for Resource Optimization and
Sustainable Development – Prof. (Dr.) Dorota Jelonek et al. (eds)
© 2024 Taylor & Francis Group, London, ISBN 978-1-032-98030-0

20

Predictive Service in the Automotive Industry: Coupling Machine Learning and IoT

Gajanan Patange
Department of Mechanical Engineering, CSPIT,
CHARUSAT, Charotar University of Science and Technology (CHARUSAT),
CHARUSAT campus, Changa, India

Nidhi Bhavsar[1]
Assistant Professor, Department of Computer engineering,
Atharva college of Engineering, Mumbai, India

Asheesh Kumar
Assistant Professor, Department of Mechanical Engineering,
Mahatma Gandhi Institute of Technology,
Hyderabad, India

Jelena Franjkovic[2]
Osijek, J. J. Strossmayer University of Osijek, Croatia

P. Satish kumar
Associate Professor, Dept. of Mechanical engineering,
School of Engineering, SR University, Warangal, India

Sorabh Lakhanpal[3]
Lovely Professional University, Phagwara, India

P. S. Ranjit[4]
Professor, Department of Mechanical Engineering,
Aditya Engineering College Surampalem, Kakinada, India

▬▬▬ **Abstract**

The fusion of machine learning and the Internet of Things (IoT) is undergoing a substantial revolution in the automotive industry. A promising use of this convergence is predictive service, which improves vehicle maintenance by providing real-time monitoring and proactive problem solving. This study examines how machine learning and the internet of things may work together to provide predictive maintenance, highlighting the potential for improved maintenance, safety, and overall driving performance. The report offers important insights into the future of automotive service by examining significant use cases, difficulties, and the influence of this technology in the automobile industry. The Internet of Things (IoT) is gaining ground and becoming more and more necessary in daily life. Recent advancements in maintenance modeling, propelled by data-driven methodologies such as machine learning and the Internet of Things, have unlocked a broad

Corresponding author: [1]nidhibhavsar2861988@gmail.com, [2]jelena.franjkovic@efos.hr, [3]sourabh.lakh@gmail.com, [4]psranjit1234@gmail.com spectrum of potential applications. Maintaining functional safety while reducing maintenance costs has always been a primary objective in the automotive industry. In order to do this, predictive maintenance (PdM) is a crucial tactic. With so much operating data available from contemporary cars, ML and IoT are a fantastic alternative for PdM. The increase of articles in this field demonstrates how important such study is. As a result, we use an application to survey, categorise, and assess papers.

▬▬▬ Keywords

Machine learning, IoT, Predictive maintenance, Automotive industries

1. INTRODUCTION

Technology breakthroughs, notably the fusion of Machine Learning and the Internet of Things (IoT), are driving a revolution in the automotive industry. With predictive service standing out as a game-changer in car maintenance, this confluence provides creative answers to age-old problems. In conventional vehicle maintenance, problems are dealt with reactively, frequently following their involvement in breakdowns or accidents. Predictive service, however, offers proactive and real-time monitoring of a vehicle's status thanks to IoT sensors and Machine Learning algorithms. It not only anticipates possible problems but also makes it easier to perform preventative maintenance, thereby boosting safety and convenience for car owners. Data-driven approaches like machine learning, which cover predictive maintenance to predictive quality and encompass monitoring of plant facilities, guarantee analytics, and safety analytics, are increasingly being used in manufacturing and mobility solutions [1]. Prognostics and health management (PHM), sometimes referred to as maintenance 4.0 or smart maintenance, is the discipline focused on the creation of methods to safeguard the soundness of components, products, and structures. It achieves this by scrutinizing, identifying, or predicting issues stemming from performance

shortcomings that might compromise safety[2]. The automotive industry can now collect enormous amounts of real-time data from multiple sensors mounted in vehicles because to the development of the Internet of Things [3].

This change in the automotive industry is especially necessary now that vehicles are getting more complicated and have more integrated systems that require precise and timely maintenance. The automobile industry may provide predictive maintenance advice and optimize vehicle service schedules by combining IoT sensors to gather real-time data from vehicles with Machine Learning algorithms to analyze this data. The transformation of the automotive sector into a more complex system is a prime illustration regarding the way machine learning (ML) and IoT have changed an industrial sector. The need for cost-effective technological solutions to ensure the longevity, effectiveness, and dependability of automobiles is increasing significantly [4], particularly in view of recent advancements in autonomous driving and the change in drive-train. The PdM of safety- and cost-relevant components using ML and IoT is a significant solution approach that is drawing more attention from researchers. This approach tries to manage the substantial system complexity while utilising the abundant data sources of the vehicles.

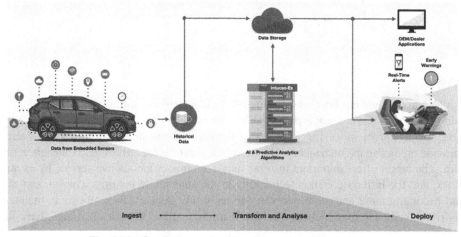

Fig. 20.1 Predictive maintenance in automotive industry

1.1 Research Objective

1. Finding IoT-based PdM for machine learning-based automotive systems and further investigating them from a machine learning viewpoint are the project's main objectives.

2. The goal of the study is to uncover unresolved issues and propose potential areas for future research in order to advance the discipline and generate new research topics.

2. LITERATURE REVIEW

Ali focuses on the most recent research and advancements in the area of sound emission signal analysis employing artificial intelligence for machine status monitors and issue diagnosis [5]. Deep Learning techniques have been discussed as a way to control systems health and monitor machine health, respectively. Bhargava supplied a collection of articles authored by several authors that covered an extensive spectrum of dependability predictions for electrical parts and a wide range of AI techniques.

Some surveys of AI vehicle applications tended to concentrate on applications for autonomous driving. The writers provide a quick overview of technical system maintenance. The poll itself is not automotive-specific, despite the fact that one of their case studies comes from the automobile industry. Deep learning (DL) for automotive systems is discussed by Falcini et al. [6], albeit they do not specifically include preventive maintenance.

Despite not focusing on automotive applications, Wu et al.'s classification of recent research can be seen as a significant foundation for our contribution [7]. The application of deep learning for machine fault diagnosis and PHM was the topic of numerous studies that examined existing trends and prospective future research directions. The two evaluations by Ahsan and Sankavaram that incorporate PdM, data-driven methodologies, and automotive applications are the works that are most comparable to our survey [8].

3. METHODOLOGY

In order to find Scopus-indexed articles that covered predictive maintenance for automotive systems utilising machine learning and IoT, we defined search parameters. This resulted in 390 publications overall.

- Regarding preventive maintenance for automotive systems based on machine learning, we aggregate and assess research articles. 65 publications in total were polled. The axes of "maintenance benefit" and "complexity" were used to divide PdM into the three subfields of statistical [9], condition-based [10] and

remaining usable life [11] in order to organise the discipline (see Fig. 20.1).

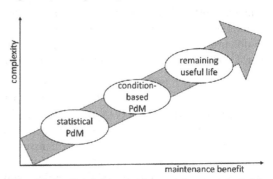

Fig. 20.2 Subfields of a few publications in PdM

- Using the exclusion criteria, we manually reviewed the list of articles in a repeatable manner. Additionally, we only selected works with five or more citations. As a result, 40 papers were written.

- Even though the number of citations is a commonly used metric to assess the worth of a piece of work, it introduces a bias in favour of older works because newly published works are more likely to have fewer citations. Without respect to the number of citations, we also picked 6 papers with publication dates in 2020 or after.

- We looked at result and grouped the findings by main use case. We carried out an individual targeted search to shed more light on these use cases. We manually picked 19 papers from this search that are designated.

4. RESULTS AND DISCUSSION

To gain further information, a quick bibliometric study was carried out. The fact that ML-enabled PdM for automotive systems has grown quickly since 2018 (see Fig. 20.3), despite the ML field's development, highlights the industry's expanding significance [12].

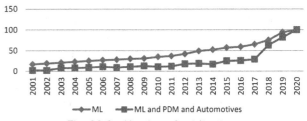

Fig. 20.3 Number of publications

The number of papers on general machine learning (blue) and ML-based automotive system preventative maintenance (green) is depicted in Fig. 20.1. The publications have converted as percentages in 2020, and the upper right corner displays the absolute values for 2020 to facilitate comparisons. Since 2018, there has been a growth in the usage of ML in automotive systems, and the field is expanding.

Utilising real-world data allows for the evaluation of developed solutions already being used. Unfortunately, because recognition involves a lot of effort and specific knowledge, real-world data is sometimes only partially annotated or not at all.

Some articles were subjected to many levels of supervision since they looked at a variety of methodologies. Even though certain unsupervised methods don't need labels, supervised or semi-supervised techniques often result in more trustworthy outcomes from Table 20.1. Even if the models were developed on unlabeled data, testing models still needs at least tagged data.

Table 20.1 The subcategories of PdM and machine learning are used to categorise survey methodologies

	Supervised learning	Unsupervised learning	Semi supervised learning
RUL	6	3	0
Statistical PDM	4	5	2
Condition based PDM	31	3	6

Predictive maintenance in the automotive industry has also been transformed by the confluence of machine learning and the internet of things. Modern cars generate an enormous quantity of operating data, including sensor readings, diagnostic data, and real-time performance metrics, which has opened up a multitude of possibilities for enhancing vehicle maintenance and lowering unforeseen breakdowns. In this conversation, we examine the value of applying ML and IoT to the context of preventative maintenance for modern cars and the expanding corpus of research in this area.

- A data-driven era for predictive maintenance in the automotive sector has arrived thanks to ML and IoT. The ongoing flow of data from vehicles enables proactive problem-solving, prompt maintenance, and a decrease in the likelihood of expensive failures.

- By analyzing this data for patterns and abnormalities, machine learning algorithms can help identify the parts or systems that need maintenance. This helps to boost vehicle dependability, which is important for both consumer satisfaction and the reputation of automobile businesses' brands.

- PdM enabled by ML and IoT results in substantial cost savings for both car owners and manufacturers. The entire cost of owning a car can be decreased by avoiding breakdowns and performing maintenance as needed rather than on a set timetable.

5. CONCLUSION

Predictive service also has broad ramifications for auto security. Early identification of possible problems enables prompt repairs or replacements, reducing the safety concerns brought on by mechanical failures. Additionally, it increases comfort for car owners by bringing them peace of mind and minimizing the inconveniences associated with unforeseen breakdowns. Predictive service deployment is not without its difficulties, though. To protect against potential attacks, cybersecurity measures must be strong and privacy concerns related to the data acquired by IoT devices addressed. Additionally, to fully realize the potential of predictive service, the automotive sector must ensure interoperability and standardization. In this paper, we reviewed and categorised recent developments in PdM for ML-IoT enabled automotive systems. This was done by using a repeatable research process, predetermined inclusion and exclusion criteria, and a systematic literature search on Scopus. 62 papers were produced as a result, which were analysed and categorised in accordance with the following criteria: the relevant predictive maintenance categories, which were predetermined based on the complexity and benefit of maintenance, (a) the relevant use cases, (b) the machine learning techniques used, (c) the ML tasks, and (d) the relevant use cases. In addition to the poll, we performed a bibliometric review to identify important books, contemporary authors, often cited use cases, and frequently. Predictive service is a testament to the ability of cutting-edge technologies to improve safety, effectiveness, and convenience for both car makers and consumers as the automobile industry continues to develop. Future vehicle maintenance could be proactive, data-driven, and user-centered thanks to the convergence of machine learning and the internet of things. This study underlines the value of predictive service for the automobile industry and the continual innovation and cooperation that are required to fully realize its potential in future work.

References

1. The Manufacturer. Annual manufacturing report 2020. Technical report, Hennik Group; 2020.
2. Pecht MG, Kang M. Prognostics and health management of electronics: fundamentals, machine learning, and the internet of things. Wiley - IEEE, Wiley; 2018.
3. Munirathinam, S. (2020). Industry 4.0: Industrial internet of things (IIOT). In Advances in computers (Vol. 117, No. 1, pp. 129–164). Elsevier.
4. Becker Jan, Helmle Michael, Pink Oliver. System architecture and safety requirements for automated driving. In: Automated driving: safer and more efficient future driving. Cham: Springer International Publishing; 2017, p. 265–83. http://dx.doi.org/10.1007/978-3-319-31895-0_11.

5. Ali Yasir Hassan. Artificial intelligence application in machine condition monitoring and fault diagnosis. In: Artificial intelligence. Rijeka: Intech Open; 2018, http://dx.doi.org/10.5772/intechopen.74932.

6. Falcini F, Lami G, Costanza A. Deep learning in automotive software. IEEE Softw 2017;34(03):56–63. http://dx.doi.org/10.1109/MS.2017.79.

7. Wu Shaomin, Wu D, Peng R. Machine learning approaches in reliability and maintenance: classifications of recent literature. In: 2020 Asia-Pacific International Symposium on Advanced Reliability and Maintenance Modeling. 2020, p. 1–5.

8. Ahsan M, Stoyanov S, Bailey C. Prognostics of automotive electronics with data driven approach: A review. In: 39th international spring seminar on electronics technology. 2016, p. 279–84. http://dx.doi.org/10.1109/ISSE.2016.7563205.

9. Manpreet Singh Bhatia, Alok Aggarwal, Narendra Kumar. (2020). Speech-To-Text Conversion Using GRU And One Hot Vector Encodings. Pal Arch's Journal of Archaeology of Egypt / Egyptology, 17(9), 8513–8524. Retrieved from https://archives.palarch.nl/index.php/jae/article/view/5796

Note: All the figures and table in this chapter were made by the author.

*Recent Trends in Engineering and Science for Resource Optimization and
Sustainable Development – Prof. (Dr.) Dorota Jelonek et al. (eds)
© 2024 Taylor & Francis Group, London, ISBN 978-1-032-98030-0*

21

Assessing the Effects of 5Gon Electronics Engineering

Ramapati Mishra

Director, Institute of Engineering and Technology,
Dr. RML Avadh University, Ayodhya, India

N.M.G. Kumar[2]

Professor, Dept of EEE, Mohan Babu University,
Sree Vidyanikethan Engineering College,
A. Rangam, India

Atul Singla[3]

Lovely Professional University,
Phagwara

Ramesh Gullapally[4]

Associate Professor, Department of ECE,
St. Martin's Engineering College,
Secunderabad, India

▬▬▬ Abstract

A new era of communication has begun with the introduction of 5G technology, which promises unheard-before speeds, dependability, and minimal latency. This study investigates the significant impacts of 5G on the discipline of electronics engineering. It looks into how the Internet of Things (IoT), communications, and electrical device development are being influenced by 5G. The study looks at how 5G could change the field of electronics engineering as well as the opportunities and difficulties it presents. This article aims to explore the impact of 5G technology on the field of Electronics Engineering. Through research and analysis, the potential implications of this revolutionary new technology for the industry are assessed. 5G technology provides immense benefit in the form of faster speeds, greater coverage, and low latency, thereby allowing for the emergence of new, more efficient technologies that may revolutionize the Electronics Engineering industry in a multitude of ways. The article evaluates the potential implications of 5G technology on the development, implementation, and widespread use of Electronic Engineering related products. Furthermore, the article discusses the opportunities, associated challenges, and considerations to ensure successful implementation of 5G technology in the field of Electronics Engineering.

Corresponding authors: [1]leszek.ziora@pcz.pl, [2]nmgkumar@gmail.com, [3]atul.singla12@gmail.com, [4]rameshreddyece@smec.ac.in

DOI: 10.1201/9781003596721-21

■■■■■■ **Keywords**

5G, Electronics engineering, Networks

1. INTRODUCTION

In the new age of fifth-generation (5G) communications, there are many technological advances that have been made to provide a more enhanced user experience for both communication and internet sources. One of the most important implications of 5G technology is its impact on electronics engineering. With increased bandwidth, improved latency, and enhanced wireless communication capabilities, 5G technology has the potential to revolutionize the field of electronics engineering. This paper will explore the impact of 5G technology on electronics engineering, looking at the current state of the field and the potential implications for the future Morocho-Cayamcela et al, (2019). To begin with, 5G technology is an immensely complex field. It uses a variety of wireless communications technologies and cutting-edge software applications to provide users with new capabilities.

The launch of 5G technology represents a critical turning point in the telecoms industry and promises to transform how we connect, communicate, and use technology. The implications of this game-changing technology on electronics engineering are significant and diverse as 5G networks continue to spread around the globe. With its impressive data rates, low latency, and improved dependability, 5G has created new opportunities for the design of electronic systems and gadgets. It is a driving force behind advancements in electronics engineering that have made it possible to develop cutting-edge technologies that were formerly thought to be unattainable.

Fig. 21.1 Impact of 5G

One of the most interesting aspects of 5G technology is its potential impact on electronics engineering. The increased connection speeds, enhanced signalling capabilities, and low latency of 5G technology can be used to improve the performance of electronic components such as chips and circuits. Furthermore, the low power consumption and high-signal-to-noise ratio of 5G networks can also be used to reduce power consumption of electronics components Ge et al, (2016). The impact of 5G technology on electronics engineering can be seen in a variety of applications, ranging from the development of autonomous vehicles to the development of medical devices. In the case of autonomous vehicles, 5G technology can be used to improve sensor processing, giving the vehicle the ability to react quickly to environmental changes. Similarly, 5G technology can be used in medical devices such as pacemakers to improve their accuracy and reliability.

The development of 5G technology has also spurred the development of new hardware and software solutions for electronics engineering Guidotti et al, (2019). These solutions are designed to take advantage of the increased connection speeds and improved signal processing offered by 5G technology. Examples of this include the development of new chips and circuits with improved signal processing capabilities, as well as new software solutions such as 5G-ready cores and 5G-enabled applications.

Finally, the impact of 5G technology on electronics engineering can also be seen in emerging fields such as the Internet of Things (IoT). IoT is a concept that involves connecting several devices to each other via a wireless network. 5G technology can be used to improve the connection speed and signal quality of these devices, as well as to reduce power consumption. This could potentially lead to a more efficient, secure, and cost-effective way of connecting various devices to each other Kanellos et al, (2016). In conclusion, 5G technology is having a major impact on the field of electronics engineering. Its increased connection speeds, improved signal processing, and low power consumption are enabling the development of new hardware and software solutions that could revolutionize the field. As the technology continues to improve and become more ubiquitous, the implications for electronics engineering will become more pronounced.

1.1 Objective of the Study

1. This study intends to investigate the various ways that 5G is affecting the discipline of electronics engineering.
2. It investigates how 5G will affect the development of the Internet of Things (IoT), communication networks, and electronic device design.

3. This study aims to offer insights into how electronics engineering is changing in response to this groundbreaking technology by exploring the potential and difficulties presented by 5G.

2. LITERATURE REVIEW

In general, 5G technology contributes to developing advanced, highly efficient networks for the electronics engineering industry (Aggarwal et al., 2020). It further enhances the speed of digital communication, while providing reliable and cost-effective services (Bhatia et al., 2020). Through support from heterogeneous networks such as terrestrial radio access solutions and satellite-based connections, 5G networks can increase the number of connected users by up to 1000x (Qadasi et al., 2018).

Furthermore, 5G technology has also opened the door to innovations like edge computing and virtualization. By bringing the needed processing speed and throughput, edge computing helps in reducing the load on both the local networks and the cloud infrastructure (Venkatapathy et al., 2019). On the other hand, virtualization of resources provides flexibility to assign specific resources on demand which can further reduce the operating cost of the system and keep the machines up-to-date (Grabinska et al., 2020).

In conclusion, 5G technology provides diverse opportunities for the electronics engineering community and can be used to realize vast automation possibilities. The sector is expected to benefit from higher data rates, low latency, and improved scalability while reducing the cost of mobile communications (Wieczorkowski et al., 2021). The findings from the literature review demonstrate that 5G technology is very instrumental in transforming the electronics engineering sector by heralding an era of unprecedented innovation.

3. RESEARCH METHODOLOGY

The research objective of this study is to assess the effects of 5G on electronics engineering. To address this research objective, the following methodology will be employed. The study will use a descriptive research design to assess the current effects of 5G on electronics engineering. This research design is appropriate, as it allows for the collection, analysis, and interpretation of the current effects that 5G has on electronics engineering. Both primary and secondary data will be collected for this research. Primary data will be collected through survey questionnaires that will be distributed to electronics engineers in different parts of the world. The survey will be designed to ask questions about the current state of electronics engineering and how it is being impacted by 5G technology. If there is a need to collect

additional primary data, interviews or focus groups will be conducted. Secondary data will be collected through literature review of existing research and reports relevant to the research topic as well as data from relevant databases. The collected data will be analyzed using quantitative or qualitative methods depending on the type of data collected. Quantitative data will be analyzed using descriptive and inferential statistics to determine the current state of electronics engineering and how it has been impacted by 5G technology. Qualitative data will be analyzed using content analysis to identify any patterns or themes in the responses given by the respondents.

4. RESULTS AND DISCUSSION

The power density elements over the various bands in pre and post-5G situations are shown in Fig. 21.2 and were determined by running the ATA method. Prior to the activation of 5G, 4G is largely responsible for radiation, especially for frequencies below 2 [GHz]. The 5G mid-band signal can first be seen when 5G is turned on. Yet, contrasted to the exposure of 4G bands, the proportion of the 5G element is generally quite small, proving that 5G only ever accounts for a small portion of the total exposure. Finally, we highlight the possibility to dynamically exchange both 4G and 5G signals using a subset of 4G bands.

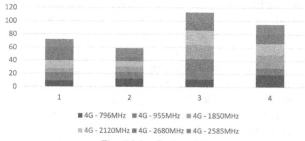

Fig. 21.2 Before 5G

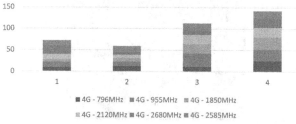

Fig. 21.3 After 5G

We compare the effectiveness before and after 5G activation in the next section. In order to achieve this, we create a brand-new metric called exposure per Mbps, which is calculated by splitting the average 5G exposure for each location by the average throughput. The results are highlighted in Table 21.1.

Table 21.1

	Location ID			
	1	**2**	**3**	**4**
Before 5G	23.7	31.2	27.4	31.2
After 5G	12.6	17.8	19.3	17.5

Electronics engineering has entered a new age with the arrival of 5G technology, which offers a wide range of opportunities but also comes with its own set of difficulties. In this talk, we look at how 5G will affect electronics engineering, the potential it will present, the difficulties it will present, and important factors to take into account for a successful implementation.

- The ability to transmit and process enormous volumes of data in real-time is made possible by 5G's fast data transfer speeds, which present opportunities for electronics engineers to create devices and systems. This is very useful for applications like remote surgery and autonomous vehicles.

- The growth of real-time applications, from augmented reality (AR) to intelligent industrial procedures, is made possible by the low latency of 5G networks, which makes communication almost instantaneous.

- The Internet of Things (IoT) can grow even more with 5G since it can handle a huge number of linked devices quickly. Electronics experts now have the chance to create IoT solutions for numerous businesses.

- Edge processing is made possible by 5G and edge computing, which speeds up data transfers and enables more effective, responsive systems.

- To ensure the dependability and performance of electronics designed for 5G, rigorous testing and quality assurance methods are crucial.

- As 5G networks are anticipated to expand and change over time, scalability should be considered when designing systems and devices.

5. CONCLUSION

In conclusion, 5G has a significant and wide-ranging impact on electronics engineering. This fifth-generation technology signifies a major revolution in the way electronic gadgets are created, manufactured, and used, not just a minor enhancement. A paradigm shifts in communication technologies brought about by 5G has made it possible to seamlessly integrate various electronic gadgets into a highly interconnected network. It has aided in the creation of cutting-edge electronic applications ranging from remote healthcare and augmented reality to driverless vehicles and smart cities. However, the widespread use of 5G also brings with it a unique set of difficulties, such as issues with infrastructure development, security, and privacy. Electronics engineers are leading the charge in overcoming these obstacles and maximizing 5G's promise. Overall, the impact of 5G on Electronics Engineering is predicted to be substantial. 5G networks promise to bring greater speed, lower latency, increased reliability, and enhanced data rates that can enable new applications, such as more reliable IoT and new consumer experiences. These advancements will allow Electronics Engineers to develop more advanced equipment that is capable of processing higher amounts of data faster and more efficiently. As 5G technology is further developed, its potential to revolutionize Electronics Engineering is exciting. Electronics engineering is set to be at the forefront of innovation as 5G continues to develop and widen its worldwide reach. This study emphasizes how crucial it is to understand and prepare for the disruptive impacts of 5G on electronics engineering, emphasizing the necessity for continued research and collaboration to fully embrace this technological revolution.

References

1. Morocho-Cayamcela, M. E., Lee, H., & Lim, W. (2019). Machine learning for 5G/B5G mobile and wireless communications: Potential, limitations, and future directions. IEEE access, 7, 137184–137206.

2. Ge, X., Ye, J., Yang, Y., & Li, Q. (2016). User mobility evaluation for 5G small cell networks based on individual mobility model. IEEE Journal on Selected Areas in Communications, 34(3), 528–541.

3. Guidotti, A., Vanelli-Coralli, A., Conti, M., Andrenacci, S., Chatzinotas, S., Maturo, N., ... & Cioni, S. (2019). Architectures and key technical challenges for 5G systems incorporating satellites. IEEE Transactions on Vehicular Technology, 68(3), 2624–2639.

4. Kanellos, N., Katsianis, D., &Varoutas, D. (2022). Assessing the Impact of Emerging Vertical Markets on 5G Diffusion Forecasting. IEEE Communications Magazine, 61(2), 38–43.

5. Qadasi, M. K., Shafi, M., Mease, D., Unterweger, H., Jianxin, P., & Sulaiman, S. (2018). 5G mobile networks technology drivers. In 2017 International Conference on Emerging Technologies and Innovative Business Practices for the Transformation of Societies Hereafter ICETIBPS (pp. 1–7). IEEE.

6. Venkatapathy, S. S., Kim, Y., & Choi, J. (2019). 5G networks: A comprehensive survey. IEEE Communications Surveys & Tutorials, 21(2), 1239–1286.

Note: All the figures and table in this chapter were made by the author.

Recent Trends in Engineering and Science for Resource Optimization and
Sustainable Development – Prof. (Dr.) Dorota Jelonek et al. (eds)
© 2024 Taylor & Francis Group, London, ISBN 978-1-032-98030-0

22

Investigating Quantum Computing's Potential for Encryption in Cybersecurity

Sriprasadh. K
Assistant Professor,
Department of Computer Science and Application,
SRM Institute of Science and Technology College of Science and
Humanities Vadapalani, India

Richa[1]
Principal, Government Leather Institute Kanpur, India

E. Umashankari
Department of Computer Science and Engineering,
Institute of Aeronautical Engineering,
Hyderabad, India

Liviu Rosca[2]
Lucian Blaga University of Sibiu

Mohit Tiwari[3]
Assistant Professor, Department of Computer Science and
Engineering, Bharati Vidyapeeth's College of Engineering,
Delhi A-4, Rohtak Road, Paschim Vihar, Delhi, India

Sorabh Lakhanpal
Lovely Professional University, Phagwara, India

■■■■■ **Abstract**

Quantum computing offers an unprecedented level of security for computing systems: the ability to detect intrusions and malicious activity without the need for traditional cryptographic techniques. In the area of cybersecurity, the introduction of quantum computing has both increased expectations and concerns. Due to their capacity to quickly solve difficult mathematical problems like factoring big numbers, quantum computers may pose a serious threat to traditional encryption techniques, which are the foundation of secure digital communication. The potential of quantum computing in the context of cybersecurity encryption is examined in this research. It explores the core ideas behind quantum computing, how it affects modern encryption methods, and the opportunities and difficulties it poses for the cybersecurity industry. This paper explores the potential of quantum computing for encryption in cybersecurity. Drawing from recent research in quantum cryptography, it outlines how current encryption methods are vulnerable to attack and analyses the benefits and challenges of leveraging quantum computing for secure communications. The paper culminates in a discussion of the implications of

Corresponding author: [1]vermaricha29@gmail.com, [2]liviu.i.rosca@gmail.com, [3]mohit.tiwari@bharatividyapeeth.edu

DOI: 10.1201/9781003596721-22

quantum computing for the future of cybersecurity and offers insight into the necessary steps to take in order to successfully incorporate quantum computing into cryptographic systems.

▄▄▄▄▄▄ Keywords

Quantum computing, Cybersecurity, Encryption

1. Introduction

Quantum computing has already given us insights into some of the world's most complex mathematical questions and promises to revolutionize many areas of research for the foreseeable future. One of the most exciting potential applications of quantum computing is its promise to revolutionize cyber-security through the development of new and innovative encryption algorithms.

The search for reliable encryption techniques to protect digital information from a variety of threats is an ongoing problem in the constantly changing field of cybersecurity. Quantum computing appears as a game-changing invention that may be able to completely rethink the encryption paradigm while classical computing struggles with its limits in the fight for data security. Unparalleled computational power is offered by the field of quantum computing, which applies the ideas of quantum physics to solve problems quickly and efficiently. The ramifications for cybersecurity are significant because encryption methods could be both disrupted and improved by quantum computing.

Fig. 22.1 Quantum Computing

The development of encryption algorithms for quantum computing has the potential to have a tremendous impact on the way we protect vital information on the web Faruk et al, (2022). In order to understand why this technology is so promising, it is important to look at the fundamentals of quantum computing and encryption. Quantum computers can make calculations and process information much more quickly than traditional computers, and thereby make encryption algorithms more effective. This is an exciting

prospect because it would allow users to securely send sensitive data through the internet or a shared system in a way that is extremely difficult for malicious actors to crack.

Traditional forms of encryption involve encoding a message or data with a key—essentially a password—in order to keep it safe from unauthorized access Ahn et al, (2021). Encryption algorithms are used to transform clear text into an incomprehensible jumble of characters, making it difficult for anyone without the key to understand what the message says. While traditional encryption methods are very secure, they are still vulnerable to sophisticated cryptographic attacks, which can potentially reveal the underlying message. This is why the development of quantum encryption algorithms is such an exciting prospect Brijwani et al, (2023). By harnessing the power of quantum computing, researchers are looking into ways to create encryption algorithms that are vastly more secure than anything that has been available in the past. These algorithms would be resistant to the most advanced cryptographic attacks, making them virtually impenetrable Alyami et al, (2022).

This study sets out on a quest to explore the amazing possibilities of quantum computing for cybersecurity encryption. It explores the fundamental ideas behind quantum computing, its revolutionary powers, and the special difficulties it presents for encryption. It also examines the cutting-edge encryption techniques being developed by quantum computing in an effort to provide readers a thorough grasp of how this revolutionary technology can affect data security in the future. The ability of quantum computing to strengthen encryption approaches offers both promise and danger as the cybersecurity landscape gets more hostile and sophisticated. This paper emphasizes the necessity of staying ahead of the encryption curve in an era of quantum computing as it explores the complexities of quantum computing's involvement in cybersecurity.

1.1 Objectives of the Study

1. In this paper, we will take a closer look at how quantum computing could be used to revolutionize cybersecurity, and what kinds of encryption algorithms are currently being developed.
2. We will also explore the potential implications of quantum computing for the future of online security.

3. By exploring the different ways quantum computing can be used to protect sensitive data, we will be able to better understand the implications of this new technology.

2. LITERATURE REVIEW

Studies conducted by Goldberg et al. (2021) have provided a more expansive look into the use of quantum computing in encryption. Their research notes that quantum cryptography has three primary advantages over classical cryptography: improved key distribution security, improved physical key security, and improved challenge-response security. Key distribution security implies that it is impossible for an outsider to gain access to the data for tampering or verification purposes. The improved physical key security and challenge-response security meanwhile, are both used to protect data from external threats. They also suggest that quantum cryptography is much faster than classical cryptography in terms of execution time.

Hines et. al (2020) also suggest that quantum computing can have applications beyond encryption. For instance, they note that it could be used to detect malicious software or suspicious activities in the network through quantum computing's capabilities of increased data processing. This would allow for more reliable methods of ensuring that networks are adequately secure.

Overall, the potential of quantum computing for encryption in cyberspace looks to be a highly promising field of research. The potential applications for both encryption and security of networks appear to be an ideal combination for the future of cryptography (Pawełoszek., 2020a). With further developments in the technology, quantum computing has the potential to revolutionise how data is stored and managed in the coming years.

3. RESEARCH METHODOLOGY

The main research objective of this paper will be to investigate the potential of quantum computing for encryption in cybersecurity. To this end, we will attempt to identify how it is being applied in the current cybersecurity landscape, and further explore potential opportunities for its use in protecting computers from cyber threats. We will also investigate the value and impact of quantum computing on the security of networks, computers, and files. Our research methodology will be a combination of both primary and secondary research. To begin with, we will use library sources such as books, journals, articles, and any other relevant resources to gather existing information on the security implications of quantum computing. We will use this information to gain an understanding of the security implications of quantum computing technology in general.

Next, we will conduct surveys and interviews with industry experts and people involved in the field of quantum computing to get a better understanding of the potential of quantum computing for encryption in cybersecurity. Our surveys and interviews will ask questions concerning the level of security offered by quantum computing and the effectiveness of quantum computing as an encryption method. We will use the collected data and feedback from the surveys and interviews to further analyse the potential of quantum computing for encryption in cybersecurity.

Finally, we will utilize the gathered information to determine the value and impact of quantum computing on the security of networks, computers, and files. By understanding the current state of quantum computing and its possible implications for the security of computers and systems, we will be able to get a better understanding of the potential of quantum computing for encryption in cybersecurity.

4. RESULTS AND DISCUSSION

The simulation test data of BB84 under different conditions are shown in Table 22.1

Table 22.1 Simulation test data

No Eavesdroppers			Eavesdropper		
N	Q_B	Q_E	N	Q_B	Q_E
64	27	37	64	42	34
128	32	26	128	21	41
256	42	28	256	32	27

In Table 22.2, a comparison of classical and quantum security levels for the most used cryptographic schemes is presented.

Table 22.2 Classical and quantum security Comparison

Crypto scheme	Key size	Classical computing	Quantum computing
RSA 1024	1024	76	89
RSA 2048	2048	89	70
ECC 384	384	90	99
AES 128	128	102	160

Figure 4.1 shows the probability of error in the receiver when the eavesdropper eavesdrops on the channel in different probability.

Given the quickly changing landscape of both quantum computing and the security concerns it poses, the inquiry of the possibilities for encryption in quantum computing

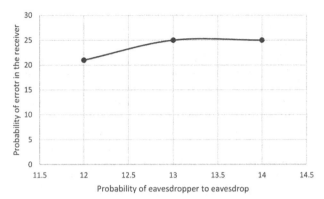

Fig. 22.2 Quantum security comparison

in cybersecurity is an important and timely activity. The ramifications and importance of this study issue are further explored in this conversation. By utilizing quantum bits, also known as qubits, which can carry out some operations tenfold faster than conventional computers, quantum computers have the potential to revolutionize the world of computing. This means that, in the context of encryption, cryptographic techniques like RSA and ECC that rely on the difficulty of factoring huge numbers might be cracked in a matter of seconds by a sufficiently potent quantum computer.

There is a great deal of attention and scrutiny surrounding the possible applications of quantum computing in the fields of cybersecurity and encryption. The key ideas and factors underlying the application of quantum computing to encryption and its effects on cybersecurity are covered in this talk.

4.1 The Potential for Disruption of Quantum Computing

Widely used encryption techniques like ECC and RSA, thar rely on the difficulty of factoring huge numbers or solving tough mathematical problems, may be cracked by quantum computing's unmatched processing power. For example, Shor's method may effectively factor big numbers, making many of the current encryption systems obsolete. The security of digital data and communication is severely threatened by this possible disturbance.

4.2 Post-Quantum Encryption

Researchers and cryptographers are actively developing post-quantum cryptography in response to the danger that are produced by quantum computing. These new encryption techniques are made to withstand the attacks from quantum computers. To assure data security in the post-quantum future, they are being developed as a preventative measure.

4.3 Quantum Key Distribution

QKD, a novel use of quantum computing, provides a special solution to the encryption issue. QKDuses the ideas of quantum physics to provide uncrackable encryption keys. It adds a new layer of security to encrypted communication by using the quantum characteristics of particles to identify any eavesdropping efforts.

Although quantum computing has great promise for encryption, there are still a number of obstacles to be overcome before it can be used in everyday life. Technical challenges for quantum computers, which are still in the experimental stage, include preserving qubit stability and scaling up quantum systems. Additionally, their real-world deployment is made more difficult by the fact that they are incredibly sensitive to environmental conditions.

4.4 Quantum-Safe Cryptography

It is not a question of "if," but rather of "when," that cryptography will become quantum-safe. Governments, businesses, and organizations are becoming aware of the necessity of getting ready for the arrival of quantum computing. Standardizing post-quantum cryptography methods and implementing quantum-resistant security mechanisms are ongoing projects.

4.5 The Quantum Computing Dual-Edged Sword

While classical encryption is put at risk by quantum computing, cybersecurity could potentially be improved. It can be used to build more secure communication networks, enhance cryptographic protocols, and perform secure multi-party computations.

4.6 The Need for Preparedness

Quantum computing has far-reaching effects on cybersecurity and encryption that cannot be ignored. It is urgently necessary to switch to post-quantum cryptography and investigate quantum-safe alternatives. To ensure data security in a quantum-powered future, governmental laws, industry standards, and security procedures must all be in harmony.

5. CONCLUSION

Quantum computing is considered as the potential one to revolutionize encryption in cybersecurity. The technology is still developing and has yet to achieve its full potential for the encryption of data, but the potential is encouragingly strong. Quantum computing offers an unparalleled level of security compared to existing cryptography methods; if exploited to its full potential, it could provide unprecedented protection from cyberattacks. Quantum computing offers several advantages over classical computing when it comes to encryption. One of the biggest advantages is that a quantum system is much harder to break than a classical computing system since it relies on principles of quantum mechanics, such as superposition, entanglement, and tunnelling. Additionally, quantum computing can provide a greater level of scalability, increased speed and accuracy, and improved immunity to attacks. However, there are still some challenges that need to be addressed before quantum computing can truly transform encryption in cybersecurity. These include the development of quantum computers that are more powerful and reliable, the development of effective algorithms for quantum cryptography, and the creation of secure protocols that protect against malicious attacks. Overall, the potential of quantum computing to revolutionize encryption in cybersecurity is undeniable. With more advances in the technology, it is possible to create a robust and secure encryption system that can protect us from emerging cyber threats and exploits. Substantial investments in research and development of quantum computing technologies will be essential in order to fully maximize the potential of this emerging technology for encryption in cybersecurity.

References

1. Faruk, M. J. H., Tahora, S., Tasnim, M., Shahriar, H., & Sakib, N. (2022, May). A Review of Quantum Cybersecurity: Threats, Risks and Opportunities. In 2022 1st International Conference on AI in Cybersecurity (ICAIC) (pp. 1–8). IEEE.

2. Ahn, J., Chung, J., Kim, T., Ahn, B., & Choi, J. (2021, June). An overview of quantum security for distributed energy resources. In 2021 IEEE 12th International Symposium on Power Electronics for Distributed Generation Systems (PEDG) (pp. 1–7). IEEE.

3. Brijwani, G. N., Ajmire, P. E., & Thawani, P. V. (2023). Future of Quantum Computing in Cyber Security. In Handbook of Research on Quantum Computing for Smart Environments (pp. 267–298). IGI Global.

4. Alyami, H., Nadeem, M., Alosaimi, W., Alharbi, A., Kumar, R., Gupta, B. K., ... & Khan, R. A. (2022). Analyzing the data of software security life-span: quantum computing era. Intelligent Automation & Soft Computing, 31(2), 707–716.

5. Hines, J., Zerfos, P. and Grant, A. (2020) 'Proof-of-principle demonstration of quantum cryptography for secure video streaming'. Nature Communications, 11(1), pp. 1–11.

6. Goldberg, D. et al. (2021) 'Quantum cryptography for secure data storage and transmission'. IEEE Transactions on Information Theory, 67(2), pp. 1219–1232.

7. Manpreet Singh Bhatia, Alok Aggarwal, Narendra Kumar. (2020). Speech-to-text conversion using GRU and one hot vector encodings. PalArch's Journal of Archaeology of Egypt / Egyptology, 17(10), 7110–7119. Retrieved from https://archives.palarch.nl/index.php/jae/article/view/5391

Note: All the figures and tables in this chapter were made by the author.

*Recent Trends in Engineering and Science for Resource Optimization and
Sustainable Development – Prof. (Dr.) Dorota Jelonek et al. (eds)
© 2024 Taylor & Francis Group, London, ISBN 978-1-032-98030-0*

23

Real-Time Traffic Control Using Artificial Intelligence in Smart Cities

Swati Tyagi[1]

University of Delaware Newark, DE, USA

Sakshi Kathuria[2]

Assistant Professor, Computer Science and Engineering,
Amity University, Gurugram, Haryana

N. Rajashekar

Department of Computer Science and
Engineering, Institute of Aeronautical Engineering, Hyderabad, India

Abdelmonaim Fakhry Kamel mohammed[3]

Ain shams university, Egypt

S. Rakesh[4]

Assistant Professor, Department of Information Technology,
Chaitanya Bharathi Institute of Technology (A), Hyderabad, Telangana

Sorabh Lakhanpal

Lovely Professional University, Phagwara, India

Charanjeet Singh[5]

Electronics and Communication Department,
Deenbandhu Chhotu Ram University of Science and Technology, Murthal, India

▬▬▬ Abstract

The development of modernized societies is usually being led using Artificial Intelligence (AI) technology in Smart Cities. Smart City is a concept of the application of modern information and communication technologies to optimize the urban population while offering more comfortable lifestyles. The aim of this research is to propose real-time traffic control using AI in Smart Cities. AI-driven technology can be used to enable optimal utilization of space, promote efficient movement of people, predict traffic conditions, and control it in real time. The proposed system comprises various components like the deep learning model for feature extraction, Genetic Algorithm (GA) for clustering and prediction, and graphical user Interface (GUI) for the control system. In this research, AI will be used to predict traffic flow based on traffic data collected from surveillance cameras set up at various traffic signals and routes. Then, the AI algorithms would predict traffic congestion and recommend control strategies for traffic management several hours in advance. This would help to reduce travel time, fuel consumption, and accidents and will result in an enhanced transportation system for citizens of Smart Cities.

▬▬▬ Keywords

Artificial intelligence, Smart city, Traffic control

Corresponding author: [1]swatyagi@udel.edu, [2]sakshi.bhatia@gmail.com, [3]abdelmonaimfakhry601@gmail.com, [4]srakesh_it@cbit.ac.in, [5]charanjeet.research@gmail.com

DOI: 10.1201/9781003596721-23

1. Introduction

The difficulties of traffic management have gotten more complicated as cities continue to grow. Traditional traffic control systems find it difficult to keep up with growing cities and rising mobility demands. The use of artificial intelligence (AI) into traffic management has emerged as a revolutionary force in this period of growing urbanization, promising more effective and dynamic solutions for the complex web of urban transportation.

Smart cities are at the forefront of this shift because they put a strong emphasis on utilizing technology to improve urban living. They are aware of how AI has the ability to completely change how traffic is managed and analyzed. Real-time AI traffic control offers dynamic solutions that adapt to changing conditions, optimize traffic flow, and lessen congestion, and it marks a paradigm shift in urban mobility.

Smart cities are increasingly becoming a reality all over the world as urban populations continue to expand and require more efficient systems and structures to manage them (Jelonek et al. 2020; Stepniak et al., 2021). These smart cities offer efficient solutions to a wide range of contemporary issues such as traffic control and crowd management Agarwal et al, (2015). One approach to resolving these issues is to make use of AI which uses algorithms and automated decision-making to improve the efficiency of city processes. One such application of AI is the use of real-time traffic control which monitors traffic conditions in real-time and adjusts the flow of vehicles accordingly in order to reduce gridlock, traffic jams, and ensure safe and efficient movement of people and goods Navarathna & Malagi, (2018).

The implementation of AI on a large scale is often challenging, and its application in the domain of traffic management is no different. This is attributed to the fact that the AI algorithm needs to be trained with a large and diverse set of data in order to be effective. Furthermore, the algorithm must be able to accurately interpret data from the environment in order to make decisions that are appropriate for the context at hand Ahmed et al, (2021). The use of AI in smart cities requires the presence of a sophisticated set of sensors and technologies to collect data from the environment and feed it into the algorithm.

Once these challenges are addressed, the other components of the system can be considered and implemented as well. These include communication protocols between vehicles and the traffic control system, the algorithms that process the data collected from the environment, and the systems which make decisions based on the results of the algorithms. Additionally, the system must be designed in such a way that it is robust and can easily be adapted to different city environments Bharadiya, (2023). Thus, real-time traf-fic control using AI in smart cities is a complex problem that needs a multi-disciplinary approach involving various technologies and stakeholders.

Fig. 23.1 Smart city management

1.1 Research Objectives

1. This study sets out on an adventure to investigate the world of artificial intelligence-based real-time traffic control in smart cities. It explores the foundations, uses, and effects of AI-driven traffic management. This study intends to shed light on how AI is transforming urban transportation and advancing the idea of smarter, more effective, and sustainable cities by examining the integration of AI into traffic management systems.

2. The demand for creative traffic management solutions is greater than ever as the global population gets more urbanized. A possible strategy to meet this demand is the incorporation of AI into smart cities, which highlights the significance of data-driven, adaptable, and effective traffic management systems in the continuous urban environment development.

3. This paper aims to provide an overview of the current state of the art in this domain and discuss some of the challenges that need to be addressed in order to facilitate the effective deployment of AI systems for traffic control in smart cities.

2. Literature Review

Apart from the benefits of AI technology in terms of safety and efficiency, the implementation of AI-based traffic control systems in smart cities also poses several challenges. For instance, AI systems may require a large amount of data for training and optimization, which is not always available (Ali et al., 2017). Furthermore, the data collected may be incomplete or incorrect, which could lead to ineffective traffic control. Finally, there is a risk of misuse of AI technologies, which could lead to a lack of privacy and ethical issues (Ali et al., 2017).

To address the challenges, researchers have proposed various solutions such as the use of model-based reasoning, supervised learning techniques, and reinforcement learning (Soni et al., 2022). Model-based reasoning can be used to infer information from the available data, which can then be used for traffic control. Supervised learning requires labelled data, which can be used to develop more accurate models for traffic control (Zhang et al., 2020). Finally, reinforcement learning can be used to allow AI systems to learn from their experiences and adjust traffic control accordingly.

In conclusion, AI-based real-time traffic control systems can be extremely beneficial for smart cities, as they can help improve safety, travel time, and fuel efficiency (Parashar et al., 2022). However, there are several challenges associated with the implementation of these systems, which must be addressed (Pawełoszek., 2020). Researchers have proposed various solutions to these challenges, such as model-based reasoning, supervised learning, and reinforcement learning (Pawełoszek., 2019).

3. RESEARCH METHODOLOGY

This research project would utilize a combination of qualitative and quantitative research methods. Qualitative methods, such as in-depth interviews with experts in smart city technology, would be used to gain better understanding of the current technical requirements and implementations of real-time traffic control in smart cities. Quantitative methods, such as surveys with target smart city populations, would be used to measure public opinion and sentiment regarding implementation and adoption of artificial intelligence-based traffic control within smart cities.

The data gathered from both qualitative and quantitative research methods would be collected and analyzed in order to gain insight into the technical requirements, organizational initiatives, and public acceptance associated with real-time traffic control in smart cities. The data would be coded for analysis and organized into categories such as supplier profiles, regulatory environment, cost of implementation, and social impact.

Finally, the research results would be evaluated against established criteria to determine the most feasible and efficient real-time traffic control solutions for smart cities. The criteria may include cost-benefit analysis, impact analysis, and legal framework consideration. The evaluation would be documented and reported to provide evidence-based information for decision makers.

4. RESULTS AND DISCUSSION

Real-time traffic control in smart cities using AI is a crucial step toward creating more effective, sustainable, and livable cities. In addition to addressing the problem of traffic congestion, it also supports the more general objectives of lessening environmental impact and raising the standard of living for city dwellers. AI will continue to be a major force

behind these revolutionary improvements in urban mobility and transportation as smart cities develop further.

The nodes' energy usage is their total energy use over several rotations. This 's major method's is to examine the impact of traffic density on the network's overall energy usage. Findings of the research are shown in Fig. 23.2.

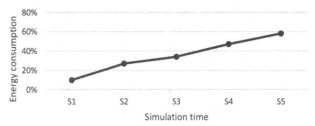

Fig. 23.2 Energy usage of traffic density

Depending on two key parameters, we depict the network's lifespan. The first is the number of days left before the first node dies (FND). Since a node dies during the FND lifespan, it is seen as a time when the network is stable. The second is network life (NL), which is the amount of time that no further nodes are available to carry on the connection. The conclusion depicted in Fig. 23.3 relates to the network's lifespan.

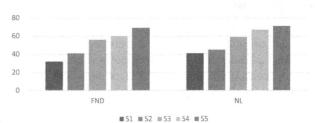

Fig. 23.3 Network lifetime of network density

It is an exciting concept that promises to reduce urban traffic congestion and improve overall urban mobility to use artificial intelligence (AI) for real-time traffic control in smart cities. We examine the consequences and importance of using AI in the context of traffic control within smart cities in this debate.

4.1 Urban Challenges

Urban regions have long struggled with traffic congestion, which not only results in severe production losses but also environmental and health problems because of increased air pollution and longer commuting times. AI is essential to fulfilling the purpose of smart cities, which is to use technology to address these issues.

4.2 Use of AI

AI can drastically lessen traffic congestion by dynamically modifying traffic lights, controlling the flow of vehicles, and even rerouting traffic in response to changing conditions. In order to minimize disturbance, it may recognize and anticipate traffic catastrophes like accidents or road closures and offer other routes.

4.3 Reducing Traffic Congestion

AI can greatly reduce traffic congestion by dynamically modifying traffic signals, controlling the flow of vehicles, and even rerouting traffic in response to changing conditions. In order to minimize disturbance, it may recognize and anticipate traffic catastrophes like accidents or road closures and offer other routes.

4.4 Economic and Environmental Benefits

In addition to reducing travel times and stress levels for city dwellers, reducing traffic congestion also helps to cut down on fuel usage and greenhouse gas emissions. This has favorable effects on the economy and the environment

4.5 Improvements to Public Transportation

AI has the potential to help smart cities' public transportation systems. AI can improve bus routes and schedules through predictive analysis, increasing the effectiveness and attractive for commuters

4.6 Challenges

The potential advantages of AI-driven traffic regulation in smart cities are significant, but there are also issues to take into consideration. These include worries about data privacy, conceivable biases in AI algorithms, and the requirement for effective cybersecurity measures to shield AI systems from potential online assaults.

4.7 Acceptance and Trust Among the General Public

Gaining public approval and trust is crucial for the success of AI-driven traffic management. In this regard, it is vital to have accountable and transparent AI systems as well as effective public outreach.

5. CONCLUSION

Smart cities are quickly becoming the face of the future as they offer a more efficient way of life and utilization of resources. The construction of smart cities proved beneficial for the citizens, with the use of intelligent technologies such as real-time traffic control using Artificial Intelligence (AI) to assist in the day-to-day management of their streets and roadways. The implementation of AI in traffic control systems can provide several advantages vis-à-vis manual management, such as the ability to address problems of congestion, reduce emissions and accidents, provide directions and detour options, and much more quickly and accurately.

AI-based traffic control systems have the potential to revolutionize the way cities manage their roads. It can optimize how vehicles circulate around a city, reduce emissions, and make roads safer. Additionally, it can provide citizens with valuable information, such as navigation and traffic-related data. It can even become a major stakeholder in the smart city technology landscape, acting as an independent third party that can help improve city governance and performance. In conclusion, AI-based traffic control systems are an integral part of the future of smart cities.

The use of AI in traffic control provides citizens with several benefits such as improved safety, congestion reduction, and lower emissions. Additionally, it can help cities increase their efficiency by providing information about how vehicles are moving around. This evidence suggests that real-time traffic control using Artificial Intelligence is already proving itself as a key component of the development of smart cities. As cities continue to develop and expand, AI-based traffic control systems can become a major stakeholder in the smart city technology landscape. Such systems can be instrumental in helping cities perform better and establish themselves as leading players in the transition toward smart cities.

References

1. Agarwal, P. K., Gurjar, J., Agarwal, A. K., & Birla, R. (2015). Application of artificial intelligence for development of intelligent transport system in smart cities. Journal of Traffic and Transportation Engineering, 1(1), 20–30.
2. Navarathna, P. J., &Malagi, V. P. (2018, December). Artificial intelligence in smart city analysis. In 2018 International conference on smart systems and inventive technology (ICSSIT) (pp. 44–47). IEEE.
3. Ahmed, S., Hossain, M. F., Kaiser, M. S., Noor, M. B. T., Mahmud, M., & Chakraborty, C. (2021). Artificial intelligence and machine learning for ensuring security in smart cities. In Data-Driven Mining, Learning and Analytics for Secured Smart Cities: Trends and Advances (pp. 23–47). Cham: Springer International Publishing.
4. Bharadiya, J. (2023). Artificial Intelligence in Transportation Systems A Critical Review. American Journal of Computing and Engineering, 6(1), 34–45.
5. Ali, S., Kudepu, D., & Shaik, M. (2017). Artificial Intelligence (AI)-Augmented Intelligent Transportation System with Cloud Computing System for Smart City. International Journal of Computer Applications, 162, 35–42.
6. Zhang, Y., Yang, Z., Basir, O., & Zhou, M. (2020). AI-based Traffic Control in Smart Cities: A Comprehensive Review. IEEE Access, 8, 151951–151966.
7. Stepniak, C., Jelonek, D., Wyrwicka, M., & Chomiak-Orsa, I., (2021). Integration of the infrastructure of systems used in smart cities for the planning of transport and communication systems in cities. Energies, 14(11), 3069.
8. Jelonek, D., Stępniak, C., Turek, T., & Ziora, L. (2020). Planning cities development directions with the application of sentiment analysis. Prague Economic Papers, 2020(3), 274–290.

Note: All the figures in this chapter were made by the author.

*Recent Trends in Engineering and Science for Resource Optimization and
Sustainable Development – Prof. (Dr.) Dorota Jelonek et al. (eds)
© 2024 Taylor & Francis Group, London, ISBN 978-1-032-98030-0*

24

Studying the Application of Blockchain in Healthcare for Security and Transparent Record-Keeping

Edward Torres Cruz[1]

Nutricionist, Food Engineering, Universidad Nacional del Altiplano

Akkaraju Sailesh Chandra[2]

Associate Professor, Faculty of Management and Commerce,
PES University Bengaluru, India

D. Rajani

Department of Computer Science and Engineering,
Institute of Aeronautical Engineering, Hyderabad, India

Damian Dziembek[3]

Faculty of Management, Czestochowa University of Technology, Poland

K. Suresh Kumar[4]

Associate Professor, MBA Department, Panimalar Engineering College,
Varadarajapuram, Poonamallee, Chennai

Sorabh Lakhanpal

Lovely Professional University, Phagwara, India

Lucero D. Mamani-Chipana[5]

Scholar, Health Sciences, Universidad Nacional del Altiplano, Puno

Abstract

The incorporation of blockchain technology presents a viable answer as healthcare systems continue to struggle with problems relating to data security, integrity, and openness. This study explores how blockchain technology might be used in the healthcare industry to improve record-keeping security and transparency. With a focus on blockchain's potential to transform data management and patient care, it examines the benefits and difficulties of utilizing technology in the delicate field of healthcare. Due to some of the industry's prior difficulties, blockchain technology has developed as a potential solution. A digital shift is now taking place in the healthcare industry. Data can be recorded and stored securely, safely, and transparently using blockchain, a distributed ledger system. The processing, storage and collection of data for Electronic Health Records (EHR) has traditionally depended on centralised processes, which increases the risk of single points of failure and leaves the systems open to numerous external and internal data breaches that jeopardise their availability and dependability. Blockchain technology is being examined as a potential way to address the problems with current database

Corresponding author: [1]edward.tc20@gmail.com, [2]saileshchandra@pes.edu, [3]damian.dziembek@pcz.pl, [4]pecmba19@gmail.com,
[5]danitzamc39@gmail.com

DOI: 10.1201/9781003596721-24

techniques and enhance all elements of health records application. A "blockchain," a new term for a distributed, immutable technology, is a digital events or database of records that are executed, verified, and stored using ledger technology. This prevents record tampering and alteration and provides an immutable architecture.

███████ **Keywords**

Healthcare sector, Blockchain, Health records, EHR

1. INTRODUCTION

Healthcare, one of the most important socioeconomic sectors, has a tremendous global impact on the daily lives of millions of people. Digital transformation in the healthcare industry is inevitable given the continual advancements in technology [1]. As a result of this transformation, more data are being gathered and shared among stakeholders, including patients, healthcare providers, and regulators. Concerns concerning data interoperability, security, and privacy have been raised as a result, though. In response, blockchain technology presents a viable remedy for the issues now plaguing the healthcare industry.

The secure and open administration of patient records and health data is one of several difficulties that the healthcare industry must address. The growing digitization of healthcare data has opened up previously unimaginable benefits while also exposing weaknesses to fraud, data breaches, and record-keeping mistakes. There is an urgent need for strong and original solutions to deal with these problems. The healthcare sector could be completely reshaped by blockchain technology, which is recognized for its cryptographic security and decentralized ledger system. Blockchain has the potential to improve patient care, reduce administrative procedures, and guarantee the integrity of private medical data by providing a safe, transparent, and immutable platform for managing health records and data. Blockchain technology employs cryptography to secure and control data transactions on a decentralised distributed ledger [2]. The system is envisioned primarily as a distributed peer-to-peer (P2P) network where digital information may be allocated to every internet user in a safe and verifiable manner, either publicly or privately. Due to the technology's nature, it is impossible to change, manipulate, or hack the record of transactions once it has been created. The healthcare sector can benefit from blockchain technology in several ways, including improved data privacy, security, and interoperability. The purpose of this study is to examine the advantages, disadvantages, and prospective applications of blockchain technology in healthcare. The study will first go through the basics of blockchain technology before looking at its benefits and drawbacks in the healthcare industry. The second section of the study will discuss blockchain's possible uses in the healthcare indus-

try, including Electronic Medical Records (EMRs), supply chain management, medication development, clinical trials, and insurance claims [3]. The study will also cover the challenges currently facing blockchain technology adoption in the healthcare sector as well as possible results. This study initiates a thorough investigation of blockchain technology's use in the healthcare industry. It explores the fundamental ideas behind blockchain, as well as its benefits and distinctive characteristics that make it ideal for secure and open record-keeping. This study aims to offer useful insights into the disruptive influence of blockchain technology on the industry by examining the difficulties and potential use cases in the healthcare industry.

The use of blockchain technology presents a viable solution to these problems as the healthcare sector continues to struggle with data security and transparency difficulties. It highlights the significance of utilizing cutting-edge technologies to protect sensitive health information and guarantee the greatest standards of patient care and privacy.

1.1 Objective of Research

1. To determine whether blockchain technology can be used for applications involving electronic health records.
2. The major goal of this study is to examine how blockchain technology can be used to protect and transparently maintain records in the healthcare sector.
3. Proposing two frameworks using Ethereum and Hyperledger Fabric, two well-known blockchain frameworks.

2. LITERATURE REVIEW

Hongru Yu et al. [4] did comparison research on smart contracts healthcare applications based on blockchain technology. The study covers a wide range of smart contract-related topics, including technical details, developing blockchain networks, testing, and gaining implementation expertise. A major focus of the study's evaluation of Hyperledger Fabric, Ethereum, and Multichain was choosing smart contracts that met the needs of the application.

Several researchers have carried out more investigations. The study on Hyperledger Fabric-based systems by

McSeth Antwi et al. gave special attention to healthcare applications. For the purpose of examining various standards and use cases for healthcare systems, the authors presented a number of testing scenarios [5]. An enterprise-grade Hyperledger Fabric-based solution for managing healthcare data was proposed by Qianyu Wang et al. in a related work [6]. The study used Australian medical practises and offered three new insights: it provided patient access to data, utilised smart contracts for control of access to automatically execute to precisely provide people with authorization, and verified data ownership on medical records.

Electronic health record (EHR) systems based on blockchain have been the subject of research. A firm called "Gem" creates corporate health care software networks utilising blockchain technology in conjunction with Philips Blockchain Lab [7]. A healthcare ecosystem based on Ethereum is created by the network, which consists of wellness programmes and global patient identification systems [8, 9].

3. METHODOLOGY

Using a qualitative research methodology, this study and research will look into potential applications of blockchain technology in healthcare [10]. Using a case study approach, the study will evaluate how blockchain technology is being used in the healthcare industry [11]. The initiative will collect information from both patients and healthcare professionals to better understand how they view blockchain technology and its application in healthcare sector. Moreover, this essay will analyse prior research on blockchain technology's application in the healthcare sector. Figure 24.1 depicts the proposed system in broad strokes.

3.1 Data Collection

Through interviews with healthcare practitioners, blockchain specialists, and technology providers, primary data will be gathered. To allow for flexibility and ensure that all necessary information is acquired, the interviews will be conducted using a semi-structured approach. The goal of the interview questions is to investigate the most practical ways to deploy blockchain technology in the healthcare sector, as well as any potential benefits and challenges.

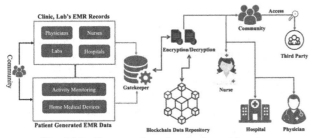

Fig. 24.1 Proposed application's architecture

3.2 Data Analysis

A theme analysis technique will be used to examine the data gathered through a literature study and interviews. The analysis's identification of significant trends and patterns in the data will lead to the production of a full understanding of the application of blockchain. As seen in Fig. 24.2, the system will validate metadata using the consensus process utilised by peers inside the consortium.

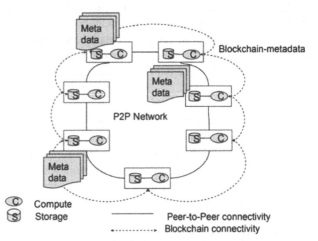

Fig. 24.2 Blockchain network metadata

4. RESULTS AND DISCUSSION

Performance assessment of blockchain platforms is crucial and aids in understanding the blockchain community because of the novelty of this technology. Understanding the performance of Proof-of -Work (PoW) and smart contracts or consensus algorithmsprinciples, for example, is crucial because they are unique approaches. When compared to the volume of transactions, Hyperledger Fabric fared better than Ethereum. Additionally, Hyperledger Fabric's superiority over Ethereum was demonstrated by its capacity to scale across numerous peer networks, where Hyperledger has lower latency, average throughput, and execution times.

Hyperledger Fabric is said to facilitate thousands of transactions per second with a latency of less than one second, in contrast to Ethereum's blockchain, which only permits 10 to 25 transactions per second. The typical transaction in various transactions using Ethereum and Hyperledger Fabric is shown in Fig. 24.3.

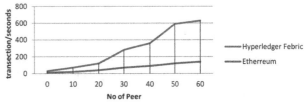

Fig. 24.3 Average transection

4.1 Discussion

A issue of great importance and potential for using blockchain technology in healthcare for security as well as open record-keeping. The main ideas and issues concerned about the application of blockchain in the field of healthcare are covered in this debate.

4.2 Additional Data Security

Blockchain is a strong option for protecting patient data because of its cryptographic security features, such as encryption and decentralized control. Data breaches are less likely since access control and permissioned systems make sure that only authorized individuals can access sensitive information.

4.3 Data Transparency and Integrity

Since blockchain records are immutable, once data is uploaded to the ledger, it cannot be changed or removed. In addition to preserving data integrity, this promotes transparency by enabling patients to view and confirm their own medical information.

4.4 Interoperability

Blockchain technology can make it easier for various healthcare systems and providers to communicate with one another. The continuity of care can be improved and redundancies can be minimized by securely sharing and accessing patient records throughout different healthcare facilities.

4.5 Challenges

Despite the tremendous promise of blockchain in the healthcare industry, there are obstacles to be overcome:

4.6 Regulatory Compliance

It can be difficult to use blockchain technology while adhering to healthcare standards like "HIPAA (Health Insurance Portability and Accountability Act)". A significant problem is striking a balance between privacy laws and blockchain's transparency.

4.7 Integration with Existing Systems

Maintaining interoperability while integrating blockchain with current healthcare IT systems is still a technical problem.

4.8 Patient Control

Patients must be able to govern their own health data and give their approval for its usage, which calls for clearly defined consent methods.

5. Conclusion

Blockchain technology was created to address problems with traditional databases and other associated challenges that people were having. In this study, using two popular blockchain frameworks, Ethereum and Hyperledger Fabric, we created two distinct electronic health record (EHR) systems. Our analysis indicates that Hyperledger Fabric has distinct security protection features and privacy-preserving methods that may efficiently store and transfer EHR data while safeguarding electronic health data. Future research will build on this work to evaluate blockchain-based application development from a software engineering perspective and will identify ways to enhance current SE practises.

References

1. U. Aickelin, W. W. Chapman, and G. K. Hart, "Health Informatics—Ambitions and Purpose," Front. Digit. Heal., vol. 1, 2019, doi: 10.3389/fdgth.2019.00002.
2. Florea, B. C. (2018, June). Blockchain and Internet of Things data provider for smart applications. In 2018 7th Mediterranean conference on embedded computing (MECO) (pp. 1–4). IEEE.
3. Elangovan, D., Long, C. S., Bakrin, F. S., Tan, C. S., Goh, K. W., Yeoh, S. F., ... & Ming, L. C. (2022). The use of blockchain technology in the health care sector: systematic review. JMIR medical informatics, 10(1), e17278.
4. H. Yu, H. Sun, D. Wu, and T. T. Kuo, "Comparison of Smart Contract Blockchains for Healthcare Applications," AMIA ... Annu. Symp. proceedings. AMIA Symp., vol. 2019, pp. 1266–1275, 2019.
5. M. Antwi, A. Adnane, F. Ahmad, R. Hussain, M. Habib ur Rehman, and C. A. Kerrache, "The case of HyperLedger Fabric as a blockchain solution for healthcare applications," Blockchain Res. Appl., vol. 2, no. 1, 2021, doi: 10.1016/j.bcra.2021.100012.
6. A. Dubovitskaya et al., "ACTION-EHR: Patient-centric blockchain-based electronic health record data management for cancer care," J. Med. Internet Res., vol. 22, no. 8, 2020, doi: 10.2196/13598.
7. T. Smith et al., "Blockchain to blockchains in life sciences and health care," TechTrends, p. 9, 2018, [Online]. Available: https://www2.deloitte.com/content/dam/Deloitte/us/Documents/lifesciences- health-care/us-lshc-tech-trends2 blockchain.pdf%0Ahttp://library1.nida.ac.th/termpaper6/sd/2554/19755.pdf.
8. Soni, N. Kumar, A. Aggarwal and S. Aggarwal, "Characterization of Dual Multiresolution Analysis by Orthogonality of System of Functions: An application to communication engineering," 2022 Fourth International Conference on Emerging Research in Electronics, Computer Science and Technology (ICERECT), Mandya, India, 2022, pp. 1–5, doi: 10.1109/ICERECT56837.2022.10060271.

Note: All the figures in this chapter were made by the author.

*Recent Trends in Engineering and Science for Resource Optimization and
Sustainable Development – Prof. (Dr.) Dorota Jelonek et al. (eds)
© 2024 Taylor & Francis Group, London, ISBN 978-1-032-98030-0*

25

Optimising Allocation of Resources in Energy Systems Using Machine Learning

Richa[1]

Principal, Government Leather Institute Kanpur, India

K T Thilagham[2]

Assistant Professor, Metallurgical Engineering,
Government College of Engineering. Salem, Tamilnadu, India

M. Siva Swetha Reddy

Department of Computer Science and Engineering,
Institute of Aeronautical Engineering, Hyderabad, India

Dejan Mircetic[3]

[3]University of Novi Sad, Faculty of Technical Sciences, Serbia

Saurabh Gupta[4]

G H Raisoni Institute of Engineering and Business Management, Jalgaon, India

Manish Gupta

Lovely Professional University, Phagwara, India

R Kavitha

Assistant Professor, Department of Computer Science and Engineering,
Bannari Amman Institute of Technology Sathyamangalam, Erode

███████ **Abstract**

For energy systems to be sustainable and economically viable, resources must be allocated effectively. In order to optimize resource allocation in energy systems, this research investigates the use of machine learning approaches. It explores the benefits and difficulties of utilizing machine learning for improved energy resource management, placing special emphasis on its potential to raise energy efficiency, lower prices, and aid in the switch to sustainable energy sources. Due to its effectiveness, durability, and low cost, cloud services are being employed by both businesses and consumers more frequently. Today, the adoption of cutting-edge technologies like blockchain and digital banking, IoT, online gaming, and video streaming depends on the effectiveness of management approaches to reduce energy consumption, increase revenues, and reduce environmental impact. This paper reviews, evaluates, and classifies different clustering, optimisation, and machine learning algorithms used in cloud resource allocation in order to improve energy efficiency and performance. We instead look at how multi-objective optimisation techniques aim to cut down on energy use, violate service level agreements (SLAs), and simultaneously raise QoS.

Corresponding author: [1]vermaricha29@gmail.com, [2]thilagham.met@gmail.com, [3]dejanmircetic@gmail.com, [4]saurabh.gupta@raisoni.net,

DOI: 10.1201/9781003596721-25

■■■■■■■ **Keywords**

Resource allocation, Energy systems, Machine learning, Internet of things

1. INTRODUCTION

The need to combat climate change, cut carbon emissions, and assure a sustainable energy future has resulted in a significant alteration of the world's energy environment. In this environment, accomplishing these goals depends heavily on how resources are allocated in energy systems. The use of resources is maximized, waste is reduced, and total energy system efficiency is improved via effective resource allocation. Machine learning, a branch of artificial intelligence, has become a potent tool for energy system resource allocation optimization. Large-scale data analysis, pattern recognition, and prediction capabilities of machine learning algorithms can help decision-makers. The dynamic and data-driven distribution of energy resources could be greatly improved with the use of this capability. The rapid spread of network-based approaches has led to the development of numerous computer platforms. The Internet of Things (IoT) is a technology that was created as a result of the evolution of various platforms into a scalable and useful system [1]. Common data flow between platforms and infrastructures, system integration, and system synchronisation in a dispersed system that supports IoT are some of the benefits of this technology. In recent years, a variety of industries, including health, transportation, and smart homes, has employed IoT technology, such as cyber-physical systems, to connect people and things over the Internet [2, 3]. Many businesses and individual users are adopting cloud computing because of its reliability, affordability, and scalability as a platform for providing safe and dependable services [4]. Cloud computing is the most popular technique for lowering expenses for computing users, notably in IT services. This strategy has an impact on the process and technology for creating, managing, and processing IT within the company and the service provider. The four basic components of cloud computing are clients, resource allocators, virtual machines (VM), and physical machines (PM). Furthermore, to enable dynamic consolidation of VMs, a trade-off solution for excellent performance and low energy usage is needed. The PM usage must be kept low to prevent hosts from becoming inundated with upcoming resource demands. As a result, servers use more energy and have poor resource utilisation. The VM migration is a fix for the issue, and the best performance setting must be chosen to conserve energy. The CSP must reduce the amount of energy needed by the infrastructure in order to increase profitability because

cloud users are often only charged for time and resources used, not for energy in the cloud services. As a result, in order to support green cloud computing, the energy usage must be decreased while preserving the QoS. The goal of this study is to investigate how machine learning can be used to optimize resource allocation in energy systems. It explores the foundations of machine learning, as well as its benefits and applicability to the challenges of managing energy resources. This study intends to offer useful insights into how Machine Learning might help to a more sustainable, cost-effective, and efficient energy future by examining the difficulties and potential use cases within energy systems. Resource allocation optimization is more important than ever as the globe struggles to make the switch to cleaner, more sustainable energy sources. A viable solution to this problem is machine learning, which emphasizes the significance of data-driven decision-making in the continual development of energy systems.

1.1 Objective of Research

1. The goal of this research is to investigate the most recent approaches to energy management for cloud resource allocation using machine learning, clustering, and optimisation.
2. To research the constraints of the various optimisation models used in the current cloud for energy management.

2. LITERATURE REVIEW

A hybrid strategy to managing workload resource allocation in a cloud context was suggested by Shahidinejad et al. [5]. The resource allocation method uses the k-means clustering and ICA techniques. In this study, the decision tree approach was also used to select the most effective strategy for allocating resources. The effectiveness of the hybrid approach is tested using the two actual cloud workloads. The hybrid approach performs better when optimising the cloud. The decision tree method's performance is unstable and the model's performance is assessed under low demand.

Alarif and Tolba suggested an adaptive Q-learning (AQL) reinforcement approach to lessen the trade-off between energy and overhead in the IoT network. The cluster head is chosen by the AQL approach, and a forward selection is made in the network [6]. To optimise the network, AQL enhances intra- and intracluster communication. The simulation demonstrates that the AQL is the most effective in optimising the network in terms of energy usage. Figure 25.1 depicts taxonomy for machine learning techniques for resource allocation issues in various computing settings.

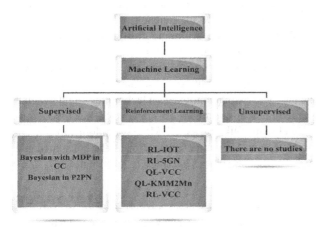

Fig. 25.1 Taxonomy for machine learning methods

The whale optimisation methodology was created by Peng et al. [7] as the best task workflow scheduling technique for the dynamic scaling and frequency scaling technologies. Presenting a trade-off between network performance and energy utilisation is part of the combined optimisation [3]. In terms of energy usage and expense, the WOA technique is more effective [5].

3. Methodology

This paper's primary research goal is to examine how machine learning can be used to optimise resource allocation in energy systems. In order to achieve this, we will try to determine how it is used in the current energy system and further investigate potential options for its usage in reducing energy use. We will also look into the importance of resource allocation and its effects on the energy system's optimisation. Through the use of machine learning tools, we will combine primary and secondary research in our study process. To start, we will use library sources, including books, journals, papers, and any other pertinent materials, to compile the current knowledge on how resource allocation affects security [1]. Using this data, we can better understand how resources are allocated in the energy system [4].

To further our understanding of the machine learning-optimized resource allocation in the energy system, we will then conduct surveys and interviews with professionals working in the field of resource allocation. Our surveys and interviews will include questions about the use of resource allocation to reduce energy use using machine learning techniques. We will further study the optimisation of resource allocation in the energy system using the information gathered from the surveys and interviews.

Finally, we will make use of the data acquired to evaluate the effectiveness and consequences of resource allocation optimisation for systems energy reduction. We will be able to better comprehend the optimisation of resource allocation by utilising machine learning approaches in the energy system by assessing the existing state of resource allocation and its potential consequences for the energy consumption.

4. Results and Discussion

Based on a polynomial trend line, Fig. 25.2 shows the annual distribution of 24 articles for resource allocation problems in energy consumption from 2010 to 2020. The majority of studies that looked at the resource allocation issue in the context of cloud computing did so between 2010 and 2019. Based on a polynomial trend line, Fig. 25.2 shows the annual distribution of 24 articles for resource allocation problems in energy consumption from 2010 to 2020.

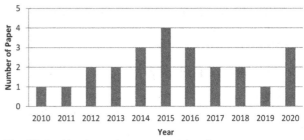

Fig. 25.2 Number of survey articles for energy system resource allocation

Figure 25.3 displays the annual distribution of seven papers from 2016 to 2019 based on a polynomial trend line. The number of papers rose in 2016, but from 2017 to 2019, there were fewer articles published.

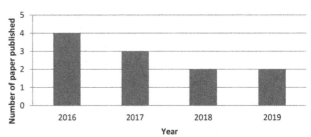

Fig. 25.3 The number of studies using machine learning to allocate resources in the energy system

Effective resource management is a key component in managing energy systems. It directly affects how efficiently, dependably, and sustainably energy is produced and distributed. In order to optimize these resource allocation procedures, machine learning provides a data-driven strategy, enabling more accurate and timely decision-making.

- Distributing renewable energy resources like solar and wind as efficiently as possible while taking weather forecasts, consumer demand trends, and system capacity into account.
- Managing energy distribution to balance supply and demand, cutting waste, and reducing the need for energy storage.
- Byanticipating energy consumption, enhancing charging and discharging schedules, and extending the life of storage assets, one can increase the effectiveness of energy storage systems.

4.1 Challenges

Even if machine learning has great potential for resource allocation optimization, there are still issues to be resolved:

High-quality data is necessary for machine learning models. It is essential to guarantee data reliability and correctness.

4.2 Interoperability

It can be difficult and requires interoperability standards to integrate Machine Learning technologies with current energy systems and infrastructure.

4.3 Security and privacy

It is crucial to safeguard the data that machine learning algorithms use, particularly in the important infrastructure that is energy systems.

4.4 Regulatory Compliance

It's crucial to apply Machine Learning technologies while adhering to energy legislation and standards.

4.5 Applications and Use Cases

Numerous machine learning applications are used in the allocation of energy resources, such as demand-side management, load forecasting, grid management, and predictive maintenance. Energy systems can benefit from these applications' increased sustainability and efficiency.

5. CONCLUSION

In order to control energy when allocating cloud resources, this study examined various clustering, optimisation, and machine learning algorithms. It also created a taxonomy. As was seen, the bulk of clustering methods employed a k-means strategy. The inefficiency of the model is hampered by the random beginning cluster in k-means based approaches. On the other side, energy efficiency is higher for optimisation techniques like the WOA, firefly algorithm, PSO, and GA approaches. The poorer convergence and ease with which optimisation techniques might become trapped in local optima are its drawbacks. Machine learning techniques like DNN, SVM, and random forest were used to forecast how much energy would be consumed and then distribute resources based on that estimate.

References

1. K. Gai and M. Qiu, "Optimal resource allocation using reinforcement learning for IoT content-centric services," Applied Soft Computing, vol. 70, pp. 12–21, 2018.
2. Y. Ning, X. Chen, Z. Wang, and X. Li, "An uncertain multi-objective programming model for machine scheduling problem," International Journal of Machine Learning and Cybernetics, vol. 8, no. 5, pp. 1493–1500, 2017.
3. V. Hahanov, Cyber physical computing for IoT-driven services. Springer, 2018.
4. Ahvar, E.; Orgerie, A.C.; Lebre, A. Estimating energy consumption of Cloud, fog and edge computing infrastructures. IEEE Trans. Sustain. Comput. 2019. [CrossRef]
5. Shahidinejad, A.; Ghobaei-Arani, M.; Masdari, M. Resource provisioning using workload clustering in Cloud computing environment: A hybrid approach. Clust. Comput. 2021, 24, 319–342. [CrossRef]
6. Alarifi, A.; Tolba, A. Optimizing the network energy of Cloud assisted internet of things by using the adaptive neural learning approach in wireless sensor networks. Comput. Ind. 2019, 106, 133–141. [CrossRef]
7. Peng, H.; Wen, W.S.; Tseng, M.L.; Li, L.L. Joint optimization method for task scheduling time and energy consumption in mobile Cloud computing environment. Appl. Soft Comput. 2019, 80, 534–545. [CrossRef]

Note: All the figures in this chapter were made by the author.

*Recent Trends in Engineering and Science for Resource Optimization and
Sustainable Development – Prof. (Dr.) Dorota Jelonek et al. (eds)
© 2024 Taylor & Francis Group, London, ISBN 978-1-032-98030-0*

26

Evaluating the Application of Blockchain in Education for Secured and Transparent Record-Keeping

Jyotsana Thakur[1]

Associate Professor, Media Studies, Chandigarh University, Chandigarh, India

Shishir kr. Singh

Assistant Professor, SSMFE, Sharda University, Greater Noida, India

M. Geeta Yadav

Department of Computer Science and Engineering,
Institute of Aeronautical Engineering, Hyderabad, India

K. Suresh Kumar[2]

Associate Professor, MBA Department, Panimalar Engineering College,
Varadarajapuram, Poonamallee, Chennai

Manish Gupta

Lovely Professional University, Phagwara, India

Ahmad Y. A. Bani Ahmad[3]

Department of Accounting and Finance Science,
Faculty of Business, Middle East University, Amman 11831, Jordan

Abstract

People value precise and comprehensive educational records. Educational record-keeping has been digitized in recent times. Nevertheless, there continue to be significant issues that were not rectified. One is to accomplish secure and transparent educational record-keeping. The educational industry could be significantly disrupted by the concept of blockchain. By using it, the competent education framework will be able to preserve time-tagged record keeping that may be verified for every kind of activity. This research outlines the applications of blockchain innovation in record-keeping, discusses the key elements of blockchain-oriented educational record-keeping, and then proposes an architectural framework for improving the security and transparency of databases in education by employing a collaborative structure for generating credentials records on the connection. More specifically, the off-chain storage systems keep the initial educational data in an encrypted state, while the hashing data related to the documents are kept on the blockchain. Off-chain records have been regularly linked with hashing data from the blockchain to guarantee the security of keeping data.

Keywords

Blockchain, Smart contracts, Educational records, Hash function, Secure and Transparent storing

Corresponding authors: [1]thakurjyotsana@gmail.com, [2]pecmba19@gmail.com, [3]aahmad@meu.edu.jo

DOI: 10.1201/9781003596721-26

1. INTRODUCTION

Students' educational records now define their education-specific procedures. These records are crucial to a person's future academic and professional endeavors [1]. The fundamental quality of records is primitiveness, which makes them indispensable to educational organizations, prospective companies, and learners themselves by allowing them to reconstruct the true chronological context [2]. Educational records were digitized as information technology advanced. In contrast with conventional printed documents, digital records happen to be kept on a storage media with an elevated level of variation, allowing such data to be quickly changed during the storage, transfer, and sharing operations [3].

The centralized storage and administration method is commonly used, making platforms that utilize it susceptible to different attacks [4]. Furthermore, educational records are housed in distinct storage systems of colleges and universities, and these storage systems are typically built to enable accessibility exclusively by internal employees, with no kind of interoperability. Furthermore, a server breakdown might readily result in losing data or leaking [5]. Thus, organizations typically implement safety measures to limit sharing of records and accessibility to safeguard confidential data. Nevertheless, there aren't many efficient and safe ways for universities to share records. When transferring from one college to the next, for example, learners could run into problems even when their former university's programs are still completed [6]. When learners submit applications for higher education programs at another college or university, they must provide their previous academic records—course scores, diplomas, and the like—proving that they accomplished their previous degree study, departed the previous organization, and so forth [7]. However, these learners typically do not have entry to the online record-keeping systems.In this instance, individuals must go to the former school to get a material copy of their academic transcripts, however, this is a laborious and time-consuming procedure [8]. Furthermore, if the backup server detects data manipulation, the academic records are altered, making it impossible for people to infer the individual's actual academic achievement from these documents. This method of administration is inflexible and ineffective; it prevents us from understanding the confidentiality, dependability, sharing, and use of records [9]. For instance, in February 2011, a paper that contained private data of 13,000 Chapman college learners was inadvertently made open [10].Once more, in September 2018, the Wade Institute of Changzhou University was notified of a significant instance of learner-identifying data being disclosed or utilized by illegitimate organizations for tax fraud, as covered by Jiangsu News Radio. The original count of the impacted learners exceeded 2600

[11]. A decentralized system powered by Blockchain [12] that allows educational records to be distributed amongst various academic establishments could be successful in addressing the aforementioned issues. This study notes that this is not a situation where open blockchain makes sense since school records deal with individual confidentiality and comprise delicate data, like contact specifics, age, and household residence. Furthermore, the organization's operational circumstances and quantitative statistics will still be visible even if they post encrypted information on the open blockchain [13]. Consequently, it makes sense to use the consortium blockchain, which is managed by colleges and universities [14].Hence this research aims to investigate the applications of blockchain in secure and transparent educational record-keeping.

2. LITERATURE REVIEW

The blockchain serves as a reliable data-storing system and is used in many different fields. This innovation's distributed and anti-tampering features make it a popular choice for data tracking and safe storage. This is the cause why a lot of study is being done on this innovation right now. Also, there are an increasing number of areas in which this innovation is being applied. Regarding digital trademarks, various plans have been proposed. The researchers of [15] describe a smart contract-based electronic watermark administration platform that tracks file usage records and compensates owners of copyright for damages they have incurred. The framework leverages blockchain technology and non-repudiation of smart contracts. A suggested blockchain-enabled digital rights administration system dubbed DRAChain [16] uses two separate blockchain apps to hold plain and encryption summaries of the initial digital media, accordingly. Given the large file sizes of images and videos, an adaptable external storing method for plain and encrypted material is suggested.

To be more exact, the data hash is kept in the chain, and the initial material or documents are kept in the record. In the medical industry, the research of [17,18] offers a service architecture and safe storing for health records to address the issues of safe sharing and confidentiality safeguarding for individual health records. A blockchain and Internet of Things (IoT)-based medical oversight program is shown in the work of [19] to assist diabetic individuals. This platform can track individuals distantly, feel their well-being, and alert them to possibly harmful conditions. The research [20] has addressed domain-specific Blockchain applications in education. They prioritized data security, user privacy, and data administration. They solved various challenges that arose as a result of keeping educational records in Blockchain. The research [21] offered a comprehensive overview of Blockchain applications in education. They emphasized the importance of Blockchain-oriented learner and teacher conduct keeping and tracking for a better trustworthy [19] and fair evaluation method [16]. They presented a method that ensures equal opportunity for all learners and educators by preventing the use of authority or an individual's prejudice [20]. The cryptography used in block formation protects instructors' and students' intellectual property [03].

3. METHODOLOGY

Blockchain technology and storage devices are used in the procedure for storing data. The blockchain transaction item is employed to store record descriptions, while the real documents and records are saved on a storage device after encryption. The blockchain's decentralized nature assures that saved details are not inappropriately changed. For optimal security, the initial records saved in the storage device have been checked with the data on the blockchain regularly.

Tampered records kept on the storage system for an extended period of period pose a security concern. An anti-tampering checking method is incorporated into the system for identifying tampering in duration, anchored off-chain records using the hash recorded in the network with this process.

Based on the suggested plan, a trial platform is created and put into action. In particular, our computational system is the Alibaba Cloud system, which we selected because of its exceptional scaling and dependability. There are now roughly twelve Ethereum nodes in use to keep the system running. Ethereum is employed to create a decentralized setting in this trial network. Table 1 describes the examination conditions for both program and hardware.

4. RESULT AND DISCUSSION

The configuration of the technology is shown in Table 26.1.

Table 26.1 Configuration

Parameter	Values
CPU	i7-8700K CPU @ 3.20 GHz
RAM	8 GB
Bandwidth	1,000 M
Hard disk	250GB
OS	CentOS 7.3
Encryption algorithm	RSA or ECC
Hash algorithm	SHA-256
No. of nodes	12
Up duration	1800

The use of digital assets depending on smart contracts and the blockchain could help with effectiveness. Users must en-

ter the appropriate spaces in the required display data compared to the requirements when distributing digital assets.

Table 26.2 and Fig. 26.1 depict the average duration it takes the blockchain system to mine a block beneath various data tampering success probabilities. In this application, the tampering-reduced PoW blockchain system produces a block in 526 ms on average.

Table 26.2 Average block duration beneath different data tampering success probabilities

Data Tampering (target-bits)	Average Block Duration (ms)
14	498
15	500
16	1650
17	2300
18	5800

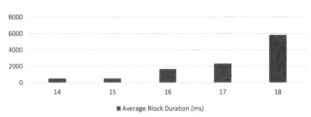

Fig. 26.1 Average block duration beneath different data tampering success probabilities

The load test outcomes of the suggested framework are shown in Table 26.3 and Fig. 26.2. As shown in the graph, response duration grows dramatically as the number of virtual institutions rises. More specifically, as the quantity of virtual institutions increases from 200-800, the minimum, average, and maximum response times all rise. In general, the highest pace of response remains within 10ms.

Table 26.3 The load test outcomes of the suggested framework

Number of Virtual Institutions	Response Time (ms)		
	Minimum	Average	Maximum
200	1.8	3	4.3
400	1.9	4	6.2
600	3.9	4.7	8.3
800	6.3	7.8	9.4

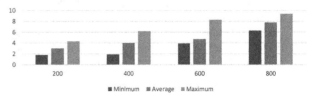

Fig. 26.2 The load test outcomes of the suggested framework

5. Conclusion

Blockchain has been employed in a variety of secure applications, including record-keeping to act as a successful data architecture. This study proposes a secure and transparent record-keeping system constructed around blockchain to address the demand for the security of educational information. A collaboration link between universities can assure the reliability and security of the records in our suggestion. To protect the security and transparency of blockchain nodes, a dispersed establishment verification technique is presented. Integrating Blockchain and Storage results in secure and transparent record-keeping. Smart contracts have been implemented for storing records to accomplish cross-institutional storage of educational data, and the authorizations are governed by smart contracts on the blockchain. Ultimately, to safeguard records in the storage system, an anti-tampering verification technique is used. In concept, the suggested strategy provides greater security, effectiveness, and trustworthiness, yet additional investigation is required.

Reference

1. Amanchukwu, R. N., &Ololube, N. P. (2015). Excellent school records behavior for effective management of educational systems. *Human Resource Management Research*, 5(1), 12–17.
2. Cheng, E. C. (2018). Managing records and archives in a Hong Kong school: a case study. *Records Management Journal*, 28(2), 204–216.
3. Han, M., Li, Z., He, J., Wu, D., Xie, Y., & Baba, A. (2018, September). A novel blockchain-based education records verification solution. In *Proceedings of the 19th annual SIG conference on information technology education* (pp. 178–183).
4. Allahmagani, K. (2014). Records management in government secondary schools: the case study of Kaduna North Local Government, Kaduna, and Kaduna State, Nigeria. *Journal of Humanities and Social Science*, 19(1), 55–60.
5. Alammary, A., Alhazmi, S., Almasri, M., & Gillani, S. (2019). Blockchain-based applications in education: A systematic review. *Applied Sciences*, 9(12), 2400.
6. Alabi, A. O. (2017). Records keeping for effective administration of secondary schools.
7. Akinloye, G. M., Adu, E. O., & Ojo, O. A. (2017). Record keeping management practices and legal issues in the school system. *The Anthropologist*, 28(3), 197–207.
8. Adebowale, O. F., & Osuji, S. N. (2008). Record-keeping practices of primary school teachers in Ondo state: implications for successful implementation of the Universal Basic Education Programme in Nigeria. *Journal of Educational Policy*.
9. Ajao, R. L., & Olawale, S. S. (2020). Record keeping as a measure of effective administration in secondary schools: Planning strategies. *Journal of Contemporary Issues in Educational Planning and Administration*, 5(2), 1.
10. Oyewole, B. K. (2015). Towards improving record keeping for effective school administration in Ekiti State, Nigeria. *International Journal of Educational Foundations and Management*, 9(1), 93â.
11. Odeniyi, O. A., & Adeyanju, A. S. (2020). Assessment of school record management in secondary schools in Federal Capital Territory. *Open Journal of Educational Development (ISSN: 2734–2050)*, 1(1), 54-65.
12. Sharples, M., & Domingue, J. (2016). The blockchain and kudos: A distributed system for the educational record, reputation, and reward. In *Adaptive and Adaptable Learning: 11th European Conference on Technology Enhanced Learning, EC-TEL 2016, Lyon, France, September 13-16, 2016, Proceedings 11* (pp. 490-496). Springer International Publishing.
13. Alnafrah, I., & Mouselli, S. (2021). Revitalizing blockchain technology potentials for smooth academic records management and verification in low-income countries. *International Journal of Educational Development*, 85, 102460.
14. Raimundo, R., & Rosário, A. (2021). Blockchain system in higher education. *European Journal of Investigation in Health, Psychology and Education*, 11(1), 276–293.
15. Zhao, B., Fang, L., Zhang, H., Ge, C., Meng, W., Liu, L., & Su, C. (2019). Y-DWMS: A digital watermark management system based on smart contracts. *Sensors*, 19(14), 3091.
16. Ma, Z., Jiang, M., Gao, H., & Wang, Z. (2018). Blockchain for digital rights management. *Future Generation Computer Systems*, 89, 746-764.
17. Chen, Y., Ding, S., Xu, Z., Zheng, H., & Yang, S. (2019). Blockchain-based medical records secure storage and medical service framework. *Journal of Medical Systems*, 43, 1–9.
18. Shae, Z., & Tsai, J. J. (2017, June). On the design of a blockchain platform for clinical trial and precision medicine. In *2017 IEEE 37th International Conference on Distributed Computing Systems (ICDCS)* (pp. 1972–1980). IEEE.
19. Kuvshinov, K., Nikiforov, I., Mostovoy, J., Mukhutdinov, D., Andreev, K., &Podtelkin, V. (2018). Disciplina: Blockchain for education. *Yellow Paper. URL: https://disciplina.io/yellow paper. pdf*.
20. Fernández-Caramés, T. M., Froiz-Míguez, I., Blanco-Novoa, O., & Fraga-Lamas, P. (2019). Enabling the internet of mobile crowdsourcing health things: A mobile fog computing, blockchain, and IoT-based continuous glucose monitoring system for diabetes mellitus research and care. *Sensors*, 19(15), 3319.
21. Chen, G., Xu, B., Lu, M., & Chen, N. S. (2018). Exploring blockchain technology and its potential applications for education. *Smart Learning Environments*, 5(1), 1–10.
22. R. Vettriselvan, C. Vijai, J. D. Patel, S. Kumar.R, P. Sharma and N. Kumar, Blockchain Embraces Supply Chain Optimization by Enhancing Transparency and Traceability from Production to Delivery," 2024 International Conference on Trends in Quantum Computing and Emerging Business Technologies, Pune, India, 2024, pp. 1-6, doi: 10.1109/TQCEBT59414.2024.10545308.

Note: All the figures and tables in this chapter were made by the author.

Recent Trends in Engineering and Science for Resource Optimization and
Sustainable Development – Prof. (Dr.) Dorota Jelonek et al. (eds)
© 2024 Taylor & Francis Group, London, ISBN 978-1-032-98030-0

27

Investigating Quantum Computing's Potential for Environmental Engineering Optimisation

Syed Omar Ballari[1]

Associate Professor and Head, Civil Engineering,
Guru Nanak Institutions Technical Campus Khanapur, India

Premendra J. Bansod[2]

Associate Professor, Mechanical Engineering,
Savitribai Phule Pune University Maharashtra, India

K. Mayuri

Department of Computer Science and Engineering,
Institute of Aeronautical Engineering, Hyderabad, India

Rakesh Chandrashekar[3]

Department of Mechanical Engineering,
New Horizon College of Engineering, Bangalore

Manish Gupta

Lovely Professional University, Phagwara, India

Prithu Sarkar

Assistant Professor Grade II,
Amity School of Communication, Amity University, Kolkata

Abstract

The objective of this work is to investigate the potential of quantum computing to environmental energy structure optimization issues and to address a few of the difficulties that quantum computers encounter, as well as methods for overcoming them. The fundamental ideas underpinning quantum computation, as well as their distinguishing features when compared to their traditional equivalents, are additionally addressed. Environment energy structure optimization financial issues of the marine engineering structure for energy structure construction can be addressed by employing both traditional algorithms carried out on traditional CPU-oriented devices and quantum algorithms realized on quantum computing equipment. Their engineering outcomes are detailed. Furthermore, this work discusses the constraints of cutting-edge quantum computers as well as their enormous potential for influencing the area of environment energy structure optimization.

Keywords

Quantum computing, Environment, Energy structures, and Optimization

Corresponding author: [1]ballarisyedomar@gmail.com, [2]premendra.bansod@gmail.com, [3]rakesh26817@gmail.com

DOI: 10.1201/9781003596721-27

1. INTRODUCTION

With the increasing need for energy and the requirements for safeguarding the environment, there has been a surge in fascination with energy structure engineering. The novel sources of energy are additionally being encompassed by energy structures, making optimal resource administration and regulation a crucial problem in harnessing the latest innovations [1]. The expense of investing in these innovations cannot be justified unless resources are used optimally. As a result, optimization instruments and algorithms are an appropriate means of addressing complicated energy structure issues in this area [2]. Models like energy organizing, energy demand-supply, forecast, sustainable energy, reducing emissions, and optimization methods must be studied to properly allocate accessible energy supplies [3]. Other domains where optimization approaches have found broad use include energy production planning and scheduled structures, location, and challenges with transport. New optimization models that include thorough energy transformation modeling and total systemic optimization a significant instruments for lowering life-cycle costs and evaluating optimal layouts for tiny grid energy facilities [4]. Simulation-based investigation of massive broad optimization for such composite energy facilities using energy administration methodologies has also been carried out. For such structures, service optimization methodologies that take system dependability into account can be used [5]. Meta-heuristic techniques have increased in popularity because of their capacity to resolve complicated optimization challenges where conventional approaches fail [6]. In the worst scenario, as the complication of certain challenges increases, exponential computing duration may be required to determine the optimal. For instance, in the context of energy grid systems, the number of energy plants that make up the grid determines how complicated the system is. Statistics from the US Energy Statistics Management indicate that there has been a steady increase in the number of energy facilities in the country over the past ten years; as a result, it is anticipated that the complication of energy systems will continue to expand. As a result, there is a requirement for novel instruments and methods suitable for resolving such complicated energy system optimization challenges while providing good responses in realistic run durations [7].

Quantum computing offers an innovative method for solving a few of the majority of complicated issues while providing a significant speed benefit over traditional computers [8]. This is demonstrated by Shor's quantum algorithm for factorization, which is substantially quicker than the standard conventional algorithm [9], and Grover's quantum search algorithm, which may search enormous records in the period square root of its dimension. Complicated energy system optimization challenges can now be tackled in quantum computing because of recent breakthroughs in hardware technologies and quantum methods.

Critical techniques that have been the subject of ongoing research include procedure insertion, superstructure optimization, and their uses in the synthesizing and ecological architecture of energy infrastructure [10]. Because of their substantial complication, these techniques rely on traditional optimization problem-solvers, which may be numerically costly and don't always yield a solution, to resolve the developed issue. Substantial computing attempt is required when doing multiple objectives on huge-scale green and sustainable energy networks using conventional optimization techniques [11]. As the magnitude of the issue rises, physical modeling and mixed-energy system optimization demand exponential computing duration [12]. Financial, environmentally friendly, and socially accountable shale gas power stations must take into consideration the best possible layout and multiple-scale decision-making [13]. Hence this research aims to investigate the quantum computing potential of the optimization of the environmental energy management approach of the marine engineering structure.

2. LITERATURE REVIEW

Conventional optimization techniques used to accomplish multiple goals on massive sustainable energy sources need a significant amount of computation [14]. Physical modeling and optimization of composite energy facilities demand exponential computing duration as the complexity of the challenge grows [15]. It is critical to account for the optimum layout and multiple scale choices for shale gas energy facilities that are cost-effective, environmentally, and socially accountable. Modeling and optimizing such shale gas power plants beneath numerous sorts of uncertainty is a computing problem [16]. Breakdown of complicated massive layouts and syntheses optimization of energy system challenges may end up in layered optimization challenges that are easier compared to the initial ones but need significantly more computer resources [17]. Customized methods for a single-issue category can surpass the most effective accessible conventional methods on occasion, but a general response method is far preferable. While predictable techniques are intractable for massive cases of complicated challenges, approximated approaches must be investigated. Such approximated methods are inherent in quantum computing devices [18]. Due to its unique characteristics, which set it apart from conventional machines in terms of operation and performance, quantum computing has grown in popularity. This investigation creates hybrid approaches and methods for solving massive mixed-integer programming issues that effectively combine the power of predictable algorithms with methods for quality assurance [19].

In [2], a particle swarm optimization (PSO) algorithm is employed to optimize the capacity setup of the hybrid energy storage structure, taking into account the power shifts of the microgrid's direct current bus and the amount of storage proportion for every preservation component [4], ensuring that the organized storage capacity for energy satisfies the microgrid's operational demands [7].

3. METHODOLOGY

As a result of the powerful worldwide optimal performance discovering competence, simplicity, and generalization, as well as its ease of execution and excellent durability when faced with complicated nonlinear issues, the quantum particle swarm optimization (QPSO) exhibits excellent accuracy and rapid convergence for resolving particular engineering issues. As a result, in this work, the QPSO is employed to optimize the energy management goal functions to satisfy the demand for loads while striving to minimize operating expenses and maximize the utilization of environmentally friendly energy. The output power of the environmentally friendly energy production component is chosen as the decision parameter for optimization when the QPSO algorithm is applied to the energy management issue related to the marine engineering structure [3]. To address the aforementioned issues, the quantum computing concept was developed to boost population diversity; quantum smart computing incorporates the ideas of quantum computation with smart computing, employs the features of quantum parallel computation, and transports out algorithm engineering in a manner that adheres to the conduct features of quantum physics, successfully compensating for certain drawbacks of conventional smart computation. The optimization concept of the artificial bee colony (ABC) algorithm is employed in this work to integrate quantum computing and the ABC algorithm in application to the issue of the optimum utilization of a marine engineering energy structure.

4. RESULT AND DISCUSSION

Real-time is a gauge of an environmental energy management structure's effectiveness, so the algorithm's convergence speed is crucial. Table 27.1 and Fig. 27.2 display the most effective convergence times for every algorithm when used with the suggested optimal planning structure beneath the same circumstances.

Table 27.2 demonstrates that, following the optimization of the environmental energy management approach of the

Table 27.1 Computation duration of the distinct optimization algorithm

Optimization algorithm	Duration (s)
PSO	0.0169
QPSO	0.0078
SQP- (Sequential quadratic programming)	1.354
ABC	1.403

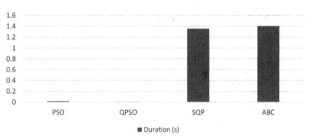

Fig. 27.1 Computation duration of the distinct optimization algorithm

Table 27.2 The outcomes of fuel and operating expenses

	Fuel Expenses	Overall Expenses
Optimization	3.71×10^5	3.87×10^5
No optimization	4.83×10^5	4.99×10^5
Deiseal functioning alone	5.24×10^5	5.32×10^5

marine engineering structure, both fuel expenses and the overall operating expense of the structure have been noticeably decreased, and the financial performance of the marine functioning has been enhanced through simulation analyses for the entire voyage.

The hash function of quantum computing is shown in Table 27.3 and Fig. 27.2.

Table 27.3 The outcomes of the hash function of quantum computing

Data	Hash function of quantum computing
2	90
4	82
6	80
8	75
10	70

5. CONCLUSION

This work provides an investigation into the fresh and arising area of quantum computing as well as certain pertinent potentials to environmental energy structure optimization. The types of issues that can be solved with quantum devic-

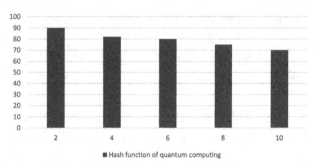

Fig. 27.2 The outcomes of the hash function of quantum computing

es and the quantum algorithms that may be employed for optimization are additionally addressed. The examination of the stated optimization issues indicates that, while an issue space discretization proximity produces an optimal approach, its probability of being the most effective alternative modification. Certain of the difficulties that quantum machines encounter in equipment structures, precision, and error prevention are additionally addressed. Although a quantum benefit might be viewed in certain circumstances, conventional algorithms tailored for that specific case have the potential to surpass a quantum machine. Quantum computing continues to be in the early phases of development and possesses many miles to travel in comparison to its much more evolved conventional counterpart. Several potentials will require the execution of both conventional and quantum assets in the coming years, so attempts must be madeto exploit the potential of quantum computing structures for massive, complicated environmental energy structure optimization issues.

Reference

1. Frangopoulos, C. A. (2018). Recent developments and trends in optimization of energy systems. *Energy*, *164*, 1011–1020.
2. Dincer, I., Rosen, M. A., & Ahmadi, P. (2017). *Optimization of energy systems*. John Wiley & Sons.
3. Biamonte, J., Wittek, P., Pancotti, N., Rebentrost, P., Wiebe, N., & Lloyd, S. (2017). Quantum machine learning. *Nature*, *549*(7671), 195–202.
4. Shayesteh, E., Yu, J., & Hilber, P. (2018). Maintenance optimization of power systems with renewable energy sources integrated. *Energy*, *149*, 577–586.
5. Xiao, H., & Cao, M. (2020). Balancing the demand and supply of a power grid system via reliability modeling and maintenance optimization. *Energy*, *210*, 118470.
6. Al-Shahri, O. A., Ismail, F. B., Hannan, M. A., Lipu, M. H., Al-Shetwi, A. Q., Begum, R. A., ... & Soujeri, E. (2021). Solar photovoltaic energy optimization methods, challeng-

es, and issues: A comprehensive review. *Journal of Cleaner Production*, *284*, 125465.
7. Mitsos, A., Asprion, N., Floudas, C. A., Bortz, M., Baldea, M., Bonvin, D., ... & Schäfer, P. (2018). Challenges in process optimization for new feedstocks and energy sources. *Computers & Chemical Engineering*, *113*, 209–221.
8. Lasemi, M. A., Arabkoohsar, A., Hajizadeh, A., & Mohammadi-Ivatloo, B. (2022). A comprehensive review of optimization challenges of smart energy hubs under uncertainty factors. *Renewable and Sustainable Energy Reviews*, *160*, 112320.
9. Mavroeidis, V., Vishi, K., Zych, M. D., & Jøsang, A. (2018). The impact of quantum computing on present cryptography. *arXiv preprint arXiv:1804.00200*.
10. Ajagekar, A., Humble, T., & You, F. (2020). Quantum computing-based hybrid solution strategies for large-scale discrete-continuous optimization problems. *Computers & Chemical Engineering*, *132*, 106630.
11. Gong, J., & You, F. (2015). Sustainable design and synthesis of energy systems. *Current Opinion in Chemical Engineering*, *10*, 77–86.
12. Cuisinier, E., Bourasseau, C., Ruby, A., Lemaire, P., & Penz, B. (2021). Techno-economic planning of local energy systems through optimization models: a survey of current methods. *International Journal of Energy Research*, *45*(4), 4888–4931.
13. Bhandari, B., Lee, K. T., Lee, G. Y., Cho, Y. M., & Ahn, S. H. (2015). Optimization of hybrid renewable energy power systems: A review. *International journal of precision engineering and manufacturing-green technology*, *2*, 99–112.
14. Wu, T., Shi, X., Liao, L., Zhou, C., Zhou, H., & Su, Y. (2019). A capacity configuration control strategy to alleviate power fluctuation of hybrid energy storage system based on improved particle swarm optimization. *Energies*, *12*(4), 642.
15. Gao, J., & You, F. (2017). Design and optimization of shale gas energy systems: Overview, research challenges, and future directions. *Computers & Chemical Engineering*, *106*, 699–718.
16. Gyongyosi, L. (2020). Objective function estimation for solving optimization problems in gate-model quantum computers. *Scientific Reports*, *10*(1), 14220.
17. Oke, D., Majozi, T., Mukherjee, R., Sengupta, D., & El-Halwagi, M. M. (2018). Simultaneous energy and water optimization in shale exploration. *Processes*, *6*(7), 86.
18. Garcia, D. J., & You, F. (2016). The water-energy-food nexus and process systems engineering: A new focus. *Computers & Chemical Engineering*, *91*, 49–67.
19. Gyongyosi, L., & Imre, S. (2019). Quantum circuit design for objective function maximization in gate-model quantum computers. *Quantum Information Processing*, *18*, 1–33.

Note: All the figures and tables in this chapter were made by the author.

Recent Trends in Engineering and Science for Resource Optimization and
Sustainable Development – Prof. (Dr.) Dorota Jelonek et al. (eds)
© 2024 Taylor & Francis Gro

28

Using AI and Blockchain to Create Secure and Transparent Energy Records

Fred Torres-Cruz[1]

Professor, Department of Computer and Statistics Engineering,
Universidad Nacional del Altiplano de Puno, P.O. Box 291, Puno Perú

Jordan Piero Borda Colque[2]

Student, Department of Computer and Statistics Engineering,
Universidad Nacional del Altiplano de Puno, P.O. Box 291, Puno Perú

E. Krishna Rao Patro

Department of Computer Science and Engineering,
Institute of Aeronautical Engineering, Hyderabad, India

Ashvini Chaudhari Bhongade[3]

Associate Professor, PhD in Computer Science, NMIMS University NGASCE

Ginni Nijhawan

Lovely Professional University, Phagwara, India

Chinnem Rama Mohan[4]

Associate Professor, Department of Computer Science and Engineering,
Narayana Engineering College, Nellore 524004, Andhra Pradesh, India

■ Abstract

The basis of this study addresses issues encountered during data distribution in traditional VENs-Vehicular Energy Networks, such as insufficient security, invasion of individual details, dearth of confidence among car owners, and so on. In this research, a blockchain-oriented announcement platform for VENs is presented to enable trustworthy and secure energy data distribution in the suggested system. The data supplied by the cars are kept in the IPFS-Interplanetary File System of artificial intelligence, which is integrated with the roadside unit. This guarantees that expenses of storage and energy data accessibility have been reduced. Furthermore, the confidentiality of vehicle owners is protected by disguising the true names of the automobiles. Furthermore, the hash values of the energy data record in IPFS have been recorded in BC on the third tier. Multiple simulations show that creating a hash of energy data record on the BC system reduces the computation duration by 15-18% and storage overhead by 80-85%, accordingly, when contrasted with putting actual energy data on the system.

■ Keywords

Artificial intelligence, Blockchain, Cuckoo filter, Vehicular energy network, Energy data record

Corresponding authors: [1]ftorres@unap.edu.pe, [2]jbordac@est.unap.edu.pe, [3]ashvinirc@gmail.com, [4]ramamohanchinnem@gmail.com

DOI: 10.1201/9781003596721-28

1. INTRODUCTION

Individuals have been moving on an extensive basis from rural to urban settings over the last few years to have exposure to the needs of existence. On the contrary, the majority have benefited greatly from the movement in many ways. However, it has also resulted in several grave problems, such as a lack of healthcare premises, a decline in work prospects, a depletion of natural assets, and so forth It has led to a significant shift away from outdated and old methods and toward the newest ones [1]. Furthermore, the globe is moving toward the creation of smart systems and the eventual replacement of old systems due to the widespread dependence of people on energy. Numerous problems, including a mismatch between supplies and demands, widespread power outages and power cuts, a sharp rise in electricity rates, etc., have also been brought on by the enormous demand for energy [2]. The environmentally friendly and ecological smart city is a viable solution to address the aforementioned problems. Smart residences, grids, hospitals, transit, etc. are all part of the smart town.

Vehicles in Vehicular Energy Networks (VENs) transmit critical energy data regarding road and traffic situations to alert other networking users about possible risks such as roadside collisions, severe weather, and so on, via messages known as announcement data [3]. Announcement messages are data given by vehicles that include pertinent energy data such as crashes on the roads, road blockages, natural disasters, and so on. These signs are vital since they direct arriving vehicles to make correct decisions and adjust their path consequently, saving both period and energy [4].

Additionally, electric vehicles (EVs) are used in VENs to transport energy between locations. Nevertheless, this category also includes other communication protocols. Moreover, as we move forward, vehicles in VENs communicate with other network members using signals known as notification signals critical data about traffic and road circumstances to warn them of possible risks such as severe weather, roadway incidents, and so on [5]. Significant data such as details regarding traffic incidents, road closures, natural disasters, etc., is sent by notification messages, which are signals sent by cars. These signals are crucial since they enable approaching cars to make the right choices early on and alter their course to conserve duration and fuel [6].

They are, nevertheless, vulnerable to a variety of dangers such as the dissemination of inaccurate energy data, Single Point of Fault (SPoF), the lack of confidentiality protection, and so on [7]. To solve these difficulties in VENs, multiple approaches based on Blockchain (BC) have been suggested [8]. The major challenges encountered with centralized VENs, such as an absence of confidence and security, are addressed by utilizing BC [9]. Because of its multiple intriguing qualities such as openness, tamperproof, and data immutability, both academics and sectors have adopted BC as a Distributed Ledger Technology to a stunning degree in recent years. To hold transactional data, the blocks in BC are cryptographically connected collectively.

Hence this research aims to create blockchain-based secure and transparent energy data records using artificial intelligence (AI). In this research energy data is distributed and recorded in the Interplanetary File System (IPFS), which is controlled by AI, ensuring storage expense reduction and energy data availability. The suggested architecture guarantees easy data sharing and lowers delay, storage, and processing energy needs. Despite their numerous advantages, BC-based VENs have several possible limitations that could reduce their effectiveness and cast doubt on their viability [10]. Furthermore, an OBU is a limited resources system that is unable to carry out a consensual method that requires a lot of computing, such as Proof-of-Work (PoW). Furthermore, the cars are unable to hold a full version of the distributed record, which is constantly expanding, because of their restricted storage.

2. LITERATURE REVIEW

Due to a multitude of problems, including a single source of breakdown and an absence of confidence, the older and centralized networks are becoming less and less relevant each day. Individuals are using more modern, decentralized structures. In addition to offering many advantages, the massive transformation in human resources from conventional to cutting-edge methods also presents significant risks. Additionally, it is necessary to solve the drawbacks of conventional energy trade plans, such as a shortage of confidence, a compromise of security and confidentiality, demands that are dangerous or unapproved, etc. Alternatively, to address the concerns of delay, network expenses, safety, and anonymity, the researchers of [11] offer BC-based alternatives. Effective trade in energy is made possible in VENs by these mechanisms. Yet, there are also problems with the suggested solutions, such as the EVs' tendency to generate a lot of queries, which adds to the computing cost of the platforms. In addition, there are traffic bottlenecks and roadside fatalities brought on by the abrupt rise in the number of cars [12]. Furthermore, the progressive loss of important data brought on by the automobiles' concurrent generation and transfer of signals contributes to traffic jams and accidents. The suggested work proposes a BC-based paradigm that guarantees effective signal distribution together with effective data storing, user confidentiality conservation, and reputation-based incentivization. This framework is driven by [13] and [14] and keeps certain difficulties in mind.

The authors of [15] presented a BC-based supply chain management paradigm that includes the Internet of Things (IoT) to maintain confidence between organizations. The suggested approach allows for simple energy data sharing while reducing processing power, recording, and latency needs. Nevertheless, despite their obvious benefits, BC-based VENs have several possible issues that restrict their efficiency and call their viability into doubt [16].

3. METHODOLOGY

The research introduces a hybrid blockchain (BC)-oriented notification distribution framework designed to facilitate the secure and efficient sharing of energy data in Vehicular Energy Networks (VENs), while safeguarding user confidentiality. This proposed architecture is structured into three distinct layers: the announcement data broadcast layer, the recording layer, and the blockchain layer. In the first layer, a certificate authority (CA) is responsible for registering vehicles within the system. Once registered, vehicles can interact with each other, exchanging energy-related data within the network. Moving on to the second layer, Road Side Units (RSUs) are integrated to oversee and monitor the overall system operations. The vehicles transmit their energy data to the RSUs, which act as intermediaries in processing and storing the data. To enhance the efficiency of data storage, Inter Planetary File System (IPFS) is utilized in conjunction with the RSUs. The data received by the RSUs is routed to IPFS for storage, significantly reducing the costs associated with traditional storage solutions while maintaining data availability. IPFS breaks the data into smaller segments and generates a hash for each segment. This hash serves as a reference for the data and is stored for future retrieval. In the third layer, the generated IPFS data hashes are securely stored on the Ethereum blockchain, ensuring the integrity and immutability of the data. Additionally, a cuckoo filter (CF) is implemented at this layer to store vehicle reputation values, providing a mechanism to track and record the trustworthiness of individual vehicles. The information related to the reputation values is stored within the CF, while private transaction data links to the IPFS hash, which is securely recorded on the Ethereum blockchain. This layered approach ensures not only the security and privacy of energy data but also the efficient management of storage and reputation tracking within VENs.

4. RESULT AND DISCUSSION

Table 28.1 and Fig. 28.1 indicate the smart contract operations' gas consumption.

Table 28.1 Gas consumption of various operations employed in smart contracts

Operations	Gas consumption (Wei)
Authorization	65000
Recording	90000
Response operation	55000
Reputation	50000
Incentive	62000

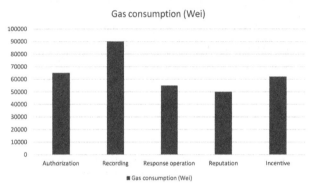

Fig. 28.1 Gas consumption of various operations employed in smart contracts

Table 28.2 and Fig. 28.2 compare the evaluation of energy data size and its accompanying hash values of the expenses of storage in bytes. The figure's increased size is because of the accumulation of hash data. On the contrary, as the number of energy records rises, so does the quantity of the real data.

Table 28.2 Storage expenses comparison of actual energy data and hash when utilizing IPFS

| Number of records | Storage expenses (bytes) | |
	Actual energy data	Hash
100	498	200
150	700	210
200	750	200
250	1100	240
300	1300	300
350	1497	320
400	1500	330
450	1700	335
500	2200	390

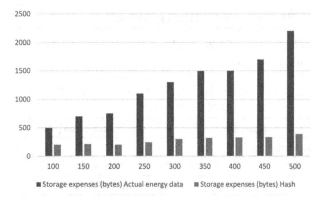

Fig. 28.2   Storage expenses comparison of actual energy data and hash when utilizing IPFS

Figure 28.3 depicts a contrast between data upload and data retrieval times for various data sizes. It should be observed that when data size rises, the rise in data upload duration outweighs the rise in data retrieve duration.

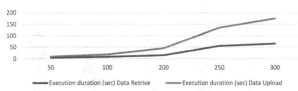

Fig. 28.3   A contrast between data upload and data retrieval times for various data sizes

5. CONCLUSION

The research suggested proposes a BC-oriented data announcement platform for effective data transmission in VENs. The suggested architecture has three levels: a data broadcast layer, a recording layer, and a BC layer. Vehicle registration takes place in the initial stratum. The data is entered and recorded in AI-oriented IPFS, which is integrated with the RSUs, in the subsequent layer. IPFS creates hash values for energy data upon recording, which are sent and recorded in the Ethereum BC, which is implemented on the third tier. Furthermore, CFs are utilized for recording the expected patterns in vehicle esteem scores. Additionally, in the suggested paradigm, users are given incentives to share honest feedback. Furthermore, comprehensive simulations show that when recording hashes of energy data on the BC system versus recording actual data on the system, the computational duration of the suggested method is lowered by 15-18% and the storage overhead is decreased by 80-85%, accordingly. The findings of the security evaluations show that the framework is resistant to both smart contract flaws and cyber assaults such as the double spending assault.

Reference

1. Ashfaq, T., Khalid, M. I., Ali, G., Affendi, M. E., Iqbal, J., Hussain, S., ... & Mateen, A. (2022). An efficient and secure energy trading approach with machine learning technique and consortium blockchain. *Sensors*, 22(19), 7263.
2. Jiang, P., Van Fan, Y., &Klemeš, J. J. (2021). Impacts of COVID-19 on energy demand and consumption: Challenges, lessons, and emerging opportunities. *Applied Energy*, 285, 116441.
3. Javed, M. U., Javaid, N., Malik, M. W., Akbar, M., Samuel, O., Yahaya, A. S., & Othman, J. B. (2022). Blockchain-based secure, efficient, and coordinated energy trading and data sharing between electric vehicles. *Cluster Computing*, 1–29.
4. GTT Wireless. (2022). DSRC vs C-V2X: Comparing the Connected Vehicles Technologies—GTT Wireless. Accessed: Jul. 23, 2022. [Online]. Available: https://gttwireless.com/dsrc-vs-c-v2x-comparing-theconnected-vehicles technologies/
5. Saputra, Y. M., Hoang, D. T., Nguyen, D. N., Dutkiewicz, E., Mueck, M. D., &Srikanteswara, S. (2019, December). Energy demand prediction with federated learning for electric vehicle networks. In *2019 IEEE Global Communications Conference (GLOBECOM)* (pp. 1-6). IEEE.
6. Mao, J., Hong, D., Ren, R., & Li, X. (2020). The effect of marine power generation technology on the evolution of energy demand for new energy vehicles. *Journal of Coastal Research*, 103(SI), 1006–1009.
7. Samuel, O., Javaid, N., Almogren, A., Javed, M. U., Qasim, U., & Radwan, A. (2022). A secure energy trading system for electric vehicles in smart communities using blockchain. *Sustainable Cities and Society*, 79, 103678.
8. Luo, Q., Zhou, Y., Hou, W., & Peng, L. (2022). A hierarchical blockchain architecture-based V2G market trading system. *Applied Energy*, 307, 118167.
9. Wang, K., Wang, J., Huang, L., Yuan, Y., Wu, G., Xing, H., ... & Jiang, X. (2022). A comprehensive review on the prediction of ship energy consumption and pollution gas emissions. *Ocean Engineering*, 266, 112826.
10. Tomar, A., & Tripathi, S. (2022). BCAV: Blockchain-based certificateless authentication system for the vehicular network. *Peer-to-Peer Networking and Applications*, 15(3), 1733–1756.
11. Al-Rakhami, M. S., & Al-Mashari, M. (2021). A blockchain-based trust model for the Internet of Things supply chain management. *Sensors*, 21(5), 1759.
12. Wang, Y., Su, Z., & Zhang, N. (2019). BSIS: Blockchain-based secure incentive scheme for energy delivery in vehicular energy networks. *IEEE transactions on industrial informatics*, 15(6), 3620–3631.
13. Baza, M., Sherif, A., Mahmoud, M. M., Bakiras, S., Alasmary, W., Abdallah, M., & Lin, X. (2021). Privacy-preserving blockchain-based energy trading schemes for electric vehicles. *IEEE Transactions on Vehicular Technology*, 70(9), 9369–9384.

14. Ayaz, F., Sheng, Z., Tian, D., & Guan, Y. L. (2021). A blockchain-based federated learning for message dissemination in vehicular networks. *IEEE Transactions on Vehicular Technology*, *71*(2), 1927–1940.

15. Shrestha, R., Bajracharya, R., Shrestha, A. P., & Nam, S. Y. (2020). A new type of blockchain for secure message exchange in VANET. *Digital communications and networks*, *6*(2), 177–186.

16. Pawełoszek I. (2018), Ontological Support for Process-Oriented Competency Management, Lecture Notes in Business Information Processing nr 311, Information Technology for Management. Ongoing Research and Development (ed.) Ziemba E., Springer, Cham pp. 41–60

17. Khalid, A., Iftikhar, M. S., Almogren, A., Khalid, R., Afzal, M. K., & Javaid, N. (2021). A blockchain-based incentive provisioning scheme for traffic event validation and information storage in VANETs. *Information Processing & Management*, *58*(2), 102464.

Note: All the figures and table in this chapter were made by the author.

Recent Trends in Engineering and Science for Resource Optimization and
Sustainable Development – Prof. (Dr.) Dorota Jelonek et al. (eds)
© 2024 Taylor & Francis Group, London, ISBN 978-1-032-98030-0

29

Studying the Application of Blockchain in Real Estate for Secure and Efficient Record-Keeping

Sudha Arogya Mary Chinthamani[1]

Assistant Professor, Department of Management Studies,
Saveetha Engineering College Chennai

S. Vidya[2]

Assistant Professor, Department of Commerce,
B. S. Abdur Rahman Crescent Institute of Science and Technology,
Vandalur, India

C. Praveen Kumar

Department of Computer Science and Engineering,
Institute of Aeronautical Engineering, Hyderabad, Telangana

Rakesh Chandrashekar[3]

Department of Mechanical Engineering,
New Horizon College of Engineering, Bangalore

Ginni Nijhawan

Lovely Professional University, Phagwara, India

Karu Lal[4]

Integration Engineer, Ohio National Financial Services, USA

▬▬▬▬ Abstract

Real estate sectors are presently dominant and constitute a significant portion of the nation's gross domestic product. Nevertheless, real estate deals can be forged through forged documents or operations. Numerous changes have been suggested, yet distributed ledger innovation has the potential to solve the majority of these problems. As Blockchain was initially developed to keep a financial record, its application can be expanded to include any decentralized computing structure, such as electronic record-keeping and administration structures. According to recent research, the application of Blockchain in the real estate field is still comparatively novel and unknown. As a result, in this paper, a blockchain-oriented framework for developing real-estate record-keeping is suggested. The suggested model was implemented in the Ethereum blockchain system, and the results demonstrated that it supported the development of secure, and efficient record-keeping.

▬▬▬▬ Keywords

Blockchain, Real estate, Record-keeping, Ethereum, and Smart contracts

Corresponding authors: [1]cmsudhaarogyamary@saveetha.ac.in, [2]vidyadinesh2003@gmail.com, [3]rakesh26817@gmail.com, [4]karu021984@gmail.com

DOI: 10.1201/9781003596721-29

1. INTRODUCTION

Real estate is among the most important subsidiaries of any government. Real estate records real(land) assets, primarily the location, (historical) possession, worth, and application [1]. The real estate record-keeping additionally holds data on the tangible, geographical, and topographic features of real estate. As a result, real estate records are critical from a societal, economic, and political standpoint [2]. Real estate records ought to be available for a long time, transparent, and by applicable laws [3]. Without effective governance, the structure may be prone to corruption, economic insecurity, and inaccuracy.

Material and legal statistics contained in real-estate record keeping are frequently regarded as reliable and appropriate. Human error, data transport, and even misuse of data can all result in manipulated real-estate record-keeping data[4]. In real-estate record keeping, operations related to data openness, validation, and mutation are often managed by a centralized power [5].This centralization may lead to a rise in the frequency of misuse and recklessness. A blockchain is an immutable ledger of statistics that is time-tagged and maintained by a group of unaffiliated computers. A work named "Bitcoin: A peer-to-peer digital currency system," written in 2008[6] under the pseudonym Satoshi Nakamoto, described the blockchain system. Blockchain is made up of points in a distributed system that use digital ledger technological advances to keep records of transactions between peers. Other parties are involved in the validation of land record updates, which can be removed for real estate transfer deals using blockchain [7]. Hence this research aims to investigate the application of blockchain for secure and efficient real estate record-keeping.

2. LITERATURE REVIEW

Several investigations have implemented Blockchain innovation into land administration systems. For instance, in [8], a conceptual model of a land management system built around the public blockchain codebase of Litecoin has been addressed, which will help to improve scalability, confidentiality, and interoperability. The system has been suggested with just one kind of transaction in mind: the transmission of worth from one user to a different one. The research [9] suggested a blockchain-oriented Land Administration System that would improve openness, reliability, security, and responsibility. They emphasized the primary characteristics of blockchain, such as immutability, openness, decentralization, and smart contracts, and traced them into the land management electronic record-keeping structure. They depicted the key characteristics of blockchain technology—transparency, immutability, decentralization, and smart contracts—in the land administration method's elec-

tronic record-keeping framework. This investigation was unable to explain the origin block's creation, nevertheless. Additionally, the PoW system's potential majority assault remained unresolved. Blockchain innovation has also been adopted for record administration in a few other industries, such as financial services, shipment of goods, electronic voting, and the maintenance of individual records in digital health systems.A blockchain-enabled electronic voting structure, for instance, was presented by [10] to lessen database tampering and preserve data integrity. This paper presents a data recording method for electronic voting that makes use of permissioned blockchain systems. Safety and openness were guaranteed by the blockchain protocol used to record voting results from each polling location [11].

3. METHODOLOGY

To create an efficient framework, the suggested approach should closely mimic the existing real estate record-keeping. The structure has been suggested with three issues in mind. To begin, the real estate record-keeping's key functions involve upgrading land-based data, checking land data, submitting a mutation demand, and completing the ownership transmit procedure. Secondly, there are the real estate record-keeping's other parties, which include the land owner, buyers, the governing body (who serves as the handles the real estate record-keeping), and the administration. Thirdly, blockchain technological advances characteristics. In this study, a local web application has been established employing Ganache to serve as the regional blockchain depending on the suggested structure. The computational power and execution duration needed to carry out every operation of the suggested real estate record-keeping are additionally covered in the below section. A block is created when a smart contract is deployed in Ethereum and any occurrence is triggered [12, 13]. Mining that blocks and adding it to the chain after validation requires a certain computing capacity [14]. The computing power on the Ethereum structure is calculated in terms of petrol fees. In addition, certain Ether digital money is needed [15].

4. RESULT AND DISCUSSION

According to Table 29.1 and Fig. 29.1, employing the contract uses the most computing power.

Table 29.1 Gas employed for every operation

Operations in Ethereum Blockchain	Gas Employed
Deploy migration	284908
Deploy real estate record-keeping	1128986
Create land	133390
Purchase land	66527
Land for sell	32687

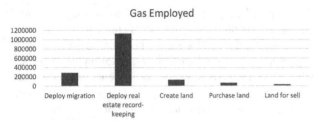

Fig. 29.1 Gas employed for every operation

The following histogram in Fig. 29.2 depicts the ether expenses associated with the same operations.

Table 29.2 Ether employed for every operation

Operations in Ethereum Blockchain	Cost of Ethereum
Deploy migration	0.00569816
Deploy real estate record-keeping	0.02257972
Create land	0.002668
Purchase land	0.001331
Land for sell	0.0006654

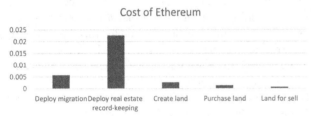

Fig. 29.2 Ether employed for every operation

Table 29.3 and Fig. 29.3 depict the execution duration of the land for sale, land creation, and land purchase operations in various instances of testing in milliseconds. The test versus time graph clearly shows that buying land requires the most duration, approximately 900 ms. The second longest execution time is spent on creating land, which takes approximately 300-400 ms. The land then selects the shortest duration of execution for the sell functioning.

Table 29.3 Executions of every operation

Test count	Execution duration (ms)		
	Land for sale	Land creation	Land purchase
1	101	398	1000
2	100	300	900
3	100	350	1100
4	100	340	898
5	110	300	880
6	150	360	920
7	110	299	900
8	140	300	910
9	140	310	810
10	130	315	920

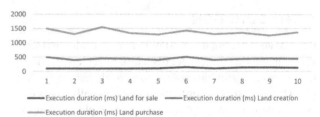

Fig. 29.3 Executions of every operation

In this study, a regional web program was developed utilizing Ganache to serve as the regional blockchain, depending on the suggested architecture. The previous part also covers the computing energy and implementation duration needed to carry out every operation of the suggested real-estate record-keeping. According to the simulation outcomes, the suggested real-estate record-keeping platform depending on blockchain technology would have the subsequent advantages: Real-estate record-keeping is provided with a plethora of security features by the suggested structure. The suggested framework's consensus mechanism makes certain that no erroneous blocks are uploaded to the blockchain, and any manipulation will result in a sequence of modification operations that promptly notify the system. Furthermore, the suggested system guarantees that individuals can access just their data because it produces a secret key for every end-user.

Secondly, a PoW consensus method is used on the Ethereum blockchain system to mimic the suggested architecture and guarantee coherence among the involved nodes [16]. As a result, every modification to the blockchain is first validated and then upgraded employing the PoW in every regional version of the blockchain. As an outcome, the suggested method's data continues to be reliable and consistent. Third, blockchain satisfies the primary requirements of land administration systems by improving confidence, data trustworthiness, and multi-factor validation [17, 18]. It also functions as an effective archive system. Once more, the suggested architecture maintains alteration logs as links in a chain that further guarantees the framework's reliability.

Fourth, mining each block requires a lot less processing energy than publishing the smart contract. Because the architecture permits a small number of nodes to act as the miner hindrance, mining requires less processing energy. A significant amount of processing energy is required for the contract's implementation, yet this only happens once during the system's lifespan. For instance, Fig. 29.3 illustrates how less computing power was used for the create and, buy land (mutation), and selling land functions. In conclusion, certain processes, such as publishing a notice of sale for land and generating a novel land block, demonstrated a mean implementation duration of fewer than 400 milliseconds. Once more, the land acquisition (mutation) shows a comparably greater implementation duration of approximately a second (1000 milliseconds), which is nevertheless regarded as reasonably quicker. The platform's adoption among end users may rise as a result of these characteristics. As a result, the suggested design has excellent user outreach and adaptability.

5. CONCLUSION

In this research, a blockchain-oriented structure for real estate record-keeping is suggested. The operation encompasses (a) upgrading land-based data, (b) checking real estate data, (c) submitting a mutation demand, and (d) the land-ownership transmit procedure. The suggested structure is tested using simulation. Utilizing the outcomes of the simulations, it was possible to create a secure, and efficient real estate record-keeping using the suggested structure. Though there are a couple of current investigations focusing on blockchain-oriented real estate record-keeping, these investigations did not take into account every potential real estate record-keeping function and were not assessed using a simulation structure. Looking at all of the factors, it is possible to conclude that the suggested structure is distinctive and contributes to the planning and development of effective real estate record-keeping. Nevertheless, the structure's performance in real-time transactions has to be examined. Our future efforts will be directed in these paths.

Reference

1. Yadav, A. S., Agrawal, S., & Kushwaha, D. S. (2022). Distributed Ledger Technology-based land transaction system with trusted nodes consensus mechanism. *Journal of King Saud University-Computer and Information Sciences*, *34*(8), 6414–6424.
2. Singh, N., & Vardhan, M. (2019) Digital ledger technology-based real estate transaction mechanism and its block size assessment. *International Journal of Blockchains and Cryptocurrencies*, *1*(1), 67–84.
3. Thakur, V., Doja, M. N., Dwivedi, Y. K., Ahmad, T., & Khadanga, G. (2020). Land records on blockchain for implementation of land titling in India. *International Journal of Information Management*, *52*, 101940.
4. Lemieux, V., Hofman, D., Batista, D., & Joo, A. (2019). Blockchain technology &and record keeping. *ARMA International Educational Foundation*.
5. Wiatt, R. G. (2020). The new management of recordkeeping. *Journal of Corporate Accounting & Finance*, *31*(2), 13–20.
6. Lemieux, V. L. (2017, November). Blockchain and distributed ledgers as trusted record keeping systems. In *Future Technologies Conference (FTC)* (Vol. 2017).
7. Crosby, M., Pattanayak, P., Verma, S., & Kalyanaraman, V. (2016). Blockchain technology: Beyond Bitcoin. *Applied Innovation*, *2*(6–10), 71.
8. Patil, V. T., Acharya, A., & Shyamasundar, R. K. (2018). Landcoin: a land management system using a Bitcoin blockchain protocol. In *Proceedings of the Symposium on Distributed Ledger Technology*.
9. Hanifatunnisa, R., & Rahardjo, B. (2017, October). Blockchain-based e-voting recording system design. In *2017 11th International Conference on Telecommunication Systems Services and Applications (TSSA)* (pp. 1–6). IEEE.
10. Bulut, R., Kantarcı, A., Keskin, S., & Bahtiyar, Ş. (2019, September). Blockchain-based electronic voting system for elections in Turkey. In *2019 4th International Conference on Computer Science and Engineering (UBMK)* (pp. 183–188). IEEE.
11. Stefanović, M., Pržulj, D., Ristić, S., Stefanović, D., & Vukmanović, M. (2018). Blockchain and land administration: Possible applications and limitations. In *Proc. EBM* (pp. 1–8).
12. Pawełoszek, I., Kumar, N., & Solanki, U. (2022). Artificial intelligence, digital technologies, and the future of law. Futurity Economics & Law, 2(2), 24–33. https://doi.org/10.57125/FEL.2022.06.25.03
13. Lemieux, V. L. (2017, December). A typology of blockchain record keeping solutions and some reflections on their implications for the future of archival preservation. In *2017 IEEE International Conference on big data (Big Data)* (pp. 2271–2278). IEEE.
14. Mylrea, M., & Gourisetti, S. N. G. (2017, September). Blockchain for smart grid resilience: Exchanging distributed energy at speed, scale and security. In *2017 Resilience Week (RWS)* (pp. 18–23). IEEE.

Note: All the figures and tables in this chapter were made by the author.

Recent Trends in Engineering and Science for Resource Optimization and Sustainable Development – Prof. (Dr.) Dorota Jelonek et al. (eds)
© 2024 Taylor & Francis Group, London, ISBN 978-1-032-98030-0

30

Electronic Health Records with Secure Sharing of Information Based on Blockchain

Dyuti Banerjee[1]

Assistant Professor, Artificial Intelligence and
Data Science department, K L University, Hyderabad, India

Syed Mujib Rahaman

Research Scholar, Paher University,
Udaipur, Rajasthan

CH. Srividhya

Department of Computer Science and Engineering,
Institute of Aeronautical Engineering, Hyderabad, India

G. Poshamallu[2]

Assistant Professor, ECE, St Martin's engineering college Secunderabad,
Telangana

Ginni Nijhawan

Lovely Professional University, Phagwara, India

P. Kiranmayee[3]

Assistant Professor, Department of ECE,
St. Martin's Engineering College Secunderabad, India

▬▬▬ Abstract

Electronic health records (EHRs) are important parts of extremely sensitive confidential information in healthcare that must be shared often among their peers. Blockchain offers a shared, immutable, and open tradition of every transaction, allowing developers to create tasks that are secure, responsible, and transparent. This presents a once-in-a-lifetime chance to use blockchain to create a secure and trustworthy EHR information administration and sharing structure. As a result, we provide a blockchain-oriented information-sharing system that adequately tackles the access regulation difficulties related to confidential information kept in the cloud by leveraging the blockchain's immutability and embedded independence. According to the results of the system assessment, our approach is portable, adaptable, and effective. The proposed research could substantially decrease the turnaround duration for EMR sharing, enhance healthcare decision-making, and transactions, and lower total expenses.

▬▬▬ Keywords

Blockchain, Electronic health records, Information-sharing, and Cloud computing

Corresponding author: [1]dbanerjee@kluniversity.in, [2]gaddi.poshamallu421@gmail.com, [3]kiranmayee239@gmail.com

DOI: 10.1201/9781003596721-30

1. INTRODUCTION

Smart innovations have transformed the design and manufacturing industries, like automobiles, computation, electronics, aviation and defense, and virtual reality (VR) and augmented reality (AR). These innovations include the Internet of Things (IoT), Artificial Intelligence (AI), VR, and Machine Learning. There is no exception to this rule when it comes to the medical systems used by hospitals and physicians. They have grown increasingly potent and beneficial as time passed [1]. Additionally, since smart systems have grown more adept at managing massive data collections in real-time, quicker disease detection and diagnosis are now possible. Treatment recommendations and evaluations are now automatic.

Openness and interaction between patients and medical professionals are also improved by the usage of blockchain technology. Delivery of patient medical records could now be tailored by the consumer because of the introduction of Healthcare 3.0 and Web 3.0. Customized and optimized encounters were made possible by the simplification and customization of user displays [2]. Electronic health records (EHRs) are vital yet extremely confidential information for medical diagnosis and therapy in medical settings that must be distributed and shared regularly among peers like medical professionals, insurance firms, pharmacy shops, investigators, families of patients, and others. Similar to this, EHR platforms [3] started to appear, integrating stand-alone non-networked structures, such as social networking pathways, for storing patient data. This simplified the process of exchanging medical data between physicians or over connected pathways, such as social networking sites, by employing EHR platforms. Additionally improved was patient-provider communication and interaction [4]. The transition of EHR to cloud-based systems has made it easier for medical and research organizations to share medical statistics, allowing for a quicker and easier interchange than was previously feasible [5]. Working together has several benefits, including better demographic health administration and the ability to analyze and obtain fresh ideas for therapies that are already present. However, there are serious potential hazards with the procedure that cannot be disregarded [6].

To cope with the numerous regulations and structures that regulate its sharing, the statistics are typically kept in cloud archives that are secured with conventional access management [7]. Due to their frequent breaches, these procedures have several drawbacks. The statistics are protected in three ways: initially, by the charges that will be imposed in the incident of a violation; secondly, users have no govern over the statistics once they evacuate the sharing institution; and thirdly, there are no technological obstacles that prevent

the merging of statistics shared from various sources, so dealing with anonymized collections of data still carries the danger of re-identifying certain people [8]. One of two issues typically plagues current EHR sharing programs: the initial one permits adequate records exchange but has inadequate control. The organization that distributes the data effectively forfeits control over it. Because there are multiple manners to refine data collection and re-identify people even in cases where anonymization was done before sharing, this presents issues with confidentiality [9]. The 2nd one is still not very good at sharing, but it does have excellent control over the statistics. This undermines the idea that sharing should make it easier to extract more significance from the collections of data that are already accessible. Strictly regulated management of access to the sharing cloud repository is thus essential [10].

3 phases are often involved when establishing access management in structures: permission, authentication, and recognition. Access management guarantees accountability by enabling the tracking of which user performed which activity within an organization, in addition to allowing only authorized users to access the platform [11]. Security managers select which user or consumers can access a certain piece of data in conventional access management platforms. Having said that, present networks are more susceptible to unauthorized invasions and hacking. Blockchain innovation will ensure data security, regulate confidential information, and ease medical information administration for patients and other participants in the healthcare sector by offering a tool for establishing agreement among dispersed organizations without depending on a single confidence party [12]. Hence this research aims to investigate HER with secure information sharing using blockchain.

2. LITERATURE REVIEW

The use of Blockchain innovation in healthcare has become a pattern in this sector, particularly for securing the sharing and handling of healthcare information. The proposed approach of [13] employs Blockchain to enable mobile users and suppliers of services to securely share information while maintaining confidentiality.

The research [14] suggested a Blockchain-based decentralized EHR records administration structure. Their method ensures immutable records, simple access, accountability, authorization, sharing of information, and the security of highly sensitive healthcare information [15, 16]. It is additionally flexible to the platforms of present data storage suppliers, ensuring interoperability [17]. Interoperability necessitates massive amounts of information sharing and transmission [18]. This could heighten the risks of copying, disseminating, or interfering with data. The study [19] proposed a safe user-focused method to enable access management and confidentiality via channel construction method while designing a blockchain-based mobile-based EHR sharing system. After [20] suggested a blockchain-based EHR sharing structure, off- and on-chain validation for the platform's storing safety was created. In their investigation of safeguarding information solutions for the healthcare industry, [2] suggested storage administration techniques that support data administration. [13] suggested blockchain-based medical records for patients, enhancing consensus processes to accomplish better data safety and confidentiality inside the system. A secured EHR sharing platform employing encryption based on attributes has been suggested by [12]. To guarantee the accuracy and accountability of medical data, they employed smart contracts. Employing blockchain technology, [20] established an attribute-based signing technique for multiple users in the administration of EHR. Through the use of a decentralized strategy for more confidentiality, they hoped to boost system safety with this attribute-based computation.

3. METHODOLOGY

This section introduces the suggested strategy for EHR information sharing depending on blockchain-based systems. As a result, a framework of a blockchain-based EHR information-sharing structure is suggested. In networked systems, different techniques and setups for block transactions are used. A shared symmetric key and a private key in the system suggested allowing the EHR to be dispersed to different blockchain network respondents. The dimension of the individual frameworks in the block is shown in Table 30.1.

Table 30.1 Format of block and size of the component

Structure Name	Size
Block size	4
Block header	80
Block format	4
Transactions	578
Transaction counter	9
Transaction lock duration	4

Evaluations depending on determined presumptions obtained from various vendor transactions per unit duration were simulated to estimate the size and scaling of the blockchain networking in this framework. For clarity, present vendor networked system transactions were stated, and the number of users on this platform was used to overstate the overall size of the blockchain in the blockchain-ori-

ented information-sharing structure for a particular time frame.

4. RESULT AND DISCUSSION

Table 30.2 and Fig. 30.1 display the server's turnaround duration.

Table 30.2 Execution duration of different operations

Operations	Average execution duration (ms)
Add Medica Facility	14
Update Medical Facility	17
Add Medical Facility Permission	13
Add Patient Record	18
Delete Medical Facility	16
Delete Patient Record	17
Get Medical Facility Access	4

Average execution duration (ms)

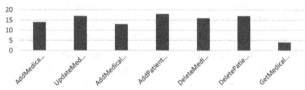

Fig. 30.1 Execution duration of different operations

Table 30.3 depicts an overstated quantity of user transactions per second. The outcomes displayed in the table beneath have an overall size in bytes that corresponds to blocks produced by this platform. Megabytes (MB), gigabytes (GB), terabytes (TB), and petabytes (PB) are the size of the information sizes produced. The number of transactions during a given period is displayed in the transaction section. The extent of the blockchain system's dimension, as time passes, is represented by the information created per session concerning the layout block.

Table 30.3 Determined blockchain network development

Transactions	Per 10 years	Per year	Per day	Per second
2000	387.9 TB	38.79 TB	108.84 GB	1.29 MB
10,000	1.95 PB	194.59 TB	545.91 GB	6.47 MB
15,000	2.92 PB	292.03 TB	819.28 GB	9.71 MB
500,000	97.4 PB	9.74 PB	26.68 TB	323.77 MB
2,100,000	409.6 PB	40.96 PB	112.22 TB	1.33 GB

Table 30.4 shows the execution expenses of the different operations.

Table 30.4 The execution expenses of the different operations

Operations	Gas	Used Price
Add Medica Facility	45296	0.06569
Update Medical Facility	58576	0.008494
Add Medical Facility Permission	25857	0.00383
Add Patient Record	55428	0.0082
Delete Medical Facility	15245	0.00225
Delete Patient Record	18091	0.00268
Get Medical Facility Access	0	0

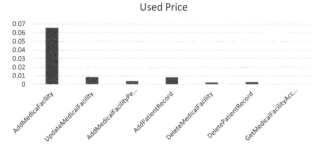

Fig. 30.2 The execution expenses of the different operations

5. CONCLUSION

In this work, we suggested a blockchain-oriented process for facilitating access between individuals and a collection of shared (sensitive) information. In comparison to the Bitcoin blockchain system, we built an adaptable (redesigned to enable faster transactions) and thin blockchain to show the effectiveness of the architecture, which enables secure information sharing while protecting the confidentiality of information. The suggested framework fails to completely investigate interaction and authorization methods and algorithms between organizations. It could be fascinating to expand on this research in subsequent studies by thoroughly investigating these. We indicate that the framework outlined in this work is the highest level of a powered by blockchain access regulate structure that is currently being implemented and tested. This study desires to undertake an experiential investigation in the future to enhance the effectiveness of the structure and gather qualitative information for additional research.

Reference

1. Jin, H., Luo, Y., Li, P., & Mathew, J. (2019). A review of secure and privacy-preserving medical data sharing. *IEEE Access*, *7*, 61656–61669.

2. Ivan, D. (2016, August). Moving toward a blockchain-based method for the secure storage of patient records. In *ONC/NIST Use of Blockchain for Healthcare and Research Workshop. Gaithersburg, Maryland, United States: ONC/NIST* (pp. 1–11). sn.

3. Fan, K., Wang, S., Ren, Y., Li, H., & Yang, Y. (2018). Medblock: Efficient and secure medical data sharing via blockchain. *Journal of Medical Systems*, *42*, 1–11.

4. Zhang, A., & Lin, X. (2018). Towards secure and privacy-preserving data sharing in e-health systems via consortium blockchain. *Journal of Medical Systems*, *42*(8), 140.

5. Cao, Y., Sun, Y., & Min, J. (2020). RETRACTED: Hybrid blockchain–based privacy-preserving electronic medical records sharing scheme across medical information control system. *Measurement and Control*, *53*(7–8), 1286–1299.

6. Xia, Q., Sifah, E. B., Smahi, A., Amofa, S., & Zhang, X. (2017). BBDS: Blockchain-based data sharing for electronic medical records in cloud environments. *Information*, *8*(2), 44.

7. Yang, J. J., Li, J. Q., & Niu, Y. (2015). A hybrid solution for privacy-preserving medical data sharing in the cloud environment. *Future Generation computer systems*, *43*, 74–86.

8. Cheng, X., Chen, F., Xie, D., Sun, H., & Huang, C. (2020). Design of a secure medical data sharing scheme based on blockchain. *Journal of Medical Systems*, *44*(2), 52.

9. Dove, E. S., & Phillips, M. (2015). Privacy law, data sharing policies, and medical data: a comparative perspective. *Medical data privacy handbook*, 639–678.

10. Scheibner, J., Raisaro, J. L., Troncoso-Pastoriza, J. R., Ienca, M., Fellay, J., Vayena, E., & Hubaux, J. P. (2021). Revolutionizing medical data sharing using advanced privacy-enhancing technologies: technical, legal, and ethical synthesis. *Journal of medical Internet research*, *23*(2), e25120.

11. Wang, R., Tsai, W. T., He, J., Liu, C., Li, Q., & Deng, E. (2018, December). A medical data-sharing platform based on permissioned blockchains. In *Proceedings of the 2018 International Conference on Blockchain Technology and Application* (pp. 12–16).

12. Xia, Q. I., Sifah, E. B., Asamoah, K. O., Gao, J., Du, X., & Guizani, M. (2017). MeDShare: Trustless medical data sharing among cloud service providers via blockchain. *IEEE Access*, *5*, 14757–14767.

13. Zyskind, G., & Nathan, O. (2015, May). Decentralizing privacy: Using blockchain to protect personal data. In *2015 IEEE Security and Privacy Workshops* (pp. 180–184). IEEE.

14. Azaria, A., Ekblaw, A., Vieira, T., & Lippman, A. (2016, August). Medrec: Using blockchain for medical data access and permission management. In *2016 2nd International Conference on Open and Big Data (OBD)* (pp. 25–30). IEEE.

15. Vinay Singh, Alok Aggarwal, Narendra Kumar, A. K. Saini. (2020). A Novel Approach for Pre-Validation, Auto Resiliency & Alert Notification for SVN To Git Migration Using IoT Devices. PalArch's Journal of Archaeology of Egypt / Egyptology, 17(9), 7131–7145. Retrieved from https://archives.palarch.nl/index.php/jae/article/view/5394

16. Dagher, G. G., Mohler, J., Milojkovic, M., & Marella, P. B. (2018). Ancile: Privacy-preserving framework for access control and interoperability of electronic health records using blockchain technology. *Sustainable cities and society*, *39*, 283–297.

17. Daraghmi, E. Y., Daraghmi, Y. A., & Yuan, S. M. (2019). MedChain: A design of a blockchain-based system for medical records access and permissions management. *IEEE Access*, 7, 164595–164613.

18. Dubovitskaya, A., Xu, Z., Ryu, S., Schumacher, M., & Wang, F. (2017). Secure and trustable electronic medical records sharing using blockchain. In *AMIA annual symposium proceedings* (Vol. 2017, p. 650). American Medical Informatics Association.

19. Liu, H., Crespo, R. G., & Martínez, O. S. (2020, July). Enhancing privacy and data security across healthcare applications using blockchain and distributed ledger concepts. In *Healthcare* (Vol. 8, No. 3, p. 243). MDPI.

20. Al Omar, A., Jamil, A. K., Khandakar, A., Uzzal, A. R., Bosri, R., Mansoor, N., & Rahman, M. S. (2021). A transparent and privacy-preserving healthcare platform with novel smart contracts for smart cities. *Ieee Access*, *9*, 90738–90749.

Note: All the figures and tables in this chapter were made by the author.

Recent Trends in Engineering and Science for Resource Optimization and Sustainable Development – Prof. (Dr.) Dorota Jelonek et al. (eds)
© *2024 Taylor & Francis Group, London, ISBN 978-1-032-98030-0*

31

Investigating the Factors of ML in Electronic Business

Richa[1]
Principal, Government Leather Institute Kanpur, India

C. Ramya
Assistant Professor, Electrical and Electronics Engineering,
SNS College of Technology, Coimbatore

Gopu Srilekha
Department of Computer Science and Engineering,
Institute of Aeronautical Engineering, Hyderabad, India

Ganesh Waghmare[2]
Associate Professor, MIT College of Management,
MIT Art, Design and Technology University, Pune, India

Amandeep Nagpal
Lovely Professional University, Phagwara, India

Ms.T Thangam, M.E.[3]
Associate professor, Department of ECE, PSNA college of Engineering and Technology,
Muthanampatti, Kothandaraman Nager, Dindigul, Tamilnadu, India

Tomasz Turek[4]
Faculty of Management,
Czestochowa University of Technology, Poland

Abstract

The research's primary objective is to detect the essential factors of using Machine learning (ML) approaches to enhance (E-business) electronic business in the firm. The studies concentrated on the key factors of ML that impacted the prediction of demand, utilization of ML in actions of purchase, developing enhanced customer engagement, and supporting total cross-selling of commodities to enhance E-business and accomplish development. The deployment of novel and enhanced ML technologies has allowed the organization to enjoy greater benefits, such as precise forecasts of consumer needs, improved engagement, and commodity cross-selling.

Keywords

Machine learning, Electronic business, Consumer engagement, and Cross-selling

Corresponding authors: [1]vermaricha29@gmail.com, [2]gntilu@gmail.com, [3]thangam7280@gmail.com, [4]leszek.ziora@pcz.pl

DOI: 10.1201/9781003596721-31

1. INTRODUCTION

To begin, advancements in artificial intelligence (AI) are widely regarded as key factors in improving the efficiency of numerous businesses, including the web business and electronic business (E-commerce) [1].

Machine learning (ML) is a component of AI that employs critical methods and instruments to gather and evaluate enormous quantities of data, apply predictive algorithms to forecast the need for goods and offerings, and allow the growth of effective assets that allow businesses to react successfully to the increasing demands of their customers [2]. Because of the creation of these innovations, e-commerce is increasing at an exponential rate. The creation and rebalancing of both qualitative and statistical characteristics between the production and usage of goods and assistance is primarily the responsibility of business and trade [3]. For the commodities to eventually enter the fingers of the customers, this involves a variety of operations, including the purchase of raw supplies, their preparation, shipping, storing, and merging additional operations [4]. In the past, the majority of businesses operated as brick-and-mortar stores, where things were sold from an established place or through distant selling, where shipments were sent by carrier, postal service, and so forth. Nevertheless, manufacturers are increasingly aiming to offer their goods and services web to clients all over the globe due to the development of innovations such as the web, smartphone programs, and other tools. To analyze the use of various innovations, such as ML, statistical analysis, and the Internet of Things, to improve functioning effectiveness and accomplish long-lasting development, a novel corporate framework has to be developed [5].The novel structure, which was put forth by [6], allows for the description of retail tasks depending on data systems. The paradigm focuses on crucial key information, and other facets revolve around it, including data that adds significance and other functioning decision-making in the e-commerce industry.

Businesses that use ML are capable of estimating demand depending on a variety of criteria, allowing them to increase customer interaction, purchasing actions, and cross-selling [7].The primary objective of the research is to examine the essential factors of ML in improving E-business.

2. LITERATURE REVIEW

Forecasting not just what clients are interested in purchasing, but additionally the cost and the likelihood that they will make quotations, requires ML to play a significant role. Since pricing and pricing strategies have a significant impact on sales, marketing experts should focus on investigating this area. Therefore, determining the best method

for AI to determine which costs are ideal and if trading in currencies ought to be made available is a crucial subject for subsequent studies [8]. Electronic salesmen who employed ML to better comprehend their clients' actions enhanced their e-commerce advertising strategy and offered new services including customization, adaptation, and referrals [9].Russell [10] asserts that ML is a fantastic instrument for online advertisers to (i) increase market revenues; (ii) comprehend consumer activity; (iii) enhance customer contentment; (iv) draw in more clients; and (v) reduce car buy denials. ML also drives e-commerce and has an impact on decision-making [11]. When ML is put into autonomous machinery, the robots are expected to play a key role in the lives of consumers as suppliers, partners, grandkids, or pet representatives. Aside from the formerly recognized UV problems, some study indicates that interacting with robots incorporated with AI produces pain and compensating conduct. It is critical to identify if customers have adverse views about the robots included in ML and if these views may be addressed as time passes [05].

3. METHODOLOGY

This part includes providing the methodology of the study as well as the additional factors that the researchers used to conduct the investigation. The present setting plays a part in comprehending the function of ML in improving E-business in substantial firms; for this investigation, the investigators used an exploratory approach since the concern of the study is novel and changing as different firms are currently contemplating using ML in their E-business framework for data analysis and framing methods for improving their procedures [08], revenue [10], and earnings [11]. Furthermore, this investigation can provide additional backing for a past study that was conducted, as well as assistance for forthcoming studies. The main objective of the investigation is to obtain details from supervisors in sectors that are mainly present in E-business, so the investigators seek to employ a survey for gathering the data. The essential data originates using a Likert 5-point size, which will allow the investigators to successfully collect responses from the participants. The investigators distributed approximately 180 surveys, with only 155 submitted.

4. RESULT AND DISCUSSION

According to Table 31.1, roughly 75.5 percent of the participants claimed that methods and instruments for ML had been utilized by their organization in the E-business procedure, while 24.5 percent claimed that such instruments were not presently used. Figure 31.1 depicts the application of ML across different organizations.

Table 31.1 Outcomes of implementing ML in E-business

Particulars	Frequency	Percentage
Yes	117	75.5
No	38	24.5
Overall	155	100

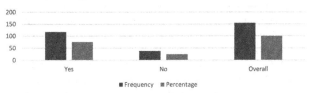

Fig. 31.1 Outcomes of implementing ML in E-business

Correlation assessment offers an in-depth examination of the degree of correlation between the dependent variable—improving e-business in the organization—and the independent variables—such as demand forecasting, customer purchase action, engagement levels, and product cross-selling. Table 31.2 demonstrates how ML-driven purchase action helps the organization's e-business to flourish, with a type of connection of 0.89. Similarly, a greater degree of connection is observed between product cross-selling and improving e-commerce, with approximately 0.86 correlation. Furthermore, it can be stated that each of the factors has a significant and beneficial impact on the dependent factor because all of the intent factors have a larger correlation with the dependent factors.

Table 31.2 Outcomes of correlation assessment

	Customer engagement	Demand forecasting	Purchase action	Cross-selling product	Improving E-business
Customer engagement	1	0.83**	0.84**	0.86**	0.83**
Demand forecasting	0.83**	1	0.81**	0.84**	0.82**
Purchase action	0.84**	0.81**	1	0.84**	0.89**
Cross-selling product	0.86**	0.84**	0.84**	1	0.86**
Improving E-business	0.83**	0.82**	0.89**	0.86**	1

** = Correlation coefficient

According to Table 31.3 and Fig. 31.2 evaluation, the Cronbach alpha level for all concepts is greater than + 0.700, indicating that the framework is accurate and reliable and may be employed in additional research.

In Table 31.4 and Fig. 31.3, the (p) significant value for 3 significant independent factors, such as demand forecast-

Table 31.3 Accuracy of cronbach alpha framework

Factors	Cronbach alpha
Customer engagement	0.862
Demand forecasting	0.859
Purchase action	0.811
Cross-selling product	0.766
Improving E-business	0.715

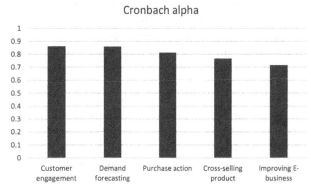

Fig. 31.2 Accuracy of cronbach alpha framework

Table 31.4 Outcomes of regression weighs

Dependent factor	Independent factor	Value	S. E	C.R	p-score	Results
E-business	Customer engagement	-0.006	0.10	-0.058	0.954	Rejected
E-business	Demand forecasting	0.074	0.09	0.752	0.002	Accepted
E-business	Purchase action	0.442	0.10	4.042	0.000	Accepted
E-business	Cross-selling product	0.543	0.14	3.718	0.000	Accepted

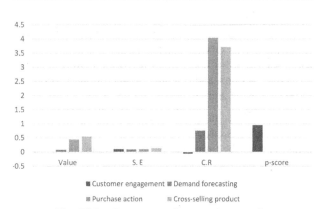

Fig. 31.3 Outcomes of regression weighs

ing, purchasing habits, and product cross-selling, is less than 0.05. In contrast, the hypothesis for consumer engagement is denied. As a result, it could be argued that ML has an important effect on enhancing E-business by anticipating the demand for various products and assistance, determining customer purchase conduct, and backing in cross-selling for future growth and expansion [11].

This indicates that businesses incorporating ML methods are highly likely to maintain development and can realize raised production and results. The research reveals that the type of connection between ML methods and improving E-business is more favorable in the environment. According to the research, it is observed that ML methods for demand forecasting, customer action evaluation associated with purchases, and product cross-selling have assisted in the enhancement of e-business. Businesses across various fields are now attempting to comprehend client requirements, forecast demand, and leverage the current product accordance to boost revenue and earnings. This is made possible by the SEM framework, which makes it possible to compare factor evaluation and regression models. This promotes the business's expansion and makes it possible for the parties to effectively realize their aims.

5. CONCLUSION

Companies are currently attempting to sell their goods and offerings online and reach clients all over the globe, due to technological advancements like the web, mobile applications, and other mediums. Firms employ ML to estimate requirements depending on an array of criteria to improve customer communication, purchasing trends, and customer management. Customers nowadays buy things online using smartphones, the web, and associated technologies. As an outcome, businesses increasingly employ e-commerce to capitalize on potential markets. According to the findings, all intended factors connect significantly with the dependent factors, implying that every factor has a substantial and beneficial impact on the factor that is dependent. Furthermore, ML possesses an important impact on the growth of electronic commerce by anticipating the need for an array of items and offerings, monitoring client purchasing actions, and enabling product minimization for future expansion and growth.

The use of ML techniques for a range of sectors, including financial services, production supply chains, and manufacturing, is what the future holds. According to some, ML is among the most useful technical instruments for helping organizations maximize efficiency and production. Additionally, ML technologies are being utilized to develop smart groups, structures, and towns that will aid in preserving the environment for future generations.

Reference

1. Nadikattu, R. R. (2016). The emerging role of artificial intelligence in modern society. *International Journal of Creative Research Thoughts*.
2. Silver, D., Schrittwieser, J., Simonyan, K., Antonoglou, I., Huang, A., Guez, A., ... & Hassabis, D. (2017). Mastering the game of go without human knowledge. *nature*, *550*(7676), 354–359.
3. Soni, N., Singh, N., Kapoor, A., & Sharma, E. K. (2018). Low-resolution image recognition using cloud hopfield neural network. In *Progress in Advanced Computing and Intelligent Engineering: Proceedings of ICACIE 2016, Volume 1* (pp. 39–46). Springer Singapore.
4. Schütte, R. (2017). Information Systems for Retail Companies: Challenges in the Era of Digitization. In *Advanced Information Systems Engineering: 29th International Conference, CAiSE 2017, Essen, Germany, June 12-16, 2017, Proceedings 29* (pp. 13–25). Springer International Publishing.
5. Sustrova, T. (2016). A suitable artificial intelligence model for inventory level optimization. *Trendy ekonomiky a Managementu*, *10*(25), 48.
6. Zangeneh, E., Rahmati, M., &Mohsenzadeh, Y. (2020). Low-resolution face recognition using a two-branch deep convolutional neural network architecture. *Expert Systems with Applications*, *139*, 112854.
7. Leung, K. H., Choy, K. L., Siu, P. K., Ho, G. T., Lam, H. Y., & Lee, C. K. (2018). A B2C e-commerce intelligent system for re-engineering the e-order fulfillment process. *Expert Systems with Applications*, *91*, 386–401.
8. Chui, M. (2017). Artificial intelligence is the next digital frontier. *McKinsey and Company Global Institute*, *47*(3.6).
9. Russell, S. J., & Norvig, P. (2010). *Artificial intelligence is a modern approach*. London.
10. Chen, C. C. V., & Chen, C. J. (2017). The role of customer participation in enhancing repurchase intention. *Management decision*.
11. Ferrario, A., Loi, M., &Viganò, E. (2020). In AI we trust incrementally: A multi-layer model of trust to analyze human-artificial intelligence interactions. *Philosophy & Technology*, *33*, 523–539.

Recent Trends in Engineering and Science for Resource Optimization and Sustainable Development – Prof. (Dr.) Dorota Jelonek et al. (eds)
© 2024 Taylor & Francis Group, London, ISBN 978-1-032-98030-0

32

Blockchain Technology Application in Electronic Engineering

Srinivasarao Dharmireddi[1]

Principal Architect, MasterCard, Cybersecurity,
St. Louis, Missouri, USA

Richa[2]

Principal,
Government Leather Institute Kanpur, India

Kilaru Aswini

Department of Computer Science and Engineering,
Institute of Aeronautical Engineering, Hyderabad, India

Shobha Kulshrestha[3]

Assistant professor in physics,
Department of nims University Jaipur, rajasthan

Amandeep Nagpal

Lovely Professional University, Phagwara, India

Sreekanth Dekkati[4]

Assistant Vice President (System Administrator),
MUFG Bank, New York, USA

Abstract

Electronic engineering is now pervasive, playing an indispensable role in our daily routines and vital facilities like the electrical grid, communications via satellite, and public transit. The safety of software operating on these platforms has gotten a lot of focus over the past few years. Nevertheless, hardware has been considered to be dependable and trustworthy without thoroughly investigating any weaknesses in the electronics supply chain. With the rising globalization of the electronics sector, ensuring the safety and reliability of electronics has grown difficult. A unique IoT meter monitors monthly use and transfers the data to a decentralized application maintained on the blockchain in the suggested blockchain-oriented consumer electronics information sharing and secure transaction architecture. This decentralized technology will create the bill and offer incentives to authentic consumers. Lastly, end-to-end lag and throughput have been utilized to assess the efficacy of the suggested technique.

Keywords

Blockchain, Decentralized application, Electronic engineering, Electronic supply chain, and Information sharing

Corresponding authors: [1]Srinivasarao.dharmireddi@gmail.com, [2]vermaricha29@gmail.com, [3]shobha011986@gmail.com, [4]sreekanthd041987@gmail.com

DOI: 10.1201/9781003596721-32

1. INTRODUCTION

Electronic engineering provides numerous options in fields like telecommunications, electronics, generation of energy, supply & dispersion, robotics, transportation, defense, mining, and assistance with consulting [1]. Each of these topics can be tackled from the standpoint of the quality of staff or as a commercial potential. Numerous more business concepts can be launched, but they all require a certain early financial investment.

As a result, consumer electronics like computers, mobile devices, and certain electronic healthcare equipment are increasingly visible and utilized in daily life. Because of multiple fundamental weaknesses, dangers, and assaults, the protection of programs, firmware, and communication paths has garnered an extensive amount of focus [2]. Blockchain is a data storage technology that renders it hard or difficult to breach, fraud, or change the network [3]. A blockchain is an extensive structure of computing devices that replicates and disperses a digital record of every transaction [4]. A blockchain is an accessible, dispersed, decentralized electronic record that is used to record operations across several systems in a manner that makes it impossible to make modifications later on without impacting the agreement of the system and every block that came before them [5]. Compared to conventional techniques of storing data on external systems or servers, which are frequently overseen by a single entity, blockchain offers better safety for the transmission of consumer electronics data and safe financial transactions [6]. Users don't have to expect credit card firms or financial institutions to approve transactions before proceeding. The decentralized nature of the blockchain makes it resistant to assaults since there isn't a single point of collapse, and once data is stored on the blockchain, it is unable to be altered [7]. Since blockchain enables numerous people to communicate data without a central power, it is frequently employed to safeguard data sharing [8]. This renders it an excellent tool for transferring data since it is open unchangeable, and safe.

Blockchain [9] systems may tackle the authenticity of information and secrecy constraints, and they can be utilized for electronic money transactions or commodities transport. A similar method can be used to create a trustworthy supply chain for electronic devices. Throughout the consumer electronics supply chain, the suggested blockchain-based structure provides assurance and transparency.

2. LITERATURE REVIEW

Blockchain is now recognized as a key strategy for decentralized cryptocurrency platforms. Both scientific and electronic sectors have expressed strong attention. Most modern blockchain-oriented structures, nevertheless, merely manage transactions between participants [10]. The application of blockchain has a significant edge in consumer electronics. Blockchain provides greater security for consumer gadgets, sharing of information, and secure transactions [11].

3. METHODOLOGY

Cryptography ensures the confidentiality and integrity of data by encrypting communication so that only the intended recipient can decrypt it. Typically, encryption protocols such as GPG (GNU Privacy Guard) or PGP (Pretty Good Privacy) are used for this purpose. These protocols utilize a public-key cipher to securely exchange a symmetric cipher key. Once the key is shared, the actual data is encrypted using this key and transmitted to the receiver. The receiver then decrypts the data with the same key. By encrypting the data in blocks and ensuring its resistance to tampering, cryptography eliminates the need for trust in a centralized authority. Blockchain technology, being decentralized, enhances security by removing a single point of control, thereby reducing the risk of fraud and manipulation. The use of cryptography within blockchain adds an additional layer of security, enabling the creation of a transparent and secure transaction platform. Blockchain facilitates fast, secure transactions without the need for intermediaries, such as banks. This not only reduces transaction costs but also increases accessibility for all users. Moreover, the immutable and permanent record of every transaction on the blockchain makes it harder for illicit activities, such as money laundering, to occur, as it simplifies tracking and prosecuting offenders. Smart contracts, a key feature of blockchain platforms like Ethereum, enable automated operations that involve "gas," which is the cost paid in ether (Ethereum's native token) to execute transactions. The more complex the business logic of a smart contract, the more gas it consumes, as the contract's bytecode increases with the size of its source code. To mitigate high gas fees, the proposed solution executes transactions from an account to a smart contract using an Ethereum stable token, ensuring more efficient and cost-effective operations.

4. RESULT AND DISCUSSION

Verification and maintenance are required in the real-world electronics supply chain to ensure electronic equipment quality. All critical data regarding an electronic part is stored in the suggested system. End-to-end lag and throughput measures are used to assess the suggested blockchain-oriented consumer electronic information-sharing performance and secure transaction architecture. The investigation depends on two frameworks: the suggested paradigm with a blockchain technology and the present one [11] without a blockchain. The present design uses a smart energy gauge connected to the IoT that uses a GSM device to track energy consumption. Through this approach, this study obtains current data on the quantity of energy consumed independent of blockchain. When contrasted with the current method, the end-to-end lag is negligible.

The transaction cost that the transmitter has to cover to the mining groups in the shape of gas and the number of operations that the mining groups should satisfy define how delayed Ethereum transfers are from the time of genesis to distribution. As a result, end-to-end lag reduces as gas prices continue to rise. This suggests that supplying higher gas expenses would result in superior service levels, implying that transactions involving consumer electronic information sharing and a secured transaction system are more likely to be validated with less interruption. As a result, in terms of end-to-end latency, the suggested blockchain-oriented Consumer Electronic Information Sharing and Secured Transaction Infrastructure lag is significantly lower than the present technique, as shown in Table 32.1 and Fig. 32.1.

Table 32.1 End-to-end lag

Lag (Ms)	Workload (MI)	
	Lag with Blockchain	Lag without Blockchain
200,000	200,000	210,000
400,000	400,000	415,000
600,000	600,000	615,000
800,000	800,000	815,000
1,000,000	1,000,000	1,015,000
1,200,000	1,200,000	1,215,000

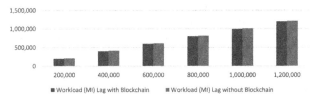

Fig. 32.1 End-to-end lag

End-to-end throughput is a critical aspect that may impact the system's overall performance. When a transaction should be validated under the suggested structure, it should initially be broadcasted to every node, and then the answers from those nodes should be gathered to create an agreement depending on a majority decision as shown in Fig. 32.2.

Because of this, the recommended framework—which has a designated system capability—significantly lowers the system's processing duration and boosts its overall throughput. The suggested approach performs better than the current paradigm [8] in terms of end-to-end throughput when evaluating total workload versus. throughput.

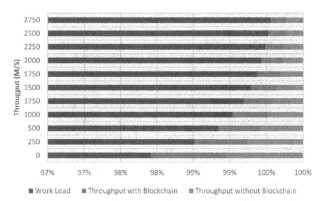

Fig. 32.2 End-to-end throughput

Furthermore, the performance of the suggested approach is assessed by contrasting the standard information-sharing technique with the blockchain-based information-sharing procedure for consumer electronics as shown in Table 32.2 and Fig. 32.3.

Table 32.2 Comparison of access regulation

	Duration of Overhead (ms)	
	Access control of the standard method	Access control of the blockchain-based method
1	1	3
2	4.5	6
3	5.5	8
4	9.5	12.5
5	13.5	16
6	17	20

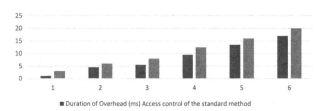

Fig. 32.3 Comparison of access regulation

As shown in Table 32.2 and Fig. 32.3, the blockchain-based procedure will require more computing and network connection overhead than the current way [17] of data exchange. This is due to transaction-related consensus lag and the long and laborious encoding process required by the blockchain-based platform.

In the worst-case scenario, the extra expense resulting from the blockchain-based solution for the data-stored function is anticipated to be 50 ms. The data request procedure and access regulation approach contribute 10 ms to the overall time expense.

5. CONCLUSION

We suggest a blockchain-oriented architecture for tracking the reliability of the electronics supply chain in this article. We additionally examine each entity's position in the suggested secure electronics supply chain. With increased electronic information sharing and monetary transactions, handling and safeguarding these electronic interactions is vital while maintaining a dispersed and shared record structure. Blockchain innovation may be the greatest solution for keeping an accurate and transparent record of goods. Using blockchain innovation to tackle current challenges can assist a variety of applications, including those utilized in energy manufacturing, transmission, dispersion, and use, those with information sharing, and secure transactions. Blockchain innovation can potentially be utilized to help design novel approaches. An innovative IoT meter collects monthly usage data and transfers it to a decentralized application. As a component of the structure suggested for blockchain-oriented consumer electronics sharing of information and safe transactions, the information is then recorded in the blockchain. This decentralized platform will generate a charge and reward clients for using the service. Lastly, the suggested model's efficacy was assessed using end-to-end lag and throughput. In comparison to the current paradigm, it showed low lag and great throughput [7].

Future work would involve developing a gamification structure that provides an unbiased, permanent ledger with an established structure for certifying investments, a compensation structure depending on measurable accomplishments, and a global, verified evaluation of these accomplishments for authorization by the company's key players

Reference

1. Damian, C., Lazăr, I., Vişoiu, D. G., Romanescu, Ş., & Alboaie, L. (2019, October). Applying blockchain technologies in the funding of electrical engineering industry applications. In *2019 International Conference on Electromechanical and Energy Systems (SIELMEN)* (pp. 1–5). IEEE.
2. Xu, X., Rahman, F., Shakya, B., Vassilev, A., Forte, D., & Tehranipoor, M. (2019). Electronics supply chain integrity enabled by blockchain. *ACM Transactions on Design Automation of Electronic Systems (TODAES)*, 24(3), 1–25.
3. Huh, S., Cho, S., & Kim, S. (2017, February). Managing IoT devices using a blockchain platform. In *2017 19th International Conference on Advanced Communication Technology (ICACT)* (pp. 464–467). IEEE.
4. Hlaing, K. M., & Nyaung, D. E. (2019, November). Electricity billing system using Ethereum and Firebase. In *2019 International Conference on Advanced Information Technologies (ICAIT)* (pp. 217–221). IEEE.
5. Pee, S. J., Nans, J. H., & Jans, J. W. (2018). A simple blockchain-based peer-to-peer water trading system leveraging smart contracts. In *Proceedings on the International Conference on Internet Computing (ICOMP)* (pp. 63–68). The Steering Committee of The World Congress in Computer Science, Computer Engineering and Applied Computing (WorldComp).
6. Gür, A. Ö., Öksüzer, Ş., & Karaarslan, E. (2019, April). Blockchain-based metering and billing system proposal with privacy protection for the electric network. In *2019 7th International Istanbul Smart Grids and Cities congress and Fair (ICSG)* (pp. 204–208). Ieee.
7. Albrecht, S., Reichert, S., Schmid, J., Strüker, J., Neumann, D., & Fridgen, G. (2018). Dynamics of blockchain implementation case study from the energy sector.
8. Adeyemi, A., Yan, M., Shahidehpour, M., Botero, C., Guerra, A. V., Gurung, N., ... & Paaso, A. (2020). Blockchain technology applications in power distribution systems. *The Electricity Journal*, 33(8), 106817.
9. Pilkington, M. (2016). Blockchain technology: principles and applications. In *Research handbook on digital transformations* (pp. 225–253). Edward Elgar Publishing.
10. Liang, W., Tang, M., Long, J., Peng, X., Xu, J., & Li, K. C. (2019). A secure fabric blockchain-based data transmission technique for industrial Internet-of-Things. *IEEE Transactions on Industrial Informatics*, 15(6), 3582–3592.
11. Thapliyal, H. (2017). Internet of things-based consumer electronics: Reviewing existing consumer electronic devices, systems, and platforms and exploring new research paradigms. *IEEE Consumer Electronics Magazine*, 7(1), 66–67.
12. Manpreet Singh Bhatia, Alok Aggarwal, Narendra Kumar. (2020). Speech-to-text conversion using GRU and one hot vector encoding. PalArch's Journal of Archaeology of Egypt / Egyptology, 17(10), 7110–7119. Retrieved from https://archives.palarch.nl/index.php/jae/article/view/5391
13. Wu, A., Zhang, Y., Zheng, X., Guo, R., Zhao, Q., & Zheng, D. (2019). Efficient and privacy-preserving traceable attribute-based encryption in blockchain. *Annals of Telecommunications*, 74, 401–411.
14. Deny, J., Narasimha, A. B., Reddy, R. G. V., & Sathish, S. (2021, May). Electricity Monitoring And Auto Bill Generation Using IOT. In *2021 3rd International Conference on Signal Processing and Communication (ICPSC)* (pp. 695–698). IEEE.
15. R. Vettriselvan, C. Vijai, J. D. Patel, S. Kumar, R, P. Sharma and N. Kumar, "Blockchain Embraces Supply Chain Optimization by Enhancing Transparency and Traceability from Production to Delivery," *2024 International Conference on Trends in Quantum Computing and Emerging Business Technologies*, Pune, India, 2024, pp. 1-6, doi: 10.1109/TQCEBT59414.2024.10545308.

Note: All the figures and tables in this chapter were made by the author.

Recent Trends in Engineering and Science for Resource Optimization and Sustainable Development – Prof. (Dr.) Dorota Jelonek et al. (eds)
© 2024 Taylor & Francis Group, London, ISBN 978-1-032-98030-0

33

An Empirical Research of AI Approaches in Electronic Engineering

V. Muralidharan[1]

Assistant Professor, Department of ECE, Dr. N.G.P. Institute of Technology, Coimbatore, Tamil Nadu

Kantilal Pitambar Rane[2]

Department of Electronics and Telecommunications Engineering, Bharati Vidyapeeth College of engineering, Navi Mumbai, India

K. Rashmi

Department of Computer Science and Engineering, Institute of Aeronautical Engineering, Hyderabad, India

Rakesh Chandrashekar[3]

Department of Mechanical Engineering, New Horizon College of Engineering, Bangalore

Amandeep Nagpal

Lovely Professional University, Phagwara, India

Purnendu Bikash Acharjee

Associate Professor, Computer Science, CHRIST University, Bangaluru, India

Abstract

The function of artificial intelligence (AI) in electronic engineering has been suggested to overcome the issue of an elevated structure error rate in electronic engineering. Using LPWAN innovation in AI as an instance, an innovative structure is presented to increase the safety of the Internet of Things wireless communication infrastructure. The business, processing of information, terminal accessibility, and communication innovation layers make up the majority of the framework. The empirical findings indicate that the structure's setup transmits four types of data transfer directions every 30 seconds, and the receiver port constantly gathers the aforementioned command information for 4 hours, contrasted to the conventional framework (15.6 percent), and the package loss rate is 4 percent, significantly enhancing the system's throughput and processing performance. The developed framework is more stable, has a lower bit error rate, and may assist wireless communication effectively. It has been demonstrated that the novel design, using LPWAN innovation as an instance, has a big total capacity as well as a good performance of its electronic engineering connection. The structure number error rate is considerably lowered, and the signal structure number is highly accurate as an outcome.

Keywords

Artificial intelligence, Electronic engineering, Wireless communication, and Low power wide area network

Corresponding authors: [1]muralivlsi5@gmail.com, [2]kantiprane@rediffmail.com, [3]rakesh26817@gmail.com

DOI: 10.1201/9781003596721-33

1. INTRODUCTION

In the realm of electronic engineering, classic commercial equipment function has steadily evolved into contemporary electronic mechanical functioning, which has evolved into intelligence [1]. Artificial intelligence (AI) innovation is now extensively utilized, and it may significantly boost output, economic efficiency, and company competition [2]. Many industries are already benefiting from the latest advancements in high-speed data transfer communications ideal for AI uses, including the healthcare industry, the smart schooling industry, and the economy. It must be emphasized that AI has been employed in numerous contexts and methods to expand and improve a broad spectrum of sectors, like financial institutions and financial transactions, education, manufacturing lines, manufacturing, retailing and e-commerce, and medical [3].

Specifically, AI-based interactions in E-healthcare allow individuals or medical professionals in various places to interact and advantage of one another. For instance, a doctor using telerobotics can operate on a patient even if they are in separate countries [4]. Societal productivity increases in tandem with the ongoing growth of the societal economy and the ongoing raising of individual levels of living. Within the discipline of electronic engineering, the functioning of conventional manufacturing equipment has progressively evolved into contemporary electronic mechanic functioning, which has been evolving toward smart. The application of AI is extensively employed today and has the potential to increase production to some degree, boost financial effectiveness, and make businesses more competitive [5].

An increasing number of initiatives are utilizing AI innovation to achieve automated manufacturing and intellect, which increases labor effectiveness and speeds up production to satisfy the demands of the rapidly evolving periods. Computing and electronic data innovation are the two fundamental components of AI innovation. It also satisfies the standards of logic and social psychology [6]. The best quality of AI is its capacity to reason and carry out activities that maximize the likelihood of accomplishing a given objective. Within the field of AI, machine learning pertains to the idea of computing devices acquiring knowledge from and adjusting to fresh statistics without the need for human intervention [7].

To fulfill the criteria of the periods' rapid growth, an increasing number of projects are utilizing AI innovation to realize automation and intelligence of manufacturing, increase the pace of production, and increase productivity at work [8]. AI is mostly a combination of electronic data technology and computing innovation. Hence this research aims to investigate the AI approaches in electronic engineering for medical uses.

2. LITERATURE REVIEW

Deep learning frameworks, natural language creation, analyzing information, and robotics have all generated significant advances in AI in the past decade [9]. The Internet of Things (IoTs) is a key component of the contemporary era of information technology, as well as a crucial phase in the "information" age's evolution.

The research [10] has recognized the benefits of low power wide area network (LPWAN) innovation, and numerous companies are starting to engage and support the industrialization of different LPWAN specifications like Sigfox [11] and Lora [12]. Furthermore, numerous other technological standards are being developed and commercialized continually [13, 14]. Furthermore, numerous other technological standards are also undergoing ongoing research and commercialization [15]. According to [16], certain types of machinery, like street lighting, well-encompassing water and gas sensors in city systems, and outside meteorological, hydrological, and mountainous data-gathering machinery, have a tiny data transfer volume and a low transfer rate. If these devices use the high-speed network that existing telecommunications companies supply, they will not just take up network space and other assets, but they will also waste a lot of money and bandwidth. Furthermore, the strain that excessive energy use places on the need for a long-lasting source of energy will also be a challenging issue to resolve. The study [17] Because of the challenges of connecting equipment dispersed across long distances to a central communication system, the setup of LPWAN systems is severely restricted, particularly globally. According to [18], an AI innovation platform can include a variety of equipment, including a control system, standardized source, electronic multimeter, and more. By incorporating various devices, technology based on AI can be made more intelligent and automated, its growth can be continuously accelerated, and its worth may be demonstrated.

3. METHODOLOGY

Sound transfer assistance and data collecting assistance make up the majority of the framework proposed in this work. A data acquisition gadget, a data interface, and a recording terminal make up the data acquisition assistance. The functioning of the data structure, which may acquire the communication information gathered through the acquisition equipment for an extended period, uses AI-based LPWAN architecture. The data interface's maximal data processing capability is 1280 megabytes, and there are a variety of data access connections, making it appropriate for use with different kinds of acquisition equipment. The LPWAN's specifications for technology are shown in Table 33.1.

Table 33.1   The LPWAN's specifications

Parameter	LPWAN
Broadband	500 Hz to125 KHz
Maximum transmission power	20dbm
Peak data rate	280 bps to 50 Kbps
Spectrum	Unauthorized
Modulation mode	CSS

4. RESULT AND DISCUSSION

In the electronic interactions of the IoTs, the data analyzing unit primarily performs the functions of storing information and categorization. Outside data signals must be provided to the data line to validate the data analysis unit. In this article, information impulses will be sent by FPD parts to the data analysis unit, which will then analyze the information. The sent data will subsequently be examined by an XJSGH-3 oscilloscope. Examine the waves of the data [19].Numerous data sets are sent out by the FPD part, which then contrasts the frames gathered for this research with those produced by the conventional framework. Table 33.2 displays the data that the contrasting endpoint obtained. The evaluation comparison data contains the standard set count, error set scale, and unmarked error set count based on the real scenario. The data analysis unit in this study can function steadily in the LPWAN-based electronic interaction system of the IoTs, as demonstrated by the data comparative findings.

Table 33.2 and Fig. 33.1 display the information that the comparison terminal received. The data comparison findings show that the data processing component in this study can operate reliably in the LPWAN-based electronic communication architecture of the IoTs.

Table 33.2   Outcomes of recelved structure

Number kind of structure	The convention-al framework success rate (%)	Success rate (%)	Outcomes of test comparison
Conventional structure number	98.1	100	Matching
Error structure number	98.5	100	No data and Prompt error
Unlocated error structure number	97.1	100	No marked error structures and no data

The components of an electronic interaction interface examination are an output port, data analyzing unit, reception port, and oscilloscope. Set up the data terminal, take out the preserved statistics, decipher the message, and transmit it to the information-obtaining terminal. Following

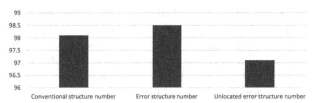

Fig. 33.1   Outcomes of received structure

that, analyze and process the data using the data processor. Producing the outcomes using the oscilloscope. Determine if the data waveform curves are typical. Lastly, determine the set number and examine the various phases of the set number.

Testing with the above-described test framework. Initially, set up the data terminal. Next, take the signal data out of the memory, decode it, and send it to the data-obtaining port. After that, process the data using the data processor. Lastly, use an oscilloscope to see if the data waveform matches the standard waveform. Lastly, determine the set number from the data pattern and examine its various stages. The port-accessible data processing capacities of the paradigm and the conventional paradigm are contrasted to confirm the design's stability and efficacy. The study comparison is displayed in Table 33.3 and Fig. 33.2. The framework proposed in this work operates steadily concerning the access port's information processing capability, possesses a lower bit error rate, and is more capable of assisting with IoTs' wireless communication depending on LPWAN innovation, as per the contrast findings shown in the figure. This paper's architecture demonstrates faster management of time than conventional methods. The results suggest that

Table 33.3   Comparison of error rate assessment

Handling time (s)	Structure value error rate (%)	
	Conventional Framework	Framework of this paper (%)
0.0	24	13
0.2	28	16
0.4	34	18
0.6	40	23
0.8	48	22
1.0	56	25

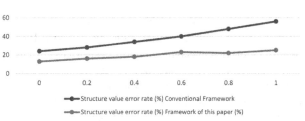

Structure value error rate (%) Conventional Framework

Structure value error rate (%) Framework of this paper (%)

Fig. 33.2   Comparison of error rate assessment

this advancement in the IoTs' electronic communication innovation may allow distant functioning while using little power and incurring little operating expenses. To satisfy the need for larger quality electronic interaction, this study explores and layout an IoT electronic interaction innovation structure depending on LPWAN innovation. The objective is to integrate LPWAN innovation using IoT electronic interaction innovation. The visual method of data transfer measurement used in Fig. 33.3 demonstrates that the structure has a lower error rate than the old design.

To confirm the stability and efficacy of the data transfer arrangement, several external characteristics, including temperature and illumination, have been monitored during the stability evaluation. Together with the examinations mentioned above, the structure's stability is also examined, mostly through the measurement of the package loss pace. 4 data transmission orders—"temperature, illumination, airflow, and humidity"—are delivered for the system's setup every thirty seconds. The recipient port records the data from these instructions for four hours. The dependability of a mesh unit route is evaluated by the package loss pace. The number of packages transmitted out divided by the overall number of packages not obtained yields this measure. Bad packages in computer networks are often disregarded beyond bit mistakes as most researchers focus on data integrity. When contrasted to the package loss pace of the conventional buffering method (15.3%), which is four percent, the package loss pace can significantly increase the system's throughput and analyzing effectiveness. The IoTs electronic interaction innovation depending on LPWAN innovation has the benefits of inadequate energy usage, excessive transmission pace, and excellent stability, it may be determined after data analyzing unit validation, the port accessibility examination, and analyzing capacity evaluation.

5. CONCLUSION

Steady and dependable framework assistance is required for the steady realization of IoT electronic communication innovations. Depending on the results, it is possible to deduce that this innovation in IoTs' electronic communication systems may allow lengthy functioning while consuming minimal power and pricing little to operate. This research investigates and develops an IoTs electronic communication innovation architecture depending on LPWAN innovation, to merge LPWAN and IoTs electronic communication innovation to meet the requirements of individuals for better electronic communication. The novel structure described in this work depends on LPWAN innovation and enhances conventional communication reception signal channels and diverse setups, addresses numerous issues in the basic layer of

structure layout, and assures the safety and confidentiality of user data transfer. This design, when accompanied by empirical validation and assessment, can reliably back the IoT electronic communication platform. Nevertheless, doubts exist about how this change will and must take place. Before adoption, 'soft' issues including ensuring the greatest grade engineering practice and evidential norms, medico-legal necessities for making choices, and fair dispersion of advantages should be tackled.

In the last few parts, the process for determining dynamics equivalence depending on coherence has been thoroughly covered. Tables 33.1 and 33.2 provide a thorough evaluation of generator aggregating and coherence detection. Apart from the conventional coherency-oriented techniques, future research has been conducted to enhance the technique's suitability for actual tasks.

Reference

1. Ström, P., Kartasalo, K., Olsson, H., Solorzano, L., Delahunt, B., Berney, D. M., ... & Eklund, M. (2020). Artificial intelligence for diagnosis and grading of prostate cancer in biopsies: a population-based, diagnostic study. *The Lancet Oncology*, 21(2), 222–232.
2. Suhel, S. F., Shukla, V. K., Vyas, S., & Mishra, V. P. (2020, June). Conversation to automation in banking through chatbot using artificial machine intelligence language. In *2020 8th International Conference on Reliability, Infocom Technologies and Optimization (Trends and Future Directions) (ICRITO)* (pp. 611–618). IEEE.
3. Sodhi, G. K., Kaur, S., Gaba, G. S., Kansal, L., Sharma, A., & Dhiman, G. (2022). COVID-19: Role of robotics, artificial intelligence, and machine learning during the pandemic. *Current Medical Imaging*, 18(2), 124–134.
4. Kumar, P., Kansal, L., Gaba, G. S., Mounir, M., Sharma, A., & Singh, P. K. (2021). Impact of peak to average power ratio reduction techniques on Generalized Frequency Division Multiplexing for 5th generation systems. *Computers and Electrical Engineering*, 95, 107386.
5. Şimşir, Ş., & Taşpınar, N. (2020). Cumulative symbol optimization–based partial transmit sequence technique for PAPR reduction in low complexity GFDM system. *Transactions on Emerging Telecommunications Technologies*, 31(6), e3801.
6. Al Harthi, N., Zhang, Z., Kim, D., & Choi, S. (2021). Peak-to-average power ratio reduction method based on partial transmit sequence and discrete Fourier transform spreading. *Electronics*, 10(6), 642.
7. Licato, J., & Zhang, Z. (2019). Correction to: evaluating representational systems in artificial intelligence. *Artificial Intelligence Review*, 52, 2743–2743.
8. Li, Z. Y., Yu, H. L., Shan, B. L., Zou, D. X., & Li, S. Y. (2020). Code design for run-length control in visible light communication. *Frontiers of Information Technology & Electronic Engineering*, 21(9), 1397–1411.

9. Kishor, A., Chakraborty, C., &Jeberson, W. (2021). A novel fog computing approach for minimization of latency in healthcare using machine learning.

10. Çiçek, S., Kocamaz, U. E., & Uyaroğlu, Y. (2019). Secure chaotic communication with a jerk chaotic system using the sliding mode control method and its real circuit implementation. *Iranian Journal of Science and Technology, Transactions of Electrical Engineering, 43*, 687–698.

11. T. Parashar, K. Joshi, R. R. N, D. Verma, N. Kumar and K. S. Krishna, "Skin Disease Detection using Deep Learning," 2022 11th International Conference on System Modeling & Advancement in Research Trends (SMART), Moradabad, India, 2022, pp. 1380–1384, doi: 10.1109/SMART55829.2022.10047465.

12. Jelonek D., Pawełoszek I., Stępniak C., Turek T. (2015), The Role of Spatial Decision Support System in Regional Business Spatial Community, Applied Mechanics and Materials, Vol.795, IT Systems and Decisions in Business and Industry Practice s.107–114.

13. Cicek, S., Kocamaz, U. E., & Uyaroğlu, Y. (2018). Secure communication with a chaotic system owning logic element. *AEU-International Journal of Electronics and Communications, 88*, 52–62.

14. Singh, P. P., Singh, J. P., & Roy, B. K. (2016). SMC-based synchronization and anti-synchronization of chaotic systems for secure communication and analog circuit realization. *Int. J. Control Theory Appl, 9*(39), 171–183.

15. Yang, Z., Pan, X., Zhang, Q., & Chen, Z. (2021). Finite-time formation control for first-order multi-agent systems with region constraints. *Frontiers of Information Technology & Electronic Engineering, 22*(1), 134–140.

16. Hu, X., Zhang, Z., & Li, C. (2021). Consensus of multi-agent systems with dynamic join characteristics under impulsive control. *Frontiers of Information Technology & Electronic Engineering, 22*(1), 120–133.

17. Gupta, S., Vyas, S., & Sharma, K. P. (2020, March). A survey on security for IoT via machine learning. In *2020 International conference on computer science, engineering and applications (ICCSEA)* (pp. 1–5). IEEE.

Note: All the figures and tables in this chapter were made by the author.

Recent Trends in Engineering and Science for Resource Optimization and
Sustainable Development – Prof. (Dr.) Dorota Jelonek et al. (eds)
© 2024 Taylor & Francis Group, London, ISBN 978-1-032-98030-0

34

Assessing 5G's Implication on Electronic Engineering

B. Suneela[1]

Department of ECE, Associate Professor,
Malla Reddy Engineering College, Secunderabad, India

S. Sharmila[2]

Assistant Professor, Department of EEE,
SNS college of Technology, Coimbatore, India

Ala Harika

Department of Computer Science and Engineering,
Institute of Aeronautical Engineering, Hyderabad, India

Asha V[3]

Department of Computer Applications,
New Horizon College of Engineering, Bangalore

Navdeep Singh

Lovely Professional University, Phagwara, India

Charanjeet Singh[4]

Assistant Professor, Electronics and Communication Department,
Deenbandhu Chhotu Ram University of Science and
Technology, Murthal, India

■■■■■ Abstract

The influence of 5G on present and prospective applications will raise the need for electronic devices. Massive quantities of data with high speeds may be carried utilizing the bandwidth range of 5G, generating significant advantages and several damages. Particularly significant damage to the group could result if the primary visual data conduits, like photographs and films, are fraudulently assaulted and published to the web, where they may swiftly expand. As a result, we offer a distinctive structure as an electronic forensics instrument for safeguarding electronic devices. It is constructed depending on deep learning and may identify assaults through classification. The suggested system's data-collecting effectiveness, resilience, and detection performance are all improved when contrasted with standard techniques and supported by our studies. Furthermore, with the help of 5G, the suggested system enables the provision of outstanding real-time forensics operations on electronic devices like mobile phones and computers, bringing enormous practical advantages.

■■■■■ Keywords

5th generation, Electronic devices, Digital forensics, Deep learning, and Detection performance

Corresponding author: [1]sunilareddyk@gmail.com, [2]sharmilaurfriend@gmail.com, [3]asha.gurudath1@gmail.com, [4]charanjeet.research@gmail.com

DOI: 10.1201/9781003596721-34

1. INTRODUCTION

Interfering with electronic multimedia data seemed to be a significant issue for average people who lacked knowledge. Nevertheless, because of current scientific advances, several advanced programs have been devised to make it far simpler to harm images and movies. Images and clips, on the contrary, continue to be trusted forms of data in the modern world. As a result, the validity and authenticity of these visual data resources should be ensured. Numerous investigators have historically dedicated attention to digital forensics to attain this objective [1]. Without a doubt, 5G networks will increase the precision and effectiveness of electronic monitoring and assessment. In the areas of engineering machinery, cell phones, vehicles, gadgets, household goods, machinery for industry, etc., 5G uses have grown extensively. The primary value chains, including antennas, radiofrequency, connectivity chips, and communication components, will experience fresh life because of the fifth-generation innovation [2]. The electronic manufacturing sector will soon be able to operate smoothly and continuously in a variety of situations thanks to 5G networks, which will boost working circumstances for employees, reduce employee effort on assembly lines, and greatly increase the manufacturing procedure's management [3]. Under these conditions, a growing number of devious forgeries could be broadcast on the internet more rapidly than ever before. When dealing with enormous quantities of data, conventional forensics techniques might suffer from decreased efficiency [4]. If the system is developed on 5G Het-Nets, this could pose a significant threat to the group. As a result, in this research, we present a unique deep learning (DL)--based architecture that can be used as a universal digital forensics instrument for safeguarding multimedia information stored on electronic devices.

2. LITERATURE REVIEW

The convolutional neural network (CNN) is a popular DL approach. As a result, CNNs can tackle several forensics challenges that previously baffled academics [5]. Every year, various novel algorithms for digital forensics are developed. Since each assault on multimedia data might depart distinct hand prints, these forensics techniques are usually used to distinguish between assaulted and undisturbed material by empirical pattern categorization. CNNs are also trained as forensics sensors with this overall aim in mind [6]. The research [7] presented the initial forensics investigation utilizing the CNN framework in 2015 [8]. The suggested approach can determine if media filtering was used on the assaulted photos [9]. Before this study, experts questioned CNN's capacity to detect assaulted images by tracing pixel-wise variations [7].

DL has also had a significant impact on visual forensics. Numerous deep neural networks, in particular, have been developed to identify false faces produced by deep-fake [6]. CNNs have additionally been used to solve a variety of other visual forensics challenges, including dual MPEG contraction, tamper identification, and furthermore. In general, DL is the key element that propels digital forensics study to new heights. It might be suggested as an approach to the majority of digital forensics challenges [5].

3. METHODOLOGY

The main component of our suggested methodology is a CNN version that may be used as an electronic device forensics detector. Initially, our framework has a smart framework with numerous vision components. Every part is made up of several convolutional, batch normalization, and activation layers. As a result, the extraction of features may be performed on an entire picture with great resolution, offering a wealth of data for the study. Following analysis of these vision components, many feature vectors may be provided accessible. To overcome this excess, our system includes numerous full connection layers that link all neurons for categorization. As a result, unlike prior models that were implemented on a regional disc, our system is hosted on the cloud for testing and training using cloud computing. To analyze their data, users can immediately access the framework using 5G Het-Nets. The detector can also use the information uploaded for training. Furthermore, with the help of 5G Het-Nets, the framework may autonomously gather smart electronic gadgets in the IoE. The detector may be maintained in an ongoing process of training with this theoretically limitless quantity of acquired data, continually being upgraded against any novel sort of assault.

4. RESULT AND DISCUSSION

Initially, we tested how well the CNN approach performed in terms of forensics depending on the volume of training data. Table 34.1 and Fig. 34.1 show the framework's

Table 34.1 Detection efficiency under different no. of training data

Size	Detection accuracy		
	Sharpening	Resampling	Compression
500	0.5	0.49	0.52
1000	0.61	0.55	0.69
2000	0.68	0.58	0.7
5000	0.81	0.74	0.76
10000	0.9	0.86	0.84
50000	0.99	0.93	0.91

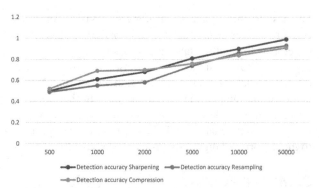

Fig. 34.1 Detection efficiency under different no. of training data

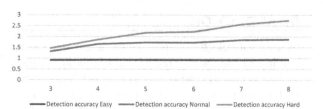

Fig. 34.2 Detection efficiency under different no. of modules

detection performance against multiple assaults with various training data densities. As more diverse data are gathered for training, the resilience of the detector is improved, allowing for a detection accuracy of about 95 percent to be attained. This outcome demonstrates the validity of the suggested architecture in 5G Het-Nets.

A further study was carried out to assess the models at varying stages of sophistication. The extent of the framework is represented by the number of units used in the CNN framework. These systems of varied architectural degrees were used to address image forensic categorization tasks of varied complexity. The simplest challenge was to develop a binary classification algorithm for sharpness identification. A typical challenge was to train the CNN algorithm to recognize five distinct image-based assaults. The most demanding task asked the system to distinguish between photos that had been assaulted by 11 distinct alterations. Each model was trained using enough data from the identical database. The classification efficiency is presented in Table 34.2 and Fig. 34.2.

Table 34.2 Detection efficiency under different no. of modules

Number of modules	Detection accuracy		
	Easy	Normal	Hard
3	0.92	0.4	0.15
4	0.93	0.73	0.2
5	0.93	0.8	0.45
6	0.93	0.8	0.5
7	0.93	0.92	0.72
8	0.94	0.93	0.87

The number of units affects the amount of sophistication. When the assignment was simple, perhaps a system with three tiers could do well. Furthermore, raising the number of units had no favorable influence on identification accuracy. When the number of units was minimal, raising

the extent of the framework significantly improved the detection rate at the standard degree of complexity. When the number of units was considerable, nevertheless, we did not see any enhancements as the consequence of a deeper framework. The extent of the CNN framework has a significant impact on classification results for the most difficult job. We can execute systems with up to 8 units due to the abundance of computing power. The depth constraint can, nevertheless, be lifted for the cloud-based forensics instrument, which has infinite computing power. Furthermore, in actuality, there are over 11 different sorts of attacks. To identify all conceivable assaults on multimedia information, a considerably broader system with a more advanced structure must be used. A cloud-based architecture can help us create useful forensics instruments for consumers. Furthermore, this system can be maintained in an ongoing process of training with better training effectiveness to guarantee that it is ready for any novel assaults.

Lastly, we ran tests to assess systems with no constraints. As previously stated, because of the significant computing load and constrained training information quantity, present CNN algorithms are constrained since they must employ techniques like pooling to prevent overfitting yet experience performance losses. In this paper, we use two techniques for predicting 5 assaults on photos. The primary distinction between the 2 systems is that one requires featured map pooling while one of them does not. Table 34.3 and Fig. 34.3 show the performance of the two approaches to the training epochs. Nevertheless, after more training, the framework can obtain an accuracy close to 98 percent when the constraints are lifted.

Table 34.3 Detection efficiency about epochs for frameworks with and without constraints

Epoch	Detection accuracy	
	Model with limitation	Model without limitation
20	88	75
40	90	90
60	90	98
80	90	99
100	91	99

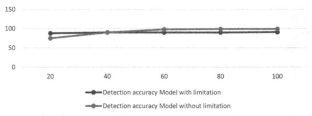

Fig. 34.3 Detection efficiency about epochs for frameworks with and without constraints

Convergence rate and overfitting are unimportant in the creation of CNN algorithms when 5G HetNets have access to enough data and computing power. As a result, the suggested framework can offer reliable forensics services. Additionally, the framework can employ 5G networks to instantly communicate the detection outcomes to end customers' electronic devices. The architecture is the best option for the security of data in the years to come because of these characteristics.

5. CONCLUSION

A DL-based architecture is suggested as a forensics instrument for electronic devices in this study. To offer customers real-time forensics offerings, it is presumable that the suggested architecture will function with 5G networks. The suggested structure, in contrast to the majority of current frameworks, is made to increase its usefulness as a forensics instrument for protecting end-user electronic data on electronic devices. Furthermore, it is built with cutting-edge technologies to overcome any possible challenges caused by 5G networks. A high level of functionality is guaranteed by the DL system's excellent detection capability and the 5G Het-Nets' high data transmission speed. Despite the suggested framework's strong performance, there are a few possible drawbacks as well. Even though the suggested model might not be ideal, we still think it is a suitable approach to multimedia data security in 5G systems. Its possible problems pale in comparison to the immense value it provides to electronic devices as a real-time realistic forensics instrument.

Reference

1. Javed, A. R., Jalil, Z., Zehra, W., Gadekallu, T. R., Suh, D. Y., & Piran, M. J. (2021). A comprehensive survey on digital video forensics: Taxonomy, challenges, and future directions. *Engineering Applications of Artificial Intelligence*, *106*, 104456.
2. Ikram, M., Sultan, K., Lateef, M. F., & Alqadami, A. S. (2022). A road towards 6G communication—A review of 5G antennas, arrays, and wearable devices. *Electronics*, *11*(1), 169.
3. Sodhro, A. H., Awad, A. I., van de Beek, J., & Nikolakopoulos, G. (2022). Intelligent authentication of 5G healthcare devices: A survey. *Internet of Things*, 100610.
4. Camacho, F., Cárdenas, C., & Muñoz, D. (2018). Emerging technologies and research challenges for intelligent transportation systems: 5G, HetNets, and SDN. *International Journal on Interactive Design and Manufacturing (IJIDeM)*, *12*, 327–335.
5. Wang, Q., & Zhang, R. (2016). Double JPEG compression forensics based on a convolutional neural network. *EURASIP Journal on Information Security*, *2016*(1), 1–12.
6. Verde, S., Bondi, L., Bestagini, P., Milani, S., Calvagno, G., & Tubaro, S. (2018, October). Video codec forensics based on convolutional neural networks. In *2018 25th IEEE International Conference on Image Processing (ICIP)* (pp. 530–534). IEEE.
7. Chen, J., Kang, X., Liu, Y., & Wang, Z. J. (2015). Median filtering forensics based on convolutional neural networks. *IEEE Signal Processing Letters*, *22*(11), 1849–1853.
8. Paweloszek I., Approach to Analysis and Assessment of ERP System. A Software Vendor\'s Perspective, Proceedings of the 2015 Federated Conference on Computer Science and Information Systems 2015, M. Ganzha, L. Maciaszek, M. Paprzycki, Annals of Computer Science and Information Systems\", 1415–1426, IEEE\", DOI:10.15439/2015F251, URL:http://dx.doi.org/10.15439/2015F251
9. Soni, N. Kumar, V. Kumar and A. Aggarwal, "Biorthogonality Collection of Finite System of Functions in Multiresolution Analysis on L2(K)," 2022 10th International Conference on Reliability, Infocom Technologies and Optimization (Trends and Future Directions) (ICRITO), Noida, India, 2022, pp. 1–5, doi: 10.1109/ICRITO56286.2022.9964791.

Note: All the figures and tables in this chapter were made by the author.

Recent Trends in Engineering and Science for Resource Optimization and Sustainable Development – Prof. (Dr.) Dorota Jelonek et al. (eds)
© 2024 Taylor & Francis Group, London, ISBN 978-1-032-98030-0

35

Understanding the Potential Advantage of Big Data Analytics for Healthcare Organizations

Swati Tyagi[1]

University of Delaware Newark, DE, USA

K. K. Ramachandran[2]

Director/ Professor: Management/Commerce/International Business,
DR G R D College of Science, India

M. Kalaiarasi

Department of Computer Science and Engineering,
Institute of Aeronautical Engineering,
Hyderabad, India

Asha V[3]

Department of Computer Applications,
New Horizon College of Engineering, Bangalore

Navdeep Singh

Lovely Professional University, Phagwara, India

Parashuram Shankar Vadar

Assistant Professor, Yashwantrao Chavan School of
Rural Development, Shivaji University,
Kolhapur, Maharashtra, India

Abstract

Despite computer researchers' claims that big data innovations offer enormous potential for the healthcare industry, current studies have directed little consideration to the study of its commercial significance. To address this gap in understanding, this study suggests an organization analytics-based company worth architecture in which we employ resource-based concept (RBC) and IT potential developing to connect big data design elements to a big data advantages structure via big data analytics (BDA) potential. We used content evaluation as a study method to determine the reliability of the theoretical framework. The collection included 109 case narratives from 63 different healthcare organizations. Our findings revealed five BDA potentials and their fundamental elements, and also 3 path-to-value networks.

Keywords

Big data analytics, Healthcare, Potentials, Big data architecture, and Commercial significance

Corresponding author: [1]swatyagi@udel.edu, [2]dr.k.k.ramachandran@gmail.com, [3]asha.gurudath1@gmail.com

DOI: 10.1201/9781003596721-35

1. INTRODUCTION

In the United States, the medical field is struggling to convert the value of information technology (IT) to commercial significance due to issues with IT, including insufficient medical system connectivity and poor healthcare data administration. Healthcare organizations are looking for efficient IT artifacts that can assist them in pooling assets to provide a top-notch encounter for patients, boost productivity, and perhaps even develop brand-new, more efficient based on data company structures [1].

The potential use of big data analytics (BDA) represents an intriguing development. BDA, which emerged from enterprise intelligence and decision aid structures, enables healthcare organizations to analyze a massive amount, diversity, and velocity of information throughout an extensive spectrum of healthcare to aid in research-based choices and actions taken [2]. These advanced BDA solutions simplify medical data fusion and offer cutting-edge company knowledge to assist healthcare organizations in satisfying patients' requirements and future market patterns, hence improving the standard of treatment and economic success [3]. Hence this research aims to investigate the instances of potential advantages of BDA in healthcare.

2. LITERATURE REVIEW

In research, multiple descriptions of BDA's potential have been created. BDA skills are obtained generally from BDAs, which entails the potential to handle a large number of different information sources to enable users to perform statistical analysis and response [4]. According to [5], BDA's potential for optimizing a company's commercial significance should include speed to insight [6], or the capacity to change unprocessed data into useful knowledge [7], as well as widespread use [8], or the potential to employ BDA in management (Jelonek, 2017), in gaining a competitive edge (Jelonek, Stępniak, Ziora, 2019) and throughout the corporation [9].

3. METHODOLOGY

We used content evaluation as a research method for evaluating the framework's reliability in Fig. 35.1. Content evaluation is defined as "a technique used in research for subjective assessment of textual data contents via a systematic categorization procedure of coding and recognizing topics or trends." Learning from breakthrough narratives and seeing optimal practices is an improved approach to gaining a greater knowledge of how organizations' IT expenditures pay off. As a result, it is suitable to collect secondary data from practical use scenarios to gain an understanding of how BDA capacities and advantages will evolve.

Several investigations have used secondary data (for example, instance resources) to expand on investigating the commercial benefits of a particular data platform. In present investigations, where no BDA structures are accessible, induction content evaluation is especially ideal for our attempt to develop classes and subclasses inductively using instances and analyze their connections. Our technique entails analyzing assertions from instance supplies, as well as investigating the different BDA capacity classes and the results that result from these capacities. These assertions include technical approach explanations, business benefits, the functions of certain big data technologies, and how they can be utilized in medical treatments or tasks. We considered assertions in the instance materials' textual to be proof of backing for the trends and relationships of structures in our framework. In contemporary research where no BDA structures are accessible, inductive material assessment is especially relevant for our attempt to develop groups and subgroups inductively using instances and analyze their linkages. Our technique entails analyzing comments from the case documents and investigating the different BDA potential groups as well as the outputs that result from these potentials. These assertions include technical options, commercial advantages, the functions of a particular big data method, and the way they are utilized in health care operations. We regarded remarks in the case

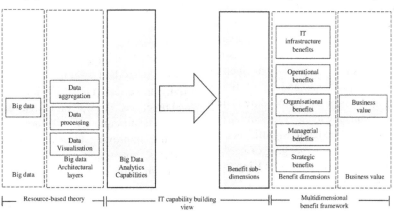

Fig. 35.1 BDA business value framework

resources' content as proof of backing for the system's trends and relationships of structures. The data collection includes 109 case narratives for 63 firms that are especially relevant to the healthcare sector.

4. RESULT AND DISCUSSION

The statistical potential was the main BDA potential discovered (marked as a component of 74 chains), accompanied by pace to traceability, interoperability, predictive analytics, and speed to decision potential. Splitting apart 5 BDA potentials, illustrated in Table 35.1, which demonstrates the frequency of instances in the case resources for every potential, we discover that the potential to evaluate semi-structured and unorganized statistics is referenced the most frequently.

Table 35.1 Outcomes of splitting apart 5 BDA potentials in healthcare

BDA potentials	Structures	No. of instances
Traceability	Historical laboratory outcomes, present patient pills, and immunization history are all available.	27
Interoperability potential	Statistics from other healthcare facilities and data sources should be incorporated.	40
Predictive analytics potential	Allow for the prediction of patient actions.	46
Speed to decision potential	Alert clinicians of major concerns automatically.	51
Analytical potential	Conduct huge multidisciplinary investigations to glean key information from massive amounts of medical information.	74

In medical services, unorganized and semi-structured statistics relate to information that is unable to be saved in standard relational files into preset data structures, like XML-driven electronic healthcare records (EHRs), diagnostic shots, medicinal records, and test outcomes. Previous study indicates that having the capacity to analyze unorganized data is critical to the effectiveness of big data in medical environments, as unorganized data accounts for eighty percent of medical statistics. The primary distinction between BDA administration platforms and conventional administration platforms in the evaluation procedure is that the former has a distinctive capacity to analyze semi-structured or unorganized information, revealing significant correlation trends that were previously challenging or hard to identify. Leeds University Hospital in the United Kingdom, for example, analyzes around one million unorganized patient records per month and has found 30 separate situa-

Fig. 35.2 Outcomes of splitting apart 5 BDA potentials in healthcare

tions where there is a possibility for development in either expenses or operational processes by leveraging natural language processing (NLP). Leeds can enhance effectiveness and minimize expenses by detecting expensive medical care like unneeded extra diagnosis examinations and therapies using this unorganized information's analytical potential. The capacity to make judgments quickly helps medical facilities to autonomously create warmings or warnings and send them to physicians promptly, as well as create clinical summaries and show them utilizing graphical dashboards/systems. Documents created by BDA machines differ from transitory IT systems in that they allow for the evaluation of historical and current operating environments at all organizational levels. Visual updates are typically produced after the near-real-time analysis of information and presented on medical functionality displays to aid wellness researchers in recognizing developing healthcare concerns like surgical mistakes, potential patient security problems, and suitable prescription usage. In 46 instances, prediction analytics competence was demonstrated. The majority of incidences, nevertheless, are documented by IT suppliers. The instances taken from research papers show that the utilization of prediction intelligence in the medical sector is still in its infancy. The instances illustrated the difficulties in creating a viable prediction framework in the absence of a huge amount of good databases. In a similar vein, [10] finds that the complexity of customizing legacy medical data platforms for prediction systems restricts prediction standards. They additionally claim that predictive systems may fail to react to modifications in EHRs, requiring IT professionals to manually update the prediction algorithms, lowering speed and production.

The outcomes of potential advantages of BDA in the healthcare sector are shown in Table 35.2 and Fig. 35.3. This conclusion indicates that when BDAs are deployed in a firm, it has dual potential. It not only enhances IT productivity and effectiveness, but it also helps to optimize clinical functions. The outcomes also show that the aspects with the highest frequency in the area of operational advantages include improving the standard and accuracy of clinical choices, processing an extensive amount of medical records in milliseconds, and enabling early intervention before the illness deteriorates. The factors most noted in the field of IT structure advantage include reducing system re-

Table 35.2 The outcomes of potential advantages of BDA in the healthcare sector

Advantages of BDA potentials	Structures	No. of instances
Strategic advantages	Offer an extensive perspective of therapy provision to address potential demands.	7
Managerial advantages	Offer excellent decision-support knowledge regarding every aspect of the healthcare environment to members of the executive committee and the heads of departments.	11
Organizational advantages	Deliver a consistent, organized, and consenting encounter for patients throughout each of its institutions.	37
IT infrastructure advantages	Procedure standardization across diverse healthcare IT platforms	55
Operational advantages	Enhance the standard and precision of clinical choices	128

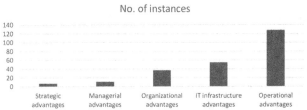

No. of instances

Fig. 35.3 The outcomes of potential advantages of BDA in the healthcare sector

dundancy, avoiding needless IT expenses, and transferring data swiftly among medical IT platforms. This conclusion indicates that when BDA is deployed in an organization, it has a dual potential. It not only boosts IT productivity and effectiveness, but it also aids in the optimization of healthcare functions.

5. CONCLUSION

Finally, using the RBC and potential constructions perspectives, this investigation demonstrates how BDA design improves a healthcare firm's BDA skills, which could assist in the development of commercial significance. We discovered that the methods for achieving commercial significance do not appear to be governed by a single component, but rather by several BDA potential characteristics. The approach was validated by analyzing secondary information from big data situations in the healthcare domain. Instead of utilizing a single example to analyze our data collection, we utilized the number of instances of structures as markers for uncovering the noteworthy paths leading BDAs to generate commercial significance. As a result, we feel that this investigation provides distinctive perspectives into the conceptual and administrative consequences of the BDA investigation. Despite the previously mentioned insights and consequences, our study has a shortcoming. One issue in the medical services sector is that the use of IT typically falls behind other sectors, which is among the primary explanations that examples are difficult to find. Despite attempts to identify instances from multiple places, the bulk of the instances discovered for this research were from suppliers. As a result, there is a possible bias, because suppliers typically only publicize their success experiences. On the contrary, if supplier situations do not reveal a few of the path-to-value chains, it is a wonderful signal for regions that need work. Various sectors have distinct requirements or aims when it comes to employing big data innovations. Because we focused on medicine in this research, the outcomes are industry-specific. The reasoning of the framework can be applied to different sectors in future studies. Various BDA potential, methods, advantages, and results may emerge. Lastly, in light of these possibilities in the future, we think that the big data study stream, with an emphasis on strategy, has significant potential to assist in increasing the number of investigations into big data's technical and management viewpoints.

Reference

1. Kamble, S. S., Gunasekaran, A., Goswami, M., & Manda, J. (2018). A systematic perspective on the applications of big data analytics in healthcare management. *International Journal of Healthcare Management*.
2. Khan, I. H., & Javaid, M. (2021). Big data applications in the medical field: A literature review. *Journal of Industrial Integration and Management*, 6(01), 53–69.
3. Wang, Y., Kung, L., Ting, C., & Byrd, T. A. (2015, January). Beyond a technical perspective: understanding big data capabilities in health care. In *2015 48th Hawaii International Conference on System Sciences* (pp. 3044–3053). IEEE.
4. Wang, Y., Kung, L., & Byrd, T. A. (2018). Big data analytics: Understanding its capabilities and potential benefits for healthcare organizations. *Technological forecasting and social change*, 126, 3–13.
5. Dash, S., Shakyawar, S. K., Sharma, M., & Kaushik, S. (2019). Big data in healthcare: management, analysis, and prospects. *Journal of Big Data*, 6(1), 1–25.
6. Anderson, J. E., & Chang, D. C. (2015). Using electronic health records for surgical quality improvement in the era of big data. *JAMA surgery*, 150(1), 24–29.
7. Jelonek, D. (2017). Big Data Analytics in the Management of Business, MATEC Web of Conferences, 125, 04021.
8. Jelonek, D., Stępniak, C., Ziora, L. (2019). The meaning of big data in the support of managerial decisions in contemporary organizations: Review of selected research, Advances in Intelligent Systems and Computing, 886, 361–368.

Note: All the figures and tables in this chapter were made by the author.

Recent Trends in Engineering and Science for Resource Optimization and
Sustainable Development – Prof. (Dr.) Dorota Jelonek et al. (eds)
© 2024 Taylor & Francis Group, London, ISBN 978-1-032-98030-0

36

Integrating Organisational Theory, Resource-Based View, and Big Data Culture to Improve Manufacturing Performance

Parihar Suresh Dahake[1]

Assistant Professor, Department of Management Technology,
Shri Ramdeobaba College of Engineering and Management, Nagpur

Mini Jain[2]

Assistant Professor, Institute of Business Management,
GLA University, Mathura, Uttar Pradesh

B. Anusha

Department of Computer Science and Engineering,
Institute of Aeronautical Engineering, Hyderabad, India

Sabbineni Poojitha[3]

Assistant Professor, School of Management and Commerce,
Mallareddy University Hyderabad, India

Navdeep Singh

Lovely Professional University, Phagwara, India

Anjali Sharma[4]

Assistant Professor, IIMT University, Meerut

Abstract

Big data were at the cutting edge of functioning and managerial manufacturing study. The impact of big data on manufacturing and functional performance has been described in research, yet material was scarce on the function of outside organizational demands on the assets of an organization to establish capabilities of big data. To fill this void, this work uses the resource-based view of the company's organizational theory, and organizational culture to establish and evaluate a framework that outlines the significance of assets in developing potential, competencies, and big data culture, in addition to boosting expense and performance in operations. We set our study ideas to evaluation with 195 evaluations collected utilizing a pre-tested questionnaire.

Keywords

Big data, Organizational theory, Big data culture, Resource-based view, and Manufacturing performance

Corresponding authors: [1]parihar_dahake@rediffmail.com, [2]minijain06@gmail.com, [3]sabbinenipoojitha@gmail.com, [4]anjali.shail@gmail.com

DOI: 10.1201/9781003596721-36

1. INTRODUCTION

Big data has garnered enormous interest among professionals and academics due to its capacity to revolutionize entire business operations. By collecting, handling, and analyzing 5V data-based measurements, a company can produce company knowledge, improve performance, and beat its rivals. The capacity to acquire, evaluate, and manage massive amounts of data with the help of strong systems of information - which is big data culture - to enhance the performance of manufacturing organizations has piqued the attention of research and businesses alike [1].The capacity to acquire, analyze, and manage massive amounts of data with the help of a strong data structure - that is, big data prediction evaluation (BDPE) - to enhance the functioning of production companies has piqued the attention of academics and businesses alike.

Big data and predictive assessment are commonly defined in the field of information systems as an organizational competence for handling enormous amounts and types of data at the velocity needed to get useful knowledge, allowing organizations to achieve a competitive edge [2]. The phrase 'big data' is frequently used to characterize large, complicated, and real-time information that necessitates advanced administration, mathematical, and statistical approaches to acquire administrative information. The prediction analytics mathematical frameworks attempt to forecast future behavior on the premise that what has occurred in the past will occur again shortly.

As a result, big data and predictive assessment can be vital instruments for realizing yield gains, especially in any manufacturing context with operational difficulty, variation in processes, and capacity limits. As a result, we synthesized different viewpoints in this research to offer a deeper comprehension of the way manufacturing organizations with a particular asset profile choose actions in the face of external challenges [3]. To conceptually support the results of our experiments, we combine organizational theory, Resource-based view (RBV), and BDC, since each paradigm may clarify the firsthand performance consequences of big data on organizational performance on its basis.

2. LITERATURE REVIEW

According to [4], manufacturing organizations can prevent costly activities like extra manufacturing, revenue losses, and overstock by integrating richer, current data into managerial choices and finding improved options promptly.

According to [5], from an RBV viewpoint, big data-based organizational assets can be used to create a distinct edge when they satisfy the conditions of value [6], rarity [7], imprecise imitability [8], and non-substitutability [9].

3. METHODOLOGY

This research obtained information utilizing a survey-oriented tool and relevant metrics from current studies to evaluate the study's hypotheses. On 5-point Likert ratings, the sizes have been determined. Additionally, because of the scale of the organizations and the Indian setting in which we obtained the data, we were unable to obtain objective statistics from the data we collected. Administration's opinions of production, revenue, market share, and consumer contentment in comparison to competition are examples of subjective performance measurements. In the present research, objective assessments were hard to collect since respondents did not desire to divulge this delicate data and instead limited themselves to filling out perception ratings. Also, because of the scale of the companies and the Indian setting in which we obtained the data, we were unable to obtain objective statistics from our collection. Expanding marketplaces lack the same amount of data accessibility as advanced nations (for example, the United States or the United Kingdom). We pre-tested the tool in two phases before finalizing it for gathering data, following the best procedure described by [10]. Initially, we asked 8 expert academics to review the survey's phrasing for uncertainty, simplicity, and the suitability of the measurements employed to operationalize every construct.

After expert feedback, we revised the format of the queries in the survey tool. We emailed the study questionnaire to ten top executives and advisors in charge of their firms' data analytics divisions. These executives and advisors were tasked with reviewing the evaluation tool carefully for framework, accessibility, confusion, and thoroughness. The advisors' and executives' recommendations were incorporated into the ultimate study tool.

The availability of data in new markets is lower than in developed markets. This research pre-tested the tool in two phases before finalizing it for gathering information, which is standard procedure. For statistical analysis in this research, we employed partial least squares (PLS).

H1: Coercive pressure (CP) exerts a considerable effect on tangible resource (TR) choice.

H2: CP has a considerable effect on human skill selection (HS).

H3: Normative pressure (NP) exerts a considerable influence on the choice of TR.

H4: NP exerts a major influence on the development of HS.

H5: Mimetic pressure (MP) influences the choice of TR.

H6: MP positively moderates HS.

H7: Big data culture (BDC) positively moderates cost performance (Cost_P).

H8: BDC influences operational performance (OP) positively.

Previous research specified the sampling frames slightly. In general, previous research was quite comparable in terms of data collection methodologies. In all of the investigations, data were gathered by interviewing senior executives. The greater part of the study relied on cross-sectional information. The simultaneous gathering of data on CP, NP, TR, human abilities, BDC, and MP may result in a possible issue of simultaneity; that is, the causal relationship between autonomous external and inherent structures cannot be identified definitively. The sampling processes accompanied well-defined trends. Sampling was either narrowly confined to a single sector or extensively spanned across sectors.There is no dataset resource from earlier studies on the same topic and environment as this investigation. As a result, initial information was required. As a result, essential aspects of the research included identifying the target audience, designing a suitable sampling for the research, and ensuring that replies were free of prejudice by utilizing reliable samples. We employed a cross-sectional e-mail study of industrial enterprises from across India. The first collection comprised 375 enterprises drawn from CII Center for Manufacturing datasets and confirmed using a dataset. We used an altered form of the entire layout of the test procedure to boost our reaction level. Each survey form was delivered to an important responder.Participants are required to be operational leaders in operating administration, purchasing/procurement, manufacturing, or quality control to participate. We accompanied inquiries to the respondents' offices after every questionnaire contained a cover letter. Given India's particular cultural and societal environment, we think that this layout is appropriate for study. As previously noted, financial transactions in India are heavily focused on human connections, and reward methods may not produce better results.

4. RESULT AND DISCUSSION

Table 36.1 and Fig. 36.1 illustrate the impact of the indicator factors, Cronbach's alpha, average variance extracted (AVE), and scale composite reliability coefficients (SCR). This demonstrates that our conceptions have sufficient convergence validity.

PLS does not make any assumptions about the multivariate typical distribution. As a result, typical parametric-oriented significance testing procedures are ineffective. For calculating standard errors (SEs) and the importance of variable estimations, PLS employs a bootstrap technique. Table 36.2 and Fig. 36.2 show the path coefficients of PLS and p-sores. The path coefficients computed have been identified as standard beta coefficients.

Table 36.1 The impact of the indicators of, SCR, Cronbach's alpha, and AVE

Structures	Cronbach's alpha	SCR	AVE
NP	0.93	0.75	0.51
BDC	0.92	0.75	0.51
OP	0.73	0.78	0.54
Cost_P	0.68	0.80	0.50
CP	0.93	0.81	0.59
TR	0.92	0.95	0.72
HS	0.92	0.96	0.76
MP	0.92	0.96	0.86

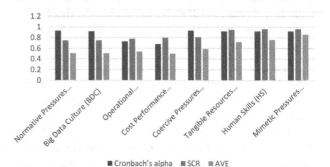

■ Cronbach's alpha ■ SCR ■ AVE

Fig. 36.1 The impact of the indicator of, SCR, Cronbach's alpha, and AVE

Table 36.2 Structure determinants

Hypothesis	Impact on	Impact of	β	p-score	Outcome
H1	TR	CP	0.28	<0.001	Accepted
H2	HS	CP	0.05	0.25	Rejected
H3	TR	NP	0.44	<0.001	Accepted
H4	HS	NP	0.27	<0.001	Accepted
H5	TR	MP	0.40	<0.001	Accepted
H6	HS	MP	0.24	<0.001	Accepted
H7	Cost_P	BDC*TR	0.32	<0.001	Accepted
H8	OP	BDC*HS	0.45	<0.001	Accepted

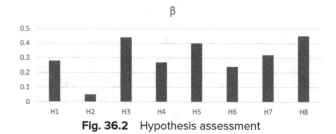

Fig. 36.2 Hypothesis assessment

Following that, we looked at the study model's explanation ability depending on the explained variation (R2) of the endogenous variables (Table 36.3).

Table 36.3 R2, and F2, and Impact Size

	Q²	R²	F²				
			HS	TR	BDPE	Cost-P	OP
MP			0.10	0.27			
NP			0.12	0.27			
CP			0.21	0.17			
HS	0.26	0.23			0.21		
TR	0.57	0.58			0.19		
BDPE	0.25	0.26				0.32	0.36
Cost_P	0.36	0.38					
OP	0.41	0.41					

To calculate the impact size of every predictor, we employed Cohen's f2 equation. Cohen (1988) classified f2 scores as big, medium-sized, and tiny.

To assess the framework's prediction potential, Stone-Geisser's Q2 for endogenous variables are TR, HS, BDC, Cost_P, and OP, all of which are more than zero, showing adequate prediction significance.

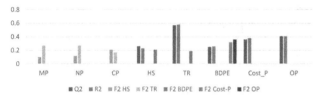

Fig. 36.3 R2, and F2, and Impact Size

5. CONCLUSION

The research results show that, when it comes to big data usage, the organizational demands of manufacturing organizations have a direct impact on the inner utilization of resources and, ultimately, big data implementation. The rapid rush in attention to big data in manufacturing performance and operational administration research prompted this investigation. To tackle the current constraints of RBV, we suggested a conceptual structure rooted in organizational theory and RBV, and we empirically investigated the way choosing resources influenced by 3 parts of organizational stresses may develop big data abilities, which consequently may assist in accomplishing manufacturing performance. The subsequent are the constraints of our investigation. To begin, while the RBV has received considerable focus, we contend that it is contextually insensitive. We define context indifference as RBV's inability to recognize the circumstances under which assets or skills can be most valued. As a result, in the years to come, the present system can be modified by employing the contingency concept to account for the way external and internal variables affect manufacturing performance.

Reference

1. Dubey, R., Gunasekaran, A., & Childe, S. J. (2018). Big data analytics capability in supply chain agility: The moderating effect of organizational flexibility. *Management Decision.*
2. Aydiner, A. S., Tatoglu, E., Bayraktar, E., Zaim, S., & Delen, D. (2019). Business analytics and firm performance: The mediating role of business process performance. *Journal of Business Research, 96*, 228–237.
3. Zhong, R. Y., Lan, S., Xu, C., Dai, Q., & Huang, G. Q. (2016). Visualization of RFID-enabled shopfloor logistics Big Data in Cloud Manufacturing. *The International Journal of Advanced Manufacturing Technology, 84*, 5–16.
4. Srinivasan, R., & Swink, M. (2018). An investigation of visibility and flexibility as complements to supply chain analytics: An organizational information processing theory perspective. *Production and Operations Management, 27*(10), 1849–1867.
5. Braganza, A., Brooks, L., Nepelski, D., Ali, M., & Moro, R. (2017). Resource management in big data initiatives: Processes and dynamic capabilities. *Journal of Business Research, 70*, 328–337.
6. Dillman, D. A. (2011). *Mail and Internet surveys: The tailored design method--2007 Update with new Internet, visual, and mixed-mode guide.* John Wiley & Sons.
7. Cohen, J. (1988). Statistical power analysis.

Note: All the figures and tables in this chapter were made by the author.

Recent Trends in Engineering and Science for Resource Optimization and Sustainable Development – Prof. (Dr.) Dorota Jelonek et al. (eds)
© 2024 Taylor & Francis Group, London, ISBN 978-1-032-98030-0

37

Research on Big Data for "Industry 4.0" Cyber-Physical Systems

Ashutosh Singh[1]

Assistant professor, Institute of Business Management,
GLA University, Mathura, India

Nitesh Kumar Saxena[2]

Asst Professor, MBA, INVERTIS UNIVERSITY,
Bareilly, Uttar Pradesh

J. Alekhya

Department of Computer Science and Engineering,
Institute of Aeronautical Engineering,
Hyderabad, India

Uma Reddy[3]

Department of Artificial Intelligence and Machine Learning,
New Horizon College of Engineering,
Bangalore, India

H Pal Thethi

Lovely Professional University, Phagwara, India

Purnendu Bikash Acharjee[4]

Associate Professor, Computer Science, CHRIST University, Bangaluru, India

Abstract

The objective of the revolution known as Industry 4.0 seeks to optimize goods creation based on consumer requirements, specifications for quality, and financial viability. Big data collected by the Internet of Things (IoT)-based commercial Cyber-Physical Systems (CPS) plays an essential part in boosting platform operation efficiency to promote throughput with improved consumer encounters in Industry 4.0. This study shows big databases derived from IoT-based Optical-Wireless CPS (OWCPSs) for optimizing the functioning of maintenance networks in the electronics-manufacturing Industry 4.0. This research collected and analyzed big databases including five parameters: data delivery, delay, overload, throughput, and package error percentage in OWCPSs. The information gathered is important for optimizing the functioning of service systems in the production of electronic goods Industry 4.0.

Keywords

Big data, Industry 4.0, Internet of things, and Cyber-physical systems

Corresponding author: [1]assingh86@gmail.com, [2]niteshsaxena2005@gmail.com, [3]nvumareddy409@gmail.com, [4]pbacharyaa@gmail.com

DOI: 10.1201/9781003596721-37

1. INTRODUCTION

Industry 4.0 arose from a German government-supported initiative for modern manufacturing ideas in 2011 and has since become an increasingly popular idea. From the standpoint of consumers, Industry 4.0 enables them to obtain personalized goods since producers may constantly rearrange manufacturing facilities depending on customer requests gathered through a digital medium [1]. Despite the enormous promise for personalization and efficient use of resources in Industry 4.0, numerous hurdles persist. Cyber-Physical System (CPS) is a particularly pertinent word for Industry 4.0 [2].

On one side, CPS has industrial uses and is seen as an essential part of Industry 4.0, yet it could likewise be employed in various sectors like medical services, public transit, and the armed forces [3]. Furthermore, with the latest developments in lower-cost detectors, superior information acquisition structures, and more rapid communication technologies in the industry 4.0 CPS, there has been an increasing number of interrelated physical structures that are constantly producing an immense quantity of information for analysis, known as Big Data [4]. Hence this research aims to investigate big databases for Industry 4.0 in CPS

2. LITERATURE REVIEW

The research [5] presented a 5-layer CPS framework, beginning with gathering data, transformation, and evaluation and progressing to intervention and adaptive behavior [6]. Numerous current initiatives in the manufacturing sector use big data from various gadgets and platforms to enhance both efficiency [7] and quality [8]. As a result, using Big Data approaches becomes vital for an additional adaptable, smart, and robust CPS in Industry 4.0 [9].

3. METHODOLOGY

In this study, a collection of optic wireless CPS has been dynamically installed in various structures in interior manufacturing of electronics commercial settings. The total quantity of optic sensor nodes that comply with the optical layer's specification and operate at wavelengths ranging from 7000-300nm is assigned to 100. On the contrary, the optic wireless CPS that adheres to the tangible layer norm has a configuration of 450. In the setup, sites outfitted with both electronic and optic communication methods function as portal head points, aggregating data collected from adjacent nodes and transmitting it to the co-bot using optic communication methods. The big database provided in this paper is critical for real-time assessments of the electronics manufacturing procedure in an Industry 4.0 electronics manufacturing environment.

4. RESULT AND DISCUSSION

Table 37.1 and Fig. 37.1 list the big databases linked to the data delivery proportion in OWCSPs. The data delivery proportion of OWRP in the first 1000 rounds is substantial, approximately 99.95%, contrasted with 93.15% in CARP.

Table 37.1 Outcomes of big data collection for data delivery proportion in OWCSPs

No. of rounds	OWRP (%)	CARP (%)	DCFBR (%)
1000	0.00999	0.00931	0.00909
2000	0.00996	0.00931	0.00888
3000	0.009981	0.00934	0.00872
4000	0.00998	0.00927	0.00865
5000	0.00995	0.0093	0.00856

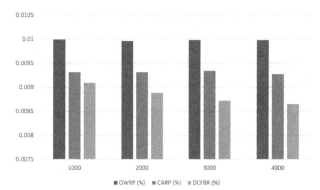

Fig. 37.1 Outcomes of big data collection for data delivery proportion in OWCSPs

Table 37.2 and Fig. 37.2 present the databases linked to the OWCSP delay. The big databases acquired show that the delay ratio (DR) of OWRP having a node count is low, approximately 30ms, as opposed to 63ms in CARP.

Table 37.2 Outcomes of big data collection for the delay in OWCSPs

No. of rounds	OWRP (%)	CARP (%)	DCFBR (%)
100	0.0029.5	0.0063	0.00616
200	0.00477	0.00984	0.00894
300	0.00658	0.0169	0.0136
400	0.00854	0.023	0.0186
500	0.0117	0.0290	0.0266

Table 37.3 and Figure display the records associated with OWCSP controlling congestion (CC). The collected big databases show that the (CC) ratio (CM) of OWRP having node counts is approximately 99.8 percent, whereas CARP has a CC of 98.5 percent.

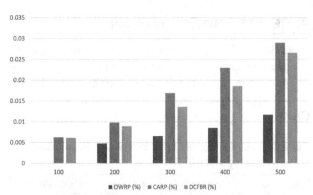

Fig. 37.2 Outcomes of big data collection for the delay in OWCSPs

Table 37.3 Outcomes of big data collection for CC in OWCSP

No. of rounds	OWRP (%)	CARP (%)	DCFBR (%)
100	0.00997	0.00987	0.00983
200	0.00994	0.00961	0.00956
300	0.00986	0.00911	0.00919
400	0.00986	0.0087	0.00859
500	0.00974	0.0086	0.0082

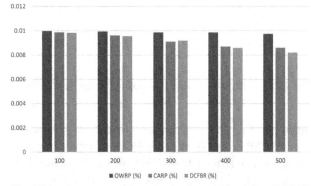

Fig. 37.3 Outcomes of big data collection for CC in OWCSP

Table 37.4 and Fig. 37.4 display the databases linked to OWCSP throughput. The acquired big databases reveal that the throughput significance (TP) of OWRP having node concentration is roughly 99.2%, whereas CARP has 91.2%.

Table 37.4 Outcomes of big data collection for throughput in OWCSP

No. of rounds	OWRP (%)	CARP (%)	DCFBR (%)
1000	0.00991	0.00915	0.0087
2000	0.00990	0.00914	0.00875
3000	0.00988	0.00902	0.00874
4000	0.00989	0.00903	0.00874
5000	0.00988	0.00916	0.00871

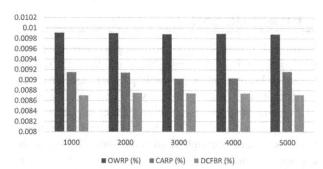

Fig. 37.4 Outcomes of big data collection for throughput in OWCSP

Table 37.5 and Fig. 37.5 display the records relating to the package error percentage (PEP) in OWCSPs. According to the acquired big databases, the PEP ratio of OWRP having a node count between 1 and 100 is inadequate, approximately 0.2 percent, as opposed to 0.35 percent in CARP and 0.39 percent in DCFBR.

Table 37.5 Outcomes of big data collection for PEP in OWCSP

No. of rounds	OWRP (%)	CARP (%)	DCFBR (%)
100	0.0019	0.0035	0.0038
200	0.0033	0.0046	0.0061
300	0.0037	0.0063	0.0097
400	0.0046	0.0109	0.015
500	0.0058	0.016	0.0262

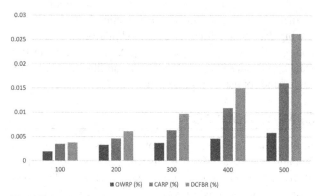

Fig. 37.5 Outcomes of big data collection for PEP in OWCSP

5. CONCLUSION

ACPS is a framework that combines data, computing, interaction, and management. Electronic manufacturing has the potential to increase performance in both production operations and manufacturing scheduling. CPSs, technology for the web, parts as data carriers, and comprehensive security and safety, encompassing confidentiality and understanding safeguarding, are the four key theoretical concepts of Industry 4.0. Big Data, CPSs, portable, cloud computing, and the IoT, among other purposes, possess an enormous effect on Industry 4.0.

Reference

1. Lu, Y. (2017). Industry 4.0: A survey on technologies, applications, and open research issues. *Journal of Industrial Information Integration*, *6*, 1–10.
2. Chen, H. (2017). Applications of the cyber-physical system: a literature review. *Journal of Industrial Integration and Management*, *2*(03), 1750012.
3. Lee, J., Bagheri, B., & Kao, H. A. (2015). A cyber-physical systems architecture for industry 4.0-based manufacturing systems. *Manufacturing letters*, *3*, 18–23.
4. Kao, H. A., Jin, W., Siegel, D., & Lee, J. (2015). A cyber-physical interface for automation systems—methodology and examples. *Machines*, *3*(2), 93–106.
5. Lee, J., Ardakani, H. D., Yang, S., & Bagheri, B. (2015). Industrial big data analytics and cyber-physical systems for future maintenance &and service innovation. *Procedia cirp*, *38*, 3–7.

Note: All the figures and tables in this chapter were made by the author.

Recent Trends in Engineering and Science for Resource Optimization and Sustainable Development – Prof. (Dr.) Dorota Jelonek et al. (eds)
© 2024 Taylor & Francis Group, London, ISBN 978-1-032-98030-0

38

HealthCare and Big Data: Management, Evaluation, and Prospects

Manisha Goswami[1]

Assistant Professor, Institute of Business Management,
GLA University, Mathura, Uttar Pradesh

Sunil Bajeja

Associate Professor, Faculty of Computer Applications,
Marwadi University, Rajkot, Gujarat, India

S. Aswini

Department of Computer Science and Engineering,
Institute of Aeronautical Engineering, Hyderabad, Telangana

P. Krishna Priya[2]

Assistant Professor, KL Business School,
KL Education Foundation, Vaddeswaram, India

H Pal Thethi

Lovely Professional University, Phagwara, India

Vaishnavi M

Assistant professor, CSE, Bannari Amman Institute of Technology,
Sathyamangalam, Erode

▄▬▬▬ Abstract

The healthcare database built around big data has grown fast in the past few decades and is being applied to healthcare data to discover critical health patterns and enable prompt preventative treatment. The purpose of this study is to evaluate organizational impediments to deploying big data-based healthcare management and prospects. The suggested framework can give hospital management projections and consequences that can help them remove organizational impediments when implementing a big data-driven healthcare data structure into their healthcare delivery network. The findings may have implications for improving the efficacy as well as the standard of the healthcare data platform depending on big data in the healthcare business. Management can create efficient methods to tackle issues with suitable emphases if they comprehend the order of essential resistant elements.

▄▬▬▬ Keywords

Healthcare, Big data, Medical big data, Organizational issues, and Adoption method

Corresponding author: [1]manisha.goswami@gla.ac.in, [2]parvathanenikrishnapriya@gmail.com

DOI: 10.1201/9781003596721-38

1. INTRODUCTION

The healthcare organization is information-intensive and might benefit from engaging dynamic big data systems with novel technology and instruments for improving patient treatment and operations [1]. Daily, the healthcare field maintains a large volume of statistics from medical and functional computer structures like Electronic Medical Records (EMR) and Lab Data Library Services (LDMS). Professionals are building novel tools to aid healthcare players in increasing prospects for better value [2].

EMRs that employ big data techniques for significant illness evaluations and the efficacy of epidemiology analysis might be considered to be groundbreaking in healthcare data management. Despite efforts to build effective big data platforms, several healthcare organizations have experienced initial failures while implementing these novel systems [3]. The purpose of this research is to determine and evaluate the organizational issues that impede healthcare organizations from establishing an effective big data structure, as well as to provide management with strategic answers to these issues.

2. LITERATURE REVIEW

Multiple investigations have added to our comprehension of big data in healthcare in various manners. The study [4] is an evaluation of research on the significance of big data in medical services. The investigations of [5] offer an overall description through an evaluation of instances in the realm of health analytics, focusing on specific components of the discipline. Employing the healthcare industry as an instance investigation, [6] investigated the links between big data analytics potential, IT-enabled conversion practices [7], and advantages [8].

3. MATERIALS AND METHOD

Using interviews with specialists in the field, this investigation intends to evaluate the issues of big data management in healthcare organizations. Interviews with professionals provide numerous advantages, including a better knowledge of interviewers [9], a faster hiring and planning procedure, rich data collection, and direct connection with interviewees [10]. This study conducted 32 interviews with key professionals in the healthcare sector, comprising physicians, medical personnel, and professionals, for this study. This study conducted thirty-two interviews with key professionals in the medical field, comprising specialists, medical personnel, and academics, for this study. 31 specialists from significant and famous healthcare facilities in Taiwan, including Taipei Vets Memorial Hospital in Taipei, China Medical School Hospitals in Taichung,

and National Cheng Kung School Hospital in Tainan, are among the attendees. Because of the geographical separation, a single interview with a big data professional and scholar at the College of Toronto in Canada is carried out over a telephone conversation. The interviews last 45 to 60 minutes and are held at the organizations where the respondents operate. Following that, we assess the respondent's interaction and synthesize the primary concepts that they communicated. There are normally 5 major barrier aspects accessible, including experience, functioning, control, material, and accessibility to market restrictions. Respondents indicate multiple, although distinct, resistive elements in every aspect. These variables are ultimately classified into 4 categories, each of which contains the most often stated variables for each aspect. Following that, the analytical network procedure (ANP) approach is used to examine these size and resistivity variables.

4. RESULT AND DISCUSSION

The mass of components in this study is calculated using an ANP. Every aspect has a single primary element. The ideal mass is the super-matrix of element loads shown in Table 38.1 and Fig. 38.1. Lastly, the super-matrix of component relative scores can be obtained. The weighted levels of the 5 factors are multiplied by the weight level of every variable based on the super-matrix outcomes.

Table 38.1 The outcomes of super-matrix element

Components	Principal elements	Scores
Market access issues (MAI)	Incentives and bonuses are restricted.	0.210
Operation issues (OI)	Obtaining data on patients is hard.	0.206
Issues of resource (IOR)	Inadequate assistance and demanding work	0.202
Expertise issues (EI)	Communication breakdowns and cross-domain analyses	0.192
Regulation issues (RI)	Data usage and access are restricted.	0.189

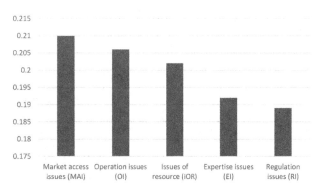

Fig. 38.1 The outcomes of the super-matrix element

ANP is employed in this investigation to evaluate comparative weighted scores. Table 38.2 and Fig. 38.2 display the outcomes of the healthcare big data relevance index. Regarding S_{vk}, physicians possess a maximum score of 0.257, while professionals possess a minimum level of 0.144. Physicians have the greatest Q_{vk} value of 0.282, while intellectuals have a minimum level of 0.171.

Table 38.2 The outcomes of the healthcare big data relevance index

Components	Scores	Physicians	Healthcare staffs	Professionals
Market access issues (MAI)	0.210	0.243	0.205	0.15
Operation issues (OI)	0.205	0.282	0.197	0.171
Issues of resource (IOR)	0.201	0.238	0.209	0.135
Expertise issues (EI)	0.194	0.253	0.2	0.133
Regulation issues (RI)	0.191	0.269	0.18	0.127
S_{vk}		0.257	0.198	0.144
Q_{vk}		0.282	0.209	0.171

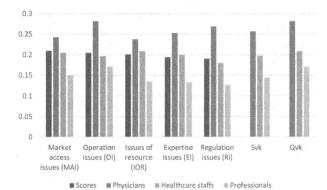

Fig. 38.2 The outcomes of the healthcare big data relevance index

Table 38.3 and Fig. 38.3 stratify the respondent's R_{vk} at $\mu = 0.5$, verifying the method order. The findings demonstrate that, when contrasted with physicians and medical personnel, professionals are more concerned about healthcare big data management and offer the most credible recommendations for removing issues to the introduction of big data platforms in healthcare organizations.

Table 38.3 The outcomes of the order of strategies

$\mu=0.5$	Physician	Healthcare staff	Professionals
R_{vk}	0.269	0.203	0.157
$1-R_{vk}$	0.731	0.797	0.843
Order	3	2	1

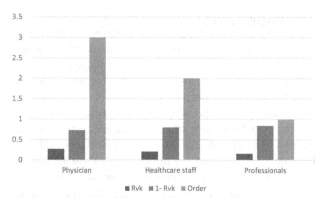

Fig. 38.3 The outcomes of the order of strategies

The outcomes of the ANP and VIKOR studies in this research reveal that the primary obstacles to healthcare big data utilization are OI and MAI. The biggest OI stems from gathering information and excellence, which can be created by the non-parametric framework for healthcare big data, leading to hurdles generated by concealed nodes in the data system and exponential development in computing challenges. As a result, the administration and its departments must continue supporting and handling data collecting and preservation. Furthermore, limitations on value-added applications for healthcare big data are a substantial MAI. The objectives of big data initiatives established by administrations and medical companies are comparable. Nevertheless, there are several impediments to accomplishing meaningful real-world uses. A robust IT framework, assessment instruments, procedures, and connections, and an equilibrium between healthcare big data advantages and safeguarding patients are required for efficient utilization of healthcare big data. As a result, the criteria and laws for healthcare big data usage, exchange of data, moral confidentiality, and administration will become critical in determining if healthcare big data can provide actual advantages.

This study also presents the perspectives of three groups of specialists on healthcare big data: doctors, medical personnel, and researchers. The researchers are regarded as experts capable of providing the most strategic and dependable advice to healthcare organizations seeking an effective utilization of big data technologies. Although doctors and medical personnel employed directly within healthcare facilities are widely regarded as having a solid understanding of the internal evolution of healthcare organizations, several plausible consequences underpin this judgment. Initially, doctors and medical professionals from healthcare facilities expressed a lack of confidence in healthcare big data throughout the interview procedure. According to them, healthcare big data has more significant drawbacks than conventional data-gathering methods. Since they operate directly with the big data framework, they can sense the challenges and high demands when the structure is novel to them. As result, doctors and medical personnel, who are considered "insiders" in the issue, can be expected to give fewer insightful ideas than researchers, who are considered "outsiders" or "observers" of the whole situation.

For several decades, researchers majoring in big data platforms have undertaken various important investigation projects and gained real-world expertise at major big data enterprises. These researchers have a good basis of technological and conceptual skills, allowing them to offer helpful recommendations regarding impediments to healthcare big data. According to the inquiries, all of the researchers had previous expertise working with numerous colleges and institutions in Taiwan and Canada. As a result, they see issues not just from an "outsider" viewpoint, but additionally from the viewpoint of a professional or investigator who is deeply involved in the creation and deployment of big-data platforms in healthcare facilities. As a result, academics are thought to have a precise understanding of the situation in general and to prevent the makers of such structures from succumbing to the hurdles.

5. CONCLUSION

This study employs the innovation-resistant concept to discover adopter opposition to healthcare technologies and to elucidate the challenges to the effective implementation of big data platforms in healthcare settings. In this study, we first identify the barriers to big data establishment in healthcare organizations through discussions with commercial professionals. By assessing the interdependence between all recognized aspects, this strategy can assist with reaching optimal judgments. The ANP technique is then used to promote element modeling as a system of variables and options classified. By assessing the interactions between all detected aspects, this strategy can assist in making optimal judgments. The optimum answer (favorable optimum solutions) and unfavorable optimum answer are then defined using VIKOR. The so-called ideal answer is the optimum alternate in terms of all assessment variables, while the bad optimum answer is the least favorable option in terms of assessment variables. A variety of physicians, medical personnel, and professionals have identified numerous impediments as to why numerous healthcare organizations are failing to implement healthcare big data platforms. Human experience, resource allocations, functional processes, rules & regulations, and marketplace entry potential are all tightly linked. Deployment of big data technology is not possible until healthcare organizations remove these restrictions. The study's results additionally indicate that when tackling growth difficulties,

management ought to give close consideration to the importance-driven pattern of the identified issues to take the most explicit and expeditious route.

Despite the positive consequences of this work, scientists should consider various limits for future studies. Initially, the study environment is limited since a great deal of data is gathered solely in Taiwan. Although Taiwan's medical sector is recognized as highly regarded and developed, more study in broad terms is required to ensure the accuracy and practicality of data in a general medical situation. Secondly, due to the difficulties in accessing medical specialists, this study lacks fairness among the subject groups. Specifically, while 23 professionals make up the majority of the respondents, healthcare workers and academics account for only 6 and 3 people, accordingly. As a result, if the number of respondents in every category had been comparable, the contrasts between the material of every category disclosed would have been significant and acceptable. Lastly, only three types of professionals take part in the study procedure: doctors, healthcare workers, and academics. We are confident that other key players, like scientists, administrations, and possibly individuals, can contribute critical data and thoughts to the growth of healthcare big data. Future studies can establish a foundation to comprehend such parties' perspectives and build novel useful barrier structures.

Reference

1. Hong, L., Luo, M., Wang, R., Lu, P., Lu, W., & Lu, L. (2018). Big data in health care: Applications and challenges. *Data and information management*, 2(3), 175–197.
2. Harerimana, G., Jang, B., Kim, J. W., & Park, H. K. (2018). Health big data analytics: A technology survey. *Ieee Access*, 6, 65661–65678.
3. Dash, S., Shakyawar, S. K., Sharma, M., & Kaushik, S. (2019). Big data in healthcare: management, analysis, and prospects. *Journal of Big Data*, 6(1), 1–25.
4. Baro, E., Degoul, S., Beuscart, R., &Chazard, E. (2015). Toward a literature-driven definition of big data in healthcare. *BioMed research international*, 2015.
5. Chen, M., Hao, Y., Hwang, K., Wang, L., & Wang, L. (2017). Disease prediction by machine learning over big data from healthcare communities. *Ieee Access*, 5, 8869–8879.
6. Wang, Y., Kung, L., Wang, W. Y. C., & Cegielski, C. G. (2018). An integrated big data analytics-enabled transformation model: Application to health care. *Information & Management*, 55(1), 64–79.
7. Manpreet Singh Bhatia, Alok Aggarwal, Narendra Kumar. (2020). SPEECH-TO-TEXT CONVERSION USING GRU AND ONE HOT VECTOR ENCODING. PalArch's Journal of Archaeology of Egypt / Egyptology, 17(9), 8513–8524. Retrieved from https://archives.palarch.nl/index.php/jae/article/view/5796

Note: All the figures and tables in this chapter were made by the author.

Recent Trends in Engineering and Science for Resource Optimization and Sustainable Development – Prof. (Dr.) Dorota Jelonek et al. (eds)
© 2024 Taylor & Francis Group, London, ISBN 978-1-032-98030-0

39

Wireless Ad hoc Networks Coherent Time-Based Collaborative MAC Protocol

Ahmad Y. A. Bani Ahmad[1]

Department of Accounting and Finance, Faculty of Business, Middle East University, Amman 11831, Jordan,Applied Science Research Center, Applied Science Private University, Jordan

Surekha M

JSS Academy of Technical education, Noida. Uttar Pradesh

K. Praveena

Department of Computer Science and Engineering, Institute of Aeronautical Engineering, Hyderabad, Telangana

Gur Sharan Kant[2]

Assistant Professor, Department of Computer science, Sir Chhotu Ram Institute of Engg. & Technology, CCS University Meerut, India

H Pal Thethi

Lovely Professional University, Phagwara, India

D. Suganthi[3]

Assistant Professor, (SG), Department of Computer Science, Saveetha College of Liberal Arts and Sciences, Saveetha Institute of Medical and Technical Sciences, Saveetha Nagar, Thandalam, Chennai

■■■■■■ **Abstract**

Collaborative communication is an efficient way to improve transmitting efficiency in wireless ad hoc networks (WANET). The quick motion of automobiles, on the other hand, causes regular shifts in network architecture and diminishes the likelihood of effective transfer of information on the medium access control (MAC) tier. We suggest a time-based collaborative MAC protocol to enhance collaboration effectiveness and multiple access effectiveness in wireless ad hoc networks. This decreases allocation overhead while also increasing channel use. Both conceptual assessment and experimental findings reveal that the suggested protocol outperforms standard approaches in terms of throughput.

■■■■■■ **Keywords**

Collaborative communication, Medium access control, Coherence time, and Wireless ad hoc networks

Corresponding authors: [1]aahmad@meu.edu.jo, [2]gskant9319@gmail.com, [3]suganthiphd@gmail.com

DOI: 10.1201/9781003596721-39

1. INTRODUCTION

The growing need for better throughput and reduced latency in WANET prompted substantial study into innovative approaches, computations, and technology. The concept of "Collaborative Communication" in ad hoc networks is one such notable involvement [1]. Collaborative communication takes advantage of the wireless channel's broadcasting characteristics and employs a geographical variety of separate pathways to alleviate channel inefficiencies, increase network throughput capability, and minimize retransmission delay. The physical (PHY) level was the primary focus of initial studies on collaborative communication strategies. Channel state information (CSI) is commonly considered to be employed to pick relays [2]. In this study, we concentrate on collaborative transfer at the MAC level.

The MAC protocol is continually employed in WANET to deal with the channel-sharing issue for numerous nodes, which is crucial to effectively store and reuse channel resources and therefore increase network performance [3]. To overcome the issue of efficient relay selection and obtain increased cooperation efficiency, we present a collaborative MAC protocol involving time-based selection (TBS-CMAC) for WANET.

2. LITERATURE REVIEW

In the following contention stage, relays use their information packages to transmit and accessible recipients use the k-round contention resolution (k-CR) procedure suggested in [4] to choose the optimum relay with the greatest speed of data from itself to its receiver for package piggyback transfers, decreasing contention expenses and enhancing collaboration effectiveness.

The cooperative MAC-aggregation (Coop-MACA) norms, the rapid collaboration distinguished MAC norms with package piggyback (RCD-CMAC), and the priority-distinguished cooperative MAC norms with contention resolution (CRP-CMAC) are examples of collaborative MAC standards that use the relay contention technique [5].

3. METHODOLOGY

In a WANET, every node has been assigned at random and shares a single wireless symmetric channel [7]. Every node uses a halfway-duplex transceiver with an established transmitting power for exchanging every package on the channel [8] and continuously detects the network if it has no packages to transmit [9, 10]. Furthermore, since the suggested protocol is completely dispersed and just one

node is permitted to send packages at all periods, no time synchronization is necessary at the physical level. To increase multiple access efficiency, the TBS-CMAC protocol employs an efficient relay selection technique and a package piggyback technique. The protocol is divided into three phases: reservation, relay choice, and information package transfer.

4. RESULT AND DISCUSSION

The fundamental parameter of simulation is shown in Table 39.1.

Table 39.1 The fundamental parameter of the simulation

Parameter	Score
PHY header	192 bits
MAC header	272 bits
Package Duration	0.512 s
RTS	160 bits
Retry limit	6
Slot duration	20 µs
CTS/ACK/RTH/RTR	112 bits

The saturated throughput evaluation of the TBS-CMAC and Coop-MACA protocols for varied counts of nodes having a package duration L_{PKT} of 1024 bytes is shown in Table 39.2 and Fig. 39.1. The picture also depicts the TBS-CMAC protocol's conceptual evaluation and simulation outcomes. The graph also indicates that as N rises, so does the saturated throughput of every method. This is because as N increases, the amount of created information packages for every node with a certain overall provided load drops, and hence the collision risk lowers.

Table 39.2 Outcomes of saturated throughout beneath various N

No. of nodes	CMAC assessment	CMAC simulation	Coop-MACA
10	2.6	2.7	2.5
20	2.9	2.9	2.6
30	3.1	3.11	2.70
40	3.12	3.10	2.71
50	3.12	3.9	2.70
60	3.11	3.9	2.69
70	3.10	3.9	2.6
80	3.10	3.9	2.68
90	3.10	3.9	2.68
100	3.10	3.8	2.67

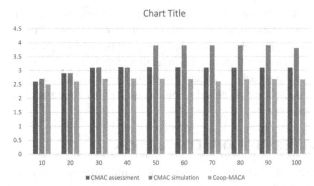

Fig. 39.1 Outcomes of saturated throughout beneath various N

The saturated throughput assessment of the CMAC and CoopMACA simulation protocols for varied counts of nodes with a package size LPKT of 1024 bytes is shown in Fig. 39.1. The illustration also depicts the CMAC protocol's conceptual evaluation and simulation outcomes. The figure shows that the Markov chain-based conceptual evaluation appropriately examines protocol functionality, which is supported by simulated findings. The graph also indicates that as N rises, so does the saturated throughput of all methods. It happens as N increases, the amount of created data packages for every node with a certain overall provided load drop, and hence the likelihood of a collision likelihood lowers. Furthermore, several more sites can act as relays to obtain better cooperative advantage and enhance the likelihood of package piggyback transfers, so enhancing network performance even further. The CMAC protocol beats the other 2 protocols, as seen in Fig. 39.1. As indicated by the findings in Fig. 39.1, when N = 50, the CMAC protocol accomplishes 15.5 percent and 51.7 percent higher saturation throughput than the Coop-MACA technique, accordingly. This is because every phase of its relay-choosing process utilizes smaller mini-slots for choosing the best relay, and its CRP terminates if the top-performing relays in the PDP do not possess data packages to transmit, therefore lowering contention cost and crash likelihood.

Table 39.3 and Fig. 39.2 depict the throughput effectiveness of the TBS-CMAC and Coop-MACA protocols when the package size L_{PKT} = 1024 bytes and package arrival speed are varied. N is fixed to 50 in this case. The graph demonstrates that when L_{PKT} increases, so do the throughputs of every protocol. This is because as L_{PKT} increases, additional bits are conveyed in limited package transmits reserving overhead.

Figure 39.2 demonstrates that the TBS-CMAC protocol has the highest saturation throughput due to its powerful 3-phase relay choosing plan, which is capable of choosing the most suitable relay in an ever-changing setting, and its

Table 39.3 Outcomes of throughput beneath various L_{PKT} and package arrival rate

Package arrival rate	TBS-CMAC	Coop-MACA
1	2.48	2.3
5	3.12	2.7
10	3.3	2.7
15	3.3	2.7

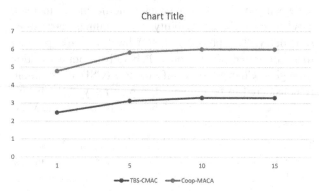

Fig. 39.2 Outcomes of throughput beneath various L_{PKT} and package arrival rate

package piggyback transfer plan, which efficiently minimizes reserving overhead. The Coop-MACA protocol continues to have the least favorable efficiency since it has a lengthy contention period during the relay choosing stage and fails to employ an effective k-CR strategy or package piggyback technique.

The effect of M and k on the TBS-CMAC throughput norms with N = 100 and L_{PKT} = 1024 bytes is shown in Table 39.4 and Fig. 39.3. The figure demonstrates that the simulation findings are marginally less than the quantitative outcomes, which is due to the simulation outcomes having a longer average length of CRP beneath the identical contention variables of CRP.

Table 39.4 Outcomes of the saturated throughput for differing M and k

No. of rounds	Simulation throughput (Mbps)			
	2	3	4	5
1	1.3	1.31	1.32	1.33
2	1.31	1.4	1.55	2.3
3	1.32	1.9	2.55	3.0
4	1.51	2.7	3.2	3.21
5	1.9	3.1	3.25	3.25

The effect of M and k on the saturated throughput of the TBS-CMAC protocol with N = 100 and LPKT = 1024 bytes is shown in Fig. 39.3. The illustration indicates that

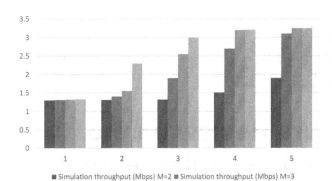

■ Simulation throughput (Mbps) M=2 ■ Simulation throughput (Mbps) M=3
■ Simulation throughput (Mbps) M=4 ■ Simulation throughput (Mbps) M=5

Fig. 39.3 Outcomes of the saturated throughput for differing M and k

the simulation outcomes are a bit less than the statistical outcomes. This is due to the simulation outcomes having a longer mean length of CRP compared to the data-driven outcomes in previous studies, which raises contention expenses and reduces throughput beneath the identical contention variables as CRP. As k grows for the same M, saturated throughput initially rises and then marginally declines. This is because as k increases, the likelihood of choosing a distinctive relay increases, and the package piggyback method can boost saturated throughput. Nevertheless, if k keeps rising while pSR is sufficiently high, the CRP uses additional mini-slots without raising the number of package piggyback transfers, raising collaboration overhead and leading to lower saturated throughput. The image also shows that when k = 5 and M = 4, the saturated throughput of the TBS-CMAC protocol is greatest for every k and M level.

5. CONCLUSION

In this research, we presented a collaborative MAC protocol with time-based selection (TBS-CMAC) for WANET. This protocol employs effective rate distinction, prioritized distinction, and k-round contention resolution techniques to quickly and precisely identify the appropriate relay, hence enhancing collaboration performance. In the meantime, a package piggyback method has been developed to minimize reserve costs and speed up information package deliveries.

Experimental findings verified conceptual studies of the chance of a single relay winning in the k-CR procedure and the saturated throughput of the suggested protocol. When N = 50 and L_{PKT} = 1024 bytes, the TBS-CMAC protocol increased saturated throughput by 15.5 percent and 51.7 percent, respectively, when contrasted to the Coop-MACA protocol.

Reference

1. Zhou, M., Han, L., Lu, H., & Fu, C. (2020). Distributed collaborative intrusion detection system for vehicular Ad Hoc networks based on invariant. *Computer Networks*, *172*, 107174.
2. Gu, C., Xu, H., Yao, N., Jiang, S., Zheng, Z., Feng, R., & Xu, Y. (2022). Enabling a MAC Protocol with Self-Localization Function to Solve Hidden and Exposed Terminal Problems in Wireless Ad Hoc Networks. *IEICE Transactions on Fundamentals of Electronics, Communications, and Computer Sciences*, *105*(4), 613–621.
3. Akande, D. O., & Salleh, M. F. M. (2020). A multi-objective target-oriented cooperative MAC protocol for wireless ad-hoc networks with energy harvesting. *IEEE Access*, *8*, 25310–25325.
4. Geng, K., Gao, Q., Fei, L., & Xiong, H. (2017). Relay selection in cooperative communication systems over continuous time-varying fading channels. *Chinese Journal of Aeronautics*, *30*(1), 391–398.
5. Liu, K., Wu, S., Huang, B., Liu, F., & Xu, Z. (2016). A power-optimized cooperative MAC protocol for lifetime extension in wireless sensor networks. *Sensors*, *16*(10), 1630.
6. Vinay Singh, Alok Aggarwal, Narendra Kumar, A. K. Saini. (2020). A Novel Approach for Pre-Validation, Auto Resiliency & Alert Notification for SVN To Git Migration Using IoT Devices. PalArch's Journal of Archaeology of Egypt / Egyptology, 17(9), 7131–7145. Retrieved from https://archives.palarch.nl/index.php/jae/article/view/5394
7. Alok Aggarwal, Smita Agarwal, Narendra Kumar. (2020). VANILLA Framework for Model Driven Re-Engineering of Declarative User Interface. PalArch's Journal of Archaeology of Egypt / Egyptology, 17(9), 7120–7130. Retrieved from https://archives.palarch.nl/index.php/jae/article/view/5392

Note: All the figures and tables in this chapter were made by the author.

Recent Trends in Engineering and Science for Resource Optimization and
Sustainable Development – Prof. (Dr.) Dorota Jelonek et al. (eds)
© 2024 Taylor & Francis Group, London, ISBN 978-1-032-98030-0

40

A Preliminary Literature Review on the Rules and Application of the Internet of Things in Modern Businesses

Budhi Sagar Mishra[1]
Assistant Professor, Department of Management,
L. N. Mishra College of Business Management, Muzaffarpur, Bihar

Sammaiah Buhukya[2]
Associate Professor, Department of MBA, Omega PG College

G. Lalitha
Department of Computer Science and Engineering,
Institute of Aeronautical Engineering, Hyderabad, India

Anand Kopare[3]
Associate Professor, Management,
Atlas SkillTech University Mumbai

Gaurav Sethi
Lovely Professional University, Phagwara, India

Ahmad Y. A. Bani Ahmad[4]
Department of Accounting and Finance Science,
Faculty of Business, Middle East University, Amman 11831, Jordan

■■■■■■■ **Abstract**

The Internet of Things (IoT) technology is rapidly changing modern business operations and the usage of devices as it connects and links these devices to the Internet. This paper aims to cover the advantages, rules, and processes of applying IoT technology to modern businesses. Through a review of relevant literature, this paper will provide an overview of the application of the Internet of Things solution to business enterprises. In particular, the advantages and disadvantages of IoT in current business operations, the regulations and guidelines for utilizing it, and the applications of IoT technology for a multitude of purposes will be discussed. In addition, this paper will further explore and evaluate the various issues that may arise from its implementation. Moreover, this paper will suggest several strategies and solutions to resolve problems associated with the application of the Internet of Things in modern enterprises. This literature review is expected to equip businesses and organizations with the necessary know-how and insights to effectively deploy and manage IoT technology in their operations.

■■■■■■■ **Keywords**

Internet of things, Modern business, Business operations, Modern, Strategies

Corresponding author: [1]budhi_mlpm73@rediffmail.com, [2]sammanayak@gmail.com, [3]anand.kopare@atlasuniversity.edu.in, [4]aahmad@meu.edu.jo

DOI: 10.1201/9781003596721-40

1. INTRODUCTION

The Internet of Things (IoT) represents a net of linked tangible gadgets, automobiles, constructions, as well as additional items. These objects can be controlled and monitored using technology such as sensors, actuators, and communication networks. The IoT can improve efficiency and productivity by connecting devices and systems that would not ordinarily be connected (Rose et.al, 2015).

Implementing automation and tracking circumstances may assist to decrease expenses and increase security. The IoT is an interconnected system of tangible things and software that allows them to communicate with one another and with their surroundings (Li et.al, 2015). IoT allows gadgets to autonomously gather and share data, allowing for real-time tracking and monitoring of items & actions. The Internet of Things can be utilized to boost efficiency and production while also improving safety and protection.

The IoT is a system composed of tangible gadgets, cars, and other things that are linked together to allow interaction and information sharing. These items might range from smart home appliances to commercial machinery (Laghari et.al, 2021). By 2020, the Internet of Things is estimated to have 100 billion gadgets. The IoT is a web of networked devices that gather and transmit data using detectors and communication standards. The Internet of Things can be utilized for tracking items like the consumption of energy, traffic patterns, and weather patterns (Villamil et.al, 2020). The Internet of Things is additionally available for managing equipment and cars. The IoT is a net of tangible and software objects that connect to share information. The devices can include everything from cars to smart homes to factories (Nord et.al, 2019). The IoT can help us to better understand and control our environment and devices and can also help us to automate processes. The IOT is a network of physical and digital devices that are interconnected to share data (Habibzadeh et.al, 2019). These devices can include everything from cars to factories to homes. The IoT can help us to improve our lives by making things like transportation easier, providing us with more accurate information, and helping us to save energy (Nguyen et.al, 2021).

The IoT is an emerging technology revolutionizing a variety of industries including retail, healthcare, and automotive. With the ability to connect physical "things" to the Internet, companies around the world are leveraging sensors and connected devices to automate processes, offer customers greater convenience, and reduce costs Hui et al, (2016). This introductory literature review seeks to explore how IoT can be used in modern businesses. Investments in projects relating to this technology are growing rapidly,

as businesses engage in new and innovative ways to stay competitive. The review will include researching regulations around the usage of IoT, their various applications, and their beneficial outcomes for businesses.

This paper will also explore the challenges that arise from incorporating IoTs into organizational operations and how businesses can mitigate these risks. The development of the Internet of Things has yielded a great deal of growth in the technology sector over the last few decades Boston et al, (2019). Companies ranging from small startups to large organizations have adopted ways to apply this technology to gain a competitive advantage. While the application of IoT has been limited in certain industries, the potential of IoT remains largely untapped, a potential thatbusinesses can use to gain insight into customer behavior, reduce operating costs, and facilitate more efficient operations.

The growth of the IoT has involved rapid advances in both hardware and software, leading to major changes in the way that companies interact with customers and other stakeholders. IoT, as a technology, has already made a profound impact on businesses in terms of cost savings, increased productivity, and better customer engagement Kumar et al, (2019). To further understand this impact, and to gain insights into how businesses can best leverage this technology, it is necessary to explore the regulations surrounding its usage, the various applications of IoT, the challenges associated with its implementation, and the potential solutions to these challenges. The literature review will begin by exploring the existing regulations around the use of IoT technology, particularly those established by governments, international organizations, and industry standards boards. It will then move on to research the diverse applications of IoT, particularly its use in various industries. It will also review the potential risks and benefits associated with the usage of the technology, including the potential for data privacy and security issues.

Additionally, the literature review will seek to identify the major initiatives that companies are undertaking when utilizing this new technology, to highlight the best practices and common pitfalls of the usage of IoT in businesses. This literature review will provide the grounds for further research on existing best practices and regulations around the use of IoT in modern businesses Stam, (2016). It is intended to provide an overview of the current landscape of this technology, as well as an introduction to the potential applications and implications for businesses. By reviewing current literature on the application and usage of IoT, the review will provide the foundation for academics and industry stakeholders to further explore the possibilities of successful and profitable IoT implementations in modern businesses.

2. Literature Review

The IoT encompasses a variety of technology advancements, such as sensors, RFID tags, and machine-to-machine communication (Pawełoszek., 2013; Pawełoszek., 2013a). It can be used to monitor and manage processes, maintain services, improve security, and facilitate data collection and analysis. It also enables businesses to remotely manage operations and optimize processes. By creating connections between people and processes, it can create greater efficiency and facilitate better decision-making (Scase and Haigh 2018).

In addition, businesses must also consider how the IoT can be leveraged to remain competitive in the digital age (Agarwal et al., 2020). This includes exploring opportunities to develop new products and services, optimize customer experiences, and increase efficiency in operations (Bhatia et al., 2020). Focused research and experimentation are key to successfully implementing the IoT in modern businesses (Zhang et al. 2018).

3. Research Methodology

This research study aims to investigate the rules and application of the Internet of Things (IoT) in modern businesses. The research methodology for this research study will involve both qualitative and quantitative approaches. To begin, a systematic literature review will be conducted on published journal articles, books, white papers, and other relevant sources. This will be an extensive process that will analyze past works and research on the topic to gain insight into the applicable rules and application of IoT in modern businesses. After this review, a survey questionnaire will be designed to obtain a further quantitative perspective on the same. The survey respondents will include business decision-makers and IT managers to provide industry-wide insight on the subject. Furthermore, multiple interviews with experts on the topic of IoT in businesses will be conducted to form the basis of an in-depth qualitative analysis. Lastly, a comparative case study between different businesses would be useful to contrast the different levels of IoT use in various sectors. The research methodology outlined should allow for a comprehensive analysis of the research topic, cover both qualitative and quantitative angles of the same, and provide a structured approach to the investigation.

4. Results and Discussion

According to Fig. 40.1, 54% of businesses in America are deploying IoT technology, with a higher percentage of 67% in Europe. The figures are comparatively lower in the APAC region at 32%.Through the implementation of various

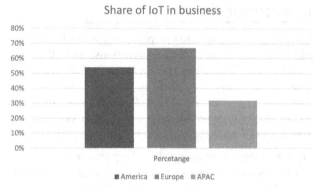

Fig. 40.1 Share of IoT in business

rules and applications, IoT has enabled businesses to streamline their processes and achieve innovative solutions with data-driven decisions. By applying these rules and applications to their business operations, companies can take advantage of the power of the IoT to stay competitive in today's market.

The potential benefits of IoT for businesses include:

1. *Improved Efficiency and Productivity:* IoT can automate manual processes and enable real-time insights into product performance, which can lead to increased efficiency and productivity.

2. *Remote Monitoring and Management of Assets:* IoT can enable remote monitoring and management of assets, such as vehicles, plants, and factories.

One of the key benefits of IoT is that it can help businesses reduce costs by automating processes and managing resources more efficiently. For example, a business could use IoT to automate the collection of data from sensors in its facilities, which would then allow it to better monitor and manage its resources. Additionally, IoT could be used to manage customer interactions and transactions, helping businesses to reduce costs associated with customer service and transactions. The IoT is a network of physical devices, vehicles, buildings, and other objects that are embedded with electronics and software to enable them to collect and share data. Figure 4.2 shows the use of machine learning in modern business.

The role of IoT in business during COVID-19 (Peter et.al, 2022):

The IoT technology helped during the COVID-19 pandemic by helping to track the virus and dispense the right amount of antiviral medication to those who needed it. The IoT helped during the COVID-19 pandemic by providing real-time information about the virus. By monitoring the virus, companies were able to adjust their safety protocols. Additionally, by tracking the location of patients, healthcare providers were able to provide better care.

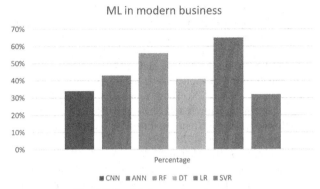

Fig. 40.2 ML in modern business

The IoT played an important role during the COVID-19 pandemic. By tracking the spread of the virus, IoT devices helped to identify and isolate infected individuals, and to provide real-time updates on the pandemic's progress.

In addition, IoT devices enabled the timely distribution of vaccines and medication to infected individuals and helped to track their health and progress. By monitoring vital signs and other data, IoT devices helped to identify potential health risks and to provide early warning signals of potential outbreaks.

Overall, the IoT played an important role in helping to prevent the spread of the COVID-19 pandemic. By tracking the virus's progress and by providing real-time updates, IoT devices helped to protect both individuals and the wider community.

IoT technology was used to help during the COVID-19 pandemic. By using sensors to monitor vital signs, IoT technology was able to help healthcare providers identify when patients were starting to experience symptoms and track their progress. This information was then used to provide treatment recommendations and help patients stay as healthy as possible.

IoT was used extensively during the COVID-19 pandemic. By using sensors to track the location and movement of patients, health officials were able to rapidly identify and isolate infected individuals. In addition, IoT enabled the collection and analysis of data from medical devices and other sensors to help diagnose and track the virus. By using this information, health officials were able to rapidly respond to outbreaks and provide care to those affected. The COVID-19 pandemic was a global outbreak of coronavirus that began on June 12, 2019, and ended on September 25, 2019. During the pandemic, IoT devices were used to help identify and track cases, monitor patient health, and provide critical information to public health officials.

One of the ways IoT devices were used during the COVID-19 pandemic was to help identify and track cases.

IoT devices were used to collect data from patients and their caregivers, which was then used to identify outbreaks and track the spread of the virus. This information was then used to provide public health officials with critical information about the pandemic.

Another way IoT devices were used during the COVID-19 pandemic was to monitor patient health. IoT devices were used to collect data about patients' conditions and symptoms. This data was then used to monitor the health of patients and provide public health officials with information about the progress of the pandemic.

Overall, IoT devices were used to help during the COVID-19 pandemic. IoT devices allowed public health officials to track the spread of the virus, monitor patient health, and provide critical information to the public.

5. CONCLUSION

The Internet of Things (IoT) is rapidly transforming the world of modern business. It has revolutionized how businesses operate and has set the standard for efficient, innovative, and profitable operations within organizations. Through the implementation of various rules and applications, IoT has enabled businesses to streamline their processes and achieve innovative solutions with data-driven decisions. By applying these rules and applications to their business operations, companies can take advantage of the power of the IoT to stay competitive in today's market. With the continuous evolution of IoT technology, businesses have the potential to reach even greater goals, making the IoT an essential tool for future success.It is clear from the results of the review that IoT has the potential to revolutionize business operations and offer innovative approaches to creating value for organizations. Companies must carefully consider their approach to IoT, as it can be instrumental in gaining a competitive advantage. While the implementation of IoT technology into a business structure does come with inherent risks and challenges, which must be managed, the potential benefits for organizations that are willing to integrate IoT solutions are significant.

References

1. Rose, K., Eldridge, S., & Chapin, L. (2015). The internet of things: An overview. *The Internet Society (ISOC)*, *80*, 1–50.
2. Li, S., Xu, L. D., & Zhao, S. (2015). The internet of things: a survey. *Information Systems Frontiers*, *17*, 243–259.
3. Laghari, A. A., Wu, K., Laghari, R. A., Ali, M., & Khan, A. A. (2021). A review and state of the art of the Internet of Things (IoT). *Archives of Computational Methods in Engineering*, 1–19.
4. Villamil, S., Hernández, C., & Tarazona, G. (2020). An overview of the Internet of Things. *Telkomnika (Telecom-*

munication Computing Electronics and Control), 18(5), 2320–2327.

5. Nord, J. H., Koohang, A., &Paliszkiewicz, J. (2019). The Internet of Things: Review and theoretical framework. *Expert Systems with Applications, 133*, 97–108.

6. Habibzadeh, H., Dinesh, K., Shishvan, O. R., Boggio-Dandry, A., Sharma, G., &Soyata, T. (2019). A survey of healthcare Internet of Things (IoT): A clinical perspective. *IEEE Internet of Things Journal, 7*(1), 53–71.

7. Nguyen, D. C., Ding, M., Pathirana, P. N., Seneviratne, A., Li, J., Niyato, D., ... & Poor, H. V. (2021). 6G Internet of Things: A comprehensive survey. *IEEE Internet of Things Journal, 9*(1), 359–383.

8. Hui, Y., Kranenburg, R. and Peters, S. (2016). Internet of Things: Enabling Technologies, Platforms, and Use Cases. In Encyclopedia of Database Systems, pp. 2508–2514. Springer,

9. Boston, MA. Katancik, R. and Clark, T. (2019). The Internet of Things: A Primer for Business Leaders. Journal of Business & Economics Research, 17(2), pp. 117–121.

10. Kumar, N., Sahu, P., Chowdhury, S. and Himani, S. (2019). Internet of Things: A Revolution in the Technology-Driven Businesses. International Journal of Pharmaceutical and Clinical Research, 11(8), pp. 85–91.

11. Stam, F. (2017). The Uniqueness of the Internet of Things and Its Impact on Business. International Journal of Research in Business Studies and Management, 4(1), pp. 19–25.

12. Scase, R. and Haigh, R. (2018). 'The Internet of Things, Machine-to-machine Communication and Smart Objects.' Annals of Information Systems, 9, pp. 63–84.

13. Zhang, C., Zhang, X., Peng, Y. and Zhang, J. 2018. 'From the Internet of Things to Business Intelligence: A Literature Review of Applications and Enabling Technologies.' IEEE Access, vol. 6, pp. 4672–4693.

14. Manpreet Singh Bhatia, Alok Aggarwal, Narendra Kumar. (2020). Smart Traffic Light System to Control Traffic Congestion. PalArch's Journal of Archaeology of Egypt / Egyptology, 17(9), 7093–7109. Retrieved from https://archives.palarch.nl/index.php/jae/article/view/5389 (Original work published December 30, 2020)

15. Peter, O., Swain, S., Muduli, K., & Ramasamy, A. (2022). IoT in combating COVID-19 pandemics: lessons for developing countries. *Assessing COVID-19 and other pandemics and epidemics using computational modeling and data analysis*, 113–131.

Note: All the figures in this chapter were made by the author.

Recent Trends in Engineering and Science for Resource Optimization and
Sustainable Development – Prof. (Dr.) Dorota Jelonek et al. (eds)
© 2024 Taylor & Francis Group, London, ISBN 978-1-032-98030-0

41

Monitoring the Development of the IoT Concept in Various Application Domains

A. B. Mishra[1]

Associate Professor,
International Institute of Management Studies,
Pune, Maharashtra

Sarita Rani Panda[2]

Lecturer in ECE for Diploma

G. Anitha

Department of Computer Science and Engineering,
Institute of Aeronautical Engineering, Hyderabad, Telangana

Uma Reddy[3]

Department of Artificial Intelligence and Machine Learning,
New Horizon College of Engineering, Bangalore, India

Gaurav Sethi

Lovely Professional University, Phagwara, India

Purnendu Bikash Acharjee[4]

Associate Professor, Computer Science, CHRIST University, Bangaluru, India

▬▬▬ Abstract

For several decades, the concept and technology of combining actuators and sensors into a system to monitor and operate tangible structures distantly was understood and developed. Nevertheless, slightly over a decade back, the notion of the Internet of Things (IoT) emerged and was utilized to merge such techniques into a prevalent architecture. The study outlines and addresses IoT conceptual structures suggested as part of continuing standardization attempts, layout problems regarding IoT hardware and software parts, and delegates of IoT application domains like healthcare, smart cities, the farming industry, and nano-scale uses. The research verifies the argument that an agreement on the precise scope of the IoTs will likely be formed, as enabling innovation evolves and novel application domains have been presented. Current modifications, nevertheless, are a bit muted, and their variants on application domains have been distinct, with statistics and information technologies serving a significant part in the IoT environment.

▬▬▬ Keywords

Internet of things, Application domains, IoT applications, and Conceptual structures

Corresponding authors: [1]amishra.iims@gmail.com, [2]vesankri@gmail.com, [3]nvumareddy409@gmail.com, [4]pbacharyaa@gmail.com

DOI: 10.1201/9781003596721-41

1. INTRODUCTION

In recent times, the Internet of Things (IoT) has been at the forefront. It is considered to be among the century's disruptive innovations [1] and, thus far, has piqued the interest of the public, the private sector, and academia as a means of technically improving tasks, novel company models, goods, and offerings, and as a rich source of study subjects and concepts. Nevertheless, getting big data from IoT-based OWSNs with high precision is difficult because of shifting items, impediments, line-of-sight, and non-line-of-sight challenges in an electronics-producing 4.0 system. Numerous organizations, institutions, businesses, and even nations have recognized its significance and the possible advantages that may be acquired from the IoT, prompting them to embark on strategic initiatives and activities aimed at developing and profiting from this subject [2]. Hence this study aims to evaluate and monitor the development of IoT concepts in various application domains.

2. LITERATURE REVIEW

The study [3] investigated important IoT problems and QoS requirements, including accessibility and dependability, movement, performance and administration, scaling and interoperability, and safety and confidentiality.

The study [4] identified significant IIoT problems as tolerating faults, operational security, data delay and scaling, mixed importance, and scalable and safe real-time cooperation. According to [5], the primary obstacles in IoT development encountered by organizations are data management and mining, safety, and confidentiality. [6] concentrated on the safety and confidentiality problems in IIoT, as well as its susceptibility to various cyberattacks.

3. METHODOLOGY

The development of the IoT conceptual framework is critical for promoting worldwide IoT adoption and the development of IoT applications and services [7]. Recently, a variety of architectural proposals has been presented, encompassing not just globally recognized organizations [8], but additionally research and educational institutes, businesses [9], key players, and civic groups [10]. Commercial perception additionally serves as a critical component in IoT implementation. The IoT protocol layer's fundamental needs encompass low-power interaction protocols as well as an IP-oriented, safe, and dependable communication layer. Standardized organizations are tackling challenges such as protocol layer interoperability and the openness of open protocols, interactions, and designs for IoT. This endeavor will result in the development of an IoT-specific protocol layer.

In the study, the transmitting power was set to 0 dBm. Every Open-mote in a TSCH (Time-Slotted Channel Hopping) network is synchronized, and the duration is separated into duration. Each open mote adheres to a communication plan, which constitutes a cell matrix. The number of cells needed is determined by the IoT application, and a trade-off between delay, dependability, and energy consumption is necessary. The number of nodes delivering packages at the same duration is three, five, and seven, accordingly. We carried out an extensive testing operation. 1000 packages have been produced for each assessment. The content is 20 bytes in size. We tried three alternative inter-package periods: 1000, 1500, and 3000 ms.

4. RESULT AND DISCUSSION

Because of space constraints, we provide the outcomes in the context of Package Loss Rate (PLR) for five distinct testing situations, as shown in Tables 41.1–41.5. The outcomes reveal that, regardless of the presence of disturbance and a dynamic setting with moving individuals, PLR remains at a suitable level. PLR is additionally influenced by the number of motes sent at the same period; it is greatest when all seven motes have been transmitting packages to the root (Table 41.3).

Table 41.1 Outcomes of PLR (Node = 3 and inter package duration= 3000ms)

Node	Packages received	Packages scented	Packages duplicate	Regular packages	PLR (%)
4F	1071	1000	95	976	2.4
31	995	1000	15	980	2
6F	1067	1000	72	995	0.5

Fig. 41.1 Outcomes of PLR (Node = 3 and inter package duration= 3000ms)

Table 41.2 Outcomes of PLR (Node = 5 and inter package duration= 3000ms)

Node	Packages received	Packages scented	Packages duplicate	Regular packages	PLR (%)
4F	853	1000	57	786	20.4
31	1001	1000	13	988	1.2
6F	1100	1000	107	993	0.7
B4	1028	1000	51	977	2.3
54	1033	1000	62	971	2.9

PLR (%)

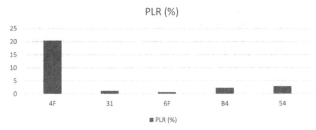

Fig. 41.2 Outcomes of PLR (Node= 5 and inter package duration = 3000 ms)

Table 41.3 Outcomes of PLR (Node = 7 and inter package duration = 3000ms)

Node	Packages received	Packages scented	Packages duplicate	Regular packages	PLR (%)
4F	894	1000	62	832	16.8
31	414	1000	8	406	59.4
6F	1080	1000	92	998	1.2
B4	1046	1000	53	993	0.7
54	1066	1000	74	992	0.8
E3	699	1000	0	699	30.1
5A	776	1000	58	718	28.2

PLR (%)

Fig. 41.3 Outcomes of PLR (Node = 7 and inter package duration = 3000ms)

Table 41.4 Outcomes of PLR (Node = 3 and inter package duration = 1500ms)

Node	Packages received	Packages scented	Packages duplicate	Regular packages	PLR (%)
4F	882	1000	48	834	16.6
31	972	1000	16	956	4.4
6F	1040	1000	75	965	3.5

PLR (%)

Fig. 41.4 Outcomes of PLR (Node = 3 and inter package duration = 1500ms)

Table 41.5 Outcomes of PLR (Node = 5 and inter package duration = 1500ms)

Node	Packages received	Packages scented	Packages duplicate	Regular packages	PLR (%)
4F	955	1000	50	905	9.5
31	741	1000	7	734	26.6
6F	1034	1000	62	972	2.8
B4	1003	1000	48	955	4.5
54	984	1000	50	934	6.6

PLR (%)

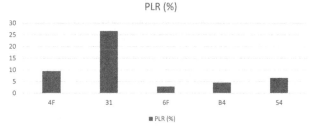

Fig. 41.5 Outcomes of PLR (Node = 5 and inter package duration = 1500 ms)

5. CONCLUSION

IoT applications have been many, various, and deeply embedded in our daily lives and actions. This work provides a discussion of current standardization initiatives, IoT software, and hardware layout problems, and tackles the needs of important IoT application domains for potential IoT situations. We report the outcomes of monitoring the Open-Mote hardware platform's functionality for commercial IoT applications, including the initial findings achieved by the development of 2 standards for potential Commercial IoT applications.

Reference

1. Hassan, Q. F., & Madani, S. A. (Eds.). (2017). Internet of things: Challenges, advances, and applications.
2. Shahid, N., & Aneja, S. (2017, February). Internet of Things: Vision, application areas and research challenges. In *2017 International Conference on I-SMAC (IoT in Social, Mobile, Analytics, and Cloud) (I-SMAC)* (pp. 583–587). IEEE.
3. Al-Fuqaha, A., Guizani, M., Mohammadi, M., Aledhari, M., & Ayyash, M. (2015). Internet of Things: A survey on enabling technologies, protocols, and applications. *IEEE communications surveys & tutorials*, *17*(4), 2347–2376.
4. Breivold, H. P., & Sandström, K. (2015, December). Internet of things for industrial automation--challenges and technical solutions. In *2015 IEEE International Conference on Data Science and Data Intensive Systems* (pp. 532–539). IEEE.

5. Lee, I., & Lee, K. (2015). The Internet of Things (IoT): Applications, investments, and challenges for enterprises. *Business Horizons, 58*(4), 431–440.

6. Sadeghi, A. R., Wachsmann, C., & Waidner, M. (2015, June). Security and privacy challenges in industrial Internet of Things. In *Proceedings of the 52nd annual design automation conference* (pp. 1–6).

7. Manpreet Singh Bhatia, Alok Aggarwal, Narendra Kumar. (2020). Speech-to-text conversion using GRU and one hot vector encoding. PalArch's Journal of Archaeology of Egypt / Egyptology, 17(10), 7110–7119. Retrieved from https://archives.palarch.nl/index.php/jae/article/view/5391

Note: All the figures and tables in this chapter were made by the author.

Recent Trends in Engineering and Science for Resource Optimization and
Sustainable Development – Prof. (Dr.) Dorota Jelonek et al. (eds)
© 2024 Taylor & Francis Group, London, ISBN 978-1-032-98030-0

42

Using IOT and ITV to Connect, Monitor, and Automate Common Areas Used by Households

Shailendra Kumar Rai[1]

Research Scholar, School of Commerce and Management,
IIMT University Meerut Uttar Pradesh, India

Kush Bhushanwar[2]

Assistant Professor, Information Technology,
Medicaps University, Indore

M. Hari Krishna

Department of Computer Science and Engineering,
Institute of Aeronautical Engineering, Hyderabad, Telangana

Kavi Bhushan[3]

Assistant Professor, Department of Computer science,
Sir Chhotu Ram Institute of Engg. & Technology,
CCS University Meerut (U.P), India

Gaurav Sethi

Lovely Professional University, Phagwara, India

Patel Yogesh kumar Jethabhai[4]

Assistant Professor, Faculty of Computer Science,
Shri C.J. Patel College of Computer Studies, Sankalchand Patel University

■■■■■■ **Abstract**

The Internet of Things (IoT) seeks to embed detectors and actuators in everyday things, maximizing miniaturization and lowering the financial expense of these hardware parts. The objective is to connect these parts to the web via wireless and fixed connections, producing real-time data that can be saved for subsequent analysis. Interactive TV (iTV), on the contrary, mixes regular television with interaction comparable to that of the web and personal computers. This paper describes an architecture that comprises and combines a wireless sensor network, an IoT system, and a real-world interactive television application. It was monitored in a household area to offer real-time data to enhance the residents' standard of existence. Furthermore, it includes the capability of analyzing this data to build procedures to minimize the usage of energy, hence promoting sustainability and assisting in the effective application of current assets. The suggested architecture can be used as the foundation for any deployment with comparable features.

Corresponding authors: [1]rais6316@gmail.com, [2]kush.bhushanwar@gmail.com, [3]kavybhushan@gmail.com, [4]yjpatel.fcs@spu.ac.in

DOI: 10.1201/9781003596721-42

■■■■■■■ **Keywords**

Internet of things, Interactive TV, Wireless sensor networks, Smart city, and Households

1. INTRODUCTION

As per research conducted by the UN's Population Trends Observation [1,] over 60 million individuals migrate to metropolitan regions every year, indicating approximately a million per week. Metropolis will house seventy percent of the global populace by 2050. This is an issue that will necessitate the transformation of towns to make them habitable and environmentally friendly, while also taking into account the vital function of information technology (IT) and communication in turning modern towns into "smart cities" [2].

A Smart City is defined as an urban area that uses IT and communication in an integrated and effective manner. This difficulty intends to create ideas for its physical structures that will allow users to communicate with a variety of city features.This is an electronic structure on which a complicated network of numerous drivers is created (the ones that receive data are referred to as detectors, and those that behave in response to incidents are identified as actuators), controlling a large amount of data and permitting you to offer certain services and data to citizens more effectively. The ultimate goals are to boost resource efficiency, improve the standard of service, uncover new requirements, deliver real-time data to citizens, and establish an environmentally friendly route for socioeconomic growth.

The application of information technologies to address urban issues may be traced to an intense desire for long-term viability a societal need for current data, and the development of novel Internet-based innovations like smartphones, the semantic web, cloud computation, or the Internet of Things (IoT) [1]. The most recent architectural patterns in Smart Cities are monitored on bioclimatic urbanization. The households in this area depend on a smart layout that enables for seamless utilization of natural energies while preventing traditional energy use, with associated savings, and is built using bioclimatic standards, with an emphasis on producing an environmentally friendly ventilation and heating structure for the building [2]. The data gathered is delivered to a centralized machine that is accountable for computing the outcomes, allowing for continuous monitoring of the function and data verification for the specified duration [3]. Hence this research aims to investigate IoT and interactive television (ITV) to connect, monitor, and automate households' common areas.

2. LITERATURE REVIEW

An IoT system creates a large amount of data that should be saved and analyzed; yet, these gadgets' storing and processing capacities are frequently constrained. As a result, the capacity of IoT gadgets to communicate data independently via conventional web protocols allows the storing of such information in outside storage platforms that offer durability and analysis skills. These characteristics are regarded as intermediary layers or middleware in IoT design, which is a software layer between tangible gadgets and software that allows the programmer to choose the specifics for different fundamental innovations and contributes to the quick prototyping of service-based uses. The apps that utilize data preserved in the intermediary layer would be formed in the 3rd tier. Furthermore, interaction with electronic gadgets can be bidirectional, which means that specific details given to these gadgets may trigger them to activate if certain criteria are met.

The setting up of monitors and their internet accessibility was effectively resolved in this effort, despite certain hurdles [4]. On the contrary, iTV (which symbolizes the progression of conventional television) provides an innovative platform for the development of applications for consumers, utilities, and activities. In this situation, we offer a new way to regulate households from the convenience of your house chair. As is the case with us, iTV has the issue of locating specifications that enable the additional potential it can provide [5] as well as instructions and resources for the creation of novel applications.

3. METHODOLOGY

The technologies presented in this work are proposed as the foundation for the construction of an IoT system that enables real-time monitoring of shared spaces in a group of nearby residents, with iTV[6] serving as an agent of contact. This design has been suggested to have three distinct layers which is shown in Fig. 42.1. First, a physical layer is composed of actuators [7], detectors [8], and control components equipped with interaction devices [9]. Secondly, there is a logical layer made up of storage devices and computer applications for analyzing information, and third, there is an illustration or application level that enables data visualization and comprehension across multiple systems.This study has taken into consideration the following

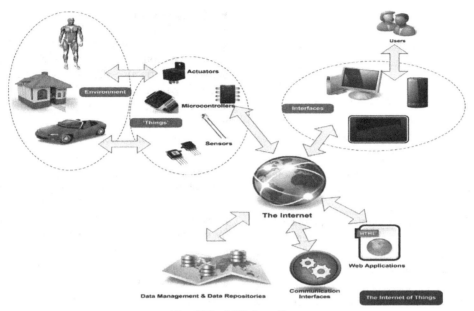

Fig. 42.1 IoT interactions

factors in the layout of the suggested architecture, adhering to the guidelines:

1. *The incorporation of a diverse infrastructure:* To provide the operation of the system in a typical manner, the diversity of the gadgets and systems employed must be homogenized.

2. *IoT-based infrastructure:* a wide range of gadgets must be incorporated into the framework to enable consistent entry to and from it. To assist the creation of applications, it is also required to openly disclose all possible possibilities of identification or performance using well-defined endpoints.

3. *Services Cloud:* By using cloud computing's capacities, application creators must be able to access instruments and operations that allow them to set up apps using pre-existing software artifacts as much as feasible and create novel features out of pre-existing ones.

4. RESULT AND DISCUSSION

Thing-Speak, a cloud IoT framework for data storage and processing, was used to build the logical tier. Smart-Residents' middle tier is made up of 2 channels, which include various areas where the information produced by the detectors is assigned. Table 42.1 shows the relationship between channels and areas.

Smart-residents interface for a given detector (Temperature) presented in Table 42.2 and Fig. 42.2, via vertical selections from panels and icons on the map, whereby the program gives additional statistical information on each of

Table 42.1 The relationship between channels and areas

Channel	Area	Detector
CH1_Smart-Resident	1	Temperature
	2	Air standard
	3	Rain
	4	Humidity
	5	Mobility
	6	Temperature of water
CH2_Smart-Resident	1	Container tier of rubbish
	2	Container cap of rubbish
	3	Consumption of electricity
	4	Noise
	5	Smoke
	6	Pressure of atmosphere

Table 42.2 Smart-residents interface for a given detector (Temperature)

Date	Temperature (°C)
03:00	26
06:00	23
09:00	25

the observations. Filter settings like the variety of dates and duration scales can be specified via the form. The graphic depicts a temperature progression graph with a duration range of 10 minutes.

Smart-Residents' basic offering to its consumers is the provision of current information in various forms that may

Temperature (°C)

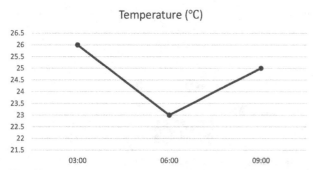

Fig. 42.2 Smart-residents interface for a given detector (Temperature)

be checked via TV as the main interface component. The user understands in a moment if the paddle tennis field has been utilized at any particular duration, the temperature of the pool or the air outside, if the trash in a specific region is complete, or the disturbances rates in a region, which certainly increases the standard of life for residents. Furthermore, modifications to all of these aspects can be evaluated by employing other periods or dates, enabling a more detailed examination of a specified end objective. Table 42.3 and Fig. 42.3 depict the variations in temperature for a typical day from 08:00-18:00 H in 10-minute periods, utilizing graphics to demonstrate the advantages of cloud computation. Smart-Residents incorporates this feature within the application.

Table 42.3 Outcomes of temperature alteration

Date	Temperature (°C)
09:00	6
12:00	13
15:00	16
18:00	14

Temperature (°C)

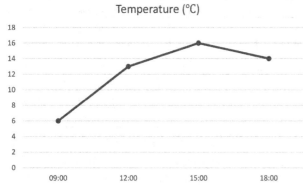

Fig. 42.3 Outcomes of temperature alteration

5. CONCLUSION

This serves as the foundation for the development of complicated environments in which various agents connect to deliver more effective offerings, typically inside Smart Cities. TV, on the contrary, has developed into ITV, with tools suitable for operating applications and connected to the web, enabling viewers to experience further amenities beyond the material offered by traditional TV, which depends on one-way interaction. Smart-Residents offers a structure for monitoring and automated processes in households by integrating frameworks resulting in the Future web: on the contrary, IoTs, a framework for the gathering, transfer, preservation, and computation of information produced by detectors; and, on the other hand, ITV, which enables access to gadget data across uses, with simple user connections tailored to the gadget's specifics. TVs employed as access devices enable fresh offerings to be accessed via an appliance that has historically been utilized for TV show usage, but only while using its computational potential. As a result, TV constitutes an increasingly more engaged part of the household, an instrument through which we connect to our surroundings. The suggested Smart-Residents layout is expandable and adaptable, as it is feasible to alter the scale or layout of the physical layer to meet novel needs without impacting the level of service.

Reference

1. Singh, K. J., & Kapoor, D. S. (2017). Create your Internet of things: A survey of IoT platforms. *IEEE Consumer Electronics Magazine*, 6(2), 57–68.
2. Zhou, W., Jia, Y., Yao, Y., Zhu, L., Guan, L., Mao, Y., ... & Zhang, Y. (2018). Discovering and understanding the security hazards in the interactions between IoT devices, mobile apps, and clouds on smart home platforms. *arXiv preprint arXiv:1811.03241*.
3. Doan, T. T., Safavi-Naini, R., Li, S., Avizheh, S., & Fong, P. W. (2018, August). Towards a resilient smart home. In *Proceedings of the 2018 workshop on IoT security and privacy* (pp. 15–21).
4. Chen, N., Xiao, C., Pu, F., Wang, X., Wang, C., Wang, Z., & Gong, J. (2015). Cyber-physical geographical information service-enabled control of diverse in-situ sensors. *Sensors*, 15(2), 2565–2592.
5. Beyer, S. M., Mullins, B. E., Graham, S. R., & Bindewald, J. M. (2018). Pattern-of-life modeling in smart homes. *IEEE Internet of Things Journal*, 5(6), 5317–5325.
6. K. M. Sahu, Soni, N. Kumar and A. Aggarwal, "σ-Convergence of Fourier series & its Conjugate series," 2022 5th International Conference on Multimedia, Signal Processing and Communication Technologies (IMPACT), Aligarh, India, 2022, pp. 1–6, doi: 10.1109/IMPACT55510.2022.10029267.

Note: All the figures and tables in this chapter were made by the author.

Recent Trends in Engineering and Science for Resource Optimization and Sustainable Development – Prof. (Dr.) Dorota Jelonek et al. (eds)
© 2024 Taylor & Francis Group, London, ISBN 978-1-032-98030-0

43

Concerns in IoT Environments: Adoption, Architecture, and Innovation of Enterprise IoT Systems

Tushar A. Champaneria[1]

Research Scholar, U & P U. Patel Department of Computer Engineering,
CHARUSAT University

Prof. Kamlesh Vasantrao Patil[2]

Assistant Professor, Department of Information Technology,
Bharati Vidyapeeth College of Engineering for Women Pune 43, Maharashtra

Sunkara Yamini

Department of Computer Science and Engineering,
Institute of Aeronautical Engineering, Hyderabad, Telangana

Uma Reddy[3]

Department of Artificial Intelligence and Machine Learning,
New Horizon College of Engineering, Bangalore

Lavish Kansal

Lovely Professional University, Phagwara, India

Purnendu Bikash Acharjee[4]

Associate Professor, Computer Science, CHRIST University, Bangaluru, India

Abstract

The Internet of Things (IoT) has received a lot of interest in recent times. IoT depicts the upcoming internet and is defined as an environment of linked gadgets, computational processes, and other items that collaborate to transmit information or data with greater ease and economic advantages. Nevertheless, because of the presence of numerous concerns, IoT adoption, architecture, and innovation continue concerns. As a result, the purpose of this study was to identify and analyze the concerns in the adoption, architecture, and innovation of IoT systems in construction enterprises in the Indian environment. The research analysis and professional comments have been employed to identify the barriers to IoT adoption, architecture, and innovation. This research may assist professionals and policymakers in addressing barriers to successful IoT adoption and spread. At last, findings and potential research possibilities are provided.

Keywords

Internet of things, Environment, Smart system

Corresponding authors: [1]tchampaneria@acm.org, [2]kamlesh.patil@bharatividyapeeth.edu, [3]nvumareddy409@gmail.com, [4]pbacharyaa@gmail.com

DOI: 10.1201/9781003596721-43

1. INTRODUCTION

The Internet of Things (IoT) is a novel innovation architecture intended as a worldwide network of interconnected equipment and objects. The IoT is recognized to be among the most significant fields of future innovation and is garnering widespread interest from a variety of businesses [1]. When connected gadgets can interact with one another and combine with inventory management systems, assistance systems, business information tools, and business analytics, the full potential of IoT for organizations may be realized [2].

The adoption of novel innovations is fraught with concerns, which can be divided into three categories: the process of adoption, the absence of adoption, and the absence of knowledge and experience [3]. This research intends to examine construction enterprises' awareness of IoT adoption and relevance; it then analyses concerns about adopting IoT in construction enterprises', and ultimately determines the main concerns of adopting it in the construction enterprises.

2. LITERATURE REVIEW

IoT has been widely adopted in a variety of enterprises, including consumers, commerce, and infrastructure [4]. Due to the complicated nature of construction enterprises and the substantial likelihood of failure, it can be hard to adapt and adopt novel innovations in construction enterprises. Despite these concerns, IoT has been applied in construction enterprises. It has also been utilized for tracking construction efficiency during catastrophes, as well as providing real-time security alerts and risk identification. [5] studied an extensive variety of IoT uses in buildings, including smart cities [6], housing [7], and smart transportation architecture [8].

3. METHODOLOGY

An IoT analytical research was conducted to determine the concerns of adopting IoT in construction enterprises. The concerns were extracted by examining present patterns connected to IoT and determining the concerns; also, preventing the recurrence of typical phrases by conceptually picking a standard term that describes the concerns. The questionnaire is divided into three sections, the initial of which emphasizes the demographic information of the respondents in terms of firm type, present designation, level of education, and years of expertise in the enterprises. While the subsequent half emphasizes respondents' understanding and consciousness of IoT, the final piece emphasizes the concerns [9]. A Likert scale is used to evaluate the concerns highlighted in the research. During this stage, information is gathered by contacting 132 construction enterprise individuals to complete a questionnaire survey. The process of random sampling is frequently employed in construction investigation, in which a sample is drawn at random from a demographic with a non-zero likelihood.

4. RESULT AND DISCUSSION

The demographic profile of the research is shown in Table 43.1 and Fig. 43.1–43.4.

Table 43.1 Demographic profile

Parameters	Percentage (%)
Categories of enterprises	
Private	72
Governmental	23
Others	5
Respondents' designation	
Civil engineer	31
Project manager	21
Planning Engineer	7
QA/QC engineer	8
Architect	17
Company director	12
Others	4
Respondents' qualification	
PhD	2
Masters	4
Bachelors	71
Diploma	21
Expertise years in construction enterprises	
0-10 yrs.	42
11-20 yrs.	31
21-30 yrs.	23
31 and more	5

■ Private ■ Governmental ■ Others

Fig. 43.1 Company categories

Table 43.2 and Fig. 43.5 depict the concerns about IoT adoption in the construction enterprise, as well as their

Fig. 43.2 Designation of participants

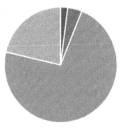

Fig. 43.3 Respondents' qualification

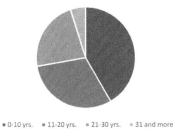

Fig. 43.4 Expertise years in construction enterprises

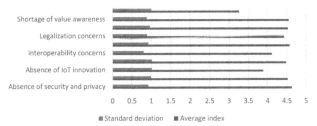

Fig. 43.5 Evaluation of concerns of IoT environment in the adoption of IoT systems in construction enterprises

Table 43.2 Evaluation of concerns of IoT environment in the adoption of IoT systems in construction enterprises

Classification	Concerns about the adoption of IoT in construction enterprises	Average index	Standard deviation
Innovation	Absence of security and privacy	4.63	0.93
	Absence of robustness in networking	4.52	0.99
	Absence of IoT innovation	3.89	1.02
	Big data Concerns	4.48	1.02
	Interoperability concerns	4.11	0.810
Administrative and legalization	Absence of written guidelines	4.57	0.93
	Legalization concerns	4.43	0.90
Awareness	Poor IoT adoption	4.53	0.97
	Shortage of value awareness	4.55	0.89
	Extra funds are required to obtain IoT innovations.	3.26	1.01

level of security in which instruments are vulnerable to attacks daily. While the absence of written guidelines is the second most significant concern, with an AI of 4.57, this is because it is a novel innovation that calls for more policies and standardizations. The third dominating concern, with an AI of 4.55, is a shortage of value awareness, while the 4th dominant concern, with an AI of 4.53, is poor IoT adoption, which refers to the incorrect selection of techniques utilized to adopt IoT in the enterprise. With an AI of 4.52, the fifth most prominent concern is an absence of robustness in networking.

5. CONCLUSION

As a result, there are numerous positive effects to using IoT in construction enterprises, and these perks can be connected to enhancing enterprises' effectiveness, security, and performance, and all of this is accomplished by transmitting an effective means of gathering data and knowledge in real-time. The research investigated the level of awareness towards a shift and adoption of IoT systems in construction enterprises and also recognized and evaluated the concerns of using IoT in construction. The analytical investigation identified issues that were divided into four categories. The research discovered the most prevalent concerns to the adoption of IoT in construction enterprises,

appraisal depending on an examination of data acquired via a questionnaire survey. According to the findings, the most prevalent concern of IoT adoption is an absence of security and privacy, with a total score of 4.63. As a result, this conclusion is warranted, which views protecting internet-connected gadgets to be many concerns to keep an elevated which are an absence of security and privacy, a shortage of defined norms, a shortage of value awareness, the incorrect adoption of IoT, and a shortage of resilience in connection. Additional studies on drafting standardization and network safety are recommended by the investigation to build more successful items.

Reference

1. Riggins, F. J., & Wamba, S. F. (2015, January). Research directions on the adoption, usage, and impact of the Internet of Things through the use of big data analytics. In *2015 48th Hawaii international conference on System Sciences* (pp. 1531–1540). IEEE.

2. Mahmud, S. H., Assan, L., & Islam, R. (2018). Potentials of Internet of Things (IoT) in Malaysian construction industry. *Annals of Emerging Technologies in Computing (AETiC), Print ISSN*, 2516–0281.

3. Zhong, R. Y., Peng, Y., Xue, F., Fang, J., Zou, W., Luo, H., ... & Huang, G. Q. (2017). Prefabricated construction enabled by the Internet of Things. *Automation in Construction, 76*, 59–70.

4. Perera, C., Liu, C. H., & Jayawardena, S. (2015). The emerging Internet of things marketplace from an industrial perspective: A survey. *IEEE Transactions on Emerging Topics in Computing, 3*(4), 585–598.

5. Chandanshive, V. B., & Kazi, A. M. (2017, March). Application of Internet of Things in Civil Engineering Construction Projects State of the Art. In *Proceedings of the 11th INDIACom, 4th International Conference on Computing for Sustainable Global Development* (Vol. 4, pp. 1836–1839).

6. Soni, N. Kumar, V. Kumar and A. Aggarwal, "Biorthogonality Collection of Finite System of Functions in Multiresolution Analysis on L2(K)," 2022 10th International Conference on Reliability, Infocom Technologies and Optimization (Trends and Future Directions) (ICRITO), Noida, India, 2022, pp. 1–5, doi: 10.1109/ICRITO56286.2022.9964791.

Note: All the figures and tables in this chapter were made by the author.

*Recent Trends in Engineering and Science for Resource Optimization and
Sustainable Development – Prof. (Dr.) Dorota Jelonek et al. (eds)
© 2024 Taylor & Francis Group, London, ISBN 978-1-032-98030-0*

44

Future Battlefield System Using Graph Database and Internet of Things (IoT)

Chetan shelke[1]

Associate Professor, Alliance University Banglore

Harish Chowdhary[2]

School of doctoral Studies and research,
National Forensic Sciences University,
Sector 9, Gandhinagar, Gujarat

Priyanka Gupta

Department of Computer Science and Engineering,
Institute of Aeronautical Engineering,
Hyderabad, Telangana

Swathi B[3]

Department of Information Science Engineering,
New Horizon College of Engineering, Bangalore

Lavish Kansal

Lovely Professional University, Phagwara, India 144433

Purnendu Bikash Acharjee[4]

Associate Professor, CHRIST University, Pune, India

Abstract

The Internet of Things (IoT) concept is rapidly evolving and is expected to influence each field of the computational realm. These advances have an impact on any nation's defence force. The defense industry's solution mostly depends on detectors and their installations. The major goal of sensory statistics is to provide information that may be used for strategic choices and evaluation in future battling fields. Each piece of statistics, from documenting a soldier's essential health metrics to its ammunition, weapons, and position circumstance, has a function and is especially important to the strategic commander stationed in the control unit. This research proposes an innovative approach that combines the IoTs with the growing graph database to produce a contextual consciousness regarding each characteristic of the personnel on the battlefield. We show a projected future battlefield application condition in which we explore the graph database for contextual consciousness patterns to gain a strategic benefit over our competitors.

Keywords

Internet of things, Future battlefields, Detectors, Graph database, and Contextual consciousness

Corresponding authors: [1]Chetan.shelke@alliance.edu.in, [2]harish.phdcs21@nfsu.ac.in, [3]baswarajuswati@gmail.com, [4]pbacharyaa@gmail.com

DOI: 10.1201/9781003596721-44

1. INTRODUCTION

The term "Internet of Things" (IoT) describes a weakly connected platform made up of several uniform and diverse gadgets with networking, processing, and detecting features [1].

These gadgets can only do so much computation. As intelligent habits become more popular, there has been a sharp increase in the number of smart gadgets available. The globe is getting closer to being linked. There are numerous instances, including smart transit networks' smart patrolling technologies and hospital-based smart healthcare surveillance networks, among many others. IoT uses span a broad spectrum of sectors, such as production, smart transit, wireless internet, automated knowledge labor, farming procedure surveillance, medical surveillance, and more. The ability to link and compute data is greatly enhanced by the power of microcomputers and sensing statistics. When sensor-oriented IoT items are involved, interaction and sharing of information become essential for any structure. Sharing data with contextual emphases is possible [2]. Millions of outdated gadgets could become electronic gadgets thanks to the widespread characteristics of sensor-based system gadgets in the IoTs. A theoretical and interacting representation of detectors, gadgets, routes, structures, and data interaction with clouds is shown in Fig. 44.1. This study recognizes that information can be gathered in combat and stored on cloud storage [3]. Essentially, the research's processes start after the cloud data is collected. Analyzing this data can provide valuable insights into the evolving nature of the battlefield. The utilization of modern information, such as graphical records, can enhance the portrayal of the commander's ability to gain a comprehensive understanding of the situation [4]. For the commander to constantly be conscious of the circumstances that could arise, operational knowledge must be connected to his mentality.

Fig. 44.1 IoT concepts

The strategic control of future battlefield situations is a difficult challenge to solve. There constitutes a mission set for the assaulting troops in a situation like this. In a battle situation, the current military establishment and other regulating units encounter vast numbers of gadgets such as fighter planes, aircraft, commercial aircraft, etc [5]. Each of these parties desires the greatest possible share of the available airspace. In the event of fights, the scenario deteriorates when allied troops fire artillery and shells at the opposing side [6].

This technology can alert battlefield structures of the status of their defences on the battlefield. The database core is built on the cloud and stores current information from flight schedules, flying resources, and other both dynamic and static firing troops. Therefore, an innovative information technology (IT) conceptual change is taking place in methods that combine the cloud and the IoT [7] to make strategic future battlefield technology successful. Hence this research aims to investigate future battlefields using graph databases and IoT.

2. LITERATURE REVIEW

A Strategic Battlefield Area (SBA) is a strategic area in which several elements influence the direction of the war and battlefield objects. Weapons and warriors are valuable assets on the battlefield, where the net of war is woven. Conceptual doctrine definitions and patterns for fighting any conflict are presented in [8]. As the technique is novel, there are various challenges in linking diverse and homogenous gadgets. To be specific, an IoT-oriented platform needs mobility assistance, geo-distribution, and low delay. Latest technology solutions, like fog computation, have been presented to allow computation immediately at the network's edge, which can bring fresh activities and applications to billions of linked gadgets [9]. A novel model where statistics can be graphically displayed is found in graph databases [10]. Graphs have been utilized in different quantitative issues to hold and display statistics. With a graph database, finding and navigating the graph is considerably simpler, making it easier to retrieve data than with a relational database. An innovative method for saving data in the form of nodes and edges is to use graph databases. Nodes are where the data is stored. There is a graphic relationship between these nodes. The connection that separates nodes is referred to as an edge. The links between linked nodes are frequently displayed by these edges. These graph-based architectures are useful since they can be employed in cases where the characteristics of a specific item of data hold greater significance than the data resource. For instance, the soldier is the primary node in Fig. 44.2, but it also has other crucial characteristics like guns, gear, ammo, and physiological sensing variables. This is easily convertible to a graphical database. This makes it possible to model the data organically. For some kinds of systems,

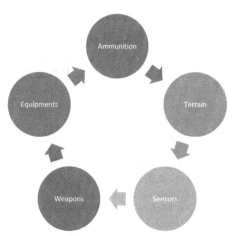

Fig. 44.2 Soldiers and features

database administration is required [11]. The databases can be categorized into 4 groups.

3. METHODOLOGY

In this research, we attempt to give a built-in IoT system in any reputable wireless network system (WNS) linked to a strategic base where warriors may be monitored for contextual consciousness. Body characteristics such as body temperature [12], heart rate detector [13], blood flow [14], sugar levels, ECG levels, device condition, weapon condition, and ammunition condition are all monitored and transmitted to strategic base stations (SBS) [15]. Employing a graph database, we have been concerned with the total machinery and health consciousness of deploying warriors from the strategic commander's perspective. The commander can make strategic opinions on their warriors' trends by utilizing the effective perspectives of the questions of graph databases. On the battlefield, vital indicators such as body specifications, machinery, weapon, and ammunition must be checked. Results can be transferred across portals using normal cloud services. We wish to emphasize that a graph database may be utilized for graphical

questions illustrating the links between information and its related edges.

4. RESULT AND DISCUSSION

This research took a simulated strategic location and established a segment of 12 warriors, which is strategically known as a Segment of warriors. This group of warriors is now quite near the opposing border and is engaged in a furious struggle. Aside from strategic-stage actions, we offer the strategic commander with question-orientated outcomes of graphical databases over warriors. Ammunition, body variables, weaponry, and machinery are the critical characteristics we are operating with. The parameters of minimal requirements (Table 44.1):

Table 44.1 Minimal requirements

Parameters	Minimal requirement
Memory	2 GB
Disk	10 GB SATA
CPU	Intel core i3
File system	Ext4
Operating system	Linux, Windows Server 2012
Java	Open JDKB
Structure	X86

In terms of memory needs, this research can observe that the requirements have been stringent. It additionally needs a lot of computing resources. With the above-mentioned prerequisites, we can quickly run sophisticated searches with graphically attractive outcomes, allowing the strategic commander to make contextual opinions about shifting battlefield scenarios. A simulator fed statistics from 12 warriors into the framework, which is kept in the Neo4j database. We are running numerous searches, and the graphical outcomes are shown in the image beneath. Table 44.2 shows the statistics that are being supplied to the platform.

Table 44.2 Graphical database of Warrior's segments

Unit_ID	Alpha 1	Alpha 1	Charlie 1	Romeo 1	Strike 1	Bite 1	Bite 1	Gamma_1
Mission_ID	Capture_X_ Hills, Capture_ Delta_Y	Capture_X_ Hills	Capture_X_ Hills	Capture_ Delta_Y, Capture_Z_Hills	Capture_ Delta_Y	Capture_ Delta_Y	Capture_ Delta_Y	Capture_Z_ Hills
Warrior_ID	A	B	C	D	E	F	F	H
Body temperature	98	99	100	97	98.5	99.2	99.2	97
Body sugar	110	100	90	110	100	120	120	110
Weapon	AK-47, HMG	Rocket launcher	LMG	AK 47	Rocket launcher	LMG	LMG	AK 47
Ammunition	20	25	30	20	25	30	30	20
Equipment	Radio	Radio	GPS	Radio	Radio	GPS	GPS	Radio

The data presented above pertains to twelve warriors who have distinct areas in the framework. Every warrior is assigned to a battalion within a hierarchical framework that is followed in fighting troops. Every warrior is given a unique mission id that involves taking a strategic area on the battlefield. Important body metrics are additionally included in the database. Table 44.3 additionally contains information about the warrior's weapons, ammo, and machinery. This information is saved in neo4j's graph database.

Table 44.3 Neo4j colour code

Node	Colour
Mission	Purple
Equipment	Grey
Ammunition	Yellow
Weapon	Pink
Unit	Red
Pearson	Blue

Query used to create a graphical database in Neo4j and the graphical database is shown below.

CREATE (a:Person), (z:Weapon), (n:Unit), (m:Mission), (b:Ammunition), (c:Equipment)

RETURN m,z,a,n,b,c

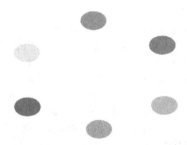

Fig. 44.3 Neo4j color code and graphical database of query 1

5. CONCLUSION

The work focuses on the usage and setup of Strategic Military Attribute Monitoring Structures (SMAMS) in the battlefield landscape using the IoT approach and graph database computation approach. It is a theoretical strategy in which the conventional method of sending the soldier's statistics from the portal to the cloud is used. We discussed an instance of several searches that are increasingly graphically intuitive and useful in strategic choice-making on the battlefield. We would develop this concept for more complicated combat situations where the complications of the

battlefield may be quickly grasped by the graphical database.

References

1. Nandalal, V., & Anand Kumar, V. (2021). Internet of Things (IoT) and Real-Time Applications. *Artificial Intelligence for COVID-19*, 195–214.
2. Goyal, S., Sharma, N., Bhushan, B., Shankar, A., & Sagayam, M. (2021). IoT-enabled technology in secured healthcare: Applications, challenges, and future directions. *Cognitive Internet of Medical Things for Smart Healthcare: Services and Applications*, 25–48.
3. Lakhwani, K., Gianey, H. K., Wireko, J. K., & Hiran, K. K. (2020). *Internet of Things (IoT): Principles, paradigms, and applications of IoT*. BPB Publications.
4. Besta, M., Gerstenberger, R., Peter, E., Fischer, M., Podstawski, M., Barthels, C., ... & Hoefler, T. (2023). Demystifying graph databases: Analysis and taxonomy of data organization, system designs, and graph queries. *ACM Computing Surveys*, *56*(2), 1–40.
5. Yushi, L., Fei, J., & Hui, Y. (2012, May). Study on application modes of military Internet of Things (MIOT). In *2012 IEEE International Conference on computer science and Automation Engineering (CSAE)* (Vol. 3, pp. 630–634). IEEE.
6. Zheng, D. E., & Carter, W. A. (2015). *Leveraging the Internet of Things for a more efficient and effective military*. Rowman & Littlefield.
7. Tortonesi, M., Morelli, A., Govoni, M., Michaelis, J., Suri, N., Stefanelli, C., & Russell, S. (2016, December). Leveraging the Internet of Things within the military network environment—Challenges and solutions. In *2016 IEEE 3rd World Forum on Internet of Things (WF-IoT)* (pp. 111–116). IEEE.
8. Suri, N., Tortonesi, M., Michaelis, J., Budulas, P., Benincasa, G., Russell, S., ... & Winkler, R. (2016, May). Analyzing the applicability of the Internet of Things to the battlefield environment. In *2016 International Conference on military communications and Information Systems (ICMCIS)* (pp. 1–8). IEEE.
9. Kassab, W. A., & Darabkh, K. A. (2020). A–Z survey of Internet of Things: Architectures, protocols, applications, recent advances, future directions, and recommendations. *Journal of Network and Computer Applications*, *163*, 102663.
10. Schneeweiss, S., Rassen, J. A., Brown, J. S., Rothman, K. J., Happe, L., Arlett, P., ... & Wang, S. V. (2019). Graphical depiction of longitudinal study designs in health care databases. *Annals of Internal Medicine*, *170*(6), 398–406.
11. Singh, D. (2023, May). Future Field Systems using Graph Database and IoT. In *2023 3rd International Conference on Advance Computing and Innovative Technologies in Engineering (ICACITE)* (pp. 2183–2186). IEEE.

Note: All the figures and tables in this chapter were made by the author.

Recent Trends in Engineering and Science for Resource Optimization and Sustainable Development – Prof. (Dr.) Dorota Jelonek et al. (eds)
© 2024 Taylor & Francis Group, London, ISBN 978-1-032-98030-0

45

Key Technologies, Protocols, and Applications for the Internet of Things

Nitesh Kumar Gupta

Research Scholar, Department of Economics,
Mahatma Gandhi Kashi Vidyapith, Varanasi, U.P

Priti Gupta

Assistant Professor, P.G. Department Of Economics,
Bhupendra Narayan Mandal University (West Campus)
P.G. Centre, Saharsa, Bihar 9

CH. Veena

Department of Computer Science and Engineering,
Institute of Aeronautical Engineering,
Hyderabad, Telangana

Chetan Shelke*

Associate professor, Alliance College of Engineering and Design,
Alliance University Bangalore

Lavish Kansal

Lovely Professional University, Phagwara, India

S. Gnana Prasanna

Assistant professor, CSE department,
St. Martin's Engineering College, India

■■■■■■ **Abstract**

The paper gives an outline of the Web of Things (IoT) with an accentuation on empowering technology, rules, and application-specific difficulties. The Internet of Things is made possible by the most recent expansions in RFID, intelligent sensors, technology for correspondence, and Web conventions. The fundamental thought is to have savvy sensors interact with one another directly and autonomously to produce a completely new set of uses. The ongoing ascent in machine-to-machine (M2M), portable, and Web innovation should be visible as the beginning of the Web of Things (IoT). Before long, it is projected that the IoT will link physical objects to improve intelligent decision-making, fusing various technologies to create new applications.

■■■■■■ **Keywords**

Protocols, Application, IoT, Key technologies, Web of things

---

*Corresponding author: Chetan.shelke@alliance.edu.in

DOI: 10.1201/9781003596721-45

1. Introduction

As a record-breaking number of physical objects are associated with the Web, the possibility of the Web of Things (IoT) is starting to take shape. Simple examples of these products include thermostat and HVAC (Heating, Ventilation, and Air Conditioning) control and surveillance systems, which enable smart homes. There are numerous specific circumstances and ventures where the IoT can have a significant effect and work on our satisfaction. A portion of these applications incorporates medical services, transportation, modern robotization, and pressing reactions to calamities of the two sorts where the human independent direction is troublesome [1].

The Internet of Things (IoT) enables physical objects to "talk" to one another, share knowledge, and organize actions. This enables objects to see, hear, think, and perform tasks. The IoT changes these items from being customary to insightful by utilizing its fundamental advancements, like omnipresent and unavoidable PCs, gadgets with implanted innovations for correspondence, organizations of sensors, Web conventions, and applications. Savvy objects and their supposed capabilities make up area explicit applications (vertical business sectors), whilst omnipresent computing and statistical analysis make up program domain independent offerings (horizontal markets) [2]. The broad IoT concept is shown in Fig. 45.1, where each domain-specific application connects with services that are not specialized to any one domain, but where within each domain the actuators and sensors directly communicate with one another.

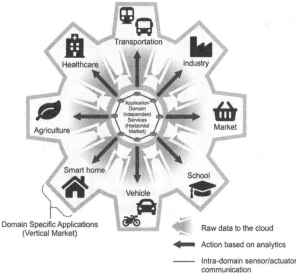

Fig. 45.1 IoT applications

The IoT can be used for a variety of applications, including monitoring and managing assets, tracking the movement of goods and vehicles, and providing users with real-time information about their surroundings. The IOT is also being used to create smart cities and other infrastructure projects and to enable connected devices to communicate with each other and with the internet using machine-to-machine (M2M) communication protocols [3]. There are many different types of IoT applications, but some of the most common are consumer applications, such as home automation and security systems; enterprise applications, such as manufacturing and supply chain management; and mobile applications, such as smartphones and tablets [4]. IoT applications are often based on sensors that collect data about conditions in the environment or on objects that are being monitored. This data can then be used to control devices or to generate alerts or notifications. IoT applications can be used to improve the efficiency and safety of operations in a variety of industries, including transportation, healthcare, and energy. They can also help to improve the customer experience by providing more accurate information about the state of systems or the location of objects. The IoT market is expected to grow from \$1.4 trillion in 2020 to \$5.8 trillion by 2025, according to Market [5]. The IOT has the potential to revolutionize a variety of industries, including transportation, manufacturing, healthcare, and energy management. It is also being used to create new applications and services, such as smart home appliances, smart city infrastructure, and autonomous vehicles [6]. The IOT is already being used to improve the efficiency of transportation systems, by allowing drivers to use mobile apps to find the best route and avoid traffic congestion. It is also being used to create smart cities, by allowing citizens to access information about traffic conditions, weather updates, and emergency alerts [7]. The IOT is also being used to improve the efficiency of healthcare systems, by allowing patients to access medical records and receive real-time updates about their health. It is also being used to create autonomous vehicles, by allowing them to navigate roads without the need for a driver [8].

2. Literature Review

Several survey papers that look at various facets of IoT technology have been released. For instance, the study by [9] examines the fundamental wired and wireless communication enabling technologies as well as the components of WSNs. The IoT architecture and the difficulties in creating and deploying IoT applications are covered by the authors in [10]. Some studies present empowering innovations and application administrations that use a centralized cloud view [11]. The authors of [12] offer an overview of the IoT for specialized clinical remote gadgets utilizing Bluetooth, NFC, and 6LoWPAN/IEEE 802.15.4 for mHealth and eHealth applications. Additionally, [13] discusses IoT enabler technologies with a focus on RFID and its possible uses. To overcome any barrier between research and application, IoT challenges are given in [14]. Some authors provide a summary of the current IETF standards and IoT-related problems [15].

Additionally, architecture standardization can be considered the foundation for the IoT to foster competition among businesses and enable them to produce high-quality goods. To address the IoT concerns, the conventional Internet infrastructure must also be changed. For instance, many underlying protocols should take the enormous number of items eager to connect to the Internet into account. The number of Internet-connected things had overtaken the number of people on the planet in 2010 [16]. To satisfy client demand for smart objects, it becomes important to use a broad addressing space (like IPv6) [17]. Because of the inborn assortment of Web-associated things and the capability to monitor and manage physical objects, security, and privacy [18] are additional crucial requirements for the IoT [19]. In addition, the IoT needs to be managed and watched over to guarantee that users receive high-quality services at reasonable prices [20].

3. METHODOLOGY

This paper's primary study goal is to examine the key IoT technologies, protocols, and applications. To achieve this, we will try to determine how it is being used in the current IoT application landscape and further investigate the chances for its use in other industries. We'll also look into the importance and effects of the various IoT technologies. Primary and secondary research will be combined in our study process. We will first use library sources, including books, journals, papers, and any other pertinent materials, to compile the most up-to-date knowledge on the use of important technologies, applications, and protocols in IoT. We will utilize this data to better understand how various IoT applications, protocols, and technologies are used in general.

To better grasp the potential of IoT at various stages, we will then undertake surveys and interviews with industry experts and participants in the current rebellion in M2M, mobile, and internet technology. The most recent developments in Internet protocols, smart sensors, technologies for communication, and RFID will be covered in our surveys and interviews. We will further examine the potential of major technologies, protocols, and application use in IoT using the information gathered and comments from the surveys and interviews. Finally, we will make use of the data acquired to assess the importance and effect of protocols and application use in IoT.

4. RESULTS AND DISCUSSION

IoT-based services have significantly increased their economic value for businesses as well. Applications in industry and healthcare are expected to have a big financial impact. With the aid of healthcare apps and related IoT-based operations like mobile health care (m-Health) and telecare, medical well-being, early detection, treatment, and tracking services can be effectively provided through electronic media. A market share projection for the most prominent IoT applications is shown in Fig. 45.2.

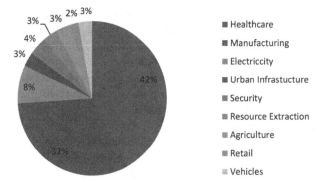

Fig. 45.2 Potential economic impact of iot applications on various sizes

IoT devices, systems, and applications must be able to share data to improve the overall efficiency and effectiveness of the IoT ecosystem. Data sharing is essential for the success of IoT devices, systems, and applications [21]. By sharing data, devices can work together to achieve common goals, and applications can learn from each other. One way to share data is through the use of APIs. APIs allow devices and applications to communicate with each other using standardized protocols. This makes it easier for developers to access and use data from different devices, and it makes it easier for users to find and use the data they need. Another way to share data is through the use of cloud services. Cloud services allow devices to share data with a remote server. This allows devices to offload the processing of data from local devices to a more powerful server. Finally, devices can share data through the use of sensors. Sensors allow devices to collect data about the environment around them. This data can be used to improve the performance of devices and the accuracy of applications.

The IEEE 802.11 standard defines a wireless LAN medium access control (MAC) and physical layer (PHY) protocol suite. IEEE 802.11 defines a basic service set (BSS) structure and provides for multiple access points (APs) in a wireless network. IEEE 802.11 also defines a management framework that allows for centralized management of the wireless network. IEEE 802.11 is a common wireless

LAN medium access control (MAC) and physical layer (PHY) protocol suite. IEEE 802.11 defines a basic service set (BSS) structure and provides for multiple access points (APs) in a wireless network. IEEE 802.11 also defines a management framework that allows for centralized management of the wireless network. IEEE 802.11 is a part of the IEEE Standards Association's 802.11 family of standards [22]. It specifies wireless networking technologies for use in personal area networks (PANs), local area networks (LANs), and wide area networks (WANs). The IEEE 802.11 standards are divided into three sub-groups: wireless local area networks (WLANs), wireless metropolitan area networks (WMANs), and wireless wide area networks (WWANs). IEEE 802.11 is an open standard, meaning that it is freely available to the public with no licensing fees. 802.11 is a wireless networking protocol that provides high-speed wireless connectivity for gadgets like laptops, tablets, smartphones, and various wireless-enabled gadgets. 802.11 standards are maintained by the IEEE. 802.11 is a wireless networking protocol that provides high-speed wireless connectivity for devices such as laptops, tablets, smartphones, and other wireless-enabled devices. 802.11 standards are maintained by the IEEE [23]. 802.11 is a popular choice for wireless networking in homes and small offices and is also used in public places such as airports and coffee shops. 802.11 is not as popular as some other wireless networking protocols, but it is still in use and is supported by most wireless routers. 802.11 is a wireless networking protocol that uses radio waves to connect devices within a network. Devices that support 802.11 can communicate with each other using radio waves, without the need for a cable or wire. 802.11 is a popular choice for wireless networking in homes and small offices and is also used in public places such as airports and coffee shops. 802.11 is not as popular as some other wireless networking protocols, but it is still in use and is supported by most wireless routers [24].

IoT administrations can have a large number of clients, thus it's interesting to know how many of those users are unique over time. Additionally, it might be expensive to store the individual IP tends to in a social data set. All things being equal, the help's neighborhood Invertible Sprout Channel (IBF) can hash the IP address of every parcel. The tradeoff between how much the IBF and the achieved precision is shown in Fig. 45.3. One thousand five hundred records in a relational table serve as the basis for comparison.

5. CONCLUSION

The newly developed concept of the Web of Things is rapidly advancing across contemporary existence fully intent on improving personal satisfaction by melding numerous

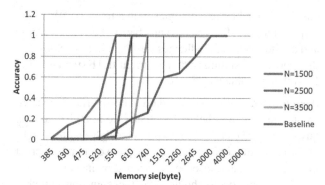

Fig. 45.3 Accuracy and IBF size are trade-offs

astute devices, advancements, and applications. Overall, the IoT would make it possible for everything in our surroundings to be automated. The basic presumptions of this theory, in addition to its associated technologies, procedures, applications, and current studies addressing numerous IoT-related concerns, were summarised in this study. To understand the basic architecture and purpose of all of the parts and protocols that comprise the IoT, academics, and professionals who have an interest in acquiring knowledge about IoT protocols and technologies should have a strong foundation from which to start.

References

1. Burhan, M., Rehman, R. A., Khan, B., & Kim, B. S. (2018). IoT elements, layered architectures, and security issues: A comprehensive survey. *sensors, 18*(9), 2796.

2. Mahmoud, R., Yousuf, T., Aloul, F., &Zualkernan, I. (2015, December). Internet of Things (IoT) security: Current status, challenges, and prospective measures. In *2015 10th international conference for Internet Technology and secured transactions (ICITST)* (pp. 336–341). IEEE.

3. El-Hajj, M., Fadlallah, A., Chamoun, M., &Serhrouchni, A. (2019). A survey of Internet of things (IoT) authentication schemes. *Sensors, 19*(5), 1141.

4. Babiceanu, R. F., & Seker, R. (2016). Big Data and virtualization for manufacturing cyber-physical systems: A survey of the current status and future outlook. *Computers in industry, 81,* 128–137.

5. Saleem, Y., Crespi, N., Rehmani, M. H., & Copeland, R. (2019). Internet of things-aided smart grid: technologies, architectures, applications, prototypes, and future research directions. *Ieee Access, 7,* 62962–63003.

6. Bansal, S., & Kumar, D. (2020). IoT ecosystem: A survey on devices, gateways, operating systems, middleware, and communication. *International Journal of Wireless Information Networks, 27,* 340–364.

7. Mashal, I., Alsaryrah, O., Chung, T. Y., Yang, C. Z., Kuo, W. H., & Agrawal, D. P. (2015). Choices for interaction with things on the Internet and underlying issues. *Ad Hoc Networks, 28,* 68–90.

8. Weyrich, M., & Ebert, C. (2015). Reference architectures for the Internet of things. *IEEE Software*, *33*(1), 112–116.

9. Lombardi, M., Pascale, F., & Santaniello, D. (2021). Internet of Things: A general overview between architectures, protocols, and applications. Information, 12(2), 87.

10. Shahid, N., & Aneja, S. (2017, February). Internet of Things: Vision, application areas and research challenges. In 2017 International Conference on I-SMAC (IoT in Social, Mobile, Analytics, and Cloud)(I-SMAC) (pp. 583–587). IEEE.

11. Kassab, W. A., &Darabkh, K. A. (2020). A–Z survey of Internet of Things: Architectures, protocols, applications, recent advances, future directions, and recommendations. Journal of Network and Computer Applications, 163, 102663.

12. Almotiri, S. H., Khan, M. A., & Alghamdi, M. A. (2016, August). Mobile health (m-health) system in the context of IoT. In 2016 IEEE 4th international conference on the future Internet of Things and Cloud workshops (FiCloudW) (pp. 39–42). IEEE.

13. Rghioui, A., &Oumnad, A. (2017). Internet of things: Surveys for measuring human activities from everywhere. International Journal of Electrical and Computer Engineering, 7(5), 2474.

14. Chernyshev, M., Baig, Z., Bello, O., &Zeadally, S. (2017). Internet of Things (IoT): Research, simulators, and testbeds. IEEE Internet of Things Journal, 5(3), 1637–1647.

15. Boubiche, D. E., Pathan, A. S. K., Lloret, J., Zhou, H., Hong, S., Amin, S. O., & Feki, M. A. (2018). Advanced industrial wireless sensor networks and intelligent IoT. IEEE Communications Magazine, 56(2), 14–15.

16. Corcoran, P. (2015). The Internet of Things: Why now, and what's next?. IEEE Consumer Electronics Magazine, 5(1), 63–68.

17. Mun, D. H., Le Dinh, M., & Kwon, Y. W. (2016, June). An assessment of Internet of things protocols for resource-constrained applications. In *2016 IEEE 40th Annual Computer Software and Applications Conference (COMPSAC)* (Vol. 1, pp. 555–560). IEEE.

18. Al-Fuqaha, A., Guizani, M., Mohammadi, M., Aledhari, M., & Ayyash, M. (2015). Internet of Things: A survey on enabling technologies, protocols, and applications. *IEEE communications surveys & tutorials*, *17*(4), 2347–2376.

19. Chahid, Y., Benabdellah, M., & Azizi, A. (2017, November). Internet of things protocols comparison, architecture, vulnerabilities, and security: State of the art. In *Proceedings of the 2nd International Conference on Computing and Wireless Communication Systems* (pp. 1–6).

20. Salman, T., & Jain, R. (2019). A survey of protocols and standards for the Internet of things. *arXiv preprint arXiv:1903.11549.*

Note: All the figures in this chapter were made by the author.

Recent Trends in Engineering and Science for Resource Optimization and
Sustainable Development – Prof. (Dr.) Dorota Jelonek et al. (eds)
© 2024 Taylor & Francis Group, London, ISBN 978-1-032-98030-0

46

Think Big with Big Data: Finding Appropriate Big Data Strategies for Corporate Cultures

E. Mythily[1]

Associate Professor & Head, Department of Commerce with Cost and
Management Accounting, PSG College of Arts & Science,
Coimbatore, India

S. S. Ramya[2]

Assistant professor, Department of Commerce,
PSG College of Arts & Science, Coimbatore, India

K. Sangeeta

Department of Computer Science and Engineering,
Institute of Aeronautical Engineering, Hyderabad, India

Swathi B[3]

Department of Information Science Engineering,
New Horizon College of Engineering, Bangalore, India

Manish Kumar

Lovely Professional University, Phagwara, India

Purnendu Bikash[4]

Acharjee, Associate Professor, CHRIST University, Pune, India

Narendra Kumar[5]

NIET, NIMS University, Jaipur, India

Abstract

The aim of this research is to learn how big data strategies (BDS) as a corporate culture might improve confidence and cooperative performance across civil and defence sectors involved in disaster relief activities. The research conceptualized a unique conceptual framework to demonstrate, employing the competitive value model (CVM), how BDS influences swift confidence (SC) and cooperative performance (CP) beneath the moderating impact of the corporate culture. The findings have four significant consequences. Initially, the BDS has a strong beneficial influence on SC and CP. Secondly, neither adaptable orientation (AO) nor regulated orientation (RO) had any effect on constructing SC. Thirdly, AO has a strong and beneficial moderating influence on the path connecting BDS and CP. As a result, RO shows a considerable negative moderating impact on the path linking BDS and CP.

Keywords

Big data, Corporate culture, Big data strategies, Predictive analytics, and Confidence

Corresponding authors: [1]mythilyparthasarathi@gmail.com, [2]ramyaspsg@gmail.com, [3]baswarajuswati@gmail.com, [4]pbacharyaa@gmail.com, [5]drnk.cse@gmail.com

DOI: 10.1201/9781003596721-46

1. INTRODUCTION

The objective of the organization is not just manufacturing, nor is it enhancing the behavioural habits of operating people, but instead a combination of the 2 areas by altering people's behavioural habits so that they can be persuaded of the corporate objectives and embrace the strategies embraced by the organization to accomplish the objective [1]. Here comes the function of corporate culture, which is meant to be capable to cultivate a feeling of relating and need to accomplish organizational objectives, as well as a feeling of faith and confidence that corporate objectives within their organization should be accomplished to guarantee the organization's achievement and accessibility to excellent results [2]. The Internet of Things (IoT) concept violates the conventional approach of segregating technology and data transmission, really achieves the connectivity between objects, and embraces the integrated idea of the web, which is extremely important [3]. If data is incorporated in the IoT interaction the number of interaction links that could be incorporated may exceed billions, providing a huge area for data transfer. With the explosive growth of machine degrees and technological communication, the use of IoT in electronic commerce, power surveillance, armed forces, national security, and additional private areas has grown more prevalent, and secure big data sharing and transfer has assumed a more significant place. Big data interaction is the primary way of data transmission, and it is critical for ensuring the security of big data interaction in the IoT [4].

Businesses may analyze user-generated data in several formats, like video, blogs, and social media data, which has various benefits for them, including increasing production and competitiveness [5]. Digital transaction data is frequently used by companies like Apple, Google, Facebook, eBay, and Amazon to improve their company operations. These businesses must regularly record transaction information such as transaction times, product pricing, purchase amounts, and client credentials to estimate market conditions, consumer behavior, trends, patterns, and other factors. The present investigation is motivated by the requirement to comprehend how big data strategies (BDS) promote confidence and cooperation among humanitarian organizations (civil & military) (Refer to Fig. 46.1). We have noticed that the terms Big Data, BDS, and BDS capacity are frequently employed concurrently in research [6]. This investigation, nevertheless, controls for their conceptual variations in viewpoint, thinking, and measurement. We propose a theory-based assessment of the functions of BDS capabilities and corporate culture in the creation of confidence and the implementation of analytics-oriented cooperation between military and sectors participating in disaster relief missions in this paper.

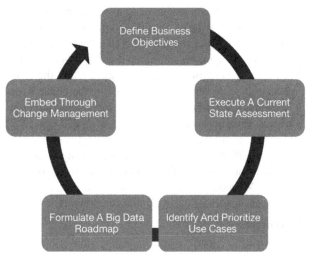

Fig. 46.1 Corporate BDS

2. LITERATURE REVIEW

According to [7], corporate culture has a significant impact on the economic and non-financial achievement of organizations, as the organization may guide people in a manner that ensures the greatest degree of accomplishment connected with the complete belief that this organization is their organization and that they should offer its greatest effort to assure its longevity, and this applies to numerous organizations.

According to [8], adaptability-oriented (AO) (i.e., growth and collection culture) and regulation-oriented (RO) [9] (i.e. logical and structured culture) cultures possess distinct impacts on the organization's conclusions of outside factors [10] and therefore impact its reactions to the environment's demands [11] and standards in distinct manners [12]. Big data and predictive assessment are commonly defined in the field of information systems as an organizational competence for handling enormous amounts and types of data at the velocity needed to get useful knowledge, allowing organizations to achieve a competitive edge [13]. The phrase 'big data' is frequently used to characterize large, complicated, and real-time information that necessitates advanced administration, mathematical, and statistical approaches to acquire administrative information. The prediction analytics mathematical frameworks attempt to forecast future behavior on the premise that what has occurred in the past will occur again shortly.

3. METHODOLOGY

A questionnaire was employed to collect data collections for the inquiry hypothesis. We conducted a pre-survey with six academics and six directors from Indian civil and military companies. We have modified our language with the assistance of their data resources to increase openness and ensure that the duration of the survey is precise. Ultimately, the survey was ready for data collection. The target sample consisted of corporations in the civil and military sectors. As a result, developing an effective survey was the most important stage in data collection. All of the concepts in our hypothetical framework are implemented as reflective structures. Finally, we gathered replies from 373 organizations. This elevated rate of response was attained by personally calling every respondent and promising them that they would be capable of reviewing the results. We next used ANOVA assessment to carry out a non-response bias assessment on each of the evaluated components.

Hypothesis

H1: BDS is positively connected with swift confidence (SC)

H2: BDS is positively connected with cooperative performance (CP)

H3: SC is positively connected with CP

H4/H5: Organizational AO significantly moderates the connection between BDS and SC/CP

H6/H7: Organizational RO negatively moderates the connection between BDS and SC/CP

4. RESULT AND DISCUSSION

We were required to evaluate the reliability of the respondents' perspectives utilizing 3 respondents from each respondent organization. Table 46.1 shows the findings of inter-rater agreement assessment employing 4 distinct techniques; the percentage technique, the ratio technique, the inter-class correlation coefficient, and the combined

Table 46.1 Inter-rater consistency outcomes

Constructs	Percentage technique (%)	Ratio technique (%)	Inter-class correlation coefficient (%)	Combined t-test
BDS	92	81	42	Not-significant
SC	91	79	36	Not-significant
CP	88	77	42	Not-significant
AO	86	76	43	Significant
RO	84	75	23	Not-significant
Temporal orientation (TO)	85	79	34	Not-significant
Interdependency (I)	87	76	28	Significant

t-test. As a result of the outcomes in Table 46.1, we concluded that the inter-rater consistency in the results was adequate.

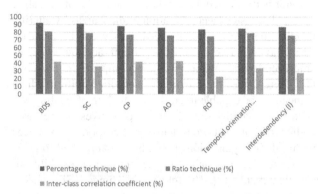

Fig. 46.2 Inter-rater consistency outcomes

As a result, we figure out that all reflecting measurements have satisfactory discriminant validity. Lastly, we look at the discriminant validity of the constructs, which is the degree to which measurements of one construct vary from measurements of different constructs in the conceptual framework. Table 46.2 demonstrates that the constructs' discriminant validity is accepted.

Table 46.2 Outcomes of discriminant validity

	BDS	SC	CP	AO	RO	TO	I
BDS	0.74						
SC	0.0	0.77					
CP	0.14	0.43	0.71				
AO	0.17	0.31	0.58	0.69			
RO	0.24	0.32	0.41	0.39	0.74		
TO	0.15	0.35	0.69	0.37	0.60	0.75	
I	0.16	0.33	0.60	0.43	0.64	0.39	0.81

Conventional parametric-oriented significance assessments are inappropriate for this investigation since PLS fails to follow a multivariate normal dispersion. PLS employs a bootstrapping process to calculate standard mistakes and the significance of parametric values. Table 46.3 and Fig. 46.3 show the PLS path coefficients and p-scores.

Table 46.3 Structural assessment

Hypothesis	Impact on	Impact of	β	p-score	Outcomes
H1	SC	BDS	0.28	***	Accepted
H2	CP	BDS	0.36	***	Accepted
H3	CP	SC	0.23	***	Accepted
H4	SC	BDS*AO	0.05	*	Not Accepted
H5	CP	BDS*AO	0.48	***	Accepted
H6	SC	BDSRO	0.06	*	Not Accepted
H7	CP	RO	-0.46	***	Accepted

*p>0.1 and ***p<0.01

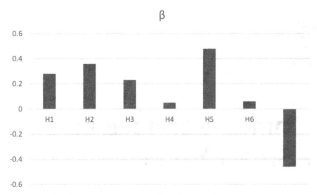

Fig. 46.3 Hypothesis assessment

5. CONCLUSION

Civilian and defence organizations often have quite different operating techniques and corporate cultures, however, there is some overlap. Secondly, effective administration of resources is essential for operation success. Among the primary drivers of disputes between civilian and defence organizations is an absence of confidence. Therefore, in this work, we employ BDS to demonstrate how an organization's data processing capabilities, when moderated by corporate culture, can foster rapid confidence and additionally enhance collaboration. These research results indicate the areas of organizational competence growth that seem to be required for complete analytics capability utilization. Also, the findings imply that the significance of complementing capability growth is dependent on a corporate culture, which governs how it functions during crises to foster confidence and increase cooperation. We anticipate that our research will help expand research on humanitarian operations administration.

Reference

1. Gorshunov, M. A., Armenakis, A. A., Feild, H. S., & Vansant, B. (2020). The Sarbanes-Oxley Act of 2002: Relationship to the magnitude of financial corruption and corrupt organizational cultures. *Journal of Management*, *21*(2), 73.
2. Murphy, W. H., Gölgeci, I., & Johnston, D. A. (2020). Power-based behaviors between supply chain partners of diverse national and organizational cultures: the crucial role of boundary spanners' cultural intelligence. *Journal of Business & Industrial Marketing*, *35*(2), 204–218.
3. Liu, L., Kong, W., Cao, Z., & Wang, J. (2017). Analysis of one certificateless encryption for secure data sharing in public clouds. *International Journal of Electronics and Information Engineering*, *6*(2), 110–115.
4. Phillips, M. (2018). International data-sharing norms: from the OECD to the General Data Protection Regulation (GDPR). *Human genetics*, *137*, 575–582.
5. Maklan, S., Peppard, J., & Klaus, P. (2015). Show me the money: Improving our understanding of how organizations generate a return from technology-led marketing change. European Journal of Marketing, 49(3/4), 561–595.
6. Wang, G., Gunasekaran, A., Ngai, E. W., & Papadopoulos, T. (2016). Big data analytics in logistics and supply chain management: Certain investigations for research and applications. *International journal of production economics*, *176*, 98–110.
7. Ogbeibu, S., Senadjki, A., Emelifeonwu, J., & Vohra, P. S. (2020). Inspiring creativity in diverse organizational cultures: An expatriate integrity dilemma. *FIIB Business Review*, *9*(1), 28–41.
8. Prasanna, S. R., & Haavisto, I. (2018). Collaboration in humanitarian supply chains: an organizational culture framework. *International Journal of Production Research*, *56*(17), 5611–5625.
9. Pawełoszek, I., Kumar, N., & Solanki, U. (2022). Artificial intelligence, digital technologies and the future of law . Futurity Economics & Law, 2(2), 24–33. https://doi.org/10.57125/FEL.2022.06.25.03
10. Manpreet Singh Bhatia, Alok Aggarwal, Narendra Kumar. (2020). SPEECH-TO-TEXT CONVERSION USING GRU AND ONE HOT VECTOR ENCODINGS. PalArch's Journal of Archaeology of Egypt / Egyptology, 17(9), 8513–8524. Retrieved from https://archives.palarch.nl/index.php/jae/article/view/5796
11. Aydiner, A. S., Tatoglu, E., Bayraktar, E., Zaim, S., & Delen, D. (2019). Business analytics and firm performance: The mediating role of business process performance. *Journal of Business Research*, *96*, 228–237.

Note: All the figures and tables in this chapter were made by the author.

Recent Trends in Engineering and Science for Resource Optimization and
Sustainable Development – Prof. (Dr.) Dorota Jelonek et al. (eds)
© 2024 Taylor & Francis Group, London, ISBN 978-1-032-98030-0

47

RFID-based Internet of Things Services and Infrastructures for Smart Homes

Radwan M. Batyha[1]

Department of Computer Science,
Irbid National University,
Irbid, Jordan

Eswararao Boddepalli

Department of Electrical and Computer Engineering,
Virginia Tech, Blacksburg, United states of America

N. M. Deepika

Department of Information Technology,
Institute of Aeronautical Engineering,
Hyderabad, Telangana

Swathi B[2]

Department of Information Science Engineering,
New Horizon College of Engineering, Bangalore

Manish Kumar

Lovely Professional University, Phagwara, India

Potaparthini Kiranmayee[3]

Assistant Professor,
Electronics and Communication Engineering,
St. Martin's Engineering College, Dhulapally,
Secunderabad, Telangana

Abstract

Radio-frequency identification (RFID) innovation has become pervasive in smart functions or Internet of Things (IoT) instances such as smart learning, homes, medical care, cities, etc. Enhancements to address problems such as security, confidentiality, and cost were developed in the research. For secured RFID interactions, the investigators employed cryptographic instruments, hashing operations, and symmetrical key encryption. Nevertheless, the degree of security continues to be insufficient. In this research, we suggested a cloud-based remote RFID authentication strategy using smart homes as an instance study. Aside from being portable, the suggested approach provides characteristics such as forward confidentiality, anonymity, and intractability. It can survive a variety of security assaults. According to our experiment findings, the suggested system paradigm and the cloud-based distant RFID identification method have been successful in offering confidentiality and safety as a component of the access management system in the smart home IoT service instance.

Corresponding authors: [1]rbatiha@inu.edu.jo, [2]baswarajuswati@gmail.com, [3]kiranmayee239@gmail.com

DOI: 10.1201/9781003596721-47

■■■■■ **Keywords**

Smart homes, Internet of things, Radio frequency identification, and Automation

1. INTRODUCTION

Machine-to-machine incorporation can occur without the need for human interaction thanks to the development of dispersed innovations and the broad adoption of intelligent handheld gadgets. It has caused innovations to combine and give rise to an innovative one called the Internet of Things (IoT) [1]. It is possible to effortlessly connect computational equipment with practical tangible assets using IoT connection. That is to say, it is possible to combine the digital and real worlds. Smart systems such as smart houses, cities, electronic government, e-healthcare, and so forth are built using architecture powered by the IoTs [2].

IoT usage realization is aided by cloud-based computing innovation [3]. IoT applications cannot be completely realized in real life without using cloud-based capabilities. IoT applications cannot be completely realized in real life without cloud computation assets [4]. Since IoT-using sensing networks generate enormous amounts of data, or "big data," cloud assets are crucial. Cloud computation systems and IoT can work together to safeguard and oversee such data easily. Thus, this study is also related to cloud computation. In actuality, cloud computing has several benefits beyond eliminating temporal and location restrictions, such as accessibility, scaling, and adaptability [5]. The IoT constitutes a notion [6] that envisions connecting the Internet to physical objects in smart places such as the modern home. By connecting dummy items to the internet, novel services might be established and utilized by objects, gadgets, and humans. The IOT research identifies Radio Frequency Identification (RFID) as among the enabling innovations for making dumb objects smart.

An RFID identification infrastructure is composed of an RFID reader, several tags, and a back-end system that uses a network connection such as a wireless area network [7] [8]. RFID readers can decode the identification data contained in tags and send it to the back-end infrastructure for operation. Hence this research aims to investigate RFID-based IoT services and infrastructure in smart homes.

2. LITERATURE REVIEW

It is crucial to assess the role that wireless connections play in the development of IoTs, as linked devices depend on wireless communications for assistance. This is especially true in the setting of developing IoT technologies.

[9] looked into various applications for 5G wireless networking. They anticipated that 5G systems would be able to handle future increases in traffic capacity. Additionally, they can support the fusion of several access innovations into a unified, seamless interface. Acceleration and a rich user encounter are only two of the numerous advantages that 5G wireless systems are predicted to offer. The IoTs are linked to several wireless networks. IPv6, WiFi, 3G, and 5G are a few of them [10]. IoT-based industries including medicine have incorporated nano interaction and wireless body-worn sensing [11]. The research [12] investigates the use of RFID with instances that combine smart homes with email, SMS, and furthermore. [13] is researching a secured infrastructure for accessing smart homes to provide IoT-based services. [14] discusses the significance of smart phones in smart home IoT use cases. The research of [15] provides RFID-based identification for medical facilities, making hospital management easier. [15] examines the challenges associated with such uses in the medical field. [16] discusses a unified structure that centers on the use of RFID for successful data dissemination. Numerous uses for RFID may be identified in the research. This involves Smith's [17] inventory control structure, RFID use in supply chain management, and the medical sector, and livestock record-keeping systems. While [18] discusses the appropriateness of RFID for different uses, [19] investigates anonymous RFID authentication.

3. METHODOLOGY

We presented a methodology for interaction among RFID tags, tag readers, backend servers, and Authenticated Cloud Servers (ACS). Every component in the secured cloud-assisted authorization procedure has a task to perform. The infrastructure is broken down into multiple types of RFID systems. Every RFID system is additionally divided into groups. A group consists of a collection of RFID tags and a couple of readers used for reading information from the tags. Every RFID network has its server, recognized as the backbone server. A network's RFID scanners and tags have been enrolled with a single backend server. The backend server provides secured passwords to tags and readers [20]. Consequently, each backside server connected to the RFID network must register with an authorized cloud service. It manages communication between backend servers. Sharing similar safety basic functions and mutual authorization is part of the interaction [21, 22]. RFID tags can travel from one group to the next. It could additionally switch to an alternate network. The suggested authorization technique is examined in conjunction with the infrastructure for overcoming various threats, attaining the requisite anonymity, and having shared authorization between tags and backbone services.

4. RESULT AND DISCUSSION

The parameters of the simulation are shown in Table 47.1.

Table 47.1 Simulation parameters

Parameter	Value
MAC type	Mac/802_11
Network interface type	Phy/Wireless_Phy
Max package	100
No. of mobile nodes	29
Routing protocol	AODV
X and Y topography dimension	1000

Table 47.2 and Fig. 47.1 clearly show that the experiment's duration and package delivery ratio (PDR) have been displayed in the vertical and horizontal axes, accordingly. The findings demonstrated a link between PDR and model duration. As the simulation duration increases, so does the PDR. A further significant finding is that the suggested infrastructure has an outstanding PDR when contrasted with competing systems.

Table 47.2 Outcomes of PDR

Simulation duration	PDR (%)			
	Certificate Authentication Scheme (CAS)	Direct Storage Authentication Scheme (DAS)	Broadcast Authentication (BA)	Proposed infrastructure
100	1	4	6	9
200	9	10	12	21
300	11	19	22	30
400	20	28	32	41
500	29	35	39	50
600	35	41	51	60

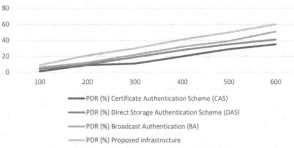

Fig. 47.1 Outcomes of PDR

Table 47.3 and Fig. 47.2 clearly show that the duration of the simulation and delay duration is given in the vertical and horizontal axes, accordingly. The findings demonstrated a link between delay duration and the duration of the simulation. The packet latency increases as the simulation duration increases. Another significant finding is that the suggested infrastructure has the shortest duration of delay when contrasted with different techniques.

Table 47.3 Outcome of delay duration usage

Simulation duration	PDR (%)			
	CAS	DAS	BA	Proposed infrastructure
100	3000	2300	2000	1998
200	3900	3700	2100	2010
300	5800	5000	4000	3050
400	6100	6000	5300	4040
500	8100	7900	7600	6000
600	9000	8100	8000	7200

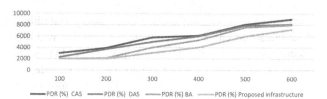

Fig. 47.2 Outcome of delay duration usage

Table 47.4 and Fig. 47.3 clearly show that simulation duration and throughput are given in the vertical and horizontal axes, accordingly. The findings demonstrated a link between simulation duration and throughput. As the simulation duration increases, so does the throughput. Another significant finding is that the suggested infrastructure outperforms previous approaches in terms of throughput.

Table 47.4 The outcomes of throughput

Simulation duration	PDR (%)			
	CAS	DAS	BA	Proposed infrastructure
100	30	22	20	18
200	39	36	24	21
300	41	39	30	28
400	50	43	39	36
500	60	50	41	39
600	64	61	58	50

5. CONCLUSION

RFID-oriented smart IoT services are proliferating at a rapid rate. Nevertheless, there is a requirement for safety

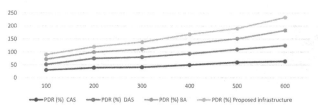

Fig. 47.3 The outcomes of throughput

advancements to IoT-based service instances in real life. Because RFID innovation is utilized to assign particular identities to items and objects in smart services, authorization is comprised of an RFID-oriented phenomenon. Secured interactions are critical in this situation. To accomplish this, we suggested cloud-based distant RFID authorization with safety characteristics like intractability, forward confidentiality, & anonymity. It is also portable and considerably decreases computing complications. A system infrastructure and authorization technique are suggested. A computerized investigation is conducted using the system infrastructure for cloud-oriented distant RFID authorization for the smart home service instance. The experimental research findings demonstrated that the suggested technique outperforms state-of-the-art techniques. In the long term, we plan to examine our method using real-world IoT experimental beds.

References

1. Khanna, A., & Kaur, S. (2020). Internet of things (IoT), applications and challenges: a comprehensive review. *Wireless Personal Communications, 114,* 1687–1762.
2. Lee, I., & Lee, K. (2015). The Internet of Things (IoT): Applications, investments, and challenges for enterprises. *Business Horizons, 58*(4), 431–440.
3. Talavera, J. M., Tobón, L. E., Gómez, J. A., Culman, M. A., Aranda, J. M., Parra, D. T., ... & Garreta, L. E. (2017). Review of IoT applications in agro-industrial and environmental fields. *Computers and Electronics in Agriculture, 142,* 283–297.
4. Zeng, X., Garg, S. K., Strazdins, P., Jayaraman, P. P., Georgakopoulos, D., & Ranjan, R. (2017). IOTSim: A simulator for analyzing IoT applications. *Journal of Systems Architecture, 72,* 93–107.
5. Kotha, H. D., & Gupta, V. M. (2018). IoT application: a survey. *Int. J. Eng. Technol, 7*(2.7), 891–896.
6. Parashar, S., Zaid, M., Vohra, N., & Kumar, S. (2018). Advanced IOT-Based Home Automation. *International Journal for Advance Research and Development, 3*(3), 113–116.
7. Alsinglawi, B., Nguyen, Q. V., Gunawardana, U., Maeder, A., & Simoff, S. J. (2017). RFID systems in healthcare settings and activity of daily living in smart homes: a review. *E-Health Telecommunication Systems and Networks,* 1–17.
8. Chávez-Santiago, R., Szydełko, M., Kliks, A., Foukalas, F., Haddad, Y., Nolan, K. E., ... & Balasingham, I. (2015). 5G: The convergence of wireless communications. *Wireless Personal Communications, 83,* 1617–1642.
9. Monserrat, J. F., Mange, G., Braun, V., Tullberg, H., Zimmermann, G., & Bulakci, Ö. (2015). METIS research advances towards the 5G mobile and wireless system definition. *EURASIP Journal on Wireless Communications and Networking, 2015*(1), 1–16.
10. Shafi, M., Molisch, A. F., Smith, P. J., Haustein, T., Zhu, P., De Silva, P., ... & Wunder, G. (2017). 5G: A tutorial overview of standards, trials, challenges, deployment, and practice. *IEEE journal on selected areas in communications, 35*(6), 1201–1221.
11. Alsinglawi, B., Elkhodr, M., Nguyen, Q. V., Gunawardana, U., Maeder, A., & Simoff, S. (2017). RFID localization for Internet of Things smart homes: a survey. *arXiv preprint arXiv:1702.02311.*
12. Shingala, K., & Patel, J. (2017). Automatic home appliances and security of smart home with RFID, SMS, Email, and real-time algorithm based on IOT. *Int. Res. J. Eng. Technol, 4*(4), 1958–1964.
13. Gope, P., Amin, R., Islam, S. H., Kumar, N., & Bhalla, V. K. (2018). Lightweight and privacy-preserving RFID authentication scheme for distributed IoT infrastructure with secure localization services for the smart city environment. *Future Generation Computer Systems, 83,* 629–637.
14. Winston, T. G., Paul, S., & Iyer, L. (2016, January). A study of privacy and security concerns on doctors' and nurses' behavioral intentions to use RFID in hospitals. In *2016 49th Hawaii International Conference on System Sciences (HICSS)* (pp. 3115–3123). IEEE.
15. Bhattacharya, M. (2015). A conceptual framework of RFID adoption in retail using Rogers stage model. *Business Process Management Journal, 21*(3), 517–540.
16. Moniem, S. A., Taha, S., & Hamza, H. S. (2017, August). An anonymous mutual authentication scheme for healthcare RFID systems. In *2017 IEEE SmartWorld, Ubiquitous Intelligence & Computing, Advanced & Trusted Computed, Scalable Computing & Communications, Cloud & Big Data Computing, Internet of People and Smart City Innovation (SmartWorld/SCALCOM/UIC/ATC/CBDCom/IOP/SCI)* (pp. 1–6). IEEE.
17. Rahman, F., Hoque, M. E., & Ahamed, S. I. (2017). Anonpri: A secure anonymous private authentication protocol for RFID systems. *Information Sciences, 379,* 195–210.
18. Gabhane, J., Thakare, S., & Craig, M. (2017). Smart homes system using Internet-of-Things: issues, solutions, and recent research directions. *International Research Journal of Engineering and Technology, 4*(5), 1965–1969.
19. Korczak, J.J., Pawełoszek, I. (2022). Constructive Approach to Students' Error Processing in E-learning Courses. In: Uskov, V.L., Howlett, R.J., Jain, L.C. (eds) Smart Education and e-Learning - Smart Pedagogy. SEEL-22 2022. Smart Innovation, Systems and Technologies, vol 305. Springer, Singapore. https://doi.org/10.1007/978-981-19-3112-3_15

Note: All the figures and tables in this chapter were made by the author.

Recent Trends in Engineering and Science for Resource Optimization and
Sustainable Development – Prof. (Dr.) Dorota Jelonek et al. (eds)
© 2024 Taylor & Francis Group, London, ISBN 978-1-032-98030-0

48

Progress and Ramifications of Using Computer Models to Solve Intelligence Test Issues

Pallavi Shetty[1]

Associate Professor, Department of MCA,
NMAM Institute of Technology, Nitte (Deemed to be University)

Eswararao Boddepalli

Department of Electrical and Computer Engineering,
Virginia Tech, Blacksburg, United states of America

B. Varasree

Department of Information Technology,
Institute of Aeronautical Engineering, Hyderabad, Telangana

Swathi B[2]

Department of Information Science Engineering,
New Horizon College of Engineering, Bangalore

Manish Kumar

Lovely Professional University, Phagwara, India

Pathan Shahnavajkhan Akabarkhan[3]

Assistant Professor, Faculty of Computer Science,
Shri C.J. Patel College of Computer Studies,
Sankalchand Patel University, India

Abstract

A few computational models of knowledge test issues were published throughout the second half of the 20th century, but in the first years of the 21st century, we have seen a growth in the number of computer systems that can complete certain intelligence test tasks. Despite this growing trend, a thorough examination of all of these works' relationships to each other and their real successes has not been done. We give a thorough account of computer models and their relationships in this paper to shed some light on these issues, with an emphasis on the assortment of knowledge test undertakings they address, the reason for the models, how general or concentrated these models are, the simulated intelligence procedures they use for each situation, their correlation with human execution, and their assessment of thing difficulty.

Keywords

AI, Computer models, Human performance, Progress

Corresponding author: [1]pallavirjsh@nitte.edu.in, [2]baswarajuswati@gmail.com, [3]sapathan.fcs@spu.ac.in

DOI: 10.1201/9781003596721-48

1. INTRODUCTION

Artificial intelligence (AI) is commonly referred to as "the empirical comprehension of the processes underpinning cognition and smart behavior and their implementation in systems." The (human-centered) idea that intellect underpins most human behavior is linked to the concept of spontaneous and AI entities. The belief that "intellect is the computing component of the capacity to accomplish objectives in the environment" was at the root of AI study.

There has been some notable advancement in AI research. For instance, Arthur Samuel introduced a self-learning program that could play draughts or checkers as early as 1959. The prediction made by Herbert Simon in 1957 that a computer would defeat the human chess champion Garry Kasparov within ten years came true in 2002. Watson, a program developed by IBM, won the Jeopardy! TV game show in 2010 Alphago is the primary simulated intelligence framework to effectively dominate the exemplary round of Go in 2016 [1]. Be that as it may, one can contemplate whether the system fundamental to these projects is equivalent to or similar to the interaction of the basic clever human way of behaving. In reality, this achievement in specialized tasks is a great example of the huge switch method in the study of AI. If we confine the shift to one specific issue, we can develop a network that outperforms humans after a few decades of study [2]. We can even integrate several specialized applications into an architecture and create an autonomous transition. For example, if we have an excellent application for checkers, great software for chess, and so on, we may create a meta-system that can recognize the type of activity that has to be performed and shift to the suitable application. Non-intelligent technologies will be capable of surviving if AI assessment depends on specified standards that have been established in advance. Game playing is a fantastic example of how a counter-reaction to this specialism is taking shape [3]. Since 2005, the functionality of game-playing platforms has been tested in game-playing contests [4] on a broad spectrum of games, a few of which are developed and not revealed to the players until the tournament. As a result, in the field of game-playing, platforms employing a huge switch strategy fare little better than platforms implementing generic game-playing methods. Figure 48.1 shows computer modes to solve intelligence test problems.

While winning games is a unique form of intellectual behavior, intellect evaluations examine the fundamental capacity for acting smartly in a variety of contexts [5]. The traditional strategy for intellect evaluation in psychology studies is to use psychometric assessments evaluating intellect [6]. Applying psychometric tests to measure the intelligence quotient (IQ) and other cognitive capacities is the traditional method of evaluating intelligence in psychology

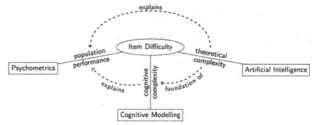

Fig. 48.1 Computer models

research [7]. These assessments are standardized so that people can be categorized as having below-average, average, or above-average intellect. The IQ test questions cover a wide range of reasoning skills, such as solving number series puzzles, spotting patterns in spatial arrangements, or comprehending verbal analogies. In the early days of AI, the traditional IQ test method of assessing human intelligence was thought to be valuable for studying cognitive processes, developing new approaches, evaluating AI systems, and even serving as the endpoint for AI research [8]. Following the early fascination of AI study in IQ assessment issues, this particular field of study faded into obscurity during the subsequent several decades. Nevertheless, studies in cognitive sciences resurrected this area of inquiry in the 1990s, and cognitive frameworks have been suggested to recreate the human cognitive procedures that occur when addressing induction deduction in IQ assessment issues [9] In 2003, 40 years following the research of Evans, Simon, and Kotovsky, machine systems completing intellect assessments resurfaced in AI. On the contrary, [10] intended to demonstrate how simple it was to create non-intelligent systems to meet IQ examinations. [11] on the contrary, sought to resurrect the function of psychological evaluations in AI, encompassing not just intelligence evaluations but also assessments of character, artistic inventiveness, and furthermore. They suggested that psychometric testing should not be discarded, but rather played an important part in defining what AI is, and recommended "psychometric artificial intelligence" (PAI) as a study path. While this technique shifts away from the traditional task-focused classification of difficulties, it is unclear if it is independent of the huge shift method, particularly when the types of activities that arise in intellect assessments are identified ahead of time. In actuality, the [12] method employed a large shift strategy. Given these opposing views on the application of intellect assessments, it is difficult to predict the impact they were or will have.

1.1 Research Objective

1. This study's goal is to evaluate the effectiveness and implications of employing computer models to address problems with intelligence tests.
2. This study's goal is to assess the comparison of human performance and a computer model.

2. LITERATURE REVIEW

2.1 Interviewing and Observation

The most traditional method of determining IQ is observation. We have been able to ascertain (at first informally but later more systematically) that a few creatures have predominant gifts than others and that a few people are more cunning than others by checking the direction of creatures and people [13]. Ethology and brain research have been framed on the perception of conduct in regular or fake situations. Observation is a very strong tool. Conversation or, more precisely, interviewing, is a more effective method of determining intelligence and other cognitive capacities in people. This is a standard procedure in the fields of psychology, education, and hiring.

2.2 Efficacy at a Given Task

Using specialized tasks is the common way in computerized reasoning to assess its frameworks all the more (particularly when contrasts are slight). For practically every general application of AI, including planning, learning, gameplay, and deductive reasoning, as well as numerous specialized applications like machine translation, driving a car, chess, and robotics; we have benchmarks and competitions [14, 15, 16]. This is analogous to how people were assessed before the development of psychometrics. Some individuals were regarded as keen since they could peruse and compose Latin, were capable chess players, or knew the names of every Pope. In any case, this didn't necessarily suggest that these people would be successful in other endeavors.

2.3 Psychometrics-Based Systematic Testing

The widespread misconception that "idiot savants" were superior to people with average intelligence was one of the factors that led to the development of psychometrics in the late 19th century [17]. The first part of the 20th century saw a consolidation of IQ tests for both children and adults in humans. The assurance of the IQ (intelligence level), which is a score accomplished on a state-sanctioned test that evaluates an individual's insight, is a common psychometric way to deal with the estimation of knowledge [18]. There are numerous cognitive tests, though, that are not utilized to estimate IQ.

3. METHODOLOGY

This study's research goal is to evaluate the effectiveness and implications of employing computer models to address problems in IQ tests. This study objective will be addressed using the following methods. The study will evaluate the present impacts of computer models in AI using a descriptive research design. The collection, analysis, and interpretation of the impacts that the computer model has on the intelligence test are all made possible by this research design, which is suitable. For this study, both primary and secondary data will be gathered. Primary data will be gathered using survey questions that will be sent out to AI all across the world. The poll will be created to include questions regarding the present state of AI and how computer models are affecting it. Interviews or focus groups will be held if more primary data needs to be collected. Through a study of prior studies and publications pertinent to the research issue, as well as information from pertinent databases, secondary data will be gathered. Depending on the kind of data that was gathered, either quantitative or qualitative methodologies will be used to analyze the data. Descriptive and inferential statistics will be used to analyze quantitative data to ascertain the efficacy and ramifications of using computer models to address issues with intelligence tests. To find any patterns or themes in the responses provided by the respondents, qualitative data will be subjected to content analysis.

4. RESULTS AND DISCUSSION

The margin for error for both machines and humans has evolved, as seen in Fig. 48.2. The highest accuracy of the human mind is only 90%, according to philosophical research. The human mind has been required to remain constant since the dawn of civilization, whereas AI does not [9, 12]. Over time, artificial intelligence has expanded its breadth of knowledge and progressed into a world of reality where it can make decisions, recommend courses of action, and even take action [1, 2].

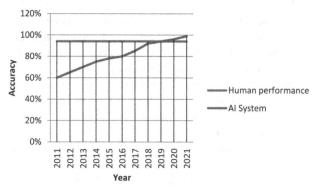

Fig. 48.2 Human performance and AI performance error comparison

In February 2015, Microsoft released the first picture categorization algorithm that outperformed humans, with an error rate of just 5.94%. Microsoft surpassed its prior mark in December 2015 with an error rate of 4.6%. Figure 48.3 shows the annual winning error rates for the ImageNet competition.

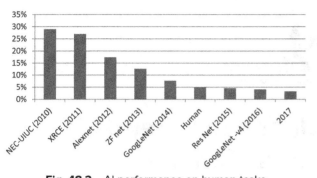

Fig. 48.3 AI performance on human tasks

AI programs are faster and, in most situations, more accurate at processing massive volumes of visual data than humans. However, when contextual information or comprehension of a complex connection is needed, computers may suddenly struggle to recognize images. The ability of AI programs to distinguish between a real man and a statue of a real man or to recognize an object from an obscured image may be problematic.

5. CONCLUSION

The motivation for this study was an evident increase in the number of studies including computational systems solving intelligence examination concerns. Half of the approximately 30 studies we examined were published after 2010. We sought to see if this growth was coincidental or caused by an increased demand for these assessments and the computational systems that solve them. When we started our research, we quickly realized that computational systems tackling intelligence examinations have a variety of reasons and applications: to develop AI through the implementation of complex issues (this is the Psychological AI method), to employ them for the assessment of AI platforms, to more effectively comprehend intellect tests and what they determine (which incorporates item challenge), and, ultimately, to more effectively comprehend what (human) intellect is.

The growth in the number of articles using computer models to solve problems related to intelligence tests served as the inspiration for this paper. At the point when we began our exploration, we immediately understood that PC models tending to knowledge tests have different objectives and applications, including propelling simulated intelligence by utilizing troublesome issues (this is the psychometric artificial intelligence approach), utilizing them to assess artificial intelligence frameworks, better comprehension insight tests and what they measure (counting thing trouble), lastly, better grasping what (human) insight is. The creation of a database or issue generator would be the most challenging objective. We are aware that many IQ tests are not accessible to the general public, and as a result, many of the approaches we have examined here have employed

alternate formulations. AI might organize these issues, keep track of how humans and computer models perform on them, and conduct competitions.

References

1. Silver, D., Huang, A., Maddison, C. J., Guez, A., Sifre, L., Van Den Driessche, G., ... & Hassabis, D. (2016). Mastering the game of Go with deep neural networks and tree search. *nature*, *529*(7587), 484–489.
2. Silver, D., Schrittwieser, J., Simonyan, K., Antonoglou, I., Huang, A., Guez, A., ... & Hassabis, D. (2017). Mastering the game of go without human knowledge. *nature*, *550*(7676), 354–359.
3. Tian, Y., & Zhu, Y. (2015). Better computer go player with neural network and long-term prediction. *arXiv preprint arXiv:1511.06410*.
4. Clark, C., & Storkey, A. (2015, June). Training deep convolutional neural networks to play go. In *International conference on machine learning* (pp. 1766–1774). PMLR.
5. Silver, D., Hubert, T., Schrittwieser, J., Antonoglou, I., Lai, M., Guez, A., ... & Hassabis, D. (2018). A general reinforcement learning algorithm that masters chess, shogi, and Go through self-play. *Science*, *362*(6419), 1140–1144.
6. Sternberg, R. J. (Ed.). (2020). The Cambridge handbook of intelligence. Cambridge University Press.
7. Goldstein, S., Princiotta, D., & Naglieri, J. A. (2015). Handbook of intelligence. *Evolutionary theory, historical perspective, and current concepts*, *10*, 978–1.
8. Darwiche, A. (2018). Human-level intelligence or animal-like abilities? *Communications of the ACM*, *61*(10), 56–67.
9. Smirni, P., & Smirni, D. (2022). Current and Potential Cognitive Development in Healthy Children: A New Approach to Raven Coloured Progressive Matrices. *Children*, *9*(4), 446.
10. Liu, Y., He, F., Zhang, H., Rao, G., Feng, Z., & Zhou, Y. (2019). How Well Do Machines Perform on IQ tests: a Comparison Study on a Large-Scale Dataset. In *IJCAI* (pp. 6110–6116).
11. Veldkamp, B. P. (2023). Trustworthy Artificial Intelligence in Psychometrics. In *Essays on Contemporary Psychometrics* (pp. 69–87). Cham: Springer International Publishing.
12. Pellert, M., Lechner, C., Wagner, C., Rammstedt, B., & Strohmaier, M. (2022). Large language models open up new opportunities and challenges for the psychometric assessment of artificial intelligence.
13. Eibl-Eibesfeldt, I. (2017). Human ethology. Routledge.
14. Bonsignorio, F., Del Pobil, A. P. D. P., & Messina, E. (2014). Fostering progress in performance evaluation and benchmarking of robotic and automation systems [tc spotlight]. IEEE Robotics & Automation Magazine, 21(1), 22–25.
15. Risi, S., & Preuss, M. (2020). From Chess and Atari to Starcraft and beyond How game AI is driving the world of AI. KI-Künstliche Intelligenz, 34, 7–17.
16. Nardi, D., Roberts, J., Veloso, M., & Fletcher, L. (2016). Robotics competitions and challenges. Springer handbook of robotics, 1759–1788.
17. Eysenck, H. (2019). The structure and measurement of intelligence. Routledge.
18. Sternberg, R. J. (2022). Intelligence. Dialogues in clinical neuroscience.

Note: All the figures in this chapter were made by the author.

Recent Trends in Engineering and Science for Resource Optimization and
Sustainable Development – Prof. (Dr.) Dorota Jelonek et al. (eds)
© 2024 Taylor & Francis Group, London, ISBN 978-1-032-98030-0

49

State-of-the-Art and Upcoming Trends in IoT-Enabled Smart Cities

Sushmita Goswami[1]

Assistant Professor, Institute of Business Management,
GLA University, Mathura, Uttar Pradesh, India

Vinay Kumar Nassa

Department of Information Communication Technology(ICT),
Tecnia Institute of Advanced Studies(Delhi),
Affiliated with Guru Gobind Singh Indraprastha University

Shaik Saddam Hussain

Department of Information Technology,
Institute of Aeronautical Engineering, Hyderabad, Telangana

Anandhi R J[2]

Department of Information Science Engineering,
New Horizon College of Engineering, Bangalore, India

Lovi Raj Gupta

Lovely Professional University, Phagwara, India

Purnendu Bikash[3]

Acharjee, Associate Professor, CHRIST University, Pune, India

■■■■■ Abstract

Modern cities' tremendous development of urbanization necessitates smart responses to pressing problems like mobility, medical care, power, and civil construction. The Internet of Things (IoT), which can use sustainable data and communication innovations, is evolving into the foundation for the upcoming trends of smart cities. To meet the demands of the expanding populace, several demands of the smart city must be taken into account. The IoT expansion has greatly generated a variety of study avenues for the smart city on the flip side of developing innovation. The suggested research proposal offers the analytic network procedure (ANP) for analyzing smart cities while maintaining in mind application instances of the smart city. In complicated circumstances when there are ambiguous options, the ANP technique performs effectively. The projected method's experimental findings demonstrate its viability for use case-based evaluation of IoT-enabled smart cities.

■■■■■ Keywords

Internet of things, Smart cities, Smart structures, Applications, and Technology

Corresponding author: [1]sushmita60485@gmail.com, [2]rjanandhee@hotmail.com, [3]pbacharyaa@gmail.com

DOI: 10.1201/9781003596721-49

1. INTRODUCTION

Because of their rapid development, modern cities are dealing with serious administration effectiveness along with the additional urban standard of life concerns [1]. For these and other contemporary difficulties in urban centres, technology for smart cities seem to offer numerous workable solutions [2]. A contemporary city that is customized in a smart and environmentally friendly manner that guarantees efficiency and sustainability is referred to as a "smart city." Combining different facilities and offerings into cohesive systems that can be observed and managed by smart gadgets is one way to accomplish this objective. The practice of building cities smartly has become more sophisticated in the past few decades. Additionally, more and more study is being done on smart city. A smart city has been completely computerized, connected, and informatized. It additionally utilizes the brains of people for independent thinking and reasoning [3, 4]. Municipalities, healthcare, property management, transit, public utilities, and public privacy and security are among the more integrated, effective, and intelligent sectors in smart cities [5]. It offers citizens improved assistance for their personal and professional lives, as well as improving the environment in which businesses can grow. Additionally, it makes government administration and operations more effective [6, 7]. The creation of a smart city has seen some early success thanks to the "smart" idea's widespread adoption. There has been an improvement in individual living circumstances. Businesses have discovered business prospects from it as executives' understanding of managerial data has gotten better and more precise [8].

On the fundamental question of "smart cities," nevertheless, researchers have differing opinions. Individuals describe and categorize "smart cities" in the context of technological advances and finances, according to studies on the topic. A smart city should, from a technological perspective, employ rational and scholarly information and communication technologies (ICT) to fully use the network's potential for bridging the technological gap and promoting the sharing of data [9]. To make decision-making processes more empirical and reasonable in the administration of towns and the environment, the societal risk evaluation of participative smart city development is applied. To make decision-making processes more empirical and reasonable in the administration of towns and the environment, the societal risk evaluation of participative smart city development is applied [10]. At the same point, it is encouraged to distribute and use natural assets, societal wealth, and human resources scientifically [11]. A smart city is defined from the standpoint of urban strategy as a structured and methodical expenditure on novel ICT

for societal growth and security. A tangible way to support urbanization is through the data of a smart city. Its main characteristic is the ability to emerge in a new manner and circumstance [12].

Internet of Things (IoT) applications are used in the smart city without the participation of humans. Various IoT gadgets are linked to one another to interact with one another for various purposes [13]. The likelihood of a breach of confidentiality and data exposure is rising as there will be a significant increase in IoT gadgets in the upcoming trends. IoT connections between billions of gadgets generate enormous volumes of data that are processed, managed, and stored in the cloud. Sending all of your data to the cloud could put safety and confidentiality in danger. Fog computation subverted the cloud computation model and served as a conduit between the cloud and the IoT [14]. Hence this research aims to investigate state-of-the-art and upcoming trends in IoT-based smart cities.

2. LITERATURE REVIEW

To outfit the IoT-oriented smart city, [15] suggested the idea of software-based connectivity architecture. The method enhances privacy preservation for completing data packages coming from divided IoT device information.

The idea of edge computation has been applied by [16] to enhance state awareness in an IoT-enabled smart city and to reduce the delay of gadgets at the edge generating data that needs to be transferred to the cloud [17]. Analyzing the massive amounts of data that are generated from the consumer's endpoint can significantly improve performance [11]. The trials' findings show that analyzing IoT unprocessed data at edge gadgets effectively provides contextual awareness for those making decisions in latency-sensitive smart cities [9, 12].

3. METHODOLOGY

Depending on the resources functioning in a tangible cyberspace context that offers communication with the system for information stream, like the web, the smart city layout can be divided into layers. Resources utilization and other gadgets perform a hugely important part in analyzing, managing, and controlling the IoT assets in the smart city. Various IoT gadgets are linked to one another and communicate with one another for various purposes. The 6 application instances from fictitious smart cities that were originally discovered have been taken into consideration for the planned investigation. Smart homes (SH), automobiles (SA), medical care (SMC), surroundings (SS), farming (SF), and monitoring (SM) are some of these application instances. Sending all of one's statistics to the cloud could put your confidentiality and safety in danger. The suggested investigation takes into account various smart city application instances that are used to evaluate smart cities. ANP's methodology has been applied to the assessment of smart cities.

4. RESULT AND DISCUSSION

To attain the objective of a smart city, the pairwise assessment calculation method was carried out for each of the options and standards. The pairwise assessment of Smart City 1's potential application instances is shown in Table 49.1. The procedure in Table 49.2 has been normalized, and Table 49.2 displays the normalization procedure. The normalized matrix is concatenated into a single matrix, creating a calculated super-matrix, following all the pairwise assessment procedures. In Table 49.3, the calculated super-matrix is displayed. For making choices, the calculated super-matrix was transformed into a limit matrix. This table was created by square-rooting the calculated matrix until all row values were the same. The limit matrix is displayed in Table 49.4. Figure 49.1 displays the order of the smart cities in comparison to the other options depending on application instances.

Table 49.1 The pairwise assessment of Smart City 1's potential application instances

	SH	SMC	SA	SM	SS	SF
SH	1	3	2	5	4	4
SMC	1/3	1	2	3	2	3
SA	½	1/2	1	3	3	2
SM	1/5	1/3	1/3	1	2	2
SS	¼	½	1/3	½	1	3
SF	1/4	1/3	1/2	1/2	1/3	1

Table 49.2 Outcomes of normalization of Table 49.1

	SH	SMC	SA	SM	SS	SF
SH	0.39	0.53	0.32	0.38	0.32	0.26
SMC	0.13	0.18	0.32	0.23	0.16	0.20
SA	0.20	0.09	0.16	0.23	0.24	0.13
SM	0.08	0.06	0.05	0.08	0.16	0.13
SS	0.10	0.09	0.05	0.04	0.08	0.20
SF	0.10	0.06	0.08	0.04	0.03	0.06

Table 49.3 Calculated super-matrix

	Application instances					
	SH	SMC	SA	SM	SS	SF
Smart City 1	0.43	0.50	0.404	0.406	0.48	0.44
Smart City 2	0.28	0.24	0.32	0.31	0.24	0.30
Smart City 3	0.15	0.15	0.17	0.15	0.14	0.16
Smart City 4	0.12	0.09	0.105	0.12	0.11	0.08

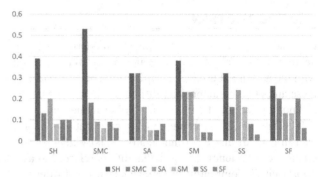

Fig. 49.1 Outcomes of normalization

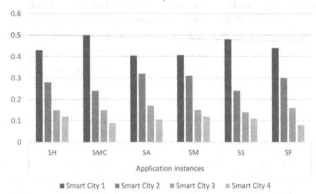

Fig. 49.2 Calculated super-matrix

Table 49.4 The outcomes of the limit matrix

	Smart City 1	Smart City 2	Smart City 3	Smart City 4
Smart City 1	0.453	0.452	0.452	0.452
Smart City 2	0.28	0.28	0.28	0.28
Smart City 3	0.161	0.160	0.160	0.160
Smart City 4	0.11	0.11	0.11	0.11

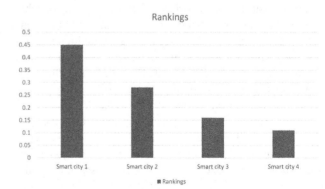

Fig. 49.3 Application instances of smart cities ranking

5. CONCLUSION

In this study, we examined the state-of-the-art of IoT investigation as it pertains to the upcoming trends of smart cities. IoT technologies are used in the smart city without the need for human intervention. IoT technology connects and communicates with one another for multiple reasons. The IoTs connect billions of technologies, generating enormous amounts of data that must be processed, managed, and stored on the cloud. Sending all data to the cloud could put your security and confidentiality at risk. For both immediate and efficient solutions to meet the demands of the expanding population, various demands of the smart city must be taken into account. The suggested investigation shows the ANP's method for assessing smart cities while maintaining in mind the application instances of the smart city. To choose a smart city, the procedure was mostly focused on the standards established for smart cities. The choice of a smart city may also have a significant impact on upcoming trends. The experimental outcomes of the suggested study demonstrate the efficacy of the method for assessing IoT-enabled smart cities depending on application scenarios.

References

1. Yin, C., Xiong, Z., Chen, H., Wang, J., Cooper, D., & David, B. (2015). A literature survey on smart cities. *Sci. China Inf. Sci.*, *58*(10), 1–18.
2. Silva, B. N., Khan, M., & Han, K. (2018). Towards sustainable smart cities: A review of trends, architectures, components, and open challenges in smart cities. *Sustainable cities and society*, *38*, 697–713.
3. Yin, C., Xiong, Z., Chen, H., Wang, J., Cooper, D., & David, B. (2015). A literature survey on smart cities. *Sci. China Inf. Sci.*, *58*(10), 1–18.
4. Lai, C. S., Jia, Y., Dong, Z., Wang, D., Tao, Y., Lai, Q. H., ... & Lai, L. L. (2020). A review of technical standards for smart cities. *Clean Technologies*, *2*(3), 290–310.
5. Kim, H., Choi, H., Kang, H., An, J., Yeom, S., & Hong, T. (2021). A systematic review of the smart energy conservation system: From smart homes to sustainable smart cities. *Renewable and sustainable energy reviews*, *140*, 110755.
6. Ghazal, T. M., Hasan, M. K., Alshurideh, M. T., Alzoubi, H. M., Ahmad, M., Akbar, S. S., ... &Akour, I. A. (2021). IoT for smart cities: Machine learning approaches in smart healthcare—A review. *Future Internet*, *13*(8), 218.
7. Ahad, M. A., Paiva, S., Tripathi, G., & Feroz, N. (2020). Enabling technologies and sustainable smart cities. *Sustainable cities and society*, *61*, 102301.
8. Javed, A. R., Shahzad, F., ur Rehman, S., Zikria, Y. B., Razzak, I., Jalil, Z., & Xu, G. (2022). Future smart cities: Requirements, emerging technologies, applications, challenges, and future aspects. *Cities*, *129*, 103794.
9. Ismagilova, E., Hughes, L., Rana, N. P., & Dwivedi, Y. K. (2020). Security, privacy and risks within smart cities: Literature review and development of a smart city interaction framework. *Information Systems Frontiers*, 1–22.
10. Ghazal, T. M., Hasan, M. K., Alzoubi, H. M., Al Hmmadi, M., Al-Dmour, N. A., Islam, S., ... & Mago, B. (2022, May). Securing smart cities using blockchain technology. In *2022 1st International Conference on AI in Cybersecurity (ICAIC)* (pp. 1–4). IEEE.
11. Guo, Y. M., Huang, Z. L., Guo, J., Li, H., Guo, X. R., &Nkeli, M. J. (2019). Bibliometric analysis on smart cities research. *Sustainability*, *11*(13), 3606.
12. Heidari, A., Navimipour, N. J., & Unal, M. (2022). Applications of ML/DL in the management of smart cities and societies based on new trends in information technologies: A systematic literature review. *Sustainable Cities and Society*, 104089.
13. Mehmood, Y., Ahmad, F., Yaqoob, I., Adnane, A., Imran, M., &Guizani, S. (2017). Internet-of-things-based smart cities: Recent advances and challenges. *IEEE Communications Magazine*, *55*(9), 16–24.
14. Zhang, C. (2020). Design and application of fog computing and Internet of Things service platform for a smart city. *Future Generation Computer Systems*, *112*, 630–640.
15. Gheisari, M., Wang, G., Khan, W. Z., & Fernández-Campusano, C. (2019). A context-aware privacy-preserving method for IoT-based smart city using software-defined networking. *Computers & Security*, *87*, 101470.
16. Hossain, S. A., Rahman, M. A., & Hossain, M. A. (2018). Edge computing framework for enabling situation awareness in IoT-based smart city. *Journal of Parallel and Distributed Computing*, *122*, 226–237.
17. Soni, N. Kumar, V. Kumar and A. Aggarwal, "Biorthogonality Collection of Finite System of Functions in Multiresolution Analysis on L2(K)," 2022 10th International Conference on Reliability, Infocom Technologies and Optimization (Trends and Future Directions) (ICRITO), Noida, India, 2022, pp. 1–5, doi: 10.1109/ICRITO56286.2022.9964791.

Note: All the figures and tables in this chapter were made by the author.

*Recent Trends in Engineering and Science for Resource Optimization and
Sustainable Development – Prof. (Dr.) Dorota Jelonek et al. (eds)
© 2024 Taylor & Francis Group, London, ISBN 978-1-032-98030-0*

50

The Tracking of (Machine) Intelligence's Evolution Using an Intelligence Catalogue

Sirisha K L S[1]

Assistant Professor,
Keshav Memorial Institute of Technology

Richa[2]

Principal, Government Leather Institute Kanpur, India

R. Anuradha

Department of Information Technology,
Institute of Aeronautical Engineering,
Hyderabad, India

Anandhi R J[3]

Department of Information Science Engineering,
New Horizon College of Engineering, Bangalore

Lovi Raj Gupta

Lovely Professional University, Phagwara, India

Purnendu Bikash Acharjee

Associate Professor, Computer Science, CHRIST University,
Bangaluru, India

Abstract

The purpose is to investigate the usage of capabilities that identify intelligence in the scientific discourse on AI, to track the evolution of the mythology around intelligence and how it appears in both people and computers across time. The form of a catalog, and covering various domains, including AI, intelligence capabilities, and related traits that are used to define intelligence were extracted from prior research in this area. Even if intelligence is still a nebulous, ill-defined term, examining and comprehending the language surrounding it could influence how we utilize it as well as how intelligent artifacts are made now and in the future.

Keywords

Machine intelligence, Human intelligence, AI, Intelligence catalog

Corresponding authors: [1]klssirisha@gmail.com, [2]vermaricha29@gmail.com, [3]rjanandhee@hotmail.com

DOI: 10.1201/9781003596721-50

1. INTRODUCTION

In recent decades, intelligence has been a hot issue across several fields, including human science, computer and information science, social science, and biology [1]. The greatest way to define intelligence is via the abilities of awareness and cognition as well as the wide application of natural laws to all living systems [2-4]. The broadest spectrum of intelligent occurrences is encompassed by this broad word. The capacity to actualize and maximize the operation of artificial systems in line with human desires and natural laws, and to activate such systems in answer to such demands, is the traditional definition of intelligence. Intelligence enables a system to respond with the necessary actions at the relevant moments in suitable ways to accomplish the intended results. This is done through an autonomous decision-making process. As far as we know, no research has been done that looks at how terminology used in intelligence research has changed over time or how that change has affected the advancement of machine or artificial intelligence (AI) and the terms used by the machine intelligence community to define it. Over time, several things have grown increasingly obvious: There is no widely accepted definition of artificial intelligence, the field is multidisciplinary, and almost every study determines intelligence through the lens of their discipline. In addition, the recent media AI hype has attempted to obstruct further understanding, fueled conflict (not just within the AI community), and diverted investigations [3].

1.1 Objective of Study

1. To explore the evolution of the scientific community's use of the vocabulary related to intelligence over time.
2. To describe the words used to describe intelligence in the scientific literature.
3. To examine how a common concept of intelligence and AI is impacted by the language of intelligence.

2. LITERATURE REVIEW

Wiener attempted to examine the distinction between people and machines [4]. He claimed that humans had unique skills for recognizing and responding to environmental changes. He thinks that human beings and machines also have control and communication networks and that this is why artificial and live systems function similarly. "Cybernetics" has a specific meaning in his book, encompassing control, feedback, communication, and interaction.

Machine intelligence may be defined as the use of automated machining in place of manual labor, machine computing in place of the human intellect, and data circulation in place of human mobility. The value of the aforementioned three features is significantly raised by complicated business processes, constrained time windows, and rising labor prices [5]. For instance, intelligent manufacturing was proposed as a way to release humans from tasks that robots can perform. There is a lot of evidence to suggest that robots can execute some jobs more effectively than people [2, 6].

The capacity for decision-making is a measure of system intelligence. A greater level, for instance, denotes that a system is capable of handling more circumstances. Five key traits—state recognition, real-time analysis, autonomous decision-making, accurate execution, and learning-based promotion—can be used to quantify the level of system intelligence [7]. As an extension of Wiener's approach, we developed five components to measure the level of intelligence of a physical object, a human consciousness, and a cyber-entity. Hu and colleagues categorized intelligent systems by the five characteristics [8].

3. METHODOLOGY

3.1 Data Collection

This study's goal is to use an intelligence catalog to trace the development of (machine) intelligence. This study aim will be addressed using the following methods. To evaluate the comparison between human intellect and machine intelligence, the study will employ a descriptive research approach. This study approach is suitable because it enables the collection, analysis, and interpretation of the intelligence catalogue-based assessment of machine intelligence. For this study, both primary and secondary data will be gathered. Survey questions will be given to both human intelligence and machine intelligence to gather primary data. The poll will be made to include questions regarding the state of machine intelligence right now and how the intelligence catalog is affecting it.

Interviews or focus groups will be held if more primary data needs to be collected [9]. Through a study of prior studies and publications pertinent to the research issue, as well as information from pertinent databases, secondary data will be gathered [1]. Depending on the kind of data that was gathered, either quantitative or qualitative methodologies will be used to analyse the data [7,2]. To assess the machine intelligence using an intelligence catalogue, quantitative data will be analyzed using descriptive and inferential statistics. Utilizing content analysis, qualitative data will be examined for any themes or patterns in the respondents' replies.

3.2 Data Pre-Processing and Assessment

The dataset was processed and examined using Python. Some of the libraries utilized are Matplotlib & NumPy, which are employed to create various charts, and tz and PyMuPDF, which are utilized for reading the written material from PDF documents (in this instance of the IJCAI articles). Pandas can be utilized to browse Excel spreadsheets and perform multiple helpful functions on single and multiple-dimensional arrays. It additionally serves for data purification, layout, and assessment.

The configurator examines two setup documents: one containing the titles of the pertinent meta data's columns and the other including the search terms or intelligence features that need to be looked for. After that, the configurator compares the statistics with the databases' contents and gets it ready for processing by the evaluation instrument's other parts. After that, the databases are cleaned up, and any superfluous material and metadata are removed. Additionally, other typical NLP pre-processing operations are carried out, such as lower-casing the search terms, removing emojis and other distinctive characters, and transforming the data to CSV files for additional processing. A handy illustration of the rest of the material has been produced. The year and rate of release of the various academic publications are then taken into consideration at the beginning of the study. Additionally, the search term instances are recorded to prevent sporadic repeats of the same item. The data is generated for the last stage and proportions about the annual number of articles are computed. Eventually, numerous charts that are prepared for display are created.

When it comes to the instance of the IJCAI articles with accessible material, the relevant assessment method is substantially simpler. There are just two parts to it: the PDFMiner and the analysis tool. The former takes the text out of the documents in PDF format and determines how frequently the relevant search terms occur. The last analyzes the data that has been processed, creates the appropriate charts, and stores the output in the designated folder. Here are also completed the handy NLP pre-processing activities. It's important to note that using the PyMuPDF library's suitable operations for processing PDF documents is considerably quicker than using different libraries (such as tex-tract, Apache Tika, or PyPDF2)—about 0.001 seconds for every PDF document.

4. Results and Discussion

Figure 50.1, which depicts the historical rise in the employment of various intelligence-related phrases in papers from AI Topics, enables a more detailed, extensive analysis of particular trends spanning more than 60 years. Observe that throughout the five AI winters (1974–1980 and 1987–1993), the term "learn" was least employed and began to get traction in the 1990s.

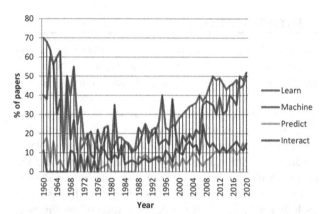

Fig. 50.1 Percentage of intelligence-related terms in the metadata of articles published on AI Topics

In Figs. 50.2 and 50.3, together with some of the capabilities, it is demonstrated how frequently various intelligence capabilities are employed in the text of IJCAI articles. Observe how certain words like predict and think have become less common over time quicker than others. They may be progressively losing their significance in the academic discussion and, consequently, in the lexicon related

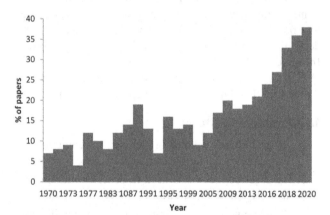

Fig. 50.2 Paper containing the word Predict

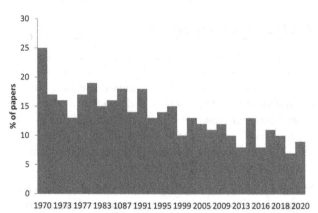

Fig. 50.3 Paper containing the word 'Think'

to machine intelligence. Winston's famous statement [13], which once regarded machine intelligence as the highest form of intelligence, may no longer be applicable given the seriousness of the "AI impact," as McCorduck mentions in her work.

They could be losing significance over time in scientific discourse and, consequently, in the lexicon around artificial intelligence. The metadata gathered from AAAI and IJCAI just extends back to 1997, while the arXiv articles go just back to 1993. However, the discipline of AI is far earlier. Although the meeting has been held since 1974 and the ECAI sessions are an excellent resource of knowledge, only a small portion of the sessions are published electronically (starting in 2000 and occurring every two years after that).

Also, this study endeavour does not take into account the numerous other AI meetings and seminars that are held annually, some of which have a long history within the AI community. For instance, the 1987–2018 NeuriIPS digital sessions are accessible online; however, this meeting is focused on particular aspects of machine learning and does not address other significant subdivisions of artificial intelligence. However, it will be taken into account in future research. For comparable seminars, data has also been gathered and is accessible; however, the time frames are extremely constrained, and the sessions cannot be downloaded. The code for data scraping may be also accessible. However, a few of these gatherings are more focused than others.

As correctly highlighted by among of the unidentified reviewers, we fail to consider how English linguistics and the significance of certain phrases have changed as time passes, particularly in the context of artificial intelligence and its subdivisions. This represents another potential constraint of our study as well as the larger investigation which encompasses it. Nevertheless, we believe that our research would help achieve that objective more effectively, particularly when taking into account semantic evaluation (a topic present work), as it would be simpler to identify the situations in which certain smart abilities and reports are employed by the academic society.

5. CONCLUSION

Although early, the methods employed for gathering, organizing, and analyzing the data, as well as the first findings that are provided in this study, form the cornerstone for a much more thorough investigation of how the language used to characterize intelligence has changed through time. However, further study is required in this area. The vast volume of intelligence research conducted outside the AI field must also be taken into account in future efforts, to sum up. Over the years, more correct use of the language on intelligence could be framed by factors such as semantic characteristics that enable the distinction of certain settings

on the one hand, and the corresponding significance of the terms employed to describe intellect on one side. Research on this subject is continuing. A comparable examination of the descriptions of artificial and human intelligence was previously started. It is not necessary to fundamentally alter those already available techniques; rather, they only require to be extended to use the novel information presented in this study.

Lastly, a large corpus of studies on intelligence from disciplines other than AI (such as neuroscience and psychology, two of the most well-known disciplines in that area of study) needs to be taken into account in further studies. Although the use of the term "intelligence" in these fields does not suffer from the same lack of consensus that the AI community cannot ignore, acknowledging their best practices and lessons learned would be beneficial to both the AI field and the discussion about intelligence both inside and outside of it.

To achieve these new goals, new data must first be gathered, examined, and prepared, such as previously published papers and their metadata in the fields of psychology, neuroscience, and related fields. These tasks shouldn't be devalued because they take up the greatest time.

References

1. Wang, L. From Intelligence Science to Intelligent Manufacturing. Engineering 2019, 5, 615–618.
2. Zhou, J.; Li, P.; Zhou, Y.;Wang, B.; Zang, J.; Meng, L. Toward New-Generation Intelligent Manufacturing. Engineering 2018, 4, 11–20.
3. Bi, Q., Goodman, K. E., Kaminsky, J., & Lessler, J. (2019). What is machine learning? A primer for the epidemiologist. American journal of epidemiology, 188(12), 2222–2239.
4. Wiener, N. (2019). Cybernetics or Control and Communication in the Animal and the Machine. MIT Press.
5. Zhou, J.; Zhou, Y.;Wang, B.; Zang, J. Human–Cyber–Physical Systems (HCPSs) in the Context of New-Generation Intelligent Manufacturing. Engineering 2019, 4, 624–636. [CrossRef]
6. Wang, B.; Zang, J.; Qu, X.; Dong, J.; Zhou, Y. Research on New-Generation Intelligent Manufacturing based on Human-Cyber- Physical Systems. Strategy. Study Chin. Acad. Eng. 2018, 20, 29–34.
7. Yang, J., Xiu, P., Sun, L., Ying, L., & Muthu, B. (2022). Social media data analytics for business decision-making system to competitive analysis. Information Processing & Management, 59(1), 102751.
8. Hu, H.; Zhao, M.; Ning, Z. Three-Body Intelligence Revolution; China Machine Press: Beijing, China, 2016.
9. Soni, N. Kumar, V. Kumar and A. Aggarwal, "Biorthogonality Collection of Finite System of Functions in Multiresolution Analysis on L2(K)," 2022 10th International Conference on Reliability, Infocom Technologies and Optimization (Trends and Future Directions) (ICRITO), Noida, India, 2022, pp. 1–5, doi: 10.1109/ICRITO56286.2022.9964791.

Note: All the figures in this chapter were made by the author.

Recent Trends in Engineering and Science for Resource Optimization and Sustainable Development – Prof. (Dr.) Dorota Jelonek et al. (eds)
© 2024 Taylor & Francis Group, London, ISBN 978-1-032-98030-0

51

Research on Theoretical Contributions and Literature-Related Tools for Big Data Analytics

M. Ravichand[1]

Professor of English, Velagapudi Ramakrishna Siddhartha Engineering,
Siddhartha Academy of Higher Education - A Deemed to be University; Vijayawada, Andhra Pradesh

Kapil Bansal[2]

Assistant Professor, Institute of legal studies and Research,
GLA University, Mathura, Uttar pradesh

G. Lohitha

Department of Information Technology,
Institute of Aeronautical Engineering, Hyderabad, Telangana

Anandhi R. J.

Department of Information Science Engineering,
New Horizon College of Engineering, Bangalore

Lovi Raj Gupta

Lovely Professional University, Phagwara, India

Patel Chaitali Mohanbhai[3]

Assistant Professor, Faculty of Computer Science,
Shri C.J. Patel College of Computer Studies, Sankalchand Patel University

Narendra Kumar[4]

NIET, NIMS University, Jaipur, India

Abstract

Big data analytics and data science are becoming increasingly important as businesses prepare to use their information assets to their advantage. Big data analytics' versatility enhances both functional and organizational effectiveness. We aim to analyze the big data research that has been published in reputable business management journals in the first phase of the study. Big data and text mining methods were used to visualize the information to identify the key themes and their relationships. After that, a classification of the research according to industry was carried out to identify the most important use cases. It was shown that consumer discretionary is the main focus of the majority of the current study, followed by public administration.

Keywords

Big data, Systematic literature review, Information management

Corresponding author: [1]ravichandenglish@gmail.com, [2]kapil.bansal@gla.ac.in, [3]cmpatel.fcs@spu.ac.in, [4]drnk.cse@gmail.com

DOI: 10.1201/9781003596721-51

1. INTRODUCTION

The Internet of Things (IoT) concept violates the conventional approach of segregating technology and data transmission, really achieves the connectivity between objects, and embraces the integrated idea of the web, which is extremely important [1]. If data is incorporated in the IoT interaction the number of interaction links that could be incorporated may exceed billions, providing a huge area for data transfer.

The objective of secure big data sharing is to enable consumers who utilize various software and hardware in various locations to access other individuals' big data and execute multiple tasks and assessments. The extent of secure big data sharing indicates an area and the nation's degree of data growth. The more advanced the degree of data growth, the greater the level of secure big data sharing. By enabling secure big data sharing, additional individuals may make better utilization of current big data assets, lowering the expenses of repetitive work and big data collecting while focusing on the creation of novel uses and integrating systems [2]. Because data contributed by various consumers can arrive from diverse sources, the big data material, design, and quality vary greatly. Secure big data sharing becomes extremely complex because the data format is unable to be transformed without losing big data. Big data flow and exchange in multiple areas and software platforms are substantially hampered as a result of unsolvable challenges like these [3]. With the explosive growth of machine degrees and technological communication, the use of IoT in electronic commerce, power surveillance, armed forces, national security, and additional private areas has grown more prevalent, and secure big data sharing and transfer has assumed a more significant place. Big data interaction is the primary way of data transmission, and it is critical for ensuring the security of big data interaction in the IoT [4].

Businesses may analyze user-generated data in several formats, like video, blogs, and social media data, which has various benefits for them, including increasing production and competitiveness [5]. Digital transaction data is frequently used by companies like Apple, Google, Facebook, eBay, and Amazon to improve their company operations. These businesses must regularly record transaction information such as transaction times, product pricing, purchase amounts, and client credentials to estimate market conditions, consumer behavior, trends, patterns, and other factors. The big data analytics review of theoretical contributions and literature-based techniques is shown in Fig. 51.1.

Most present secure big data sharing encryption systems do not take into account the interaction overhead and burden of data owners and consumers; on the contrary, conducting calculations on encrypted secure big data shared in

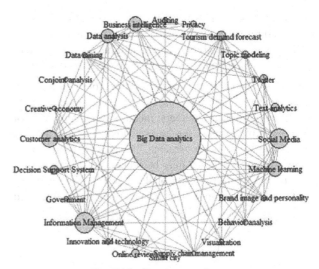

Fig. 51.1 Big data analytics

the cloud was not performed properly. The implementation of impact optimization in social media platforms, in particular, nevertheless confronts considerable obstacles [6]. Right now, secure sharing of big data is in its early stages, and not every element is complete, particularly in law and rules; there nevertheless remain significant gaps, and there have been a rising number of lawbreakers, who will gradually utilize their disadvantages to damage the community. Data-driven choices regarding production and inventory, sales forecasting, commodity price optimization, inventory optimization, supplier coordination, demand forecasting, and logistics optimization may all be made by producers using the data [7]. They can also use the data to improve their services and boost customer satisfaction. Businesses are relying more and more on information, knowledge, and evidence-based insights derived from data, claim Chowdary, Muthineni, and Singh.

Ten years ago, systems to handle such massive amounts of data were not necessary, but today's big data storage, analysis, and visualization demand scientific methodologies for businesses to make the greatest use of the information assets that are easily accessible to them. For academic certification, big data architecture with five steps has been established in the literature [8]. However, a radical departure from conventional data analysis is required for the study of large data, which has given rise to the field of data science.

2. LITERATURE REVIEW

The 5 V's of big data were created to address the drawbacks of relational database management systems, which is the main reason why big data is growing in popularity [9]. Original literature highlighted the four V's of big data. The first V, or volume, refers to the large amount of data that is frequently produced by machines and sensors. According to published research, storing such large amounts of data and dealing with its scalability is a challenging task. The velocity of data generation is represented by the second V [10].

One of the most well-known examples of high-velocity created data, where a lot of user-provided unprocessed data is routinely produced, is social media data streams. One of the most difficult tasks for data scientists is managing and processing various data streams. The third V represents a range of data types that may be looked at to provide important information. Traditional databases, for instance, contain clearly defined data formats, but semi-structured and unstructured data must now be captured as well [11].

The importance of practical insights obtained from big data is symbolized by the fourth V. Finding the important information that can be taken from the broader database's valuable data is necessary. The fifth V of big data is veracity, and it has changed since 2012. Veracity describes the degree of data uncertainty. According to studies, the veracity traits of big data include objectivity, honesty, and believability [12]. Another difficult aspect of large data analysis is the technology an analyst must utilize.

Although the researcher provided a particular solution, it has yet to be tested in practice, hence its viability is unknown [13]. As per [13], the existing situation of the medical business makes it hard to validate, maintain, and synchronize clinical data, making clinical big data exchange an exhausting endeavor. The outcome is in significant validation assets and period expenses. To address this issue, he developed a block-based health secure big data sharing paradigm that benefits from decentralization, extreme safety, collaborative preservation, and tamper tolerance. It has the potential to make secure big data sharing easier, safe, and quick, as well as improve compatibility between various kinds of healthcare organizations.

While the researcher highlighted the benefits of his suggested approach, he did not explain how to put it into healthcare big data [14]. According to [14], keyword searching for encrypted documents constitutes a basic demand of customers in a secure big data sharing platform. While typical searchable encryption approaches can give security for confidentiality, two critical challenges must still be addressed.

He devised a method known as "Verifiable Searching Encryption using Collectively Keys." In this technique, once the consumer has such a key, it may be utilized for verifying the server's safety.

3. METHODOLOGY

To evaluate the development of such literature, we first selected the database that may let us conduct searches of the published big data literature. Scopus was picked for this as one of the larger databases of academic journal articles that contained applicable articles since it is the largest database for literature with an engineering and management focus. Additionally, it offers a variety of fields for the user to look for research articles [5]. The Scopus database is utilized to locate the research publications related to this development. On March 21, 2018, we looked for pertinent studies in the Scopus database. First, we searched for big data text in the "Article Title" and "Keywords," and 27,562 articles were retrieved. After that, we restricted our search to journal papers only to discover publications with a high level of peer review. Around 6820 studies have been collected in the second round. We had "business access to the location in the third round. There were 640 fewer studies overall.

Since there are a lot of academic publications concerning big data, inclusion/exclusion criteria were employed to draw attention to big data analytics applications in certain sectors in respected academic journals [6, 7]. Studies that only discuss methods and don't address or give any answers to challenges unique to an industry were left out [8]. Publications that address issues specific to the industry and appear in journals with a grade of three or above were included. Studies that investigate how a real-world, practical application of business analytics functions were deemed eligible for inclusion.

4. RESULTS AND DISCUSSION

In Fig. 51.2, the research studies chosen for this systematic review are listed by year of publication. Only 21 of the 123 research articles were published before 2016; the remaining 123 studies were published in 2016 or later until March 2018. As shown, there is a growing emphasis on publishing big data studies in reputable publications.

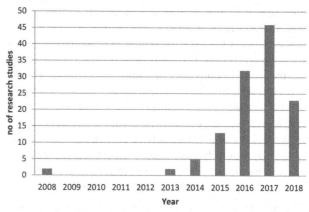

Fig. 51.2 Distribution of the research studies by year

There are 123 research papers, although 42 of them are broad in nature and target relevant fields associated with big data (such as marketing and policy), rather than focusing on a single analytics-related subject. Figure 51.3 displays the remaining 81 research papers, which span six industry categories.

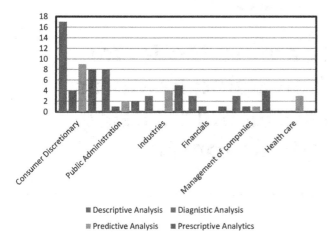

Fig. 51.3 Distribution of the papers chosen for systematic review throughout the fields of industry and the big data analytics that were followed

It is clear from Fig. 51.3 that most industries combine descriptive analytics with big data analytics. Additionally, only three case studies of diagnostic analytics usage were found. Prescriptive analytics-focused big data analytics research is rather few. It has been shown that the healthcare industry heavily relies on both big data analytics and predictive analytics.

5. Conclusion

This study aims to provide recommendations for big data analytics researchers. Understanding the prospective fields where big data has been actively employed would be the first step in such a study inquiry. This would open the door to studying possible regions that have received less attention. Based on how the tools were categorized, numerous big data services and their applications in different sectors were examined in this study report. Users can select the tools they need to implement any solution for solving their research topic. According to the literature, employing these big data tools makes big data processing computationally simpler since distributed computational methodologies are used. It is important to highlight that the currently available literature does not pay much attention to the platforms and technologies that enable the outcome realization from big data difficulties. In particular, the field of data aggregation and transmission has not yet been well investigated. The 5 V problems of big data analytics should be manageable for anyone employing these tools and technologies. Big data analytics would be able to identify and then attempt to

solve complicated problems by coupling the proper problem domains with the proper technologies.

References

1. Liu, L., Kong, W., Cao, Z., & Wang, J. (2017). Analysis of one certificateless encryption for secure data sharing in public clouds. *International Journal of Electronics and Information Engineering, 6*(2), 110–115.
2. Marwick, B., & Birch, S. E. P. (2018). A standard for the scholarly citation of archaeological data as an incentive to data sharing. *Advances in Archaeological Practice, 6*(2), 125–143.
3. Xu, C., & Li, P. (2017). pth moment exponential stability of stochastic fuzzy Cohen–Grossberg neural networks with discrete and distributed delays. *Nonlinear Analysis: Modelling and Control, 22*(4), 531–544.
4. Phillips, M. (2018). International data-sharing norms: from the OECD to the General Data Protection Regulation (GDPR). *Human genetics, 137*, 575–582.
5. Maklan, S., Peppard, J., & Klaus, P. (2015). Show me the money: Improving our understanding of how organizations generate a return from technology-led marketing change. European Journal of Marketing, 49(3/4), 561–595.
6. Liang, K., Fang, L., Wong, D. S., & Susilo, W. (2015). A ciphertext-policy attribute-based proxy re-encryption scheme for data sharing in public clouds. *Concurrency and Computation: Practice and Experience, 27*(8), 2004–2027.
7. Bradlow, E. T., Gangwar, M., Kopalle, P., & Voleti, S. (2017). The role of big data and predictive analytics in retailing. Journal of Retailing, 93(1), 79–95.
8. Hussain, M., Al-Mourad, M., Mathew, S., & Hussein, A. (2017). Mining educational data for academic accreditation: Aligning assessment with outcomes. Global Journal of Flexible Systems Management, 18(1), 51–60.
9. Jin, J., Liu, Y., Ji, P., & Liu, H. (2016). Understanding big consumer opinion data for market-driven product design. International Journal of Production Research, 54(10), 3019–3041.
10. Hammer, C., Kostroch, M. D. C., & Quiros, M. G. (2017). Big data: Potential, challenges and statistical implications. International Monetary Fund.
11. Bafna, A., Wiens, J. (2015). Automated feature learning: Mining unstructured data for useful abstractions. In 2015 IEEE International Conference on Data Mining (ICDM), (pp. 703–708). I
12. Lozano, M. G., Brynielsson, J., Franke, U., Rosell, M., Tjörnhammar, E., Varga, S., & Vlassov, V. (2020). Veracity assessment of online data. Decision Support Systems, 129, 113132.EEE.
13. Li, J., Zhang, Y., Chen, X., & Xiang, Y. (2018). Secure attribute-based data sharing for resource-limited users in cloud computing. *computers & security, 72*, 1–12.
14. Xue, T. F., Fu, Q. C., Wang, C., & Wang, X. (2017). A medical data sharing model via blockchain. *Acta Automatica Sinica, 43*(9), 1555–1562.

Note: All the figures in this chapter were made by the author.

Recent Trends in Engineering and Science for Resource Optimization and Sustainable Development – Prof. (Dr.) Dorota Jelonek et al. (eds)
© 2024 Taylor & Francis Group, London, ISBN 978-1-032-98030-0

52

Ensemble Machine Learning Model for ECG based Identification Using Features Extracted from Digital ECG Signal using Deterministic Finite Automata

Mamata Pandey[1]

Computer Science and Engineering, Birla Institute of Technology Mesra,
Ranchi, Jharkhand, India

Anup Kumar Keshri[2]

Computer Science and Engineering, Birla Institute of Technology,
Mesra, Ranchi, Jharkhand, India

Abstract

Biometric systems use some biological data such as finger prints, face or iris to identify and authorize individuals. Many security systems rely on biometric. But most of the systems used today are subjected to fraud as they are easy to mimic. This paper presents a method for identifying an individual using ECG signal. A Deterministic Finite Automaton (DFA) is designed to identify PQRST points in ECG signal. DFA accepts digital ECG signal and produce set of fiducial points P, Q, R, S and T for each cardiac cycle. These PQRST values are used to calculate a set of features which include temporal, amplitude, distance, slope, HRV and some miscellaneous features. Machine learning models are then used to identify individuals using these features. An efficient model is required that can identify an individual accurately. This paper presents a comparative study of six popular classifiers MLP, KNN, SVM (kernel: linear), SVM (kernel: RBF), Decision tree, Naïve Bayes classifiers, Boosting, Stacking and Voting techniques are used to gives better accuracy.

Keywords

ECG, DFA, PQRST, ECG biometrics, Ensemble classifier

1. INTRODUCTION

Electrocardiogram (ECG) represents electrical activity of heart. Recent researches have proved that ECG is unique and can be used for identification and authentication (Uwaechia and Ramli, 2021). ECG is symbol of life and it's hard to mimic. This trait gives ECG a very high potential as tool for next generation biometric. Features can be fiducial or non- fiducial. Several types of features are used by researchers for ECG based identification. Non- fiducial

[1]m.pandey@bitmesra.ac.in, [2]anup_keshri@bitmesra.ac.in

DOI: 10.1201/9781003596721-52

features can be statistical like mean, standard deviation (Khan et. al., 2019) or features obtained by performing various wave transforms on ECG wave or wave features (Aziz et. al., 2019). Fiducial features are calculated using amplitude and time instances of fiducial points on cardiac cycle. (Ingale et al, 2020). Figure 52.1 shows fiducial points P, Q, R, S and T on a cardiac cycle. In this paper fiducial features have been used. To calculate fiducial features, it is necessary to locate fiducial points P, Q, R, S and T on ECG signal accurately. A number of methods has been practiced by researches in past such that Kalman filter (Avendano et.al., 2020), wavelet transformation (Spicher, N. and Kukuk, M., 2020), convolutional network (Silva et. al., 2018). Many methods detect fiducial points in two phase (Patro & Kumar, 2016). In first phase R peak is determined. Other points are detected in second phase as peaks and dips by moving back and forth from the R peak in time domain (Israel, et al., 2005, Patro & Kumar, 2017). This work uses 71 features similar to those used by previous researches for classification (Mincholé, et al., 2019). They include classifiers like artificial neural networks (ANN), convolution neural networks (CNN), support vector machines (SVM), Bayesian networks, decision trees, k-nearest neighbor (KNN) and clustering method like k-mean clusters, density-based clustering (DBSCAN), fuzzy C-mean clustering (Nezamabadi, et. al., 2022). Performance of machine learning models are evaluated with metrices accuracy, recall, precision and F1-score (Flach, 2019). Accuracy represents the number of correctly classified data tuples over the total number of tuples. Accuracy may not be right metric if dataset is not balanced (Dietterich., 2000) Ensemble classifier combines a number of classifiers to predict the final output. Final output is determined by combining outputs of all constituent classifiers. Basically, there are three ensemble techniques, bagging, boosting and stacking (Zhang et al., 2022). Bagging consists of bootstrapping and aggregation. Bootstrapping is creating a number of random samples from the dataset and each sample is used to train a different model (Dietterich, 2000). Some works has experimented with ensemble classifiers like Ad boost and Gradient boost that contain a number of homogeneous base classifiers. Ada boost stands for adaptive boosting (Soui et al., 2021). In Ad boost ensemble classifier weights are assigned to constituent classifiers as well as training data

samples. In each iteration weights are adjusted to increase prediction accuracy. In gradient boosting each constituent classifier improves on its predecessor.

2. METHODOLOGY

The experiment is performed on 21 digital ECG recording samples of 19 individuals (age 20 to 55, male/female). Data has been recorded in department of Bio-engineering and Biotechnology laboratory, Birla Institute of Technology, Mesra, Ranchi, Jharkhand, India. Samples are of different durations from 30 second to 10 minutes. Sampling frequency is 250 Hz that is there are 7500 readings in 30 second. Each reading is a datapoint of form (x, y) where y is amplitude of signal at time instance x. For extracting fiducial features from ECG signal, first step is to find fiducial features P, Q, R, S and T. This work has used algorithm based on Deterministic Finite Automata (DFA) to detect all PQRST sets in ECG signal. DFA is model of computer that can identify a specific pattern in input string. ECG signal is first transformed into ECG string constitute of symbols 1. 0 and -1. Transformation is on basis of slope between two consecutive readings. Slope between two points (x1, y1) and (x2, y2) is (y2-y1)/(x2-x1). A threshold value 100 is selected after try and error. The slope is replaced by a symbol from {1,0, -1} depending on its value. If slope > 100 then symbol is 1, if slope < -100 then symbol is -1 else symbol is 0. The slope on QRS is very high in respect to the slope on P wave and T wave. QRS wave shows a very clear pattern of five or more 1's followed by five or more -1's. P is highest point before QRS. T is highest point after QRS. End of T wave is identified by three consecutive -1's. DFA is defined by 5-tuple {Q, Σ, δ, q0, F}, where, Q is set of states = {q0, q1, q2, q3, q4, q5, q6, q7, q8, q9, q10, q11, q12, q13, q14, q14, q15}, Σ is finite non-empty set of input symbols, {1,0, -1}, δ is transition function, which controls the movement from one state to another depending upon the current state and current input symbol. It is defined as δ: Q × Σ → Q given by transition diagram in Fig. 52.2. q0 is starting state and F is finite non-empty set of final states {{q0, q1, q5, q10, q12}. Initially, DFA is at state q0 and starts reading ECG string from left to right. Transition to next state is performed according to transition function δ

Fig. 52.1 PQRST fiducial points

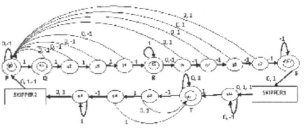

Fig. 52.2 DFA for PQRST identification

depending on current symbol and current state. The final states q0, q1, q5, q10 and q12 determine fiducial points P, Q, R, S and T respectively. It gives output as a set of tuples (Px, Py, Qx, Qy, Rx, Ry, Sx, Sy, Tx, Ty). For each 21 samples of digital ECG signal all PQRST sets are identified. The accuracy is 96%. Z-score outlier detection technique is used to detect misidentified points and corresponding tuples are removed from the set. For each PQRST set 76 fiducial features are calculated. This includes 24 amplitude features, 18 temporal features, 9 distance features, 10 slope and 6 angle features and 14 miscellaneous features. Five temporal features Px, Qx, Rx, Sx and Tx are different for each PQRST as they are time instances of fiducial points in digital ECG signal. Hence for input to machine learning only 71 features are used. 100 feature sets from each sample are merged and shuffled to create a dataset 2100 record and each record is labeled with 'pid' (person ID). 'pid' is number from 1 to 19 assigned to 19 individuals and is used as class label in classification process. Six popular classifiers MLP, KNN, SVM (linear kernel), SVM (RBF kernel), Decision tree and Naïve Byes classifiers are used to identify individuals.

3. RESULTS AND DISCUSSION

The 21 digital ECG recording samples of 19 individuals are first converted to ECG string. Figure 52.3 shows ECG string over a sample of digital ECG signal. It's clearly visible that ECG string on rising edge consists of a series of

1's and falling edge consists of series of -1's. Where slope is not sharp ECG string consists of 0's. DFA based algorithm is applied on each ECG string to identify PQRST values. Figure 52.4 shows a sample of all PQRST sets detected using DFA in a digital ECG signal. Category of fiducial features calculated using these PQRST values are listed in Table 52.1. Table 52.2 shows performance of all classifiers with ensemble classifiers with metrices accuracy, recall, precision and F1-score. Figure 52.4 depicts the comparison of all classifiers against these measures. It is very clear that ensemble classifiers perform better than single classifiers. It is observed that accuracy of KNN and SVM (kernel: RBF) is 0.965079 and 0.961905 respectively which is lower than all other classifiers. SVM (kernel: Linear), Decision Tree and Naïve Bayes classifiers give same accuracy 0.980952 but F1-score of SVM (kernel: Linear) 0.982163 is better in comparison to that of decision tree 0.981203 and Naïve Bayes 0.981019. Performance of MLP is in between with accuracy 0.977778 and F1-score 0.978125.

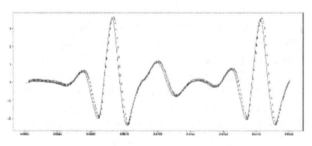

Fig. 52.3 ECG string

Table 52.1 Fiducial features

Amplitude Features	Py, Qy, Ry, Sy, Ty, PS, QS, RS, ST, PQ, PT, QR, QT, RS/QR, RS/QS, ST/QS, PQ/RS, PQ/QS, RS/QT, PQ/QT, ST/PQ, PQ/PS, ST/QT, PQ/QR	24
Temporal Features	Px, Qx, Rx, Sx, Tx, PQ, QS, QT, PT, PS, ST, QT/QS, PT/QS	13
Distance Features	PQ, RS, QS, QR, ST, ST/QS, PR, RS/QR	9
Slope Features	PQ, ST, PS, QR, QS, QT, RS, PT, PR, RT	10
Angle Features	PQR, RST, RSQ, QRS, RQS, RTS	6
Miscellaneous Features	QRS area, S angl/PQ dis, QRS perimeter, QRS area/RS^2, (R/Q) angle, QRS in radius, (R/S) angle, (R/T) angle, QRS x-centroid, R angle/QS time, (Q/T) angle, QRS y-centroid, S angle/QT time, QRS area/QR amp, RR	14
Total		76

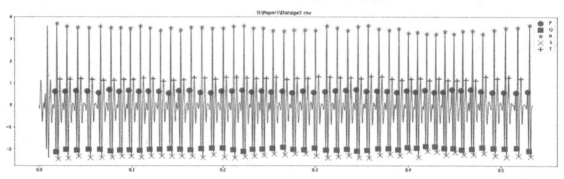

Fig. 52.4 Fiducial points detected by DFA

Table 52.2 Classifiers comparison

Model	Accuracy	Recall	Precision	F1_Score
MLP	0.977778	0.977778	0.980995	0.978125
KNN	0.965079	0.965079	0.967295	0.965191
SVM: Linear	0.980952	0.980952	0.986667	0.982163
SVM: RBF	0.961905	0.961905	0.964645	0.961823
Decision Tree	0.980952	0.980952	0.983656	0.981203
Naïve Bayes	0.980952	0.980952	0.98308	0.981019
Ad boost	0.990476	0.990476	0.992156	0.990759
Gradient Boost	0.980952	0.980952	0.983359	0.981178
Stacking	0.984126	0.984126	0.986281	0.984262
Ensemble (voting: Hard)	0.986162	0.984126	0.986281	0.984262
Ensemble (voting: soft)	0.990476	0.990476	0.991836	0.990664

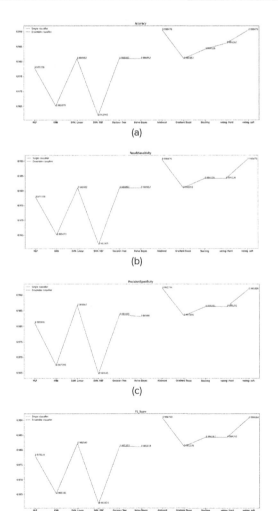

(a)

(b)

(c)

(d)

Fig. 52.5 Classifiers' performance (a) Accuracy (b) Sensitivity (c) Specificity (d) F1-Score

4. CONCLUSION

ECG is a potential tool for next-generationbiometrics. There is a need for an efficient system to identify individuals with greater accuracy. This work presents a comparison of six popular classifiers. The performance of ensemble classifiers, such as Ada-boost, gradient boost, stacking, and voting, is also analyzed. This work presents a novel DFA-based algorithm for identifying fiducial points PQRST in digital ECG signals. The proposed algorithm is applied to 21 digital ECG samples of 19 individuals. The algorithm identifies the PQRST sets for all cardiac cycles in the signal. These PQRST values has been used to compute 71 features that include amplitude, temporal, distance, slope, and miscellaneous features. A dataset is then created with an equal number of feature sets from each sample. The six most popular classifiers Multilayer perceptron (MLP), Support Vector Machine (Kernel: linear), Support Vector Machine (Kernel: RBF), K- nearest neighbor (KNN), Decision tree, and Naïve Bayes classifier are then trained and tested. Their performance is compared against metrics accuracy, precision, recall, and F1_score. It has been observed that SVM (kernel: Linear) performs best as a single classifier, with an accuracy of 0.980952 and an F1 score of 0.982163. All ensemble classifiers perform better than single classifiers. Ada-boost and Voting Classifiers (Voting: hard) give a high accuracy of 0.990476. Thus, these two are the most suitable machine-learning models for ECG-based identification.

References

1. Aziz, S., Khan, M.U., Choudhry, Z.A., Aymin, A. and Usman, A., 2019, October. ECG-based biometric authentication using empirical mode decomposition and support vector machines. In *2019 IEEE 10th Annual Information Technology, Electronics and Mobile Communication Conference (IEMCON)* (pp. 0906–0912). IEEE.
2. Dietterich, T.G., 2000. Ensemble methods in machine learning. In *Multiple Classifier Systems: First International Workshop, MCS 2000 Cagliari, Italy, June 21–23, 2000 Proceedings 1* (pp. 1–15). Springer Berlin Heidelberg.
3. Garg, G. and Garg, R., 2021. Brain tumor detection and classification based on hybrid ensemble classifier. *arXiv preprint arXiv:2101.00216.*
4. Han, J., Pei, J. and Tong, H., 2022. *Data mining: concepts and techniques*. Morgan Kauffmann.
5. Hopcroft, John E.; Motwani, Rajeev; Ullman, Jeffrey D. (2013). "Introduction to Automata Theory, Languages, and Computation" (3rd ed.). Pearson. ISBN 1292039051
6. Ingale, M., Cordeiro, R., Thentu, S., Park, Y. and Karimian, N., 2020. Ecg biometric authentication: A comparative analysis. *IEEE Access, 8*, pp. 117853-117866.

7. Israel, S.A., Irvine, J.M., Cheng, A., Wiederhold, M.D. and Wiederhold, B.K., 2005. ECG to identify individuals. *Pattern recognition*, *38*(1), pp. 133–142.

8. Keshri, A.K., Sinha, R.K., Singh, A. and Das, B.N., 2011. DFAspike: A new computational proposition for efficient recognition of epileptic spike in EEG. *Computers in biology and medicine*, *41*(7), pp. 559–564.

9. Keskes, N., Fakhfakh, S., Kanoun, O. and Derbel, N., 2022. Representativeness consideration in the selection of classification algorithms for the ECG signal quality assessment. *Biomedical Signal Processing and Control*, *76*, p. 103686.

10. Khan, M.U., Aziz, S., Iqtidar, K., Saud, A. and Azhar, Z., 2019, December. Biometric Authentication System Based on Electrocardiogram (ECG). In *2019 13th International Conference on Mathematics, Actuarial Science, Computer Science and Statistics (MACS)* (pp. 1–6). IEEE.

11. Kim, S.K., Yeun, C.Y., Damiani, E. and Lo, N.W., 2019. A machine learning framework for biometric authentication using electrocardiogram. *IEEE Access*, *7*, pp. 94858–94868.

12. Kumar Keshri, A., Kumar Sinha, R., Hatwal, R. and Nand Das, B., 2009. Epileptic spike recognition in electroencephalogram using deterministic finite automata. *Journal of medical systems*, *33*(3), pp. 173–179.

13. M. Pandey and A. K. Keshri, "QRS Detection in digital ECG signal using Deterministic Finite Automata," 2023 Fifth International Conference on Electrical, Computer and Communication Technologies (ICECCT), Erode, India, 2023, pp. 1–5, Doi: 10.1109/ICECCT56650.2023.10179620.

14. Mahmud, T., Barua, A., Begum, M., Chakma, E., Das, S. and Sharmen, N., 2023, February. An improved framework for reliable cardiovascular disease prediction using hybrid ensemble learning. In *2023 International Conference on Electrical, Computer and Communication Engineering (ECCE)* (pp. 1–6). IEEE.

15. Mincholé, A., Camps, J., Lyon, A. and Rodríguez, B., 2019. Machine learning in the electrocardiogram. *Journal of electrocardiology*, *57*, pp. S61–S64.

16. Patro K. and Kumar R. "A Novel Frequency-Time Based Approach for the Detection of Characteristic Waves in Electrocardiogram Signal", Springer India 2016 S.C. Satapathy et al. (eds.), Microelectronics, Electromagnetics and Telecommunications, Lecture Notes in Electrical Engineering 372, DOI 10.1007/978-81-322-2728-1_6

17. Patro, K.K. and Kumar, P.R., 2017. Effective feature extraction of ECG for biometric application. *Procedia computer science*, *115*, pp. 296–306.

18. Uwaechia, A.N. and Ramli, D.A., 2021. A comprehensive survey on ECG signals as new biometric modality for human authentication: Recent advances and future challenges. *IEEE Access*, *9*, pp. 97760–97802.

19. Zhang, Y., Liu, J. and Shen, W., 2022. A review of ensemble learning algorithms used in remote sensing applications. *Applied Sciences*, *12*(17), p. 8654.

Note: All the figures and tables in this chapter were made by the author.

*Recent Trends in Engineering and Science for Resource Optimization and
Sustainable Development – Prof. (Dr.) Dorota Jelonek et al. (eds)*

53

Artificial Intelligence in the Onboarding Process

Suruchi Pandey*

Professor, Symbiosis Institute of Management Studies,
Symbiosis International University

Ritu Pandey

Assistant Professor, CSA Kanpur

Shanul Gawshinde

Assistant Professor,
CSA Kanpur Symbiosis Institute of Management Studies,
Symbiosis International University, Kanpur

Abstract

The study aims to present how artificial intelligence helps onboard employees comfortable with the new system. AI in Human Resource Management procedures is changing how business hire, manage and engage employees. AI has an impact on new employees' Onboarding. AI can potentially improve employee experience; however, it needs to be more utilized in many aspects across the globe. The automation and AI implementation across business function is seeking attention from researchers and practitioners. This study is one such study presenting views of employees on their experience towards AI base onboarding process.

Keywords

Artificial intelligence, Onboarding, HR, Technology and Implementation

1. INTRODUCTION

Companies are implementing better and novel technologies such as "artificial intelligent (AI)" to assist work procedures such as recruiting and innovation facilitation which may usher in a new age of work practices. However, research has yet to be conducted to determine how these AI apps can help onboarding. As a result, this study performs a specific literature analysis on the existing onboarding processes and employs expert interviews to assess AI's increasing potential drawbacks for each activity. The study contributes to the literature by comprehensively reviewing onboarding approaches and evaluating possible AI application areas in the onboarding process.

*Corresponding author: suruchi.p@sims.edu

DOI: 10.1201/9781003596721-53

2. Literature Review

New workers are initially scared and apprehensive, which is why managers must assist them in understanding the fundamental corporate operations and culture (Bauer, 2006). According to current literature, onboarding techniques must be adapted to new workers to provide a unique experience. As a result, managers must grasp each employee's workers own personality and tailor their introduction to the same (Sharma & Stol, 2019). Onboarding entails agreeing on a shared definition of procedures, interactions and communication channels among superiors, teams and newcomers. Not unexpectedly, Onboarding has an impact on organizational retention. Multiple research articles argue for the development of an extensive onboarding process, with some emphasizing how long new employees stay with a company. Previous work advocates for the establishment of an extensive onboarding process, with some emphasizing the length of time new employees stay with a company stated by Bauer (2007),

2.1 AI Reshaping Onboarding

Artificial Intelligence (AI) technology has established itself as the new norm. Because AI now powers everything, it has transformed our approach to life (Pandey and Bahukhandi 2022) The application of AI into HRM operations is changing how businesses hire, manage and engage their employees and so are employees' perceptions about AI barging into their work life (Malik, Tripathi, Kar & Gupta, 2021)) Based on the prevailing data sets and behavioral patterns, AI allows robots to make more precise judgements than humans (Gogate and Pandey 2015). Businesses must retain staff while also hiring new, competent personnel. AI can assist in this endeavor. There is little doubt that AI technology has altered HR practices, and this shift has increased since the outbreak of Covid -19. HR managers employ AI, and it has assisted them in making less biased, accurate, and data-backed judgements throughout the onboarding process (Pattnaik & Tripathy 2022).Related work may be categorized into two key study streams. The first look at the AI implications and applications in Human Capital Management, while the following looks at how firms use AI to improve the employee experience (Pandey and Khaskhel 2019). For instance, Makarius etal (2020) focus on AI adoption in machine related operations optimization in an organization's onboarding process. Mayer Von Wolff et al. (2020) proposed chatbots to respond to inquiries; however, they also incorporated tailored training recommendations and automatic scheduling ideas with coworkers. The onboarding utilization voltages of AI should be more utilized. The aim was to understand and explore the current **AI** approaches in induction and Onboarding.

3. Research Methodology

In this research paper, primary as well as secondary methods of research were adopted. The data for the primary research was gathered through a questionnaire that had self-administered questions filled by the employees of all levels employed across different sectors in the Indian economy. Twenty-five research papers were referred to about AI's scope and increasing application for various human resource management functions, especially job onboarding. In this research, the population was also new hires going through induction and Onboarding or had recently gone through.

3.1 Selection of Participants

New hires from different departments and roles, ranging from analyst to managerial role were chosen as respondents. This study helped us to conclude the implementation of AIin induction training by the participants' answers. Purposive sampling was used and 50 respondents were reached. Out of which 41 respondents reverted to participate in this study.

3.2 Questionnaire

Questionnaires are widely recognized as a powerful data-gathering technique for answering research questions using a well-crafted set of questions. The researcher created the questionnaire and distributed it to responders via the online tool Google Forms.

4. Data Analysis and Interpretation

Analysis of understanding of AI-assisted Onboarding and Induction across the industry based on data collected from a sample of 41 responses

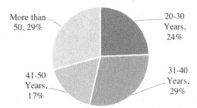

Fig. 53.1 Age of respondents

Figure 53.1 Indicates the Age of respondents who filled out the questionnaire. There was a fair participation of respondents across age categories, with 29% (31-40 Years) and (More than 50). An equitable response was collected from people 24% (20-30 Years).

Figure 53.2 Indicates the gender of respondents who filled out the questionnaire—immediate balance responses of Males and females with 54% and 46%. Getting responses

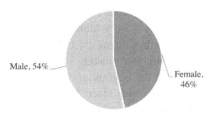

Fig. 53.2 Gender of respondents

from female professionals is a positive and healthy sign of the study's relevance.

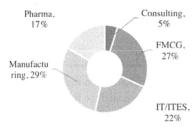

Fig. 53.3 Organization (Sector) of respondents

Figure 53.3 Indicates the Sectoral distribution of respondents who completed the questionnaire—having the maximum contribution of Manufacturing (29%) followed by FMCG (27%). Data collected across sectors shows the diversity of data that helped analyze and evaluate the data.

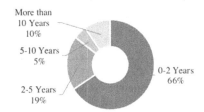

Fig. 53.4 Association with the organization

Figure 53.4 Indicates the Association of the respondents with their existing Organization in years. People with 0-2 years show the highest of 66% of respondents, followed by 2-5 years with 19%of them. Based on this data, one inference is that people tend to change their jobs quickly across sectors. Reasons can be many, but this is one of the prevailing phenomena.

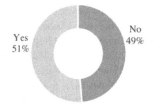

Fig. 53.5 AI assisted onboarding experience better than the traditional method

Figure 53.5 Indicates **AI** assisted on boarding organization better than the traditional method. In response, the results show a mixed reaction where 51% of the respondents agreed to the fact and said Yes, whereas 49% said No.

Based on the organization's values, interests and proactive approach, the experience scan vary across sectors. Only so many industries are undertaking **AI**-backed processes; hence, the responses can also be perception-based. However, This is a changing period, and gradually, companies are shifting towards automation and more tech-driven approaches.

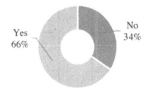

Fig. 53.6 AI assisted in making onboarding and induction n cost-effective option

Figure 53.6 Indicates **AI** assisted in making the Onboarding and induction function a cost-effective option. Responding to the same, the respondents showed Yes as 66% and No as 34%. This is again a perception-based response to the question as only a few industries across sectors have successfully implemented AI-assisted Onboarding and induction, but no doubt the organizations are planning to tend towards it; hence they are collecting literature and conducting intensive studies to collect the data. In the long run, AI-based processes will help save costs and make processes more sophisticated.

Response on the AI and its usage in onboarding process is presented in Table 53.1.

Does AI boost the efficiency of the onboarding and induction process? In response to the same maximum number of respondents Disagreed (14) with it, and 12 with Neutral opinions. Companies are adopting AI-based Onboarding and induction to match with the competitors. In response to the same, 19 responses out of 41 Agreed to it. AI assistance adversely affects privacy and security. In response, 22 respondents out of 41 Agreed to it. The transition from a traditional to an AI-assisted process takes time to adapt. AI-assisted Onboarding will enhance employees' delight and will be a helping hand for HR to handle operations. Responding to the same, 21 out of 41 respondents Agreed that it helps new joiners have a fair amount of idea about the processes and understanding of the organization, its vision, mission and values and other essential things. Information shared during Onboarding and induction helps establish an understanding of the role, department, and organization. In response to the same, 23 out of41 respondents Agreed to it as they believed that Onboarding and induction are

Table 53.1 Shows responses based on the Like rt scale on specific questions

	Indicators	Agree	Dis-agree	Neu-tral	Strongly Agree	Total
1	Does AI boost the efficiency of the Onboarding and induction process?	4	14	12	11	41
2	Companies are adopting **AI-based** onboarding and induction to match with the competitors.	19	6	3	13	41
3	**AI** assistance adversely affects privacy and security.	22	I	5	13	41
4	The transition from traditional to The al-assisted process takes time to adapt.	22	4	5	IO	41
5	AI-assisted Onboarding will enhance employees' delight and will be a helping hand for **HR** to handle operations.	21	3	5	12	41
6	Information shared during Onboarding and induction is helpful in establishing an Understanding of the role, department, and organization.	23	4	5	9	41
7	AI-assisted Onboarding has become more accessible as it can connect anyone from anywhere in the world.	16	3	7	15	41
8	AI onboarding and induction help in reducing turnover as they make employees feel valued, respected from the beginning?	18	5	5	13	41
9	Onboarding and induction help in keeping data secured and organized with a proper Time line reduce the risk of data breaches?	16	4	4	17	41
10	**AI** onboarding helps in understand Diversity and Inclusion by eliminating biases From the very beginning.	16	5	4	16	41

an integral part of the HR process and are very helpful in establishing an understanding of What and How of the organization. Al-assisted Onboarding has become more accessible as it can connect anyone from anywhere in the world. In response tothesame16outof 41responses Agreed to it as the scalability is very high and requires less time, coordination with other teams. AI onboarding and induction help reduce turnover, making employees feel valued and respected from the beginning. In response, 18 out of 41 respondents agreed that since day one, a sense of care and belongingness had been extended to the employees. Onboarding and induction help keep data secured and organized with a proper timeline to reduce the risk of data breaches. In response to the same, 17 out of 41 respondents Strongly agreed to it. One thing that can be inferred here is when traditionally the process is happening, it limits the risk of data breach and privacy.

5. CONCLUSION

In this study, authors focused on finding out how Onboarding using AI experience differs from traditional methods. The views were tied up, with 51% agreeing that the AI onboarding experience is better than traditional methods and 49% disagreeing. Although people were tied between the experience provided in AI onboarding versus traditional Onboarding, most respondents agreed that AI provides us with more advantages. Companies are moving towards AI, which would help companies at a large scale where the advantages are much more than the disadvantages. It helps the HR department keep track of its employees more efficiently and effectively.

References

1. Ritz, E., Donisi, F., Elshan, E., and Rietsche, R. (2023). Artificial Socialization? How Artificial Intelligence Applications Can Shape A New Era of Employee Onboarding Practices. In Hawaii International Conference on System Sciences (HICSS).
2. Anjana, S., Bhubaneswar, B., Sornashanthi, D., and Vijay, P. (2022). Employee Onboarding Automation. In ICT Systems and Sustainability: Proceedings of ICT4S-D2022(pp.623–631). Singapore: Springer Nature Singapore.
3. Zel, S., and Kongar, E. (2020). Transforming digital employee experience with artificial intelligence. In 2020 IEEE/ITU International Conference on Artificial Intelligence for Good (A/4G) (pp. 176–179). IEEE.
4. Waterworth, D., Sethuvenkatraman, S., and Sheng, Q.Z. (2023). Deploying data-driven applications in smart buildings: Overcoming the initial onboarding barrier using machine learning. Energy and Buildings, 279:112699.
5. MERSON, A. (2021). On boarding and Developing Employees in a Covid world. HR Future, 2021(3):10–13.
6. Ramirez, E. (2022). Artificial Intelligence in Medical Dosimetry: A Quantitative Analysis of Artificial Intelligence Adoption Among Medical Dosimetrists (Doctor AI dissertation, University of Southern California).

7. Nishad, **N.**U., and Gurav, M.D. (2019). Impacts of Artificial Intelligence in Human Resource Management. Think India Journal, 22(32): 45–47.

8. Chandar, P., Khazaeni, Y., Davis, M., Muller, M., Grasso, M., Liao, Q.V., and Geyer, W. (2017). Leveraging conversational systems to assist new hires during On boarding. In Human-Computer Interaction-INTERACT2017: 16th IFIP TC 13 International Conference, Mumbai, India, September 25-29, 2017, Proceedings, PartII16 (pp. 381–391). Springer International Publishing.

9. George, G., and Thomas, **M.** R. (2019). Integration of artificial intelligence inhuman resources. Int. J. Innova Technol. Exp/or. Eng, 9(2): 5069–5073.

10. Achchab, S., and Temsamani, Y. K. (2021). Artificial intelligence use in human resources management: strategy and operation's impact. In 2021 IEEE2nd International Conference on Pattern Recognition and Machine Learning (PRML) (pp. 311–315). IEEE.

11. On wubiko, C., and Ouazzane, K. (2019). Cyber Onboarding is 'broken'. In 2019 International Conference on Cyber Security and Protection of Digital Services (Cyber Security) (pp. 1–13). IEEE.

12. Upadhyay, A.K., and Khandelwal, K. (2018). Applying artificial intelligence: implications for recruitment. Strategic HR Review,17(5):255–258.

13. Kaur, S., and Sharma, R. (2021). Emotion Al:integratin gemotional intelligence with artificial intelligence in the digital workplace. In Innovations in Information and Communication Technologies (IICT-2020)

14. Kaushal, **N.,** Kaurav, R. P. S., Sivathanu, B., and Kaushik, **N.** (2021). Artificial intelligence and HRM: identifying future research Agenda using systematic literature review and bibliometric analysis. ManagementReviewQuarterly,1–39.

15. Nishad, **N.**U., and Gurav, M.D. (2019). Impacts of Artificial Intelligence in Human Resource Management. Think India Journal, 22(32): 45–47.

16. Pandey, S., and Bahukhandi, M.(2022) Applicants' Perception Towards the Application of AI in Recruitment Process 2022 International Conference on Interdisciplinary Research in Technology and Management, IRTM 2022 - Proceedings, 2022

17. Pandey, S. and Khaskel, P (2019) Application of AI in Human Resource Management and Gen Y"s Reaction, International Journal of Recent Technology and Engineering (IJRTE) ISSN: 2277–3878, 8 (4):10325–10331

18. Gogate, L,and Pandey, S., (2015)- Indian Journal of Science and Technology, 2015 Impact of Structured Induction on Mutually Beneficial Deployment and Talent Retention

19. Bauer, T.N., and Erdogan, B. (2011). Organizational socialization: The effective onboarding of new employees.

20. Bauer, T. N., and Elder, E. (2006). Onboarding newcomers into organizations. Presentation at the Society for Human Resource Management Annual Meeting, Washington, D.C.; Van Velsor, E., & Leslie, J. B. (1995). Why executives derail: Perspectives across time and culture. Academy of Management Executive, 8:62–72.

21. Bauer, T (2007) Onboarding new employees: Optimizing success. www.shrm.org/hr-today/trends-and-forecasting/special-reports-and-expert-views/documents/onboarding-new-employees.pdf

22. Sharma, G. and Stol, K (2019) Exploring onboarding success, organizational fit, and turnover intention of software professionals, Journal of Systems and Software159:110442

23. Malik, N., Tripathi, S., Kar, A., and Gupta, S. (2021) Impact of Artificial Intelligence on Employees working in Industry 4.0 Led Organization, International Journal of Manpower

24. Pattnaik, S. and Tripathi, S (2022) The effect of organizational justice on employee performance in the Indian Public Sector Units: the role of organizational identification, Benchmarking An International Journal

25. Meyer von Wolff, R., Hobert, S., Masuch, K., and Schumann, M., (2020) Chatbots at Digital Workplaces – A Grounded-Theory Approach for Surveying Application Areas and Objectives, *Pacific Asia Journal of the Association for Information Systems*: 12 (2): 3.

26. Makarius, E., Mukherjee, D., Fox, J., and Fox, A. (2020) Rising with the machines: A sociotechnical framework for bringing artificial intelligence into the organization, Journal of Business Research, 120: 262–273, ISSN 0148–2963

Note: All the figures and table in this chapter were made by the author.

Recent Trends in Engineering and Science for Resource Optimization and Sustainable Development – Prof. (Dr.) Dorota Jelonek et al. (eds)
© 2024 Taylor & Francis Group, London, ISBN 978-1-032-98030-0

54

A Differential Evolution Algorithm Using Constraint Handling Function for Optimized Routing in Mobile Ad Hoc Network

Anju Sharma*

Assistant Professor, BIT Mesra, Jaipur campus

Madhavi Sinha

Associate Professor, BIT Mesra, Jaipur campus

Tanvi Sinha

Project Manager, Carmichale Tompkins Property Group,
Melbourne, Australia

▬▬ Abstract

The goal of optimization is to find the best solution within the constraints placed on a chosen optimization objective function. Physical and natural principles are used to design robust optimization algorithms, which are applicable to a wide range of problems. Differential Evolution (DE) is a natural heuristic method with biological influences based on natural selection. MANET (Mobile Ad-hoc NETwork) is a network with mobile nodes and infrastructure-less. In this paper, the Differential Evolution algorithm using constraint handling function (DECHF) is proposed to solve the non-linear programming issue in MANETs, which contains integer and discrete variables. The major goal of this study is to simulate and evaluate the performance of Differential Evolution based AOMDV (DE_AOMDV) with DECHF in terms of fitness value at different iterations for different numbers of nodes. Based on simulation results and data analysis, it can be seen that for Mobile adhoc network, routing speed in DE using constraint handling function speed is faster than network convergence in DE.

▬▬ Keywords

Mobile adhoc NETworks, DECHF, Optimization, Routing strategy

1. INTRODUCTION

In MANET, there is a list of routing protocols. In MANETs, multipath routing can transfer data simultaneously and offers fault tolerance in place of single path routing. Thus, the AOMDV routing protocol is the main emphasis of this research. Adhoc On-Demand Multi-path Distance Vector, or AOMDV, is a multi-path reactive routing protocol that

*Corresponding author: anjusharma@bitmesra.ac.in

DOI: 10.1201/9781003596721-54

offers numerous loop-free paths but necessitates route discovery when an existing path fails. Each route discovery has a high overhead and latency.

2. PROBLEM DESCRIPTION

Data delivery in MANET is highly difficult since nodes are movable and their behavior between the source and destination is unpredictable. Data delivery may take a single path or several. AOMDV offers alternate routes that have redundant data to provide fault tolerance in a network routed to the destination. AOMDV can select alternate routes to reroute traffic when a link gets congested in order to lessen the load on the congested link. During route discovery in AOMDV, the source node broadcasts an RREQ packet that is disseminated throughout the network, and the destination node builds a reverse route RREP to the source node. We introduced Differential Evolution algorithm using constraint handling function (DECHF) in this research due to the high convergence of DE and to solve the non- linear programming issue in MANETs as compared to other evolutionary computation.

The DECHF mutation operation produces mutant vectors in the current population and inserts parameters for the subsequent iteration. The DECHF crossover operation combines the mutant vector with the chosen parent vector to create the vector known as the trial vector. In the selection process, the vector (trial) is compared to the vector (parent), and the performance of the better (minimum) vector is chosen for the following iteration until the optimal value is found. The performance of multi-path routing is improved by using intelligent path selection and constraint handling approach.

3. PROPOSED PROTOCOL

This part will move forward with the AOMDV using the network model as its main focus. The connected, undirected MANET model graph G (V,E), where V stands for mobile nodes and E for a collection of links (communication) between the two nearby mobile nodes within the range. Additionally, there will be a substantial delay and cost associated with message delivery through a wireless network.

Total delay (transmission) on path Pi is calculated by:

$$\Delta(P_i) = \sum_{l \in P_i(s,d)} dl \qquad (1)$$

Total cost (distance) of the path P_i is calculated by:

$$C(P_i) = \sum_{l \in P_i(s,d)} cl \qquad (2)$$

Where dl , cl is the delay and cost (distance) on the link l respectively between s (source) and d (destination) node.

DECHF technique work on two stages: 1) Route-Discovery phase, 2) Route-Recovery phase.

3.1 Route-Discovery Phase

The responsibility of this phase is to find routes between source and target. Source sends RREQ (route request) message to its adjacent nodes and receives multiple RREP (route reply) messages from multiple paths. They select best path as primary path on the basis of lower hop count.

Another paths are set as secondary paths for backup which are used when primary path beaks up. Priorities are also set on paths to choose next primary path. These priorities are set by the DE. In RREQ flooding mechanism, a node receives a route advertisement using

$$advertised \_ hop \_ count_s^d := \max \{hop \_ count_s^d\}, s \neq d$$
$$:= 0, otherwise \qquad (3)$$

The DE has components like fitness function, population initialization, selection, mutation and crossover which are to be described as.

a) Initialization

In DE, this process starts with the NP (population), D (dimensions) in search vectors. As generations (G) increases, the vectors are expected to be optimized or changed.

$$\overrightarrow{X_{i,G}} = [x_{i1,G}, x_{i2,G}, x_{i3,G} \ldots \ldots x_{iD,G}] \qquad (4)$$

Theses vectors are referred as "chromosomes". Initially, each chromosome contains more than one solution. These initial chromosomes are obtained from route discovery phase in AOMDV.

Mutation Operation: This operation is utilized in DECHF to produce a mutation vector v_i^G with individual vector (target) x_r^G, in the population. For each vector (target) x_r^G (initially taken at random) at the generation (iteration) G, mutant vector $v_i^G = \{v_{i1}^G, v_{i2}^G, \ldots \ldots v_{iD}^G\}$ can be via strategy of mutation. For every target vector x_r^G, a mutation vector v_i^{G+1} is generated for discrete according to the following:

$$v_{j,i}^{G+1} = \begin{cases} x_{r1,G} + F.(x_{r2,G} - x_{r3,G}) & if \ rand_j(0,1) < CR \wedge j = k \\ x_{ij}^G & otherwise \end{cases} \qquad (5)$$

Where randomly selected i is in range of 1 and NP, j is in the range of 1 and D, and r1, r2, r3 are in the range of 1 and NP but $r1 \neq r2 \neq r3 \neq i$, $k = (int(rand_i[0,1] \times D) + 1]$, and value of CR, and F is between 0 and 1. The randomly chosen vectors are x_{r1} , x_{r2} , and x_{r3} from the population.

b) Crossover Operation

After the previous phase, to generate a vector (trial), Eq. 7 is applied to each pair of the vector (target) x_i^G and vector (mutant) v_i^G using crossover factor (CR):

$$U_i^{G+1} = v_i^{G+1} \quad if \ rand(0,1) < CR,$$
$$= x_{ij}^G \quad else.................. \quad (6)$$

The crossover may be binomial or exponential.

c) Selection Operation

In the selection process follows the principle of Darwinian "survival of fittest" as

$$x_i^{G+1} = U_i^{G+1} \quad if \ f(U_i^{G+1}) \leq f(x_i^G)$$
$$= x_i^G, \quad otherwise \quad (7)$$

3.2 Constraint Function

After selecting feasible solution from DE, Adaptive Penalty Function is used as a Constraint function which improved objective values instead of the original objective function values

3.3 Route-recovery Phase

By using DECHF, feasible paths can be obtained from backup routing table. If primary path fails or a node detects failure in the network, protocol will recover the connection by accessing the backup paths from the backup routing table and replace the primary path by backup path to avoid reroute discovery process. In MANETs, a multipath routing protocol based on the DECHF algorithm is suggested to shorten the period when a route fails

4. RESULTS AND DISCUSSIONS

NS-2 network simulator is used to simulate behavior of network on a simulator. Figure 54.1 shows that as number

of nodes increases routing overhead may vary in DE_AOMDV. Due to high congestion, routing overhead in DE_AOMDV and DECHF increases. These routing protocols have almost routing overhead for less number of nodes. The difference between routing overhead in DE_AOMDV and DECHF increases as node density decreases. Therefore DECHF perform better routing overhead at high mobility and high congestion.

Figure 54.2 shows that as number of nodes increases packet deliver ratio may vary in DE_AOMDV and DECHF. Throughput of DE_AOMDV and DECHF decreases gradually as node density increases. These routing protocols have least difference for packet delivery ratio in less number of nodes. The packet delivery ratio in DECHF is always less than DE_AOMDV. Therefore DECHF perform better packet delivery ratio than DE_AOMDV at high mobility and high congestion.

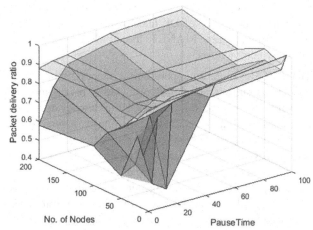

Fig. 54.2 Packet delivery ratio of DECHF (red) Vs DE_AOMDV (blue)

Figure 54.3 shows that as number of nodes increases, End-to-end delay vary in DE_AOMDV and DECHF. End-to-end

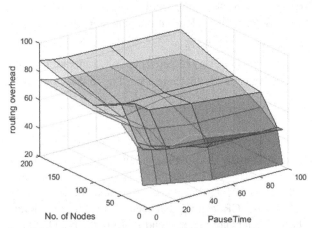

Fig. 54.1 Routing overhead of DECHF (red) Vs DE_AOMDV

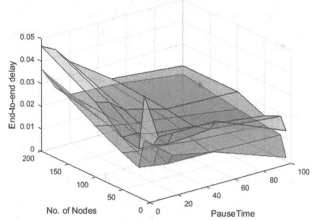

Fig. 54.3 End-to-end delay of DECHF (red) Vs DE_AOMDV (blue)

delay of DE_AOMDV and DECHF increases gradually as node density increases. The end-to-end delay in DECHF is always less than DE_AOMDV. As pause time increase the end-to-end delay of DE_AOMDV and DECHF increases. The difference between end-to-end delay in DE_AOMDV and DECHF decreases as node density increases. Therefore DECHF perform better end-to-end delay at high mobility and high congestion.

References

1. Ali, M.M. and Torn, A. (2004) 'Population set based global optimization algorithms: some modifications and numerical studies', Computer Operation Research, Vol. 31, No. 10, pp. 1703–1725.
2. A Sharma and M. Sinha, "A differential evolution-based routing algorithm for multi-path environment inmobile ad hoc network," Int. J. Hybrid Intell., vol. 1, no. 1, p. 23, 2019, doi: 10.1504/ijhi.2019.10021294.
3. A Sharma and M. Sinha, "Influence of crossover and mutation on the behavior of Genetic algorithms in Mobile Adhoc Networks," 2014 International Conference on Computing for Sustainable Global Development (INDIACom), New Delhi, India, 2014, pp. 895–899, doi: 10.1109/IndiaCom.2014.6828092.
4. Das, S. et al. (2008) 'Particle swarm optimization and differential evolution algorithms: technical analysis, applications and hybridization perspectives', studies in Computational Intelligence (SCI), Vol. 116, pp. 1–38.
5. Deb, K. (2000) 'An efficient constraint handling method for genetic algorithms', Computer Methods in Applied Mechanics and Engineering, Vol. 186, Nos. 2–4, pp. 311–338, Elsevier, Netherlands.
6. Elhoseny, M., Tharwat, A., Farouk, A. and Hassanien, A.E. (2017) 'K-coverage model based on genetic algorithm to extend WSN lifetime', IEEE Sensors Lett., August, Vol. 1, No. 4.
7. Godfrey Onwubolu , Donald Davendra (2006), "Discrete Optimization Scheduling flow shops using differential evolution algorithm" European Journal of Operational Research 171 (2006) 674–692 (ELSEVIER).
8. Goldberg, D.E. (1989) Genetic Algorithms in Search, Optimization and Machine Learning, Addison- Wesley, Reading. Gundry, S. and Kusyk, J. (2012) 'Performance evaluation of differential evolution based topology control method for autonomous MANET nodes', IEEE, pp. 228–233.
9. Lee, S-J. and Gerla, M. (2000) SMR: Split Multipath Routing with Maximally Disjoint Paths in Ad Hoc Networks, Technical report, August, Computer Science Department, University of California, Los Angeles.
10. Li, Z., Wang, R. and Bi, J. (2009) 'A multipath routing algorithm based on traffic prediction in wireless mesh networks', Intern. Conf. on Natural Computation, pp. 115–119, Tianjin, China.
11. Matre, V. and Karandikar, R. (2016) 'Multipath routing protocol for mobile ad hoc networks', Colossal Data Analysis and Networking, DOI: 10.1109/CDAN.2016.7570946.
12. Morrison, R.W. (2004) Designing Evolutionary Algorithms for Dynamic Environments, Springer-Verlag, Berlin, Germany.
13. Mueller, S. and Tsang, R.P. and Ghosal, D. (2012) 'Multipath routing in mobile ad hoc networks: issues and challenges'.
14. Paulose, N. and Paulose, N. (2016) 'Comparison of on demand routing protocols AODV with AOMDV', International Journal of Science, Engineering and Technology Research (IJSETR), Vol. 5, No. 1, pp. 181–184.
15. Perkins, C.K. and Bhagat, P. (1994) 'Highly dynamic destination-sequenced distance-vector routing (DSDV) for mobile computers', in Proceedings of ACMSIGCOMM, August, pp. 234–24.
16. Perkins, C.K. and Royer, E.M. (1999) 'Ad hoc on-demand distance vector routing', Proceedings of IEEE Workshop on Mobile Computing Systems and Application, February, pp. 90–100.
17. Qin, A.K. and Suganthan, P.N. (2005) 'Differential evolution algorithm with strategy adaptation for numerical optimization', in Proc. IEEE Congr. Evolut. Comput., September, Edinburgh, Scotland, pp. 1785–1791
18. Qin, A.K., Huang, V.L. and Suganthan, P.N. (2009) 'Differential evolution algorithm with strategy adaptation for global numerical optimization', IEEE Transactions on Evolutionary Computation, April, Vol. 13, No. 2.
19. Riad, A.M., El-minir, H.K. and El-hoseny, M. (2013) 'Article: secure routing in wireless sensor networks: a state of the art', International Journal of Computer Applications, April, Vol. 67, No. 7, pp. 7–12.

Note: All the figures in this chapter were made by the author.

Recent Trends in Engineering and Science for Resource Optimization and Sustainable Development – Prof. (Dr.) Dorota Jelonek et al. (eds)
© 2024 Taylor & Francis Group, London, ISBN 978-1-032-98030-0

55

Cyrus Mistry and Corporate Governance Mystery at TATA

Ameya Patil*
Assistant Professor, School of Business,
Dr. Vishwanath Karad MIT World Peace University, Pune, India

Rakesh Yadav
Assistant Professor, S. P. Mandali's Prin.
L.N. Welingkar Institute of Management Development & Research,
Mumbai, India

Rajeev Sengupta
Associate Professor, School of Business,
Dr. Vishwanath Karad MIT World Peace University, Pune, India

Namita Shivlal Mane
Assistant Professor, Faculty of Business Management and
Commerce, JSPM University, Wagholi, Pune, India

Jayashree Patole
Assistant Professor, Global Business School and
Research Centre, Dr. D.Y. Patil Vidyapeeth, Pimpri, Pune, India

Abstract

The research article takes a review of Cyrus Mistry's sudden removal as Chairman at Tata group. The board's explanation behind their decision and the counter-allegations made by Mistry have been discussed, along with financial performance of Tata group companies under the leadership of Cyrus Mistry. The paper ponders over a possible failure of corporate governance at Tata group, India's most trusted and ethical brand. It also suggests a few solutions to improve corporate governance standards of Indian businesses, especially the ones related to independent directors in the corporate boards.

Keywords

Corporate governance, Independent director, TATA Group, Financial performance, ROCE, EBIT

1. INTRODUCTION

Effective corporate governance is essential to solve the agency problems between shareholders and management (Jensen,1986). In this direction, shareholders elect a board of directors to oversee the firm's affairs on their behalf (Bansal & Sharma, 2019). Many professionals in the fields of finance, economics, behavioral science, law,

*Corresponding author: ameyapatil0786@gmail.com

DOI: 10.1201/9781003596721-55

and business have long been fascinated by the impact of corporate governance on company performance (Bonazzi & Islam, 2007). Given this context, a high profile exit from the most trusted brand in India does arouses a great deal of interest.

It was shocking to hear about the ouster of Cyrus Mistry, the former chairman of Tata Sons. Tata Sons Limited is the Tata Group's holding company, accounting for the majority of the group's shareholdings. It also owns the Tata name and trademarks. Traditionally, the chairman of Tata Sons has served as the chairman of the Tata Group, which has nearly 30 group companies ranging from salt to software (Dhameja & Agarwal, 2017). On October 24, 2016, the board of Tata Sons, the holding company of the Tata Group, voted to remove Cyrus Mistry as chairman. The board cited the erosion of the Group's values and of non-performance as the reasons behind this sacking. Former chairman Ratan Tata had been then named the interim chairman, and a selection team has been created to find a replacement. In addition, Nusli Wadia, who was at the time an independent director on a number of Tata Group firms, was dismissed for working with Mistry to undermine the main shareholder (Bansal & Rajkumar, 2024).

In December 2012, Ratan Tata retired, and Cyrus Mistry was selected as sixth chairperson of Tata Sons by a selection panel. Unlike his predecessors, Mistry was not a member of the Tata family. However, his family owned around 18 percent equity in Tata Sons (Goldstein, 2013). The selection committee deemed Mistry to be a responsible and qualified candidate, praising him for his capabilities and selflessness at the time of his appointment. However, there were differences in management style, leadership approach, and cultural compatibility in case of Mistry (Puliyolli, 2023), which might well be a reason for his ouster. Mistry's primary objective as chairman was to consolidate the group's finances rather than engage in significant acquisitions. He was more focused on each company's and the group's bottom line. If the Nano project or Corus Steel failed to produce a certain rate of return, they were worthless in his opinion. Mistry disposed of the loss-making and heavily indebted Corus in UK, which led to clashes with his predecessor. When Tata Steel had acquired Corus, it was widely believed in the business community that Ratan Tata had overpaid in order to secure the bid. This implies that it held substantial sentimental worth for him. Mistry made the decision to divest Tata Chemicals Ltd.'s urea division (Balabhaskaran, 2019). .Another big mistake of Mistry was getting rid of loss-making Nano project (Jhunjhunwala, 2020). Though Nano was a loss-making proposition for Tata, it was the dream project of Ratan Tata. Thus, Mistry divested a few loss-making businesses of Tata, which did made a business

sense. However, Mistry's strategy of selling off portions of the business rather than maintaining assets and expanding the company's global presence led to reversing of Ratan Tata's business endeavors (Balabhaskaran, 2019), causing dissatisfaction to the principal shareholder (Vakhariya, 2017).

This removal of TATA chairman within a span of 4 years (Bennedsen et al.,2022) is in sharp contrast to the view of company insiders, fairly of the outsider world as well, who believed that he would remain at the helm for close to three decades, going by the retirement age of above 70 for the earlier Tata bosses. For a company with a culture of consensus, the abruptness of such a sacking is indeed brutal. After the incident, Ratan Tata, the interim chairman stated that the decision was made after through and extensive considerations, and was essential for the future prosperity of Tata group. However, Mistry was not offered any opportunity to explain his case (Sudhakar, 2018). Many people think that his firing stemmed from a disagreement with promoter and main shareholder Ratan Tata (Bansal & Vajpeyi, 2022). Several inquiries on Tata Group's governance and the failed succession experiment were prompted by Mistry's departure (Jain, 2018).

On the other hand, Cyrus Mistry leveled a series of allegations against the Tata Group and his predecessor, Ratan Tata. He firstly stated that the board altered articles of association of business entity to restrict the chairman's authority (Kaur & Gupta, 2019). Further, he claimed there were fraudulent transactions amounting to Rs 22 crore in the Air Asia case, involving non-existent parties in India and Singapore, prompting an ED probe. In the Tata Capital case, a loan given to the Siva group by advice of Executive Trustee Venkatraman had turned into a non-performing asset, necessitating cleanup due to bad loans in the infrastructure sector. Mistry also criticized the group's foreign acquisition strategy, except for JLR and Tetley, stating it had burdened the group with significant debt, citing acquisitions in IHCL, the steel business, and Tata Chemicals. He highlighted the Nano project as a loss-making endeavor continued for emotional reasons and for supplying Nano gliders to an electric car entity in which Mr. Tata had a stake. Additionally, he pointed out that capital employed in legacy businesses (IHCL, Tata Motors PV, Tata Steel Europe, Tata Power Mundra, and Teleservices) had increased from Rs 132,000 crore to Rs 196,000 crore. This huge figure, which is close to group's networth, has be attributed to operational losses, interest, and capital expenditure. Mistry suggested that a realistic assessment of these businesses could result in a write-down of about Rs 118,000 crore over time. He also alleged a total lack of corporate governance, with trust-nominated directors reduced to mere postmen, failing to apply independent

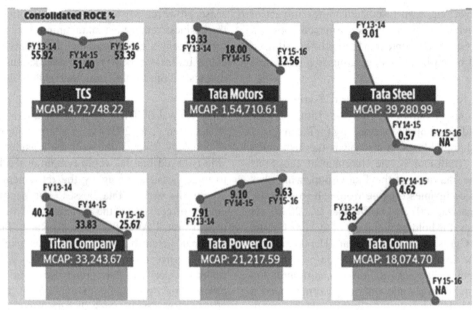

Fig. 55.1 ROCE of tata group companies during mistry's tenure

Source: Economic Times

judgment and fulfill their fiduciary duties, and accused the board of reducing him to a "lame duck chairman".

The group's trustworthiness and honesty are called into question by such vitriolic criticisms from the two sides. Both Mr.Ratan Tata, and Cyrus Mistry had put PR teams in place, in hope of a long battle ahead. Meanwhile, investors of Tata group companies lost over Rs 25,000 crore in the next two trading days following this event. The market capitalization of Tata Group listed companies, declined from Rs 851,020 crore to Rs 824,594 crore during this period.

2. MISTRY'S PERFORMANCE

If financial numbers are an evaluator of the boss's performance, we can see that major TATA companies have shown disappointing returns on capital employed (ROCE) since Mistry has taken over (Fig. 55.1). ROCE (returns on capital employed) indicates how well a company is using its funds. ROCE is calculated as Earnings before Interest and Tax (EBIT) divided by the Capital Employed in the business.

Similarly, revenues and profits of TATA Sons have eroded considerably over the past financial year, as seen from the Fig. 55.2.

However, a note has to be taken of Cyrus Mistry's letter regarding the financial performance. The letter mentioned that the group's operating cash flows have increased at a compounded annual rate of 31% during his tenure. From

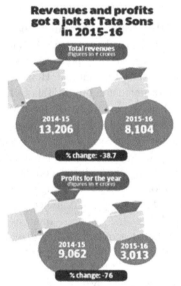

Fig. 55.2 Revenue and profits change in Tata Sons (2014–15 to 2015–16)

Source: Economic Times

2013 to 2016, the Tata Group's valuation climbed by 14.9 percent annually in rupee terms, whereas the BSE Sensex experienced an annual increase of 10.4 percent over the same period. After accounting for impairments, Tata Sons' net worth has climbed from around Rs.26,000 crore to Rs 42,000 crore. This has greatly fortified the group's balance sheet, so increasing its capacity to withstand additional shocks from the restructuring of the companies.

The graph below indicates absolute returns given by major Tata group companies since Mistry took over.

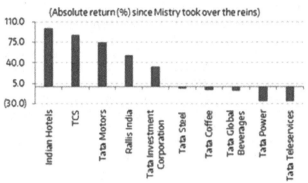

Fig. 55.3 Absolute returns generated by Tata group companies during mistry's tenure

Source: Economic Times

3. ISSUES OF MAJOR CONCERN

The absence of a well-defined process for selecting family members to serve as company heads raises worries over the impact of family dynamics on this topic. Subsequently, conflicts of authority will arise between the owner and the selected managers (Pandey, 2020). Was Cyrus Mistry removed in such a short span of time because he was the first chairman of the conglomerate not connected to the founding family? Or was it due to clash of egos with Ratan Tata? Is it the case of succession planning going wrong for TATA? If it is indeed the case,what can be the reasons for the same. Where has such a large conglomerate gone wrong when it comes to succession, and who has to be accountable? Looking from the angle of corporate governance, this responsibility lies with the board of directors (BOD). Members of BOD are meant to represent the interests of all shareholders. Still, it is a rare to see that the board will challenge the promoter of a family-owned company. Is this exactly the case that happened when in December 2012 when Mistry was appointed as the chairman of Tata Sons? Surprisingly, many of the board members who have voted to sack Mistry are the same ones who had then made appointment of Mistry. Was Mistry picked up as Chairman in 2012 just because his family was having18.5%stake in Tata group? Incidentally, majority of large Indian conglomerates are family-run business, wherein such challenges and board room conflicts are bound to exist (Pandey,2020). Excessive interference by promoter group can be seen in this case (Vij, 2017).

leadership change initiated by family patriarch Mr Ratan Tata was because of socioemotional wealth reasons, rather than financial performance. Hence, the article provides evidence that the Indian family-owned business groups consider socioemotional wealth more important than financial interests.

If Mistry's allegations (even if some of them) happen to be true, what image does that set of the most trusted brand in India? The allegations are as serious as fraudulent transactions, powerless chairman and directors, acquisitions and projects eroding shareholder's wealth, conflict of interest and the lack of corporate governance on a whole. It is indeed a fact that the largest investments by the Tatas happened under Mr. Ratan Tata from 2000 onwards (notable being Chorus and JLR) and overall, they have lost big money for shareholders. The ideal degree of director oversight of management is seen to be a key factor in influencing the company's financial performance (Bonazzi & Islam, 2007). **If this is the quality of Corporate Governance at TATA group, what standard can be expected of other Indian corporates?**

4. CONCLUSIONS AND SUGGESTIONS

The financial figures and the ensuing letter by Mistry indicates that the financial performance during Mistry's reign was not bad, contrary to the one alleged by the BOD and Ratan Tata. At the same time, even despite being chairman of Tata group, Mistry didn't really had control over the company matters due to reduction in his powers. The article suggests that family ties, rather than financial metrics, were the primary driver behind the leadership change initiated by Ratan Tata. This implies that preserving family control and influence within the Tata Group took precedence over the company's financial performance, highlighting the potential dominance of socioemotional wealth over financial interests in Indian family-owned conglomerates (Subramanian, 2022).

The case highlights a major hurdle in Indian corporate governance regarding keeping powerful shareholders, typically the founding families (promoters), in check. Their dominance on the company board weakens its independence. This lack of independence extends to the selection of independent directors, who are often chosen by the very majority shareholders they are meant to oversee. There is a lack of accountability and transparency in corporate systems with complicated board structures, and the law cannot explain this (Maria & Singh, 2021). This case necessitates a reevaluation of the legal and governance framework under which Indian corporations function (Vyas & Singh, 2022).

It is time that the issue of corporate governance is taken seriously in India and the shareholder's interests protected. More number of competent independent directors should be appointed (Shikha & Mishra, 2019) and by independent, they should be truly independent. They should act in the interest of the shareholders, apply their own judgement and not be under the undue influence of a Chairman or CEO. Also to prevent such an influence, periodical rotation of Board of Directors must take place. An adequate check must be kept on BOD and they must be held accountable for any lapses on the corporate governance front.

References

1. Balabhaskaran, P. (2019). Tata Group: Trauma of Transition. South Asian Journal of Business and Management Cases, 8(1), 32–46. https://doi.org/10.1177/2277977918803250

2. Bansal, N., & Sharma, A. K. (2019). Corporate governance and firm performance in an emerging economy context: new evidence from India. International Journal of Comparative Management, 2(2), 123–147.

3. Bansal, S., & Vajpeyi, A. (2022). The warning of an ambush: Disarming and appeasing activist shareholders. ILI Law Review.

4. Bansal, S., & Rajkumar, J. (2024). The Trilemma of Indian Independent Directors: Concerns and Directions for Reform. http://ir.nbu.ac.in/handle/123456789/5244

5. Bennedsen, M., Lu, Y. C., & Mehrotra, V. (2022). A survey of Asian family business research. Asia-Pacific Journal of Financial Studies, 51(1), 7–43. https://doi.org/10.1111/ajfs.12363

6. Bonazzi, L., & Islam, S. M. N. (2007). Agency theory and corporate governance: A study of the effectiveness of board in their monitoring of the CEO. Journal of Modelling in Management, 2(1), 7–23. https://doi.org/10.1108/17465660710733022

7. Dhameja, N. L., & Agarwal, V. (2017). Corporate Governance Structure: Issues & Challenges–Cases of Tata Sons & Infosys. Indian Journal of Industrial Relations, 53(1), 72–85.

8. Economic Times. (2016, October 29). Cyrus Mistry ouster: One more mess for Ratan Tata to sort out before his second exit. Economic Times. Retrieved from https://economictimes.indiatimes.com/news/company/corporate-trends/cyrus-mistry-ouster-one-more-mess-for-ratan-tata-to-sort-out-before-his-second-exit/articleshow/55133980.cms

9. Goldstein, A. (2013). The political economy of global business: the case of the BRICs. Global policy, 4(2), 162–172. https://doi.org/10.1111/1758-5899.12062

10. Jain, D. (2018). Who's at Fault?? Tata Sons in 2012 or Mistry in 2016: A Case Study on Tata Group. IIUM Journal of Case Studies in Management, 9(1), 1–6.

11. Jensen, M. C. (1986). Agency costs of free cash flow, corporate finance, and takeovers. The American Economic Review, 76(2), 323–329

12. Jhunjhunwala, S. (2020). Tata Sons and the Mystery of Mistry. Vikalpa, 45(3), 170–182. https://doi.org/10.1177/0256090920965420

13. Kaur, J., & Gupta, M. (2019). Removal of cyrus mistry from tata group-A study of legal provisions. *Asian Journal of Multidimensional Research (AJMR)*, 8(1), 55–74.

14. Maria, A. V., & Singh, K. D. (2021). Decoding Corporate Governance and Insolvency Related Issues in India. *Facets of Corporate Governance and Corporate Social Responsibility in India*, 69–83.

15. MC, A. B., & Rentala, S. (2018). Role of leadership and corporate governance: The case of Tata group and Infosys. *FIIB Business Review*, 7(4), 252–272.

16. Pandey, A. (2020). Tata Sons and the Mystery of Mistry. *Vikalpa*, 45(3), 183–185. https://doi.org/10.1177/0256090920973062

17. Puliyolli, D. (2023). Changes in Top Management and Shift in Strategy—A Comprehensive Study. In INDAM: Indian Academy of Management at SBM-NMIMS Mumbai (pp. 839–852). Singapore: Springer Nature Singapore. https://doi.org/10.1007/978-981-99-0197-5_53

18. Shikha, N., & Mishra, R. (2019). Corporate governance in India-battle of stakes. International Journal of Corporate Governance, 10(1), 20–41. https://doi.org/10.1504/IJCG.2019.098041

19. Sudhakar, G. P. (2018). Tata: the biggest boardroom coup. Emerald Emerging Markets Case Studies, 8(3), 1–24. https://doi.org/10.1108/EEMCS-03-2017-0041

20. Subramanian, S. (2022). Importance of Socioemotional Wealth in Indian Family Business Group: The Case of Tata Group. Paradigm, 26(2), 138–154.

21. Vij, S. (2017). Surgical Strike at Tata Group.

22. Vakhariya, S. (2017). Tata at a Crossroads. American Journal of Educational Research, 5(3), 284–295.

23. Vyas, M., & Singh, K. (2022). Tata Versus Mistry: A Boardroom Battle of Governance. Journal of Positive School Psychology, 6(10), 2339–2348.

Recent Trends in Engineering and Science for Resource Optimization and
Sustainable Development – Prof. (Dr.) Dorota Jelonek et al. (eds)
© 2024 Taylor & Francis Group, London, ISBN 978-1-032-98030-0

56

Acomparative Analysis of the Performance of Selected Banking Stockslisted in the National Stock Exchange (NSE)

Neha Nupoor[1]

Research Scholar,
Faculty of Commerce and Management,
Sarala Birla University Ranchi,

Mukesh Babu Gupta[2]

Assistant Professor,
Faculty of Commerce and Management,
Sarala Birla University Ranchi

Sandeep Kumar[3]

Dean & Associate Professor,
Faculty of Commerce and Management,
Sarala Birla University Ranchi

Abstract

A bankis anentity that gathers funds from the general public through deposits and provides loans to individuals, and businessmen, alsoto the entire nation. In recent times, the Indian securities market has emerged as a highly adaptable and efficient platform, showcasing versatility and effectiveness. The whole credit goes to RBI, which took severalmeasures to make the Indian Banking sector sturdier as well as healthier. This study is intended to examine the riskreturn dynamics, return on investment, and behavioral trends of specific Indian banks of the public and private sector viz State Bank of India, Punjab national bank, HDFC, and ICICI, over the period from 2018 to 2022. The samples are selected using the convenience sampling methodbased ontheir weightage in NIFTY, and upsurge in the banking industry. Averages and correlation statistical tools have been used to calculate the average return and risk return relationship of stocks.During the pandemic in 2020, the banking sector shares had a panic selling so the trend was negative which affected the return of this sector. Despite consistent growth in the return on investment for banking sector shares each year, the risk associated with investing in these banks appears to be greater than the returns they generate. This implies that investors in the banking sector may be taking on a higher level of risk relative to the financial gains they receive from their investments.

Keywords

Bank, Public and private sector banks, Performance, Risk and return, Return on investment

[1]nehanupoor1617@gmail.com, [2]mukesh.gupt@outlook.com, [3]sandeep.kumar@sbu.ac.in

DOI: 10.1201/9781003596721-56

1. INTRODUCTION

The economic growth of any nation is directly proportional to growth of their banking sector as bank transfers savings into investments, which generates income for the bank. Banking is an activity involving the collection of excess money with the public, which is over and above their requirements, and lending the same to the needy in the form of loans and advances. Bank of Hindustan is the first bank in India established in 1770, by the name. After that many more banks were established and therefore there was a need for one Central bank to regulate the monetary policy, manage other commercial banks, and issue currencies. Therefore, in 1935, the Reserve Bank of India was established. There were various private sector banks and over time it was observed that contribution of private sector banks in economic growth of the country remained far below the desirable and intended levels. This has led to the idea of the nationalization of banks. The nationalization of 14 banks in 1969 paved the way for the expansion of the banking industry in the country. Further in 1980 6 more banks were nationalized during the regime of Mrs. Indira Gandhi (Sengupta, A & De, 2020).

Entities that gather funds from the general public market potential of Indian banks is influenced by their social profitability, growth rate, and risk exposure. As a result, stocks from the Indian banking sector have become highly attractive, making it a prominent and leading segment in the Indian stock market through deposits and providing loans to individuals. Initially, bank stocks found a place in the BSE-100, BSE-500, and BSE-Sensex indices. The Sensex was formed in 1986 being the first benchmark index, that includes thirty blue-chip companies' stocks. NSE was established in 1992, having head office in Mumbai, Maharashtra. On the recommendation ofthe Government of India,the Pherwani committee established NSE. Being the first of its kind in the country itreplaced paper-based trading with an electronic trading system that offered a versatile trading platform to investors across the country.The National Stock Exchange (NSE) and the Bombay Stock Exchange (BSE) stand out as the two most crucial stock exchanges in India, handling the majority of share transactions. Within the NSE, there are seven major indices along with fifteen sector-specific indices. Among these, the CNX Bank Index or BANK NIFTY specifically tracks the price movements of shares from 17 listed banks, serving as a distinct benchmark for assessing the performance of the banking sector in the stock market

It's like taking a stroll through the financial history books of India! The fact that the roots of the stock market go back almost two centuries is mind-boggling. The blend of banking and cotton pressing businesses adding to the market dynamics gives it a unique flavor. It's like witnessing the early days of a market that would eventually become a vital part of the country's economic landscape. The evolution from those meagre and obscure records to the bustling stock market we know today is quite a journey. The essence of a bank's profitability lies in that interest spread—taking in money at one rate and lending it out at another. It's a classic banking model. And it seems like India is catching the eye of investors as a promising opportunity. With a growing economy, there's that potential for significant expansion. The prospect of offering various financial services to depositors adds another layer to the appeal. It's like the financial world sees India as a canvas for potential growth and prosperity. International investors are placing their bets on the Indian stock market! The rising conviction in the performance of Indian securities is attracting Foreign Institutional Investors (FIIs). The expectation that Indian stocks might outshine those in other emerging markets in the medium term is quite an endorsement. It's like the global fund managers are foreseeing a promising journey for Indian stocks, and they're eager to ride that wave of potential outperformance. It's always intriguing to see how global economic dynamics influence investment strategies. The remarkable expansion of the banking and financial services sector in India appears to be a consequence of well-planned government policies, particularly in the area of financial inclusion. It is intriguing to observe how significant investments and supportive policies at a large scale can trigger a chain reaction, propelling business growth throughout the country. The financial performance, the cost of capital, rate of return and return on equity, and net interest margin are some of the determinants of banking stock prices. Over the next decade, the banking sector is projected to create up to two million new jobs, driven by the efforts of the RBI and the Government of India to integrate financial services into rural areas. Factors such as financial performance, cost of capital, rate of return, return on equity, and net interest margin play pivotal roles in influencing banking stock prices. Looking ahead, the banking sector is anticipated to generate substantial employment growth, with projections indicating the creation of up to two million new jobs over the next decade. This surge is attributed to the concerted efforts of the Reserve Bank of India (RBI) and the Government of India in extending financial services to rural areas, reflecting a broader initiative to enhance financial inclusion. (Ramakrishnan & Toppur, 2016).

2. LITERATURE REVIEW

Berger et al., (2005), Analyzing data from Argentina in the 1990s reveal that the most robust and significant findings pertain to the impact of state ownership on bank performance. State-owned banks exhibited poor performance, indicating a static effect. Those undergoing privatization displayed particularly weak performance prior to the selection process. However, following privatization, these banks showed a substantial improvement, suggesting a dynamic effect. It's worth noting that a significant portion of the observed improvement may be attributed to the practice of transferring nonperforming loans into residual entities, thereby leaving the privatized banks with a healthier, 'good' portfolio. **Kamath & G., (2007)** in their study affirms significant variations in the performance of Indian banks across different segments, with an overall enhancement in performance observed throughout the study period. Notably, there is a noticeable inclination favoring the performance of foreign banks when compared to their domestic counterparts. **Hundal, Eskola & Chan (2019)**, the study propose a novel approach to enhance the Capital Asset Pricing Model (CAPM). Instead of using traditional market indices as the basis for estimating the beta coefficient, the suggestion is to employ economic growth as the foundation. This innovative methodology aims to shift the focus from market-specific indicators to broader economic indicators when estimating beta, potentially providing a more nuanced and relevant measure of investment risk. **Mutairi, S. & Sapuan (2020)**, in their study, found Correlating individual securities returns with market returns using Karl Pearson's coefficient of correlation was employed to understand the relationship between share prices and overall market movements. The analysis revealed that securities categorized as low-risk demonstrated better performance, particularly in the aftermath of the global financial crisis and within the Dubai market during the specified period. **Kunt, Pedraza & Ortega (2021)**, this paper investigates how announcements related to financial sector policies influenced the performance of bank stocks globally at the beginning of the COVID-19 crisis. **Mishra & Mishra (2021)**, this study examined the herding behavior of 54 stocks of banking and financial services sectors listed in the national stock exchange. **P. & H. (2023)** analysed Investments with higher perceived risk, such as stocks or speculative ventures, which have the potential for greater returns, but they also come with a higher likelihood of loss. On the other hand, lower-risk investments, like government bonds or certain fixed-income securities, tend to offer more stable returns but at the expense of lower potential gains. Balancing risk and return is a key aspect of constructing investment portfolios and making financial decisions that align with an individual's or an organization's risk appetite and financial objectives. Investors must conduct risk and return analysis to make informed choices and achieve a portfolio that reflects their unique preferences and circumstances.

3. RESEARCH METHODOLOGY

Source of Data: The study is based on secondary data collected from the National Stock Exchange Site (NSE). The data collection process is thorough and accurate, and considers using statistical and analytical tools to derive meaningful insights from the collected monthly market prices. Other sites like money control have also been considered. The yearly data has been considered for better results. To collect data for more longer period is not possible so yearly data has been collected.

Sample Size: The sample size 4 banking stocks, 2 from the public sector and 2 from private sector banks, they are top-performing banks. The four banks are SBI, PNB, HDFC and ICICI.

Method: The convenience sampling method has been used to choose the 4 banks. The statistical and financial tools are accompanied by average return, minimum, maximum, and correlation. These tools were employed to analyze and comprehend the relationship among risk and return of selected banks. The paper focused on key metrics such as average values, minimum and maximum values, return on investment, and correlations to gain insights ofthe financial performance and risk profiles of banks under consideration.

Period of the Study: The period under consideration is for 05 financial years i.e. from 2018-22.

4. OBJECTIVE AND SCOPE

Main objective of the study is to find performance of selected banking stocks on NSE index. The objective is to examine and understand the growth trajectory of nationalized banks in the Indian stock market over period from 2018 to 2022. The primary focus is on assessing return on investment (ROI) of selected bank stocks during this time frame.

Hypotheses

H0: There is no rational relationship between banking stocks' performance on the NSE index.

H1: There is a rational relationship between banking stocks' performance on the NSE index.

5. ANALYSIS AND INTERPRETATION

Interpretation of the result

Table 56.1 shows overall return of all four banks from 2018 to 2022. Here HDFC shows the highest return of 76.58%, ICICI Bank shows a return of 41.96% whereas in public sector banks, PNB shows a return of 10.24%, returns of SBI is 25.16%. In the initial years from 2018 to 2022,

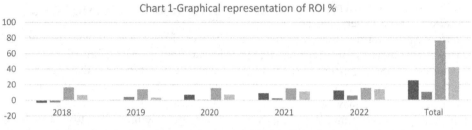

Fig. 56.1 Graphical representation of ROI %

Table 56.1 ROI % for the Banking stocks for 5 years

Year	SBI	PNB	HDFC	ICICI
2018	-3.37	-2.85	16.45	6.63
2019	0.39	4.2	14.12	3.19
2020	6.95	0.58	15.35	6.99
2021	8.86	2.41	15.27	11.21
2022	12.33	5.9	15.39	13.94
Total	25.16	10.24	76.58	41.96

Source: Prepared from the data downloaded from www.nseindia.com

Table 56.2 Consolidated calculation of ratios for 5 years

Name of bank	Average Return %	Std. dev	Average risk
SBI	5.03	2.45	3
PNB	2.1	1.45	2
HDFC	15.32	3.91	4
ICICI	8.4	2.9	3

Source: Prepared from the data downloaded from www.nseindia.com

HDFC has performed well with fluctuations in the upcoming years. Whereas the performance of ICICI is very good in 2022 and low in the following years and PNB improved after 2019 with some fluctuations in 20202, 2021 and 2022. Chart 1 shows performance of banks from 2018 to 2022, where SBI and PNB gave lowest return in 2018 and gradually it increased in the following years. In 2019 due to the emerging disease COVID-19 pandemic, the performance reduced for all the banks.Though Indian banking stocks did not perform well yet recovered later as the Indian banks' working system is controlled by RBI. Hence, resilience demonstrated by Indian banks is quickly recovering from the previous crisis suggesting a positive outlook for the future of the Indian markets. The data implies that the country has the potential to sustain a growth rate of 6-7 percent. Additionally, as a developing nation, India presents ample opportunities for further development, fostering optimism for the future performance of its markets. The combination of a robust banking sector recovery and the country's developmental scope paints a promising picture for the continued economic advancement of India.

Interpretation of result

It's important to note that while the data suggests HDFC has provided the highest average return of 15.32% over the last 5 years and PNB the lowest return of 2.1%, focusing solely on returns might not provide a comprehensive view of the investment scenario. Considering risk is crucial for a well-rounded investment analysis. In Table 56.2 above, it is mentioned that Punjab National Bank (PNB) has the lowest risk at 2%, followed by ICICI and SBI at 3%. If PNB has the lowest risk, it might not necessarily mean it's the best investment choice. Investors generally seek a balance between risk and return, aiming for an optimal risk-adjusted return. It's also crucial to consider other factors such as market conditions, the economic environment, and potential future trends. Before concluding that PNB is worth investing in due to its lower risk, investors should study all other inclusive factors like financial statement and management strength of equity, with overall economic climate. Investing decisions should be made based on a comprehensive understanding of the market and individual stocks rather than solely relying on one or two metrics like average return and risk. Additionally, seeking advice from financial experts and considering a diversified portfolio can help manage risks effectively.

Correlation

Correlation study of SBI, PNB, HDFC and ICICI versus NIFTY 50:

$$C0 = [\{N(\Sigma XY) - \Sigma X \, \Sigma Y\}/\sqrt{N\Sigma X^2 - (\Sigma X)^2} \sqrt{N\Sigma Y^2 - (\Sigma Y)^2}]$$

Where:

    X- Share price of Nifty 50

    Y- Share prices of Stocks

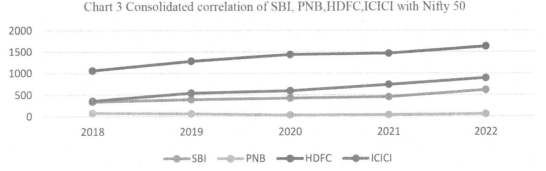

Fig. 56.2 Consolidated correlation of SBI, PNB, HDFC, ICICI with Nifty 50

Interpretation of result

In Table 56.3–56.4, the correlation coefficient between two variables is 1, it signifies a perfect positive linear relationship between them. This implies that as one variable increases, the other variable also increases proportionally, following a straight-line pattern on a scatter plot. The strength and direction of the relationship are at their maximum, indicating a perfect positive correlation. Here, correlation between Nifty and SBI shares is 1. This implies that as the Nifty index increases, the value of SBI shares also consistently increases, and when the Nifty index decreases, the value of SBI shares consistently decreases. The scatter diagram further reinforces this perfect positive relationship. Scatter diagrams visually represent the correlation between two variables. In this context, the scatter diagram for Nifty and SBI shares would show a clear upward-sloping line, indicating that as one variable (Nifty)

increases, the other variable (SBI shares) also increases. The data points would be tightly clustered around this line, suggesting a high degree of correlation. Similarly taking Nifty onthe X-axis and PNB on the Y-axis correlation between these two was negative initially and then positive to 1. This implies that as the Nifty index increases, the value of PNB shares also consistently increases, and when the Nifty index decreases, the value of PNB shares consistently decreases. The scatter diagram further reinforces this perfect positive relationship. Scatter diagrams visually represent the correlation between two variables. In this context, the scatter diagram for Nifty and SBI shares would show a clear upward-sloping line, indicating that when one variable (Nifty) accelerates, the other variable (PNB shares) also goes up. For the bank HDFC, a correlation coefficient of 0.91 indicates a linearitywith Nifty 50. Therefore here in this case, the correlation between Nifty and HDFC shares is 0.91. This implies that as the Nifty index increases, the value of HDFC shares also consistently increases, and when the Nifty index decreases, the value of HDFC shares consistently decreases. The scatter diagram further reinforces this perfect positive relationship. Scatter diagrams visually represent the correlation between two variables. In this context, the scatter diagram for Nifty and HDFC shares would show a clear upward-sloping line, indicating that as one variable (Nifty) increases, the other variable (HDFC shares) also increases. The data points would be tightly clustered around this line, suggesting a high degree of correlation. Now for ICICI bank, a correlation coefficient of 0.96 shows the linear relationship to be positive between two variables. In this case, the correlation between Nifty and ICICI shares is 0.96. This implies that as the Nifty index increases, the value of ICICI shares also consistently increases, and when the Nifty index decreases, the value of SBI shares consistently decreases. The scatter diagram further reinforces this perfect positive relationship. Scatter diagrams visually represent the correlation between two variables. In this context, the scatter diagram for Nifty and ICICI shares would show a clear upward-sloping line,

Table 56.3 Consolidated correlation of SBI, PNB, HDFC, ICICI with Nifty 50

Year	Nifty	SBI	PNB	HDFC	ICICI
2018	10900	337	78	1061	360
2019	12168	395	64	1282	544
2020	13981	425	33	1436	595
2021	17354	452	37	1462	736
2022	18105	613	57	1628	891

Source: Prepared from the data downloaded from www.nseindia.com

Table 56.4

	Correlation Matrix				
	Nifty	**SBI**	**PNB**	**HDFC**	**ICICI**
SBI	0.87407435	1			
PNB	-0.5752627	-0.293854376	1		
HDFC	0.91850905	0.90982526	-0.651402273	1	
ICICI	0.96221148	0.948807177	-0.505843208	0.968273	1

indicating that as one variable (Nifty) increases, the other variable (ICICI shares) also increases. The data points would be tightly clustered around this line, suggesting a high degree of correlation.

6. Findings

In year 2018, return of HDFC was highest to16.45% and that of PNB was lowest equal to(2.85%). In 2019 HDFC-gave highest return of 14.12% and State Bank of India was lowest to 0.39%.In 2019 deadly disease created COVID-19 createda panic sellingtherefore bank shares had a negative impact which affected return of this sector which was case in whole world. All the banks have a negative return. The SBI had the highest negative value of 0.39% and HDFC was better to 14.12 %. This proves, among the four banks, HDFC is the leading and best bank of all to invest in, it is India's leading private bank, whereas the public sector bank, PNB has some growth issues. The year 2020 showed some recovery. This was the case after the announcement of the second lockdown in June 2020Kunt, Pedraza & Ortega (2021). In 2021 and 2022 again HDFC showed the highest results among the four and PNB was the lowest of all. Certainly, based on the correlation analysis it could be described, that Nifty index and the shares of banks, with HDFC, State Bank of India (SBI), and ICICI is directly proportional. In simpler terms, when the Nifty index experiences a positive movement (increase), the shares of these banks, especially HDFC, also tend to move in a positive direction. Likewise, when the Nifty index shows a negative movement (decrease), the shares of the four banks, including PNB, tend to move in a negative direction. This correlation suggests that the performance of the banks, and particularly HDFC, SBI, and ICICI, is closely tied to the overall movements in the Nifty index. Investors and analysts often use such correlations to make predictions or assess the potential impact of market changes on specific stocks or sectors.

7. Implications and Limitations of the Study

The study is limited to Indian banking stocks which is listed in the National Stock Exchange. Further study is based only on 04 banks according to their weightage in the NIFTY 50. Further, we have done study for 5 years only. More number of Years with more data could give more relevant results. So, further studies could be done keeping these two limitations in view. However, it is important to acknowledge that, this study may encounter limitations related to the statistical analysis of its results. Despite the acknowledged limitations, the outcomes of this study hold the potential to offer valuable insights for a range of stakeholders, including investors, policymakers, and those involved in the banking industry. The primary objective is to enhance comprehension regarding the factors influencing the performance of nationalized bank stocks within the Indian stock market. The hope is that these insights will play a role in shaping future investment strategies and policy decisions within the banking sector.

References

1. Al Mutairi, A. M. S., & Sapuan, N. M. (2020). Impact of Corporate Governance on Financial Risk of Omani Non-financial Companies Listed on the Muscat Securities Market. International Journal of Psychosocial Rehabilitation, 24(02), 2121–2127. https://doi.org/10.37200/ijpr/v24i2/pr200513
2. A. P., & -, N. M. H. (2023). A Study on Risk and Return Analysis of Selected Private Sector Banks in India. International Journal For Multidisciplinary Research, 5(2). https://doi.org/10.36948/ijfmr.2023.v05i02.1880
3. Barathi Kamath, G. (2007). The intellectual capital performance of the Indian banking sector. Journal of intellectual capital, 8(1), 96–123.
4. Berger, A. N., Clarke, G. R., Cull, R., Klapper, L., & Udell, G. F. (2005). Corporate governance and bank performance: A joint analysis of the static, selection, and dynamic effects of domestic, foreign, and state ownership. Journal of Banking & Finance, 29(8-9), 2179–2221.
5. Hundal, S., Eskola, A., & Tuan, D. (2019). Risk–return relationship in the Finnish stock market in the light of Capital Asset Pricing Model (CAPM). Journal of Transnational Management, 24(4), 305–322. https://doi.org/10.1080/15475778.2019.1641394
6. Kunt, A., Pedraza, A., & Ruiz-Ortega, C. (2021)- Demirgüç-Kunt, A., Pedraza, A., & Ruiz-Ortega, C. (2021). Banking sector performance during the COVID-19 crisis. Journal of Banking & Finance, 133, 106305.
7. Mishra, P. K., & Mishra, S. K. (2023). Do banking and financial services sectors show herding behavior in Indian Stock Market amid the COVID-19 pandemic? Insights from quantile regression approach. Millennial Asia, 09763996211032356.
8. Ramakrishnan, P., &Toppur, B. (2016). A study of banking stocks in India to develop a model for prudent investment. Universal Journal of Management, 4(9), 477–487.
9. Sengupta, A., & De, S. (2020). Assessing. Performance of Banks in India Fifty Years After Nationalization. Springer.

Note: All the figures and tables in this chapter were made by the author.

*Recent Trends in Engineering and Science for Resource Optimization and
Sustainable Development – Prof. (Dr.) Dorota Jelonek et al. (eds)
© 2024 Taylor & Francis Group, London, ISBN 978-1-032-98030-0*

57

Numerical Modeling of Mangalore Port Including Wave Energy Dissipators using the Boundary Element Method

Komal Thakran[1]
Department of Mathematics, School of Basic and Applied Sciences,
K. R. Mangalam University, Gurugram, Haryana India

Rupali[2]
Department of Mathematics, School of Basic and Applied Sciences,
K. R. Mangalam University, Gurugram, Haryana India

Prachi Priya[3]
Department of Mathematics, School of Basic Sciences,
Galgotias University, Greater Noida, Gautham Budh Nagar, Uttar Pradesh, India

Prashant Kumar[4]
Department of Applied Sciences,
National Institute of Technology, Delhi, Delhi, India

Yogendra Kumar Rajoria[5]
Department of Mathematics, School of Basic and Applied Sciences,
K. R. Mangalam University, Gurugram, Haryana India

■■■■■ **Abstract**

This study presents a comprehensive numerical modeling approach for analyzing the wave dynamics in Mangalore port, with a focus on the integration of wave energy dissipators using the boundary element method (BEM). The primary objective is to assess the effectiveness of wave energy dissipators in attenuating wave-induced forces within the port environment. The study encompasses the development and implementation of a numerical model that incorporates the complex geometrical features of the Mangalore port and integrates wave energy dissipators to mitigate the impact of wave forces. The numerical model considers the interaction of incident waves with the port structures and the influence of wave energy dissipators in reducing wave heights and energy levels. Through numerical simulations and analysis, the effectiveness of different configurations of wave energy dissipatorsis evaluated, offering guidance for optimal design and placement to minimize wave-induced effects on port infrastructure and vessels.

■■■■■ **Keywords**

Boundary element method, Energy dissipaters, Mangalore port, Helmholtz equation

komalthakran8@gmail.com, [2]guptarupali1105@gmail.com, [3]prachipriya167@gmail.com, [4]prashantkumar@nitdelhi.ac.in, [5]yogendrarajo@gmail.com

DOI: 10.1201/9781003596721-57

1. Introduction

Mangalore port serves as a vital trade gateway, facilitating the import and export of a wide range of goods. It plays a crucial role in fostering international trade, and connecting India to various global markets. The port's economic impact extends beyond its immediate vicinity, contributing to job creation and livelihoods. Mangalore port boasts substantial cargo handling capabilities, its efficiency in cargo handling makes it a preferred choice for commercial shipping and trade-related activities. Its strategic location on the west coast of India positions it as a key player in transshipment and maritime trade routes. From a strategic standpoint, Mangalore port's operations are integral to India's national security interests.

The harbor resonance at Mangalore Port encompasses the interaction of waves, tides, and other hydrodynamic phenomena within the harbor, influencing various aspects of port operations, infrastructure design, vessel behavior, environmental considerations, and safety management. Understanding and effectively managing harbor resonance is essential for ensuring the safe and efficient functioning of the port while minimizing risks and environmental impacts. The numerical modeling of Mangalore port, specifically incorporating wave energy dissipaters (Xing, 2009), is a complex and crucial engineering task. One approach to this problem is the application of the boundary element method (BEM)(H. S. Lee, 2004), which offers a powerful technique for simulating wave interactions with geographic features and human-made structures. This method is particularly useful for analyzing the behavior of wave energy dissipators and their impact on the overall port environment.

Oscillations in simple shaped harbor such as rectangular and circular harbors were initially studied theoretically and experimentally in 1950s by (McNown, 1952). Various numerical methods have been developed to study the wave oscillations on harbors of complex geometry such as Boundary Element Method (BEM) (Kumar et al., 2021, 2022, 2017; Kumar & Gulshan, 2017; Jiin Jen Lee, 1971), Hybrid Finite-Element method (HFEM) (Rupali & Kumar, 2021), Finite-Element Method (FEM) (Kumar & Rupali, 2018). Researchers successfully applied the numerical studies on realistic harbor based on linear and nonlinear methods like Long beach harbor in California, USA (J J Lee et al., 1998), Marina di Carrara in Italy (Guerrini et al., 2014), Hua-Lien Harbor in Taiwan (Chen et al., 2004), Pohang New Harbor in South Korea (Kumar et al., 2016; Rupali et al., 2020, 2020). Recently, Kumar et al., (2021) used the numerical model based on BEM to obtain the distribution of wave energy within the Visakhapatnam port.

In this context, the primary goal of the numerical modeling in this study is to understand how wave energy dissipators affect the wave climate within the Mangalore port, including wave heights, wave patterns, and energy dissipation. By accurately capturing these aspects, engineers and researchers can make informed decisions regarding the design and placement of wave energy dissipators to optimize their performance and minimize potential negative effects on the port infrastructure.

2. Mathematical Formulation

2.1 Model Geometry

The mathematical model is represented in Fig. 57.1. The model geometry is divided into two regions i.e. REGION I (unbounded region or open sea region) (Ω_{UR}) and REGION II (bounded region or harbour region) (Ω_{BR}). Partially reflecting boundary is denoted by Ω_b and pseudoboundary $B_1 B_2 B_3$ is denoted by Ω_a. In unbounded region REGION I the exterior point $\vec{x}$ is approaching to the pseudoboundary $\vec{x_I}$ and similarly, the interior point $\vec{x}$ approaching to boundary point $\vec{x_o}$ and corner point $\vec{x_I}$ in bounded region REGION II as shown in the Fig. 57.1.

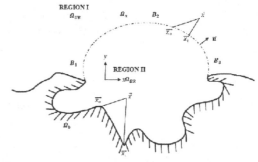

Fig. 57.1 Sketch for the model configuration with region I and II separated by an interface $B_1B_2B_3$.

Assuming the fluid to be incompressible and inviscid, and the fluid flow to be irrotational. Continuity equation for fluid is $\nabla \cdot \vec{V} = 0$. The velocity vector can be written as in terms of gradient of potential function ϕ. Thus, according to potential flow theory, ϕ will satisfy the Laplace Equation with solution given by

$$\phi(x,y,z;t) = \frac{i}{\sigma} f(x,y) Z(z) e^{-i\sigma t} \qquad (1)$$

Where σ is the angular frequency, $f(x, y)$ is the wave amplitude in direction of x and y from the sea surface, $Z(z)$ represents depth function, and $i = \sqrt{-1}$ is an imaginary number. Using Separation of variables method, Helmholtz Equation and Depth Equations are obtained which are expressed below,

$$\frac{\partial^2 f}{\partial x^2} + \frac{\partial^2 f}{\partial y^2} + k^2 f = 0 \quad \text{and} \quad \frac{d^2 Z}{dz^2} - k^2 Z = 0 \quad (2)$$

Where k represents wave number and d represents the depth of the harbor. The boundary condition at seabed is given by $\frac{d\phi}{dz}(x, y, -d, t) = 0$. Using seedbed equation, we can find the solution of Eq. (2).

2.2 Wave Function in REGION II (i.e. Harbor Region)

Green's Identity Formula is applied in REGION II and using the Hankel function of the first kind and zeroth order, the function f_2 at any point $\vec{x}(x, y)$ inside the bounded region Ω_{BR} is derived as

$$f_2(\vec{x}) = \frac{-i}{4} \int_{\Omega b} \left[f_2 \frac{\partial}{\partial n}\left(H_0^1(kr)\right) - H_0^1(kr)\frac{\partial f_2}{\partial n} \right] ds \quad (3)$$

2.3 Wave Function in REGION I (i.e. Unbounded Region)

Wave function in REGION I can be expressed as sum of an incident and a radiated wave and the boundary conditions are:

i) On the solid boundary of the harbor $\frac{\partial f_1}{\partial n} = 0$.

ii) On the pseudo boundary $B_1 B_2 B_3$ $\frac{\partial f_1}{\partial n} = -\frac{\partial f_2}{\partial n}$.

iii) The radiated wave decay at infinity $\lim_{r \to \infty} f_1 = f_{inc}$.

$$f_{rad}(\vec{x_i}) = \frac{-i}{2} \int_{B_1 B_2 B_3} \left[f_{rad}(\vec{x_0})\frac{\partial}{\partial n}\left(H_0^1(kr)\right) \right.$$

$$\left. - H_0^1(kr)\frac{\partial}{\partial n}\left(f_{rad}(\vec{x_0})\right) \right] ds(\vec{x_0}) \quad (4)$$

The function f_1 is to be evaluated in REGION I at the harbor entrance, similarly, the function f_2 is determined in REGION II at the entrance. The two solutions were matched at the entrance of the harbor and the unknown function C_j is determined.

3. CONVERGENCE ANALYSIS AND VALIDATION

Numerical convergence refers to the property of a numerical method where the solution approaches the exact solution as the computational resources are refined. By systematically evaluating convergence and quantifying errors,

engineers and researchers can ensure the reliability and accuracy of the simulated results, leading to better-informed decisions in the design, operation, and maintenance of harbor infrastructure. The application of the least squares method to analyze numerical convergence and error approximation for a rectangular harbor, resulted 1.76 order of convergence. Therefore, the numerical technique can be used to predict accurate and robust solutions for the relevant physical and mathematical problems associated with the behavior of water in the harbor.

3.1 Implementation of Numerical Scheme on Mangalore Port

Mangalore Port is located in Panambur, Mangalore in the state of Karnataka and on the west coast of India. It is 354 km north of Kochi Port and 310 km south of Mormugao Port. Plotted position of Mangalore Port on map is 12° 50' 56.5008" N, 74° 50' 13.8804" E.

The response of amplification factor is analyzed with or without dissipators at different locations in Mangalore Port by conducting the numerical simulation for incident waves. The amplification factor at Mangalore port (displayed in Fig. 57.2) can be influenced by the presence or absence of dissipators. It was observed that the peak of wave amplification is high without dissipator whereas the peak of wave amplification reduces significantly with dissipator. With the use of dissipators at different locations, the amplification of waves can be minimized. Wave height contours in Mangalore port are a vital consideration for port operations, coastal protection, and maritime safety. The wave height contours shown in Fig. 57.3 represent the spatial distribution of wave heights in the vicinity of the port, providing valuable information for port planning, design, and risk assessment.

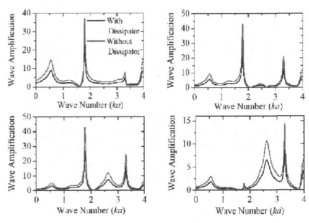

Fig. 57.2   Comparison of the wave amplification results with and without application of energy dissipaters in the port area

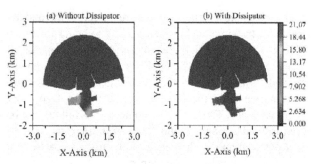

Fig. 57.3 Contour sketch of the Mangalore plot depicting the wave field distribution over the entire region of port with and without application of energy dissipaters

4. Conclusion

The findings of this study provide valuable insights into the performance of wave energy dissipators in mitigating wave forces within the Mangalore port. The study demonstrates the potential of the boundary element method as a reliable tool for assessing wave dynamics and the integration of wave energy dissipators in port environments. The implications of this research extend to the realm of coastal engineering and port infrastructure design. By elucidating the efficacy of wave energy dissipators in Mangalore port, this study contributes to the development of sustainable and resilient port structures capable of withstanding wave forces, thereby enhancing operational safety and reducing maintenance costs. In conclusion, this research underscores the significance of numerical modeling and the boundary element method in evaluating wave dynamics and the integration of wave energy dissipators within port environments.

References

1. Chen, G. Y., Chien, C. C., Su, C. H., & Tseng, H. M. (2004). Resonance induced by edge waves in Hua-Lien Harbor. *Journal of Oceanography*, *60*(6), 1035–1043.
2. Guerrini, M., Bellotti, G., Fan, Y., & Franco, L. (2014). Numerical modelling of long waves amplification at Marina di Carrara Harbour. *Applied Ocean Research*, *48*, 322–330.
3. Kumar, P., & Gulshan. (2017). Extreme Wave-Induced Oscillation in Paradip Port Under the Resonance Conditions. *Pure and Applied Geophysics*, 1–16.
4. Kumar, P., Gulshan, & Rajni. (2017). Multidirectional Random Wave Diffraction in a Real Harbor by using 3-D Boundary Element Method. *AIP Conference Proceedings 1897*, 020006-1–020006-020008.
5. Kumar, P., Priya, P., & Rajni. (2021). Boundary Element Modeling of Multiconnected Ocean Basin in Visakhapatnam Port Under the Resonance Conditions. *China Ocean Engineering*, *35*, 662–675.
6. Kumar, P., Priya, P., & Rajni. (2022). Mathematical modelling of Visakhapatnam Port utilizing the porous and non-porous breakwaters with finite depth green function. *Ocean Dynamics*, *72*, 557–576.
7. Kumar, P., & Rupali. (2018). Modeling of shallow water waves with variable bathymetry in an irregular domain by using hybrid finite element method. *Ocean Engineering*, *165*, 386–398.
8. Kumar, P., Zhang, H., Ik, K., & Yuen, D. A. (2016). Modeling wave and spectral characteristics of moored ship motion in Pohang New Harbor under the resonance conditions. *Ocean Engineering*, *119*, 101–113.
9. Lee, H. S. (2004). Boundary element modeling of multidirectional random wave diffraction by multiple rectangular submarine pits. *Engineering Analysis with Boundary Elements*, *28*, 1149–1155.
10. Lee, J J, Lai, C. P., & Li, Y. (1998). Application of computer modeling for harbor resonance studies of Long Beach & Los Angeles harbor basins. *Coastal Engineering Proceedings*, *1*(26).
11. Lee, Jiin Jen. (1971). Wave-induced oscillations in harbours of arbitrary geometry. *Journal of Fluid Mechanics*, *45*(2), 375–394.
12. McNown, J. S. (1952). Waves and seiche in idealized ports. *Gravity Waves Symposium, National Bureau of Standards Circular*, *521*, *521*, 153.
13. Rupali, & Kumar, P. (2021). Mathematical modeling of arbitrary shaped harbor with permeable and impermeable breakwaters using hybrid finite element method. *Ocean Engineering*, *221*, 108551.
14. Rupali, Kumar, P., & Rajni. (2020). Spectral wave modeling of tsunami waves in Pohang New Harbor (South Korea) and Paradip Port (India). *Ocean Dynamics*, *70*(12), 1515–1530.
15. Xing, X. (2009). Computer modeling for wave oscillation problems in harbors and coastal regions. In *University of Southern California*.
16. N. Kumar, A. Pant and R. Kumar Singh Rajput: "A Computational Study of Elastico-Viscous Flow between Two Rotating Discs of Different Transpiration for High Reynolds Number" International Journal of Engineering, vol-22(2), aug-2009, pp. 115–122. (http://www.ije.ir/article_71770.html)
17. N. Kumar, U S Rana and J. Baloni: "A Mathematical Model of Growth of Homogeneous Tumor with Delay Time" In International journal of Engineering, vol-22(1), April -2009, pp. 49–56. (http://www.ije.ir/article_71759.html)

Note: All the figures in this chapter were made by the author.

Recent Trends in Engineering and Science for Resource Optimization and Sustainable Development – Prof. (Dr.) Dorota Jelonek et al. (eds)
© 2024 Taylor & Francis Group, London, ISBN 978-1-032-98030-0

58

Review on Deterministic Mathematical Modelling Studies on COVID-19

Sadhana Gupta

Department of Mathematics (School of Basic and Applied Science),
K. R. Mangalam University, Gurugram, India

Surabhi Pandey

Indian Institute of Public, Public Health Foundation, India

Rupali*

Department of Mathematics (School of Basic and Applied Science),
K. R. Mangalam University, Gurugram, India

Yogendra Kumar Rajoria

Department of Mathematics (School of Basic and Applied Science),
K. R. Mangalam University, Gurugram, India

■■■■■ **Abstract**

We discuss various modelling studies conducted on COVID-19 in this review study, along with the conclusions drawn from each paper. We have mainly concentrated on the deterministic COVID-19studies. We have outlined some of the most important findings from the studies and also talked about some of the potential directions for further investigation into mathematical models for understandingCOVID-19.

■■■■■ **Keywords**

SARS-CoV-2, COVID-19, Reproduction number

1. INTRODUCTION AND MOTIVATION

Wuhan city experienced a pandemic brought on by a beta coronavirus in December 2019. This virus quickly spread to most of the countries from the city of origin. Due to its quick spread, WHO declaredCoV-19 as pandemic on 17th March 2020. SARS-CoV-2 is the current name for the RNA virus that causedCOVID-19. Various groups of people,

doctors, clinicians, researchers came together and worked to understand the virus and the disease. The existing techniques and therapies then were not found to be effective against the virus. At this hours of difficulty, a practical, clinically relevant, logical framework was needed to understand the disease and its dynamics. In the past mathematical models were found to be very useful in understanding the behaviour of an infection[Zeb et al., 2020]. For infections

*Corresponding author: guptarupali1105@gmail.com

DOI: 10.1201/9781003596721-58

like influenza, MERS-Cov, and HIV, mathematical models were found to be very useful to understand the dynamics of the infection and also to find out conditions under which the disease can be wiped out or continued [Dar Assi et al., 2022]. Owing to this contributions of mathematical models, disease modellers started working on developing models to understand and interpret COVID-19 disease. In the past 2 years various models on COVID-19 were developed and studies. Also to study the effectiveness of the controls like vaccination, isolation of infected individuals, quarantine etc optimal control problems were also formulated and studied for COVID-19 pandemic. Optimal control theory (OCT) is used in epidemiology to evaluate the best of the above mentioned control strategies so that the disease burden is reduced [Abakuks, 1973]. Filippov's existence theorem and Pontryagin's maximum (or minimum) principle were used in OCT to prove the existence of optimal control measures and for obtaining the optimal solutions [Schättler and Ledzewicz, 2012].

In this review study we discuss various modelling studies done on COVID-19 along with the findings from each of the paper. Our major focus will be on the deterministic studies. In the next section we have two subsections. In the first subsection we discuss the deterministic studies done on COVID-19 and in the second subsection optimal control studies done on COVID-19 is discussed. Following this we will provide the main findings from the studies that are done so far and also discuss some of the possible approaches for future research in understanding COVID-19 using mathematical models.

2. LITERATURE REVIEW

2.1 Mathematical Modelling Studies on Covid-19

Here we discuss the modelling studies on COVID-19 and also highlight the main outcomes of each of the studies. Because our interest is only on the compartment models, therefore in this review work we only discuss the deterministic modelling studies done in COVID-19. In the last two years, many deterministic models were developed to analysis & understand prevalence and behaviour of virus CoV-19 disease. In Samui et al., 2020 compartmental models are proposed to predict the rise cases in India & to understand various ways to prevent the spread of the virus. Stability analysis for the admitted equilibria is done with respect to R_0. The study reveals that transmission rate is very effective to lower R_0. In Joshi et al., 2021, the authors used the SMAART RAPID Tracker to map the spread the coronavirus across states in India. This study highlighted the requirement for a country-centric ways to monitoring and combating the COVID-19 pandemic. In Mandal et al.,

2021, a compartmental model is used to examine the conditions under which a third wave can occur in India. The authors have identified possible mechanisms by which India will weather the third wave.

A new SQIR model is developed and studied taking saturate incidence rate in Algehyne and ud Din, 2021 to study susceptible, quarantine, infected and recovered population. Lyapunov function based techniques used for establishing the global stability and the finite difference method is used to validate the theoretical findings numerically. In Ndaïrou et al., 2020, parameters were discussed which are sensitive to determine the most spreading ability of susceptible individuals. To investigate the role of isolation mathematical models containing isolation class were also developed. A mathematical model is formulated in Zeb et al., 2020 to comprehend the spreading behaviour of COVID-19 with isolation class. According tothis article, isolation of the infected person as a whole is necessary to lower the chance of COVID-19 spreading. Effective use of social distancing, lockdown and face masks was found the lower COVID-19 cases. The detailed stability and sensitivity analysis of the within-host model developed for COVID-19 can be found in Chhetri et al., 2021. In Riyapan et al., 2021 a model is developed to study the transmission of COVID-19 in Bangkok, Thailand. To analyze the transmission of the disease in Nigeria, an epidemic model considering awareness campaigns and various hospitalization plans has been developed in Musa et al., 2021. It was evident that campaigns for awareness and on time medical support including hospitalization of critical cases has played vital role in controlling the spread of virus. The model results for the countries like Italy, UK, Germany and Spain are estimated in Bärwolff, 2020 were evaluation proper data of infected people using a SIR COVID-19 model. The summary and in-depth comparisons of all the mathematical models that are developed to estimate the dynamics and spread of COVID-19 are presented in Kumr et al 2009, Al Arjani et al., 2022.

SARS-Cov-2 virus also evolved into many variants to name a few variant of concern, Alpha, Delta, and Omicron. Along with clinical research, many mathematics based models are also developed to study different variants of SARS-CoV-2 virus and its impact. Omicron model is stable when the reproduction number $R_0 < 1$, while when $R_0 \leq 1$, the model is found to be globally stable. Using PRCC method the global sensitivity analysis is done to find out the various influential variables that decreases or increases R_0. A model with age-structure and control measures such as antiviral medication, vaccine, and the influx of the omicron variant is developed and studied. Besides omicron variant delta and beta variants of SARS- CoV-2 also affected large portion of world population. Numerous studies were done to understand delta variant and also the effectiveness of control measures against this variant. In Yu et al., 2022, transmissibility of Delta, Beta, and Omicron variants are calculated using a mathematical model. It was found that the transmissibility with respect to the Omicron is comparatively higher than the other two. The dynamics and spread patterns of COVID-19 were fairly well understood by mid-2020. However, assessing the effectiveness of control strategies was equally important. Effectiveness of control strategies, including vaccination are well studied incease of COVID-19. Because the initial supply of the vaccine was limited, it took an hour to develop effective vaccination strategies.

The study concluded that screening of travellers from the countries with COVID-19 cases, can result insignificant drop in spread of the virus. The effect of new variant and vaccination impact on new infection, hospitalization, and deaths is studied using a modified SEIR model in Nakhaeizadeh et al., 2022. The results of the study suggest that focus must be on improving vaccination strategy to prevent COVID-19 spread. An age-structured model with vaccination is developed and studied in Choi et al., 2021. For lower social distance, it was found that vaccination of the age group with the highest infection rate reduced morbidity, but if social distancing was effective, vaccination of the older age group was better at reducing infection. To stop the spread of the variants and to lower the peak of a multi-strain pandemic minimum number of fully vaccinated individuals is calculated in the study. In Chhetri et al., 2022b, Chhetri et al., 2022a optimal control analysis is discussed considering different controls, such as treatment,vaccination and screening. The results from the study imply that controls are very helpful for reducing infected individuals and enhancing population health. The effectiveness of control measures like quarantine, home containment, and infected person isolation is discussed.

3. Key Finding

Today, a lot of effort has been put into using mathematical modelling as instrument to comprehend the dynamics and patterns of disease spread. Many of these models were created and studied following the COVID-19 outbreak. In this paper, we have highlighted and briefly discussed the results of the COVID-19 deterministic modelling work. Some of the key findings from these studies include the following.

- The compartment models developed mostly admitted two equilibrium states, the infection-free and the infected equilibrium states.
- The local and global stabilities of the admitted equilibrium states were found to mostly depend on

the reproduction number R_0. In most of the cases the infection-free state was found to be globally stable for the value of R_0 less than unity.

- Apart from the traditional ordinary differential equations fractional differential equation are also being used to understand and interpret disease dynamics. In these cases the tentative results of the model equations were obtained using the Laplace transform and fixed-point method.
- Many mathematics based models are also developed to understand the role of different variants (Omicron, delta, beta) of SARS-CoV-2 virus and its impact. Because of these variants rapid rise in COVID-19 cases were found in different parts of the globe.
- From the optimal control studies it was found that the different preventive measure such as vaccination, lockdown, face masks, quarantine, social distancing etc if implemented seriously and effectively, were crucial in limiting spread of the virus and enhancing public health.

4. Discussion and Conclusion

SARS-CoV-2 is the cause behind COVID-19, a contagious respiratory and vascular disease. The disease's emergence has presented a significant challenge to governments and health organizations all over the world. One of the biggest pandemics the world has ever seen, the COVID-19 has break out in the entire world. Almost everything on the planet has been affected negatively by the SARS-CoV-2 virus. The deadliest infection has a significant negative impact on social life, health, the economy, and education.

The pandemic has everyone, including researchers, policymakers, the government, and health authorities baffled. In the past, it has been demonstrated that models can help for comprehending the dynamics of infectious diseases. Because of this contributions disease modellers felt the need of mathematical models and many models were developed study the spread pattern disease. Mathematical modelling is very critical to understand the pandemic and to predict the future course under different cases.

We have cited and highlighted the key deterministic studies on COVID-19 in this review study. We may not have covered all of the deterministic works done across all nations, but we did our best to cover nearly all types of the deterministic works. This includes compartment models for understanding the stability of the admitted equilibrium states and finding sensitive parameters, compartment models with different variants of the virus, and finally the optimal control studies to study the effectiveness of various strategies.

The stabilities of the admitted equilibrium states were found to depend on R_0. In most of the cases the infection-free state was found to be globally stable for the value of R_0 less than unity. The new variants of the virus were found to be highly contagious compared to the original one and these variants were found to result in the rapid rise in COVID-19 cases globally. In optimal control studies the efficacies and roles of various available and possible strategies such as vaccination, lockdown, face masks, quarantine, social distancing etc were studied. It was found that if implemented seriously and effectively, these control measures were crucial in limiting spread the virus and enhancing public health.

References

1. Abakuks, A. (1973). An optimal isolation policy for an epidemic. Journal of Applied Probability, pages 247–262.
2. AlArjani, A., Nasseef, M. T., Kamal, S. M., Rao, B. S., Mahmud, M., and Uddin,M. S. (2022). Application of mathematical modeling in prediction of covid-19 transmission dynamics.Arabian Journal for Science and Engineering, 47(8):10163–10186.
3. Algehyne, E. A. and ud Din, R. (2021). On global dynamics of covid-19by using sqir type model under non-linear saturated incidence rate. Alexandria Engineering Journal,60(1):393–399.
4. [Bärwolff, 2020] Bärwolff, G. (2020). Mathematical modeling and simulation of the covid-19 pandemic. Systems, 8(3):24.
5. Chhetri, B., Bhagat, V. M., Vamsi, D., Ananth, V., Mandale, R., Muthusamy, S.,Sanjeevi, C. B., et al. (2021). Within-host mathematical modeling on crucial inflammatory mediatorsand drug interventions in covid-19 identifies combination therapy to be most effective and optimal.Alexandria Engineering Journal, 60(2):2491–2512.
6. Chhetri, B., Vamsi, D., Prakash, D. B., Balasubramanian, S., and Sanjeevi, C. B.(2022a). Age structured mathematical modeling studies on covid-19 with respect to combined vaccination and medical treatment strategies. Computational and Mathematical Biophysics, 10(1):281–303.
7. Chhetri, B., Vamsi, D., and Sanjeevi, C. B. (2022b). Optimal control studies onage structured modeling of covid-19 in presence of saturated medical treatment of holling type iii.Differential Equations and Dynamical Systems, pages 1–40.
8. Choi, Y., Kim, J. S., Kim, J. E., Choi, H., and Lee, C. H. (2021). Vaccination prioritization strategies for covid-19 in korea: a mathematical modeling approach. International Journal of Environmental Research and Public Health, 18(8):4240.
9. DarAssi, M. H., Shatnawi, T. A., and Safi, M. A. (2022). Mathematical analysis of a mers-cov coronavirus model. Demonstratio Mathematica, 55(1):265–276.
10. Joshi, A., Kaur, H., Krishna, L. N., Sharma, S., Sharda, G., Lohra, G., Bhatt, A., and Grover, A. (2021). Tracking covid-19 burden in india: A review using smaart rapid tracker. Online Journal of Public Health Informatics, 13(1).
11. Mandal, S., Arinaminpathy, N., Bhargava, B., and Panda, S. (2021). Plausibility of a third wave of covid-19 in india: A mathematical modelling based analysis. The Indian journal ofmedical research, 153(5–6):522.
12. Musa, S. S., Qureshi, S., Zhao, S., Yusuf, A., Mustapha, U. T., and He, D. (2021).Mathematical modeling of covid-19 epidemic with effect of awareness programs. Infectious disease modelling, 6:448–460.
13. Nakhaeizadeh, M., Chegeni, M., Adhami, M., Sharifi, H., Gohari, M. A.,Iranpour, A., Azizian, M., Mashinchi, M., Baneshi, M. R., Karamouzian, M., et al. (2022). Estimatingthe number of covid-19 cases and impact of new covid-19 variants and vaccination on the population inkerman, iran: a mathematical modeling study. Computational and Mathematical Methods in Medicine, 2022.
14. [Ndaïrou et al., 2020] Ndaïrou, F., Area, I., Nieto, J. J., and Torres, D. F. (2020). Mathematical modeling of covid-19 transmission dynamics with a case study of wuhan. Chaos, Solitons & Fractals, 135:109846.
15. Riyapan, P., Shuaib, S. E., and Intarasit, A. (2021). A mathematical model ofcovid-19 pandemic: A case study of bangkok, thailand. Computational and Mathematical Methods inMedicine, 2021.
16. Samui, P., Mondal, J., and Khajanchi, S. (2020). A mathematical model for covid-19transmission dynamics with a case study of india. Chaos, Solitons & Fractals, 140:110173.
17. [Schättler and Ledzewicz, 2012] Schättler, H. and Ledzewicz, U. (2012). The pontryagin maximum principle: From necessary conditions to the construction of an optimal solution. In Geometric Optimal Control, pages 83–194. Springer.
18. Yu, Y., Liu, Y., Zhao, S., and He, D. (2022). A simple model to estimate the transmissibility of the beta, delta, and omicron variants of sars-cov-2 in south africa. Mathematical Biosciences and Engineering, 19(10):10361–10373. Zeb, A., Alzahrani, E., Erturk, V. S., and Zaman, G. (2020). Mathematical model for coronavirus disease 2019 (covid-19) containing isolation class. BioMed research international, 2020.
19. N. Kumar, A. Pant and R. Kumar Singh Rajput: "A Computational Study of Elastico-Viscous Flow between Two Rotating Discs of Different Transpiration for High Reynolds Number" International Journal of Engineering, vol-22(2), aug-2009, pp. 115–122. (http://www.ije.ir/article_71770.html)
20. N. Kumar, U S Rana and J. Baloni: "A Mathematical Model of Growth of Homogeneous Tumor with Delay Time" In International journal of Engineering, vol-22(1), April -2009, pp. 49–56. (http://www.ije.ir/article_71759.html)

Recent Trends in Engineering and Science for Resource Optimization and Sustainable Development – Prof. (Dr.) Dorota Jelonek et al. (eds)
© *2024 Taylor & Francis Group, London, ISBN 978-1-032-98030-0*

59

Fuzzy Economic Order Quantity Models with Carbon Emission Considerations and Planned Discounts

Anubhav Pratap Singh[1],
Sahedev[2]
Department of Mathematics, S.G.R.R. (P.G) College,
Dehradun, India

Anand Chauhan[3]
Department of Mathematics, Graphic Era Deemed University,
Dehradun, India

Deo Datta Aarya[4]
Department of Mathematics, Acharya Narendra Dev College,
University of Delhi

Yogendra Kumar Rajoria[5]
Department of Mathematics, SBAS, K.R. Mangalam University,
Gurgaon, India

▬▬▬ Abstract

Nowadays, consumers are becoming increasingly conscious of greenhouse gases from industry because of global warming. This study aims to establish fuzzy economic order quantity models of inventory that take carbon emissions and quantity discounts into account. The unit purchase cost is associated with the demand rate under this scenario. Due to the presence of uncertainty, model parameters like carrying cost, purchasing cost, ordering cost, and back-ordering cost are considered by the fuzzy approach and defuzzy by the sign distance method. The validity and Specialness of the most favorable solution to the model are investigated under carbon emission regulations, and comparisons of optimal solutions among the suggested model is presented through appropriate examples. Also provided are some managerial perspectives on inventory policies and carbon reduction plans.

▬▬▬ Keywords

Carbon-emission awareness, Quantity discounts, Price-sensitive demand, Triangular fuzzy number

[1]drapsingh78@gmail.com, [2]sahdevsingh0120@gmail.com, [3]dranandchauhan83@gmail.com, [4]deodatta.aarya@gmail.com, [5]yogendrarajo@gmail.com

DOI: 10.1201/9781003596721-59

1. INTRODUCTION

Industries nowadays are striving in order to mitigate the carbon dioxide emissions connected to business production. Optimizing system parameters, including such small batches or transaction volumes, has proven to be a beneficial way to reduce emissions. The production inventory model includes two concepts: partial backordering and scheduled discounts, with greenhouse gases considered. We have taken into consideration the various parameters in fuzzy systems in the practical aspect model developed in fuzzy environments. The best order volume for a business to procure in terms of reducing stock outs such as storage expenditures, shortfall expenses, and order extra costs is referred to as the economic order quantity (EOQ). Stocking executives are responsible for determining the number of items a company must add to its stock for each lot order in order to bring down inventory stocking expenses.

Zadeh (1965) was first to introduce Fuzzy set theory and their basic component. Jain(1976) investigated a model designed for declining products which time dependent demand function under fully shortfall. Authors considered fuzzy triangular number to input data and sign distance, centroid method used as defuzzifier. Dubois and Prade (1978) were the first to introduce these operations on fuzzy data.De and Rawat (2011) extended an ordering model predicated on the idea of triangular fuzzy numbers, where there would be no shortfall. Kumar and Singh (2017) presented their research work on a fuzzy model with modified trapezoidal fuzzy numbers while Syed and Aziz (2007) presented a comprehensive review on use of the crisp and fuzzy logic methodologies to supply chain networks. Kim and Sarkar (2017) reviewed the proposal for the impact of business quality assurance on environmental sustainability and the role of ecological strategy in controlling this correlation. The source of this ambiguity, which is exacerbated by fuzziness.

Rajput et al. (2021) developed a signed distance optimization method for a medical industry-specific fuzzy EOQ model with three separate trends. Rajoria et al. (2014) Investigated and developed an ordering strategies designed for declining substances having store, under inflationary conditions.Poswal et al. (2022a) developed a model for declining products consuming stock and price reliant on demand service under shortfall and fuzziness surrounding. Mondal et al. (2022) investigated a study of the behaviour of several inventory parameters in a partially backlogged inventory model with periodic uncertainty, and items deteriorate at a specific pace under stock-in circumstances. Zhou et al. (2022) investigated the fuzzy economic models that take into consideration different fuzzy environments.

Poswal et al. (2022,b) and Poswal et al. (2022,c), Poswal et al. (2022,d) investigated and review of fuzzy EOQ models under shortages, an EOQ model for a two-parameter weibull degrading under shortages, and the production of sustainable products under a trade credit period were all explored.

2. ASSUMPTIONS

The model is under the following presumptions:

(i) The holding cost is distributed into two parts: a predetermined component, i, and a variable component, w, which generally increases with the unit sales price. As a result, the component holding cost c_{ji}, is equivalent to the component sales price; this presumption is specified as

$$h = i + wc_j \qquad (1)$$

(ii) For the all-units bulk purchases, the component purchase fee is determined as follows:

$$C(Q) = c_j, \text{ if } q_{j-1} < Q \le q_j,$$
$$c_1 > c_2 > c_3 > \dots\dots > c_j$$

(iii) Linear Demand function considered,

$$D(P) = a - b\,P, \qquad (2)$$

Where a and b are constant parameter and selling price coefficient.

(iv) Consider two assumption demand function must be greater than zero and selling price greater than the purchasing cost which satisfied following given equation

$$c_j < P < \frac{a}{b} \qquad (3)$$

(v) Shortfall permitted which is partly backlogged

(vi) Deterioration is not taking into account.

(vii) Carbon dioxide emissions take into account in this model which is the delivery frequency, warehousing capacities, sustainable effect, and emissions of carbon from outmoded substances.

We presented various variables in the diagram as follows:

$$Q^* = \text{Consumed amount}$$
$$= \text{FT D(P)} + (1 - F)\,\beta\,D(P)T \qquad (4)$$

$$DFT = \text{Inventory having maximum level}$$
$$= D(P)\,FT \qquad (5)$$

$$L = \text{level of shortages which complexly lost}$$
$$= (1 - \beta)\,D(P)\,(1 - F)T \qquad (6)$$

$$B = \text{Order Backing} = \beta(1 - F)\,T\,D(P) \qquad (7)$$

Average level of Inventory $= F^2T/2\,D(P)$

3. MATHEMATICAL MODELLING IN FUZZY SYSTEM

We consider an EOQ model (Fig. 59.1) in the beginning of the cycle (T = 0) the inventory level is maximum (DFT) the inventory level is decreased with demand rate D(P).when the inventory level becomes zero then the demand rate becomes β D(P) and the time T the back ordering quantity is $\beta(1 - F)$ T D(P) while the level of shortage is $(1 - \beta)$ D(P) $(1 - F)$T.

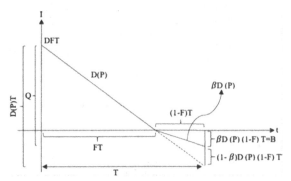

Fig. 59.1 Depicts a schematic diagram of the proposed model

In fuzzy system we take ordering cost, backordering cost, lost sales cost and holding cost and as well as demand constants take as fuzzy variable Let fuzzy ordering cost $\tilde{A} = (A_1, A_2, A_3)$ and fuzzy holding cost $\tilde{h} = (h_1, h_2, h_3)$ and fuzzy backordering cost $\tilde{\pi} = (\pi_1, \pi_2, \pi_3)$ and lost sales cost $\tilde{\pi}' = (\pi'_1, \pi'_2, \pi'_3)$ and the purchase cost $\tilde{c}_j = (c_1, c_2, c_3)$ are in the fuzzy sense. These are the triangular fuzzy numbers.

Cost of Ordering: Fixed ordering cost take as fuzzy

variables $= \dfrac{\tilde{A}}{T}$ (10)

Cost of Holding: Holding cost $= \tilde{h}\left(\dfrac{D(P)F^2T}{2}\right)$ (11)

Cost of Backordering: this cost calculated as,

$$s = \tilde{\pi}\left[\frac{\left[\beta(1-F)^2 T(a-bP)\right]}{2}\right]$$ (12)

Cost of Lost Sales: It takes as fuzzy variables and calculated as

$$= \left(P - C_j + \tilde{\pi}'\right)(1-\beta)(1-F)(a-bP)$$ (13)

Cost of Purchasing:

$$\tilde{c}_j\left(\frac{P[DFT + \beta(1-F)D(P)T]}{T}\right)$$
$$= \tilde{c}_j(a-bP)(F+\beta-\beta F)$$ (14)

The total annual profit (TAP) function:

$$P(a-bP)(F+\beta-\beta F) - \left\{\frac{\tilde{A}}{T} \oplus C_j(a-bP)(F+\beta-\beta F)\right.$$

$$+\tilde{h} \otimes \left(\frac{(a-bP)}{2}\right)F^2T + \tilde{\pi} \otimes \left(\frac{\left(\beta(1-F)^2 T(a-bP)\right)}{2}\right)$$

$$+\left(P - \tilde{c}_j \oplus \tilde{\pi}'\right) \otimes (1-\beta)(1-F)(a-bP)$$

$$+\frac{(P-P')b'F(a-bP)}{2} + \frac{e}{T} + k(a-bP)(F+\beta-\beta F)$$

$$+\left(g+b_1 a_1 c_e\right)\left(\left(\frac{(a-bP)}{2}\right)F^2T\right)\right\}$$

4. NUMERICALLY DEMONSTRATE UNDER FUZZY

In the practical aspect model established in fuzzy environments, we consider the following parameters in fuzzy systems, which are hereunder: Ordering cost $\tilde{A} = (90, 100, 110)$, backordering cost $\tilde{\pi} = (0.8, 1.0, 1.2)$, lost sale cost $\tilde{\pi}' = (1, 2, 3)$, Holding cost $\tilde{h} = (5.0, 5.2, 5.4)$, purchase cost $\tilde{c}_j = (20, 25, 30)$. We obtain the best value of TAP of proposed model in the fuzzy system. By using software of Mathematica The optimum value of TAP is given by 3044.66, optimum selling price, P is given by 81.9963, optimum time, T is given by 0.856458.

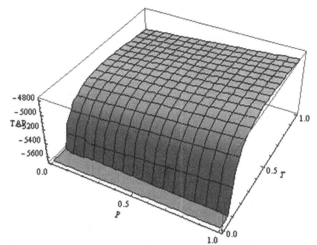

Fig. 59.2 TAP's (fuzzy system) function over time T and selling price P

5. SENSITIVITY ANALYSIS

With the aid of sensitivity analysis, an economist can determine if a change in one variable will have an impact on the outcomes of an occurrence based on a particular combination of components.

Table 59.1 Analyzing the sensitivity of various parameters

parameter	% variation of parameter	Value of parameter	T (cycle length)	P (selling price)	ATP (average total profit)
	-40%	0.48	1.4187	84.96	1893.64
F	-20%	0.64	1.1103	84.75	2315.72
	20%	0.96	0.7515	84.53	3152.26
	40%	1.12	0.6433	84.46	3568.81
	-40%	120	1.3522	58.44	483.05
a	-20%	160	1.0581	71.46	1391.92
	20%	240	0.7964	97.84	4508.67
	40%	280	0.7222	111.10	6712.20
	-40%	0.9	0.8395	129.005	6136.12
b	-20%	1.2	0.8679	101.25	3990.87
	20%	1.8	0.9352	73.55	1923.45
	40%	2.1	0.9756	65.65	1366.44

(i) If we increase the Percentage of consumption that will be satisfied from available supply (F) then cycle length (T), selling price (P) decreased while the average total profit (ATP) is increased

(ii) As increased market growth opportunity (a) is increasing then time cycle (T) is decreased while the selling price (P) is increased and also the average total profit (ATP) is increased

(iii) An increased the selling rate component b then time cycle (T) is increased while the selling (P) price and ATP is decreased.

6. CONCLUSION

In this work, the signed distance defuzzification technique was used to defuzzify the entire yearly profit function; here, we calculated the total yearly profit for both the crisp system and the fuzzy surroundings. The triangular fuzzy number was used to defuzzify the findings. We notice a small but significant rise or reduction in the values of P and T. The study's main finding is that the fuzzy system boosts annual overall profits while the selling price and time, as compared to the crisp system, decrease. We suggest that organizations use the planned discount evaluation under uncertainty provided here to assure that important sources of environmental impacts are not ignored throughout their supply chains. Such information can help firms reduce carbon and environmental emissions not only inside their own operations, but also throughout their supply chain.

References

1. Zadeh LA.Fuzzy sets. Information Control.1965; 8: 338–353.
2. Kim MS, Sarkar B. Multi-stage cleaner production process with quality improvement and lead time dependent ordering cost. J Clean Prod. 2017; 144: 572–590.
3. Jain R. Decision making in the presence of fuzzy variables. IIIE Transactions on Systems, Man and Cybernetics.1976; 17: 698–703.
4. Dubois D, Prade H. Operations on fuzzy numbers. Int. J. Systems Sci.1978; 9(6):613–626.
5. De PK, Rawat A. A fuzzy inventory model without shortages using triangular fuzzy number. Fuzzy Information & Engineering. 2011; 1:59–68.
6. Kumar P, Singh SB. Fuzzy system reliability using generalized trapezoidal intuitionistic fuzzy number with some arithmetic operations. Nonlinear Studies.2017; 24(1): 915–922
7. Syed JK, Aziz LA. Fuzzy Inventory model in fuzzy environment without shortage using triangular fuzzy number with sensitivity analysis, Int. J. Agricult. Stat. Sci.2007; 1(2):203–209.
8. Rajput N, Chauhan A, Pandey R.K. EOQ model with a discount rate of inflation and optimization with pentagonal fuzzy number, International Journal of Mathematics in Operational Research.2021; 20(2):264–280.
9. An Inventory Model with Time Dependent Demand Under Inflation and Trade Credits. In: Pant, M., Deep, K., Nagar, A., Bansal, J. (eds) Proceedings of the Third International Conference on Soft Computing for Problem Solving. Advances in Intelligent Systems and Computing,2014;vol.259:155–165. Springer, New Delhi.
10. Poswal P, Chauhan A, Boadh R, Rajoria YK, Kumar A,Khatak N.Investigation and analysis of fuzzy EOQ model for price sensitive and stock dependent demand under shortages,Materials Today: Proceedings. Volume 56, Part 1,2022a,Pages 542–548.
11. Mondal R, DasS, Chandra DasS, Shaikh AA, Bhunia AK. Pricing strategies and advance payment-based inventory model with partially backlogged shortages under interval uncertainty, International Journal of Systems Science: Operations & Logistics. 2022.
12. Zhou W, LuoD,XuZ. Review of fuzzy investment research considering modelling environment and element fusion, International Journal of Systems Science.2022; 53(9): 1958–1982.
13. Poswal P, Chauhan A, Boadh R, Rajoria YK A review on fuzzy economic order quantity model under shortage, AIP Conference Proceedings 2481, 040023, Volume 2481, Issue 1 (2022b);
14. Poswal P, Chauhan ARajoria Y K, Boadh R, Goyal A Fuzzy optimization model of two parameter Weibull deteriorating rate with quadratic demand and variable holding cost under allowable shortages,Yugoslav Journal of Operations Research, 32(4) pp. 453–470, 2022c.
15. Poswal P, Chauhan A Aarya DD, Boadh R, Rajoria Y K, GaiolaSU, Optimal strategy for remanufacturing system of sustainable products with trade credit under uncertain scenario, Materials Today: Proceedings, Volume 69, Part 2,2022d, Pages 165–173.

Note: All the figures and table in this chapter were made by the author.

*Recent Trends in Engineering and Science for Resource Optimization and
Sustainable Development – Prof. (Dr.) Dorota Jelonek et al. (eds)*
© 2024 Taylor & Francis Group, London, ISBN 978-1-032-98030-0

A Study to Review Global Regulations Regarding Mitigation of Operational Risk Associated with Crypto-Assets

Deepankar Roy[1]

Associate Professor,
Information Technology,
National Institute of Bank Management, Pune

Ashutosh Dubey[2]

Senior Lead Market Innovations,
National Payments Corporation of India, Mumbai

Sarika Lohana[3]

Senior Consultant,
G Institutions and Consultancy, Mumbai

▆▆▆▆ Abstract

The purpose of this paper is to provide clarity on the various aspects of the Crypto-assets ecosystem and to suggest possible policy options. In order to understand the major risks associated with crypto-assets as currently regulated globally, a comprehensive analysis of current crypto-asset operational guidelines has been conducted. As a result of the analysis of the existing regulations, the paper has attempted to identify gaps in the operations of crypto-asset related organizations. This has helped in understanding what gaps are and how it can be mitigated. The paper reviews the Crypto-asset associated Operational Risk Management to determine how the operational risk management associated with crypto-assets of financial institutions can be mitigated in reaction to the increasing demand for crypto-assets, cross border payments, electronic money, and cryptocurrencies.

▆▆▆▆ Keywords

Blockchain, Crypto-assets, Operational risk, Risk management

1. INTRODUCTION

Cryptocurrencies have emerged as one of the most exciting investment prospects in recent years. As cryptocurrency prices continue to rise, an increasing number of institutional investors and wealth managers are preparing to invest in cryptocurrencies in the coming years. "According to Data Bridge Market Research, by 2029, the global cryp-

[1]deepankarroynibm@gmail.com, [2]adashutosh@gmail.com, [3]drslohana@gmail.com

DOI: 10.1201/9781003596721-60

to asset management market is expected to reach USD 2801.87 million, representing a CAGR of 25.50% during the forecast period of 2022-2029" (Data Bridge Market Research, 2022). During the 2022 period, after the FTX saga (KPMG, 2022), cryptocurrencies experienced another bubble similar to that which occurred during the 21st century dotcom bubble. Heady speculation led the prices on dot com organisations to reach to very high levels before crashing. A sudden global interest in cryptocurrency and assets has made them part of everyday news across the globe. Several types of crypto-assets have been developed and adopted worldwide based on blockchain technology like non-fungible tokens, cryptocurrencies, fungible tokens, and central bank digital currencies (CBDC). 90% of the world's central banks nowadays are weighing the risks and rewards of issuing CBDC (FinTech Department Reserve Bank of India, 2022), but most of the analysis remains somewhat abstract. As central bankers contemplate entering the digital currency space race, they must weigh numerous potentially destabilizing risks. It is true that a race is underway for the future of money, currency and payments. CBDCs, or Central Bank Digital Currencies, offer a unique alternative to cryptocurrency. Unlike cryptocurrency, which is decentralized, CBDCS are issued, regulated and backed by central banks. This means that they are backed by a government and offer an additional layer of trust and security. Additionally, CBDCs can be used to facilitate transactions and payments between banks, making them a potentially more efficient and cost-effective option than cryptocurrency. Furthermore, CBDCs can also be used to make transactions more transparent and trackable, providing governments with better visibility into financial activities. Ultimately, CBDCs if generated over blockchain can be classified as Crypto asset, offer a new way for central banks to interact with their citizens, providing a secure and regulated alternative to cryptocurrency

Operational risk is the risk of loss caused by weak processes, people, or systems - or by external factors. Risks associated with financial institutions include legal risks and information technology risks. Actual and potential operational risk events are assessed for their reputational, regulatory, and operational impacts. Managing operational risk (KPMG, n.d.) is an integral part of a bank's risk management programme. Operational risk is built into all banking products, activities, processes, and systems. The paper has reviewed the development and evolution of associated operational risk types in the context of crypto-assets. The paper is divided into several sections, such as

- First section introduces crypto assets and blockchain,
- Second section gives a timeline of risk management failures in crypto-assets,
- Third section includes global initiatives to manage risks and associated risks that have been observed,
- Finally the last section gives a summary of Operational Risk Pillars associated with Crypto-assets

2. TIMELINE OF RISK MANAGEMENT FAILURES IN CRYPTO-ASSETS

"The storage problem" is the largest risk. Cryptocurrency addresses are not stored as coins, but rather as cryptographic keys. The funds associated with that address can be lost completely if the keys are disclosed or if control is lost even briefly. 850,000 bitcoins were lost when Mt. Gox crashed in January 2014. In 2016, Ethereum was launched, allowing other platforms to create their own cryptocurrencies and run smart contracts. In 2016, Cardano, Tezos, and Neo followed this model. Coin check, a Tokyo-based exchange, was robbed of $530 million in cryptocurrency in January 2018. Using a "hot wallet" (Buck, 2018), the thieves stole funds from Coin check and brought attention to the exchange's security. Over $600 million was stolen from Poly's (Gagliardoni, 2021) decentralized finance platform in August 2021.

A tweet appealed for $33 million in Tether from the project's developers. It was hacked in December 2021 (Thurman, 2021). BitMart addresses were exhausted by security firms. The Binance Smart Chain handled $96 million in cryptocurrencies. The address was identified as "BitMart Hacker" by Etherscan. By allowing unlimited approvals to an Externally Owned Account (EOA), a front-end attack on the Badger DAO (Decentralised Autonomous Organisations) in December 2021 resulted in a $120 million Bitcoin and Ethereum theft. Badger paused smart contracts after learning of the exhausting of users' addresses. After two hours and twenty minutes, the malicious transactions failed. The use of specialised key-storing hardware, such as hardware security modules (HSMs) and hardware wallets, is an efficient security measure. Hardware solutions are not flawless, however. Multi-signature wallets and compartmentalization of funds are also security controls.

Mining power centralization poses a systemic risk since it could lead to currency manipulation and blockchain manipulation. In October 2022, Binance was hacked for $570 million, one of the biggest attacks in cryptocurrency history. 2 million Binance Coins (BNB) were withdrawn and extra Binance Coins were created as a result of hacking the Binance Smart Chain network (Livni, 2022). The crypto exchange BNB has its own token. Blockchain security should be tightened because of a bug in a smart contract. There is a great deal of debate over the policy implications of crypto-assets.

3. GLOBAL INITIATIVES TO MANAGE RISK ASSOCIATED WITH CRYPTO-ASSETS

Over the past few years, there has been a marked increase in the regulatory emphasis on digital assets, and this trend is anticipated to continue. As retail and institutional adoption increased, market capitalisation increased rapidly and volatility increased. Recent high-profile failures of crypto firms, fraud, scams, and mismanagement of customer funds have undermined consumer trust. Regulators have been brought into sharp focus because of this. A swift and thorough global regulatory policy approach and supervisory structure are needed to ensure better consumer protection. The regulations are categorised into two main categories:

3.1 Category 1: New Regulations for Holding Crypto-Assets by Regulated Entities

In its second consultation on the prudential handling of crypto asset exposures in December 2022, the Basel Committee on Banking Supervision (BCBS) (Basel Committee on Banking Supervision, 2022) specifies criteria for assessing whether a bank's exposure to a crypto asset would be allocated. All rights and obligations relating to the crypto asset are clearly defined and legally enforceable. This includes settlement finality, whether it is a tokenized traditional asset or has an effective stabilization mechanism linking its value to a traditional asset. Crypto-assets should be classified on an ongoing basis under the standard into two groups, Group 1 crypto-assets: Group 1a tokenised traditional assets and Group 1b crypto-assets with effective stabilization mechanisms. Basel Framework capital requirements apply to Group 1 crypto-assets based on the risk weights of the portfolio exposures. Unbacked crypto-assets are included in Group 2. To determine which Group 2 crypto-assets can be hedged (Group 2a) and which cannot (Group 2b), hedging recognition criteria are used. Financial and non-financial risks associated with crypto-assets are identified in the December 2019 (Basel Committee on Banking Supervision, 2020) discussion paper by BCBS is provided in Table 60.1 below.

Crypto assets (including their operating networks) mitigate material risks sufficiently. Organizations and processes involved in the organization and processing of crypto assets must be regulated and supervised, or subjected to appropriate risk management procedures. "In October 2021, the Financial Action Task Force (FATF) published a risk-based approach to virtual assets and virtual asset providers"

Table 60.1 Basel committee for banking supervision (BCBS) risk classification framework for crypto-assets

Category	Risk	Description
Financial Risks	Liquidity risk	In the event that crypto-assets cannot be sold at little or no loss of value, market liquidity risk arises. During times of stress, banks that issue crypto assets and/or receive crypto asset deposits may also be at risk due to a lack of liquidity in funding
	Market Risk	The valuation and pricing of crypto-assets display a high degree of volatility, and disjointed trading platforms may hinder price discovery.
	Credit & counterparty credit risk	In the same way as traditional assets, crypto-assets which constitute legal obligations create credit and counterparty credit risks. It notes that limited historical data on crypto-assets makes it hard for banks to price the risk of lending to crypto-asset companies.
Non-Financial Risks	Cyber & operational risk	Cyber risks and operational risks are obvious concerns with crypto-assets since they are digital and not backed by physical collateral. Financial institutions are exposed to a whole new set of vulnerabilities both from a cybersecurity and governance perspective as a result of the technologies underlying crypto-assets.
	Legal & regulatory risk	Crypto-assets pose new legal and regulatory risks to firms without a robust regulatory framework. Crypto-assets are not regulated centrally, which may result in regulatory arbitrage. Furthermore, to comply with KYC, AML, and terrorist financing regulations, financial institutions will need to come up with innovative solutions as blockchain technologies ease value transfer.
	Reputational risk	Reputational risks are associated with the use of innovative coin offering and crypto-asset management technologies. Unlike traditional assets, crypto-assets are distributed, which means any negative sentiment or action of one party may have a negative impact on the entire ecosystem.
	Third party Risk	Unregulated external parties using community-driven software run most Crypto asset. In addition, financial institutions may seek out third-party developers, partners, or solution providers to enhance their product offerings. A financial institution's third-party risk increases as a result of all of these factors
	Implementation Risk	A crypto-asset's lifecycle requires internal policies and procedures to be developed both at the onset and throughout. An operational procedure, accounting treatment, and other frameworks must be in place before a crypto-asset can be adopted.

Source: https://assets.kpmg/content/dam/kpmg/ca/pdf/2020/03/basel-iv-crypto-en.pdf

(FATF, n.d.). Besides helping authorities develop regulatory and supervisory guidelines for virtual asset operations, this document also helps Virtual Asset Service Providers (VASPs) understand and act on their anti-money laundering (AML) and counter-terrorism financing (CTF) duties.

The German government was among the first to offer financial institutions legal certainty (Crypto Custody Business, n.d.) that permitted them to keep cryptocurrency assets. Regulations specify that citizens and legal bodies can buy or trade crypto assets till it is done through licensed exchanges and custodians. Companies must be licensed with the German Federal Financial Supervisory Authority (BaFin).

"The UK Financial Conduct Authority (FCA), the Bank of England, and HM Treasury make up the country's Crypto-assets Taskforce" (*Crypto assets Taskforce: Final Report*, 2018). Rules specifically designed for crypto assets by the FCA address know your customer (KYC), AML, and CTF. In order to protect VASPs, restrictions have also been made, but caution has been taken not to stifle innovation. Cryptocurrency exchanges that haven't already applied for an e-money licence must register with the FCA. Cryptocurrencies are not regarded as legal money, and taxes on them are depending on activity. The trading of cryptocurrency derivatives has been outlawed by the FCA.

3.2 Category 2: Classifying Crypto-Assets as Financial Products Already Subject to Regulation and Extending Regulation to Other Components of the Ecosystem

"The Israeli Securities Authority (*Warning to Investors Regarding Cryptocurrency Investments*, n.d.) has ruled that cryptocurrency is a security subject to Israel's Securities Laws". The public has been warned by the regulator about the dangers associated with cryptocurrency. Comparable to FATF, the Israel Money Laundering and Terror Financing Prohibition Authority has adopted a similar stance towards AML/CFT regulations. The Israel Tax Authority requires a 25% capital gains tax and classifies cryptocurrencies as assets.

The Saudi Arabian Monetary Authority (SAMA) and its Minister of Finance have issued a warning "against dealing or investing in Crypto Assets including cryptocurrencies as they are not recognised by legal entities in the kingdom" (A statement by MOF regarding dealing in virtual currencies, including cryptocurrencies that claim any relationship with the Kingdom, 2019). They are not traded by local financial institutions and are beyond the purview of the regulatory framework.

Despite overlap and divergent opinions between agencies, the regulatory environment surrounding cryptocurrency in the US is changing. The Federal Reserve Board, the Commodity Futures Trading Commission (CFTC), the Securities and Exchange Commission (SEC)—all of which are widely regarded as the most powerful regulatory bodies—as well as the Financial Crimes Enforcement Network (FinCEN) have all released divergent interpretations and guidelines. Cryptos are time and again viewed as securities by the SEC, "the CFTC (*The CFTC's Role in Monitoring Virtual Currencies*, 2020) calls bitcoin a commodity, and Treasury calls it a currency". Cryptocurrencies are defined as "a digital representation of value that functions as a medium of exchange, a unit of account, and/or a store of value" by The Internal Revenue Service (IRS) and has issued tax guidance accordingly.

The Monetary Authority of Singapore (MAS) regulates cryptocurrencies (*A Guide to Digital Token Offerings*, 2019). The Payment Services Act of 2019 regulates Traditional and cryptocurrency payments and exchanges. Issuance of digital tokens are also subject to the Securities and Futures Act.

By passing the Finance Bill 2022, the Indian legislative council approved taxation rules on virtual digital assets (VDAs) or crypto tax, as proposed in the Budget 2022-23. Crypto assets are subject to a 30% tax. The government of India is working on drafting a bill. There was a major focus on this issue at the G20 summit in 2023, which was be hosted under the leadership of India. Cryptocurrencies are prohibited as legal tender or currency in the draft Bill (*Draft Banning of Cryptocurrency & Regulation of Official Digital Currency Bill, 2019*). Additionally, cryptocurrency mining, holding, selling, dealing in, issuance, disposal or use are prohibited. The purpose of mining is to create a cryptocurrency and/or validate cryptocurrency transactions between buyers and sellers.

4. ASSOCIATED RISKS WITH CRYPTO-ASSETS

This paper has reviewed the existing risk management for Financial Institutions regarding to Crypto asset management and virtual assets. It is important to understand that operational risk deals with external factors such as unexpected events and human mistakes. It is important to categorize risk on a global level, and to understand how crypto-asset providers and service providers operate and function. Several major risks pillars as depicted in the Table 60.2 characterize crypto-assets.

Table 60.2 Summary of Operational Risk Pillars associated with Crypto-assets

S. no.	Risk	Description
1	Business Model	Different digital asset investment approaches and business models will involve different types of operational risks, including direct investment, futures trading, and stake assets to generate income. In addition to unauthorized transactional activity, inaccurate or incomplete books and records, and digital asset holdings that do not reconcile with the custodian or blockchain are examples of operational risks.
2	Technology	In the context of technology risks, logical and physical access to critical systems may be inadvertent or unauthorized, change management activities may result in system errors and reporting, and an ineffective response to extreme market conditions may be possible.
3	Custody and security	Providing services involving crypto-asset custody functions such as on boarding, deposits/withdrawals, and reconciliation should be accompanied by robust controls at every stage of the private key life cycle, including generation, distribution, storage, security, and usage as well as private rotation and destruction.
4	Market access and data	In order to maintain market data and liquidity, market data service providers have implemented controls. The key risk associated decision is whether the service consumer will connect to each decentralized exchange and blockchain separately, or will leverage an infrastructure provider to aggregate and provide an all-in-one solution for all services
5	Confidentiality and privacy	To build trust and meet stakeholder expectations, it is essential that confidentiality and privacy are maintained. The greatest risk is the loss of transaction data and data leaks
6	Compliance and tax	Crypto asset service providers must demonstrate compliance with financial industry standards and regulations, including anti-money laundering (AML), know your customer (KYC) requirements, and tax reporting.
7	Centralisation	The controls of the business model, technology decisions, operations and market decisions at the hands of limited persons (mostly owners) without any maker –checker governance.

Source: PwC Global Crypto Regulation Report 2023; https://www.bis.org/bcbs/publ/d490.pdf

5. CONCLUSION

Effective crypto operational risk management is critical for the success and sustainability of any organization operating in the cryptocurrency industry. It involves identifying, assessing, and mitigating risks associated with the technology, processes, and people involved in crypto operations. Cryptocurrencies and other crypto assets offer numerous advantages, including decentralization, transparency, and fast transactions. However, like any financial asset, they also come with inherent risks, such as market volatility, security breaches, and regulatory uncertainty. As crypto assets continue to gain mainstream acceptance and adoption, it is becoming increasingly important for institutions involved in the financial ecosystem to have a well-defined crypto asset operational risk management framework. Such a framework would help institutions to identify, assess, and mitigate risks associated with crypto asset operations, while still participating in the innovation that crypto brings. By taking a proactive approach to crypto asset operational risk management, institutions can minimize the potential impact of crypto asset-related operational incidents, maintain their reputation, and protect the interests of their stakeholders. While crypto assets offer exciting opportunities for innovation and growth, institutions must be vigilant in managing the risks associated with these assets. A well-defined crypto asset operational risk management framework is critical for institutions to participate in the crypto ecosystem with confidence, while still effectively mitigating the risks associated with crypto assets.

References

1. *A Guide to Digital Token Offerings.* (2019). Monetary Authority of Singapore. URL: https://www.mas.gov.sg/regulation/explainers/a-guide-to-digital-token-offerings
2. *A Statement By Mof Regarding Dealing In Virtual Currencies, Including Cryptocurrencies That Claim Any Relationship With The Kingdom.* (2019). Saudi Central Bank. URL: https://www.sama.gov.sa/en-US/News/pages/news 21082019.aspx
3. Basel Committee on Banking Supervision. (2020). Designing a prudential treatment for cryptoassets. In Basel Committee on Banking Supervision. Retrieved December 26, 2022, URL: https://www.bis.org/bcbs/publ/d490.pdf
4. Basel Committee on Banking Supervision. (2022). Prudential treatment of cryptoasset exposures. In Basel Committee on Banking Supervision. Retrieved December 26, 2022, URL: https://www.bis.org/bcbs/publ/d545.pdf
5. Buck, J. (2018). Coincheck: Stolen $534 Mln NEM Were Stored On Low Security Hot Wallet. Cointelegraph.URL: https://cointelegraph.com/news/coincheck-stolen-534-mln-nem-were-stored-on-low-security-hot-wallet
6. *Cryptoassets Taskforce: final report.* (2018). HM Treasury contacts. URL: https://assets.publishing.service.gov.uk/government/uploads/system/uploads/attachment_data/file/752070/cryptoassets_taskforce_final_report_final_web.pdf

7. Financial Action Task Force (FATF). (n.d.).URL: https://www.fatf-gafi.org/publications/fatfrecommendations/documents/guidance-rba-virtual-assets-2021.html

8. FinTech Department Reserve Bank of India. (2022). Concept Note on Central Bank Digital Currency. RBI. Retrieved December 26, 2022, URL: https://rbidocs.rbi.org.in/rdocs/PublicationReport/Pdfs/CONCEPTNOTEAC-B531172E0B4DFC9A6E506C2C24FFB6.PDF

9. Gagliardoni, T. (2021). The Poly Network Hack Explained. Kudelski Security Research. URL: https://research.kudelskisecurity.com/2021/08/12/the-poly-network-hack-explained/

10. KPMG. (n.d.). Beyond Basel IV: Incorporating crypto-assets into the Basel framework. In KPMG. Retrieved December 26, 2022, URL: https://assets.kpmg/content/dam/kpmg/ca/pdf/2020/03/basel-iv-crypto-en.pdf

11. KPMG. (2022). The collapse of FTX: Lessons and implications for stakeholders in the crypto industry. In KPMG. Retrieved December 26, 2022, URL https://assets.kpmg/content/dam/kpmg/cn/pdf/en/2022/11/the-collapse-of-ftx.pdf

12. Livni, E. (2022). Binance Blockchain Hit by $570 Million Hack. The New York Times. URL: https://www.nytimes.com/2022/10/07/business/binance-hack.html

13. *The CFTC's Role in Monitoring Virtual Currencies*. (2020). Commodity Futures Trading Commission. URL: https://www.cftc.gov/media/4636/VirtualCurrencyMonitoringReportFY2020/download

14. Thurman, A. (2021). Crypto Exchange BitMart Hacked With Losses Estimated at $196M. URL: https://www.coindesk.com/business/2021/12/05/crypto-exchange-bitmart-hacked-with-losses-estimated-at-196-million/

15. *Warning to Investors Regarding Cryptocurrency Investments*. (n.d.). Israel Security Authority. URL: https://www.isa.gov.il/sites/ISAEng/Pages/unregulated-investments.aspx

Recent Trends in Engineering and Science for Resource Optimization and Sustainable Development – Prof. (Dr.) Dorota Jelonek et al. (eds)
© 2024 Taylor & Francis Group, London, ISBN 978-1-032-98030-0

61

An Analysis on Blend Edge Square Multiband Microstrip Patch Antenna for Wireless Telecommunications

Nairaj Jat, Manisha Gupta

Department of Physics, University of Rajasthan,
Jaipur, India

Abstract

An antenna with higher gain, greater broadband, multiband compatibility, and compact layout sizes are needed for today's wireless and satellite networks of communication. This research paper presents a blend edge square microstrip patch antenna as well rectangular notch ground with also blend edge. This simulated antenna employs a recursive fractal geometry method that runs up to four iterations. The architectural framework of this antenna was designed and modelled using the application simulator CST Microwave Studio Suite. This antenna is created upon a FR4 base and has tiny dimensions of 40×40 mm² (L x W), h = 1.6 mm thickness, 4.3 permittivity ε_r, and loss tangent of 0.02. The intended antenna is stimulated by a micro strip feed line. It has a wider spectrum between 4 GHz and 17 GHz in the Ultra-Wide-Band (UWB)range, with s11 < -10dB, and four harmonic frequencies at 5.22 GHz, 8.4 GHz, 10.62 GHz, and 14.7 GHz. The coverage area of this antenna is determined by its impedance bandwidth, which is 13GHz. Fractal geometry is a viable method for improving impedance matching and size decreasing. As a result, antenna can work at many frequencies having maximum gain 6.5 dB. The antenna is used in satellite Internet service and WLAN networking broadcasts.

Keywords

Blend edge, Rectangular notch, Recursive fractal geometry, Square microstrip patch, UWB

1. INTRODUCTION

In contemporary wireless communication systems, there is a rising requirement for additional frequency bands because of a growing need for wireless networks. Additionally, such devices require tiny measurements and the addition of more functional bands without compromising performance (Karimkashi and Zhang, 2013; Raghavendra, Saritha and Alekhya, 2018; Parthasarathy, 2022). As a result, there is a constant need for multi-band antenna development. Therefore, a forward-looking antenna needs to be able to operate in multiple bands and have a small, straightforward framework. In order to expand the portion of a substance that can transmit electromagnetic radiation inside a specific area of surface or volume, fractal geometry adopts a fractal, or self- similar concept (Rao *et al.*, no date; Fotedar *et al.*, 2015; Bhus, 2016; Gupta and Mathur, 2018). Multilevel and space-filling shapes are referred to as

*Corresponding author: drguptamanisha@gmail.com

DOI: 10.1201/9781003596721-61

equivalent fractal antennas, and their number of iterations are what matter most. Fractal antennas are characterised by this trait as being very small, multiband or wideband, and having beneficial functions for communication over wireless networks. Fractal antennas respond differently than conventional antenna systems (Dattatreya and Naik, 2019; Dakhli and Choubani, 2020).

When contrasted with conventional develops, fractal element antennas are small and don't require any additional parts, provided the framework has the desired resonating input impedance. Fractals are intricate patterns with a repeating pattern on different sizes. As instances of this trait, the Mandelbrot established, the Lorenz attractor, and the Minkowski arc can all produce multifrequency phenomena. Sierpinski gasket, circular disc, and Parany monopole antenna exhibit log periodic resonance (Prabhakar *et al.*, no date; Rao *et al.*, 2022; Sreenivasulu *et al.*, 2022; Yadav *et al.*, 2022; Pourahmadazar, 2023).

Due to the vast number of devices utilised for military and wireless communication systems, as well as worldwide positioning satellite systems, circular polarisation is another necessity for antennas that has come into attention. For a circularly polarised antenna, there is simply no requirement for the placement of the electric field at the transmitter and receiver(Gupta and Mathur, 2018).

This endeavour involves the creation of a brand-new fractal blend edge square monopole antenna. The fabrication principle of Sierpinski gaskets and circular discs is also taken into account. A blend edge square-shaped microstrip fractal antenna featuring four successive passes is suggested in the present article. ISM and C band services are among the frequencies where the antenna echoes. The subsequent sections cover the intended antenna's antenna mathematics, current shipping, modelling, and outcomes from experiments.

2. ANALYSES OF ANTENNA METHODOLOGIES FOR DESIGN

2.1 To Find Length and Width of Patch

The rectangular-shaped radiating patch has dimensions of L × W. The formulas listed below are used to create the rectangular patch(Bhus, 2016; Raghavendra, Saritha and Alekhya, 2018): -

$$W_{width} = \frac{c}{2f_o}\sqrt{\frac{2}{C_r+1}} \tag{1}$$

$$\varepsilon_e = \frac{c_r+1}{2} + \frac{c_r-1}{2}\left[\frac{1}{\sqrt{1+\frac{12h}{W_{width}}}}\right] \tag{2}$$

$$L_{eff} = \frac{c}{2f_o\sqrt{c_r}} \tag{3}$$

$$\Delta L = 0.412h\frac{(C_{reff}+0.3)\left(\frac{W_{width}}{h}+0.264\right)}{(C_{reff}-258)\left(\frac{W_{width}}{h}+0.8\right)} \tag{4}$$

$$L = L_{eff} - 2\Delta L \tag{5}$$

2.2 Modelling for Fractal Structure

The repetitive procedure is used to create the fractal geometry on the square patch. To get fractal structure we use following equations (Raghavendra, Saritha and Alekhya, 2018): -

$$2R_n = \left(\frac{1}{3}\right)^n L \tag{6}$$

Here n is repetitive factor and R_n is radius of slotted circles for all repetitions.

For 0th iteration n = 0; 1st iteration n = 1; 2nd iteration n = 2 and for 3rd iteration n = 3.

By using equations (1) to (6) we get all dimensions of patch and slotted circles which are given in Table 61.1. We constructed a design using all data in Table 61.1 and simulated it by CST.

Table 61.1 Dimensions of composed blend edge square antenna

S. No.	Variables of blend edge square Antenna	Size of blend edge square antenna
1.	Length of substrate 'L'	40 mm
2.	Width of substrate 'W'	40 mm
3.	Size of side of square patch 'L□'	20 mm
4.	Length of ground 'L_g'	14.5 mm
5.	Width of ground 'W_g'	40 mm
6.	Length of feed 'L_{feed}'	15.5 mm
7.	Width of feed 'W_{feed}'	2 mm
8.	Blend radius 'r'	4 mm
9.	Length of rectangular slot 'L_{rg}'	1 mm
10.	Width of rectangular slot 'W_{rg}'	4 mm
11.	R_1	3.33 mm
12.	R_2	1.11 mm
13.	R_3	0.37 mm

Following figures show images of simulated square antennas:-

For simulated antennas, in 0th iteration we made a blend edges square patch of size 'L□' and rectangular slotted ground also with above two edges blend. To get proper impedance matching we use 50Ω microstrip feed line with its position shifted from below left edge of blend edges square patch. The shifting distance is 2.5 mm for getting good s11.

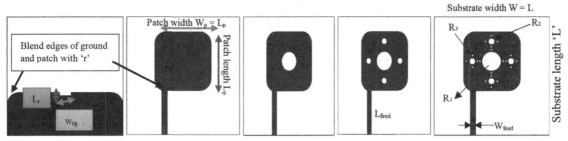

Fig. 61.1 Images (a), (b), (c), (d) and (e) of all iterations of blend edge square antenna (a) slotted ground (b) 0th iteration of antenna (c) 1st iteration of antenna (d) 2nd iteration of antenna (e) 3rd iteration of antenna

In 1st iteration we cut a circle of radius R_1 which is $\frac{1}{6}$ of size of patch whereas in 2nd iteration we cut four circles in 1st iteration of radius R_2 which is $\frac{1}{18}$ of size of patch and in 3rd iteration we removed twenty circles in 2nd iteration of radii R_3 which is $\frac{1}{54}$ of size of patch (from equation (6)). All data related to blend edges square microstrip patch antenna (BESMPA) is given below in Table 61.1.

The suggested antenna has been built on a 1.6 mm thick, FR4 dielectric platform with an apparent tangent of 0.02 and a dielectric constant of 4.3. To improve the antenna's efficiency, an inflexible substrate measuring 40 mm by 40 mm is used. Antenna features that facilitate multiband functioning are enabled by the starting point and first repetition.

3. THE OUTCOMES OF SIMULATION AND ANALYSIS

3.1 Influence of Various Substrate Materials on the Data Acquired

Figure 61.2 shows that changes in substrate material result in variations in return loss. We are all aware of how crucial the substrate is to the mechanical strength of an antenna. We achieve enhanced return loss for substrate FR-4 in Fig. 61.2. The graph demonstrates that nearly the whole frequency range has a return loss of less than -10 dB for all materials, with variance occurring solely in the value of s11 at resonant frequencies. Here, at the bottom border of the frequency spectrum, Rogers 5880 exhibits greater S11 magnitude. In image 2 shows that variation in s11 is almost same for all materials with respect to frequency except rogers 5880.

3.2 Influence of Various Hight 'h' of Substrate on the Data Acquired

Figure 61.3 demonstrates how variations in substrate thickness cause variations in return loss. We are all aware of the significance the substrate is to the mechanical durability of an antenna. In Fig. 61.3, we have improved return loss with h = 4mm. The plot illustrates that the 13 to 15 GHz band has less than -10 dB return loss for parametric heights more than 1.5 mm, and that the only value that varies with the frequency of resonance in this band is s11.

Figure 61.4 displays the way the return loss s11 fluctuates for various feed line widths. We achieve a superior s11 for W_{feed} = 2mm at 14.7 GHz frequency for the BESMP antenna's required range. With a small chance in the resonance frequency values, we get the same number of resonating frequencies for all widths. Signal attenuation and reflective power are lowered when adequate balancing takes place. The figure also shows that there is larger bandwidth for all widths.

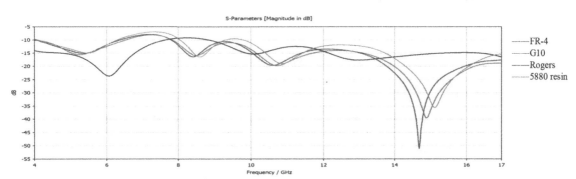

Fig. 61.2 Graph between reflection coefficient s11 v/s frequency for several materials

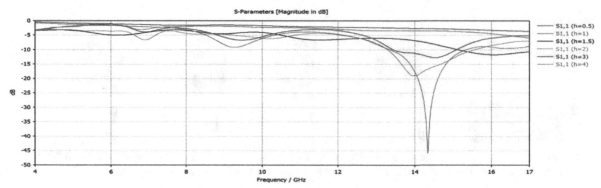

Fig. 61.3   Graph between reflection coefficient s11 v/s frequency for several hight 'h'

3.3 Influence of various width 'W_feed' of feed line on the data acquired

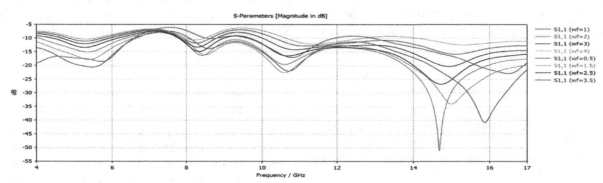

Fig. 61.4   Graph between reflection coefficient s11 v/s frequency for several width 'W_feed'.

3.4 Influence of Repetitions on the Data Acquired

Figure 61.5 also shows the return loss for all four repetitions. S11 plotted vs frequency over four iterations. By increasing the number of repeats, additional bands can be formed, and the resonant frequency shifts to the lower end of the frequency spectrum, according to a plot analysis. However, the strength of the s11 is growing as the number of repetitions rises. For all iterations we get UWB range.

VSWR

Picture 6 depicts the VSWR v/s frequency plot for all repetitions, revealing the fact that at the optimal frequency span, VSWR is less than 2 or as $1 \leq VSWR \leq 2$.

Gain

The graphical representation shows that the antenna has positive gain at frequencies that resonate. The gain increases as the number of repeats increases, according to a review of the simulation results. We obtain positive gain across the entire frequency range.

3.5 Assessment of the E-Field' Far Fields

From Fig. 61.8 we can say that all far fields radiation patterns are omnidirectional.

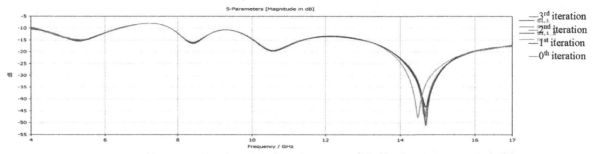

Fig. 61.5   Influence of several iterations on the reflection coefficient s11

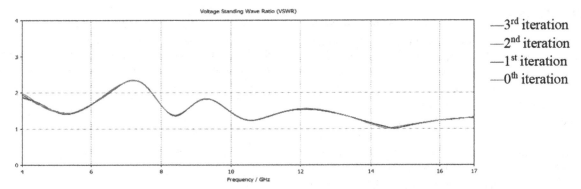

Fig. 61.6 Image shows VSWR versus frequency for all iterations

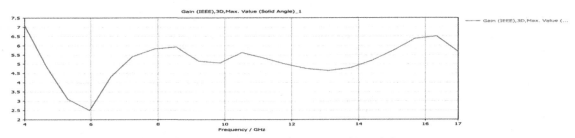

Fig. 61.7 Picture shows graph between Gain v/s frequency.

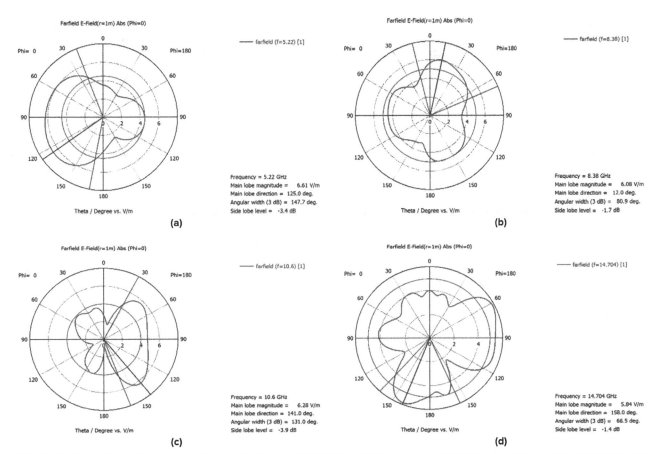

Fig. 61.8 Modelling of radiation pattern in E-Plane of BESEMPA (a) E-Plane at f = 5.22 GHz, (b) E-Plane at f = 8.38 GHz, (c) E-Plane at f = 10.6 GHz, (d) E-Plane at f = 10.6 GHz

4. CONCLUSIONS

Due to its many different variations in dimension and layout, fractal geometry delivers an impressive alternative to the multiband that can be used for many different kinds of purposes. A fractal aerial with square blend edges is recommended for use in wireless networks. It delivers a better current distribution at the boundary at mid resonant frequencies. As a result, the radiation pattern can be better controlled. By changing the ring's thickness and radius, the required frequency region and bandwidth can be achieved. The aerial length is decreased by the blend edge form compared to monopole antennas. To remove the form of circles, various radii are utilised. The VSWR, transmission pattern, and reflection rate measurements are used to assess antenna performance. Vibration frequencies that were produced through simulation are 5.22 GHz, 8.4 GHz, 10.62 GHz, and 14.7 GHz and antenna's bandwidth of approximately 13 GHz. The maximal gain of BESMPA is 6.5dB at 16.4 GHz. The chosen antenna is suitable for broadband functions such as WLAN, radar, Wi-Fi, C-band, X-band, Ku-band, and ISM band as well as other systems that work in the frequency range of 4 GHz to 17 GHz.

ACKNOWLEDGEMENT

The author would like to acknowledge CSIR for JRF.

References

1. Bhus, B. (2016) 'A Novel Design for f Circular Patch Fractal Antenna for Multiband', pp. 449–453.
2. Dakhli, N. and Choubani, F. (2020) 'Multiband L-Shaped Slot Antenna Loaded with Split Ring Resonator', *2020 International Wireless Communications and Mobile Computing, IWCMC 2020*, pp. 1087–1090. Available at: https://doi.org/10.1109/IWCMC48107.2020.9148052.
3. Dattatreya, G. and Naik, K.K. (2019) 'A low volume flexible CPW-fed elliptical-ring with split-triangular patch dual-band antenna', *International Journal of RF and Microwave Computer-Aided Engineering*, 29(8), pp. 3–11. Available at: https://doi.org/10.1002/mmce.21766.
4. Fotedar, R. *et al.* (2015) 'Performance Analysis of Microstrip Antennas Using Different Shapes of Patch at 2.4 GHz', *Proceedings - 2015 2nd IEEE International Conference on Advances in Computing and Communication Engineering, ICACCE 2015*, (3), pp. 374–377. Available at: https://doi.org/10.1109/ICACCE.2015.132.
5. Gupta, M. and Mathur, V. (2018) 'Multiband Multiple Elliptical Microstrip Patch Antenna with Circular Polarization', *Wireless Personal Communications*, 102(1), pp. 355–368. Available at: https://doi.org/10.1007/s11277-018-5843-x.
6. Karimkashi, S. and Zhang, G. (2013) 'A dual-polarized series-fed microstrip antenna array with very high polarization purity for weather measurements', *IEEE Transactions on Antennas and Propagation*, 61(10), pp. 5315–5319. Available at: https://doi.org/10.1109/TAP.2013.2273813.
7. Parthasarathy, H. (2022) *Antenna Theory, Select Topics in Signal Analysis*. Available at: https://doi.org/10.1201/9781003344957-2.
8. Pourahmadazar, J. (2023) 'Towards Fractal Antennas : A multiband Sierpinski Triangle (Gasket) Fractal Vivaldi Antenna Javad Pourahmadazar To cite this version : HAL Id : hal-03974528 Tow ards Fractal A nten- nas : A m ultiband Sier- pinski Triangle (G asket) fractal V ivaldi a'.
9. Prabhakar, D. *et al.* (no date) 'Implementation of Koch Snowflake Fractal Antenna for Multi-Band Applications', pp. 30–33.
10. Raghavendra, C., Saritha, V. and Alekhya, B. (2018) 'Design of modified sierpinski carpet fractal patch antenna for multiband applications', *IEEE International Conference on Power, Control, Signals and Instrumentation Engineering, ICPCSI 2017*, pp. 868–871. Available at: https://doi.org/10.1109/ICPCSI.2017.8391835.
11. Rao, B. *et al.* (no date) 'Multiband slotted elliptical printed antenna design and analysis', *researchgate.net*, 14(4), pp. 378–386. Available at: https://doi.org/10.26782/jmcms.2019.08.00031.
12. Rao, S.J.M. *et al.* (2022) 'Fractal segmented lotus shape planar monopole antenna for multiband applications', *Materials Today: Proceedings*, 66, pp. 3450–3456.
13. Sreenivasulu, M. *et al.* (2022) 'Minkowski Fractal Antenna with Circular DGS for Multiband Applications', *2022 IEEE Wireless Antenna and Microwave Symposium, WAMS 2022* [Preprint], (June). Available at: https://doi.org/10.1109/WAMS54719.2022.9847789.
14. Yadav, K. *et al.* (2022) 'Internet of Thing based Koch Fractal Curve Fractal Antennas for Wireless Applications', *IETE Journal of Research* [Preprint], (November). Available at: https://doi.org/10.1080/03772063.2022.2058631.

Note: All the figures and table in this chapter were made by the author.

*Recent Trends in Engineering and Science for Resource Optimization and
Sustainable Development – Prof. (Dr.) Dorota Jelonek et al. (eds)
© 2024 Taylor & Francis Group, London, ISBN 978-1-032-98030-0*

An Elliptic Shaped Patch with Triangular Slot Fractal Antenna for Wireless Communication

Kiran Githala[1] and Manisha Gupta[2]

Department of Physics, Rajasthan University, Jaipur, India

■■■■■■■ **Abstract**

This paper includes a thorough parametric analysis of the techniques utilized to improve the standard antenna's range in bandwidth for SWB applications. This proposed antenna is printed on a $36 \times 44 \times 1.6$ mm^3 FR-4 substrate. The suggested SWB antenna was developed from a regular elliptical patch antenna using a partial ground plane and a taper feeding into an embedded fractal arrangement. Having a bandwidth ratio of 12.4:1, this antenna performs over 2.41 and 30 GHz. The range of the simulated gain is 2 to 4.27 dB. The suggested antenna can be used for super-wideband and satellite applications in the S, C, X, and Ku bands.

■■■■■■■ **Keywords**

Elliptical patch, Bandwidth, Super wide band

1. INTRODUCTION

Low-profile and minimalist antennas are essential in applications such as outstanding performance of aero planes, satellites, wireless communication, and missiles but are restricted by dimensions, mass, cost, efficiency, simplicity of deployment, and aerodynamic profile. Modern communication has recently advanced to the point that long- distance transmission requires a wideband antenna. Ultra-wideband technologies, with an FCC-allocated frequency of operation range of 3.1 GHz to 10.6 GHz, can only deliver improved data rates for contemporary wireless communication over limited distances because of their poor rate of adaptation and prolonged signal collection time (2002, no date; Balani *et al.*, 2019). For long-distance communication outside, there's a strong need for a greater

transfer rate of data and increased bandwidth. The super wideband (SWB) antenna, which may be used for both indoor and outdoor communication, is a promising response to the rising need for connectivity. Narrow band technology is inferior to SWB technology in terms of benefits. It has a large bandwidth and innovative features, including a large channel capacity, better time accuracy, and better resolution, which enable it to send data across large distances. The FCC defines a SWB antenna as one that may have a bandwidth ratio of 10:1 with a -10 dB return loss (Alluri and Rangaswamy, 2020; Dhasarathan *et al.*, 2020; Sagne and Pandhare, 2022). Greater bandwidth and higher data rates are provided by SWB technology.

A lightweight monopole SWB antenna for SWB applications has been proposed in (Okan, 2020) and consists of a patch with an octagon ring form, an extra segment in its

[1]kiranch7231@gamil.com, [2]drguptamanisha@gmail.com

DOI: 10.1201/9781003596721-62

upper right corner, and a partially scooped ground surface. The constructed antenna has a bandwidth fraction of 169% and a bandwidth ratio of 12.02:1, and it is able to function in the 2.59 GHz to 31.14 GHz frequency band. The gain, however, is rather small, falling between 2 and 5 dBi. The fractured antenna in (Singhal, 2016) has a bandwidth of impedance of 4.6 GHz to 52 GHz with a star pattern inside a ring-like ground plane. In the observed frequency spectrum, it has a variable gain with an average value of 4 dBi. The ground plane with two circular-shaped grooves and a quasi-square with the correct dimensions and slopes on the patch's bottom sides is illustrated in (Oskouei *et al.*, 2019). Applications for SWB usually have this shape. The bandwidth is achieved with a varying gain of 1 to 9.47 dBi over 3 GHz and 50 GHz. A clover-structured antenna with a notched surface in the shape of a semi-elliptical letter "V" is suggested in (Palaniswamy *et al.*, 2017). It features a gain that varies from 0.7 to 5.6 dBi and a 176% range between 1.9 and 30 GHz. With a semi-circular ground, a bandwidth of 1.42 to 90 GHz, and an optimal gain of 7.67 dB, the small SWB trapezoidal design radiator suggested in (Rahman *et al.*, 2019) achieves this goal despite the cross-polar structure becoming more significant than the co-polar structure.

In this study, a detailed design and evaluation of an elliptic patch fractal SWB antenna having a 12.4:1 proportion bandwidth is discussed. The proposed antenna operates between 2.41 GHz and 30 GHz. This paper investigates and analyses several strategies for increasing bandwidth. The proposed antenna's design was inspired by a printed elliptical patch antenna because it has an omnidirectional radiation pattern and a large impedance bandwidth. The bandwidth is initially improved to some extent by using blended ground plane corners. The recommended antenna's bandwidth has also been improved by using the tapered feed approach and the fractal procedure. The recommended antenna can be designed and its many properties can be studied using CST microwaves simulator tool.

2. GEOMETRY OF ELLIPTICAL PATCH ANTENNA

The recommended SWB antenna's configuration is depicted in Fig. 62.1. The suggested antenna is designed on a FR-4 substrate with a 1.6 mm thickness, a dielectric constant of 4.3, and a tanδ value of 0.025. An elliptical patch with axes a_1 and a_2 and the fractured structure of two iterations make up the recommended antenna geometry. A circle with radius R_1 is etched in a triangle with its side length l_1 in the first iteration, while a circle with radius R_2 is etched in a triangle with its side length l_2 in the next iteration. As illustrated in Fig. 62.1(a), a feed line with taper with its length

L_f, width of feed Wf_2 at the confined side, and Wf_1 at the opposite side is used to feed material onto the top surface of the substrate.

$$K = 4\sqrt{\varepsilon_r}$$

For lower edge frequency, the outer ellipse axis of the antenna is calculated as shown below.

$$f_L = \frac{c}{\lambda} = \frac{7.2}{(L + R + p) \times K}$$

Here f_L = inferior edge frequency in GHz

p = space between the elliptical patch and the ground plane's length,

R = a_1/4, L = $2a_2$ (Gupta and Mathur, 2018).

$$\varepsilon_{eff} = \frac{S_r + 1}{2}$$

Here FR-4 substrate is used for antenna design with dimensions $36 \times 44 \times 1.6$ mm^3. FR-4 with 1.6 mm thickness gives $\varepsilon_{eff} = 1.626$; K = 1.15.

$$a_{1eff} = \left[\frac{a_1}{\pi seff} .a2 + 2ha\{\ln a + (1.41\varepsilon eff + 1.77) + \frac{1}{a_1}h.(0.286\ \varepsilon eff + 1.65) \right]^{1/2}$$

$$a_{2eff} = \left[\frac{a_2}{\pi seff} .a2 + 2ha\{\ln a2 + (1.41\varepsilon eff + 1.77) + \frac{1}{a_2}h.(0.286\ \varepsilon eff + 1.65) \right]^{1/2}$$

Here $a_{1\ eff}$ = optimal radius of major axis $a_{2\ eff}$ = optimal radius of minor axis

Table 62.1   Values for the proposed antenna's parameters

Antenna's parameter	size (units in mm)	Antenna's parameter	size (units in mm)	Antenna's parameter	size (units in mm)
W_s	36	a_1	14	Wf_2	1.6
L_s	44	a_2	10	l_1	18
hs	1.6	R_1	5.4	l_2	9
W_g	36	R_2	2.7	r	10
L_g	15	Wf_1	3	w_1	3
L_f	16	w_2	2.4		

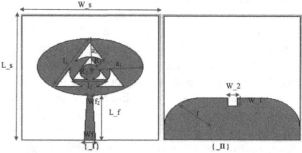

Fig. 62.1   Front and bottom views of the intended antenna's geometry

3. ANTENNA DESIGN

The elliptical patch fractal antenna's design development is depicted in Fig. 62.2. The recommended super-wide band antenna was developed from a conventional elliptical patch antenna with the use of an embedded fractal pattern. A feed line with a taper is used for this designed antenna, and Radius r is used for blending the outer edges of the partially grounded plane.

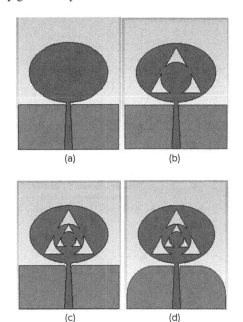

(a) (b)

(c) (d)

Fig. 62.2 The planned antenna's creation phases (a) elliptical patch antenna{I}, (b) elliptical patch with 1st iteration{II}, (c) elliptical patch with 2nd iteration{III}, (d) elliptical patch with 2nd iteration blended ground{IV}

4. RESULTS AND ANALYSIS

Multiple parameters were used to simulate the recommended antenna with CST Microwave Studio software. The following is a discussion of the simulated VSWR, gain, E_ field and H_ field radiation patterns, return loss S11 and surface current distribution:

4.1 Antenna's Return Loss

Figure 62.3 depicts return loss v/s frequency graph for different steps of designed antenna (S11 _orange for step I, S11_ green for step II, S11_ blue for step III and S11_ red for step IV). Here it has been observed from Fig. 62.3 that resonant frequency shifts towards lower as increasing the no. of iteration. Wider bandwidth (2.41 GHz to 30 GHz) with minimum return loss has been obtained for step IV antenna.

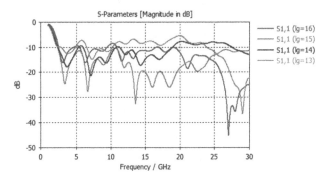

Fig. 62.3 S11 graph for different steps of designed antenna

4.2 VSWR

Figure 62.4 depicts the simulated VSWR comparison for designed antenna in step III(green) and step IV (red).

Here it can be noted that VSWR < 2 has been obtained for 2.41 GHz to 30 GHz.

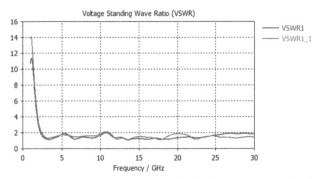

Fig. 62.4 Comparison of VSWR v/s frequency graph of designed antenna for step III and step IV

4.3 Far-field Radiation Patterns

Figure 62.5 [i– iv] shows 2D E–field and H–field patterns, respectively, at resonant frequencies of 3.34 GHz, 6.72 GHz, 13.57 GHz, and 16.12 GHz. The E—field and H—field omnidirectional radiations show that an antenna radiates in a plane.

4.4 Gain

Figure 62.6 depicts the gain versus frequency plot for designed antenna(step 4). It can noted that gain increased from 2 dB to 4.27 dB. Maximum gain of simulated antenna is 4.27 dB at 16.12GHz.

5. CONCLUSION

An elliptical patch with triangular slots and a fractal super-wide band antenna has been designed in this article. The bandwidth was enhanced by applying iterative methods to the patch and ground with several defects and rounded edges. With a bandwidth proportion of 12.4, this planned

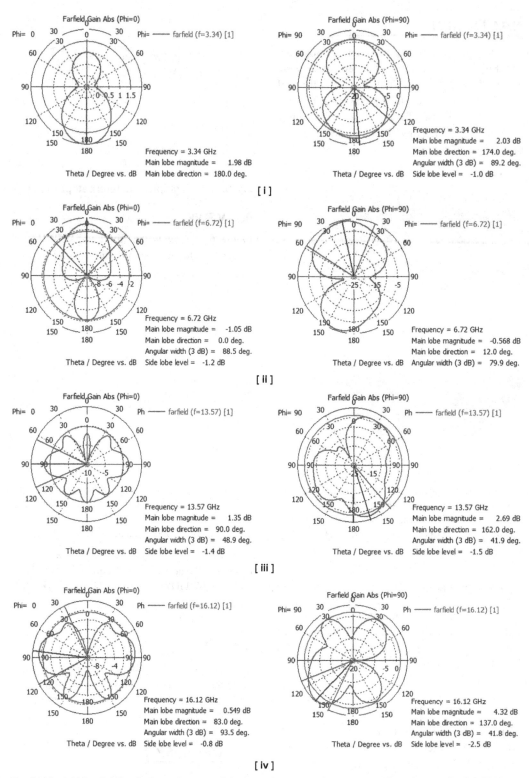

Fig. 62.5 E – field and H – field radiation patterns of designed antenna for resonant frequencies [(i) 3.34GHz, (ii) 6.72 GHz, (iii) 13.57GHz, (iv) 16.12GHz]

antenna functions over 2.41 GHz and 30 GHz. The maximum gain of the suggested antenna is 4.27 dB at 16.12 GHz. The suggested antenna is appropriate for use with the GPS, the internet of things, satellite communication (S, C, X, and Ku-band), super wide-band communication, and other wireless communication applications.

Table 62.2 Analyzing the recommended antenna in comparison with already developed antennas

Ref.	Dimensions (mm³)	Substrate material used	Frequency range	Bandwidth ratio
(Okas et al., 2018)	52 × 42 × 1.575	Rogers RT 5880	0.96-10.9 GHz	11.35:1
(Okas, Sharma and Gangwar, 2018)	52 × 46 × 1.6	Rogers RT 5880	0.95-13.6 GHz	14.31:1
(Oskouei and Mirtaheri, 2017)	30 × 30 × 1.6	FR-4	3-50 GHz	16.7:1
(Maity and Nayak, 2023)	44.5 × 47.2 × 0.8	FR-4	1.61-26.35 GHz	16.34:1
Proposed	36 × 44 × 1.6	FR-4	2.41-30 GHz	12.4:1

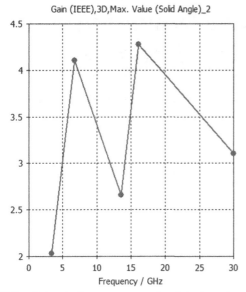

Fig. 62.6 Gain v/s frequency graph of designed antenna (step 4)

6. Acknowledgement

The author would like to acknowledge UGC for JRF.

References

1. 2002, F. (no date) 'Report and order in the commission's Rules Regarding Ultra-Wideband Transmission Systems (2002). Released by Federal Communications Commission.'
2. Alluri, S. and Rangaswamy, N. (2020) 'Compact high bandwidth dimension ratio steering-shaped super wideband antenna for future wireless communication applications', *Microwave and Optical Technology Letters*, 62(12), pp. 3985–3991. Available at: https://doi.org/10.1002/mop.32541.
3. Balani, W. *et al.* (2019) 'Design Techniques of Super-Wideband Antenna-Existing and Future Prospective', *IEEE Access*, 7, pp. 141241–141257. Available at: https://doi.org/10.1109/ACCESS.2019.2943655.
4. Dhasarathan, V. *et al.* (2020) 'Integrated bluetooth/LTE2600 superwideband monopole antenna with triple notched (Wi-MAX/WLAN/DSS) band characteristics for UWB/X/Ku band wireless network applications', *Wireless Networks*, 26(4), pp. 2845–2855. Available at: https://doi.org/10.1007/s11276-019-02230-0.
5. Gupta, M. and Mathur, V. (2018) 'Multiband Multiple Elliptical Microstrip Patch Antenna with Circular Polarization', *Wireless Personal Communications*, 102(1), pp. 355–368. Available at: https://doi.org/10.1007/s11277-018-5843-x.
6. Maity, B. and Nayak, S.K. (2023) 'A Super Wideband CPW-Fed Elliptical Slot Monopole Antenna for Wireless Applications', *Progress In Electromagnetics Research C*, 130(December 2022), pp. 139–154. Available at: https://doi.org/10.2528/PIERC22120505.
7. Okan, T. (2020) 'A compact octagonal-ring monopole antenna for super wideband applications', *Microwaveand Optical Technology Letters*, 62(3), pp. 1237–1244. Available at: https://doi.org/10.1002/mop.32117.
8. Okas, P. *et al.* (2018) 'Elliptical slot loaded partially segmented circular monopole antenna for super wideband application', *AEU - International Journal of Electronics and Communications*, 88, pp. 63–69. Available at: https://doi.org/10.1016/j.aeue.2018.03.004.
9. Okas, P., Sharma, A. and Gangwar, R.K. (2018) 'Super-wideband CPW fed modified square monopole antenna with stabilized radiation characteristics', *Microwave and Optical Technology Letters*, 60(3), pp. 568–575. Available at: https://doi.org/10.1002/mop.31006.
10. Oskouei, H.D. and Mirtaheri, A. (2017) 'A monopole super wideband microstrip antenna with band-notch rejection', *Progress in Electromagnetics Research Symposium*, 2017-Novem, pp. 2019–2024. Available at: https://doi.org/10.1109/PIERS-FALL.2017.8293470.
11. Oskouei, H.R.D. *et al.* (2019) 'A small cost-effective super ultra-wideband microstrip antenna with variable band-notch filtering and improved radiation pattern with 5g/IoT applications', *Progress In Electromagnetics Research M*, 83(May), pp. 191–202. Available at: https://doi.org/10.2528/PIERM19051802.
12. Palaniswamy, S.K. *et al.* (2017) 'Super wideband printed monopole antenna for ultra wideband applications', *International Journal of Microwave and Wireless Technologies*, 9(1), pp. 133–141. Available at: https://doi.org/10.1017/S1759078715000951.
13. Rahman, S.U. *et al.* (2019) 'Compact design of trapezoid shape monopole antenna for SWB application', *Microwave and Optical Technology Letters*, 61(8), pp. 1931–1937. Available at: https://doi.org/10.1002/mop.31805.
14. Sagne, D. and Pandhare, R.A. (2022) 'Design and Analysis of Inscribed Fractal Super Wideband Antenna for Microwave Applications', *Progress In Electromagnetics Research C*, 121(June), pp. 49–63. Available at: https://doi.org/10.2528/PIERC22030703.
15. Singh, S.S. and A.K. (no date) 'MODIFIED STAR-STAR FRACTAL (MSSF) SUPER-WIDEBAND ANTENNA', *MICROWAVE AND OPTICAL TECHNOLOGY LETTERS*, 59(March 2016), pp. 624–630. Available at: https://doi.org/10.1002/mop.30357.

Note: All the figures and tables in this chapter were made by the author.

Recent Trends in Engineering and Science for Resource Optimization and Sustainable Development – Prof. (Dr.) Dorota Jelonek et al. (eds)
© 2024 Taylor & Francis Group, London, ISBN 978-1-032-98030-0

63

Revolutionizing the Floriculture Industry with Excellence in Infrastructure Development: A Case Study of Rajasthan

Diksha Sinha[1]

Research Scholar, Department of Management,
Birla Institute of Technology, Mesra, off Campus Jaipur

Roopali Sharma[2]

Professor, Department of Management,
Birla Institute of Technology, Mesra, Off Campus Jaipur

▬▬ Abstract

Floriculture as a branch of horticulture deals with the breeding, growing, and marketing of flowers and ornamental plants. Commercialization of the Floriculture industry in India is of recent development owing to the globalization era. The paper explores the current problems and challenges faced by the floriculture exporting firms of Rajasthan. A survey is conducted through a mailed questionnaire approach involving 178 exporting firms. The questionnaire is developed on afive-point Likert scale to gauge the severity of the problems faced by the exporters of the floriculture industry. The study uses an exploratory factor analysis approach with the Varimax rotation technique to identifythe most influencing challenges faced by the exporters of the floriculture industry in Rajasthan. The analysis resulted in 7 major factors produced out of 29 variables. The result suggests that the attributes majorly contributing to the challenges in the floriculture industry are marketing-related constraints followed by production,logistics, and transport-related barriers. The study will help the decision makers in framing the best suitable schemes specifically designed to combat the growth-resistant barriers and also suggest the further scope of development for the growth and expansion of the floriculture industry.

▬▬ Keywords

Floriculture industry, Exporting firms, Exploratory factor analysis, Rajasthan, India

1. INTRODUCTION

Whenever we come across the word "Horticulture", the first thing that comes to our mind is "cultivation of fruits and vegetables" but we tend to forget one of the most integral parts of horticulture i.e. floriculture or flower farming. Floriculture is a division of horticulture that deals with the production, cultivation, and marketing of flowers and ornamental plants for commercial purposes. India has a very conducive environment for the growth of floriculture

---

[1]dikshasinha.rgsc@gmail.com, [2]roopalisharma@bitmesra.ac.in

DOI: 10.1201/9781003596721-63

products with a rich diversity of agro-climatic zones and soil textures. The floriculture industry of India comprises flowers such as roses, tuberose, gladiolus, anthurium, carnations, marigolds, etc. with foliage, tubes, branches, root cuttings, and cut flowers. According to the 3rd advanced estimate ofthe National Horticulture Board, the area under floriculture production in India was 322 thousand hectares with the production of 2295.07 thousand tonnes of loose flowers and 833.16 thousand tonnes of cut flowers during 2020-21. The cultivation of flowers for commercial purposes has been recently introduced in several states of India with Karnataka, Tamil Nadu, Madhya Pradesh, and West Bengal going ahead of other producing states. The palaces and traditional rich heritage of Rajasthan have made the state a destination for weddings and cultural festivities. This has led to a boom in the flower business despite low production. The flower business in the state accelerates during the festive and marriage seasons. Pushkar, Jaipur, Kota, and Bharatpur are the main areas for flower cultivation and Pushkar is well known for rose farming. The present farming of marigolds, roses, and other flowers like jasmine in the state is fulfilling the daily needs of temple towns within and outside the state.

The export of floriculture products from Rajasthan, as depicted in Table 63.1, has experienced a thrust after 2019-20 and exported 228.66 metric tonnes of floriculture products worth 2294.22 US thousand dollars. The analysis of the statistics shows that the export of floriculture products from Rajasthan is much less in proportion to the share of exports of other leading states. Despite having the potential for commercial production of flowers in the state, the export of floriculture products from Rajasthan is lagging due to

Table 63.1 Export of floriculture products from rajasthan (2012-2023)

Year	Quantity (in Metric tonne)	Value (in US thousand $)
2012-13	4.45	18.84
2013-14	2.00	18.68
2014-15	2.85	7.26
2015-16	0.01	0.01
2016-17	0.35	0.94
2017-18	0.00	0.00
2018-19	2.47	20.25
2019-20	12.43	57.25
2020-21	190.90	631.31
2021-22	414.46	2845.15
2022-23	228.66	2294.22

Source: Agricultural and Processed Food Products Export Development Authority (APEDA)

several constraints related to the infrastructure, production, and marketing of floriculture products. The above-identified research problem sets the background of our research question i.e. Research Question: Why the export volume of floriculture products from Rajasthan is much less in comparison to other leading floriculture exporting states?

1.1 Research Objective

The growth of any industry needs proper management of production, marketing, and infrastructure-related amenities. The major objective of the study is to find the prevailing problems and contemporary challenges related to the production, infrastructure, and marketing of floriculture products from export-oriented floriculture firms in Rajasthan. In the face to answer the above research question the following objective is determined for the study (Kaushik & Dhir, 2019):

a) To identify and explore the major factors leading to problems and challenges faced by the exporters of floriculture products in Rajasthan.

To attain the research objectives, primary data is collected from a sample of active exporters of Rajasthan registered under APEDA. The scale of the questionnaire is developed and refined through a pilot survey. After data collection and cleaning, Exploratory Factor Analysis using SPSS software (version 22) is appliedto the dataset to identify the underlying factors leading to export constraints of floriculture products in Rajasthan. The paper is further classified into six sections. Section 2 focuses on the Review of Literature, followed by Research Methodology in Section 3, Data analysis and Results in Section 4, Findings and Discussion in Section 5, and lastly Conclusions and future research directions in Section 6.

2. REVIEW OF LITERATURE

The floriculture industry in India is still lagging due to the inadequacy of prominent infrastructural and marketing facilities. The review of the literature was undertaken in light of finding the variables indicative of the challenges persistent in the floriculture and perishable products industry worldwide. Many studies have focused on the infrastructure-related aspects of agriculture. To explain the transport and logistics-related infrastructure Veerappan et al., (2020)analyzed the relationship between air freight demand and airport characteristics. Applying regression analysis, a set of statistical models was developed to estimate the air freight weight holding capacity in an airport. The findings of Sinha & Sharma, (2022) pointed out insignificant growth of floriculture exports in terms of volume in India and suggested the need for proper infrastructure facilities required to ensure the overall growth of the floriculture industry. Several studies have incorporated exploratory factor analysis to generate the factors influencing the concerned study. Patnaik & Bhowmick, (2022) determine the appropriateness for the management of appropriate technology generating four factors defining appropriateness. Magesa & Jonathan, (2022)analyzed the attributes of a compelling leader to lead Digital Transformation in a formal organization producing seven factors in five roles. Similarly, Marakova et al., (2021)identified the key sources of competitive advantage of large enterprises in Poland, by using EFA and Varimax criterion of rotation. Ojo, L. D., Oladinrin, O. T., & Obi, (2021)investigate the barriers to Environmental Management System (EMS) implementation in the Nigerian construction industry (NCI). Other studies that have derived the factors using EFA are(Aung et al., 2023; Konanahalli et al., 2022). After going through the studies we short-listed the variables that may act as a barrier to the export of floriculture products and the papers supporting the use of Exploratory Factor Analysis (EFA) for the present study were also enumerated.

3. RESEARCH METHODOLOGY

The primary objective of this study is to identify the factors contributing to problems and challenges faced by the exporters of the floriculture industry in Rajasthan. The perception of exporters towards the challenges faced in the export of floriculture products is analyzed (Mgale & Yunxian, 2021). For this purpose, a comprehensive and systematic approach was followed (Agostini et al., 2022; Ahmed & Ting, 2019; Bhat et al., 2022; Kao et al., 2020).The process initiated with scale development, sample frame, and data collection, followed by the application of Exploratory factor analysis on the final dataset. The research employed a quantitative research design in studying the constructs that define the challenges in the floriculture industry. This involved a self-reported questionnaire based on the interview and focus group discussion with the exporters, academicians, researchers, and marketing experts. The questionnaire was filled out by the respondents drawn from the exporting firms and organizations listed in the exporter's directory of APEDA.

3.1 Scale Development

Following the literature review, focus-grouped discussion, and in-depth interviews with exporters, academicians, researchers, and practitioners, a set of 32 items divided into three constructs of 12 production-related constraints, 8 infrastructure-related problems, and 12 marketing-related challenges was generated to gaugethe prominent challenges prevailing in the floriculture industry (Cook et al., 2020). The questionnaire was reviewed by experts working in the areas of marketing, international business, and en-trepreneurship. To check the validity, reliability, and suitability of the questionnaire, a pilot survey was conducted involving 51 active exporters of floriculture products in Rajasthan in December 2022. In reliability testing, we test for the internal consistency of the statements used in the Likert scale. The reliability of the instrument is tested through Cronbach's Alpha. The sample size taken for the pilot survey is 51. Using SPSS software (version 22), construct-wise reliability and overall reliability of the questionnaire are tested. The minimum value of Cronbach's alpha should be greater than 0.7 (Nunnally, 1987). The validity of the questionnaire is tested using Pearson's correlation (r) method and significance value (p-value) using SPSS software. For a two-tailed test degree of freedom is N-2. At a 95% confidence interval, the obtained value is more than the table value. For each statement, the calculated value of r is greater than the table value and the significance level (p) is less than 0.05, which indicates that the statements are correlated and converge toward finding out the answer toa similar set of questions (Sony et al., 2022). Hence, all statements of the questionnaire are valid and focus on the problems and challenges of the floriculture industry. The time gap between the pilot survey and the main survey was reduced to the possible extent to maintain the authenticity of the results. The questionnaire dictated the context and purpose of this study.

3.2 Sample Frame and Data Collection

From the exporters' directory of APEDA, all the active exporters of floriculture products from Rajasthan were considered as the population which accounted for a total of 339 exporting firms including manufacturer cum merchant (41) and merchant (298) based on monthly export return. Since the population is finite and known, Yamane's (1967:886) simplified formula for calculating sample size (Uakarn, 2021) is used. At 95% confidence level and P=0.5, the following formula is used as per Yamane:

$$n = \frac{N}{1 + N(e)^2}$$

Where n is the sample size, N is the population size, and e is the level of precision.

Data for the study were collected by mailingquestionnaires to the listed mail id of the respondents. The period of the study is between January and March 2023. A total of 339 respondents were contactedby mailing the questionnaire, out of which 181 responses were received. Then the standard deviation of each respondent's statements was calculated and theunengaged response with a standard deviationof zero was deleted from the list. Thereafter, after the process of data cleaning, 178 responses were found suitable for further analysis.

4. Data Analysis and Results

4.1 Exploratory Factor Analysis

To determine the factors contributing to problems and challenges in the floriculture industry we applied exploratory factor analysis (EFA) to validate the scale (Higuchi et al., 2020; Zickar, 2020). Factor analysis on 32 variables using principal component analysis and Varimax rotation was conducted (Ramadani et al., 2022). As a rule of thumb, the variables having factor loadings less than 0.4 and those with low communalities (<0.5) are typically dropped. In addition, factors were extracted based on eigenvalues greater than 1. An adequacy test of the dataset is considered to be the first step in factor analysis (Ojo, L. D., Oladinrin, O. T., & Obi, 2021). Therefore, the Kaiser-Meyer-Olkin measure of sampling adequacy was conducted in which a KMO index of 0.876 was obtained, which is greater than 0.6which shows the appropriateness of the sample data (Table 63.2). Bartlett's Test of Sphericity indicates that there is a substantial correlation in the data. Correlation among the variables was tested using Pearson correlation and the result suggests that the variables were significantly correlated. In the further analysis Table 63.3 shows 7 significant factors explored with Eigen values greater than 1.0 that accounted for 65.16% of the total variance explained. Table 63.4 shows the matrix of factor loadings of all 29 itemsthatare above 0.5 and the structure of the variables associated with each factor. The Cronbach's alpha calculated for each factor is greater than 0.5 which proves the reliability of each factor; hence all 29 items were considered for the study.

Table 63.2 KMO and bartlett's test

Statistics	Value
Cronbach's alpha	0.917
Number of items	29
Kaiser-Meyer-Olkin Measure of Sampling Adequacy	0.876
Bartlett's Test of Sphericity	
Approx. Chi-Square	2449.197
Degree of Freedom	406
Sig.	.000

Source: Authors' compilation

5. Findings and Discussion

The EFA findings resulted in the extraction of 7 factors out of 29 variables with Eigen values greater than 1 that accounted for 65.16% of the variance explained. As suggested by experts, we have interpreted and named the factors as per the meaning of variables associated with the factors. Table 63.4 illustrates all the factor variables leading

Table 63.3 Matrix of eigenvalues for factors

Factors	Eigen-values	Percentage of variance explained	Percentage of variance after rotation	Cumulative % of variance after rotation
F1	9.097	31.37	18.27	18.27
F2	3.071	10.588	10.956	29.225
F3	2.252	7.765	8.366	37.592
F4	1.233	4.253	7.859	45.45
F5	1.123	3.871	7.709	53.159
F6	1.096	3.78	6.952	60.111
F7	1.025	3.534	5.051	65.162

Source: Authors' compilation

to barriers and challenges to the growth of the floriculture exporting firms in Rajasthan. The first factor is "Pricing and Promotional chellenges" which include nine variables explaining the highest percentage of variance. The second factor is "Production-related challenges" which comprise seven variables. The third factor is "Logistics-related challenges"which include insufficient cold storage facilities, high transportation costs including air freight rates, and poor connectivity to markets. Fourth facor is "Absence of Floriculture board" to enhance more focused development of the floriculture trade. Fifth factor is "R&D Challenges" related to quality improvement, virus resistance, plant nutrition, tissue culture, and othet research arescrucial for the growth of any organization and industry in the long run. The sixth factor posing a challenge is "Climatic and Cost Challenges"including the cost of a farm establishment, high commercial electricity rates, high-cost farming inputs like equipment, seeds, fertilizers, etc. The seventh barrier is related to "Financial Challenges" as there is lack of financial support and improper post-harvest management facilities. All the barriers and challenges of the floriculture export industry directly or indirectly fall in the purview of production, infrastructure, and marketing infrastructure.

6. Conclusion and Future Research Directions

The study has determined significant factors contributing to the barriers and challenges faced by the exporters of the floriculture industry in Rajasthan that have been discussed in the previous section. Despite making a huge investment in marketing infrastructure-related projects, with 16 projects in Rajasthan for the rural primary market with an assistance of Rs.59.97 lakh as per Horticulture statistics 2018, the major challenges facing the industry are related to the trading and marketing of perishable product like flowers. As per the statistics of the National Horticulture Board, there are 167 cold storages in Rajasthan with a ca-

Table 63.4 Matrix of factor loadings and structure of variabl

Code	Variables	Factors 1	2	3	4	5	6	7	Alpha
M11	31. Lack of fairs/events/buyer-seller meets/trade delegations for promotion of products	0.865							.907
M7	27. Decreasing profit-margin	0.755							
M10	30. High fluctuation in prices	0.748							
M12	32. No MSP (Minimum Support Price) for floriculture produce	0.686							
M9	29. Lack of bilateral trade agreement	0.681							
M6	26. High tariff and non-tariff barriers	0.673							
M1	21. Lack of market information	0.595							
M8	28. Underutilization of existing financial and human resources	0.588							
M2	22. Lack of efficient distribution channel	0.537							
P12	12. Different quality specifications of products in different countries		0.752						.834
P11	11. Lack of quality and quantity assurance from the producer		0.732						
P6	6. Forced sale due to perishability		0.609						
P9	9. Underutilization of production capacity		0.582						
P4	4. Lack of availability and accessibility of new technology		0.544						
P10	10. Lack of information about advanced cultivation practices		0.528						
P8	8. Unaffordable greenhouse technology		0.434						
I6	18. Insufficient cold-storage facility			0.829					.753
I5	17. High Transportation costs (including air freight rates)			0.808					
I4	16. Poor Connectivity to markets			0.559					
M4	24. Unhealthy competition among players				0.728				.721
M3	23. Exploitations by the middleman				0.641				
M5	25. Absence of a separate board for floriculture products				0.573				
I7	19. Inadequate irrigation facility					0.701			.714
I8	20. Lack of established R&D labs					0.627			
I3	15. Seasonality of trade					0.622			
P1	1. Unpredictable agro-climatic condition						0.742		.751
P2	2. High production cost						0.637		
I1	13. Improper post-harvest management facility							0.886	.553
I2	14. Lack of Financial support							0.518	

Source: Authors' compilation

pacity of 561293 million tons, still, the logistics and infrastructure challenges have prevailed and pinpointed by the respondents. The government needs to ensure proper and adequate infrastructural facilities related to the production and marketing of floriculture products with the frequent organization of export promotional activities with better technical knowledge in the field for post-harvest management skills. The government should also provide technical and managerial education to the floriculturists through proper training programs with support from educational institutions. A separate Floriculture Board is the pressing need of the hour and to ensure the overall sustainable growth of the industry. The result of the analysis is based on the responses of the floriculture products exporters of Rajasthan registered under APEDA; therefore caution should be made before generalization. A more advanced study could be undertaken on a large basis covering all the states of India with a bigger sample size. In further studies, structural equation modeling (SEM) could be adopted for grouping the barriers and challenges in the industry. It is recommended to perform the study on a much larger scale, which could help the government in forming better operational plans for the floriculture industry nationwide.

References

1. Agostini, L., Nosella, A., Teshome, M. B., & Holgersson, M. (2022). What is Patent Management? The Purification and Validation of an Integrated Measurement Scale. *IEEE Transactions on Engineering Management*, 1–15. https://doi.org/10.1109/TEM.2021.3138331

2. Ahmed, S., & Ting, D. H. (2019). The shopping list in goal-directed shopping: scale development and validation. *Service Industries Journal*, 39(5–6), 319–342. https://doi.org/10.1080/02642069.2018.1532997

3. Aung, Z. M., Santoso, D. S., & Dodanwala, T. C. (2023). Effects of demotivational managerial practices on job satisfaction and job performance: Empirical evidence from Myanmar's construction industry. *Journal of Engineering and Technology Management - JET-M*, 67, 101730. https://doi.org/10.1016/j.jengtecman.2022.101730

4. Bhat, S. A., Mir, A. A., & Islam, S. B. (2022). Scale Purification and Validation: A Methodological Approach to Sustainable Online Retailing. *Vikalpa*, 47(3), 217–234. https://doi.org/10.1177/02560909221123632

5. Bohle Carbonell, K., Könings, K. D., Segers, M., & van Merriënboer, J. J. G. (2016). Measuring adaptive expertise: development and validation of an instrument. *European Journal of Work and Organizational Psychology*, 25(2), 167–180. https://doi.org/10.1080/1359432X.2015.1036858

6. Cook, R., Jones-Chick, R., Roulin, N., & O'Rourke, K. (2020). Job seekers' attitudes toward cybervetting: Scale development, validation, and platform comparison. *International Journal of Selection and Assessment*, 28(4), 383–398. https://doi.org/10.1111/ijsa.12300

7. Finney, S. J. (2007). Book Review: Exploratory and Confirmatory Factor Analysis: Understanding Concepts and Applications. *Applied Psychological Measurement*, 31(3), 245–248. https://doi.org/10.1177/0146621606290168

8. Higuchi, A., Coq-Huelva, D., Arias-Gutierrez, R., & Alfalla-Luque, R. (2020). Farmer satisfaction and cocoa cooperative performance: Evidence from Tocache, Peru. *International Food and Agribusiness Management Review*, 23(2), 217–234. https://doi.org/10.22434/IFAMR2019.0166

9. Kao, C. Y., Tsaur, S. H., & Huang, C. C. (2020). The scale development of organizational culture on customer delight. *International Journal of Contemporary Hospitality Management*, 32(10), 3067–3090. https://doi.org/10.1108/IJCHM-02-2019-0128

10. Kaushik, V., & Dhir, S. (2019). Non-conformance in apparels: exploring online fashion retail in India. *Journal of Fashion Marketing and Management*, 23(2), 257–276. https://doi.org/10.1108/JFMM-05-2018-0067

11. Konanahalli, A., Marinelli, M., & Oyedele, L. (2022). Drivers and Challenges Associated With the Implementation of Big Data Within U.K. Facilities Management Sector: An Exploratory Factor Analysis Approach. *IEEE Transactions on Engineering Management*, 69(4), 916–929. https://doi.org/10.1109/TEM.2019.2959914

12. Magesa, M. M., & Jonathan, J. (2022). Conceptualizing digital leadership characteristics for successful digital transformation: the case of Tanzania. *Information Technology for Development*, 28(4), 777–796. https://doi.org/10.1080/02681102.2021.1991872

13. Marakova, V., Wolak-Tuzimek, A., & Tuckova, Z. (2021). Corporate social responsibility as a source of competitive advantage in large enterprises. *Journal of Competitiveness*, 13(1), 113–128. https://doi.org/10.7441/joc.2021.01.07

14. Mgale, Y. J., & Yunxian, Y. (2021). Price risk perceptions and adoption of management strategies by smallholder rice farmers in Mbeya region, Tanzania. *Cogent Food and Agriculture*, 7(1). https://doi.org/10.1080/23311932.2021.1919370

15. Ojo, L. D., Oladinrin, O. T., & Obi, L. (2021). Critical Barriers to Environmental Management System Implementation in the Nigerian Construction Industry. *Environmental Management*, 68(2), 147–159. https://doi.org/10.1007/s00267-021-01473-y

16. Patnaik, J., & Bhowmick, B. (2022). Determining appropriateness for management of appropriate technology: an empirical study using factor analysis. *Technology Analysis and Strategic Management*, 34(2), 125–137. https://doi.org/10.1080/09537325.2021.1890013

17. Ramadani, V., Agarwal, S., Caputo, A., Agrawal, V., & Dixit, J. K. (2022). Sustainable competencies of social entrepreneurship for sustainable development: Exploratory analysis from a developing economy. *Business Strategy and the Environment*, 31(7), 3437–3453. https://doi.org/10.1002/bse.3093

18. Sahi, G., Gupta, M. C., & Patel, P. C. (2017). A Measure of Throughput Orientation: Scale Development and Nomological Validation. *Decision Sciences*, 48(3), 420–453. https://doi.org/10.1111/deci.12227

19. Sinha, D., & Sharma, R. (2022). Impact of National Horticulture Mission on the Growth of Indian Floriculture Industry. *International Journal of Accounting, Business and Finance*, 1(2), 1–10. https://doi.org/10.55429/ijabf.v1i2.52

20. Sony, M., Antony, J., Tortorella, G., McDermott, O., & Gutierrez, L. (2022). Determining the Critical Failure Factors for Industry 4.0: An Exploratory Sequential Mixed Method Study. *IEEE Transactions on Engineering Management*, PP, 1–15. https://doi.org/10.1109/TEM.2022.3159860

21. Uakarn, C. (2021). Sample size estimation using Yamane and Cochran and Krejcie and Morgan and Green formulas and Cohen statistical power analysis by G*power and comparisons. *Apheit International Journal*, 10(2), 76–88.

22. Veerappan, M., Sahu, P. K., Pani, A., Patil, G. R., & Sarkar, A. K. (2020). Analysing and modelling the relationship between air freight movement and airport characteristics in India. *Transportation Research Procedia*, 48(2019), 74–92. https://doi.org/10.1016/j.trpro.2020.08.007

23. Zickar, M. J. (2020). Measurement Development and Evaluation. *Annual Review of Organizational Psychology and Organizational Behavior*, 7, 213–232. https://doi.org/10.1146/annurev-orgpsych-012119-044957

*Recent Trends in Engineering and Science for Resource Optimization and
Sustainable Development – Prof. (Dr.) Dorota Jelonek et al. (eds)
© 2024 Taylor & Francis Group, London, ISBN 978-1-032-98030-0*

Satisfaction and Level of Awareness Towards Post Office Investment Schemes of Investors in Kolkata

Shailendra Dayal*

Research Scholer,
PhD Management, Usha Martin University, Ranchi

▬▬▬ Abstract

Post Office is one of the oldest institutions linked with our lives since more than 150 years. India has more than 1.50 Lac post offices with 90% in Rural India. Apart from their main job of delivering letters post offices provides a wide variety of investment avenues starting from Saving banks to Time Deposits, NSC, KVPs, PPF, Sukanya Samriddhi, SCSS, PLI/ RPLI, etc. Investments under post office schemes comes under Ministry of Finance & are backed by Sovereign guarantee. It offers more interest rates as compared to its counterparts like banks which is most prime concerns for most of investors. It is observed that people especially in cities are not using Post office as their investment options and no. of peoples using this attractive option is less. Present study analyses Satisfaction & Level of Awareness towards post office investment schemes of investors in Kolkata. Satisfaction and awareness as input variables alongwith demographic features and other factors like interest rate, market volatility, ease of operations etc. results in Investment decision of an investor as an output. Satisfaction and awareness are assessed through various questions in a questionnaire. This study is an attempt to find awareness, attitude, perception and problems faced by the respondents with post office. What type of financial product marketing is required to make the post office savings scheme more popular and effective also analysed. Most of studies are done with persons having post office accounts, in this study overall population is studied where person may or may not be investor of post office, which can give more holistic view. It connects Financial management, financial behavior, attitude with marketing management as well as psychology (behavioral finance) which makes in multi-disciplinary in nature.

▬▬▬ Keywords

Satisfaction level, Level of awareness, Investment decisions

1. INTRODUCTION

Most of Post Office investment is done in rural areas where people invests due to nearness & are comparatively less tech. savvy. Researchers have tried to analyse awareness, attitude perception and problems faced by investors in post office Most of research is done for a particular rural or urban area. In most of studies mixed group of respondents ie. Lower class, middle class and upper class considered. In our country there is expanding middle class working

*Corresponding author: shailendra.dayal@umu.ac.in, shailendra1001@gmail.com

DOI: 10.1201/9781003596721-64

in PSU and private jobs as well as retired persons forms a large chunk of investors, some of motivation for them for investment is Tax Savings, Investment security with higher returns and ease of operation. Hence analysis of the awareness, attitude and perception of urban middle class respondents towards post office saving schemes was undertaken. Most of studies are done with persons having post office accounts in this proposed study person may or may not be investor of post office schemes are studied.

2. RESEARCH OBJECTIVES

1. To analyse investors awareness towards post office schemes
2. To study investors level of satisfaction post office schemes
3. To study problems faced by investors towards post office schemes
4. To describe the features of various post office saving schemes

3. CONCEPTUAL FRAMEWORK

Fig. 64.1 Conceptual framework

4. LITERATURE REVIEW

(Thaker VK, 2021)[1] took survey in Mehsana PO, 79 respondent with non-random sampling method. He found that people opined that Agent service is very poor in post offices. An other problem is slow processing & time taken in with draw alof money.

(Shetter Rajeshwari M., 2020) [2] reported profiles of various post office saving schemes as on 31.03.2020, explained benefits of investing in various PO schemes. He concluded that schemes are best when compared to other financial services in the market. He opined that post office must introduce schemes such as advertising etc. in radios and TV Channels.

(Shanmugapriya S. & Saravanan S., 2020) [3] states post office provides an opportunity and habit for poor and rural surrounding people to invest in saving. Sample of total 150 was studied in Mettupalayam subdivision of Coimbatore

district post circle. They concluded that investors were overall satisfied with the post office financial schemes.

(Karunakaran N. & Babu Athira, 2020)[4] selected sample of 60 persons from Kasaragod district of Kerela. People were satisfied with Localities and Parking facilities of Post office most. Prompt payment, Safety and Convenience to operate were first three reasons for Post office Saving schemes better than other schemes. Most people were found aware about Post Office Saving Account (33%) and Time Deposit (25%).

(Kamala Santhana S., 2020) [5] said postal schemes give opportunity to peoples who don't approach banks. Random 75 people were selected from Thoothukudi district. Easy formalities, well being of children, and to repay borrowings were found first three factors that influenced to save. He concluded that there is no significant relationship between the age group of respondents and the level of preference in postal schemes.

(KoleyJyotirmoy, 2020)[6] A sample size of 106 persons of Singur Block was studied. They found Lack of employees in post office, delay in withdrawal and more formalities in withdrawal were major problems faced by post office investors. Positive correlations was found in respect to overall satisfaction level of customers with demographic variables like education, income and overall liquidity of post office saving schemes.

(Saranya Baby K. & Hamsalakshmi, 2018)[7] Authors explained reasons of development of post saving historical perspective postal system in India, various saving products of Post office. They said expanding middle class families with both couple working will play vital role in saving and investment and post office must target these groups through proper advertisements.

(Ravindran G & Venkatachalam V, 2016) [8] studied sample of 100 respondents of Coimbatore. They found that safety and security was first objective for investment. They also found that there is a significant relationship between age group of respondents and awareness about the post office saving schemes.

(Kalaiarasi N &Saranyadevi S, 2016)[9] Studied sample of 85customers of post office at Udumalpet out of which 75 responses were considered for analysis. Major problem faced by depositors was in closure of scheme before maturity period. No significant relationship was found between nage and level of satisfaction of investors but there was significant relation between monthly income and level of satisfaction of investors.

(Balu A & Muthumani K, 2016)[10] Stated that post office is a friendly place for local people. Post office is must in a rural country like India and it must go for repositioning strategy.

5. RESEARCH DESIGN

It is an Exploratory as well as Explanatory research. Where various aspects related to investment in post office schemes were explored through questionnaire. Convenient sampling was done by sending google form to respondents and were asked to forward the same to others who are residing at Kolkata. It consisted questions regarding demographic profile, investment profile at post office, question regarding awareness regarding post office investment schemes, satisfaction regarding post office investment, post office amenities and source of knowledge about these schemes as well as overall remarks. Based on information provided by respondents analysis is done and conclusions are drawn.

6. ANALYSIS OF DATA:
DEMOGRAPHY OF RESPONDENTS

Table 64.1 Demography of respondents

Educational Qualification	No	%
PG and above	11	22
Graduation including Professional degree like Engineering/CA	36	72
Intermediate	2	4
Matric or below	1	2
Annual Income	**No**	**%**
0-500000	7	14
1000001-1500000	14	28
1000001-1500000	12	24
1500000+	17	34
Working in	**No**	**%**
Govt job	35	70
Private job	7	14
Retired	7	14
Unemployed	1	2
Age	**No**	**%**
20-35	4	8
35-50	34	68
50-60	6	12
60+	6	12
Gender	**No**	**%**
Male	42	84
Female	8	16

Source: Author's compilation

7. ANALYSIS OF AWARENESS ABOUT POST OFFICE INVESTMENTS

Awareness of Respondents about Post office schemes was checked through 6 questions: 1 question about awareness of fact that Investment under Post office Schemes comes under Ministry of Finance and backed by Sovereign Guarantee and safer option with assurance i.e. Zero Risk to deposits i.e. no limit of Guarantee of amount invested while in Bank Govt gives guarantee upto only 5 Lac. One Question regarding awareness that Post Office Term Deposit Schemes offers 1-1.2% more interest rates as compared to bank for 5 year period for General public. 2 questions regarding Income Tax treatment of Post Office schemes and 2 questions regarding initiatives of Internet Banking and ATM of Post Office. Following are the findings:

Table 64.2 Awareness level of respondents about post office investment schemes

Awareness No of Correct Answers out of 6	No of Respondents	Percentage
0	3	6
1	3	6
2	5	10
3	11	22
4	9	18
5	9	18
6	10	20
Total	50	100

Source: Author's compilation

Observations are as follows:

1. If we consider more than 3 correct answer as more awareness then total 56% are aware of Post office investment schemes and 44% were less aware.

2. 32% respondents were unaware of the fact that that Investment under Post office Schemes are fully safe. Though total 34 respondents were aware of this fact, 11 have still no investment in any of schemes.

3. 64% respondents were unaware of the fact that that Investment under Post office provides more interest rate as compared to banks. Though total 32 respondents were aware of this fact, 10 have still no investment in any of schemes.

4. 20% were unaware about the interest earned in various schemes of Post Office needs to be added in total taxable income while 36% were unaware regarding fact that no TDS is deducted from maturity value in Post Office investment schemes.

5. 42% were unaware about fact that Post Office ATM can be used in any Bank ATM as well as online payment, while 60% respondents were unaware about internet banking facilities of post office. This unawareness regarding above facts contributes largely about non movement of persons towards post office schemes.

8. MOST VITAL FEATURE OF POST OFFICE INVESTMENT SCHEMES

Security of Investment (50%) and High Interest rates (34%) were highest rates feature among respondents which is available in post office investment schemes. Liquidity and Tax Benefits were 2% each while Other scored 12%.

9. SATISFACTION FROM POST OFFICE INVESTMENT SCHEMES

In this section only 28 respondents were analysed who invested in Post office schemes.

Table 64.3 Satisfaction from various aspects of post office investment schemes

	Satisfaction from your post office investment schemes		Satisfaction Level with Overall return of Schemes		Satisfaction Level with Liquidity of Post Office Schemes	
	No of Respondents	%	No of Respondents	%	No of Respondents	%
Most Dissatisfied	1	3.57	1	3.57	2	7.14
Dissatisfied	1	3.57	1	3.57	2	7.14
Neutral	8	28.57	8	28.57	6	21.43
Satisfied	13	46.43	15	53.57	16	57.14
Most Satisfied	5	17.86	3	10.71	2	7.14
Total	28	100	28	100	28	100

Source: Author's compilation

From Table 64.3 it is clear that those who have invested in Post Office investment schemes are more or less satisfied with overall schemes, liquidity and overall returns.

10. SATISFACTION LEVEL WITH AMENITIES OF POST OFFICE

Tables 64.4 shows that respondents are highly dissatisfied with Behavior of Dealing Staff and Infrastructure, Seating, Ambience which needs to be improved.

11. SOURCE OF INFORMATION REGARDING POST OFFICE SCHEMES

Above table shows that most respondents get information regarding Post Office Investment Schemes through their family members and friends (40% + 24% = 64%). Agents 8%, Post Office Employees 4% Newspaper advertisement/ articles14% and Others 10%. Other effective source of information needs to be increased to spread overall awareness.

12. CONCLUSIONS

Post office investment have two vital features –Safety & comparatively higher returns, still people are not going to it due to lack of awareness & dissatisfaction from post office amenities. Investors specially middle class investors can use Post office as a diversification avenue in investment portfolio. More active and innovative advertising on various platforms as well as awareness programs needs to be planned by government and post office department need to be designed to encourage people for taking up this investment, which will further channelize saving and same can be utilized for growth of economy as well as protection of investors' interest. Post office needs to work on technical upgradation, as well as skill enhancement of employees through training programs as well as improving look & seating arrangements at post office to attract people towards this investment avenue.

Table 64.4 Satisfaction level with amenities of post office

		Location		Behaviour of Dealing Staff		Infrastructure, Seating, Ambience	
Level	Weightage	No of respondents	Overall score	No of respondents	Overall score	No of respondents	Overall score
Most Dissatisfied	-2	12	-24	9	-18	16	-32
Dissatisfied	-1	8	-8	17	-17	14	-14
Neutral	0	8	0	9	0	11	0
Satisfied	1	12	12	14	14	8	8
Most Satisfied	2	10	20	1	2	1	2
Total		50	0	50	-19	50	-36

Source: Author's compilation

13. LIMITATIONS OF STUDY

1. Being financial in nature respondents may hesitate to expose their investment data.
2. Limited analysis is done due to time constraints.
3. Result of study will be subject to all limitations of the primary data.
4. Only Urban respondents were contacted.

References

1. Vinal Kishor bhai Thaker (2021), "Trouble for Investors from the Indian Post Office," *Glob. J. Commer. Manag. Perspect.*, vol. 10, no. 4, pp. 1–4, Apr. 2021.
2. Dr (Smt.) Rajeshwari M. Shetty (2021), "Financial Services From Indian Post Office: A Study," *IOSR J. Bus. Manag.*, vol. 23, no. 12, pp. 42–48, Dec. 2021.
3. Shanmugapriya S. & Saravanan S., (2020), "Rural Investor's Behaviorand Satisfaction Level of Financial Saving Schemes towards Post Office," *Int. J. Recent Technol. Eng. IJRTE*, vol. 8, no. 6, pp. 4122–4125, Mar. 2020, doi: 10.35940/ijrte.F9169.038620.
4. N. Karunakaran & A. Babu (2020), "Post office savings and attitude of rural investors in Kerala: A study from Kasaragod district," *J. Manag. Res. Anal.*, vol. 7, no. 2, pp. 60–63, Jul.2020, doi: 10.18231/j.jmra.2020.013.
5. Kamala Santhana S. (2020), "A study on Post Office Saving Schemes in Thoothukudi District," *J. Xian Univ. Archit. Technol.*, vol. XII, no. IV, pp. 1622–1632.
6. Koley Jyotirmoy (2020), "Customers' Investment Behavoir and Satisfaction towards Post Office Savings Schemes: A Study with reference to Singur Block," *Int. Journak Creat. Res. Thoughts*, vol. 8, no. 10, pp. 615–626, Oct. 2020.
7. Saranya Baby K. & Hamsalakshmi, (2018), "Performance of Indian Post Office Saving Schemes In Recent Trends.," *Int. J. Adv. Res.*, vol. 6, no. 3, pp. 998–1004, Mar. 2018, doi: 10.21474/IJAR01/6757.
8. V.V. Ravindran G (2016), "Investment Opportunities of Postal Services Sectors in India," *Int. Conference Res. Ave. Soc. Sci. Organised SNGC Coimbtore*, vol. 1, no. 3, pp. 226–229, 2016.
9. S.S. Kalaiarasi N (2016), "Depositors' satisfaction and Level of Awareness towards Post Office Savings Bank Schemes with special reference to Udumalpet," *Int. J. Innov. Res. Adv. Stud.*, vol. 3, no. 9, pp. 227–231, Aug. 2016.
10. M.K. Balu A (2016), "An Overview on Post Office as an Avenue for Savings," *Internaional J. Appl. Res.*, vol. 2, no. 8, pp. 47–52, Jul. 2016.

Note: All the figures and tables in this chapter were made by the author.

Recent Trends in Engineering and Science for Resource Optimization and Sustainable Development – Prof. (Dr.) Dorota Jelonek et al. (eds)
© 2024 Taylor & Francis Group, London, ISBN 978-1-032-98030-0

Cold Plasma Treatment of Raw Milk and Analysis of its Physicochemical and Microbial Properties

Vigyan Gadodia*,
Kiran Ahlawat, Ramavtar Jangra and Ram Prakash

Department of Physics, Indian Institute of Technology Jodhpur,
Rajasthan, India

▬▬▬ Abstract

Milk, one of the most complete nutrition mixes for human consumption, also presents itself as the most potent nutrient medium for bacterial growth. This becomes especially pronounced for a hot country like India with highly fragmented milk production, spoiling the milk before it reaches processing plants. In this work, an effort has been made to study the changes in the physical parameters of the milk, like pH, conductivity, temperature, and biological parameters of microbial load after direct treatment of milk using a Dielectric Barrier Discharge (DBD) based large volume surface plasma discharge system. The developed system is optimized for operational parameters, and raw milk has been treated at different time intervals. The pH and colour characteristics of raw milk samples have not been affected by the proposed cold plasma treatment. Nevertheless, the conductivity of the milk has significantly increased, which may help in reducing the harmful bacterial growth in the milk. The reduction in bacterial load is also reflected by the increased time for discolouration in the Methylene Blue Reduction Test (MBRT) of milk before and after treatment. Moreover, after continuous operation of the system for more than 30 minutes, there was no significant change in the raw milk temperature, which is a key requirement for milk decontamination without milk heating up using a cold plasma treatment in the developed DBD geometry. Low-temperature treatment prevents the denaturation of heat-sensitive nutrients -like Vitamins and Amino acids in milk. Further studies are being carried out for different fat-content raw milk and using some other experimental settings to increase the efficacy of treatment further.

▬▬▬ Keywords

DBD, Milk, MBRT, Non thermal plasma, pH and Conductivity

1. INTRODUCTION

Raw milk is the milk that comes out of the udder of cow/buffalo/camel/goat, etc. and has not undergone any temperature-based processing such as sterilization and pasteurization. Milk consists of water, protein, non-fat solids, lactose, vitamins, lipids, and minerals. Due to its nutritional composition, milk is considered as a functional food that

---

*Corresponding author: gadodia.1@iitj.ac.in

provides health benefits (Guetouache et al., 2014). This nutritional richness also makes raw milk an optimal media for the bacteria to grow. The unchecked bacterial growth above 8-10 °C reduces the milk nutrition and quality and also increases the acidity of milk making it harmful for human consumption. The most prominent harmful bacteria include *Staphylococcus Aureus*, *Escherichia Coli*, and a cocktail of mesophilic and thermophilic bacteria (Hickey et al., 2015). Poor cow care, improper milking, and poor cleaning of utensils further aggravate the problem increasing the initial bacterial load. Subsequently, handling and storage above 4 °C conditions make it an ideal temperature for the bacteria to grow in milk, eating up the nutrition and spoiling the milk beyond permitted standards for human consumption (Hou et al., 2015).

Heat Treatment is the most standard process to reduce bacterial contaminations by inactivating and killing the bacteria (all mesophilic and most thermophilic bacteria). The most common heat-treatment process for processing milk is pasteurization, where we have to heat the milk to a certain temperature (82-85 °C in typical Indian milk bacterial load) for 8-12 seconds and bring it back to 4 °C for packaging and storage till it reaches consumers in polypack. Another method though expensive, gaining increasing popularity is the Ultra High Temperature (UHT) method after which milk is packed in tetra pack packaging. The UHT procedure kills all bacteria cells and spores in milk by heating it to a higher temperature of 120–130 °C for 2–4 seconds. However, all the heat treatment processes have the disadvantage of denaturing heat-sensitive nutrients like vitamins and essential amino acids (Burton, n.d.).

When the temperature of the electrons is significantly higher than that of the ions and neutrals, which are typically close to room temperature, non-thermal plasma, also known as cold plasma, is formed. Non-thermal plasma has been used as a novel sterilization technology for a wide range of fields, such as in biomedical, food, and water treatment applications (López et al., 2019)(Soloshenko et al., 2000)(Ahlawat et al., 2022). Even at low temperatures, NTP or cold plasma inactivates the microorganisms faster via some physical processes like pulsed electric fields and using different active species. The plasma generates UV photons, electric fields, electrons, physical forces, charged particles, free radicals, reactive oxygen species (ROS), and reactive nitrogen species (RNS) (López et al., 2019) (Jangra et al., 2023). Non-thermal plasma has been applied to various liquid foods, including apple juice, orange juice, white grape juice, and milk (Wang et al., 2018). NTP treats both liquid and solid food, and because of its non-thermal (cold) nature, it can maintain the nutritional and functional characteristics of the food.

In this study, we have investigated the effect of non-thermal plasma on the physicochemical properties of milk, and no significant difference is observed on most of the parameters (like pH, temperature, etc.) except the conductivity. In our case, the increase in conductivity plays an important role in bacterial reduction, and the same was investigated via the MBRT test.

2. Experimental Section

2.1 Experimental Setup

A DBD-NTP-based source is designed and developed for the milk treatment, in which a dielectric layer is sandwiched co-axially between a high-voltage electrode and the ground electrode. The high voltage electrode is an aluminium rod with 12 mm diameter and 100 mm length and the grounded electrode is a wire mesh made of stainless steel (food grade) with 1 mm thickness. A glass tube with OD 16 mm, ID 12.3 mm and length 120 mm is used as the dielectric layer. The power was supplied by using a high-voltage power source (1-6 kV, 5-30 kHz, and 2 μs pulse width). The developed source is submerged in the milk and filled in a glass container. A 1000 × high voltage probe (P6015A, Tektronix) and a fast response Rogowski coil (110, Pearson) are used to visualize voltage and current waveforms through a four-channel digital oscilloscope (MDO3014, Tektronix). The complete experimental setup for milk sterilization is shown in Fig. 65.1.

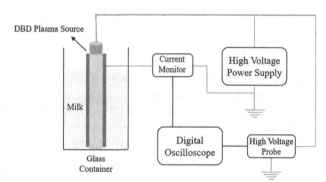

Fig. 65.1 Experimental setup for the milk sterilization

2.2 MBRT or Microbial Analysis

For the quality assessment of the milk, an MBRT test is carried out. The theory behind this test is that the milk's quality may be assessed by looking at the colour that develops after the addition of a dye, such as methylene blue. Due to the fact that methylene blue is a redox indicator and loses colour when oxygen levels are low. Due to the synthesis of reducing chemicals as a result of bacterial metabolism, the amount of oxygen in milk decreases. The amount of microorganisms in the milk and hence the metabolism of oxygen in the test tube determines how long it takes for the milk to become white again. The following process is adopted for the MBRT analysis:

1. Take 10 ml of milk sample in a sterile glass test tube
2. Adding 1 ml of standardised (0.05%) MBRT dye and mixing it well
3. Closing the tube with a sterilized vacuum tight rubber cap
4. Placing the test tube in a water bath maintained at 37°C
5. Noting the time taken for decolourization.

2.3 Investigation of Physicochemical Properties

The physical and chemical properties of the milk are also important for microbial inactivation by non-thermal plasma. Therefore, different physicochemical properties of raw milk and non-thermal (cold) plasma treated milk are analysed. The parameters studied for the non-thermal (cold) plasma-treated milk were MBRT, pH, conductivity, temperature, titratable acidity, along fat/SNF content. All the parameters were recorded immediately after the experiment and then up to 60 minutes, at an interval of 5 minutes, after the experiment. The volume of the milk for each experiment is 225 ml. The value of pH is recorded with a pre-calibrated pH meter in the pH range (0-14). Temperature and conductivity are recorded with a pre-calibrated conductivity meter. Titratable acidity and fat content are measured by using chemical processes and SNF through a density test with a standardised Lactometer. The detailed process is as follows:

1. Acidity was tested with Qualigens Phenolphthalein indicator added to the milk and titrated with Qualigens N/10 NaOH solution.

   a. 10 ml milk + 10 ml distilled water + 1 ml Phenolphthalein indicator – mix well in a glass beaker. Titration with N/10 NaOH and 0.09*ml of NaOH added.

2. Fat measurement was done as follows:

   a. 10 ml H_2SO_4 (90%) in Butyrometer + 10.75 ml Milk + 1 ml Amyl Alcohol – Cork, mix and centrifuge at a speed of 1200 rpm for 5 minutes. After that measuring the reading of fat separated.

   b. SNF is measured by using a standardised lactometer. Dip the lactometer into a measuring cylinder containing milk at 29°C, the readings in the lactometer (LR) outside the milk are taken. The formula used is

    i. CLR = LR + 3

    ii. SNF (%) = (CLR/4) + 0.29 + fat% × 0.2

3. RESULTS AND DISCUSSION

3.1 Characterization of Discharge

All the experiments are conducted in ambient conditions, and no external gas is utilized for the discharge process. The overall discharge process can turn out to be very expensive if discharges are generated using any noble gases. Figure 65.2(a) shows the typical V-I characteristic of the DBD plasma discharge at 4.5 kV/20 kHz. The electric power consumed by the developed plasma source is calculated by using the formula:

$$P = \frac{1}{T}\int_{\tau}(V \times I)dt \tag{1}$$

where V is the applied voltage, I is the discharge current, and τ denotes the period of the pulsed power source. The estimated discharge power at 4.5 kV/20 kHz is 17.64 W at

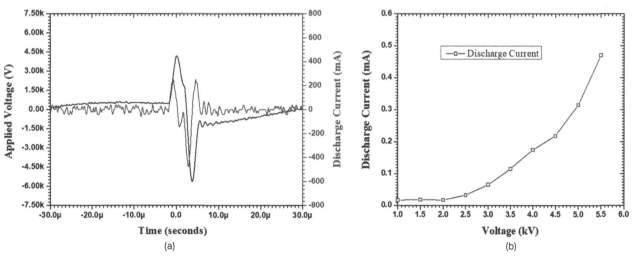

Fig. 65.2 (a) Typical V-I characteristic of the DBD plasma discharge at 4.5 kV/20 kHz, and (b) current and voltage mapping of the DBD plasma discharge

2 μsec pulse width in a pulse power source. Figure 65.2 (b) shows the current and voltage mapping of the DBD plasma discharge at different applied voltages. The current increases with increasing the applied voltage, which means that the electric field that causes the process of ionization, excitation, and recombination of suspension's molecules will increase till heating effects are significantly visible.

3.2 MBRT or Microbial Analysis

Infectious diseases (like salmonellosis, tuberculosis etc.) caused by the ingestion of multiple bacteria in contaminated milk are still a major health concern, especially for children. Several studies have focused on the microbial inactivation ability of non-thermal plasma technology (Wang et al., 2018)(Ruan et al., 2010), but to the best of our knowledge, no investigations are there on the MBRT analysis of milk treated with plasma. MBRT is especially relevant from an industry viewpoint, as it is the standard test conducted at the time of milk collection across India. Figure 65.3 shows the MBRT analysis of raw milk from the developed DBD-based non-thermal plasma source. The MBRT of raw milk is approximately 2 hours 40 minutes, after 5 minutes treatment of non-thermal plasma MBRT increases up to 3 hours 15 minutes. No significant increment is seen between 5 minutes to 10 minutes.

However, MBRT further improves to 3 hours 45 minutes when the treatment time is increased up to 15 minutes. So, a one-hour MBRT increase was achieved within just 15 minutes of operation of the cold plasma source, which is quite a significant improvement.

studies evaluate the effect of non-thermal plasma technology on physicochemical properties of milk (Nikmaram & Keener, 2022)(Wu et al., 2021). In this study, we have evaluated a number of physicochemical properties of milk like pH, conductivity, temperature, etc. The treatment time for milk is set to 10, 15, and 20 minutes. Findings show that pH, temperature, Fat/SNF, and titratable acidity of the milk remain the same in all the cases, as shown in Table 65.1, but conductivity increases rapidly for a period of time and then decreases up to its original value in a span of 45 minutes. Results are consistent with the previous studies, that the physicochemical properties of milk are not extensively affected by non-thermal plasma treatment (Nikmaram & Keener, 2022)(Manoharan et al., 2021). This increase in conductivity is generally due to the short and long-lived active species, in which short-lived active species instantly increase the conductivity up to a certain value and long-lived active species increase the time span of conductivity before it comes to its original mark. Figure 65.4 shows the conductivity of raw milk treated with non-thermal plasma at different times. The deactivation time of milk is around 45 minutes in all cases.

Table 65.1 Effect of cold plasma treatment on different physicochemical properties of milk

Treatment time	pH	Temperature (°C)	Fat/SNF (%)	Titratable acidity (g lactic acid/L)
Control	6.60±0.05	26.5±0.10	5.0/8.41±0.00	0.126±0.00
10 min	6.58±0.04	29.0±0.40	5.0/8.41±0.00	0.126±0.00
15 min	6.56±0.04	28.5±0.25	5.0/8.41±0.00	0.126±0.00
20 min	6.50±0.04	28.0±0.01	5.0/8.41±0.00	0.126±0.00

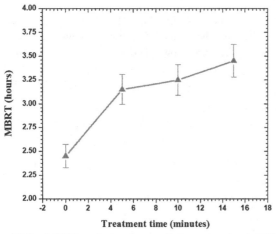

Fig. 65.3 MBRT analysis of raw milk from developed DBD-NTP source

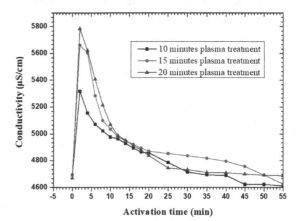

Fig. 65.4 Raw milk conductivity analysis from developed DBD-NTP source

3.3 Physicochemical Properties

Physicochemical characteristics are important parameters for the quality assessment and preservation of milk. Several

The pH value of treated milk remains the same, but conductivity changes over a period of time, which clearly indicates that in the cold plasma treatment of milk, there are an equal number of generations of acidic and basic species in the milk, and after some time (15-20 minutes) they neutralize. This increase in conductivity for a short period of time plays an important role in the microbial load reduction.

4. CONCLUSION

We infer that cold plasma technology can be utilized as a promising milk sterilization intervention for reducing the bacterial load without any extensive changes or permanent changes in the physicochemical properties of milk. The technique is environment-friendly, efficient at low temperatures, and has a low impact on the product matrix. We have investigated the effect of non-thermal plasma on the physicochemical properties of milk, and no significant difference is observed on most of the parameters except conductivity. In our case, the increase in conductivity plays an important role in bacterial reduction, and the same has been investigated via the MBRT test. This study opens a way for milk sterilization at low temperatures with minimal power consumption with a simple structural design of the system. There exists scope for further studies on designing continuous treatment systems and conjugate systems and further enhancing the MBRT improvement for immediate use in the industry. However, the obtained results are significant at this level and show the utility of cold plasma to improve the quality of milk.

ACKNOWLEDGEMENTS

The authors would like to acknowledge MSME, Govt. of India for providing financial support for this research through project no. IDEARJ000111.

References

1. Ahlawat, K., Jangra, R., Chaturvedi, S., Prakash, C., Dixit, A., Fulwani, D., Gupta, A., Jain, N., Tak, V., & Prakash, R. (2022). Photocatalytic oxidation conveyor "PCOC" system for large scale surface disinfection. *Review of Scientific Instruments*, 93(7). https://doi.org/10.1063/5.0082222

2. Burton, H. (n.d.). *Ultra-high-temperature Processing of Milk and Milk Products*.

3. Guetouache, M., Guessas, Bettache, Medjekal, & Samir. (2014). Composition and nutritional value of raw milk. *Issues in Biological Sciences and Pharmaceutical Research*, 2(10), 115–122. http://www.journalissues.org/IB-SPR/%5Cnhttp://dx.doi.org/10.15739/ibspr.005

4. Hickey, C. D., Sheehan, J. J., Wilkinson, M. G., & Auty, M. A. E. (2015). Growth and location of bacterial colonies within dairy foods using microscopy techniques: A review. *Frontiers in Microbiology*, 6(FEB), 1–8. https://doi.org/10.3389/fmicb.2015.00099

5. Hou, Q., Xu, H., Zheng, Y., Xi, X., Kwok, L. Y., Sun, Z., Zhang, H., & Zhang, W. (2015). Evaluation of bacterial contamination in raw milk, ultra-high temperature milk and infant formula using single molecule, real-time sequencing technology. *Journal of Dairy Science*, 98(12), 8464–8472. https://doi.org/10.3168/jds.2015-9886

6. Jangra, R., Ahlawat, K., Dixit, A., & Prakash, R. (2023). Efficient deactivation of aerosolized pathogens using a dielectric barrier discharge based cold - plasma detergent in environment device for good indoor air quality. *Scientific Reports*, 1–14. https://doi.org/10.1038/s41598-023-37014-2

7. López, M., Calvo, T., Prieto, M., Múgica-Vidal, R., Muro-Fraguas, I., Alba-Elías, F., & Alvarez-Ordóñez, A. (2019). A review on non-thermal atmospheric plasma for food preservation: Mode of action, determinants of effectiveness, and applications. *Frontiers in Microbiology*, 10(APR). https://doi.org/10.3389/fmicb.2019.00622

8. Manoharan, D., Stephen, J., & Radhakrishnan, M. (2021). Study on low-pressure plasma system for continuous decontamination of milk and its quality evaluation. *Journal of Food Processing and Preservation*, 45(2), 0–1. https://doi.org/10.1111/jfpp.15138

9. Nikmaram, N., & Keener, K. M. (2022). The effects of cold plasma technology on physical, nutritional, and sensory properties of milk and milk products. *Lwt*, 154, 112729. https://doi.org/10.1016/j.lwt.2021.112729

10. Ruan, R., Metzger, L., Chen, P., & Deng, S. (2010). *Nonthermal plasma pasteurization of milk using plasma technology (phase II)*. 227–231.

11. Soloshenko, I. A., Tsiolko, V. V, Khomich, V. A., Shchedrin, A. I., & Ryabtsev, A. V. (2000). *sterilization of Medical.pdf*. 26(9), 845–853.

12. Wang, R., Wang, L., Yuan, S., Li, Q., Pan, H., Cao, J., & Jiang, W. (2018). Compositional modifications of bioactive compounds and changes in the edible quality and antioxidant activity of 'Friar' plum fruit during flesh reddening at intermediate temperatures. *Food Chemistry*, 254, 26–35. https://doi.org/10.1016/j.foodchem.2018.01.169

13. Wu, X., Luo, Y., Zhao, F., Safian Murad, M., & Mu, G. (2021). Influence of dielectric barrier discharge cold plasma on physicochemical property of milk for sterilization. *Plasma Processes and Polymers*, 18(1). https://doi.org/10.1002/ppap.201900219

Note: All the figures and table in this chapter were made by the author.

Recent Trends in Engineering and Science for Resource Optimization and
Sustainable Development – Prof. (Dr.) Dorota Jelonek et al. (eds)
© 2024 Taylor & Francis Group, London, ISBN 978-1-032-98030-0

66

Analysis of Factors for usage of Electric Vehicles Taking into Account the Indian Perspective

Samridhi Bisht[1],
Chirag Sharma[2], Chaitanya Khurana[3],
Kanak Ahuja[4], Prashant Kumar[5], Sonal Khurana[6]
Vivekananda Institute of Professional Studies—Technical Campus

Abstract

As the world is moving and progressing in terms of the needs of people, the Earth is moving closer to its destruction. All the components that contribute towards the development, are also contributing towards the degradation of Earth's natural resources. Be it the conventional vehicles, which run on road everyday are a major cause of pollution in the environment. The conventional vehicles use fossil fuels, fossil fuels are being burnt excessive number resulting not only in the degradation of resources but also in the increased rate of pollution. The pollution rate has become such alarming that the people all over the world are concerned about it and are debating widely on the Net Zero Economy. As a result of such concerns, various automobile companies are primarily focusing on the Electric vehicles. Though they are being popularized all over the world, but it is not household name for the people of India. Through this paper, the author aims to identify the barriers faced by electric vehicles in gaining popularity in Delhi-NCR region. Opinion of the experts was taken by circulating a questionnaire through e-Delphi technique. For the evaluation of the same, the data collected was analysed using Interpretive Structural Modelling (ISM) technique. With the help of ISM technique, the paper highlights the factor driven and the driving factors, which further help in pinpointing the root cause for the same. The results of the analysis suggested that "Not Truly Pollution Free", "Not sustainable everywhere" and "Lesser number of choices" were considered as the driving factors and "limited recharge points" was considered as the driven factor for Electric vehicles(EVs) being unpopularized in Delhi NCR. The author aims to bring this factor into consideration and wants this shortcoming to be worked upon.

Keywords

Electric vehicles, Electric vehicle conversion industry, Interpretive structural modelling (ISM), Net zero emission, Tailpipe emissions

---

[1]samridhibisht645@gmail.com, [2]chiragsharma1703@gmail.com, [3]khuranachaitanya26@gmail.com, [4]kanakahuja2303@gmail.com, [5]prashantyadavdec@gmail.com, [6]sonal.khurana@vips.edu

DOI: 10.1201/9781003596721-66

1. INTRODUCTION

Air quality is deteriorating in smart cities, turning them into pollution hubs. Initially, people migrated to cities for a better life, but now they're returning to rural areas due to severe pollution (evirtualguru_ajaygour, 2016). Urban pollution causes numerous health issues like respiratory diseases, lung cancer, and asthma (Very Well Health, 2022). The main culprit is the use of conventional vehicles, which emit greenhouse gases from burning fossil fuels (Victor, 2016; Burning of fossil fuels, 2022). There's growing support for electric vehicles to cut down these emissions and contribute to the Net Zero Economy, which aims to balance the carbon emitted with the carbon removed from the atmosphere (Understanding Global Change, 2022; Net zero Climate, 2022). The global goal is to limit warming to 1.5°C and achieve net-zero emissions by 2050 (Ferrer, 2022; United nations, 2022).

Tailpipe emissions, comprising various harmful pollutants, come from internal combustion engines and contribute significantly to greenhouse gas levels, which can remain in the atmosphere for centuries (Tailpipe emissions, 2020; Science Learning Hub, 2020). These emissions cause smog, which accumulates near the ground, particularly during summer, leading to breathing difficulties and diseases like asthma (Krosofsky, 2021; Science Learning Hub, 2020).

Electric vehicles, which produce minimal tailpipe emissions, are gaining popularity worldwide (Iea). Despite their global uptake, their momentum in India is slow (Fady M. A Hassouna, 2020). Some Indian cities like Delhi NCR are among the world's most polluted (EPA), making them focal points for pollution studies, which consider several key factors (Climate Change Connection, 2017).

The main objective of author of this paper is to identify the key factors that are affecting the momentum of Electric vehicles and with the help of ISM Technique, to find the inter relationship between these factors.

The remaining section of the manuscript proceeds as follows:

Section 2 contains the Literature Review, which highlights the factors that are affecting its popularity

Section 3 talks about the Methodology used.

Section 4 sketches the results drawn out by this research

Section 5 talks about the implications of this research

2. LITERATURE REVIEW

As the growth of Electric Vehicles remains almost negligible as compared to the other conventional vehicle even after having such huge number of advantages in India. A literature review was conducted to identify the factors theoretically that are not enabling the people of Delhi-NCR from buying Electric vehicles in order to provide guidance to various sections of people to focus on the shortcomings and understand why they are not that effective here in India.

2.1 Scope of Electric Vehicles Globally

Several countries including France, England, Norway, the Netherlands, and India have pledged to promote the use of electric vehicles (EVs) by 2040 as part of efforts to reduce reliance on fossil fuels and curb pollution (Contributor, 2022). These commitments are supported by measures such as reducing petrol and diesel supply in cities and encouraging the shift to greener energy sources, underscoring the push towards electrification as a sustainable alternative (Sean Szymkowski, 2022).

EVs offer significant environmental benefits, such as higher energy efficiency and reduced local pollution, which are crucial in tackling urban air quality issues (Singh et al, 2020). They also provide economic advantages, including lower running and maintenance costs due to fewer moving parts and reduced fuel expenses. Furthermore, EVs benefit from tax incentives, are easier to operate, and can be

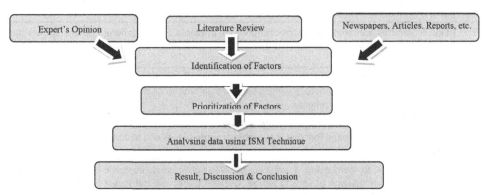

Fig. 66.1 The research flowchart of this paper

conveniently charged at home, adding to their appeal as a practical transportation solution ("11 best advantages of electric vehicles - EVs", 2022).

However, the transition to electric mobility is not without challenges. EVs still contribute indirectly to pollution through electricity production, especially in regions reliant on coal-fired power plants (World Coal Association, 2020). Also, their production involves emissions, and while they emit no tailpipe pollutants, the total environmental impact includes both direct and well-to-wheel emissions. Despite these issues, the shift towards electric vehicles appears inevitable, driven by both environmental imperatives and technological advancements, offering a cleaner, quieter alternative to conventional vehicles.

2.2 Identification of Various Factors that Affect the Hype of Electric Vehicles in India?

A literature review conducted to understand the barriers to the adoption of electric vehicles (EVs) in India, particularly in the Delhi NCR region, identified several critical factors impeding their popularity. Among the hindrances are limited recharge points, a short driving range, lengthy recharge times, fewer vehicle choices, and high initial costs. These limitations often overshadow the environmental benefits associated with EVs, leading many locals to prefer conventional vehicles (Prasanna, 2011).

The infrastructure for EVs in Delhi NCR is notably insufficient; recharge points are sparse, causing anxiety among potential users about running out of battery power far from the next charging station. Furthermore, the driving range of current EV models does not meet the expectations for long-distance travel, making them less appealing for extended journeys. Additionally, the lengthy duration required to fully charge an EV is seen as impractical for daily use.

In terms of vehicle choice and environmental impact, consumers face limited options with similar specifications and concerns over whether EVs are truly emission-free. Critics argue that while EVs are marketed as zero-emission vehicles, they still contribute to pollution through the electricity generation process and the manufacturing stages that heavily rely on fossil fuels. Initial financial outlay for EVs is also significant, deterring potential buyers despite the promise of lower lifetime operating costs (Energy Innovation, 2022; Jackie J., 2021).

These issues were thoroughly debated in a session with experts from various research fields, focusing on the perspectives and concerns of the general public in Delhi NCR. This debate helped highlight the multifaceted challenges of adopting electric vehicles in the region.

In a brainstorming session, researchers, experts, and the author collaboratively established parameters concerning the adoption of electric vehicles. Despite initial divergences in opinion, the panelists engaged in detailed discussions, breaking down each parameter into sub-factors and aligning their perspectives with those of the general public. After thorough consideration and exchange of views, the

Table 66.1 Table highlighting the major key factors affecting the growth of electric vehicles and their description

Variables	Description	References
LIMITED RECHARGE POINTS	When it comes to services like providing regular recharge points at regular intervals, it does causes inconvenience to the general public.	Imeon Energy (2022), Electric Vehicles (2022), Vikaspedia (2020)
HIGH INITIAL INVESTMENT	When it comes to choose between giving a hefty amount at once and giving the same amount in chunks, general public will prefer to give it in chunks.	India Briefing (2022), Ijert (2021), Tuhin Guha (2021)
NOT TRULY POLLUTION FREE	Many Electric vehicles do not stand by what they promise to the general public on it being pollution free. The general public does feel betrayed and do not want to invest in such products.	Daisy Simmons (2022), Riad Kherdeen (2021), Souptik Ghosh (2021),
LESSER NUMBER OF CHOICES	Electric Vehicles with almost same specifications are available. So, the general public face problem because of the limited number of choices.	Lesser known benefits of electric vehicles (2022) Blink Charging. Filip Mandys (2021)
LONGER RECHARGE TIME	A vehicle that takes longer time to get recharged will cause inconvenience to the general public. A conventional vehicle takes lesser time to fill the fuel in comparison to the Electric Vehicles.	Jeremy Laukkonen (2021), Nayan Madhav Sarode, M.T. Sarode (2020)
NOT SUITABLE EVERYWHERE	Electric vehicles are not suitable everywhere. They need fuels for the making of rechargeable battery and that fuel is available at some places and at some places, they are deficient.	Ashutosh Sinha (2021), Mukesh Malhotra (2019)
SHORT DRIVING RANGE	When it comes to purchasing a vehicle, general public sees its basic specifications like mileage or Driving Range. General public will not invest in a vehicle with short driving range, people will not invest in such vehicles.	Matthew Gindin (2020), Willet Kempton (2016)

panelists reached a consensus on all proposed parameters, demonstrating the session's effectiveness in synthesizing diverse viewpoints into a unified understanding.

3. SOLUTION METHODOLOGY

The methodology used in represented below is the form of a flow chart as shown below in Fig. 66.1.

The methodology was built on various steps, which are:

Step I: Identification of the variables.

Step II: Prioritization of the Factors

Step III: Analyzing data with the help of ISM method

3.1 Interpretive Structural Modelling Technique

Interpretive Structural Modelling is an interactive and effective way of learning process. It is an effective way of finding out the inter relationship among various variables. It summarizes the interdependency of these variables on each other in a very effective manner.

It gives out the rank of the listed variables and establishes interrelationship among those variables. It begins with listing the various variables, then a Structural Self Interaction Matrix is developed with the help of various inputs received. A Structural Self Interaction Matrix is a matrix that highlights relationship of variables in pair of twos. After that a Reachability Matrix is developed and after that, the hierarchies of the variables are formed.

Symbols to define relationships:

V → row variable influences corresponding column variable

A → row variable is influenced by corresponding column variable

X → row and corresponding column variable influence each other

O → row and corresponding column variable have no relationship

Table 66.2 Structural self interaction matrix

Variables	1	2	3	4	5	6	7
Limited Recharge Points		O	A	O	O	O	A
High Initial Investment			O	O	V	V	V
Short Driving Range				X	V	O	V
Longer Recharge Time					O	O	V
Lesser Number of Choices						V	O
Not Trully Fully Pollution Free							X
Not Suitable Everywhere							

Table 66.3 Reachability matrix

Variables	1	2	3	4	5	6	7	Driving Power
Limited Recharge Points	1	0	0	0	0	0	0	1
High Initial Investment	0	1	0	0	1	1	1	4
Short Driving Range	1	0	1	1	1	0	1	5
Longer Recharge Time	0	0	1	1	0	0	1	3
Lesser Number of Choices	0	0	0	0	1	0	0	2
Not Truly Fully Pollution Free	0	0	0	0	0	1	1	2
Not Suitable Everywhere	1	0	0	0	0	1	1	3
Dependence Power	3	1	2	2	3	4	5	

Table 66.4 Final reachability matrix

Variables	1	2	3	4	5	6	7	Driving Power
Limited Recharge Points	1	0	0	0	0	0	0	1
High Initial Investment	1	1	0	0	1	1	1	5
Short Driving Range	1	0	1	1	1	1	1	6
Longer Recharge Time	1	0	1	1	1	1	1	6
Lesser Number of Choices	1	0	0	0	1	1	1	4
Not Trully Fully Pollution Free	1	0	0	0	0	1	1	3
Not Suitable Everywhere	1	0	0	0	0	1	1	3
Dependence Power	7	1	2	2	4	6	6	

Table 66.5 Level partitioning

Elements	Reachability Set	Antecedent Set	Intersection Set	Level
1	1	1,2,3,4,5,6,7	1	1
2	2	2	2	4
3	3,4	3,4	3,4	4
4	3.4	3,4	3,4	4
5	5	2,3,4,5	5	3
6	6,7	2,3,4,5,6,7	6,7	2

Table 66.6 Level partitioning iterations

Variables	1	6	7	5	2	3	4	Driving Power	Level
1	1	0	0	0	0	0	0	1	1
6	1*	1	1	0	0	0	0	3	2
7	1	1	1	0	0	0	0	3	2
5	1*	1	1*	1	0	0	0	4	3
2	1*	1	1	1	1	0	0	5	4
3	1	1*	1	1	0	1	1	6	4
4	1*	1*	1	1*	0	1	1	6	5
Dependence Level	7	6	6	4	1	2	2		
Level	1	2	2	3	4	4	4		

Table 66.7 Conical matrix

Variables	1	6	7	5	2	3	4	Driving Power
Limited Recharge Points	1	0	0	0	0	0	0	1
Not Truly Fully Pollution Free	1*	1	1	0	0	0	0	3
Not Suitable Everywhere	1	1	1	0	0	0	0	3
Lesser Number of Choices	0	1	1*	1	0	0	0	4
High Initial Investment	0	0	0	1	1	0	0	5
Short Driving Range	0	0	0	1	0	1	1	6
Longer Recharge Time	0	0	0	1*	0	1	1	6
Dependence Level	7	6	6	4	1	2	2	
Level	1	2	2	3	4	4	4	

Table 66.8 Reduced conical matrix

Elements	Reachability Set	Antecedent Set	Intersection Set	Level
1	1	1,2,3,4,5,6,7	1	1
2	1,2,5,6,7	2	2	
3	1,3,4,5,6,7	3,4	3,4	
4	1,3,4,5,6,7	3,4	3,4	
5	1,5,6,7	2,3,4,5	5	
6	1,6,7	2,3,4,5,6,7	6,7	
7	1,6,7	2,3,4,5,6,7	6,7	
1 234				

3.2 Prioritization of the Factors

After the selection of the parameters. The panelists decided to discuss on the prioritization of the various factors. But all the panelists were never on the same ground when it came to prioritization of factors. So, it was eventually decided why not to take the viewpoints of the general public, which was considered as one of the main objectives.

A questionnaire was prepared. These variables were sent to the general public for the prioritization of the variables. The prioritization of variables is necessary in order to identify and breakdown the major variable which is in every individual's mind when it comes to buying the Electric vehicles.

3.3 Analysing the Data

After the questionnaire was sent to the general public in Delhi NCR and the data was collected. The data was cleaned and was analyzed in order to prioritize the factors that were listed. The data was finalized according the thinking of the majority of the people in Delhi NCR. After that, it was analyzed using ISM Method.

The following Table 66.2 shows the dependence of one variable on the other variable. This is a self-interaction matrix.

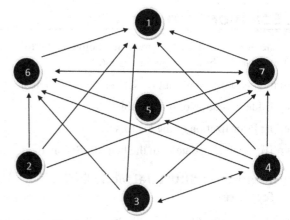

Fig. 66.2 Flowchart showing interdependency of variables among each other

4. OBSERVATION OF THE DATA ANALYSIS

The following research was designed to identify the factors that are disabling the people in Delhi NCR from buying electric vehicles.

It was found that "limited Recharge Points" is refraining people from investing in the electric vehicles. It is considered to be an important factor by the people of Delhi NCR as it is one of the services that must be available in regular intervals as a service to the products. The services play an important role and as a perk to a costumer. If a customer is not served the basic services, his/her is left unsatisfied and is forced to change his/her mind from buying a certain product.

It was found that "Longer Recharge time" and "short driving range" together as a factor are also keeping the people in Delhi NCR from investing into the electric vehicles. The people in Delhi NCR are concerned if the driving range is short and it will cost them so much time to recharge their vehicles, then, they wouldn't be able to travel a long distance. If they get restricted to travel only to short distances, then, what is it good for? They feel that after a point of time, they will ultimately switch to conventional vehicles in order to travel longer distances.

Then, it comes to the money. The people are concerned in buying cheaper goods. Since, the Electric vehicles have high initial investment as compared to the conventional vehicles. People are concerned about their money. When it comes to choose between giving a hefty amount at once and giving the same amount in chunks, general public will prefer to give it in chunks.

The factors "Lesser number of choices", "Not suitable everywhere" and "not fully pollution free" are the least bothered factors among the listed factors. Though, they need to be worked upon but they are not the need of the hour.

5. IMPLICATIONS OF THE ANALYSIS MADE

The following data analysis is made in order to identify the factors or variable that is stopping people of Delhi NCR from investing in electric vehicles. The data can be used by the various companies which are involved in selling electric vehicles. Such companies must work on these factors in order to ensure good growth of electric vehicles.

The data can also be used by the government policy makers in order to make them more sellable than the conventional vehicle. The government can take care of the fact that they are not pollution free as they promise. They can ensure a quality check and ensure that they live up to the expectations they set up for themselves as people expect them to be.

The scientists could use make use of the fact that it is not suitable everywhere and has high initial investment because of the fuel used in making rechargeable batteries. They can work on this factor and can make an alternative for the fuel that is less expensive and available almost everywhere.

And, at the end everyone must join their hands in order to make electric vehicles a successful step in reducing the pollution and tackling this problem. This is a must take step that every citizen must take to support Net zero Economy.

6. CONCLUSION AND FUTURE WORK

The main objectives of the author of this paper were to identify the barriers behind the popularity of Electric Vehicles and to establish an inter-relationship among the various key factors.

It was concluded that "Not Truly Pollution Free", "Not sustainable everywhere" and "Lesser number of choices" were considered as the driving factors and "limited recharge points" was considered as the driven factor for Electric vehicles being unpopularized in Delhi NCR.

Furthermore, on the future level, the objective of the author of this paper was to provide researchers with a challenge to look at these factors and help in suggesting with the solutions so that these challenges and barriers can be overlooked.

Also, the companies involved in the production of the Electric Vehicles can target these factors as the 'factors hindering their growth' and can work on these variables.

The author looks forward to the day when the problems people are facing in the current times no longer exists and people stand by each-other in each problem they all face together, so that, the problem can be resolved and the society becomes a better place to live in.

References

1. Burning of fossil fuels (2022) Understanding Global Change.
2. Burning of fossil fuels - Understanding Global Change (berkeley.edu)
3. Contributor, G. (2022). 8 European Countries & Their EV Policies. Retrieved 16 July 2022, from https://cleantechnica.com/2018/11/04/8-european-countries-their-ev-policies/
4. Despite pandemic shutdowns, carbon dioxide and methane surged in 2020 - Welcome to NOAA Research. (2022). Retrieved 16 July 2022, from https://research.noaa.gov/article/ArtMID/587/ArticleID/2742/Despite-pandemic-shutdowns-carbon-dioxide-and-methane-surged-in-2020#:~:text=Since%202000%2C%20the%20global%20CO2%20average%20has%20grown,from%20about%20380%20to%20450%20parts%20per%20million.
5. Eleven best advantages of electric vehicles - EVs. (2022). Retrieved 16 July 2022, from https://www.primecom.tech/blogs/news/advantages-of-electric-vehicles-evs
6. EPA. Environmental Protection Agency. Air Pollution: Current and Future Challenges | US EPA
7. evirtualguru_ajaygour (2016) Essay on "Pollution in cities" complete essay for class 10, class 12 and graduation and other classes., eVirtualGuru. Essay on "Pollution in Cities" Complete Essay for Class 10, Class 12 and Graduation and other classes. (evirtualguru.com)
8. Ferrer, M.de (2022) Global warming must be limited to 1.5°C to avoid 'climate disaster', euronews. Global warming must be limited to 1.5°C to avoid 'climate disaster' warns the latest IPCC report | Euronews
9. Greenhouse gases, facts and information. (2022). Retrieved 16 July 2022, from https://www.nationalgeographic.com/environment/article/greenhouse-gases
https://afdc.energy.gov/vehicles/electric_emissions.html
https://thecsrjournal.in/are-electric-vehicles-answer-sustainable-future/
https://www.acko.com/car-insurance/convert-your-petrol-or-diesel-cars-into-an-electric-vehicle/ https://www.acko.com/car-insurance/convert-your-petrol-or-diesel-cars-into-an-electric-vehicle/ https://www.edfenergy.com/for-home/energywise/electric-cars-and-environment
https://www.edfenergy.com/for-home/energywise/electric-cars-and-environment#:~:text=They%20emit%20fewer%20greenhouse%20gases,their%20impact%20on%20the%20environment
https://www.ijedr.org/papers/IJEDR1902084.pdf#:~:text=An%20electric%20vehicle%20conversion%20is%20the%20modification%20of,with%20Low%20Aerodynam-

ic%20drag%20increases%20efficiency%20and%20Range. https://www.indepthnews.net/index.php/sustainability/climate-action/3409-hybrid-cars-and-the-myth-of-zero-emission-driving
https://ypte.org.uk/factsheets/electric-cars/what-are-the-downsides-to-electric-cars

10. Iea Electric vehicles – analysis, IEA. https://www.iea.org/reports/electric-vehicles

11. Krosofsky, A. (2021) What emissions do cars produce? it's not just CO2, Green Matters. What Emissions Do Cars Produce? (greenmatters.com)

12. Limitations of electric vehicles | Zigwheels. (2022). Retrieved 16 July 2022, from Limitations of electric vehicles (zigwheels.my) Nations, U. (2022). Net Zero Coalition | United Nations. Retrieved 16 July 2022, from https://www.un.org/en/climatechange/net-zero-coalition

13. Pahuja, N. (2022). GHG Mitigation in India. Retrieved 16 July 2022, from https://www.wri.org/research/ghgmitigation-india

14. Tailpipe emissions (2020) Greener Cars. Tailpipe Emissions | Greener Cars

15. Victor (2016) Home, ImportantIndiacom. Vehicle Pollution: Meaning, Causes, Effects and Solution - ImportantIndia.com

Note: All the figures and tables in this chapter were made by the author.

Recent Trends in Engineering and Science for Resource Optimization and
Sustainable Development – Prof. (Dr.) Dorota Jelonek et al. (eds)
© 2024 Taylor & Francis Group, London, ISBN 978-1-032-98030-0

Exploring Internet Trends: An Empirical Study of Usage Patterns among Indian Millennials

Kunal Gaurav*

Professor, School of Business,
Dr. Vishwanath Karad MIT World Peace University,
Pune, India

Madhuri Mahato

Associate Professor,
Amity Business School, Amity University
Jharkhand, Ranchi, India

Neelam Raut, Ameya Patil

Assistant Professor, School of Business,
Dr. Vishwanath Karad MIT World Peace University,
Pune, India

Abstract

This study delves into the internet usage patterns among millennials in Hyderabad by employing exploratory factor analysis (EFA) through IBM SPSS 24.0, aimed at uncovering the underlying motivations and behaviors driving their digital interactions. The study's methodology was underpinned by the Kaiser-Meyer-Olkin (KMO) measure and Bartlett's test of sphericity, which confirmed the suitability of the data for factor analysis with a KMO value of 0.553 and a significant Bartlett's test outcome. The analysis successfully identified ten factors that collectively explained 74.52% of the variance in the dataset, with factors such as social networking, information seeking, and professional development emerging as key dimensions of internet usage. These factors highlight the diverse and multifaceted ways in which the internet meets various personal and professional needs of millennials in an urban setting. The findings underscore the necessity for digital strategies that are finely tuned to the preferences and behaviors of this demographic to enhance engagement and effectiveness. Ultimately, this study contributes to a deeper understanding of the digital landscape among young adults in rapidly urbanizing areas, providing valuable insights for policymakers, educators, and marketers aiming to connect with this influential group.

Keywords

Internet usage, Millennials, Empirical study, Exploratory factor analysis (EFA)

---

*Corresponding author: kunal.gaurav2156@gmail.com

DOI: 10.1201/9781003596721-67

1. Introduction

It's undeniable that the internet has become a transformative force in contemporary society. As a ubiquitous element of daily life, it connects countless computers globally through an extensive network. The internet not only encompasses millions of networks and search engines but also engages billions of users worldwide. With the rapid advancement of technology, it's increasingly common for individuals to remain glued to their screens, diminishing face-to-face interactions as digital engagement becomes the norm.

The influence of the internet surpasses that of traditional information sources such as television, radio, and newspapers (Roy, 2008). It extends the capabilities of telephones and isolated computers into a vast networked information frontier. The internet augments traditional tools used for gathering information, creating data visualizations, disseminating news, and facilitating communication. When used adeptly, the internet shrinks the global divide, offering access to a wealth of information, experiences, and knowledge directly to users' screens.

Significant differences in internet usage patterns can be observed between Generation X and millennials, with changes in gratification also evident across generations. Even baby boomers are actively engaged online, primarily using the internet for email and maintaining personal connections. Unfortunately, the youth are often drawn to harmful content, which contributes to the prevalence of cybercrimes.

Millennials, in particular, have witnessed the rapid expansion of the internet. Numerous studies have explored the evolution and global impact of internet usage over the past decade. This specific study focuses on understanding the uses and gratifications of the internet among millennials in Hyderabad, aiming to delineate the behavioral patterns of internet usage among this demographic.

The Uses and Gratification Theory (UGT), rooted in the 1940s studies of radio listeners, provides a framework for this analysis. Unlike other media effect theories that view individuals as passive consumers, UGT posits that people actively select media to fulfill specific needs such as entertainment, relaxation, or social interaction (Ruggiero, 2000). The theory distinguishes between different types of gratification: content gratification, which involves seeking media for entertainment or information; process gratification, which relates to the act of using the medium itself, such as browsing the internet or engaging with new technologies; and social gratification, which encompasses using media for social networking or personal development.

Interestingly, process gratifications may be more closely associated with internet addiction, while content gratifica-

tions often link users to external activities such as news consumption or event participation. An online purchase, for example, results in physical delivery, integrating virtual and real-world interactions. Similarly, using emails for communication serves a content gratification purpose, reinforcing the importance of content-driven activities.

This study primarily aims to investigate the patterns and preferences of internet use among millennials in Hyderabad, identifying key activities and exploring the underlying factors of internet gratification among this group. The objective is to discern both the predominant and minimal uses of the internet, providing insight into the digital behaviors that characterize modern youth engagement.

2. Literature Review

The internet has become an integral part of daily life, significantly impacting how people access information, communicate, and entertain themselves. As traditional media like radios, TVs, and newspapers become less dominant, the internet emerges as a versatile tool that offers a variety of services and connects users globally. Roy (2008) identified key gratifications driving Indian users' internet engagement, exploring the nuanced role of gender in these dynamics.

Gratification, defined as the feeling of pleasure derived from satisfying a desire (Chambers Dictionary), is central to understanding the psychological motivations behind media usage. A significant body of research has been conducted to analyze internet use and the factors that motivate users. For instance, Larose, Mastro, and Eastin (2001) employed Social Cognitive Theory (SCT) to examine the uses and gratifications of internet users, emphasizing self-reinforcement, self-control, and self-efficacy.

The impact of excessive internet use, particularly among children and adolescents, has also been a focal point of study. Bayraktar and Gün (2007) investigated the correlation between internet use and psychological behaviors among youths aged 13 to 15 in North Cyprus, finding significant links between internet activity, aggression, and depression.

Research has also explored internet usage across different generations and cultures. For instance, Anderson (2001) conducted an exploratory study on college students' internet usage, identifying patterns that affect academic performance, social interactions, and sleep. Similarly, Chan and Fang (2007) focused on young people in Hong Kong, examining how internet access facilitates educational and social activities.

Differences in internet usage between age groups are also evident in studies like those by Adekunmisi, Ajala, and

Iyoro (2012), who examined Nigerian undergraduates' internet habits. Their findings highlighted a daily engagement of 6-7 hours, primarily for communication and academic research. This extensive use underscores the benefits of internet access but also raises questions about its broader social and cultural impacts, as explored by Sait, Al-Tawil, Sanaullah, and Faheem Uddin (2007) in their study of internet usage in Saudi Arabia.

The changing dynamics of internet gratification are also significant. Kaye and Johnson (2004) addressed this in their study of politically interested internet users, employing uses and gratification theory to understand media interactions. Meanwhile, Trocchia and Janda (2000) looked into older individuals' internet usage, discovering distinct motivations such as affiliation with reference groups and resistance to technological changes.

The widespread influence of the internet has not only changed how individuals interact with the world but has also led to concerns about internet addiction and its consequences. Song, Larose, Eastin, and Lin (2004) explored this by analyzing internet use among students, identifying various gratifications like virtual community engagement and information seeking, and examining their link to internet dependency.

As the internet continues to evolve, it becomes crucial to understand not only the motivations behind its use but also the potential consequences for individual users and society at large. The research continues to expand, reflecting the complexity of internet interactions and the diverse uses that individuals find for this pervasive technology.

3. RESEARCH METHODOLOGY

The research conducted is both exploratory and descriptive, aimed at uncovering the patterns of internet usage among millennials in Hyderabad. This study primarily utilizes a structured questionnaire developed to collect primary data, employing a 5-point Likert scale to gauge the intensity of responses on various variables. To ensure the robustness of the findings, a reliability test was also administered on the questionnaire. Data collection was supported by both primary and secondary sources. The primary data was gathered directly through surveys distributed among a diverse group of 150 daily internet users (millennials) in Hyderabad. Secondary data was sourced from reputable databases such as EBSCO and Google Scholar, providing a comprehensive backdrop to the primary data insights. The sampling technique employed was convenience sampling, chosen for its practicality in accessing a broad cross-section of internet users within the specified geographical area. The questionnaire comprised 26 questions

focused specifically on understanding the nuances of internet utilization among the respondents. Data analysis was performed using exploratory factor analysis on the IBM SPSS 24.0 software, which facilitated the reduction of variables into a manageable number of factors (Gaurav & Ray, 2020). This analysis used varimax rotation to stabilize the data and create a simpler structure, making the interpretation of factor loadings more straightforward.

4. DATA ANALYSIS

4.1 Reliability Analysis

In this study, a reliability test was conducted to assess the consistency of the scale and the variables used in the questionnaire. The purpose of this test is to ensure that the measures employed are dependable and provide stable results over time. A structured questionnaire was crafted to gather responses on various aspects of internet usage, using a 5-point Likert scale ranging from 1 (strongly disagree) to 5 (strongly agree). The questionnaire included 26 variables and was administered to a sample of 150 respondents.

The reliability of the questionnaire was quantified using Cronbach's alpha, a statistic commonly used to evaluate the internal consistency of a survey instrument. In this case, the resulting Cronbach's alpha was 0.648, which is considered acceptable for exploratory research. Generally, a Cronbach's alpha value above 0.5 is deemed satisfactory, indicating a reasonable level of reliability. This metric is crucial as it reflects the extent to which the variables within the questionnaire are related and thus, consistently measure the underlying construct of internet usage.

4.2 Exploratory Factor Analysis

In line with Gaurav & Dheer (2018) exploratory factor analysis (EFA) conducted for this study employed IBM SPSS 24.0 to scrutinize patterns of internet usage among millennials in Hyderabad. This analytical approach was rigorous, aimed at extracting significant factors from observed variables to deepen our understanding of behaviors associated with internet usage. To ensure the validity of the factor analysis, two key statistical tests were used: the Kaiser-Meyer-Olkin (KMO) measure and Bartlett's test of sphericity. The KMO measure assesses the sampling adequacy for each variable and for the complete model, indicating the proportion of variance among variables that might be common variance. For this study, the KMO value was 0.553, considered acceptable, confirming that the sample size was suitable for the analysis. Generally, a KMO value greater than 0.4 is necessary for adequate factor analysis, with higher values indicating better suitability of the data for the analysis.

Bartlett's test of sphericity further supported the appropriateness of the data for factor analysis. This test evaluates the null hypothesis that the correlation matrix is an identity matrix, implying no correlation among variables. A significant Bartlett's test result (significance level of 0.000, below the threshold of 0.05) indicated sufficient correlation among variables, hence rejecting the null hypothesis and validating the factor analysis approach.

The factor extraction identified ten factors with eigenvalues over one, signifying that each factor captured a substantial amount of the variance in the data. Factors with eigenvalues below one were excluded as they did not contribute enough information. The variance explained by these factors ranged considerably; the first factor alone accounted for 18.571% of the total variance, marking it as the most influential. The subsequent factors explained descending amounts of variance, with the second factor at 9.550%, and the tenth factor at 4.196%. Collectively, these factors accounted for 74.52% of the total variance, indicating a significant explanatory power and reinforcing the relevance of these factors in social science research.

A varimax rotation was applied to the extracted factors, facilitating a clearer interpretation by making the output more orthogonal. This method helps in simplifying the factor structure, making it easier to discern which variables are associated with specific factors. It enhances the interpretability of the data by clarifying how different variables contribute to each factor.

4.3 Factor Description

In this study, a consolidated factor analysis was performed on the data from an exploratory factor analysis (EFA) using a cutoff of 0.4 for factor loadings, focusing on 26 variables related to internet usage among millennials in Hyderabad. This rigorous approach extracted 10 distinct factors that highlight different gratifications derived from internet use. The reliability of these factors, assessed using Cronbach's alpha, varied significantly, ranging from 0.247 to 0.863. Eight of the factors demonstrated high reliability, indicating strong internal consistency and the effective measurement of underlying attributes. However, two factors exhibited low reliability, suggesting potential areas for refinement in measurement or factor structure to better capture the nuances of internet usage behaviors.

The first factor, 'Learning and Development', extracted from the factor analysis, indicates a substantial influence with a Cronbach's Alpha of .780 and explaining 12.299% of the variance. This factor comprises six variables: it enhances creativity, inspires new initiatives, provides a sense of control, offers access to job opportunities, broadens exposure, and aids in problem-solving. This factor illustrates the in-ternet's role in facilitating learning, expediting information access, and enhancing personal and professional growth.

'Global Exposure' is the second factor, characterized by a Cronbach's Alpha of .661 and explaining 8.808% of the variance. It includes three variables: it promotes global integration, broadens outlooks, and enables sharing views worldwide. This factor highlights the internet's capacity to expand users' horizons and enable global connectivity.

The third factor, 'Social Relationships', with a Cronbach's Alpha of .680 and accounting for 8.461% of the variance, includes variables related to global chatting, introducing peer groups, and minimizing time spent in direct social interactions. This factor underscores the internet's role in fostering relationships across distances while also reflecting a preference for digital over face-to-face interactions.

'Adaptability', the fourth factor, has a Cronbach's Alpha of .604 and explains 7.844% of the variance. It is associated with user-friendly interfaces, comprehensive global knowledge access, and simplified content downloads, highlighting the internet's role in facilitating easy access to information and enhancing user engagement.

The fifth factor, 'Pastime', reflects the internet's role in leisure and relaxation, with a lower reliability (Cronbach's Alpha of .324) but explaining 7.627% of the variance. It includes spending time indoors, relaxing, and preparing for global economic integration, emphasizing the internet as a source of entertainment and personal development.

'Stress Buster', the sixth factor, has a Cronbach's Alpha of .584 and accounts for 6.842% of the variance. It includes entertainment and gaining a competitive edge, underscoring the internet's role in stress relief and providing competitive advantages through accessible information.

The seventh factor, 'Improve Work Efficiency', demonstrates the highest reliability (Cronbach's Alpha of .855) and accounts for 6.247% of the variance with the single variable of speeding up work processes, highlighting the internet's significant impact on enhancing job performance efficiency.

'Expression of Opinion', the eighth factor, has the lowest reliability (Cronbach's Alpha of .247) but explains 5.962% of the variance. It involves expressing opinions and finding inspiration, reflecting on how the internet serves as a platform for voicing views and inspiring users.

The ninth factor, 'Job Opportunities', highly reliable (Cronbach's Alpha of .863) and explaining 5.279% of the variance, emphasizes the internet as a crucial tool for job searching, highlighting its utility in career advancement.

Lastly, 'Improve Ideology', the tenth factor with a Cronbach's Alpha of .840 and explaining 5.159% of the variance, focuses on the internet's role in ideation and creative thinking, illustrating its importance in fostering innovation and broadening intellectual horizons.

5. CONCLUSION AND DISCUSSION

The findings from this exploratory factor analysis offer a comprehensive understanding of the internet usage behaviors among millennials in Hyderabad, revealing multifaceted motivations that drive their digital engagement. The study identified ten distinct factors accounting for 74.52% of the total variance in internet usage patterns, with each factor reflecting a specific dimension such as social networking, information seeking, professional development, and entertainment.

The most significant factor, accounting for 18.571% of the variance, underscores the critical role of the internet in learning and development. This suggests that the internet is not merely a tool for communication but also a significant educational resource that enhances creativity and provides access to job opportunities. The prominence of factors related to global exposure and social relationships highlights the internet's role in creating global connections and maintaining social interactions, which are particularly valued in the culturally diverse and dynamically interconnected world of today.

Interestingly, the factors with lower explained variances, such as stress relief and expression of opinions, suggest that while these aspects are relevant, they are not the primary drivers of internet usage among this demographic. This nuanced understanding helps in distinguishing between the core and peripheral uses of the internet among young adults.

This analysis provides crucial insights for policymakers, educators, and marketers to develop targeted strategies that align with the intrinsic needs and preferences of millennials. For instance, educational platforms can be optimized to cater more effectively to the learning and professional development interests of young adults, while social platforms might focus on enhancing user engagement through improved tools for global interaction and self-expression. Moreover, recognizing the diverse uses of the internet can aid in crafting more effective communication and marketing strategies that resonate with this key demographic, ultimately leading to more meaningful digital experiences.

References

1. Adekunmisi, S. R., Ajala, E. B., & Iyoro, A. O. (2013). Internet access and usage by undergraduate students: A case study of Olabisi Onabanjo University, Nigeria. *Library Philosophy and Practice* (e-journal), 848.
2. Anderson, K. (2001). Internet use among college students: An exploratory study. *Journal of American College Health*, 50(1), 21–26.
3. Bayraktar, F., & Gün, Z. (2007). Incidence and correlates of internet usage among adolescents in North Cyprus. *Cyberpsychology & Behavior*, 10(2), 191–197.
4. Chan, K., & Fang, W. (2007). Use of the internet and traditional media among young people. *Young Consumers*, 8(4), 244–256.
5. Gaurav, K., & Dheer, S. (2018). Social media usage at workplace: An empirical investigation. *Sumedha Journal of Management*, 7(1), 144–158.
6. Gaurav, K., & Ray, A. S. (2020). Customer experience and customer loyalty in Indian telecom industry-An empirical investigation. *Test Engineering and Management*, 83, 9071–9085.
7. Kaye, B. K., & Johnson, T. J. (2004). A web for all reasons: Uses and gratifications of internet components for political information. *Telematics and Informatics*, 21, 197–223.
8. LaRose, R., Mastro, D., & Eastin, M. S. (2001). Understanding Internet usage: A social-cognitive approach to uses and gratifications. *Social Science Computer Review*, 19, 395–411.
9. Roy, S. (2008). Determining uses and gratifications for Indian internet users. *CS-BIGS*, 2(2), 78–91.
10. Ruggiero, T. E. (2000). Uses and gratifications theory in the 21st century. *Mass Communication & Society*, 3(1), 3–37.
11. Sait, S. M., Al-Tawil, K. M., Syed, S., & Faheemuddin, M. (2007). Impact of Internet usage in Saudi Arabia: A social perspective. *International Journal of Information Technology and Web Engineering*, 2, 81–107.
12. Song, I., LaRose, R., Eastin, M. S., & Lin, C. A. (2004). Internet gratifications and Internet addiction: On the uses and abuses of new media. *Cyberpsychology & Behavior*, 7(4), 384–394.
13. Trocchia, P. J., & Janda, S. (2000). A phenomenological investigation of internet usage among older individuals. *Journal of Consumer Marketing*, 17(7), 605–616.

Recent Trends in Engineering and Science for Resource Optimization and Sustainable Development – Prof. (Dr.) Dorota Jelonek et al. (eds)
© *2024 Taylor & Francis Group, London, ISBN 978-1-032-98030-0*

68

Navigating the Circular Economy in Sustainable Development: A Creative Framework based on People First PPP for Waste-to-Energy Projects in India

Radha Krishna Tripathy
Research Scholar, RGIPT, Jais, UP

Kavita Srivastava
Associate Professor, RGIPT, Jais,UP

Kiran Jalem
Assistant Professor, CUJ, Ranchi, Jharkhand

Abstract

The global issue of waste management has reached critical proportions, with approximately 4.5 million tonnes generated daily worldwide and projected to soar beyond 8 million tonnes by 2050, according to the World Bank. Despite significant advancements in waste management, waste has evolved beyond a mere public concern to become a thriving business sector, exemplified by the Waste-to-Energy (WtE) industry's global proliferation.

Recognizing the need for a paradigm shift, there is a growing consensus that the response to waste must align with the principles of a "circular economy." This approach seeks to eliminate waste altogether and aligns with the commitments made by United Nations Member States in the 2030 Agenda for Sustainable Development, particularly Sustainable Development Goal 12 on responsible consumption and production.

While WtE activities initially appear as a sustainable solution, its transformation of waste into energy through state-of-the-art facilities raises concerns about their alignment with the circular economy. A framework with specific guidelines proposed by UN asserts that WtE projects have the potential to contribute to a circular economy transition by adopting circular economy approaches, particularly as energy recovery activities. However,it necessitates innovative approaches to Public-Private Partnerships (PPPs) focused on people-first principles and the support of best practices in WtE projects by both governments and the private sector.

The central thesis underscores the importance of aligning WtE initiatives with circular economy principles for a sustainable and effective waste management strategy. While this approach gained momentum in EU countries, it is still to take off in India. This paper highlights the innovative framework adopted by EU and discusses modalities and way forward for successful adoption in India.

Keywords

WtE, PPP, People first PPP, Sustainability, Circular economy, Waste management

Corresponding author: Arnav1512@gmail.com

DOI: 10.1201/9781003596721-68

1. Introduction

1.1 Is Waste to Energy Compatible with Circular Economy?

The prevailing linear economy model, characterized by a take-make-use-dispose approach, has resulted in the generation of approximately 4.5 million tonnes of waste daily worldwide. This unsustainable pattern has led to severe environmental consequences, including escalating energy consumption and a surge in global Municipal Solid Waste (MSW). As the world grapples with the repercussions of this linear model, a transition toward a circular economy has gained traction. The circular economy emphasizes maintaining the value of products, materials, and resources by minimizing waste generation and resource consumption.

This shift holds promise for addressing the pressing challenges posed by traditional landfills, especially in the face of burgeoning waste streams such as the fast fashion industry. Recycling and composting programs have become integral to urban waste management, offering economic benefits and reducing the depletion of natural resources. As a parallel response, Waste-to-Energy (WtE) processes have emerged as a viable solution, using heat to recover energy or fuels from waste materials.

The global landscape of WtE reflects over 1,200 plants, with China, the EU, Japan, the United States, South Korea, and Singapore leading in capacity. Despite its potential to contribute to the renewable energy supply, the WtE market remains oligopolistic, dominated by major players. The industry's market size, valued at USD 31 billion in 2019, is projected to witness substantial growth. However, questions arise about its compatibility with the circular economy concept.

This comprehensive overview explores the evolution from a linear to a circular economy, the challenges posed by traditional waste management, and the role of WtE in addressing the growing waste crisis. The study delves into the global dynamics of WtE, its market players, and the potential hurdles in aligning it with circular economy principles. The findings highlight the need for informed strategies and collaborative efforts between governments and private sectors to ensure sustainable waste management and promote circularity(Waste to Energy in the Age of the Circular Economy:, 2020).

The primary objective of waste management hierarchy is to establish a prioritized order that minimizes adverse environmental effects, mitigates negative public health impacts, and optimizes resource efficiency by diverting waste from landfills. Ensuring that this waste management principle remains a pivotal driver in legislative actions and policies is crucial for addressing environmental concerns.

1.2 Arguments Against and in Favour of Waste-to-Energy (WtE)

Over the years, countries have adapted their approach to the WtE industry. It is pertinent to explore the arguments both opposing and supporting WtE, as outlined in Table 68.1.

Table 68.1 Arguments against and in favor of waste-to-energy

Arguments For WtE	Arguments against WtE
WtE can be part of a holistic waste management strategy. The EU countries reduce landfilling of wastes, by a combined effort of recycling/composting and WtE.	WtE reduces recycling/composting, acting as a disincentive or even barrier to circular economy or zero waste practices. Turning unsorted and usable trash into a valuable fuel commodity means communities are less likely to choose to reduce, reuse and recycle it.
In the United States of America, counties and municipalities thatutilise WtE consistently show an increased recycling rate, in parallel to WtE practice.	WtE raises environmental concerns, exacerbating climate change, emitting toxic emissions and giving rise to air pollution.
Today's technology allows WtE projects to operate with limited to no polluting effects. WtE plants must comply with stringent environmental standards, such as the EU Industrial Emissions Directive. WtE facilities equipped with sophisticated Air Pollution Control (APC) systems have far less severe impacts on air pollution.	WtE raises public health concerns for the population, emitting carcinogenic pathogens.
WtE plants monitor their emissions continuously, and report these on site and/or online. Many WtE plants around the world are built in the middle of residential or industrial sites so as to facilitate the use of heat for district or industrial heating or cooling. Some cities, such as Brescia, Osaka, Paris, Vienna, have built WtE plants that have become tourist attractions. The most recent addition is the new WtE Plant in Copenhagen that is planned to have a roof that can be used as a ski slope.	WtE raises societal concerns and communities are opposed to them in their neighbourhoods. In some countries, popular protests have taken place over the location of WtE plants reflecting serious concerns by residents on the impact to their health.

WtE can act as a transitional step toward a more circular and sustainable development path, depending on countries' starting points. While the prominence of WtE is expected to decrease with increased reusing and recycling, it remains essential for residual mixed waste. Certain products that cannot be recycled, the need to reduce reliance on landfills, and emerging technologies are cited as reasons supporting WtE's potential contribution to circular economy principles and practices.

2. METHODOLOGY

This research draws heavily from the UNECE study (Guidelines on Promoting People-First PPP *, 2019)and research on the subject that follows the EU for drawing comparison and set guidelines for people first PPP that can be adopted globally. However, an extensive literature review on global waste management issues, including current trends, challenges, and projections were conducted to know the global aspects on the circular economy and people perception on WTE as a circular economy(Cobo et al., 2018).

The literature study explores existing research on the circular economy and its principles, especially focusing on how it aligns with waste management practices. This also investigated the historical and current landscape of Waste-to-Energy (WtE) initiatives, considering their evolution, successes, and criticisms. The literature review Identifies factors contributing to success or challenges in circularity adoption, considering cultural, economic, and regulatory differences.

The study also compares waste management strategies across different nations to identify patterns and best practices. Several international and national policies related to waste management, with a focus on regulations governing WtE projects are analysed. Also, evaluation was done considering the effectiveness of policies that promote or hinder the alignment of WtE initiatives with circular economy principles.

Interviews with key stakeholders in the waste management and WtE sectors, including government officials, industry experts, and representatives from non-governmental organizations from India were conducted to find the best practices and to know their perceptions whether the WtE can be considered within the circular economy concept. That provides firm perceptions and attitudes of various stakeholders toward circular economy principles in waste management. The interviews gather insights on the role of Public-Private Partnerships (PPPs) in enhancing circularity and people-first principles in WtE projects(Tripathy & Tyagi, 2019).

Based on the findings, the research tries to formulate policy recommendations for governments and the private sector to enhance circularity in WtE projects(Priyadarshini & Abhilash, 2020)and propose strategies for improving PPPs with a focus on people-first principles and sustainable waste management practices. This also align recommendations with the United Nations Sustainable Development Goals, particularly SDG 12 on responsible consumption and production for government of India.

By incorporating these approaches, the research paper explored the intersection of waste management, circular economy principles, and the role of WtE initiatives. The combination of quantitative data, policy analysis, and stakeholder perspectives will contribute to a nuanced understanding of the challenges and opportunities in aligning WtE projects with circular economy goals.

3. LIMITATIONS

While conducting research on the intersection of waste management, circular economy principles, and Waste-to-Energy (WtE) initiatives, several limitations were encountered. It's important to acknowledge these potential challenges to maintain transparency and ensure the accuracy and reliability of the research findings. The limitations are provided below but not limited to;

- *Incomplete or Inaccurate Data:* Availability and accuracy of waste management data in India that may be limited or unreliable.
- *Lack of Standardization:* Variations in waste reporting standards and methodologies between countries that may be a hindrance to develop a standardized practice across countries.
- *Policy Implementation Gap:* Even if policies favour circular economy principles, their effective implementation may lag, affecting the success of WtE projects.
- *Limited Generalizability:* Findings from specific WtE case studies may not be universally applicable due to variations in regional contexts, regulatory environments, and project scales.
- *Rapid Technology Changes:* The WtE industry is subject to technological advancements. The research may become outdated quickly if it does not consider the latest innovations or technological shifts.
- *Cultural and Contextual Differences:* Variations in cultural perceptions and socio-economic contexts across regions may affect the generalizability of recommendations and findings.
- *Differing Circular Economy Definitions:* The interpretation and adoption of circular economy principles can vary, impacting the assessment of alignment with WtE projects.

Apart from several limitations to the subject under research, the generalist guidelines would apply to all WtE projects irrespective of locations and technologies implied.

4. DISCUSSION

4.1 The Risk Aspects for PPP based WtE Projects for Sustainability

For any WtE projects to be long term sustainable, the financial aspects take the centre stage for investment. Thus, PPPs are a favoured developmental strategy in numerous countries across various industries, including the Waste-to-Energy (WtE) sector. In a standard PPP arrangement for WtE projects, the developer assumes responsibility for project development following the Design-Build-Own-Operate (DBOO) model. Under the DBOO model, the developer secures independent financing, oversees the construction, ownership, maintenance, and operation of the WtE facility to fulfil the contracted energy capacity requirements throughout the facility's lifespan, typically spanning 25 to 30 years. WtE facilities necessitate substantial upfront investments, and developers, along with their financiers, seek assurances from the Government agency commissioning the project. These assurances are vital for ensuring satisfactory returns on the investment over the project's duration.

In conjunction with Government incentives, WtE projects primarily rely on two revenue sources. The first source is a "gate fee/ tipping fee," levied when municipalities, businesses, or other organizations dispose of their waste at the facility. The second source involves the sale of generated energy, comprising electricity and/or heat, to local power grids. Certain by-products resulting from WtE incineration, such as bottom ash, constitute a third, albeit smaller, revenue source. The gate fee is contingent on the waste volume, while energy sales are influenced by the produced heat. This dynamic, in turn, shapes the WtE project's business model. The combustibility of waste, including plastics, paper, or wood, affects the furnace temperature and the Calorific Value (CV) produced. The mix of combustible and non-combustible waste determines the facility's revenue streams.

Moreover, safety regulations mandate that the facility is designed for a specific thermal capacity. Excessive combustible waste leads to a CV above the designated level, prompting the operator to reduce waste input and subsequently gate fees. Conversely, if the CV is too low, the facility generates less electricity than it can sell. Balancing the right CV and waste quality emerges as the primary business challenge for PPP WtE projects to optimize waste volumes and power and electricity sales.

Apart from the financial and operational related to waste characteristics, there are other risks that may impact sustainability of WtE projects in the long run. These risks need to be identified, properly scrutinized and mitigated through various measures by the stakeholders of the project(Gupta & Verma, 2020).

4.2 Ways to Minimize Risk in WtE Project

It is hard to ascertain the risks in a long-term infrastructure project, specifically for WtE projects. However, there are ways to minimize the risks if risks are identified, categorised and targeted throughout the life cycle of the project rigorously. The table below provides various ways to minimize the risks in a WtE project.

Table 68.2 Ways to minimise the risks

Risk	Mitigant
Technology	Proven technology track record, avoid First of Kind facilities
Regulatory/ Subsidy	Evaluate case by case, do not rely onanticipated regulatorypolicy, use subsidies as an enhancement, not a foundation
Project phase (Development or Construction)	Rely on project milestones, involving permitting, leasing, or other approvals agreements to help inform and devaluate risk, tie release of funding to milestone achievement
Operational	Operating agreements O&M contracts with experienced providers
Feedstock	Diversified andlong-term agreements with rated / creditworthy counterparties
Offtake	Long term agreements with rated /creditworthy counterparties with merchant exposure for upside
Exit	Structure and aggregate portfolio of projects,optimise and demonstrate performance, leverage projects to improve economic return

4.3 People-first PPPs

For the WtE project to be adopted as a model for circular economy, it must adhere to the principle of people fist. Thus, the emphasis for all WtE projects are based on people first PPP concept. The [1]Economic Commission for Europe (ECE) advocates for a comprehensive developmental model, asserting that People-first PPPs should prioritize sustainable development with "people" as the primary beneficiaries. Evaluating partnerships based on "quality infrastructure" investments, People-first PPPs aim to bring significance to "value to people" and "value to the planet." These partnerships aspire to achieve and adhere to five People-first outcomes, as detailed in Table 68.2 below.

---

[1] https://eur-lex.europa.eu/legal-content/EN/TXT/?uri=CELEX%3A02010L0075-20110106

Table 68.3 People first outcomes for PPP for WtE

Outcomes	Benchmarks
Access and equity	Provide essential services Advance affordability and universal access Improve equity and social justice Plan for long-term access and equity.
Economic effectiveness and fiscal sustainability	Avoid corruption and encourage transparent procurement Maximise economic viability and fiscal sustainability Maximise long-term financial viability Enhance employment and economic opportunities
Environmental sustainability and resilience	Reduce greenhouse gas emissions and improve energy efficiency Reduce waste and restore degraded land Reduce water consumption and wastewater discharge Protect biodiversity Assess risk and resilience for disaster management Allocate funds for resilience and disaster management Advance community-driven development
Replicability	Encourage replicability and scalability Enhance Government, industry and community capacity Support innovation and technology transfer
Stakeholder engagement	Plan for stakeholder engagement and public participation Maximise stakeholder engagement and public participation Provide transparent and quality project information Manage public grievances and end user feedback

4.4 Key Challenges for Establishing People-First Waste-to-Energy Projects

Transforming Waste-to-Energy (WtE) initiatives into "high quality" investments and "People-first" endeavours poses considerable challenges for the industry. This section delves into each of the five People-first outcomes individually, illustrating the nature of challenges under each outcome and how projects are addressing and overcoming these issues.

4.5 Increase Access and Promote Equity

Increasing access and promoting equity involve assessing whether the project facilitates access to essential services, such as energy, particularly for those previously underserved or served with lower-quality services. The challenge to this outcome is the cost factor that comes with WtE projects as these projects tend to be more expensive than alternative energy sources, making them less affordable for consumers in low- and middle-income communities. Critics argue that the industry often portrays itself as power plants rather than waste disposal facilities. While WtE projects historically contribute modestly to national energy grids, this scenario is evolving. Projects in Olsztyn, Poland, and Klaipeda, Lithuania, significantly address energy needs and aid regions in transitioning from fossil fuels. While WtE can be a major portfolio for small countries

whose energy demand is less, it can not be a significant contributor for countries that have high energy demands. In these countries, rather than energy contributor, the WtE should have been perceived as a major tool for addressing health and environmental concerns.

These may be considered as social infrastructure projects and local government may support these initiatives as is happening in India. More to Energy projects, these projects may be considered as health and environmental initiatives.

4.6 Improve Projects' Economic Effectiveness and Fiscal Sustainability

This criterion focuses on the project's contribution to quality jobs, technology, innovation, economic asset utilization, and profitability. WtE projects may have limited local economic impact in terms of quality jobs. Concerns span from the provision of well-paid jobs to knowledge transfer and benefits for low-income groups, such as informal waste-pickers. WtE plants can adversely affect local communities, particularly low-income families reliant on informal recycling. However, successful projects, like the one in Cox's Bazar, Bangladesh, engaged refugees in construction, and others, such as in Dublin, Ireland, provided jobs, training, and substantial community allocations. Moreover, WtE projects often overlook gender equality and women's empowerment, requiring increased emphasis on these aspects.

4.7 Improve Environmental Sustainability and Resilience

Environmental sustainability focuses on protecting the planet and mitigating climate change, ensuring adherence to Sustainable Development Goals (SDGs).WtE combustion releases CO_2 equivalent emissions that may impact public health. Additionally, there are concerns about recycling targets and waste hierarchy compliance. Certain WtE projects, like those in Barcelona, Spain, and Glasgow, United Kingdom, offer substantial environmental benefits, reducing CO_2 emissions and enhancing recycling. Integrated approaches, as seen in Barcelona, Glasgow, and Singapore, demonstrate excellence in circularity, promoting resilience and economic gains(Malinauskaite et al., 2017).

4.8 Replicability

Replicability emphasizes the project's scalability and transferability of technologies and programs, requiring enhanced capacities and skills development. Making the WtE model replicable necessitates extensive skills transfer, training, and local staff development, which can be costly. While WtE companies often provide training, the wrong technology selection can lead to significant losses, as seen in the Tees Valley project in the United Kingdom. Skills transfer and technology selection are critical components of achieving replicability.

4.9 Stakeholder Engagement

People-First PPPs encourage developers to engage all stakeholders affected by the project, requiring transparency and data availability for evaluation. Projects often neglect plans to engage hostile local communities, giving rise to the "not in my back yard" (NIMBY) effect. Stakeholder engagement challenges involve addressing local opposition and ensuring transparency. While some projects, like those in Trimmis, Switzerland, effectively engage local groups, others face strong opposition, emphasizing the need for comprehensive engagement strategies.

Despite challenges, People-First projects in the WtE industry can achieve significant social and environmental objectives aligned with circular economy principles. However, scaling up these efforts requires active involvement from governments and stakeholders to elevate the industry to a new level. Embracing the Circular Economy: Seven Best Practice Options for Adapting and Transforming Projects into People-First Waste-to-Energy Public-Private Partnerships.

4.10 Best Practice Options for Adapting and Transforming Projects into People-first Waste-to-Energy Public-Private Partnerships

In alignment with the People-first PPPs this section proposes seven best practice options for transforming waste-to-energy (WtE) projects into holistic, circular economy principles. Aimed at governments, the private sector, and civil society groups, these options address challenges and provide specific strategies for creating People-first Waste-to-Energy Public-Private Partnerships. WtE is undergoing evolution due to factors such as government policies against climate change, technological advancements, and circular economy strategies.

Three scenarios emerge from the experience that can be replicated:

- Scenario 1: WtE continuing as landfill in the waste hierarchy,
- Scenario 2: WtE placed at the same level as landfills,
- Scenario 3: WtE fully integrated into circular economy activities.

Placing WtE in the Scenario 1 and Scenario 2 would not solve the burgeoning problem that is associated with the humongous generation of waste daily worldwide and the increasing trend of waste generation, the solution lies in scenario 3 where the WtE is fully adopted and integrated to circular economy activities and seen as a part of the overall transformational economic transition for a country. This is a must for countries like India where generally people do not want to own the waste they generate and the traditional thought process is that waste collection, segregation and treatment is the responsibility of government.

To adopt the WtE into circular economy and make it a part of the integrated economy, the following are seven best practice options applicable to these scenarios.

Option 1: Embed circular economy visions into government policies, prioritizing waste as a resource

People-first WtE PPPs should operate with purpose-oriented business models, emphasizing contribution to the circular economy. Valuing waste: projects should prioritise efficient collection and pre-processing systems, which can prevent the loss of potentially valuable waste, and should aim at avoiding the use of land for throwing waste away. In order to promote WtE it is therefore necessary to highlight the importance of preventing waste, reusing waste products and recycling as much as possible

Encouraging new WtE technologies and processes where WtE is not common: such a programme should particularly focus on low- and middle-income countries where WtE projects are relatively rare. These are the countries where WtE has to be promoted in the place of landfills which are cheaper but dangerous for the public health and the environment. Prioritize efficient waste collection and pre-processing systems to prevent valuable waste loss. Encourage new WtE technologies in low- and middle-income countries where landfilling prevails.

Option 2: Internalize externalities, gain social acceptance, and mobilize investments. Modify the waste hierarchy to reflect circular economy challenges

The waste hierarchy should encapsulate the circular economy activities. In this context, emphasis should be given on two separate activities: resource management, and waste management. The first requires advocacy of innovations, and strong regulatory environment to enhance the smarter product use and manufacture, as well as to extend the lifespan of product cycles. Waste management should be related to maximum resource and energy recovery, not landfilling or incineration of wastes without energy recovery. Also, People-first PPPs should focus on marginalised and vulnerable groups trying to survive in an ever more dangerous world, such as refugees, first nation, etc. Advocate industrial symbiosis for maximum recycling and energy recovery. Along with this, it should address the "residual" fraction of poor-quality waste, create renewable energy from biodegradable waste in WtE processes and ensure valuable product recovery from bottom ashes and sustainable disposal of fly ashes.

Option 3: Select suitable, innovative, and less polluting technologies. Promote skills development for local economies to utilize these technologies

People-first WtE PPPs should adopt the right circular economy enhancing technologiesincluding "cleaning" the circular process by removing dangerous harmful substances andhelping the local economy with skills development to utilise these technologies.People-first WtE PPPs should operate with sophisticated Air Pollution Controlsystems, and their emissions must be lower than strict emission standards, such asthe Industrial Emissions Directive. A system of monitoring of emissions from WtE plants needs to be put in place withcentralised registers controlled by the appropriate public environmental agencies.

Option 4: Provide economic incentives and price supports. Implement fiscal incentives promoting circular economy processes in People-first WtE PPPs

People-first WtE PPPs should benefit from fiscal incentives that encourage such projects toadopt circular economy processes and move upwards in the waste hierarchy. Governments should increase the landfill tax and should consider a credit for WtE for renewable energy production, e.g. feed in tariffs or the issuance of tradable green certificate with a guaranteed minimum market value for capacity installed. It should aim for results-based financing, e.g. environmental impact bonds, should be considered to address the construction, operation, and counterparty risks in WtE investments(Wang et al., 2020).

Option 5: Identify good partners and monitor their performance. Partner with enterprises displaying compatible WtE technologies and circular economy values

Innovative and cost effective local Technology should be given preference over high ended technology that is not suitable for local context. Local innovations can be employed to make it people and environment friendly. The local government should facilitate the spread of innovative technologies from successful companies to state and local urban bodies for successful implementation of the same. A scoring methodology can be employed to find best alternative in such cases. Local financial institutions can be targeted for financing such projects to promote green economies and green investment opportunities. However, caution should be exercised to inadequate procurement frameworks that lead to transparency issues and poor governance (Patil & Laishram, 2016).

Option 6: Establish transparent, open procurement processes with a zero-tolerance approach to corruption. Ensure People-first WtE PPPs adhere to circular economy values

The local government may Implement transparent and open procurement processes with a zero-tolerance approach to corruption. It should comply with the ECE Standard on a Zero Tolerance Approach or develop its own localised approach to zero corruption in PPP Procurement with Stakeholder and Community Engagement in each level of participation.

Option 7: Enhance local participation, including women's empowerment and vulnerable groups. Establish a "social contract" with stakeholders and ensure strong community engagement

The concept of social contract(Biygautane et al., 2019) is a new phenomenon and would attract the stakeholders in a big way by proposing the benefits that would accrue to them in establishing WtE projects that would come with health and environmental benefits. In this contract, efforts should be made to make the local people partners in such contracts and they can be involved in the design, construction, and operation of WtE plants according to their skills and calibre. They may be included to promote civil engineering projects benefiting the community, such as land restoration and green areas.

4.11 Sample Project Format to Ascertain to the People First PPP Framework

- Description of the Project
- Compliance to the people first PPP framework (framework can be customised to host country)

How the project actually implements the People-first approach and focuses on one or more of the five following People-first outcomes.

Outcome 1: Increase access to essential services and lessen social inequality and injustice.

For example:

- Did the project consider the needs of the socially and economically vulnerable?
- Did the project increase access to essential services to the previously underserved?
- Does the project contribute to eliminating socioeconomic inequalities and gender inequalities?

Outcome 2: Enhance resilience and responsibility towards environmental sustainability.

For example:

- Does the infrastructure or facility improve environmental sustainability: By cutting greenhouse gas emissions? Reducing loss or waste? Decreasing the use of water and energy? Were the facilities built respecting environmental standards?
- Is the facility resilient against climate change threats?
- Is there an increase in the quality of the service provided?

Outcome 3: Improve economic effectiveness and sustainability

For example:

- Does the project achieve value for money and fiscal sustainability?
- Do the projects promote local decent and sustainable employment? annual growth of local income?
- Does it advance women's economic empowerment or take into account the differentiated needs of women and men?
- Does the project empower local business communities, economically marginalized communities, vulnerable groups?
- Is there an improvement in operational efficiency? Does it reduce costs?

Outcome 4: Promote replicability and the development of further projects

For example:

- Can the project be repeated and/or scaled up?
- Did the project build capacities of the local staff and the governments to deliver similar projects?
- Did it provide training to local workforces for the transfer of skills?

Outcome 5: Fully involve all stakeholders in the projects

For example:

- Were all stakeholders -directly involved in the PPP project or directly or indirectly affected by it- consulted on the selection, design and impact of the project? Did special groups who have played a limited role to date were integrated as well?
- Were the interests of affected communities protected?

4.12 The Indian Scenario: Adopting best Practices for People First PPP in Waste to Energy Projects for Sustainable Development

According to the Solid Waste Management Rules, 2016 of Govt of India, municipal bodies have to ensure that recyclables are routed through appropriate vendors and only the segregated, non-recyclable, high-calorific fractions are sent to a WTE plant or for RDF (refuse-derived fuel) production, co-processing in cement plants or to a thermal power plant. Under SBM, the Government of India will reimburse 100 per cent of the cost of preparing the detailed project report (DPR) as per the unit cost and norms set down by the National Advisory Review Committee (NARC). The State High Powered Committee (HPC) will authorise institutes of national repute for appraisal of the DPRs for projects recommended by urban local bodies. Besides this, the Central government's grants/VGF can be used for WTE projects, either up-front or as generation-based incentive for the power generated for a given period of time. The Central government incentive for SWM projects will be in the form of a maximum of 35 per cent grant/VGF for each project. The Mission also says that states will contribute a minimum of 25 per cent funds for SWM projects to match the 75 per cent Central share (10 per cent in the case of states in India's northeast and special category states). Niti Ayog also emphasises on WtE projects for waste reduction and waste minimisation in India and promotes to develop a waste to energy market for commercial exploitation in India. This is going to be a big opportunity for private players entering to the waste sector in India(Annepu, 2013).

India is also not lagging behind in implementing the best practices for people first PPP in its waste to Energy projects. The GOI has taken several initiatives to adopt such best practices in its policy framework.

The best practices as envisaged can be summarised below in Table 68.4.

5. CONCLUSION

The pursuit of a circular economy, characterized by a perpetual and optimal material cycle with no waste, remains an aspirational goal. While technological limitations and entrenched human behaviour patterns present challenges, the responsibility to sustainably manage materials that become waste is undeniable. Waste-to-Energy (WtE) emerges as a crucial player in this narrative, representing a transitional technology in the journey toward a circular economy.

In the current context, the WtE industry is positioned as a transitional technology, offering a unique subplot in the broader transition narrative. The industry has the potential to ascend the waste hierarchy and evolve into a key player in the circular economy's future. Realizing this potential necessitates the establishment of an enabling environment that supports both the circular economy and WtE. Governments and stakeholders, including new projects, are urged to adopt the seven best practice options proposed in this document, setting the stage for a sustainable and circular

Table 68.4   Best practices in indian scenario

Best Practices	Indian Scenario
Option 1: Embed circular economy visions into government policies, prioritizing waste as a resource.	The objective of the WtE programme launched by Govt of India is to support the setting up of Waste to Energy projects for generation of Biogas/ BioCNG/ Power/ producer or syngas from urban, industrial and agricultural wastes/residues.The programme provides Central Financial Assistance (CFA) to project developers and service charges to implementing/inspection agencies in respect of successful commissioning of Waste to Energy plants for generation of Biogas, Bio-CNG/enriched Biogas/Compressed Biogas, Power/ generation of producer or syngas.
Option 2: Internalize externalities, gain social acceptance, and mobilize investments. Modify the waste hierarchy to reflect circular economy challenges	Efforts are underway to internalize the cost associated with environment , health in specific projects. There are project driven. Social acceptance increases after incorporating the cost and the society benefits from the investment. The project cost increases, however, there is a increased acceptance from the society that increases viability of the project. Certain projects in Andhra Pradesh under competitive bidding mode take these factors into account while bidding.
Option 3: Select suitable, innovative, and less polluting technologies. Promote skills development for local economies to utilize these technologies.	WtE technology is being chosen after due consideration of waste availability and local conditions so that maximum number of locals can be employed in operation and maintenance of the plant.
Option 4: Provide economic incentives and price supports. Implement fiscal incentives promoting circular economy processes in People-first WtE PPPs.	Economic incentives are provided through central financial assistance, tax breaks and putting the WtE under Renewable Energy category. WtE projects are provided with preferential tariff in India.
Option 5: Identify good partners and monitor their performance. Partner with enterprises displaying compatible WtE technologies and circular economy values.	Companies with relevant experience and domain expertise in handling waste is considered for build, operate and maintain these projects. Some organisations that have good presence in WtE India are A2Z infrastructure, Essel Infra,Ramkygroup,Jindal,SPML infra etc.
Option 6: Establish transparent, open procurement processes with a zero-tolerance approach to corruption. Ensure People-first WtE PPPs adhere to circular economy values.	For procurement, now the focus is shifted to competitive bidding for establishment of a clean, transparent process.
Option 7: Enhance local participation, including women's empowerment and vulnerable groups. Establish a "social contract" with stakeholders and ensure strong community engagement.	Social contract is ensured by employing rag pickers in the WtE facility and employing the local people with various operations of the WtE plant. Majorly women are employed for segregation, sorting and other related works that provides for a stable livelihood for them.

future. The guidelines are necessitated for a better project planning and execution with less hindrance from the local population.

In the aftermath of these guidelines, a series of recommendations are put forth to further solidify the integration of WtE into the circular economy framework such as encourage discussions on the WtE Guidelines and its best practices among governments, businesses, and civil society. Seek insights from experienced governments and those in the early stages of WtE engagement.Disseminating the guidelines to be adopted in lower hierarchy such as municipalities and panchayats to develop such framework for better evaluation of projects and monitoring. It will encourage collaborative efforts for the widespread adoption of people first PPP in WtE industry.

The UNECE model(UNECE study, 2019) that employs a framework for people first PPP gathered several project information based on the specific framework to determine whether the project can be qualified as a people first PPP

project. The same can be done in the Indian context as a test case and the existing as well as planned WtE projects can be asked to follow the framework to develop an index that qualify for the people fist WtE projects adhering the guidelines to circular economy and the results need to be shared with the local government and stakeholders for project acceptance and to mitigate the NIMBY syndrome.

Efforts need to be undertaken to prepare stepwise guidance on how the WtE industry can maximize its contribution to the transition to a circular economy. The project promoters along with local urban bodies that enter into a specific agreement can offer practical insights and strategies for sustainable and circular practices in the industry. The government of India that provides several fiscal incentives such as central financial assistance, tax exemptions, putting WtE projects under RE basket can map their fiscal incentives to the people first PPP framework and adherence to its principles. This step would encourage WtE project promoters to create such a framework at the planning state and assign numbers to their projects so as to benefit from the government schemes. This will not only help the stakeholders with quality project but also provide them hard numbers to quantify the benefits that is hard to ascertain.

As an incentive and penalty mechanism for such WtE projects Indian government should go for landfill tax, provide credit line for WtE projects under renewable energy production, may consider feed in tariff for such projects. As we navigate the complex landscape of waste management and energy generation, the holistic adoption of these guidelines promises to steer the WtE industry toward a future where circularity is not just an aspiration but a tangible reality(Malinauskaite et al., 2017). Through collaboration, innovation, and a commitment to People-first principles, the WtE industry can play a pivotal role in the circular economy's evolution.

References

1. Annepu, R. (2013). A Billion Reasons for Waste to Energy in India. *Waste Management World.*
2. Biygautane, M., Neesham, C., & Al-Yahya, K. O. (2019). Institutional entrepreneurship and infrastructure public-private partnership (PPP): Unpacking the role of social actors in implementing PPP projects. *International Journal of Project Management,* 37(1). https://doi.org/10.1016/j.ijproman.2018.12.005
3. Cobo, S., Dominguez-Ramos, A., &Irabien, A. (2018). From linear to circular integrated waste management systems: a review of methodological approaches. *Resources, Conservation and Recycling,* 135, 279–295. https://doi.org/10.1016/j.resconrec.2017.08.003
4. *Guidelines on Promoting People-first Public-Private Partnerships Waste-to-Energy Projects for the Circular Economy* *. (n.d.). https://data.worldbank.org/indicator/EN.ATM.CO2E.KT
5. Gupta, P. K., & Verma, H. (2020). Risk perception in PPP infrastructure project financing in India. *Journal of Financial Management of Property and Construction,* 25(3). https://doi.org/10.1108/JFMPC-07-2019-0060
6. Malinauskaite, J., Jouhara, H., Czajczyńska, D., Stanchev, P., Katsou, E., Rostkowski, P., Thorne, R. J., Colón, J., Ponsá, S., Al-Mansour, F., Anguilano, L., Krzyżyńska, R., López, I. C., A.Vlasopoulos, & Spencer, N. (2017). Municipal solid waste management and waste-to-energy in the context of a circular economy and energy recycling in Europe. *Energy,* 141. https://doi.org/10.1016/j.energy.2017.11.128
7. Patil, N. A., & Laishram, B. S. (2016). Sustainability of Indian PPP procurement process. *Built Environment Project and Asset Management,* 6(5). https://doi.org/10.1108/bepam-09-2015-0043
8. Priyadarshini, P., & Abhilash, P. C. (2020). Circular economy practices within energy and waste management sectors of India: A meta-analysis. *Bioresource Technology.* https://doi.org/10.1016/j.biortech.2020.123018
9. Team, P. (2019). *Circulated by the UNECE secretariat as received from the Project Team leader. April.*
10. Tripathy, R. K., & Tyagi, V. (2019). PPP: A Catalyst for Sustainable Infrastructure Development in India. *The Management Accountant Journal,* 54(7). https://doi.org/10.33516/maj.v54i7.46-50p
11. Wang, Y., Wang, Y., Wu, X., & Li, J. (2020). Exploring the risk factors of infrastructure PPP projects for sustainable delivery: A social network perspective. *Sustainability (Switzerland),* 12(10). https://doi.org/10.3390/su12104152
12. *Waste to Energy in the Age of the Circular Economy:* (2020). https://doi.org/10.22617/TIM200330-2

Note: All the tables in this chapter were made by the author.

Recent Trends in Engineering and Science for Resource Optimization and
Sustainable Development – Prof. (Dr.) Dorota Jelonek et al. (eds)
© 2024 Taylor & Francis Group, London, ISBN 978-1-032-98030-0

Parametric Study of Degree of Dissociation in a Low Pressure ECR based Hydrogen Plasma Source using Optical Emission Spectroscopy

**Surajit Kuila[1],
Ramesh Narayanan[2] and Debaprasad Sahu[3]**
Plasma-Laboratory, Department of Energy Science and Engineering,
Indian Institute of Technology Delhi, New Delhi, Indian

▬▬ Abstract

In this study, Hydrogen plasma is produced in a novel portable Compact Electron Cyclotron Resonance Plasma Source (CEPS) which had been designed, developed and patented at Indian Institute of Technology Delhi and optical emission spectroscopy is utilized to measure the intensity of Balmer-α and Balmer-β lines to calculate the DoD. Here parametric study of DoD of hydrogen plasma is reported. Gas pressure is varied from 2-10mTorr while the input power is maintained in the range 300-600 W. Two ECR field topologies are used to investigate its effect on Dissociation. The DoD is found to be always higher in case of single ring magnet configuration. At 2mTorr Pressure and 300 W power 12% DoD is obtained for single ring magnet configuration, on the other hand 7% DoD is found for three steps ring magnet configuration. Also, DoD is found to be decreasing with increase of working gas pressure and input power.

▬▬ Keywords

ECR plasma source, Optical emission spectroscopy, Degree of dissociation

1. INTRODUCTION

Electron Cyclotron Resonance (ECR) Plasmas are highly regarded for their remarkable density and wide range of applications in scientific research and industrial sectors. These applications encompass diverse fields such as the semiconductor industry, biomaterial surface treatment, large area coating, and propulsion [1] etc. ECR plasma is a microwave discharge plasma in which electron cyclotron resonance is achieved by superimposing DC space varying magnetic field. Compared to electrode-based DC and radio frequency (RF) plasma sources, ECR-generated plasmas offer several advantages, including efficient energy coupling, simpler reaction vessels without internal electrodes, adaptability to different geometries, absence of high DC voltages, potentially lower power consumption, and availability of affordable microwave sources operating at 2.45 GHz [2].

One of the oldest, most widely used, and powerful non-invasive diagnostic tools for plasma characterization is

---

[1]surajitkuila10@gmail.com, [2]rams@dese.iitd.ac.in, [3]dpsahu@ces.iitd.ac.in

Optical Emission Spectroscopy (OES). It enables the determination of internal plasma parameters such as electron temperature (T_e), gas temperature (T_g), and electron density (n_e) [3]. Spatially resolved OES measurements are employed to map plasma density, while time-resolved measurements aid in studying the dynamics of pulsed plasmas [4]. Moreover, OES is also utilized to assess more complex plasma parameters, including the degree of dissociation, electron energy distribution function (EEDF), and optical emission cross-section [5].

The accurate determination of absolute concentrations of plasma radicals in molecular discharges, such as the density of atomic hydrogen (H) in H_2 plasma, is crucial for understanding surface processes and is therefore of great interest in various plasma applications [6]. The degree of dissociation (DoD) refers to the proportion of dissociated molecules relative to all other molecules, based on their respective number densities. If n_H is the number density of atomic hydrogen in the ground state 1^2S_2 and n_{H2} is the number density of hydrogen molecules in its electronics ground state $X^1\Sigma^+_g$, then the degree of dissociation is defined as [7]

$$DoD = \frac{n_H}{n_H + 2n_{H_2}} = \frac{n_H/n_{H_2}}{2 + n_H/n_{H_2}} \quad (1)$$

At the thermodynamics equilibrium the degree of dissociation is a known function of temperature and pressure but for low temperature plasma calculation of DOD is usually extremely difficult. Several standardized experimental techniques exist to determine DOD, including chemical and calorimetric methods, measurement of Lyman-alpha lines and molecular continuum, laser-induced fluorescence, resonance- enhanced multiphoton ionization, coherent anti-Stokes Raman scattering, actinometry, and measurement of Balmer-alpha and Balmer-beta lines [8].

This study specifically focuses on the diagnostic analysis of low-pressure hydrogen plasma using optical emission

spectroscopy (OES) in a novel Electron Cyclotron Resonance (ECR) source. The goal is to determine the degree of dissociation of hydrogen plasma under different gas discharge conditions by measuring the intensity ratio of Balmer-alpha and beta lines and the effect of magnetic field on DoD. Section 2 describes the experimental setup. In section 3 the methodology has been presented. Section 4 illustrates experimental results and discussion and finally conclusion is presented in section 5.

2. EXPERIMENT SETUP

The schematic diagram of experimental setup is shown in Fig. 69.1, where CEPS [9] is connected at one end of the cylindrical extension chamber of length 550mm and diameter 150mm and on the other end an Ocean HR 2 optical spectrometer (FWHM=1nm, 1µs integration time and SNR=380:1) is combined using a collective lens and a multimode optical fiber. The DC space varying magnetic field is produced by three ring shape permanent magnet (NdFeB) in the plasma source chamber. A turbo-molecular pump backed by a mechanical pump is used to evacuate the chamber to base pressure of the order of mTorr. The microwave power is provided by a 2.45 GHz magnetron. The magnetron is connected to a waveguide system, which comprises an isolator, a three-stub tuner and a rectangular ($TE_{1,0}$) to circular ($TE_{1,1}$) Mode converter

3. THEORY

The Balmer series (Hα, Hβ, Hγ, ...) represents the spontaneous emission of hydrogen atoms undergoing transitions between upper states with principal quantum numbers of n=3, 4, 5, and so on, to the lower states with n=2. The intensity of these spectral lines is directly related to the population density of the excited energy levels.

Therefore, by analyzing these lines and applying an appropriate excitation and deexcitation model, it becomes possi-

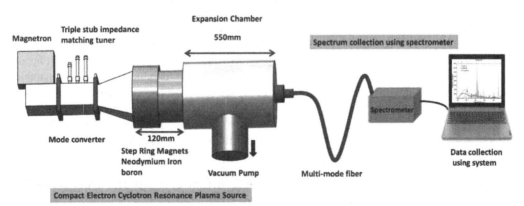

Fig. 69.1 Experimental setup, an Optical emission spectroscopy has been combined with CEPS

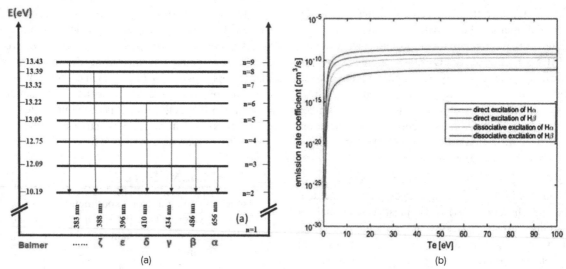

Fig. 69.2 (a) Energy diagram of the Balmer series transitions of the hydrogen atom, (b) Graph representing Emission rate coefficient of hydrogen atom.

ble to determine the degree of dissociation. In a wide range of plasma conditions, two dominant processes significantly contribute to populating the excited energy levels of hydrogen atoms.

The first process is the dissociative electron impact excitation, described by the reaction:

$$H_2(X^1\Sigma_g^+) + e \rightarrow H^* + H(1S_{1/2}) + e \qquad (2)$$

The second process involves direct electron impact excitation of hydrogen atoms from the ground state:

$$H(1S_{1/2}) + e \rightarrow H^* + e \qquad (3)$$

Considering these two processes and using the balance equation, the intensity of Balmer lines can be written as [6].

$$I_{n \rightarrow 2} = \frac{n_H}{n_{H_2}} K_{dir}^{em}(T|n \rightarrow 2) + K_{dis}^{em}(T|n \rightarrow 2) \qquad (4)$$

Where, $K_{dir}^{em}(T|n \rightarrow 2)$ and $K_{dis}^{em}(T|n \rightarrow 2)$ are the emission rate coefficients for direct excitation and dissociative excitation, respectively. For Balmer-α and Balmer-β lines, intensity can be abbreviated as $I_{H\alpha} = I_{(3 \rightarrow 2)}$ and $I_{H\beta} = I_{(4 \rightarrow 2)}$, respectively.

By taking ratio of $I_{H\alpha}$ and $I_{H\beta}$ the ratio of $\left(\dfrac{n_H}{n_{H_2}}\right)$ can be written as

$$\frac{n_H}{n_{H_2}} - \frac{I_{H\beta} K_{dis}^{em}(T|3 \rightarrow 2) - I_{H\alpha} K_{dis}^{em}(T|4 \rightarrow 2)}{I_{H\alpha} K_{dir}^{em}(T|4 \rightarrow 2) - I_{H\beta} K_{dir}^{em}(T|3 \rightarrow 2)} \qquad (5)$$

In order to calculate the DOD, the emission rate coefficients $K_{dir}^{em}(T|n \rightarrow 2)$ and $K_{dis}^{em}(T|n \rightarrow 2)$ are taken from the reference [7] as shown in Fig. 69.2(b)

4. EXPERIMENTAL RESULTS AND DISCUSSION

4.1 Effect of Gas Pressure and Input Power on DoD

Figure 69.3(a) showcases the spectrum of hydrogen plasma recorded at a power of 300 watts and a pressure of 2 millitorr. Notably, the Balmer alpha, beta, and gamma lines exhibit clear visibility in the spectrum. Figures 69.3(a) and 69.3(b) illustrate the variation of the degree of dissociation (DoD) with pressure and Power, respectively.

Interestingly, our observations reveal a negative correlation between DOD and power, indicating a decrease in dissociation with higher input power. Similarly, another negative correlation is observed between DoD and pressure, implying a decrease in dissociation as pressure rises. When pressure is increased, it reduces the electron temperature and excitation coefficients are temperature dependent as consequence DoD decreases with pressure. At the specific experimental conditions of 2 millitorr pressure and 300 watts power, the observed ratio of hydrogen atoms to molecules falls within the range of 0.06 to 0.16. It is noteworthy that similar studies conducted under comparable experimental settings have reported ratios ranging from 0.005 to 0.1 [10]. Thus, these results highlight the significant dissociation of hydrogen molecules into hydrogen atoms within plasma source, exceeding the dissociation levels observed in most previous research.

4.2 Effect of Magnetic Field on DoD

To investigate the impact of the magnetic field on DoD, two distinct ECR field topologies were created using two

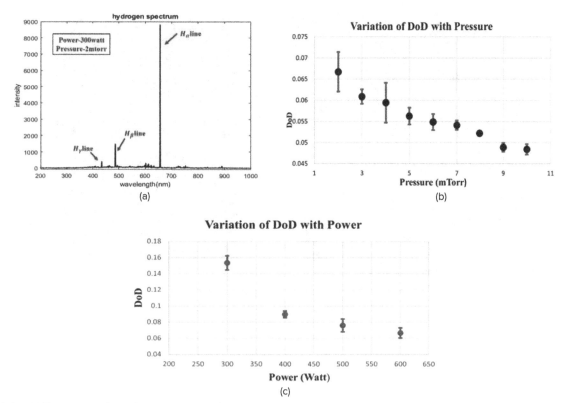

Fig. 69.3 (a) Experimental results representing Spectrum of Hydrogen plasma. (b) Variation of DoD with Pressure for 300 W input power. (c) Variation of DoD with power

types of magnets: a three-step ring magnet (CEPS) and a single ring magnet. These are depicted in Figs. 69.4(a) and 69.4(b) respectively. The simulated ECR field topologies are shown in Figs. 69.4(c) and 69.4(d); these were developed using the COMSOL Multiphysics tool.

The primary distinction between the two topologies is in their resonance zones. The three-step ring magnet has only one Electron Cyclotron Resonance zone with a magnetic field intensity of 875G located within the plasma source section. In contrast, the single ring magnet features two resonance zones. Due to the presence of these two zones, electrons are confined more strongly. As a result, the DoD for the single ring magnet consistently surpasses that of the three-step ring magnet. The results of the experiment are presented in Fig. 69.4(e).

5. CONCLUSION

In conclusion, Optical Emission Spectroscopy (OES) proves to be a valuable tool for determining the Degree of Dissociation (DoD) by analyzing the intensity ratio of the Balmer, alpha and beta lines. The findings of this study demonstrate that, like the energy efficiency of the Compact Electron Plasma Source (CEPS), the degree of dissociation achieved in this novel ECR plasma source

surpasses that of inductively coupled and capacitor coupled discharge plasma sources. However, for enhanced accuracy and precision, it is recommended to perform intensity measurements along the radial direction of the plasma. This approach enables a more comprehensive understanding of the spatial distribution and variation of dissociation within the plasma source, thereby facilitating further optimization and control of the dissociation processes.

References

1. Chu, P. K., Chen, J. Y., Wang, L. P., & Huang, N. (2002). Plasma -surface modification of biomaterials. *Materials Science and Engineering: R: Reports*, *36*(5-6), 143–206.
2. Espinho, S., Felizardo, E., & Tatarova, E. (2016). Vacuum ultraviolet emission from hydrogen microwave plasmas driven by surface waves. *Plasma Sources Science and Technology*, *25*(5), 055010.
3. Hanna, A. R., & Fisher, E. R. (2020). Investigating recent developments and applications of optical plasma spectroscopy: A review. *Journal of Vacuum Science & Technology A: Vacuum, Surfaces, and Films*, *38*(2), 020806.
4. Boffard, J. B., Lin, C. C., & DeJoseph Jr, C. A. (2004). Application of excitation cross sections to optical plasma diagnostics. *Journal of Physics D: Applied Physics*, *37*(12), R143.
5. Su, P. H., Zhu, Y. M., & Yang, S. (2008). Using OES to measure distribution of energetic electron in multi-needle-

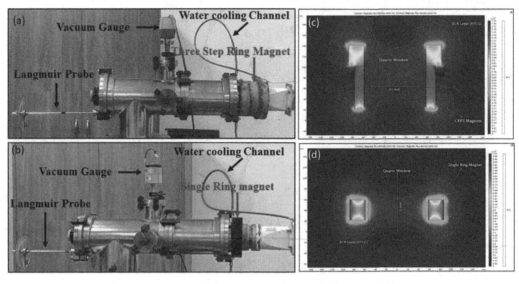

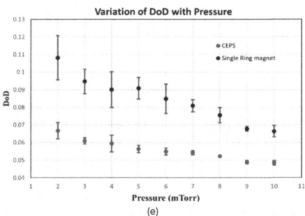

Fig. 69.4 Two different magnetic configurations and their field; (a) and (b) chamber with three step ring magnet and Single ring magnet; (c) and (d) COMSOL- Multiphysics Simulated magnetic field distribution. (e) Variation of DoD with Pressure for Three Steps Ring Magnet and Single Ring Magnet

to-plate corona discharge. *Journal of Electrostatics*, 66(3-4), 193–196.

6. Lavrov, B. P., Pipa, A. V., & Röpcke, J. (2006). On determination of the degree of dissociation of hydrogen in non-equilibrium plasmas by means of emission spectroscopy: I. The collision-radiative model and numerical experiments. *Plasma Sources Science and Technology*, 15(1), 135.

7. Pipa, A. V. (2004). On determination of the degree of dissociation of hydrogen in non-equilibrium plasmas by means of emission spectroscopy.

8. Dang, J. J., Chung, K. J., & Hwang, Y. S. (2016). A simple spectroscopic method to determine the degree of dissociation in hydrogen plasmas with wide-range spectrometer. *Review of Scientific Instruments*, 87(5), 053503.

9. Verma, A., Ganguli, A., Narayanan, R., Tarey, R. D., & Sahu, D. Compact ECR Plasma Source: its Physics and Application.

10. Abdel-Rahman, M., Schulz-Von Der Gathen, V., Gans, T., Niemi, K., & Döbele, H. F. (2006). *Plasma Sources Science and Technology*, 15(4), 620.

Note: All the figures in this chapter were made by the author.

*Recent Trends in Engineering and Science for Resource Optimization and
Sustainable Development – Prof. (Dr.) Dorota Jelonek et al. (eds)*
© 2024 Taylor & Francis Group, London, ISBN 978-1-032-98030-0

70

An Extensive Examination of ML Approaches for Predicting Heart Disease Insights: A Comprehensive Survey

Vaibhav C. Gandhi[1]

Computer Engineering Department,
Madhuben & Bhanubhai Patel Institute of Technology (MBIT),
Charutar Vidya Mandal University,
(CVM University) Anand,
Gujarat - India

Dhaval Mehta[2]

School of Engineering and Technology,
Navrachana University, Vadodara,
Gujarat – India

■■■■■■■ **Abstract**

This study aims to use a set of patient features to identify if a patient has a cardiac ailment. The goal of our study is to improve diagnostic accuracy while also saving medical institutions' human resources. In our project, we detect heart sickness using many methods such as random forest, logistic regression, SVM, naive Bayes, and artificial neural networks. Of these algorithms, Random Forest has the highest accuracy rate (95.6%). Medical facilities all throughout the world provide data on a range of health-related issues. By utilizing various machine learning techniques, this data may be utilized to provide meaningful findings. However, the amount of data being collected is massive and occasionally noisy. Even though these numbers are sometimes too big for human comprehension, machine learning techniques may be used to study them successfully. Consequently, it has recently been demonstrated that these techniques are highly effective in precisely determining the presence or absence of heart-related disorders. The use of information technology in the healthcare profession is growing by the day to help doctors make better decisions. It aids doctors and physicians in disease management, pharmaceutical development, and the identification of patterns and linkages among diagnosis data. Current models for estimating cardiovascular risk are failing to identify many individuals who could benefit from preventative care, while others receive unnecessary treatment. Through the utilization of complex relationships among risk variables, it offers a chance to improve accuracy. We aim to see whether they can aid in the prediction of cardiovascular risk.

■■■■■■■ **Keywords**

Machine learning, Healthcare, Heart disease, ANN, SVM, Naïve bayes, Decision tree, Regression, XG boost, Python

[1]Vaibhavgandhi2424@gmail.com, [2]Dhruvin.83@gmail.com

DOI: 10.1201/9781003596721-70

1. INTRODUCTION

Over 12 million personalities worldwide authorization away from heart illness each day. Heart disease is sometimes referred to as the "silent killer" meanwhile it slayed a person without showing any indications. In high-risk persons, preliminary identification of heart infection is critical for making existence fluctuations. This study explores purposes to envisage forthcoming heart illness by analysing persevering information and applying ML procedures to control whether or not they have heart illness. Heart disease comes in a lot of different forms. Heart disease is most commonly caused by constriction or blockage of the coronary arteries, which are the blood channels supplying the heart with blood. The medical word for this ailment, which develops over time, is coronary artery sickness. It is the leading cause of sentiment attacks. Other types of cardiac abnormalities can occur such as difficulties with the heart's valves or a heart that does not pump efficiently, leading to heart failure. Disease is a condition that some people are born with. Has the patient ever experienced a cardiac ailment or been told that he or she has one? If that's the case, what should you do? Gather information on the diagnosis, including when it was made, who made it, how it was made, and what was done about it.

Heart Attack: To various people, the word "heart attack" signifies different things. The majority of patients with myocardial infarction now spend at least a week in the hospital. Patients who claim to have been treated for only a day or two or who claim to have been discharged from the emergency room did not usually have an acute infarction. More documents are required

Coronary Artery Disease: Another time, specifics about the judgement are required.

Valvular Disease: Typically, a history of mitral prolapse, commonly known as a "leaky valve," is acquired. More data is needed to support this diagnosis because some persons are diagnosed with mitral prolapse based only on their personality, symptoms (usually palpitations or chest pain), and a systolic murmur (frequently an innocent flow murmur).

Heart Murmur: One important piece of historical information is the age at when the muttering was first noticed. It's also important to have a history of any physical limitations the patient had at the time of diagnosis.

Rheumatic Fever: It is important to proceed cautiously while treating this frequently misunderstood illness. The intricacies of the sickness need to be looked into. Many patients claim that when they experienced a fever, sore throat, and heart murmur as children, their parents informed them they had rheumatic fever.

Enlarged Heart: This is frequently a radiologic indication that is very broad. This history is of limited use because "enlarged heart" can signify dilatation, hypertrophy, or faulty x-ray technique.

Heart Failure: Shortness of breath with effort is frequently mistaken as congestive heart failure, especially in the elderly. Because congestive failure is more of a symptom complex (similar to fever) than a definitive diagnosis, this claim has to be investigated further.

Ectopy: Have you informed the patient about any extra or missed heartbeats? Was this based on an irregular pulse (which might be deceiving) or an EGG.

Hypertension: Is there a history of high BP in the patient? What method was used to diagnose this? Were any medications provided to you?

History of Heart Surgery: Throughout a physical examination, search for suitable surgical scars. What procedure was performed if the patient denies cardiac surgery and a thoracotomy scar is found?

History of Other Vascular Disease: This is important because there is a link between peripheral arterial disease and coronary artery disease. These symptoms alone cannot confirm a diagnosis because many individuals complain of "cold feet," "poor wound healing," and leg cramps.

The fundamental objective of the research is:

- To determine the most accurate and reliable ML methodology for predicting heart diseases risks, via rigorous ml classifier models and analysis.
- A comprehensive comparison requires the assessment of various factors, including age, fever, risk level, systolic and diastolic blood pressure along with heart rate.

The objective of our study is to improve the sensitivity and accuracy of risk assessment during heart disease by including several characteristics in our prediction models. This will allow for prompt intervention and individualized treatment.

The second section of the offered article discusses the commendable achievements, successes, and approaches used in similar investigations and research endeavors. In Section 3, the research technique and processes that were employed during the investigation are outlined in detail. Next, Section 4 provides a thorough summary of the work that has been done, including particulars, results from the experiments, and assessment criteria. Finally, Section 5 represents the conclusion of the finished work. Reviewing the study's results and outlining potential directions for more research along the same topic issue are its last sections.

2. Related Work

Because heart disease prediction is a hard endeavor, as the author indicated in their article "Heart Disease Prediction Using ML" (1), to avoid the risks associated with the procedure, it must be automated and the patient must be informed well in advance. Using data mining methods such as.

We have included the many criteria, including a person's age, gender, slope, and others, in the Fig. 70.1 below (1 - Graphical representation of heart disease). That is basically how different algorithms, such as Nave Bays, Decision Trees, Logistic Regression, etc., are implemented.

The suggested research rates a patient's probability of developing heart disease using Naive Bayes, Decision Trees, Logistic Regression, and Random Forest. The timing of a disease's detection determines how accurately an ailment is treated overall. This study aims to identify those who may have heart illness or not.

2.1 Pre-Processing and Data Collection

Out of the four datasets that comprise the Heart Disease Dataset, only the UCI Cleveland dataset was used. The Decision Tree model is shown as a flowchart, with the center node reflecting dataset attributes and the surrounding branches representing the outcome. The basic assumption and most significant aspect in establishing a classification is the independence between the dataset's characteristics. For the first time, the Heart Disease Dataset was utilized to train the Decision Tree algorithm. The dataset's characteristics are independent, which is the most important assumption in generating a classification.

As author conversed through their article [2], heart disease has become one of the ecosphere's major causes of decease. It is required to foresee the opportunity of emerging heart ailment grounded on convinced appearances. The feature selection approaches could be useful techniques for lowering diagnosis costs. Coronary Emotion Disease (CHD) is a category of CVD that disturbs the arteries that provide the bloodstream with oxygen and plasma. CHD is caused by fatty material accumulation in the coronary arteries, which is known as atherosclerosis. When an artery ruptures or narrows, blood flow to the heart muscles is cut off, resulting in a heart attack. Heart catastrophe, also acknowledged by way of congestive sentiment fiasco, is a dangerous ailment which can affect the right, left, or both sides of the heart and cannot repair correctly. Some of the modifiable risk factors include smoking, obesity, a poor diet, high BP, cholesterol, diabetes and stress.

Ayantan Dandapath, M Karthik Raja, and V. V. Ramalingam: "heart disease prediction using machine learning techniques" [3]. This is a permitted paper thing, which means it can be used, distributed, and reproduced in any medium without restriction as long as the original action

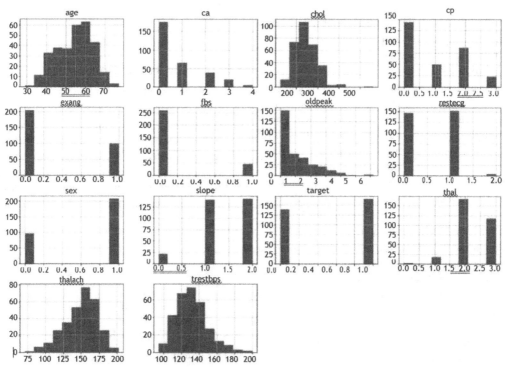

Fig. 70.1 Graphical representation of heart disease

is correctly cited. The user has requested that the down-loaded file be improved. In the human body, the heart is an essential organ. If it stops working properly, the brain and other bodily organs would get locked, and the person will die in a matter of minutes. According to World Health Organization (WHO) research, heart ill- nesses have become India's main cause of death, killing an estimated 1.7 million people in 2016. From 2005 to 2015, India is estimated to have lost up to $237 billion owing to heart and cardiovascular ailments. Medical organizations acquire a lot of data, which may be easily analyzed using ML algorithms. This study calculates the accuracy of ML techniques for forecasting heart sickness, as well as KNN, Decision Tree, Linear Regression, and SVM, using data from the UCI repository (SVM). The Anaconda (Jupiter) notebook is an excellent choice since it includes a number of libraries and header files that make the process more precise and accurate. The article will be presented. ML is a very effective testing method that is based on working out and challenging. It is a subset of AI in which robots are programmed to imitate human abilities [4]. For the main infections that are linked to the heart, cardiac prediction is required for this purpose. ML systems are taught how to process and use data through training. We calculate the accuracy of four distinct ML algorithms in this research and determine which one is the best. To evaluate the algorithms, we used data such as Cholesterol, BP, Age, Sex, and Age. The first section of this paper contains an overview of ML and heart disease.

Random Forest is used to estimate heart illness. The author suggested the Bagging Technique to reduce coronary heart infection misclassification and a Random Forest is used to make predictions. Harshit Jindal and his colleagues in 2021 updated and improved the article, which can be found online. Cardiovascular illnesses are a broad term that encompass a wide range of heart-related issues. This research has supplied us with useful information that can support us in estimating the prognosis of heart disease patients. Because it could predict indicators of heart disease in a single individual, the suggested model's strength was rather amazing [5]. K-Nearest Neighbour, Nave-Bayes, and SVM are just a few of the statistics mining and ML approaches that can help forecast cardiac dis- ease (SVM). If it does not work properly, the intellect and other body parts will stop operating and the individual will perish within minutes. Cardiovascular disease is today's number one killer, yet detecting the problem is extremely challenging [6]. The author can anticipate the occurrence of heart infection using ML procedures such as SVM, KNN, and Random Forest. It is unassuming to foresee the attendance or absenteeism of cardio vascular illnesses with this method. The current system entails a variety of tests, as well as the use of algorithms, questionnaires, and surveys. Data mining and ML techniques are critical in disease prediction. K Nearest Neighbor is an algorithm that is used to detect cardiovascular illness. The collection of measures used to detect cardiovascular disease does not yield good results, resulting in a lack of knowledge which could lead to an erroneous diagnosis.

3. METHODOLOGY

The study's methodology—which covers data collecting, preprocessing, model construction, and assessment methods—is explained in this section. These methods were used to risk factor analysis and prediction.

3.1 KNN Regression

KNN regression is a non-parametric approach for approaching acquaintances amongst self-regulating variables and continuous results by aver- aging data in the same neighborhood in a logical way. The neighborhood size must be determined by the analyst, or cross-validation can be used to identify the size that minimizes the mean-squared error [7,8,9]. The accuracy, macro, and weighted averages for several metrics, such as precision, recall, f2 score, and support, are shown in the below Fig. 70.2 - Accuracy table.

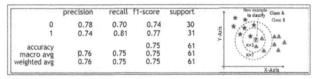

Fig. 70.2 knn accuracy table

3.2 Logistic Regression

This is a numerical investigation process for forecasting a record price grounded on past facts' customary explanations. This model aims to examine the relationship between one or more existing independent variables to predict a dependent data variable [8,9]. The Fig. 70.3 - Accuracy Table for the Logistic Regression, displays the accuracy, macro, and weighted averages for a number of measures, including precision, recall, f2 score, and support.

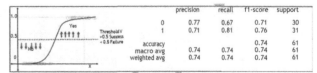

Fig. 70.3 LR accuracy table

3.3 Support Vector Machines

SVM are well-known in classification problems. Less research has been done on the use of SVMs in regression, though. These kinds of models are known as support vector regression approaches (SVR). In this article, I'll discuss the benefits of SVR against alternative regression models, go over the math underlying the technique, and give an example that uses the Boston Housing Price dataset [9]. Support Vector Machines (SVMs) are widely renowned for their prediction capabilities in classification problems. The below Fig. 70.4 - Accuracy table, displays the accuracy, macro, and weighted averages for a number of metrics, including precision, recall, f2 score, and support.

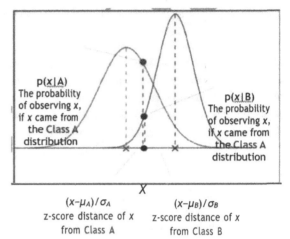

	precision	recall	f1-score	support
0	0.80	0.67	0.73	30
1	0.72	0.84	0.78	31
accuracy			0.75	61
macro avg	0.76	0.75	0.75	61
weighted avg	0.76	0.75	0.75	61

Fig. 70.4   LR accuracy table

3.4 Naive Bayes Classifier

This is a probabilistic ML-based classification model. The formula lies at the sentiment of the classifier. We can analyze the probabilities of A fashionable. if B has already happened by using Bayes' theorem. In this circumstance, the predictors/features are expected to be independent. As a product, it is designated as naïve [7,8,9]. We can determine the likelihood that A will occur if B has already occurred using the equation demonstrated by the Naive Bayes Classifier in above equation 15.5 bayes theorem.

$$P(A|B) = \frac{P(A|B)P(A)}{P(B)} \qquad (1)$$

In Fig. 70.5 and 70.6 where for the Naive Bayes Classifier Accuracy Table shows the accuracy, macro, and weighted averages for a variety of metrics, including precision, recall, f2 score, and support. The likelihood that x will be seen if x is drawn from the Class A distribution. p(x|B) The likelihood that x will be observed if x comes from the Class A distribution.

3.5 Decision Tree

Groups may be able to use this kind of cataloging technique for both numerical and category data. A tree-shaped graph's data is easy to create and analyze. Three nodes form the basis of the model research. All subsequent nodes are constructed upon the root node, which is the primary node [12, 13]. The leaf node displays the results of each test, while the inside node controls a number of properties. Using the most significant components as a guide, this method divides the data into two or more related groups. Each feature's entropy is determined, and the data is divided into predictors that have the lowest entropy and the largest information gain.

	precision	recall	f1-score	support
0	0.79	0.73	0.76	30
1	0.76	0.81	0.78	31
accuracy	0.77	0.83	0.77	61
macro avg	0.77	0.77	0.77	61
Weighted avg	0.77	0.77	0.77	61

Fig. 70.5   Naïve bayes accuracy table

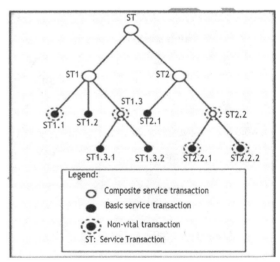

Fig. 70.6   Naive bayes classifier

This approach separates the data into two or more related categories based on the most significant components, as seen in the decision tree representation shown in (Fig. 70.7 - Decision tree classfier). Additionally, the decision tree algorithm's accuracy is displayed in Figs. 70.8 decision tree

Fig. 70.7   Decision tree classfier

	precision	recall	f1-score	support
0	0.68	0.70	0.69	30
1	0.70	0.68	0.69	31
accuracy	0.69	0.69	0.69	61
weighted avg	0.69	0.69	0.69	61

Fig. 70.8   Decision tree accuracy table

accuracy. Once the entropy of each feature is calculated, the data is split up into predictors with the highest information gain and the lowest entropy.

3.6 Random Forest

An algorithm is a formula that is used to solve problems. A supervised arranging style is the random forest approach. In order to create a forest, several trees are used. In this process, every tree emits a class expectation, and the class that receives the most votes becomes the prediction for the model. An increase in tree count improves the accuracy of a random forest classifier. It may be used for both regression and classification, despite its superior performance in classification and ability to overcome missing data. Furthermore, the findings are unexplained because large data sets and more trees are required to create predictions [13,14]. For this Fig. 70.9, Precision, recall, f2 score, and support are among the metrics represented in this random forest Accuracy Table, which also displays the accuracy, macro, and weighted averages for each [7,8,9].

	precision	recall	f1-score	support
0	0.88	0.70	0.78	30
1	0.76	0.90	0.82	31
accuracy			0.80	61
macro avg	0.82	0.80	0.80	61
weighted avg	0.81	0.80	0.80	61

Fig. 70.9 RF accuracy table

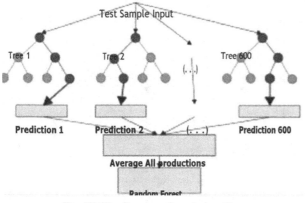

Fig. 70.10 Random forest classifier

3.7 XGBoost Classifier

Models may be utilized as regressors or classifiers with the scikit-learn framework's XGBoost wrapper class. Below in this figure presents the model for classification is called an XGB Classifier. We may construct it and then test it against our training data. To fit models, the scikit-learn API and the model are utilized [15]. We might develop it, test it against our training data, and refine it. As demonstrated in

the accuracy tables in the Fig. 70.11 accuracy figure for XGBoost classifier, the scikit-learn API and the model are used to fit the models.

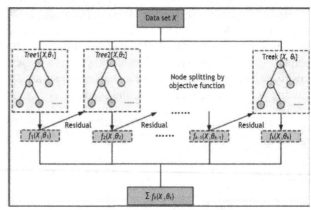

Fig. 70.11 Model XGB classifier

	precision	recall	f1-score	support
0	0.69	0.67	0.68	30
1	0.69	0.71	0.70	31
accuracy			0.69	61
macro avg	0.69	0.69	0.69	61
weighted avg	0.69	0.69	0.69	61

Fig. 70.12 Accuracy table

4. PROPOSED WORK

4.1 Dataset Details: UCI Heart illness

This dataset, which is grounded on the UCI Heart Illness Statistics Customary, has 303 cases. Conferring to UCI, "this database has 76 traits, but all printed investigate mentions to by means of a subdivision of 14 these." People have complete feature withdrawal and we diminished 76 geographies to 14. We used feature extraction to reduce the number of features from 76 to 14, anticipating that having too many features would lead to too much noise. We need to identify a few elements of the creative dataset in order to increase the relevance of the features thus,

UCI: age (in eons):

Sex: sex (male = 1; female = 0) cp: kind of chest pain

The most prevalent kind of angina is angina pectoris. Angina atypical (value 1) Treetop asymptomatic: passive blood pressure Value 2: discomfort that is not angular Value 3: passive blood pressure and asymptomatic treetops chol: milligrams per deciliter of cholesterol serum concentration (fasting blood sugar more than 120 mg/dl) fbs: more than 120 mg/dl (False = 0; true = 1) • The goal is heart disease (0 = no, 1 = yes).

4.2 Prototype Model

This Fig. 70.13 proposed system the primary idea behind this suggested system is to gather the available data set and apply various attribute selection processes to the data. Then, this process data is utilized for the best outcome of the heart decision prediction, and its accuracy measurement is shown.

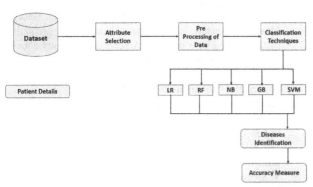

Fig. 70.13 Proposed system

4.3 Experimental Analysis

In this section, the following analysis shows the experimental results, and provide insights about the implications of our analysis for the Heart diseases Risk. Experiment configuration is as follows:

- **Hardware:** AMD Ryzen 7, 1.90 GHz (16 GB RAM)
- **Software:** Google Colab CPU, T4 GPU
- **Libraries:** Matplotlib, Seaborn, Scikit-learn
- **Architectures used:** ML Classifier

Here, Fig. 70.14 above illustrates the system implementation result that represents the programming that is basically implemented in the python programming of the system that we propose. Ultimately, our outcomes surpass those attained by the aforementioned techniques.

5. Discussion and Conclusion

The creation of a method for precisely forecasting cardiac problems is the main objective of the study. Machine learning tools such as ski learn, naïve bayes, and the LR method can be used to predict cardiac disease. The long-term aim of this article is to use new methods and processes to predict cardiac ailments early on. Given the increasing number of fatalities from heart disease, it is critical now more than ever to develop a system that can accurately and effectively predict heart illness. The accuracy ratings of the Naive Bayes, logistic regression, random forest, and decision tree algorithms for cardiac illness prediction are examined in this study. Based on the assumptions of this study, the RF algorithm is the most effective algorithm for

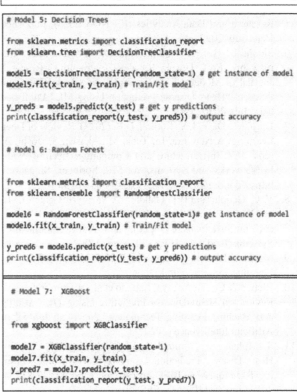

Fig. 70.14 System implementation result

identifying heart disease, with an accuracy score of 95.16 percent.

References

1. Apurb Rajdhan Student CSE, Milan Sai Student CSE, Avi Agarwal Student CSE Dundigalla Ravi Student CSE, Dr. Poonam Ghuli Associate Professor R V College of Engineering Bengaluru, India International Journal of Engineer-

ing Research & Technology (IJERT) "Heart Disease Prediction using ML" Published Date: April 2020

2. N. Satish Chandra Reddy, Song Shue Nee, Lim Zhi Min, Chew Xin Ying International Journal of Innovative Computing ISSN "Classification and Feature Selection Approaches by ML Techniques: Heart Disease Prediction" Published date: May 2019

3. V.V. Ramalingam, Ayantan Dandapath, M Karthik Raja, International Journal of Engineering & Technology · March 2018 "Heart disease prediction using ML techniques, Published Date: March 2018

4. Archana Singh Computer Science and Engineering Madan Mohan Malaviya University of Technology, Rakesh Kumar Computer Science and Engineering Madan Mohan Malaviya University of Technology, 2020 International Conference on Electrical and Electronics Engineering - "Heart Disease Prediction Using ML gorithms" (ICE3-2020)

5. Harshit Jindal, Sarthak Agrawal, Rishabh Khera, Rachna Jain, Preeti Nagrath "Heart disease prediction using ML algorithms", International Conference on Computational Research and Data Analytics (ICCRDA 2020), Published Date: July 2020

6. Siddhesh Iyer, Shivkumar Thevar, Priyamurgan Guruswamy, Prof. Ujwala Ravale - International Research Journal of Modernization in Engineering Technology and Science, "Heart Disease Prediction Using ML" Published date: July 2020.

7. Darji, M., Dave, J.A., Rathod, D.B. (2023). Review of Deep Learning: A New Era. In: Tuba, M., Akashe, S., Joshi, A. (eds) ICT Infrastructure and Computing. Lecture Notes in Networks and Systems, vol 520. Springer, Singapore. https://doi.org/10.1007/978-981-19-5331-6_33

8. V. C. Gandhi and P. P. Gandhi, "A Survey-Insights of ML and DL in Health Domain," in In2022 International Conference on Sustainable Computing and Data Communication Systems (ICSCDS), IEEE, 2022, pp. 239–246.

9. Machine Learning, Tom M. Mitchell Building Machine Learning Systems with Python,

10. Richert & Coelho Yahaya, Lamido et al. "A Comprehensive Review on Heart Disease Prediction Using Data Mining and Machine Learning Techniques." American Journal of Artificial Intelligence (2020): n. pag.

11. S. Mohan, C. Thirumalai and G. Srivastava, "Effective Heart Disease Prediction Using Hybrid Machine Learning Techniques," in IEEE Access, vol. 7, pp. 81542-81554, 2019, doi: 10.1109/ACCESS.2019.2923707.

12. V. Sharma, S. Yadav and M. Gupta, "Heart Disease Prediction using Machine Learning Techniques," 2020 2nd International Conference on Advances in Computing, Communication Control and Networking (ICACCCN), Greater Noida, India, 2020, pp. 177-181.

13. Rani, P., Kumar, R., Ahmed, N.M.O.S. et al. A decision support system for heart disease prediction based upon machine learning. J Reliable Intell Environ 7, 263–275 (2021). https://doi.org/10.1007/s40860-021-00133-6

14. Ekta Maini, Bondu Venkateswarlu, Baljeet Maini, Dheeraj Marwaha, Machine learning–based heart disease prediction system for Indian population: An exploratory study done in South India, Medical Journal Armed Forces India, Volume 77, Issue 3, 2021, Pages 302-311, ISSN 0377-1237, https://doi.org/10.1016/j.mjafi.2020.10.013.

15. Md Manjurul Ahsan, Zahed Siddique, Machine learning-based heart disease diagnosis: A systematic literature review, Artificial Intelligence in Medicine, Volume 128, 2022, 102289, ISSN 0933-3657, https://doi.org/10.1016/j.artmed.2022.102289.

16. V. Sharma, S. Yadav and M. Gupta, "Heart Disease Prediction using Machine Learning Techniques," 2020 2nd International Conference on Advances in Computing, Communication Control and Networking (ICACCCN), Greater Noida, India, 2020, pp. 177-181.

17. M. Ali, I. S. Amiri, K. Ahmed, F. M. Bui, J. M. W. Quinn, and M. A. Moni, "Heart disease prediction using supervised machine learning algorithms: Performance analysis and comparison," Computers in biology and medicine, Sep. 01, 2021. https://www.sciencedirect.com/science/article/abs/pii/S0010482521004662

18. D. E. Salhi, A. Tari, and T. Kechadi, "Using Machine Learning for Heart Disease Prediction," *Lecture notes in networks and systems*, Jan. 01, 2021. https://link.springer.com/chapter/10.1007/978-3-030-69418-0_7

19. T. Parashar, K. Joshi, R. R. N, D. Verma, N. Kumar and K. S. Krishna, "Skin Disease Detection using Deep Learning," *2022 11th International Conference on System Modeling & Advancement in Research Trends (SMART)*, Moradabad, India, 2022, pp. 1380-1384, doi: 10.1109/SMART55829.2022.10047465.

Note: All the figures in this chapter were made by the author.

Recent Trends in Engineering and Science for Resource Optimization and Sustainable Development – Prof. (Dr.) Dorota Jelonek et al. (eds)
© 2024 Taylor & Francis Group, London, ISBN 978-1-032-98030-0

71

Scheduling of Cyber Security Activities in Indian Context

Rekha Sharma*

Research Scholar, Dept. of CS and Informatics,
University of Kota, Kota, Rajasthan

Reena Dadhich

Professor, Dept. of CS and Informatics,
University of Kota, Kota, Rajasthan

Abstract

This paper aims to study about cyber security issues in India. It has been shown with facts that number of cybercrimes is increasing exponentially in India. It has seen that there has been hike in number of cybercrimes during and after pandemic. This paper also discussed about cyber attacks and its different types. It also discussed about the status of cyber security in India and measures taken by authorities for cyber security. This paper also considered about initiatives taken by Government of India to prevent their citizens. It has been discussed that what security measures can be done to improve cyber security war in India. This paper suggests the ways through which cooperation and collaboration of Government with the PPP (Public Private Partnership) can reduce the chances of cybercrimes in India. With the support of law enforcement cybercriminals can be detected and punished.

Keywords

Cyber security, Cyber attack, Cyber war, IT Act, Cybercrime

1. INTRODUCTION

Cyber security is a major issue in today's world specially in developing countries like India. From past few years it has been increasing so rapidly. Cyber security is the process to prevent computers, servers, mobile devices, electronic system and network from malicious Cyber- attack. The main aim of these cyber attacks are accessing, changing or destroying sensitive information. These incidents include SQL injection, phishing, website intrusion, defacement, virus and Denial of service (DoS) attack. The Global users of Internet have been increasing ever since it is adopted. The latest statistics shows that the increase in number of Internet users is growing exponentially. In 2022[28], the number of internet users are 4.95 billion which is 6.2 percent increase from 4.6 billion in previous years. India has delivered largest absolute growth. According to the Economics Time 2019, India has reported quarterly increase of 32 million Internet subscribers which is 5 percent increase and annual growth of 143 million subscribers which is

*Corresponding author: rekhasharma.csscholar@uok.ac.in

DOI: 10.1201/9781003596721-71

increase of 29 percent. The dependency of users and organizations has been increased on cyberspace which has resulted in increased cybercrime. Due to lack of awareness in Indian society it can be seen a bounce in number of cybercrimes. Law enforcement officers are not able to deal cybercrime due to not having proper technical training and other requisite expertise. India can succeed in combating the problem of cyber-crime by adopting working together of technological measures [22] and trained human resources. The government of India has taken some cyber security steps but still more expansive and aggressive measures are [24] required to meet the rising challenges. It may applicable to a variety of context like mobile computing and business application. In this paper various types of Cyber-attacks are discussed. Cyber security steps taken by Indian government are discussed along with improvement needed. Cyber-security awareness tips and tricks can be useful for preventing Internet users from Cybercrimes.

2. STATUS OF CYBERCRIMES IN INDIA

2.1 Types of Cyber Threats

Cybercrime

It include single actors or groups targeting systems [1] for financial gain or to cause disruption. Any unlawful act or behaviors that occur through the medium of computer system, mobile and Internet is described as Cybercrime. A variety of illegal acts are performed by the technical expert over the Internet. Some of the newly emerging cybercrimes are cyber-stalking, cyberterrorism, email- spoofing, e-mail bombing, cyber pornography, cyber defamation.

Cyber-attack

It involves politically gathering of information for unlawful activities. It has been found that more than 50 percent [8] of the organizations evidently affected by major five cyber threats such as Denial of service (DOS), phishing, malware, spear phishing and ransomware. Phishing: It is meant to stealing user's private information such as credit card numbers, bank details, social accounts credentials and PINs of debit cards etc. Type of cybercrime where hackers try to obtain key personal data such as social security no, Aadhar details, credit card or other related to impersonate someone and gain benefit with his/her nameand release sensitive data until a sum of money is paid to attackers. Maze is a common type of ransom ware attack. Identity theft happens when hacker try to obtain someone's personal datasuch as [11] social security no, Aadhar details, bank details etc.

Cyberterrorism

It is intended to undetermined electronic systems to cause panic or fear. The word [7] Terrorism implies illegal utilization of power, or hatred against people in order to threaten their administration or its resident. It can be performed through cyber space. It is very important issue of present society.

2.2 Role of Social Media in Cyber Security

Nowadays social media has become everyone's lifestyle. We use it to stay in contact with our family and friends [7]. We can share music, videos, pictures as well as important events of ourlives. Social media can also be used to spread incorrect information. Intruder can also utilize social media for cyber- crime.

2.3 Data Showing Hike in Number of Cybercrime

In India Number of cybercrimes [23] has been increased drastically as reported in 2021 compared to the previous years. Karnataka and Uttar Pradesh have highest share. It was estimated that after 2017, India collectively lost 18 billion U.S Dollars due to Cybercrimes. India has been victim of critical phishing attack worth 171 million U. S dollar on Union Bank of India in 2016. Covid-test lab results of thousands of Indian citizens were also leaked from Government websites in 2021. The main reason for hike in number of cybercrimes after 2017 (Fig. 71.1) is the launch of recent government initiatives such as a dedicated Cybercrime online portal for reporting cybercrimes online. Before 2017 no such effective measures were taken torecord number of cases. After the launch of Cybercrime Reporting Portal by Ministry of Homeaffairs Cybercrime can be registered online due to which number of cases has been increased.

2.4 Effect of COVID-19

Cybercriminals have used pandemics for their phishing campaign. During pandemic everyone was at their home to prevent COVID spread. To access their basic needs everyone was dependent on Internet get their essentials online. This was best utilized by cyber criminals to forge [9] global users. It has been reported that possibility of occurrence of ransomware has increased after Pandemics. Earlier in 2019, casualty to a ransomware bout occur every 14 seconds. It has been decreased to 11 seconds in 2021. The total cost of ransomware will increase above $20 billion worldwide.

2.5 Case Study

E-commerce sites are providing goods and services through online mode. It has also provided merchants to collaborate with such E-commerce sites to create virtual online shops. It provides facilities to select and order product online. Advertisement of such online shops are displayed at popular social networking websites. Such advertisement attracts

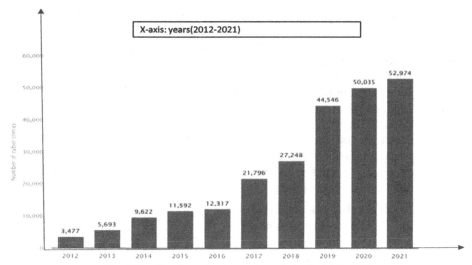

X-axis: years(2012-2021)

Fig. 71.1 Number of cyber crimes in india 2012-2021

Source: Author

users to perform shopping with unauthentic shops. These advertisements are sometime fake. One such case study has been discussed here. Once a user clicks on the advertisement shown on their Social Networking profile. It gets redirected to online shop where products at their shop are at unbelievable lower prices. User ordered one product and asked for Cash on delivery but was not allowed. User paid online for the product but no confirmation mail has been received. It was a fake online store which had received payments for not delivered product. Thus, user become victim here. Therefore, Cyber security should be implemented to prevent users from such cyberattack.

3. BENEFITS OF USING CYBER SECURITY IN INDIA

India is playing an important role in IT field therefore it has to play a significant role in implementing cyber security. In June 2021, the United Nations ITU Global Cybersecurity Agenda has placed India at 10th rank with a score of 97.5 in Global Cyber security Index (GCI)2020. India has jumped 37 positions according to the Cyber Security Ranking of Countries from its previous rank of 47. In India most of the IT industries has advanced cyber security protection and expert teams to handle security breaches, leaks and other attacks [33]. Good cyber security implementations can help protecting citizens from cybercrimes and keep safe their personal information.

In India a large number of banks are victim to cyber-attack [2]. The major cyber-crime in Indianbanking system is due to Phishing, Identity Theft and malware attack. It not only leaks valuableand sensitive information but also [20] causes heavy financial losses. Proper setting and maintenance of firewall can prevent the banking system from [19] cyber-

attack. Banks should take different security measures to prevent from such cyber-attacks. Penetration test is carried out to test security of the network and infrastructure of bank. It can be done to find out vulnerability in the system and identify security breaches. Secure Socket Layer is used to prevent Backend applications in Banks. Network must be protected through firewall setting and multiple layered protection framework should be proposed. Password too should be strong and be changed at a fixed interval of time. Passwords should be managed and stored in proper locked and encrypted way.

4. CYBER SECURITY INITIATIVES IN INDIA

In India cyber security concern has increased and it has already started working towards implementing improved cyber security policy by 2022. Various security [7] measures had beentaken to improve cyber security in India. It can be divided into further categories like policies,tools, legal framework etc;

1. Policy - National Cyber Security policy, 2013
2. Cyber law-

4.1 National Cyber Security Policy-2013

The Government of India took the first think up step towards cyber security in 2013. Ministry of Communication and Information Technology, Department of Electronics and Information Technology [2] created National Cyber Security Policy, 2013(NCSP-2013). The Policy is focused to build a secure and robust cyberspace for Indian citizens, businesses and the Government Infrastructure. Its main objective is to protect cyberspace information and infrastructure, build capabilities to prevent and respond to cyber-attacks, and minimize damages through coordinated efforts of institutional structures, people and technology. The objectives of the policy include creating a secure cyber ecosystem, compliance with global security standards, strengthen the regulatory framework, creating round the clock mechanisms for gathering intelligence and effective response. National Critical Information Infrastructure Protection Centre is built up [4] for 24×7 protection of critical information infrastructure, research and development for security technologies. It creates a 500,000 strong cyber security workforce, to provide fiscal benefits to businesses for adopting cyber security practices, to build public private partnerships for cooperative cyber security efforts.

The Prime Minister's Office created the National Cyber Security Coordinator in 2014 [13]. Ministry of Electronics and Information Technology issued several instructions in response to intrusions by infamous hacker group [14] 'Legion'. It includes National Payment Corporation of India (NPCI) to audit the financial sector, review and strengthening of the IT act. Agencies that deal with cyber security in India [3] are National Technical Research Organization, National Intelligence Grid, and the National Information Board. India's first chief information security officer (CISO) was appointed in 2016 with the aim of Improving cyber security and subsequently asked to appoint Central Information Security Officers to all the ministries.

Cyber Swachhta Kendra (Botnet Cleaning and Malware Analysis Centre): -To battle violation of cyber security. In 2017 Government of India's Computer Emergency Response Team (CERT-in) launched 'Cyber Swachhta Kendra' (Botnet Cleaning and Malware Analysis Centre) a new desktop and mobile security solution for cyber security. Ministry of Electronics and Information Technology's Digital India initiative, will détect botnet infections in India and prevent further infections by notifying, enable cleaning and securing systems of end-users. It is main function is to analyze BOTs/malware characteristics, provides information and enables citizens to remove BOTs/malware and to create awareness among citizens to secure their data,computers, mobile devices. National Cyber Security Policy in India steps in direction of securecyber ecosystem in India through the Cyber Swachhta Kendra.

The Cyber Swachhta Kendra operates with close coordination and collaboration with Internet service Provider and product/antivirus companies. It will notify to the end user regarding infection and provide them assistance to clean their system. It also provides assistance to the industry and academia to detect bot infected systems. The Centre aims to increase awareness regarding botnet, malware infection among common users. It also provides assistance for measures to be taken to prevent malware infections and secure their devices [15]. Definition of computer system and punishment of cyber offences.

4.2 Security and Protective Tools offered by Government of India

- **USB Pratirodh** – It is used to focus at controlling the unauthorized usage of removable USB storage media devices like pen drives [16], external hard drives and USB supported mass storage devices.

- **Samvid** – It is a desktop-based Application Whitelisting solution for Windows operating system. It allows only preapproved set of executable files for execution and protects desktops from malicious applications from executing.

- **M-Kavach** – is a device for security of Android mobile devices has also been developed. It provides protection against issues related to malware that steal personal data & credentials, misuse Wi-Fi and Bluetooth resources, lost or stolen mobile device, spamSMSs, premium- rate SMS and unwanted / unsolicited [1] incoming calls.

- **JSGuard** – is a tool which serves as a browser extension which detects and defends malicious HTML & Java Script attacks made through the web browser based on Heuristics. It alerts the user when he visits malicious web pages and provides a detailed analysis threat report of the web page.

Digital India Campaign: It is a flagship program started in 1st July 2015 by Government of India. It is originated by national e-Governance Project [18]. The most significant aspect of this program is information security and implementation of IT policy which support e-Governance.

National Cybercrime Reporting Portal: Ministry of Home Affairs launched the National cybercrime reporting portal [27] on 30th Aug 2019. It is a centralized mechanism for the citizens where anyone can report online all the types of cybercrime incidents. It has a special focus on Cybercrimes against women & children. According to the data 317439 cybercrime incidents and 5771 FIRs have been registered till 28, Feb 2021.

5. Proposed Recommendation to Combat with Cyberwars

India has proposed few security measures to prevent cyberattacks [25] still it needed to work more. In this research paper some ideas have been discussed which can deal with the current trends of Cybercrime.

5.1 Special Wing Development

There is a requirement to create [5] a Directorate or special wing in NSCS (National Security Council Secretariat) and National Security Advisor (NSA) in India. Some objectives of this council will be raising of Cyber commands, to development of Human Resource by providing proper training to young IT professionals to [12] deal with cybercrime. It can implement Public Private Partnership with the IT industries in India. International Cyber law is today's demand for the world. There is a need for coordination, planning, understanding and synergy of efforts amongst all civil, military, intelligence, law enforcement and educational organizations responsible for cyber security. Research and development can also be used to create tools which ensure cyber security and privacy Functioning and Software design of social networks to ensure 'security and privacy', and emphasis on 'malware detection'.

5.2 Collaboration with Industry Partners

Public Private Partnerships is an important strategy under the National Cyber Security Policy (NCS). Antivirus company Quick Heal is providing a free bot removal Tool under the Cyber Swachhta Kendra Initiative. Cisco and Ministry of Electronics and Information Technology's Indian Computer Emergency Response Team (CERT-In) have signed a Memorandum of Understanding (MoU) under which they will develop threat intelligence-sharing programme. Professionals from Cisco and CERT- In will work collectively to prevent digital threats and develop and encompass new ways to improve cybersecurity. Indian Computer Emergency Response Team (CERT-In) was launched in 2004 by Ministry of Electronics and Information Technology, Government of India. It is a national nodal agency for responding Computer security incidents.

5.3 International Cooperation Initiatives

The Government of India has tied up cyber security collaborations with countries such as the USA, European Union and Malaysia [21]. The U.K. has agreed to assist in developing the proposed National Cyber Crime Coordination Centre in India. The shared principles of the U.S.-India Cyber Relationship Framework [17] provide for the recognition of the leading rolefor governments in cyber security matters relating to national security. The areas of corporation provide among other things that both countries agree to share and implement cybersecurity best practices, share cyber threat information [26] on a real- time basis, develop joint mechanisms to mitigate cyberthreats, promote cooperation between law enforcement agencies and improve their capacity through joint training programs, encourage participation in the field of cybersecurity research, and strengthening critical Internet infrastructure in India [6].

6. CONCLUSION

On the basis of the conclusion and findings, India has implemented so many cyber security measures but still, it needs some more improvements to deal with latest trends. Government of India also took some legal framework to prevent Indian citizens from such Cybercrimes. Now it is the time to accept some amendments in such legal framework. It can be done by international cooperation and collaboration with Industries. It can also be prevented by Cyberawareness so that both technical and non-technical users can prevent themselves from cybercrimes. In future some tools are developed to prevent cybercrime with collaboration with Government and IT industries in India.

References

1. Saroj.Mehta, Vikram.Singh, "A STUDY OF AWARENESS ABOUT CYBERLAWS IN THE INDIAN SOCIETY ", *International Journal of Computing and Business Research (IJCBR)*,Volume 4 Issue 1, page no 9, January 2013.
2. Suman Acharya, Sujata Joshi," Impact of cyber-attacks on Banking Institutions in India: Astudy of safety mechanisms and preventive measures", *Palarch's Journal of Archaeology of Egypt/Egyptology* PJAEE, 17 (6) (2020).
3. S. Goel, "Cyber-Crime: A Growing Threat to Indian Banking Sector," 3rd International Conference on Recent Innovative in Science, Technology, Management and Environment,New Delhi,2016, [Online]. Available: http://data.conferenceworld.in/IFUNA18DEC16/P13-20.pdf.
4. International Institute of Strategic Studies (June, 28, 2021)., "Cyber Capabilities andNational Power: A Net Assessment [online]", Available at: https://www.iiss.org/blogs/research-paper/2021/06/cyber-capabilities-national-power (Accessed on July 10, 2021).
5. Lieutenant.General.Nitin.Kumar.Kohli,AVSM,VSM. "Challenges and prospects of Cyber Security in the Indian context", *Journal of the United Service Institution of India*, vol CXLN No 600, April, 2015.
6. Sunil. Chaudhary, vasileios. Gkioulos, Sokratis. Katsikas, "Developing metrics to assess the effectiveness of cybersecurity awareness program", *Journal of CyberSecurity*, 2022, 1–19.
7. Rohit. Kalakuntla, Anvesh. Babu, Ranjith. Reddy," Cyber Security", *Holistica*, vol 10, Issue 2, 2019, pp 115–128.
8. Jatin.Patil," Cyber Laws in India: An overview", *Indian Journal of Law and LegalResearch*, vol.IV, Issue 1, 23 Feb 2022.
9. Mrs.Ashwini. Sheth, Mr. Sachine.Bhosale, Mr. Farish. Kurupar, "Research paper on cybersecurity", *Contemporary Research in India*, April.2021.
10. M.M.Chaturvedi, Mp.Gupta, Jaijit.Bhattacharya. (2009, January). "Cyber security Infrastructure in India: A Study", *Research Gate* Available:- (PDF) Cyber Security Infrastructure in India: A Study (researchgate.net).
11. David.Churchill, "Security and Visions of the criminality and social change in Victorian and Edwardian", *The British Journal of criminology* .vol 56, 857–876, August 2015.
12. Kaspersky (2022*). What is Cyber-security? [online]*. Available: What is Cyber Security? l Definition, Types, and User Protection (kaspersky.co.in)

13. J.Satanarayana, *National Cyber Security Policy 2013(NCSP-2013)*",Ministry of Communication and Information Technology, July 02 2013.Available:National Cyber Security Policy (1).pdf (meity.gov.in).

14. Surabhi.Agarwal.(2016, Dec. 13). *IT Minister orders measures to strengthen India's cybersecurity, The Economics Times [online].* Available: https://economictimes. indiatimes.com/tech/internet/it-minister-orders-measures-to-strengthen-indias-cyber- security/articleshow/55963728. cm

15. *Cyber Swachhta Kendra, Botnet cleaning and malware analysis Centre[online].*Available: http://www.cyberswachhtakendra.gov.in/about.html.

16. Ridhima.Kedia. (2017, Aug. 17). Cyber Security Initiatives by Government of India[online]. Available: Cyber Security Initiatives by the Government of India - iPleaders.

17. "Framework for the U.S.-India Cyber Relationship", The White House, Office of the Press Secretary, U.S.

18. Rashmi. Anand et al." Transforming information security governance in India: (A SAP- LAP based case study of security, IT policy and e-governance), *Information and computersecurity,* vol 26 no.1, pp. 58–90, March.2018.

19. Jin. Manghui. Tu, Tae. Hoon. Kim et al. "Game based Cyber security training for High school students", *Proceedings of the 49th ACM Technical Symposium on Computer Science Education,*Feb 2018,Pages 68–73.

20. Keith S.Jones, Akbar.Siami.Namin et al." The Core Cyber-Defence Knowledge, Skills, and Abilities That Cybersecurity Students Should Learn in School: Results from Interviews with Cybersecurity Professionals", *ACM Transactions on Computing Education,* vol-18, Issue 3, September 2018, pp 1–12.

21. Allen.Parrish, John.Impagliazzo, "Global perspectives on cybersecurity education", *Proceedings of the 23rd Annual ACM Conference on Innovation and Technology in Computer Science Education,* July. 2018.

22. H. Arora, T. Manglani, G. Bakshi and S. Choudhary, "Cyber Security Challenges and Trends on Recent Technologies," 2022 6th International Conference on Computing Methodologies and Communication (ICCMC), 2022, pp. 115–118.

23. Tanushree.Basuroy, "Number of Cyber crimes reported in India 2012-2021",October. 2022 Available: https://www. statista.com/statistics/309435/india-cyber-crime-it-act/

24. Mallika, V. Deep and P. Sharma, "Analysis and Impact of Cyber Security Threats in Indiausing Mazarbot Case Study," 2018 International Conference on Computational Techniques, Electronics and Mechanical Systems (CTEMS), 2018, pp. 499–503.

25. Rohit. Chivukula, T. Jaya Lakshmi, L. Ranganadha Reddy Kandula and K. Alla, "A Study of Cyber Security Issues and Challenges," *2021 IEEE Bombay Section Signature Conference (IBSSC),* 2021, pp. 1–5.

26. Shivansh. Sharma and Mukul. Khadke, "Network Security: A Major Challenge in India," *2018 4th International Conference on Computing Communication and Automation (ICCCA),* 2018, pp. 1–5.

27. Shri. G. Kishan. Reddy (2021, Mar, 09), "National Cyber Crime Reporting Portal", Available: https://pib.gov.in/Pressreleaseshare.aspx?PRID=1703509#:~:text=Ministry%20 of%20Home%20Affairs%20operationalized,crimes%20 against%20women%20and%20c hildren.

28. "Number of Internet Users worldwide(2012-2022)", Available: https://www.oberlo.in/statistics/how-many-people-use-internet

Recent Trends in Engineering and Science for Resource Optimization and
Sustainable Development – Prof. (Dr.) Dorota Jelonek et al. (eds)
© 2024 Taylor & Francis Group, London, ISBN 978-1-032-98030-0

72

Role of Human Resource Management In Improvement of Employee Performance

Shaik Rehana Banu[1]

Post Doctoral Fellowship, Department of Business Management,
Lincoln University, College Malaysia

VVYR. Thulasi

Associate Professor, School of Management Sciences,
Nalla Narasimha Reddy Education Society's Group of Institutions,
Hyderabad

Rita Biswas

LLB, MBA (HR), Ph.D (Management), Senior Facilitator,
Regenesys Business School

Shailendra Kumar Rai[2]

Research Scholar, School of Commerce and Management,
IIMT University Meerut, Uttar Pradesh

Priyanka Rana[3]

Associate Professor, School of Commerce and Management,
IIMT University Meerut, Uttar Pradesh

P. Sasikala[4]

Associate Professor and Head,
Costume Design and Fashion, Kongunadu Arts and Science College
(Affiliated to Bharathiar University)

◼◼◼◼ **Abstract**

Human resource management (HRM) plays a pivotal role in enhancing employee performance within organizations. This abstract delves into the multifaceted responsibilities of HRM and its significant contributions to fostering a conducive work environment that cultivates high levels of employee engagement, motivation, and productivity.Firstly, HRM serves as the strategic architect in aligning organizational goals with the development of human capital. Through effective recruitment and selection processes, HRM ensures the acquisition of skilled individuals whose competencies match the requirements of the job roles, thereby laying the foundation for improved employee performance. Moreover, HRM facilitates the continuous development of employees through various training and development initiatives. By offering opportunities for skill enhancement, career advancement, and personal growth, HRM not only equips employees with the necessary tools to excel in their roles but also instills a sense of loyalty and commitment towards the organization.Additionally, HRM plays a pivot-

Corresponding authors: [1]drshaikrehhanabanu@gmail.com, [2]rais6316@gmail.com, [3]ranaphd81@gmail.com, [4]sasidrbharathi@gmail.com

DOI: 10.1201/9781003596721-72

al role in fostering a positive work culture and nurturing employee relationships. Through initiatives such as performance feedback mechanisms, recognition programs, and employee assistance programs, HRM fosters a sense of belongingness and wellbeing among employees, thereby enhancing their morale and job satisfaction, which are essential determinants of performance. Furthermore, HRM acts as a mediator in resolving conflicts and addressing grievances, thereby mitigating potential barriers to performance such as interpersonal conflicts or unfair treatment. By promoting fairness, transparency, and equity in organizational practices, HRM ensures a conducive work environment where employees feel valued and respected, thus motivating them to perform at their best. In conclusion, the role of HRM in improving employee performance is multifaceted and indispensable. By strategically managing human capital, fostering a culture of learning and development, nurturing employee relationships, and promoting fairness and transparency, HRM lays the groundwork for creating high-performing organizations that thrive in today's competitive landscape.

▣▣▣▣▣ Keywords

Human resource management, Employee performance

1. INTRODUCTION

Any company needs human resource management (HRM), but multinational corporations (MNCs) particularly need it as they must oversee a diversified workforce that comes from a variety of cultural backgrounds and locations Indarapu et al., (2023). HRM promotes cross-cultural understanding and helps establish a climate at work where other cultures, viewpoints, and ideas may be welcomed and encouraged by hiring the right individuals. In MNCs, effective HRM contributes to the protection and well-being of employees. HRM makes sure that workers receive fair treatment, that their workplace is secure, and that they are abiding by labour laws in all the nations in which they conduct business.

For multinational corporations (MNCs) to get a competitive edge and guarantee a secure and efficient work environment, human resource management is essential. For the company to maximise the resources at its disposal, HRM helps to ensure that the proper people—employees and suppliers—are placed in the correct positions with the necessary abilities. Today's task management is far more complicated and wide-ranging due to the rise of multinational corporations with several overseas branches Chowdhury et al., (2023). By enabling MNCs to monitor and assess international project progress in real-time, these cloud-based solutions also assist to avoid expensive delays and unanticipated disputes more effectively. Furthermore, MNCs may more easily scale their task management procedures without adding more staff when they use cloud-based solutions. AI-driven solutions, in contrast to manual human monitoring, may quickly take preventative action in anticipation of future problems before they materialise. Distributed resource planning, cloud computing, and clever AI technologies enable multinational corporations to overcome obstacles and stay one step ahead of their competitors Dutta & Kannan Poyil, (2023).

There is great potential for using ML in human resource management to provide solutions that go beyond traditional methods. The potential of ML approaches to improve personnel management and increase organisational effectiveness is examined in this article Boudreau & Marler (2017). HR professionals can make data-driven choices, optimise workflows, and cultivate a more involved and productive staff by utilising ML, which makes use of predictive analytics, natural language processing, and other sophisticated algorithms.

Talent acquisition is one of the most important components of HRM, and ML puts predictive analytics at the centre of this procedure. Large datasets may be analysed using predictive models to find trends and patterns, which can assist businesses in hiring the best candidates. ML algorithms are able to estimate future talent demands, correlate individuals with job criteria, and predict individual success by analysing recruiting data from the past. This lowers turnover rates by speeding up the hiring process and guaranteeing a more precise and effective applicant selection. According to Fallucchi et al. (2020), ML may be a critical tool for forecasting staff retention and measuring employee engagement. ML algorithms are able to determine the variables that impact work satisfaction and possible attrition risks by evaluating employee data, including productivity metrics, feedback, and sentiment analysis from communication channels. By taking a proactive stance, HR professionals may better involve employees and lower turnover by implementing focused solutions like individualised growth plans or employee well-being programmes.

When it comes to measuring employee engagement and forecasting retention, ML may be quite important. ML algorithms are able to determine the variables that impact work satisfaction and possible attrition risks by evaluating employee data, including performance metrics, feedback, and sentiment analysis from communication channels.

With this proactive strategy, HR managers may better retain employees and lower turnover by implementing focused solutions like employee well-being programmes or personalised development plansMeddeb (2021). By taking into account a wide range of elements outside of standard measurements, ML helps to provide a more nuanced picture of employee performance. These might include a person's preferred method of learning, career path, and aptitude for a given task. Through the utilisation of this data, organisations may create customised development programmes that address the distinct requirements of every staff member, promoting ongoing progress and skill improvement.

Complex datasets may be expertly analysed by ML algorithms to yield insights that guide strategic personnel planning. These models help HR managers make well-informed decisions on organisational restructuring, training initiatives, and resource allocation by taking into account factors including market trends, skill shortages, and demographic changes. In a corporate climate that is evolving quickly, organisations may remain ahead of the curve by taking a proactive strategy. In summary, a new era in employee management is being ushered in by the use of ML techniques into human resource management. Each of these applications will be examined in more detail in this study, along with their advantages, difficulties, and possible effects on organisational results. HR managers can drive strategic decision-making, obtain actionable insights, and establish a workplace that is both efficient and flexible enough to meet the changing needs of the modern corporate environment by leveraging the potential of ML.

2. LITERATURE REVIEW

Based on recent research in both domains, the goal of this study is to begin the process of creating a theoretical framework that explains HRM's function in effective CI. To achieve this, components of the CI Maturity Model as well as a framework that illustrates the function of HRM in innovation form the basis for investigating the ways in which sets of HRM practices employed at various stages of the CI implementation process can promote organisational growth and improve operational performance Jorgensen et al., (2008). This research mostly contributes to theory, but the structure has application as well because it highlights key connections amongst HRM practices and behaviours required for effective CI. This study concludes with a summary of a preliminary test of the structure in an empirical environment, along with various directions for further study.

The purpose of this study was to investigate how HRM practices affect worker performance at the Malaysian Skills Institute (MSI). It looked at the variables that affect hiring, selection, and pay in relation to worker performance at MSI Al-Qudah et al. (2014). There were forty responders in the research, which included MSI workers. In order to fulfil the goals of the investigation, the investigator created and disseminated a survey, and gathered and examined the information utilising SPSS. Based on the correlation analysis and descriptive statistics, an overall analysis was carried out. The findings showed a substantial correlation between hiring, selection, and pay practices and MSI employee performance. The report included suggestions for enhancing MSI's hiring, selection, and pay practices.

Studies confirm that without placing a high priority on human resource development and management (HRM), a high-performance organisation (HPO) cannot exist (HRD). Nevertheless, there has not always been a good fit between HRM and HRD. Richman (2015) says. The progression of HRD from its foundation in human knowledge transference to HRM and current HRD operations shows that organisations today need to concentrate employee development on flexibility, creativity, and competence due to increased environmental, social, and political pressures. The review that follows demonstrates how crucial HRM and HRD are to organisational leadership. Additionally, the literature reviewed establishes a strong correlation between the qualities of an HPO and the competencies imparted via efficient HRM and HRD cooperation.

Researchers and managers alike are becoming more and more interested in how businesses might achieve ambidextrous learning, or concurrently explore new knowledge domains and take use of existing ones. By presenting and evaluating how high-involvement HR strategies impact the social environment that influences the firm's ambidextrous learning and subsequent performance, we want to introduce HRM into this arena in this research (Prieto & Pilar Pérez Santana, 2012). High-involvement HR practices are favourably correlated with the social atmosphere, which in turn promotes ambidextrous learning and enhanced performance, according to field research of 198 Spanish organisations.

Several strategies, tactics, or instruments from the fields of quality management and human resources management may be used to boost output to enhance output quality, volume of manufactured goods produced, and production capacity Blaga, (2020). To increase production capacity, the volume of products produced, and ultimately their quality, the paper will discuss the use of quality tools in human resources management in the automotive industry. Specifically, it will discuss how staff motivation is impacted by the use and application of quality tools in the field of manufacturing electrical and electronic equipment for motor vehicles. In addition to highlighting the Quality Management systems employed by the organisation, the article also discusses a number of issues pertaining to human motivation and active participation in continuous improvement procedures aimed at boosting production efficiency.

3. RESEARCH METHODOLOGY

The advantages of ML for human resource management (HRM) are discussed in this study, with a focus on how it might reveal previously undiscovered insights from large datasets. This option is the subject of the investigation. The application of sophisticated analytics may yield important recommendations for workforce planning, resource allocation, and the creation of a culture that supports lifelong learning. Furthermore, by automating repetitive HR tasks using machine learning, HR professionals can focus on important initiatives that need the application of human expertise. The integration of ML into HRM procedures has a transformative prospect to enhance several aspects of HRM. Although there are many opportunities to improve decision-making, encourage diversity, and streamline processes, there are also concerns about bias, privacy, and ethics that must be properly addressed. Organisations which recognise and effectively manage these challenges will have a considerable competitive edge in the quickly changing field of modern HRM.

It involves converting unprocessed data into a format that ML algorithms can use. This includes scaling numerical qualities, eliminating outliers, converting parameters, and cleaning and organising data. For any ML task, it is the process of choosing the optimal subset of features from a feature pool. It aims to pinpoint the key elements that can optimise a predictive model's efficiency. It not only makes the model simpler, but it also increases the model's precision. It's critical to take data quality factors like correctness, accuracy, integrity, completeness, dependability, and timeliness into account when automating and improving HR procedures. Data validation procedures, such as verifying that data are consistent across numerous sources, checking for null values, and checking for correct data types, should also be used to validate the quality of the data. To protect the privacy and confidentiality of employee data, data security and privacy should also be considered.

4. RESULTS

The performance metrics for various machine learning models are presented in Table 72.1, indicating their effectiveness in each task. Logistic Regression demonstrates a solid overall performance with an accuracy of 85%, boasting high precision at 88% and a balanced recall of 80%. Random Forest outperforms others with an accuracy of

Table 72.1 Performance metrics of classification models

Model	Accuracy	Precision	Recall	F1 Score
Logistic Regression	0.85	0.88	0.80	0.84
Random Forest	0.90	0.92	0.88	0.90
Neural Network	0.88	0.92	0.85	0.87
Support Vector Machine	0.87	0.89	0.84	0.86

90%, excelling in precision (92%) and recall (88%), resulting in a commendable F1 score of 90%. The Neural Network achieves an 88% accuracy, showcasing strong precision (92%) and recall (85%). Support Vector Machine maintains a competitive performance with an accuracy of 87%, demonstrating balanced precision (89%) and recall (84%), resulting in an F1 score of 86%.

The Table 72.2 provides a snapshot of employee performance evaluation, comparing predicted and actual performance scores. Each employee is identified by an Employee ID, with corresponding Performance, Predicted Performance, and Actual Performance scores. Noteworthy trends include Employee 004 outperforming predictions with an impressive 95, while Employee 003 slightly un-

Table 72.2 Employee performance comparison: predicted vs. actual

Employee ID	Performance	Predicted Performance	Actual Performance
001	75	78	80
002	85	88	90
003	60	65	62
004	92	91	95
005	78	80	82
006	70	72	68
007	88	86	89
008	95	94	96

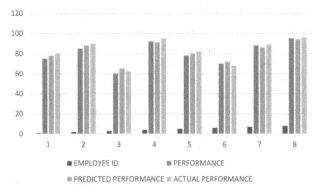

Fig. 72.1 Employee performance comparison: predicted vs. actual

derperformed. Overall, the team exhibits a mix of meeting, exceeding, and falling short of predictions, highlighting the challenges of forecasting individual performance. The data suggests a need for fine-tuning performance assessment methods to enhance accuracy and better align with employees' actual achievements.

5. CONCLUSION

In conclusion, Human Resource Management (HRM) plays a pivotal role in enhancing employee performance within an organization. Through various strategic initiatives and practices, HRM serves as the backbone for cultivating a positive work environment, fostering employee engagement, and maximizing individual and collective productivity.

One of the primary functions of HRM in improving employee performance is recruitment and selection. By employing effective hiring strategies, HRM ensures that the right talent is brought on board, aligning skills and competencies with organizational objectives. Moreover, HRM facilitates onboarding processes that help new employees integrate smoothly into the organizational culture, setting the stage for success from the outset.

Furthermore, HRM is instrumental in providing ongoing training and development opportunities. By investing in employees' professional growth, HRM not only enhances their skills and knowledge but also boosts their morale and motivation. Continuous learning opportunities empower employees to perform better in their roles and adapt to evolving industry trends and technologies.Additionally, HRM plays a crucial role in performance management systems. Through regular feedback, goal setting, and performance evaluations, HRM provides employees with clear expectations and benchmarks for success. This facilitates a culture of accountability and drives continuous improvement among employees.

Moreover, HRM is responsible for fostering a positive work culture and employee well-being. By implementing policies and programs that prioritize work-life balance, diversity and inclusion, and employee recognition, HRM creates an environment where employees feel valued, supported, and motivated to excel.Furthermore, HRM acts as a mediator in resolving conflicts and addressing employee grievances. By providing channels for open communication and conflict resolution, HRM ensures that potential barriers to performance are promptly addressed, allowing employees to focus on their roles effectively.In essence, the role of HRM in improving employee performance cannot be overstated. By strategically managing the organization's

most asset—its people—HRM drives productivity, innovation, and ultimately, organizational success. As businesses continue to navigate an increasingly competitive landscape, the importance of effective HRM practices in optimizing employee performance will only grow in significance.

References

1. Jorgensen, F., Hyland, P., & Busk Kofoed, L. (2008). Examining the role of human resource management in continuous improvement. *International Journal of Technology Management*, *42*(1-2), 127–142.
2. Al-Qudah, H. M. A., Osman, A., & Al-Qudah, H. M. (2014). The effect of human resources management practices on employee performance. *International journal of scientific & technology research*, *3*(9), 129–134.
3. Richman, N. (2015). Human resource management and human resource development: Evolution and contributions. *Creighton journal of interdisciplinary leadership*, *1*(2), 120–129.
4. Prieto, I. M., & Pilar Pérez Santana, M. (2012). Building ambidexterity: The role of human resource practices in the performance of firms from Spain. *Human Resource Management*, *51*(2), 189–211.
5. Blaga, P. (2020). The importance of human resources in the continuous improvement of the production quality. *Procedia manufacturing*, *46*, 287–293.
6. Indarapu, S. R. K., Vodithala, S., Kumar, N., Kiran, S., Reddy, S. N., &Dorthi, K. (2023). Exploring human resource management intelligence practices using machine learning models. *The Journal of High Technology Management Research*, *34*(2), 100466.
7. Chowdhury, S., Joel-Edgar, S., Dey, P. K., Bhattacharya, S., & Kharlamov, A. (2023). Embedding transparency in artificial intelligence machine learning models: managerial implications on predicting and explaining employee turnover. *The International Journal of Human Resource Management*, *34*(14), 2732–2764.
8. Dutta, D., & Kannan Poyil, A. (2023). The machine/human agentic impact on practices in learning and development: a study across MSME, NGO and MNC organizations. *Personnel Review*.
9. Marler, J. H., & Boudreau, J. W. (2017). An evidence-based review of HR Analytics. *The International Journal of Human Resource Management*, *28*(1), 3–26.
10. Fallucchi, F., Coladangelo, M., Giuliano, R., & William De Luca, E. (2020). Predicting employee attrition using machine learning techniques. *Computers*, *9*(4), 86.
11. Meddeb, E. (2021). The Human Resource Management challenge of predicting employee turnover using machine learning and system dynamics. In *BIR Workshops* (pp. 184–196).

Note: All the tables and figure in this chapter were made by the author.

Recent Trends in Engineering and Science for Resource Optimization and
Sustainable Development – Prof. (Dr.) Dorota Jelonek et al. (eds)
© 2024 Taylor & Francis Group, London, ISBN 978-1-032-98030-0

73

Customer Satisfaction and Service Quality with Effective Practices of Human Resource Management

Nivedita Pandey[1]

Assistant Professor, NIMS school of Business,
NIMS University

S. Shyam Sundar[2]

Assistant Professor (SG),
School of Commerce and Management,
Mohan Babu University, Tirupati

Brijesh Goswami[3]

Assistant Professor, Institute of Business management,
GLA University, Mathura

Somanchi Hari Krishna[4]

Associate Professor, Department of Business Management,
Vignana Bharathi Institute of Technology, Aushpur Village,
Ghatkesar Mandal, Medchal Malkajigiri Dist, India,
Telangana state

Deepti Sharma[5]

Associate Professor, School of Management,
OP Jindal University, Raigarh, Chattisgarh

Shilpa Tandon[6]

Assistant Professor, Management,
Asian Business School, Noida

▬▬▬ Abstract

Artificial intelligence (AI) has the ability to revolutionize society and alter the function of human resource management (HRM) procedures. Key competencies, business procedures like knowledge management, and customer outcomes like satisfaction and perceptions of service quality are all areas where AI is clearly being felt. Despite the fact that research has examined the connections among HRM practices and the success of various firms, it is still quite uncommon for research to examine the processes that may be the catalyst for such links.The objectives of this research are to enhance consumer satisfaction and service quality, and optimize the HRM process.A backpropagation neural network (BPNN) is used to

[1]nniv1986@gmail.com, [2]Shyamsundardgl@gmail.com, [3]brijesh.goswami@gla.ac.in, [4]harikrishnasomanchi@gmail.com, [5]deepti.sharma@opju.ac.in, [6]shilpaatandon@gmail.com

DOI: 10.1201/9781003596721-73

create a salary prediction model (SPM) based on AI digitizing technology. The Nesterov and Adaptive Moment Estimation (Nadam) algorithms are then combined to optimize the model. Findings reveal that the Nadam optimization algorithm has the best optimization effect and the fastest convergence time when compared to other optimization algorithms. The results imply that AI digitization technology based HRM practices are beneficial in improving customer satisfaction and service caliber.

▆▆▆▆ Keywords

Service quality, Human resource management, Customer satisfaction, AI, BPNN

1. INTRODUCTION

Artificial intelligence (AI) technology has given rise to a new class of labor known as "human intelligence" or "artificial intelligence," which is now essential for companies to thrive in a dynamic marketplace. AI is an interdisciplinary field of study that aims to mimic cognitive functions and human behavior. Artificial intelligence (AI) is being used more and more in business management decision-making to support managers with their mundane and repetitive daily responsibilities (Iodice et al., 2021).

In corporate organizations, customer satisfaction (CS) has been the main focus of almost all marketing management initiatives. Furthermore, as firms shift their focus to customer satisfaction, the emphasis on computer science (CS) has permeated other functional domains. In the corporate sector, human resource management (HRM) is crucial to attaining corporate social responsibility (CS).As a result, HRM now needs a specific set of techniques to build CS throughout the company. Nevertheless, not much research has been done to fully examine HRM techniques for CS and evaluate how they help CS in businesses.

HRM is a fundamental component or capability for each organisation, yet it is particularly significant for organizations in the hotel business that depend on offering client support and others related administrations as a feature of their USP. HRM rehearses that further develop service quality, consumer satisfaction, and performance is a practical way for hotel to remain cutthroat. In order to enhance the feasibility of the AI-based HRM system, this research develops a BPNN-based salary prediction model (SPM). The program is designed to anticipate a candidate's contracted wage based on information from their resume. These findings can serve as a guide for HRM systems built on data mining technologies in an effort to raise client satisfaction and service standards (Zahoor, A., and Khan, D. 2022). Given the prior, the current examination researches the connections between HRM, service quality, and consumer satisfaction in the Indian hotel business. The sections that follow provide an analysis of a selection of literature on HRM-service quality, HRM-customer satisfaction, and related literature. The third portion discusses the research approach. The results are covered in the fourth section. The paper concludes with suggestions for more studies in the fifth section.

2. REVIEW OF LITERATURE

2.1 Service Quality and HRM Practices

Quality improvement in service firms needs to focus on hiring, training, development, and compensation package, as many HR studies have previously stressed (Yurtseven, G., & Muluk, Z. 2016, Tiago et al., 2020). It's also true that in labor-intensive industries, a company can gain a competitive advantage by making efficient use of its people resources. In order for contract workers at hotels to provide excellent service, they should be provided with the appropriate training. Past HRM investigation has discovered that representative assessments of business HR practice and client evaluations of authoritative assistance adequacy in the help area are emphatically connected (Ahmed et al., 2020). Drawing from the recently referenced data, it tends to be guessed that HR approaches and service quality affiliations will reasonably improve hotel representatives' capacity to all the more actually satisfy visitor demands, honor commitments, and boost service delivery efficiency.

2.2 Customer Satisfaction and Service Quality

The substantial correlation among customer happiness and quality of service has been well-documented by the research community (Harnjo et al., 2021). Customer satisfaction is a multifaceted feeling that is influenced by various environmental and personal factors, as well as the caliber of the product or service received. A key component of success in the service sector and the hotel industry is customer satisfaction. According to Nunkoo et al., 2020, it is the outcome of specific service operations as well as the caliber of the overall service encounter. Additionally, as it promotes customer retention and the expansion of market share, firms have long sought to satisfy their customers (Belias et al., 2022).As a result, happy clients have the ability to stick around and recommend the business to others in addition to using its services again. Furthermore, satisfied customers could be more forgiving of any service lapses. The culmination of these elements is the company's financial performance.

2.3 Connection among Customer Satisfaction, Service Excellence, and Human Resource Management

The performance of service organizations, the provision of high-quality services, and customer satisfaction have all been said to be significantly impacted by HRM practices by numerous researches (Najam et al., 2020). The results of these studies may be seen as skewed because they focused mostly on individual HRM practices rather than other important variables like customer satisfaction and service quality. Several academic studies have recognized how critical it is to comprehend how consumers, staff members, and business success interact. As per the findings of Kurdi et al., 2020, contented customers lead to happy and motivated staff, and happy customers spend more, boosting the earnings and revenues of the company.As a result, it is determined that further research is necessary in order to accurately determine the relationship between several variables, including customer happiness, HRM practices, and service quality.

3. METHODOLOGY

3.1 BPNN or BP Neural Network

The Error back propagation learning serves as the primary learning basis for BPNN, and it is this process that continuously modifies and corrects the network's thresholds and weights [24]. The primary role of BPNN is to realize the mapping function from input to output, which qualifies it for use in the resolution of a wide range of intricate internal issues and the realization of complex nonlinear mappings. An input layer, a hidden layer, and an output layer make up a network. The concealed layer is the one with the most layers among them. Each layer's neurons function independently within BPNN. In two layers, there are only unidirectional connections between neighboring neurons. The input layer forwards the signal to the hidden layer, which processes it according to its function before sending the data to the output layer. To obtain the final output result, the (e function transforms the output layer's data.

3.2 SPM Employing BPNN

An example of a multi-input single-output mapping is the pay forecast. There are fourteen neurons in the network's input layer and one in the output layer, based on the format of the data used in the demand analysis. As seen below, there is no definitive way for calculating the quantity of neurons in the buried layer.

$$n_h = \sqrt{n_i + n_0} + k \tag{1}$$

Although the activation function can be either linear or nonlinear in theory, a nonlinear activation function is more useful in situations where it is unclear if the problem is linear. One of the most popular activation functions at the moment is the sigmoid function, which is calculated as

$$f(x) = \frac{1}{1 - e^{-x}} \tag{2}$$

The loss function serves as a criterion for assessing the convergence of the training learning model and is mostly used to quantify the error between the expected and actual values of the network. The quadratic mean square-error function E, which is provided as, is the most commonly used loss function in regression prediction.

$$E = \frac{1}{2}(y - z)^2 \tag{3}$$

3.3 SPM Optimization

A benchmark While BPNN has a robust learning capacity and a straightforward structure; it also has several drawbacks, like a sluggish convergence speed. It is simple for a gradient descent search to enter the parameter space's local minimum value. Therefore, BPNN needs to be tuned for practical use [26]. The back propagation update of BPNN employs the gradient descent approach; nevertheless, the result is not optimal and the iteration pace is slow.

Therefore, a hybrid optimization strategy is used to maximize gradient descent by combining extra momentum and adaptive learning rate. Among them, hybrid techniques such as Adam's adaptive moment estimation and Nesterov's mix of Adam and Adaptive moment estimation are frequently employed.

4. RESULTS AND DISCUSSION

The training results of the Adam and Nadam optimization algorithms are contrasted with those of the Root-Mean-Square Prop (RMSProp), Adaptive gradient (Adagrad), Nesterov Accelerated Gradient (NAG), and Stochastic Gradient Descent (SGD). Figure 73.1 displays the comparison findings of the convergence speeds of several optimization techniques.

When compared to SGD and NAG optimization algorithms in Fig. 73.1, the hybrid optimization algorithms Adam and Nathan exhibit superior update stability in terms of con-

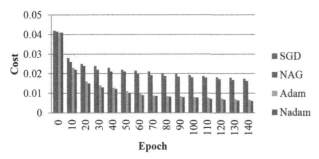

Fig. 73.1 Comparison of optimization techniques' rates of convergence

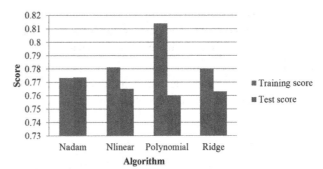

Fig. 73.3 Salary regression forecasting algorithm comparison

vergence speed. Additionally, the method performs better and converges faster. In general, the hybrid optimization algorithm Nadam converges more quickly than the hybrid optimization method Adam. As seen in Fig. 73.2, the training results of several optimization methods are compared in order to more accurately assess the performance of different optimization algorithms.

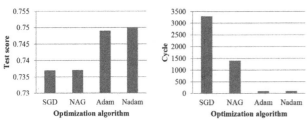

Fig. 73.2 Outcomes of different optimization algorithms' training

Comparing the hybrid optimization methods Adam and Nadam to the SGD, NAG, Adagrad, and RMSProp optimization procedures, Figure 73.2 demonstrates that they have the largest end prediction score and the shortest training duration. Among them, the Adam optimization algorithm's final prediction score is 0.7605, and the training time is 195 s. The Nadam optimization algorithm's final anticipated result score is 0.7706, and the training time is 188 seconds. The Nadam optimization e4ect has the fastest convergence speed and is the best. It seems sense that the BPNN-based SPM can be optimized using the Nadam method.

The effectiveness of several machine learning regression techniques is contrasted with that of the optimized BPNN-based SPM hybrid approach. Ten experiments were conducted for each method, and the best test results are noted as indicated in Fig. 73.3.

Figure 73.3 compares different techniques with the BPNN-based SPM optimized using Adam's hybrid approach. The test set score of 0.77 and the training set score of 0.78 are extremely close. It demonstrates the superior performance of the BPNN-based SPM enhanced by the Nadam hybrid algorithm. Figure 73.4 displays the predicted fitting effect

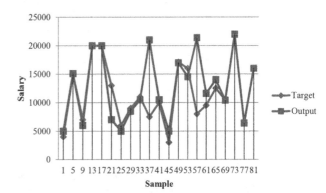

Fig. 73.4 Fitting impact of BPNN-based Nadam-optimized SPM

of the BPNN-based SPM following the Nadam hybrid algorithm's optimization.

Figure 73.4 clearly illustrates the superior fitting effect of the BPNN-based SPM following Nadam tuning. The wage data is standardized and de-normalized during model training, which could lead to certain computation errors. The error falls within the controlled range and does not significantly affect the model's outcome. It is confirmed that the Nadam-optimized BPNN-based SPM performs well in both learning and prediction outcomes, with an accuracy rate of up to 81.6%. The wage prediction model has some practicality and reference value because the outcomes validate its validity, which raises customer happiness and service quality.

5. CONCLUSION

Conversely, traditional HRM systems are not able to predict future development based on present data or analyze data correlations enough. A BPNN-based SPM is created in this study to raise customer satisfaction and enhance the HRM system's service quality. The model is optimized using the hybrid optimization method Nadam, and the BPNN-based SPM is assessed using experimental simulation. Findings demonstrated that, in comparison to other algorithms, the

BPNN-based SPM improved by the Nadam hybrid algorithm has a high score of 0.7735 on the training set and 0.7732 on the test set, which are extremely close.

Furthermore, the Adam hybrid algorithm-optimized BPNN-based SPM performs better in the learning process and in predicting customer happiness and service quality. As a result, the findings have some applicability and significance.

References

1. Iodice, F., Romoli, M., Giometto, B., Clerico, M., Tedeschi, G., Bonavita, S., ... & Digital Technologies, Web and Social Media Study Group of the Italian Society of Neurology. (2021). Stroke and digital technology: a wake-up call from COVID-19 pandemic. *Neurological Sciences, 42*, 805–809.

2. Zahoor, A., & Khan, D. (2022). Frontline service employees and customer engagement: some further insights. *IIM Ranchi journal of management studies, 1*(2), 175–190.

3. Yurtseven, G., & Muluk, Z. (2016). The practice of service quality: Hotel case. *Global Journal of Business, Economics And Management: Current Issues, 6*(1), 20–34.

4. Tiago, F., Borges-Tiago, T., & Couto, J. (2020). Human resources role in hospitality service quality. In *Strategic Innovative Marketing and Tourism: 8th ICSIMAT, Northern Aegean, Greece, 2019* (pp. 81–89). Springer International Publishing.

5. Ahmed, U., Kura, K. M., Umrani, W. A., & Pahi, M. H. (2020). Modelling the link between developmental human resource practices and work engagement: The moderation role of service climate. *Global Business Review, 21*(1), 31–53.

6. Harnjo, E., Rudy, R., Simamora, J., Hutabarat, L. R., & Juliana, J. (2021). Identifying customer behavior in hospitality to deliver quality service and customer satisfaction. *Journal Of Industrial Engineering & Management Research, 2*(4), 107–113.

7. Nunkoo, R., Teeroovengadum, V., Ringle, C. M., & Sunnassee, V. (2020). Service quality and customer satisfaction: The moderating effects of hotel star rating. *International Journal of Hospitality Management, 91*, 102414.

8. Belias, D., Rossidis, I., Papademetriou, C., & Mantas, C. (2022). Job satisfaction as affected by types of leadership: A case study of Greek tourism sector. *Journal of Quality Assurance in Hospitality & Tourism, 23*(2), 299–317.

9. Najam, U., Ishaque, S., Irshad, S., Salik, Q. U. A., Khakwani, M. S., & Liaquat, M. (2020). A link between human resource management practices and customer satisfaction: A moderated mediation model. *SAGE Open, 10*(4), 2158244020968785.

10. Kurdi, B., Alshurideh, M., &Alnaser, A. (2020). The impact of employee satisfaction on customer satisfaction: Theoretical and empirical underpinning. *Management Science Letters, 10*(15), 3561–3570.

Note: All the figures in this chapter were made by the author.

*Recent Trends in Engineering and Science for Resource Optimization and
Sustainable Development – Prof. (Dr.) Dorota Jelonek et al. (eds)
© 2024 Taylor & Francis Group, London, ISBN 978-1-032-98030-0*

74

Role of Digital Transformation and Strategies of Marketing Management

Abhishek Vaishnava[1]

Assistant Professor NIMS Institute of Travel & Tourism Management,
NIMS university

Kamal Kishor Pandey[2]

Associate Professor, TMIMT-College of Management,
Teerthanker Mahaveer University, Moradabad

Royal Tamang[3]

Assistant professor, Commerce and Management,
Centre for Distance and Online Education, Sikkim Manipal University
[2]Sikkim Manipal University, 5th Mile Tadong, Sikkim

Atul Singla

Lovely Professional University, Phagwara, India

Nilesh Anute[4]

Associate Professor, ASM'S Institute of Business Management
and Research, Pune

Kavita Tandon

Associate Professor, Amity University, Noida, UP

Abstract

The process of digital transformation has become an essential requirement for businesses in the modern corporate landscape, particularly in the ever-changing marketing industry. Despite the fact that the digital marketing strategy has gained popularity in the sales industry, there is still a dearth of study on the subject. Fuzzy system theory is used in this inquiry to maximize brand digital marketing under security management according to the machine learning classification approach. This study collects data from brand positioning, brand marketing, and customer sensitivity in order to forecast and assess future data findings. After that, it solves the correlation function using the machine learning method of classification model by utilizing the classification algorithm and the data that was gathered. Out of the four digital marketing strategies, network marketing offers the highest percentage of benefits to the company (70%) and the best amount of brand publicity (43%), according to the trial's final results. Under the network marketing strategy, clients' brand sensitivity reaches 49% at the same time.

Corresponding authors: [1]abhishek.vaishnava@nimsuniversity.org, [2]kamalpandey7719@gmail.com, [3]Royal.tamang@smudoe.edu.in, [4]nileshanute@gmail.com

DOI: 10.1201/9781003596721-74

■■■■■■■ **Keywords**

Marketing management, Digital transformation, Brand positioning and promotion, Customer sensitivity, ML

1. INTRODUCTION

In an environment of fierce and ever-changing market rivalry, many businesses are turning to digital transformation (DT) as a new strategy to obtain a competitive edge. Numerous companies have implemented digital transformation, which has improved their productivity (Downes, L., & Nunes, P. 2013) and commercial performance (Sundaram et al., 2020), increasing their superiority (Curraj, E. 2018).

Digital transformation is a disruptive and progressive process that involves integrating digital technology into every area of a business. It has the potential to radically alter how firms operate and provide value to their consumers. It combines a realistic approach with a fundamental rethinking of the way that current corporate structures are structured. As a result, businesses of all sizes, big or little, have been integrating digital transformation into their operations to keep up with the quickly changing digital landscape and to compete in the market (Kane et al., 2015).

Digital transformation has an impact on many company functions, and marketing management is not a single exception. The emergence of digital technology has resulted in a notable transformation in the development and implementation of marketing strategies (Rathore 2018, Gurbaxani& Dunkle 2019). As a result of this shift, companies have had to rewrite their policies, which have altered the conventional wisdom around marketing management (Porter and Heppelmann, 2014).

The implementation of digital transformation has permeated every facet of peoples' daily lives and profoundly influenced the trajectory of society. Machines will leave a lasting impression on human lifestyles and marketing management as technology advances. Financially speaking, machine learning techniques not only improve people's quality of life on a daily basis, but they also bring about large cash rewards for them (Yamamura et al., 2017). People need to focus more on network security management and control issues in order to achieve financial rewards.

2. REVIEW OF LITERATURE

2.1 Digital Transformation's Impact on Marketing

Even if everyone was happy about the Millennium, it was hard to predict 20 years ago that marketing would change in such a big way, especially in terms of the distribution and communication of products and services (Kotler et al., 2021). The rules governing everyday life and corporate procedures were completely broken by advances in information technologies and mobile devices.

Other brands' digital marketing initiatives are likewise influenced by the well-liked hardware products of technology giants. New fields of study in digital marketing are being brought about by the advancements in artificial intelligence, data mining, driverless car technology, and the internet of things (IoT). People have praised Amazon Go for its cashier-less shopping experience, which offers a highly advanced shopping experience. It has also demonstrated how new age marketing technology, such as face recognition technology, can alter consumers' everyday shopping habits (Parise et al., 2016).

2.2 Digital Transformation

Digitalization has impacted nearly every department in goods and service firms and has been one of the most talked-about subjects over the past several years (DEMİRBAS and YURT 2022). Additionally, it is now a necessary prerequisite for modern institutions that offer many business advantages, making it a competitive weapon. Trade laws and corporate procedures have changed as a result of digital revolution. As a result, businesses that are not digitalized will find it difficult to compete in the market today. These days, digitalization is turning into a survival strategy. As a result, organizations should budget for investments in digital transformation (Li, 2020).

Digitalization ought to extend to ongoing operations as well as those in the future. Even if it can be difficult to change, the outdated work practices need to be up dated. Furthermore, some procedures could not be competitive since they are difficult to modify for the digital age.

2.3 Machine Learning

The field of machine learning focuses on enhancing performance by investigating ideas and techniques. These methods enable machines to learn and acquire skills in a manner similar to that of humans. Starting with the objective of building a learning machine from prior experience and using it in order to categorize or forecast unknown data, machine learning (ML) aims to continuously construct and enhance the machine as it goes. One of the most promising areas of artificial intelligence is ML, which is based on the biological process of teach (Leung et al., 2015). The creation of scalable computer programs is the aim of the ML community. It is anticipated that computers will be able to assist humans in analyzing sizable and intricate data collections (Buchanan et al., 2018). The need for machine learning systems is rising as a result of machine learning's success in a variety of applications.

3. METHODOLOGY

3.1 Algorithm for Machine Learning Classification Creating Models

Supervised learning revolves around a machine learning classification method. A ML classification approach aims to learn by predicting the class for each incoming sample based on the sample model's preexisting class label. The input space in a ML classification process is the collection of all possible input values. The output space is the collection of all potential output values. The input and output spaces can be any subset of finite components or Euclidean space. Generally speaking, the output space will be much smaller than the input space. We have specified an output quantity, Y, and an input quantity, X. Each distinct input unit serves as a sample, and sample attribute vectors capture all of the sample's attributes and data. The attribute space is the region that contains the attribute vectors.

A sample feature is associated with each dimension in the sample feature space. A vector of input attributes is commonly written as

$$x_j = (x_j^1, x_j^2, \dots x_j^n)^N \qquad (1)$$

Using training examples, the ML classification approach assesses the test instances' memory category and trains the model or objective function. The collection of teaching examples is typically written as

$$T = \{(x_1, y_1), (x_2, y_2), \dots (x_N, y_N)\} \qquad (2)$$

3.2 Prediction Model

First, a training sample set is provided; the category value and space of the output sample are represented by yj ∈ y, whereas the feature and feature space of the input sample are expressed by the values xj ∈ x ⊆ Rn. According to the joint probability, the training sample data and the test sample data F(x, y) are distributed independently. Y= f(x) and the conditional probability distribution P(Y|X) can be used to represent the mapping model that the learning system created throughout the learning process based on the training sample set. The prediction system predicts the input sample data x_{N+1}, and the model's probability distribution for it is

$$y_{N+1} = argmax_{y_{N+1}} P(Y|X) \qquad (3)$$

Decision function is:

$$y_{N+1} = f(x_{N+1}) \qquad (4)$$

4. EMPIRICAL RESULTS

A brand's revenue is mostly determined by the level of attention it receives, which in turn influences consumers' awareness of and sensitivity to the brand. In order to enhance brand awareness, the PR model is quite important. In order to identify more profitable publicity strategies for the business, we examine the advantages of the various brand publicity techniques in this section, as well as the sensitivity of the target audience. Figure 74.1 displays the level of brand publicity, the sensitivity of consumers to the brand, and the sales income generated by the brand under various sales models.

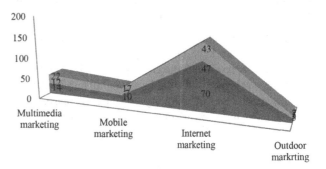

Fig. 74.1 Impact analysis diagram for sales models

Online marketing generates the highest revenue for the brand out of the four digital marketing modalities, at 70%, as evidenced by the pertinent statistics on sensitivity, brand income, and exposure level displayed in Fig. 74.1. With 43%, it has the best brand publicity degree as well. Conversely, 47% of network marketing users are brand sensitive. When it comes to customer sensitivity, brand publicity, and predictability, the outdoor strategy does not perform as well as other marketing strategies. Its brand promotion is 7%, its customer sensitivity is 5%, and its brand revenue is 5%.

In order to forecast the total number of brand consumers under the digital marketing model in the coming days, the number of customers who will make a purchase, the sensitivity of the customer, and the brand income, we study and compute the pertinent brand data within five days. The final results are shown in Table 74.1 and are based on the data collected by the ML classification approach in this article as well as the classification strategy used to forecast and evaluate the data set.

Table 74.1 Prediction data for the classification method

Days	Total Buyer	Revenue of brand (%)	Total customer	Sensitivity	Non sensitivity
Day one	274326	58	474576	.53	.47
Day two	153285	49	333275	.4	.6
Day three	95054	33	175436	.37	.63
Day four	610365	72	633678	.70	.30
Day five	412356	40	573286	.43	.57

The brand will have a very objective number of consumers in the next seven days, with the highest amount occurring on weekends, according on the forecast findings in Table 74.1. Based on the prior examination of client age groupings, it is evident that the reason for the highest volume of business on weekends is that the majority of consumers are either employees or students who are on vacation. To make the data more visually appealing, we turned the data in the table into a graph, as seen in Fig. 74.2.

The quantity of brand buyers and customer sensitivity will peak over the weekend in the upcoming five days. It follows that by appropriately promoting its goods or announcing some weekend events, the brand can obtain greater results, and the level of publicity will also increase. Table 74.2 displays the comparison results of various techniques utilizing SVM, Navie Bayes classification, machine learning classification algorithm, etc.

Table 74.2 Different techniques comparison results

Techniques	Forecasted total buyers	Predicted total consumer	Predicted revenue (%)	Accuracy
SVM	410316	573250	39	.92
NB classification method	410276	573200	45	.91
Machine learning classification method	410356	573286	41	.95

5. CONCLUSION

To have a better grasp of the dynamic character of digital marketing and the cause-and-effect factors influencing its utilization, and how digital marketing affects the variables being studied, this study has conducted a thorough investigation into the complex relationship between DT and management strategies. The study examined how different technological integrations influenced the factors of digital transformation that shaped modern marketing strategies. This study processes data linked to digital sales using classification algorithms and analyzes brand digital sales within the framework of organisation using machine learning techniques. This research examines the quantity of consumers, level of brand exposure, and brand income in a digital marketing campaign that lasts seven days. Using a machine learning classification algorithm model, the brand's associated data for the following seven days is forecasted. We may obtain the data projection for brand digital marketing for the following five days by repeatedly computing the trial data. Organizations may position themselves at the forefront of innovation and ensure sustainable growth and flexibility in the dynamic field of digital marketing by adopting and carefully implementing these findings.

This study examines digital marketing using information from online sales. Despite certain advancements in research, several deficiencies still require additional development. This paper's data collecting approach prioritizes pertinent consumer traits over those that are not useful, necessitating the use of other techniques. Additionally, the data processing in this study is restricted by memory, so it can only streamline the process to that extent. A substantial number of tests and a thorough comparison of the efficacy of various procedures are required in order to give more exact and comprehensive data and outcomes.

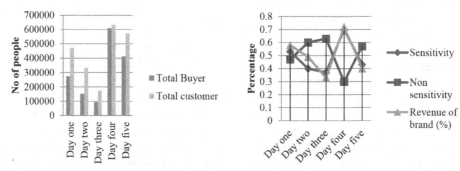

Fig. 74.2 Prediction of the next five days

References

1. Downes, L., & Nunes, P. (2013). Big bang disruption. *Harvard business review*, 44–56.

2. Sundaram, R., Sharma, D. R., & Shakya, D. A. (2020). Digital transformation of business models: A systematic review of impact on revenue and supply chain. *International Journal of Management, 11*(5).

3. Curraj, E. (2018). Business digitalization of SMEs in Albania: Innovative approaches and their impact on performance.

4. Kane, G. C., Palmer, D., Phillips, A. N., Kiron, D., & Buckley, N. (2015). Strategy, not technology, drives digital transformation. *MIT Sloan Management Review*.

5. Rathore, B. (2018). Green Strategy: Exploring the Intersection of Marketing and Sustainability in the 21st Century. *Eduzone: International Peer Reviewed/Refereed Multidisciplinary Journal, 7*(2), 83–90.

6. Gurbaxani, V., & Dunkle, D. (2019). Gearing up for successful digital transformation. *MIS Q. Executive, 18*(3), 6.

7. Porter, M. E., &Heppelmann, J. E. (2014). How smart, connected products are transforming competition. *Harvard business review, 92*(11), 64–88.

8. Yamamura, S., Fan, L., & Suzuki, Y. (2017). Assessment of urban energy performance through integration of BIM and GIS for smart city planning. *Procedia engineering, 180*, 1462–1472.

9. Kotler, P., Kartajaya, H., & Setiawan, I. (2021). *Marketing 5.0: Technology for humanity*. John Wiley & Sons.

10. Parise, S., Guinan, P. J., & Kafka, R. (2016). Solving the crisis of immediacy: How digital technology can transform the customer experience. *Business Horizons, 59*(4), 411–420.

11. DEMİRBAŞ, E., & YURT, C. (2022). DIGITAL TRANSFORMATION AND MARKETING: AN INTEGRATIVE CONCEPTUAL STUDY ON STRATEGIC DIGITAL MARKETING MANAGEMENT AND SOCIAL MEDIA MARKETING. *Research on Economics and Administration and Social Sciences*, 41.

12. Li, F. (2020). Leading digital transformation: three emerging approaches for managing the transition. *International Journal of Operations & Production Management, 40*(6), 809–817.

13. Leung, M. K., Delong, A., Alipanahi, B., & Frey, B. J. (2015). Machine learning in genomic medicine: a review of computational problems and data sets. *Proceedings of the IEEE, 104*(1), 176–197.

14. Buchanan, L., Kelly, B., Yeatman, H., & Kariippanon, K. (2018). The effects of digital marketing of unhealthy commodities on young people: a systematic review. *Nutrients, 10*(2), 148.

Note: All the figures and tables in this chapter were made by the author.

Recent Trends in Engineering and Science for Resource Optimization and Sustainable Development – Prof. (Dr.) Dorota Jelonek et al. (eds)
© 2024 Taylor & Francis Group, London, ISBN 978-1-032-98030-0

75

Marketing Management of Production and Sales Activities in the Enterprises

Smruti Ranjan Das[1]
Assistant Professor, Department of Commerce,
School of Social, Financial & Human Sciences, KIIT University,
Bhubaneswar, Odisha

Anandhi Damodaraswamy[2]
ASSISTANT PROFESSOR, Department of Commerce with
Computer Applications, PSG College of Arts & Science,
Coimbatore, Tamilnadu, India

Amol Murgai[3]
Professor in the Department of M.B.A, S.N.J.B. College of Engineering,
Chandwad, Dist. Nashik, Maharashtra

Uma Reddy
Department of Artificial Intelligence and Machine Learning,
New Horizon College of Engineering, Bangalore

Ratnakar Lande[4]
Assistant Professor, Mechanical Engineering,
G. H. Raisoni Institute of Engineering and Technology, Nagpur,
Nagpur, Maharashtra, India RTMNU, Nagpur

Ginni Nijhawan
Lovely Professional University, Phagwara, India

■■■■ Abstract

In today's corporate landscape, the key to sustainability is controlling the retail market. A lot of businesses rely a lot on past performance information and sales trends forecasted from product demand. The impact on business of these projections' correctness is substantial. The application of data mining to production and sales forecasting in retail sales and demand prediction is discussed in this paper. This paper carries out a rigorous investigation and review of understandable machine learning (ML) categorization algorithms to improve product sales forecasts. Big data and accuracy in product sales prediction pose challenges for standard statistical forecasting and prediction approaches. This work presents a comparative comparison of different machine learning techniques, including as LSTM, random forest, naïve bays, support vector machine (SVM), and additional tree regression, with the goal of developing a prediction model and accurately estimating potential sales. The results are shown in terms of the accuracy and dependability of the various prediction algorithms that were employed.

Corresponding authors: [1]smrutiranjan.dasfcm@kiit.ac.in, [2]anandhipappu2@gmail.com, [3]murgai.arcoe@snjb.org, [4]ratnakar1977@gmail.com

DOI: 10.1201/9781003596721-75

■■■■■■ **Keywords**

Marketing management, Sales, Production, Machine learning, Accuracy, Prediction

1. INTRODUCTION

One of the most important business metrics is customer happiness. Retailers constantly strive to meet customer expectations as much as they can while also seeking for ways to boost profits through strategic investments in sales. Forecasting product sales so aids retailers in making more profit on an absolute investment basis.

Retail is one of the most significant and quickly growing commercial fields in data science because of its massive amount of data and myriad optimization issues, like finding the optimal prices, discounts, suggestions, and stock levels. Numerous data analysis techniques can be used to overcome these issues. Predicting commodities sales in the dynamic and ever-changing business environment of today can be exceedingly challenging. A small number of improvements in sales forecasting could assist retailers in lowering operating expenses and increasing revenue. Additionally, it might increase client satisfaction (Jain et al., 2015).

Businesses are primarily focused on their target market. Thus, it is noteworthy that the organization was able to accomplish this goal through the use of a forecasting system. Analyzing data from numerous sources, including market trends, customer behavior, and other factors, is a necessary step in the forecasting process. Additionally, this study will assist the businesses in efficiently managing their financial resources. The forecasting method can be used to determine how much of a product will be sold in a specific amount of time, as well as to predict future demand for goods or services.

Researchers have dubbed the current era the "AI Revolution" due to the pervasive effects of machine learning (ML) and artificial intelligence (AI) on all spheres of society (Makridakis, 2017). Research across disciplines has been stimulated by this AI revolution. Such procedures have had a big influence and been a major source of innovation in the business sector. Even though they are relevant, words like artificial intelligence and machine learning may appear like foreign languages to many marketing academics and practitioners (Conick, 2017). This paper seeks to change the status quo by analyzing the critical roles that machine learning (ML) and artificial intelligence (AI) can play as research methods in the marketing sector.

Many commercial problems necessitate making decisions based on a substantial amount of historical data, and be-cause human decision-makers find it difficult to integrate this data, their choices are prone to a range of biases. In order to help businesses and organizations address the issue of gathering data and using the knowledge contained within it to predict sales, machine learning algorithms and data mining have been used recently. Businesses can increase their chances of making more informed decisions by using predictive algorithms correctly and putting in place a strong data mining infrastructure. Therefore, it is believed that sales forecasting is crucial for businesses planning to expand into new areas, introduce new services or goods, or experience rapid development. The conceptual framework needed to carry out this research is established using the following road map.

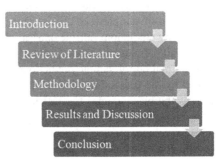

Fig. 75.1 Research structure

1.1 Objective of Study

1. To use machine learning algorithms and data mining approaches to help determine the product class that needs to be sold so that business owners can make better decisions about purchasing inventory.

2. To determine which categorization machine learning algorithms work best on the provided data by obtaining a comparative study of various algorithms in accordance with the CRIPS_DM methodology.

2. REVIEW OF LITERATURE

2.1 Marketing Strategy

Based on the requirements and previous performance of particular target consumers, this kind of business activity is mostly planned and executed for the market. It collects data on sales volume, purchasing power, and industry prestige value (Silver et al., 2017). It also includes comprehensive strategies like public relations, pricing, promotion, and channel. Through marketing, companies can help customers accept and value the products, services, and benefits they offer. Its main objectives are to improve the buy and usage experience, provide customers with the essential goods or services raise consumer approval of products, and concurrently establish a brand. Due to the dynamic and ever-changing nature of this process, decision-makers must constantly build, modify, and alter plans in response to market developments. Apply the 4P principle in accordance with the actual situation; we should carefully select markets, optimize the strategic policy, increase competitiveness, and strive to get the best economic results in a large environment at the lowest possible input costs (Tingting et al., 2018).

2.2 Machine Learning's Applicability to Marketing

According to Dzyabura and Yoganarasimhan (2018), there are some distinctions between the two. First, while causal econometric techniques seek to derive the best unbiased estimators, machine learning techniques concentrate on producing the best out-of-sample predictions. Thus, when it comes to generating predictions outside of a sample, causal inference techniques frequently perform poorly. This occurs because optimal out-of-sample predictions are not always produced by the best unbiased estimator. This issue is referred to as the bias variance tradeoff in the machine learning research. Secondly, unlike econometric techniques, a lot of machine learning approaches were created without any preconceived notions about how the results seen in the data were produced. Third, ML techniques have the ability to handle a very high number of variables and choose which ones should be kept in the study and which ones should be eliminated, in contrast to a number of empirical methods utilized in marketing. Finally, ML can be a potent strategy for handling marketing scalability. In order to attain scale and efficiency a goal that is becoming more and more crucial for a variety of real-time marketing problems—many machine learning algorithms use feature selection and optimization.

2.3 Production and Sales Prediction

Predicting sales and production brings great value to the company since it addresses the future. Businesses in the private sector can feel more confident about their upcoming investments when forecasts are reliable and precise (Cheriyan et al., 2018). In a similar vein, a precise projection enhances the market standing and adds value to publicly listed enterprises. Forecasting is also used by companies in the industrial sector to plan and control their production cycle, so reducing excess, eliminating waste, and satisfying demand (Wang et al., 1019). It is impossible to over-

estimate the significance of forecasting and projecting the future for growth and profit maximization in today's business environment since wise business judgments necessitate both prediction and decision analysis. Thus, the ability of corporate executives to accurately forecast future sales and client demands is critical to making excellent business decisions. Therefore, achieving a flawless forecast has a significant impact on sales volume, profit margin, storage, and consumer attraction (Arif et al., 2019; Lin et al., 2016). In order to increase the precision and effectiveness of forecasts, data mining can be a useful tool for analysts to extract information and hidden patterns from massive datasets that are generated over time.

3. METHODOLOGY

3.1 Proposed Method

Data mining needs a standard methodology that may assist in transforming organizational or business challenges into activities related to data mining, suggest appropriate data transformations and procedures, and offer ways to evaluate the viability of discoveries and document the information. Some of these worries were allayed by the Cross Industry Standard Process for Data Mining (CRISP-DM), which established a process model that offers a framework for the implementation of data mining projects that is not dependent on the technologies or industry. Data mining analysts find the CRISP-DM methodology helpful in a number of ways. The approach offers direction, aids in project planning, and offers direction on every task or phase of the procedure for beginners. Experienced analysts also make use of checklists specific to each position to make sure nothing important has been overlooked. Nonetheless, the CRISP-DM methodology's ability to communicate and log results is its most important feature. A cohesive and successful project can be formed by assembling a variety of instruments and individuals with a range of specialties and skills. The Fig. 75.2 below displays the CRISP-DM model.

3.2 Data Collection

The process of understanding data starts with the first data collection and continues with getting to know the data to learn about consistency concerns, gain some early understanding of the data, or identify intriguing subsets to draw inferences about hidden information. Data gathered from an off-license retail store's sales was utilized in this project. The collection, which is divided into categories such as wine/champaign and spirits/liquor, offers statistics on more than 55 different product types that were sold at the store in January. There are 14 columns and 15765 rows in the data that was gathered.

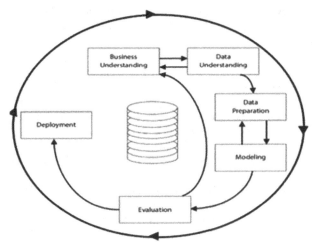

Fig. 75.2 CRISP-DM model

3.3 Preparing Data

The data preparation process includes all of the procedures necessary to create the final data, which are supplied into the modeling tool or tools from the original raw data. Procedures for data processing don't have to be finished in a certain order or several times. Selecting tables, document and attribute selection, data cleansing, creating new attributes, and data conversion for modeling tooling are some of these responsibilities. To construct the final dataset for the forecast, it simply involves properly analyzing the data that was collected. The year, month, supplier, and item code columns were removed after the first data preparation and cleaning, and the item type was changed to a binomial.

3.4 Modelling Techniques

In this step, various modeling approaches are selected, put into practice, and their parameters are adjusted to the best possible levels. Four different classification algorithms are combined in our algorithm: SVM, Random Forest, Naïve Bayes, and LSTM.

- **Neural networks with both short- and long-term memory**

  A unique RNN is the LSTM model. It addresses the issue of gradient expansion and disappearance, which traditional RNN is unable to prevent, by introducing memory units and making various modifications to the neuronal architecture of RNNs. The network architecture of the LSTM is modified to incorporate the gradient learning technique. Additionally, a processor including three gates—input, output, and forgetting—is incorporated to determine the utility of the information.

- **Naïve Bayes**

  Based on the Bayes theorem, the Naive Bayes approach works under the supposition that every pair of features is independent of one another. Spam filtering and document categorization are only two examples of the real-world applications in which naive Bayes classifiers excel. To forecast the required parameters, this method needs a little amount of training data. Comparing naive Bayes classifiers to more advanced techniques reveals how fast they are. We use Baye's Theorem to determine the likelihood that an event will occur given that another event has already occurred.

- **Random Forest**

  Different tree grades make up the random forest classifier. To categorize the input vector in the most common class, each tree casts a unit vote. A random vector sampled independently of the induction vector is used for each classification. Random Forest is a computationally efficient method that works well with big datasets. It has been applied to numerous disciplines in recent research projects and practical applications. Bagging is typically more dependable than single classifiers. However, on sometimes, it performs far less accurately than boosting algorithms.

- **Support Vector Machine**

  Throughout the past 40 years, the SVM methodology has become more and more popular as a rapid and efficient statistical learning method. A support vector machine is a machine learning supervised technique that maximizes the range of differences between two groups. It is commonly used as a classification tool. SVM can be applied to problems involving regression or classification by utilizing a kernel trick technique to alter the data. In summary, the SVM basically followed the same values, but with minor deviations.

4. EMPIRICAL RESULTS

Classification Precision and Accuracy in each class, which show how many predictions from each class can be compared to examples of that class, are typically used to measure the efficiency of classification algorithms. Choosing the ideal features for the model is the first stage. This is accomplished by using the univariate method of select the best methodology to an automated feature selection process, with the f-score serving as the scoring parameter. The performance of each of the four methods is displayed in Table 75.1 Fig. 75.3. When compared to other algorithms, the SVM algorithm fared better. It has a 97.5% recall rate, a 96.5% f1-score, and 96.5% accuracy and precision stands at 96.5%. Following with an precision of 81.6%, an F1-score of 92.3%, and a recall of 99.2% was the RF method. The accuracy of the RF algorithms is 85.8%.

Table 75.1 Performance comparison of the four algorithms

Algorithm	Precision	Accuracy	F1 Score	Recall
SVM	95.6%	96.8%	97.1%	97.2%
LSTM	60.1%	66%	75.3%	97.6%
Navie Bayes	55.6%	58.3%	72%	96.4%
Random Forest	81.6%	85.8	92.3%	99.2%

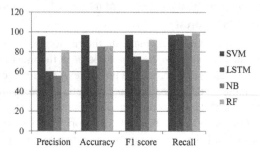

Fig. 75.3 Performance measures of algorithms

5. CONCLUSION

The project's research has led to the conclusion that businesses require product sales forecast systems to handle enormous amounts of data, and that the speed and accuracy of data processing technologies play a critical role in business decision-making. The machine learning techniques discussed in this article can offer a useful mechanism for data tweaking and decision-making. Businesses must arm themselves with specific methodologies in order to make predictions about the sales and manufacture of their products in order to maximize profit, taking into account the many forms of consumer behavior. For this study, about 15,500 data instance records were gathered in order to do the first algorithm comparison. However, because of the extended implementation duration and the difficulty of managing such a big collection of data, some of the records were discarded throughout the analysis and data processing method phase. Further analysis was not acceptable for the fields and attributes utilized in this investigation. It was a significant challenge encountered during the research.

However, by presenting practical ML approaches that would solve the issue, we carefully considered the impact of our work.

References

1. Jain, A., Menon, M. N., & Chandra, S. (2015). Sales forecasting for retail chains. *San Diego, California: UC San Diego Jacobs School of Engineering.*
2. Makridakis, S. (2017). The forthcoming Artificial Intelligence (AI) revolution: Its impact on society and firms. *Futures, 90,* 46–60.
3. Conick, H. (2017). The past, present and future of AI in marketing. *Marketing News, 51*(1), 26–35.
4. Silver, D., Schrittwieser, J., Simonyan, K., Antonoglou, I., Huang, A., Guez, A., ... & Hassabis, D. (2017). Mastering the game of go without human knowledge. *nature, 550*(7676), 354–359.
5. Tingting, L., Wendong, Z., & Guangyi, L. (2018). Research progress of text classification based on deep learning. *Power Information and Communication Technology, 16*(3).
6. Dzyabura, D., & Yoganarasimhan, H. (2018). 11. machine learning and marketing. *Handbook of Marketing Analytics: Methods and Applications in Marketing Management, Public Policy, and Litigation Support,* 255.
7. Cheriyan, S., Ibrahim, S., Mohanan, S., & Treesa, S. (2018, August). Intelligent sales prediction using machine learning techniques. In *2018 International Conference on Computing, Electronics & Communications Engineering (iCCECE)* (pp. 53–58). IEEE.
8. Wang, J., Liu, G. Q., & Liu, L. (2019, March). A selection of advanced technologies for demand forecasting in the retail industry. In *2019 IEEE 4th International Conference on Big Data Analytics (ICBDA)* (pp. 317–320). IEEE.
9. Arif, M. A. I., Sany, S. I., Nahin, F. I., & Rabby, A. S. A. (2019, November). Comparison study: Product demand forecasting with machine learning for shop. In *2019 8th International Conference System Modeling and Advancement in Research Trends (SMART)* (pp. 171–176). IEEE.
10. Lin, K. Y., & Tsai, J. J. (2016, April). A Deep Learning-Based Customer Forecasting Tool. In *2016 IEEE Second International Conference on Multimedia Big Data (BigMM)* (pp. 198–205). IEEE.

Note: All the figures and table in this chapter were made by the author.

*Recent Trends in Engineering and Science for Resource Optimization and
Sustainable Development – Prof. (Dr.) Dorota Jelonek et al. (eds)
© 2024 Taylor & Francis Group, London, ISBN 978-1-032-98030-0*

76

Perspectives of Green Ecology: Development of Marketing Management System in Digital Economy

Varalakshmi Dandu[1]

Assistant professor (a), School of Management Studies,
JNTUA, ANANTHAPURAMU, Andhra Pradesh,

Priti Gupta[2]

Assistant Professor, P.G. Department of Economics,
Bhupendra Narayan Mandal University (West Campus),
P.G. Centre, Saharsa, Bihar

Sachin Tripathi[3]

Symbiosis Law School, Nagpur,
Symbiosis International (Deemed) University (SIU),
WATHODA, Nagpur, Maharashtra, India

Pavana Kumari. H[4]

Faculty, JAIN Group of Institutions, Bengaluru

Nilesh Anute[5]

Associate Professor,
ASM'S Institute of Business Management and Research, Pune

Ankit Gupta[6]

Associate Professor, Department of Management,
ITM Gwalior (M.P) India

▬▬▬ Abstract

The pace of enterprise development is accelerating in the setting of the digital economy. Many businesses currently use pretty conventional marketing strategies. These management strategies are not only inefficient, but they also have shortcomings including closed marketing data, low digitization, information asymmetry, and insufficient real-time information. To address these problems, this research recommends managing corporate marketing scientifically through the application of digital economic techniques. This study primarily uses the mean clustering approach to examine the functioning of the organization marketing management system in order to design a system that can grow smoothly in a green environment. We do this by applying the B/S paradigm. The outcomes of the study demonstrated that the organization's marketing

[1]varalakshmi.jntua@gmail.com, [2]prity.gupta024@gmail.com, [3]sachintripathi@slsnagpur.edu.in, [4]pavanajagadish77@gmail.com, [5]nileshanute@gmail.com, [6]prof.ankit7@gmail.com

DOI: 10.1201/9781003596721-76

management system had some utility from a green viewpoint in the context of the digital economy, which can aid in the system's healthy development.

■■■■■■■ **Keywords**

Green ecology, Marketing management, Organisation, Digital economy

1. INTRODUCTION

As defined by McKinsey, the term "digital economy" refers to an economy that is built around digital technologies. As Internet technologies advance and become more widely used in the economy, China's GDP could increase to 23% and Russia's to 35%. In the US, the expected increase in value due to the introduction of digital technologies could reach 1.7-2.3 trillion US dollars by 2026 (Belassi, 2013). The concept of the digital economy is often quite straightforward: it is the economy built around digital technologies.

The global economic environment has seen a dramatic upheaval in the 21st century, primarily due to the widespread use of digital technologies. The digital economy is characterized by a widespread utilization of electronic instruments, data-driven judgment, and the pervasiveness of the internet which has become a powerful force that is changing markets, companies, and consumer behavior (Li et al., 2020). The job of marketing has changed in this digital age to become not only important but revolutionary. Marketing, which was formerly limited to conventional media, has experienced a significant transformation, adjusting and flourishing in the digital space.

The facilities and equipment of the management system are outdated, and the management mechanism is somewhat inflexible. There is minimal informatization and the managerial approach is out of date. Many businesses have seen a steady decrease in their competitiveness as a result of these. In line with organisation developments, this study looks into how green development affects organization marketing. In the framework of the digital economy, it also examines the effect of marketing and potential financial gains that companies may experience by implementing a marketing management system.

1.1 Research Objective

The goal of this investigation is to develop a more effective organisation marketing management system that will help businesses address issues related to low organisation informatization, outdated technology, and a failure to understand the demands of business customers.

2. LITERATURE REVIEW

2.1 Marketing Management and Digital Economy

In the digital economy, marketing management is no longer limited to conventional media like radio, print, and television (Kaewkhum, 2020). It has spread throughout a wide range of digital channels, including social media, email, content, and internet marketing, in light of the digital economy. Marketers employ digital platforms and tools to reach and engage with a wider audience of customers. Data-driven judgment plays a major role in marketing management in the digital economy (Awan et al., 2021). Marketers gather a tone of information about rival activity, market trends, and consumer behavior. After then, this data is examined to help with campaign optimization, resource allocation, and decision-making related to marketing.

2.2 Digital Economy

From the information economy, the digital economy evolved as a more advanced stage (Li et al., 2023). A deeper, more comprehensive, and more advanced approach to economic and social growth will be brought about by the digital economy as a result of the increased adoption, application, and penetration of digital technology in the economy (Jia et al., 2023). The improvement of digital technology prompted a quick development of the data and communication area, making it the quickest developing vital arising area in the economy and society. Consequently, a deeper understanding of the digital economy is needed. The introduction of new action plans is associated with the growth of the digital economy. The new plan of action in the digital economy, rather than the customary corporate working model, relies upon the constant progression of ICT innovation. Distance is no longer a barrier to commercial development. These facilitate the expansion of the globalized market and, to some extent, lower transaction costs (Dudko, 2020).

2.3 Success of Organizational Procedure In Digital Economic System

In the digital economy, the real world modifies the behavior of the virtual, and the machines are controlled by the world of subtle technologies (Kostakis, 2016). These characteristics are what give rise to new kinds of markets and societies. The digital economy's technological underpinnings provide insights from the fourth industrial revolution. Artificial intelligence, disseminated information, blockchain innovation, the IoT, mining offices, enormous information and distributed storage, computerized stages, 3D and hence 4D printing are a couple of them. Various frameworks' innovative plans are used to resolve specific issues.

The fourth industrial revolution has led to the digital economy which is based on the information economy can be defined as the information economy's continuation in a new form following an unprecedented and disruptive technological breakthrough (Melnyk et al., 2019). Their use alters decision-making and cognitive processes significantly, which affects economic behavior, corporate structure and operations, productivity, and the functioning of the overall economic system.

3. METHODOLOGY

3.1 Research Model

A growing number of companies are reevaluating traditional concepts in light of the digital economy in response to the national push for green ecology and to keep up with current developments. To help organizational law proceed logically, a marketing management system that is more in line with the growth of the business is designed. The B/S mode is currently the organizational marketing management system's primary and most frequently utilized mode.

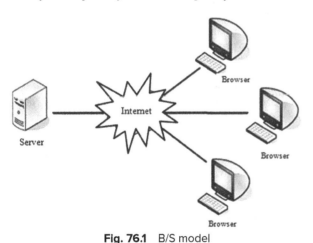

Fig. 76.1   B/S model

3.2 Used Method

K-means clustering algorithm groups similar customer samples together; cosine similarity, Manhattan distance, Euclidean distance, and other similarity measures are commonly employed. Segmenting consumer value using

the K-means clustering algorithm can improve the system more than data mining technology can.

The Euclidean distance formula is expressed as follows:

$$h(k_i, k_j) = \sqrt{\sum_x^n (k_{ix} - k_{jx})^2} \qquad (1)$$

The Manhattan distance formula expression is:

$$h(i, j) = |k_{i1} - k_{j1}| + |k_{i2} - k_{j2}| + \cdots + |k_{in} - k_{jn}| \qquad (2)$$

3.3 Database System

The database, hardware, software, and personnel are the four components that make up a basic database system. Any management system has to have a database. The primary purpose of the database is to hold the data that the system generates or uses while handling information [26]. The term "hardware" refers to the various hardware components that make up the system, such as external and storage devices. The software that runs the data in the database is referred to as software. The software performs some business operations by adding, removing, altering, and querying database data. The four categories of staff members are database administrators, system analysts, application staff, and end users.

4. RESULTS AND DISCUSSION

The current examination conducts accomplishment assessments utilizing an organization's management data set to affirm the viability of the organization's marketing management framework in the setting of the digital economy from a green environment perspective. The presentation worth of an organisation management system in the traditional B/S and C/S modes is analyzed in this research. In addition, it evaluates the accuracy of the initial data center points after randomly extracting them from the organization management database in a variety of modes. Table 76.1 shows how accurate the analysis of the dataset was at various points in the company's marketing management framework.

Table 76.1 shows how the dataset's validity is determined using the starting random data. The B/S model's mean precision is 87.93%, whereas the C/S model's mean accuracy is 75.89%. Compared to the conventional C/S model, the mean accuracy of the B/S model is 12.04% greater.

Utilizing a digital environment, a K-means clustering algorithm was used to divide the company's customer value, potential customers, customer happiness, retention, and customer turnover. Accuracy of the consumer segmentation both after and before the K-means clustering method was evaluated. The findings of the before and after comparison are displayed in Fig. 76.2.

Table 76.1 Result of C/S and B/S Models regarding accuracy

Model	First Data Location	Average Accuracy (%)
C/S Model	128534	75.89
	39673422	
	243677	
	41678933	
	2436579	
B/S Model	36732	87.93

■ Before (%) ■ After (%)

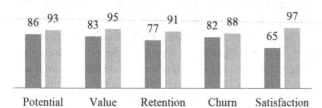

Fig. 76.2 Accuracy of consumer segmentation both prior to and following the application of K-means clustering

After comparing the organizational marketing management system with B/S architecture, Figure 76.3 shows the changes in real-time information acquisition by the level of system informatization, organization personnel, data processing capabilities, the sharing of marketing data, and the security of system information.

■ Before(%) ■ After(%)

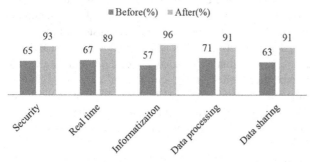

Fig. 76.3 Modifications to the organization's functionality both after and before the B/S model

5. CONCLUSION

This study looked at the development of organizational marketing management systems from a green ecology standpoint within the framework of the digital economy. The research project started out by looking at the digital economy. Next, in accordance with data mining, a significant portion of the system's client data was examined, and the results were arranged into a database. The system database was eventually acquired. In the testing portion of the article, the accuracy of the client segmentation within the system was assessed, and the system's overall functioning was assessed in combination with the K-means clustering technique. The system's analysis of potential customers, customer value, churn, and satisfaction was shown to be more accurate than it had ever been. A B/S model examination of the improvements in the business's functionality before and after the enterprise marketing management system was deployed revealed that the upgraded enterprise marketing system had an unmatched degree of informatization. The enterprise marketing management system improved the company's financial results in the setting of the digital economy. In addition to analyzing the system's operation, the study looked at how the K-means clustering algorithm affected the system's individual components and how much it helped businesses adapt to the digital economy.

References

1. Belassi, W. (2013). The impact of organizational culture on the success of new product development projects: A theoretical framework of the missing link. *The Journal of International Management Studies, 8*(2), 124–133.
2. Li, K., Kim, D. J., Lang, K. R., Kauffman, R. J., & Naldi, M. (2020). How should we understand the digital economy in Asia? Critical assessment and research agenda. *Electronic commerce research and applications, 44*, 101004.
3. Kaewkhum, N. (2020). Television industry and its role in the new media landscape under the system of digital economy.
4. Awan, U., Shamim, S., Khan, Z., Zia, N. U., Shariq, S. M., & Khan, M. N. (2021). Big data analytics capability and decision-making: The role of data-driven insight on circular economy performance. *Technological Forecasting and Social Change, 168*, 120766.
5. Li, X., Wang, J., & Yang, C. (2023). Risk prediction in financial management of listed companies based on optimized BP neural network under digital economy. *Neural Computing and Applications, 35*(3), 2045–2058.
6. Jia, D., Zhang, H., & Han, X. (2023). Construction of Enterprise Marketing Management System in Digital Economic Environment from the Perspective of Green Ecology. *Sustainability, 15*(2), 1299.
7. Dudko, P. M. (2020). Management of Brands' portfolio Positioning of Domestic Enterprises. *Management, 31*(1), 67–76.
8. Kostakis, V., Roos, A., & Bauwens, M. (2016). Towards a political ecology of the digital economy: Socio-environmental implications of two competing value models. *Environmental Innovation and Societal Transitions, 18*, 82–100.
9. Melnyk, L., Dehtyarova, I., Kubatko, O., Karintseva, O., & Derykolenko, A. (2019). Disruptive technologies for the transition of digital economies towards sustainability. Економічнийчасопис-*XXI*, (9–10), 22–30.

Note: All the figures and table in this chapter were made by the author.

Recent Trends in Engineering and Science for Resource Optimization and
Sustainable Development – Prof. (Dr.) Dorota Jelonek et al. (eds)
© 2024 Taylor & Francis Group, London, ISBN 978-1-032-98030-0

77

Use of Artificial Intelligence for Optimization of Economic and Sustainable Aspects of Construction Projects

Manisha[1]

Assistant Professor, Computer Science Department,
Agrawal P.G. College, Jaipur, Rajasthan

Shashikumara S R[2]

Assistant Professor, Civil Engineering,
JSS Academy of Technical Education

Abhishek R[3]

Assistant Professor, Civil Engineering,
JSS Academy of Technical Education

Sachin K C[4]

Research Scholar, Department of Civil Engineering,
RV College of Engineering, Visvesvaraya Technological University,
Belagavi

Deepak Tulsiram Patil[5]

Department of Business and Accountancy,
Lincoln University College, Petaling Jaya, Malaysia

Amiya Bhaumik[6]

Department of Business and Accountancy,
Lincoln University College, Petaling Jaya, Malaysia

■■■■■ **Abstract**

The built environment is greatly influenced by construction projects, which also have a big effect on the environment and global economies. The choices taken throughout a project's planning and implementation phases can have a lasting effect on its economic and environmental performance. Artificial intelligence (AI) models have proven to be capable of managing challenging, erratic, and dynamic tasks. Project delays are the primary problem that the construction industry faces because of the inherent complexity and unpredictability of the work involved in building. Therefore, it is essential to carefully consider these factors and make decisions that are consistent with the goals of sustainable development. Economic criteria and sustainable development can be balanced and optimized by using a hybrid artificial intelligence model known as an

[1]Manishay191@gmail.com, [2]shashikumarsr@jssateb.ac.in, [3]abhishekr@jssateb.ac.in, [4]kcsachin333@gmail.com, [5]dtpatil@lincoln.edu.my, [6]amiya@lincoln.edu.my

DOI: 10.1201/9781003596721-77

SVM classifier with genetic algorithm optimization, which forecasts project delays in construction project scheduling. Statistical performance measure indices are used to validate the suggested SVM-GA against the traditional SVM model. All things considered, the suggested methodology demonstrated a strong and trustworthy method for project delay prediction that supports sustainability and monitoring of construction project management.

▬▬▬▬ Keywords

Construction projects, Artificial intelligence, Supper vector machine (SVM), Genetic algorithm, and Sustainability

1. INTRODUCTION

Due to the many issues the construction sector has, which have impeded its development, its productivity levels are extremely low in contrast to other industries like production (Ahmed et al., 2019). Compared to projects in most other industries, construction projects typically have a longer lifespan and are more complicated. The construction industry is really one of the least digitalized in the world, according to Monteiro et al. (2017), and the majority of stakeholders are aware of the sector's history of resistance to change.

The success of a project is directly impacted by the particular obstacles that face project managers in the construction industry. These difficulties stem from the fact that every project has different staff, equipment, locations, and logistics, in addition to other variables like economics and cost fluctuations (Rafiei et al., 2018). These may raise the level of uncertainty in project planning and execution, increasing the risk of overspending, project delays, and conflicts amongst contractors, staff, and customers. Furthermore, the conventional project management techniques employed by modern construction firms mostly rely on the project managers' knowledge, and data is manually gathered through decentralized storage in a range of non-digital formats (You, Z., & Wu, C. 2019).

A construction schedule is one of the most important elements for creating an optimization model for construction projects. It facilitates cash flow analysis, resource balancing, and the preparation of several variants that are subject to optimization in addition to planning and directing the construction process (by demonstrating the linkages between activities and their durations). Construction schedule optimization is a complex problem that falls under the category of NP-difficult problems, making it difficult to solve. This indicates that as an issue gets bigger, the amount of time needed to solve it grows exponentially (Rosłon et al., 2018).

Delay risk is one of the biggest problems that construction companies face. Any occurrence or action that increases the amount of time needed to finish the project according to the terms of the contract is considered a delay. Project performance suffers as a result of delays, which also increase expenses and decrease production. Its impact extends to the owner, consultant, and contractor in terms of litigation, dispute, and arbitration.

2. LITERATURE REVIEWS

2.1 Sustainability in Construction

The idea of meeting both current and future demands without jeopardizing the interests of others is the foundation of project sustainability. The concept of the three foundations of sustainability—the economic, environmental, and social goals—is a generally acknowledged perspective on sustainability (Ranjbari et al., 2021). The following explains each of their objectives.

- **Economic sustainability**

  Economic sustainability's main objectives are to provide value, generate financial profits, and provide a return on investment relative to the resources used; all the while reducing costs (Trianni et al., 2017). Effective project management, adherence to rules and guidelines, and efficient risk assessment and mitigation are a few instances of this.

- **Sustainability of the Environment**

  The objective of environmental sustainability, which goes beyond business alone, is to preserve and enhance the environment while reducing the adverse effects of operations. This entails using ecologically friendly materials, cutting back on material use, and decreasing energy consumption (Hong et al., 2021).

- **Social Sustainability**

  The preservation and enhancement of human life quality is the aim of social sustainability. Clients, staff members, contractors, and anybody else who might be impacted by the project's completion are all included in this. Enhancing diversity in the workplace, enhancing training and development, enhancing health and well-being, and making a positive impact on society are some ways to achieve this (Fatourehchi et al., 2020).

2.2 Artificial Intelligence

AI is defined as "the study and growth of computer systems intended to carry out tasks normally requiring human intelligence, such as recognition of images, decision making, and language translation." Reinforcement learning (RL), gaining from named datasets for the expected result and info, unaided learning for coordinating unlabeled information, and planning from conditions to activities to expand reward are a portion of the learning methodologies (Abioye et al., 2021). A couple of instances of AI methods are ANN, multivariate-linear regression (MLR), random forests, support vector machine, decision trees, logistic regression (LR) and K-means, Bayesian interface (BI). Before granting a contract, studies used an artificial neural network (ANN) for prequalification. Through learning, the ANN instructs the network. Before putting the model into use, validation of the network is required after training to ensure that the ANN model can be trusted to handle the challenges. An ANN network consists of three layers, each with a unique configuration of neurons: the input layer, any hidden layers, and the output layer. Some datasets serve as the input layers' sources of signals. The system's critical computational capability is provided by hidden layers, which are present in the leading phases of ANN.

Natural language processing (NLP) transforms spoken language into machine-readable text for use in social networking, e-commerce, customer service, education, entertainment, finance, and healthcare applications (Hagiwara, M. 2021). Processing typewritten reports and documentation can be computed in relation to construction project management, and NLP can be used to learn new things. One example is the analysis of accident data in the building industry to identify risk factors for incidents (Baker et al., 2020).

Finding the optimal strategy to minimize or maximize an objective function while taking predetermined restrictions into consideration is the problem of intelligent optimization (Pan, Y., & Zhang, L. 2021). The genetic algorithm (GA) is an illustration of this kind of algorithm. This can be applied to system results optimization to enhance model performance in construction project management decision-making.

3. METHODOLOGY

3.1 Support Vector Machines

Throughout the past 40 years, the SVM methodology has become more and more popular as a rapid and efficient statistical learning method. A support vector machine is a machine learning supervised technique that maximizes the range of differences between two groups. It is commonly used as a classification tool. SVM can be applied to prob-lems involving regression or classification by utilizing a kernel trick technique to alter the data. In summary, the SVM basically followed the same values, but with minor deviations. The SVM can be elucidated with a clear comprehension of the three fundamental ideas that form the basis of SVM models, as depicted in Fig. 77.1:

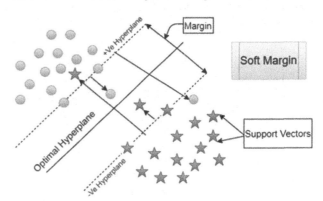

Fig. 77.1 SVM concept

The research used SVM compared them to a hybrid SVM model that was bagged and tuned using a genetic algorithm. The bootstrapping method statistical resampling method, which generates a range of training sets and prepares the SVM model to build an ensemble, is the foundation of the bagging strategy. To create an ideal SVM model, a large number of kernels and parameters must be adjusted. Typically, this is done by utilizing a trial-and-error approach to determine the parameters. The individual chromosome is assigned by the implemented GA using a combination of three real value variables. A randomly generated chromosomal population is used to start the GA. The cross-validated correlation coefficient result demonstrated that the hybrid SVM models optimized by GA and bagging performed better than the simple SVM.

3.2 Collection of Data

Sixty-five completed projects with varying degrees of schedule overrun were included in the collated data. In the US, these projects were carried out. Historical project records that were examined to determine the risk delay in construction projects were among the data that were gathered. Contract paperwork, specifications, records of modification orders, and schedule baselines were among these materials.

3.3 Performance Measure

Class performance and overall performance metrics were used to assess the anticipated model's effectiveness. Preciseness, sensitivity, and specificity were used to gauge class performance. The prediction model's overall performance was assessed using accuracy, and classification error.

4. EMPIRICAL RESULTS

The identification of the root cause of the delay issues was accomplished through analysis of the collected data based on 65. Figure 77.2 displays the characteristics of the collected data as well as the breakdown of the sources of delays in the construction project.

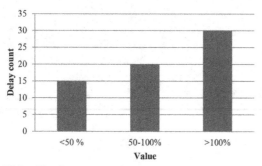

Fig. 77.2 The frequency of delays in construction activities

Figure 77.3 displays the number of projects with a 50%–100% delay, a 50% delay, and a >100% delay determined by the data supplied. The numbers are 15 (23%), 20 (30.8%), and 30 (46.2%), respectively. It is evident that a sizable portion of projects fall into the category of delays exceeding 100%. During the training and testing stages, the hybrid SVM-GA and SVM models' performance was assessed. The computation results for precision, sensitivity, specificity, accuracy and classification error are shown in Tables 77.1 and Figure 77.3.

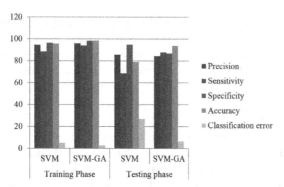

Fig. 77.3 Bar graph of performance measure of both models

Table 77.1 Evaluation of two classifiers according to testing and training phase performance metrics

Performance measure	Training Phase		Testing phase	
	SVM	SVM-GA	SVM	SVM-GA
Precision	94.6	95.9	85.65	84.43
Sensitivity	88.49	93.98	68.6	87.75
Specificity	96.48	98.4	94.89	86.8
Accuracy	95.75	98.54	79	93.78
Classification error	5.13	2.72	27	6.56

According to performance measures for both overall and class performance, Tables 77.1 and Fig. 77.3 compare the SVM-GA and SVM models. According to these findings, the SVM-GA model achieved training-phase accuracy and classification error values of 98.54% and 2.72%, respectively, whereas the SVM model yielded accuracy values of 95.75% and 5.13%, respectively. It is evident that both models performed well during the training phase. SVM-GA performed better than the SVM model, according to testing phase findings. The SVM-GA classification error and accuracy numbers that are given are 6.56% and 93.78%, respectively.

5. CONCLUSION

Artificial Intelligence presents a fresh method to overcome barriers and increase efficiency, which has the potential to change a number of areas. AI can solve the productivity problem as well as a host of other issues plaguing the building industry. According to the causes of delays, the current study offered an assessment technique that can forecast the extent of delays in building projects. To accommodate the complex and ever-changing data in the construction industry, a hybrid SVM-GA model was created. Performance measure indices were used to compare and assess the SVM-GA model against the traditional SVM model. The SVM-GA model outperformed the SVM model, according to the analysis's findings. The accuracy and classification error provided by the SVM-GA were 93.78% and 6.56%, respectively. These outcomes demonstrate how well the model can manage the complexity and nonlinearity of data in the construction industry. The genetic algorithms ability to solve issues with many solutions was also demonstrated by the results.

References

1. Ahmed, M., Qureshi, M. N., Mallick, J., & Ben Kahla, N. (2019). Selection of sustainable supplementary concrete materials using OSM-AHP-TOPSIS approach. *Advances in Materials Science and Engineering, 2019*.
2. Monteiro, P. J., Miller, S. A., & Horvath, A. (2017). Towards sustainable concrete. *Nature materials, 16*(7), 698-699.
3. Rafiei, M. H., & Adeli, H. (2018). Novel machine-learning model for estimating construction costs considering economic variables and indexes. *Journal of construction engineering and management, 144*(12), 04018106.
4. You, Z., & Wu, C. (2019). A framework for data-driven informatization of the construction company. *Advanced Engineering Informatics, 39*, 269-277.
5. Rosłon, J., Książek-Nowak, M., & Nowak, P. (2020). Schedules optimization with the use of value engineering and NPV maximization. *Sustainability, 12*(18), 7454.

6. Ranjbari, M., Esfandabadi, Z. S., Zanetti, M. C., Scagnelli, S. D., Siebers, P. O., Aghbashlo, M., ... & Tabatabaei, M. (2021). Three pillars of sustainability in the wake of COVID-19: A systematic review and future research agenda for sustainable development. *Journal of Cleaner Production, 297*, 126660.

7. Trianni, A., Cagno, E., & Neri, A. (2017). Modelling barriers to the adoption of industrial sustainability measures. *Journal of Cleaner Production, 168*, 1482-1504.

8. Hong, J., Kang, H., An, J., Choi, J., Hong, T., Park, H. S., & Lee, D. E. (2021). Towards environmental sustainability in the local community: Future insights for managing the hazardous pollutants at construction sites. *Journal of Hazardous Materials, 403*, 123804.

9. Fatourehchi, D., &Zarghami, E. (2020). Social sustainability assessment framework for managing sustainable construction in residential buildings. *Journal of building engineering, 32*, 101761.

10. Abioye, S. O., Oyedele, L. O., Akanbi, L., Ajayi, A., Delgado, J. M. D., Bilal, M., ... & Ahmed, A. (2021). Artificial intelligence in the construction industry: A review of present status, opportunities and future challenges. *Journal of Building Engineering, 44*, 103299.

11. Hagiwara, M. (2021). *Real-world natural language processing: practical applications with deep learning.* Simon and Schuster.

12. Baker, H., Hallowell, M. R., & Tixier, A. J. P. (2020). Automatically learning construction injury precursors from text. *Automation in Construction, 118*, 103145.

13. Pan, Y., & Zhang, L. (2021). Roles of artificial intelligence in construction engineering and management: A critical review and future trends. *Automation in Construction, 122*, 103517.

14. T. Parashar, K. Joshi, R. R. N, D. Verma, N. Kumar and K. S. Krishna, "Skin Disease Detection using Deep Learning," 2022 11th International Conference on System Modeling & Advancement in Research Trends (SMART), Moradabad, India, 2022, pp. 1380-1384, doi: 10.1109/SMART55829.2022.10047465.

Note: All the figures and table in this chapter were made by the author.

Recent Trends in Engineering and Science for Resource Optimization and Sustainable Development – Prof. (Dr.) Dorota Jelonek et al. (eds)
© 2024 Taylor & Francis Group, London, ISBN 978-1-032-98030-0

78

Application of Blockchain Technologies for Use of Cryptocurrency in Fees Payment Portal

Rashi Saxena[1]
Assistant Professor, Department of AIMLE,
Gokaraju Rangaraju Institute of Engineering &
Technology, Hyderabad

Simmi Madaan[2]
Assistant professor, Faculty of Engineering & Technology
SGT University, Gurugram

S. Neelima[3]
Associate Professor and Head, School of Management,
Department of BBA CA, KPR College of Arts Science and Research,
Arasur, Coimbatore

Uma Reddy
Department of Artificial Intelligence and Machine Learning,
New Horizon College of Engineering, Bangalore

Navdeep Singh
Lovely Professional University, Phagwara, India

Vipul Vikas Masal[4]
Principal, Computer Engineering,
Smt. Geeta D. Tatkare Polytechnic, Roha, MSBTE,
Mumbai, Maharastra

Abstract

Blockchain technology, which is known for being safe and decentralized, has attracted a lot of interest from a variety of sectors. The creative use of blockchain technology to integrate cryptocurrencies into fee payment websites for educational institutions is examined in this research study. This study intends to evaluate the viability, advantages, and difficulties linked to incorporating cryptocurrency as a means of payment for academic fees by utilizing the transparency, permanence, and effectiveness of blockchain technology. at order to pinpoint problems and opportunities for development, the research technique entails a thorough examination of the state of fee payment procedures at educational institutions. The design and development of a powered by blockchain fee payment gateway will next be the subject of the study, with an emphasis

Corresponding author: [1]rashisaxena.cse@gmail.com, [2]simmi.madaan@srmuniversity.ac.in, [3]neelima.s@kprcas.ac.in, [4]mvipulv01@gmail.com

DOI: 10.1201/9781003596721-78

on the incorporation of cryptocurrencies like Bitcoin, Ethereum, and other pertinent tokens. This work gives a safe entry point with continuous transaction flow. Bitcoin is used to make the payment, which is then sent to the admin's wallet. Blockchain technology is utilized to store student data and course enrollment information, ensuring privacy and security. The information is kept in blocks that are unchangeable and make it simple to identify the person who changed the information. This work is being created with the intention of using cryptocurrencies as a payment mechanism in the future.

▬▬▬ Keywords

Blockchain, Cryptocurrency, Fees payment portal

1. INTRODUCTION

The financial environment is not an exception to the revolutionary changes brought about by the widespread adoption of blockchain technology in recent years. As decentralized digital assets, cryptocurrencies have become well-known as a cutting-edge type of money that facilitates financial transactions with efficiency, security, and transparency. The field of fee payment portals, especially in educational institutions, is one interesting area where blockchain technology is being applied with remarkable success. Conventional fee payment systems sometimes face difficulties including inefficiency, security issues, and delays. For companies trying to streamline and modernize their financial operations, processing bitcoin payments through blockchain technology seems like a good solution. This study explores the possible advantages, difficulties, and ramifications for consumers and educational institutions as it dives into the usage of blockchain technologies to enable digital currency transactions within fees payment portals. Blockchain is a decentralized ledger technology that powers cryptocurrency, which is best represented by the original Bitcoin and several altcoins Balakrishnan et al., (2023). Over a network of computers, the blockchain is an immutable, distributed ledger that keeps track of and validates transactions. Blockchain's decentralized architecture improves the privacy of financial transactions, guarantees transparency, and does away with the need for middlemen. Learning institutions are increasingly thinking about integrating blockchain technology to enable the smooth usage of cryptocurrencies on fee payments, taking advantage of these built-in benefits. There are several different driving forces underlying this paradigm change. First off, transactions based on blockchain technology provide almost immediate processing times, eliminating the hold-ups caused by conventional payment systems. When it comes to paying academic fees, this quickness is especially important because pupils as well as schools frequently place a high value on punctuality. Second, a strong defense versus fraud and illegal access is offered by the security mechanisms built into blockchain technologyLundqvist et al., (2017).

This takes care of the worries about how vulnerable sensitive financial data is in traditional payment methods. International students particularly benefit from the usage of cryptocurrencies for fee payment websites because they enable cross-border transactions. Students studying abroad face obstacles since traditional banking systems frequently charge exorbitant fees and take a long time to process cross-border payments. Payments with cryptocurrencies, made possible by blockchain technology, provide a more economical and effective option that promotes financial inclusion. Notwithstanding the possible benefits, there are drawbacks to using blockchain technology in fee payment websites for cryptocurrency payments Lu et al., (2021). Notable obstacles that must be overcome for widespread use include price volatility, regulatory uncertainty, and the requirement for user education. To clarify the implications and things to consider when educational institutions are negotiating this innovative financial environment, this research aims to thoroughly examine the subtleties of implementing blockchain innovations in relation to digital currency transactions for academic fees.

2. LITERATURE REVIEW

2.1 Using Blockchain to Process Online Payments

Cryptocurrencies are virtual currencies that are used in e-commerce and banking, among other applications. It is distributed among peers via a decentralized public ledger known as Blockchain. Specialized internet markets allow you to swap cryptocurrencies for fiat money. They have a fluctuating pace of trade with the major worldwide monetary standards, including the US dollar, the English pound, and the euro. There is very little of them left (Martucci 2018). The Blockchain, which is an expert record for digital forms of money, monitors all exchanges and movement including those monetary standards along with proprietor subtleties. Miners are in charge of each node of the software network that powers cryptocurrencies. Each node stores identical copies of the Blockchain. By performing cryptographic tasks, the excavators recover exchange records, affirm their credibility, and make new blocks. The blocks are irreversible and are affixed to the previous blocks in the Blockchain. A confidential key is expected by every digital currency proprietor to trade their units and check their personality (Jaag and Bach 2016). A disseminated record comprised of each and every cryptographic money exchange that has been finished inside an organization is known as a blockchain. It goes about as the organization's only wellspring of truth. A blockchain is comprised of a progression of blocks, or information bundles. A few exchanges make up a block. The beginning block is the name of the underlying block. There are a few hubs in a blockchain framework, and everyone has its own duplicate of the data set. The nodes speak with each other to get a consensus on adding a blockchain. Consensus building is the process of reaching this understanding. A chain is added and becomes part of the distributed database name when it is entered into the ledger across all nodes. A transaction record is irreversibly recorded to the Blockchain and cannot be deleted. This characteristic is known as immutability. In a blockchain, there are two models that govern the ability to conduct transactions: permissioned and permission-less. To conduct transactions in a permissioned blockchain, users must have been enrolled in the blockchain. As a result, there is more systemic trust. The Blockchain architecture eliminates intermediaries, which promotes data security.

2.2 Main Security Problems with Blockchain Technology for Online Payments

Vyas and Lunagaria (2014) and "Distributed Ledger Technology in Payment, Clearing and Settlement" (2017) both mention the possibility of security problems while using Blockchain for online payments. These risks include:

- *Resilience and reliability:* To ensure dependability and uninterrupted operation, multiple nodes are offered. On the other hand, this might potentially provide bad actors more opportunities to get access, jeopardizing the ledger's security and integrity.
- As technology develops further, the instruments used in cryptography today may become antiquated and useless. Intrusions and security risks may arise from the integration of decentralized ledgers into the current infrastructure.
- When a bitcoin miner or group of miners acquires computing power larger than 50% and has the capability to reverse transactions, it is known as a ">50%" attack.
- Spending double happens when a hacker conducts multiple transactions using a single coin.
- Selfish mining, in whicha user group's earningsexceedstheirmining capacity.

Blockchain payment solutions include security risks and issues just like any other payment mechanism. The following is an overview of the security issues (Lin and Liao 2017):

- *Wallet software hacking attempts:* Users of cryptocurrencies utilize wallets to manage the currency they possess. For this reason, they need to have an offline backup and be encrypted. Attacks known as distributed denial of service (DDoS) might potentially affect online wallets.
- *Time jacking attacks:* In these attacks, a hacker changes a node's network time counter, tricking the node into accepting a false blockchain. The outcome is double-spending and wasted computing resources.
- *">50%" Attack:* One of the largest threats to a Blockchain network is when an individual or group of individuals seizes control of more than 50% of the processing power utilized for mining. They achieve this by finding the "nonce" value of a block. Subsequently, they have the ability to halt the mining of valid blocks and execute, modify, and reverse transactions.
- Two transactions with identical input will not be accepted by peers on the blockchain. The fraud transactions will be confirmed and the initial transaction won't be confirmed because they will only validate the first one that reaches them, Kim (2022).
- *Selfish mining:* It enables a big enough pool to make more money than it can through mining. Sincere miners will have to make unnecessary calculations as a result of the attacking miners. The self-centered miner will split their Blockchain covertly,keep each block private, and profit more.
- *Fork Issues:* A fork occurs when there is a disagreement over the version agreement of a decentralized node during software upgrades. This is a significant issue since it touches on many different Blockchain applications. The consensus mechanism in nodes also varies when a new version of the Blockchain software is implemented. Old and new nodes, issues occur when they try to reach a consensus. The consensus or agreement between the new and old nodes is incompatibleAdewole et al., (2020).

3. METHODOLOGY

This section contains a discussion of the strategies that were applied to this work. Additionally, this system employs Dapps and smart contracts on blockchain to store student data Balakrishnan et al., (2023).

3.1 Proposed Architecture

Figure 78.1 depicts the suggested system architecture's work. The students can use this technology to construct a

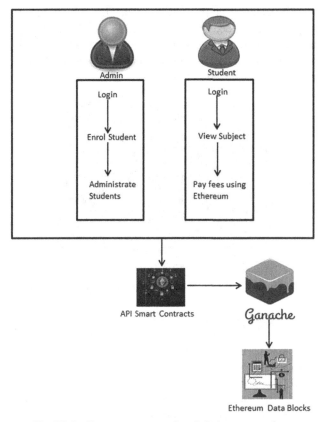

Fig. 78.1 Payment system for digital currency fees

secure payment site using cryptocurrencies. The student data is kept in a public ledger and the payment is incorruptible; authentication is required through an agreement. Students can use this work to pay their educational expenses with any kind of digital currency. In this study, smart contracts are used to maintain and retrieve data on a blockchain. The genesis block with administrator access is delivered, and the admin inputs the student data and generates the blocks. The admin, student, and transaction modules are the three primary parts that comprise the system architecture. Students are enrolled and their personal data is stored in blocks in the admin module, which is the first layer of the architecture. The "student module" sub layer enables students to view their fee information and semester receipts. This layer can only be read. A secure blockchain consensus process is used to make digital payments, and the transaction component is the last layer of the architecture used by the system to achieve the main objective of the project. The admin, student, and transaction modules are the three primary parts that comprise the system architecture. The administration module, the student module, and the transaction module make up the three tiers of the architecture.

3.2 Blockchain

Blockchain is a dispersed, encrypted ledger that serves as a public repository for unchangeable, incorruptible information. Blockchain is a public ledger that is distributed. Every trade or digital event in the general ledger requires the approval of more than 50% of the network's users in order to be verified. First-time participants can agree on how a particular transaction or virtual event can happen using the Blockchain without the need for a central authority. The owner's digital signature authenticates every transaction in this ledger and guards against tampering Lundqvist et al., (2017).

3.3 Smart Contract

On a blockchain platform, smart contracts have been programs that only execute when certain criteria are met. They are generally employed to put agreements into action among participants, providing them with instant results and avoiding time wastage or the need for an intermediary. When certain requirements are satisfied, smart contracts employ computerized algorithms that carry on the workflow by expediting the subsequent task (Lin and Liao 2017).

3.4 DApps

The part of the application that interacts with the blockchain and manages every user's state on the network is called a DApp. The basic principles of a decentralized application are shown via smart contracts. DApps' user interface is the same as that of any other standard website or application for mobile devices. A decentralized application's back end is designed to support all business logic, while its front end is designed to reflect what consumers would seeSekar et al., (2022).

3.5 Digital Money

An unreadable string representing an ounce of currency is called a cryptocurrency. It serves as a safe record of transactions, including buying, selling, and transferring, and is monitored and controlled by a peer-to-peer network called a blockchain. Cryptocurrencies are not issued by nations or other financial entities, in contrast to fiat money, because they are decentralized Adewole, (2020).

3.6 Gateway for Payment with Cryptocurrency

A gateway for cryptocurrency payments is a processor for digital currencies, much like how bank credit cards are obtained and payment processors and gateways are utilized. By using cryptocurrency gateways, you may take digital payments and get paid in fiat money right away. Payment gateways are companies that take on the perceived risk of bitcoin payments by using their wallets to process transactions between retailers and their clients.

4. EMPIRICAL RESULTS

Figure 78.2 displays the QR code's data format, which contains digitalsignature1 information and the value obtained from the online fee portal using their address to digitally sign the message "address=1ZwGYT6dBOsegTWu-1FYRQbXKcF6kX98kKdKKfp, amount = 850, timestamp = Fri, 23 Dec 2023 15:25:09GMT,m_tx_id=7436286." To facilitate the customer's simple payment, the Institution's online store displays a QR code with the digitalsignature1 parameters as well as M_Address, Amount, Date and time, and M_TX_ID.

```
{
"address":"1ZwGYT6dBOsegTWu1FYRQbXKcF6kX
98kKdKKfp",
"amount":"850",
"timestamp":"Sat ,23 Dec 2023 15:25:09 GMT",
"m_tx_id":"7436286",
"digitalsignature1":"Hy+Ngvvo3bfkXSuh3L6d0gLif
T9y/cZYVNWP/54gaToVapQu1vRSm1QvdtwNHw
5LuBCkLz1rchH3mfh+rUb4UAg="
}
```

Fig. 78.2 Information from QR codes in JSON format

To purchase a course or syllabus using the institution's online fee payment gateway, the student fills out the enrolment form. The student hits the "QR Code Scan" option in order to read the QR code on the online fee payment portal screen. The QR code will be scanned and analyzed in order to obtain the values of M_address, Value, Date and time, M_TX_ID, and digitalsignature1. Additionally, the program creates digitalsignature2 by using the student's private key and digitalsignature1 in the signmessage() method when the student clicks the "Payment" button, as seen in Table 78.1. Subsequently, the blockchain system receives

Table 78.1 Information in the digital signature2 and QR code

Element	Size in bytes	Details
M_Address	36	obtained by use of the QR code scan
Fees	String	obtained by use of the QR code scan
Date and Time	30	obtained by use of the QR code scan
M_TX_ID	8	obtained by use of the QR code scan
digitalsignature1	90	obtained by use of the QR code scan
digitalsignature2	90	Created using the function signmessage (student's address, digitalsignature1)

the parameters from the student application, which include M_Address, Value, Date and time, M_TX_ID, digitalsignature1 and digitalsignature2. The data structure of the payment details that the consumer sends to the blockchain platform is displayed in Fig. 78.3.

```
{ "address":"
1ZwGYT6dBOsegTWu1FYRQbXKcF6kX98kKdKKfp
",
"amount":"100",
"timestamp":" Sat ,23 Dec 2023 15:25:09 GMT",
"m_tx_id":"7436286", ",
"digitalsignature1":"Hy+Ngvvo3bfkXSuh3L6d0g
LifT9y/cZYVNWP/54gaToVapQu1vRSm1QvdtwN
Hw5LuBCkLz1rchH3mfh+rUb4UAg="
"digitalsignature2":"ICsWAZLREsxiq9hNk/dFtpdf
EBXtpOjPwKAsbWIIQ5/wdp2DKfdSWPPSb7F85y
TOYa9zsTrSX/TDsXKMMc4qRYg=" }
```

Fig. 78.3 Payment details for students in JSON format

5. CONCLUSION

The student fees payment portal made possible by this work gives a safe entry point with continuous transaction flow. Bitcoin is used to make the payment, which is then sent to the admin's wallet. Blockchain technology is utilized to store student data and course enrolment information, ensuring privacy and security. The information is kept in blocks that are unchangeable and make it simple to identify the person who changed the information. This work is being created with the intention of using cryptocurrencies as a payment mechanism in the future. Blockchain technology is now an emerging payment method. This leads to the more effective method of storing, retrieving, and changing data in smart contract storage. An array of technological, legislative, and user-related factors is expected to impact the outcomes of utilizing blockchain-based technologies enabling digital currency transactions in fees payment portals. Policymakers, stakeholders, and educational institutions attempting to negotiate the changing terrain of financial transactions having the education sector will find great value in a thorough examination.

References

1. Adewole, K., Saxena, N., & Bhadauria, S. (2020). Application of cryptocurrencies using Blockchain for e-commerce online payment. In *Blockchain for Cybersecurity and Privacy* (pp. 263–305). CRC Press.
2. Balakrishnan, S., Naulegari, J., Devi, A., Dharanyadevi, P., &Narashiman, D. (2023, November). Digital Currency Fees Payment Portal Using Blockchain Technology. In *2023 International Conference on System, Computation, Automation and Networking (ICSCAN)* (pp. 1–6). IEEE.

3. Jaag, C., & Bach, C. (2017). *Blockchain technology and cryptocurrencies: Opportunities for postal financial services* (pp. 205–221). Springer International Publishing.

4. Kim, S. I., & Kim, S. H. (2022). E-commerce payment model using blockchain. *Journal of Ambient Intelligence and Humanized Computing*, *13*(3), 1673–1685.

5. Lin, I. C., & Liao, T. C. (2017). A survey of blockchain security issues and challenges. *Int. J. Netw. Secur.*, *19*(5), 653–659.

6. Lu, W., Wu, L., Zhao, R., Li, X., &Xue, F. (2021). Blockchain technology for governmental supervision of construction work: Learning from digital currency electronic payment systems. *Journal of construction engineering and management*, *147*(10), 04021122.

7. Lundqvist, T., De Blanche, A., &Andersson, H. R. H. (2017, June).Thing-to-thing electricity micro payments using blockchain technology. In *2017 Global Internet of Things Summit (GIoTS)* (pp. 1–6).IEEE.

8. Martucci, B. (2019). What Is Cryptocurrency–How It Works, History & Bitcoin Alternatives. *Martucci. [Electronic resource]. URL: https://www. moneycrashers. com/cryptocurrency-history-bitcoin-alternatives/(accessed 02.09. 2021).*

9. Sekar, S., Solayappan, A., Srimathi, J., Raja, S., Durga, S., Manoharan, P., ...&Tunze, G. B. (2022). Autonomous transaction model for e-commerce management using blockchain technology. *International Journal of Information Technology and Web Engineering (IJITWE)*, *17*(1), 1–14.

10. Vyas, C. A., &Lunagaria, M. (2014). Security concerns and issues for bitcoin. *Int. J. Comput. Appl*, 10–12.

Note: All the figures and table in this chapter were made by the author.

Recent Trends in Engineering and Science for Resource Optimization and Sustainable Development – Prof. (Dr.) Dorota Jelonek et al. (eds)
© 2024 Taylor & Francis Group, London, ISBN 978-1-032-98030-0

79

Integration of Social Media and Machine Learning to Enhance Cryptocurrency Price Forecasting

Amol Dhakne[1]

Associate Professor, Computer Engineering,
D Y Patil International University,
Akurdi, Pune

Mohammed Faez Hasan[2]

Assistant Professor of Finance,
Finance and Banking department, Kerbala University, Iraq

Vijilius Helena Raj

Department of Applied Sciences,
New Horizon College of Engineering, Bangalore, India

Manish Gupta

Lovely Professional University, Phagwara, India

Sateesh Nagavarapu[3]

Professor, CSE, Malla Reddy Engineering College for Women,
Secunderabad, Telangana

Kunal D Gaikwad

Associate Professor, Department of Electronics,
MGSMS Arts Science and Commerce College,
Chopda, Dist, Jalgaon

Abstract

Accurate price forecasting is becoming more and more important for stakeholders and investors as the market for digital currencies continues to show dynamic and frequently unpredictable behavior. In order to improve bitcoin price forecasts, this study combines machine learning methods with data from social media. By utilizing the extensive and up-to-date data accessible via social media sites, this research seeks to pinpoint trends, attitudes, and indications that enhance the accuracy of bitcoin price forecasts. When we examined the models' performance both with and without the use of Twitter sentiment data, we saw that adding the sentiment feature led to consistently higher results, with TwitterRoBERTa-based sentiment providing an average F1 score of 0.88. With an F1 score of 0.94, an optimized Multi Modal Fusion predictor utilizing

Corresponding author: [1]dhakne.amol5@gmail.com, [2]mohammed.faiz@uokerbala.edu.iq, [3]drnskcse@gmail.com

DOI: 10.1201/9781003596721-79

sentiment based on Twitter-RoBERTa was the highest performing model. Through the demonstration of the possibilities of social media analysis of sentiment, onchain data integration, along with the practical use of a Multi Modal Fusion model, this study makes significant improvements to the field of financial forecasting by enhancing the precision and resilience of machine learning (ML) algorithms for predicting market trends and offering traders, brokers, and investors a useful tool for making well-informed decisions.

■■■■■■■ Keywords

Cryptocurrency, Machine learning, Price forecasting, Social media

1. INTRODUCTION

Many virtual currencies have experienced significant rises in value and acceptability since Satoshi Nakamoto'sgroundbreaking introduction of Bitcoin. For example, two of the biggest competitors were Ethereum and Litecoin. Ethereum's rise was fuelled by its smart contract capabilities, while Litecoin's emphasis on improving transaction speed as well as capacity made it a preferred option for regular users. The current state of cryptocurrencies boasts hundreds of coins, majority of which exhibit volatility and a dearth of noteworthy initiatives Härdle et al., (2020). Their popularity gained through social media word-of-mouth influences their volatility and worth. For instance, people may become interested in investing as a result of the enthusiasm these internet phenomena create, especially when it comes to certain coins and viral memes associated with them. Due in great part to exchange platforms' creative marketing approaches, the bitcoin business has experienced exceptional success in recent years Hashemi et al., (2020). With the help of these tactics, which include user-friendly interfaces and instructive materials, investing in cryptocurrencies has become more accessible, spurring market expansion. Furthermore, well-known individuals have used their expansive social media following to publicly support or oppose cryptocurrencies in tweets and speeches, which has increased public interest in the space and caused market volatility Kraaijeveld and De, (2020). This research looks at the factors that influence price swings in the constantly evolving cryptocurrency market, with a focus on market dynamics and social media content in particular. The method we use involves forecasting the fluctuations in cryptocurrency values by examining two substantial datasets: one with market data about specific cryptocurrencies, like prices and trading volumes, and the other with posts from social media on these coins. Correct integration and evaluation of these datasets were achieved by a range of ML techniques and text analytics. When comparing risk-free investments to machine learning projections, which anticipate stock Bianchi et al., (2021), risk in equity Valencia et al., (2019), or mortgage Abraham et al., (2018) premiums, several studies have shown that investors can reap significant economic rewards. Specifically, machine learning techniques often outperform popular regression-based strategies; often even doubling their efficacy this is especially true when using decision trees with neural networks (Bali et al., 2023; Zhou et al., 2023)

Large datasets can be used to extract nonlinear patterns by machine learning approaches, which do not require any prior knowledge of the data. But even conventional machine learning techniques, such neural networks, support vector machines, and multilayer perceptron (MLP) (Hajek et al., 2023), have shortcomings such as over fitting and underutilize the potential to find high-level concealed trends in sequential data pertaining to cryptocurrencies. Deep learning-based forecasting models, which can outperform conventional machine learning techniques, have been utilized to solve these issues (Cui et al., 2022). This conclusion is supported by a recent study by Murray et al. (2023), which shows that in terms of forecasting error, long short-term memory (LSTM) perform better than a variety of other statistical and machine learning techniques. Additional investigation has concentrated on the ability of ensemble learning approaches to lower variance and bias through the integration of several weak learning models (Aggarwal et al., 2020). Gradient boosting outperforms the widely used random forest strategy in terms of accuracy and durability, according to highly acclaimed research by Sun et al. (2020), and that ensemble learning models for forecasting perform better than individual machine learning models.

2. LITERATURE REVIEW

The sentiment included in tweets has been used in the past to try and forecast changes in the price of bitcoin. While their information was labeled utilizing a web-based text sentiment API, Colianni et al., (2015) showed 90% exactness in predicting cost swings utilizing similar directed learning strategies. How much their model looked like the web-based composed opinion Programming interface, as opposed to the exactness concerning forecasting cost developments, was the means by which their precision was estimated. Essentially, utilizing 2.27 million tweets and profound learning calculations at a fundamentally higher recurrence over a time of like clockwork, Stenqvist and Lonno, (2017) had the option to expect changes in the cost of bitcoin with 79% exactness. The typical size of the cost percent rises and percent declines that the models were guaging was not analyzed, and neither of these methodologies utilized information that was straightforwardly named in light of cost varieties. There have also been attempts at more traditional methods of forecasting utilizing cryptocurrency price data from the past. Using supervised learning techniques, Hegazy and Mumford, (2016) were able to forecast the actual price with 57% accuracy. Jiang and Liang, (2017) managed a bitcoin investment that produced price forecasts by using deep reinforcement learning. In terms of portfolio value, they saw a 10x rise. Finally, Bayesian regression was employed by Shah and Zhang, (2014) to double the amount they invested over a sixty-day period. To identify patterns not evident in pricing history data, none of these techniques used press or social media data.

Due to its extreme volatility and abundance of speculative activity, the bitcoin market has drawn increasing attention. For traders, investors, and other financial industry participants, being able to predict bitcoin values accurately has been essential in this regard. One area of exploration that has emerged to help bitcoin cost figures is the utilization of profound learning and AI strategies. Strategies like long-short-term memory (LSTM) neural networks have shown a lot of promise for solving problems with time series and finding nonlinear patterns in price fluctuations (Fischer et al., 2022). Some exploration has utilized the arbitrary timberland and direct relapse ways to deal with survey the

model's viability and the exchanging methodologies' practicability. Livieris et al. (2020) have proposed novel RNN-based models, like GRU, LSTM, and bi-LSTM, to produce exact expectations of bitcoin values. Trade judgments have been bolstered by these methods' ability to pinpoint intricate processes and produce precise results. Other research has looked into how analyzing text on publicly available data might improve the capacity to forecast cryptocurrency prices. Particular research has shown that sentiment analysis on social media posts might provide insightful information about the processes behind price changes. Tools like VADER, Hutto et al., (2014) as well as TextBlob, each of which have shown their efficacy in several similar publications, have been primarily used for the sentiment analysis of tweets pertaining to cryptocurrencies. According to Kim et al., (2022) using on-chain data—that is, data that comes straight from the blockchain—can give important insights into how bitcoin prices fluctuate. Combining change event detection algorithms with on-chain data analysis has also improved the accuracy of bitcoin price projections and helped investors make better judgments.

3. METHODOLOGY

Figure 79.1 below provides an explanation of the research methodology. Together with a Multi Modal Fusion Model, we trained five classification models: Support Vector Machine (SVM), K-Nearest Neighbors (KNN), Logistic Regression (LR), XGBoost and Naïve Bayes (NB). We also determined the sentiment of Tweets using VADAR and Twitter-RoBERTa, Bhatt et al., (2023).

The pipeline shown in Fig. 79.1 represents the experimental procedure utilized in this investigation. We aim to ex-

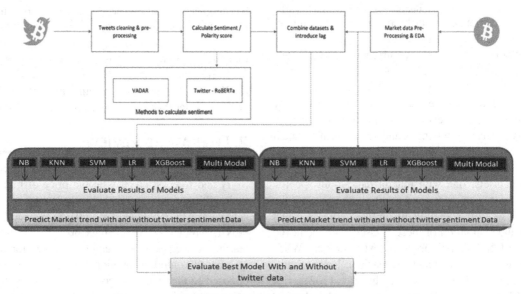

Fig. 79.1 Workflow of the study

plore if adding sentiment data from Twitter enhances the models' performance in forecasting market trends. Therefore, while evaluating regardless of Twitter sentiment, we use the same models and evaluation methodology, as illustrated in Fig. 79.1.

3.1 Dataset

This study makes use of three primary datasets: historical Bitcoin tweets, historical Bitcoin market data, and historical on-chain Blockchain data specifically related to Bitcoin. These datasets were all acquired from the Kaggle repository. Only tweets on Bitcoin from that particular time period are included in the filtered Twitter data.

3.2 Creating Models

The tweets underwent data processing and cleaning procedures. Consequently, the first stage involved removing duplicate entries. Afterwards, terms that started with @ + username—user mentions—were eliminated. The referenced URLs (such as "https") were also eliminated because the emotion of the tweet cannot be determined from this information. Additionally, since stop words (such as "a," "it," "the," and so forth) are categorized as low-level information, they were eliminated to free up the models to concentrate on the crucial data. Eliminating the stop words can also result in a smaller dataset and a faster run time. Punctuations were also eliminated. Lemmatization and word stemming were the next steps. We then used VADAR and Twitter-RoBERTa to forecast the sentiment. A selection of the cleaned tweets and the anticipated sentiments from the new dataset are shown in Table 79.1.

Table 79.1 Dataset of sentiment

Calendar	Tweets with lemmatization	Sentiments
12-08-2023	[huge, opportunity, billion, bitcoin]	0.032178
13-08-2023	[greatest, nft, stocks, btc]	0.066674
14-08-2023	[Bitcoin, anticipate, surge]	0.068656

Regardless of whether the closing price of today or tomorrow is higher or lower, the objective is to predict the direction of the market. We lag each feature by one day in this research because we hypothesized that the features (including sentiment) that were used to forecast market trend gained effect after one day.

$$next\ day\ close_n = close_{n+1} \qquad (1)$$

$$trend_n = next\ day\ close_n - close_n \qquad (2)$$

3.3 Fusion of Multiple Modalities

Utilizing a deep learning methodology, we integrated sentiment evaluation of Twitter with data on onchain parame-

ters, including the number of transactions, difficulty, as well as payments count. Using an LSTM model, our method is unusual in that it can extract and combine several kinds of data, each with distinct temporal qualities and characteristics. Both the longer-term patterns of the on-chain characteristics data and the short-term trends of the sentiment assessment are well-represented by the LSTM model. In particular, our model has two distinct branch offices: one for the modality of sentiment analysis and another for the modality of on-chain attributes. With a thick layer for sentiment analysis and an LSTM layer for on-chain properties, each branch handled its particular modality independently. The prediction was then generated by concatenating the outcomes of each branch, adding another dense layer, and then adding a final sigmoid activated layer.

3.4 Classification

The train_test_split() function of the Sklearn library was utilized to separate the dataset into training and testing sets. With the 'random_state' option set to 0, training and testing consumed up to 80% and 20% of the data, respectively. Moreover, adjusting the hyperparameters of a classification model is one of the most important phases in the process. In this work, we examined hyperparameters tuning using five different classifiers: Support Vector Machine (SVM), Gaussian Naïve Bayes, K-NearestNeighbors (KNN), Logistic Regression (LR), and XGBoost. We employed a number of methods for fine-tuning hyperparameters, such as Bayes Search CV, Randomized Search CV, and GridSearchCV.

4. EMPIRICAL RESULTS

The purpose of this research was to investigate whether adding Twitter data to our algorithms improves their performance in predicting the direction of the Bitcoin market. In order to anticipate the Bitcoin market trend combining the VADER including Twitter-roBERTa emotion analysis models, we analyzed the efficacy of five alternative models for classification as well as a Multi Modal Fusion model. Two distinct datasets—one with sentiment data and the other without—were used to train and evaluate the algorithms. The F1 values for each model, both with and without sentiment data, are displayed in Table 79.2. It is clear that employing Twitter-RoBERTa for a sentiment model yields consistently superior results. The outcomes of comparing the models' performances are shown in Fig. 79.2.

The findings demonstrate that most models perform better when sentiment data is included. The model that was most successful in forecasting market trends based on the available data was Multi-Modal Fusion with Twitter-roBERTa, which received the highest F1 score of 0.940. With a score

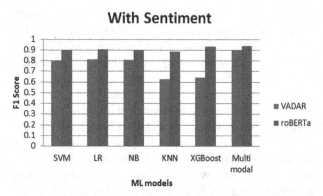
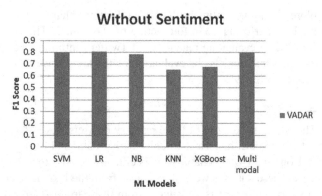

Fig. 79.2 An assessment and comparison of the models' performance

Table 79.2 Modelling outcomes: The scores of F1

ML Models	Model of Sentiment	With Sentiments	W/O Sentiments
SVM	VADAR	0.805	0.802
	roBERTa	0.903	
LR	VADAR	0.813	0.806
	roBERTa	0.905	
Gaussian NB	VADAR	0.806	0.783
	roBERTa	0.900	
KNN	VADAR	0.624	0.653
	roBERTa	0.887	
XGBoost	VADAR	0.642	0.675
	roBERTa	0.933	
Multi modal	VADAR	0.900	0.8
	roBERTa	0.940	

of F1 of 0.933, XGBoost using Twitter-roBERTa analysis of sentiment also yielded remarkable results. With sentiment data, however, models like Gaussian Naïve Bayes and KNN exhibited lower F1 scores. The results of all models declined when sentiment data was removed, indicating that sentiment data can be helpful in forecasting market changes. Based on these findings, it appears that XGBoost combined with roBERTa sentiments analysis is the most effective model for the task at hand. Logistic Regression combined with roBERTa sentiments assessment also yielded competitive results.

5. CONCLUSION

To sum up, combining social media with machine learning offers a viable way to improve the prediction of bitcoin prices. The findings of the research suggest that social media networks, which are real-time sources of sentiment and information, may have a significant influence on the dynamics of the cryptocurrency market. Through the use of machine learning algorithms, important insights can be gleaned from the massive volume of unstructured data produced on these platforms, especially in the areas of NL processing and forecasting. Despite a Multi Modal Fusion method appearing as the most successful, this research offers insightful information about how to integrate sentiment data from Twitter to enhance machine learning (ML) models' effectiveness at predicting market movements. These findings are notable given the increasing importance of online social networking data in the finance industry and the potential benefits of using this data in prediction models.

References

1. Aggarwal, D., Chandrasekaran, S., &Annamalai, B. (2020). A complete empirical ensemble mode decomposition and support vector machine-based approach to predict Bitcoin prices. *Journal of behavioral and experimental finance, 27,* 100335.
2. Bali, T. G., Beckmeyer, H., Moerke, M., &Weigert, F. (2023).Option return predictability with machine learning and big data. *The Review of Financial Studies, 36*(9), 3548–3602.
3. Bhatt, S., Ghazanfar, M., &Amirhosseini, M. (2023). Machine learning based cryptocurrency price prediction using historical data and social media sentiment. *Computer Science & Information Technology (CS & IT), 13*(10), 1–11.
4. Bianchi, D., Büchner, M., &Tamoni, A. (2021).Bond risk premiums with machine learning. *The Review of Financial Studies, 34*(2), 1046–1089.
5. Cakici, N., Fieberg, C., Metko, D., &Zaremba, A. (2024). Do anomalies really predict market returns? New data and new evidence. *Review of Finance, 28*(1), 1–44.
6. Colianni, S., Rosales, S., &Signorotti, M. (2015). Algorithmic trading of cryptocurrency based on Twitter sentiment analysis. *CS229 Project, 1*(5), 1–4.
7. Cui, T., Ding, S., Jin, H., & Zhang, Y. (2023). Portfolio constructions in cryptocurrency market: A CVaR-based deep reinforcement learning approach. *Economic Modelling, 119,* 106078.
8. Fleischer, J. P., von Laszewski, G., Theran, C., & Parra Bautista, Y. J. (2022).Time Series Analysis of Cryptocurrency

Prices Using Long Short-Term Memory. *Algorithms*, *15*(7), 230.

9. Gu, S., Kelly, B., &Xiu, D. (2020). Empirical asset pricing via machine learning. *The Review of Financial Studies*, *33*(5), 2223–2273.

10. Hajek, P., Hikkerova, L., &Sahut, J. M. (2023). How well do investor sentiment and ensemble learning predict Bitcoin prices?. *Research in International Business and Finance*, *64*, 101836.

11. Härdle, W. K., Harvey, C. R., &Reule, R. C. (2020).Understanding cryptocurrencies. *Journal of Financial Econometrics*, *18*(2), 181–208.

12. HashemiJoo, M., Nishikawa, Y., &Dandapani, K. (2020). Cryptocurrency, a successful application of blockchain technology. *Managerial Finance*, *46*(6), 715–733.

13. Hegazy, K., & Mumford, S. (2016). Comparitive automated bitcoin trading strategies. *CS229 Project*, *27*, 1–6.

14. Hutto, C., & Gilbert, E. (2014, May). Vader: A parsimonious rule-based model for sentiment analysis of social media text. In *Proceedings of the international AAAI conference on web and social media* (Vol. 8, No. 1, pp. 216–225).

15. Jiang, Z., & Liang, J. (2017, September). Cryptocurrency portfolio management with deep reinforcement learning.In *2017 Intelligent systems conference (IntelliSys)* (pp. 905–913). IEEE.

16. Kim, G., Shin, D. H., Choi, J. G., & Lim, S. (2022). A deep learning-based cryptocurrency price prediction model that uses on-chain data. *IEEE Access*, *10*, 56232–56248.

17. Kraaijeveld, O., & De Smedt, J. (2020). The predictive power of public Twitter sentiment for forecasting cryptocurrency prices. *Journal of International Financial Markets, Institutions and Money*, *65*, 101188.

18. Livieris, I. E., Pintelas, E., Stavroyiannis, S., &Pintelas, P. (2020). Ensemble deep learning models for forecasting cryptocurrency time-series. *Algorithms*, *13*(5), 121.

19. Murray, K., Rossi, A., Carraro, D., &Visentin, A. (2023). On Forecasting Cryptocurrency Prices: A Comparison of Machine Learning, Deep Learning, and Ensembles. *Forecasting*, *5*(1), 196–209.

20. Shah, D., & Zhang, K. (2014, September). Bayesian regression and Bitcoin.In *2014 52nd annual Allerton conference on communication, control, and computing (Allerton)* (pp. 409–414). IEEE.

21. Stenqvist, E., &Lönnö, J. (2017). Predicting Bitcoin price fluctuation with Twitter sentiment analysis.

22. Sun, X., Liu, M., & Sima, Z. (2020). A novel cryptocurrency price trend forecasting model based on Light GBM. *Finance Research Letters*, *32*, 101084.

23. Zhou, X., Zhou, H., & Long, H. (2023). Forecasting the equity premium: Do deep neural network models work?. *Modern Finance*, *1*(1), 1–11.

Note: All the figures and tables in this chapter were made by the author.

Recent Trends in Engineering and Science for Resource Optimization and
Sustainable Development – Prof. (Dr.) Dorota Jelonek et al. (eds)
© 2024 Taylor & Francis Group, London, ISBN 978-1-032-98030-0

80

Decision Support System Using ChatGPT in Human Resource Management

P. P. Halkarnikar[1]

Associate Professor, Department of Computer Engineering,
Dr. D. Y. Patil Institute of Engineering, Management and Research,
Akurdi, Pune

M. Guru Vimal Kumar[2]

Associate Professor, Department of Computer Science and
Engineering, Vel Tech Rangarajan Dr. Sagunthala R&D Institute of
Science and Technology

Simmi Madaan[3]

Assistant professor, Faculty of Engineering & Technology,
SGT University, Gurugram

Sundarapandiyan Natarajan[4]

[4]Professor and Head, Department of Management Studies,
Adithya Institute of Technology, Coimbatore

Melanie Lourens[5]

Deputy Dean Faculty of Management Sciences,
Durban University of Technology, South Africa

Punamkumar Hinge[6]

Associate Professor, Management,
Indus business school, Pune, Maharashtra

▬▬▬ Abstract

An organisation can run successfully with human resource management which is considered to be an significant aspect. Efficient and effective HR Management helps the organisation for optimal performance and to achieve their goals by involving every team member of that organisation. Even though HR Management faces huge number of challenges such as technological environment changes, business changes expectations and diverse need of the members as well as large team management. The main objective of the study to use ChatGPT as decision support system in human resource management. The research study involves methods for interpreting and analysing ChatGPT as the effective decision support tool for its great potentiality transparency effectiveness and efficiency of HR processes. ChatGPT is an artificial intelligence model that provide virtual assistance in the form of text responses employee support process performance management employee

[1]pp_halkarnikar@rediffmail.com, [2]guruvimal09@gmail.com, [3]simmi.madaan@srmuniversity.ac.in, [4]nt_sundar@yahoo.com, [5]melaniel@dut.ac.za, [6]punamkumar.hinge@gmail.com

DOI: 10.1201/9781003596721-80

development and providing assistance in recruitment process. ChatGPT has more insights and valuable data about the work environment as well as employees to provide informed editions to the HR Managers. The accuracy of the ChatGPT as decision support system in human resource management can be analysed with KNN and CNN algorithm.

▬▬▬▬ Keywords

ChatGPT, Decision support system, Human resource management, Organization

1. INTRODUCTION

The success of an corporate industries or an achievement of organisation depends on the efficient and improved human resource management (Gadzaliet al., 2023). The efficient and effective use of HRM practises helps in significant enhancement in the ability of the organisation for attaining its objectives, effective team members as well as maximisation of productivity (Wahyoedi et al., 2023).Furthermore, human resource management faces few complex situations and challenges that can encompass the oversight of small or large sized teams to address expectations and recommends of the individuals belong to the team and adoption of transformational swift in technological and commercial landscape. The technological advancements include artificial intelligence and machine learning technologies that is huge number of potential breakthroughs which helps in the human resource management process effectively and efficiently (Sutrisno et al., 2023). The recent advancement in technologies is chatbot system development which has the ability to process and engage in text conversations using natural language processing with humans as one to one process (Harahap et al., 2023). ChatGPT Is referred as chat generator pre trained transformer is belongs to the category of artificial intelligence driven chat bots that may possess considerable number of promises in the application of decision support system is the safer hands for human resource management.

The artificial intelligence model ChatGPT involves extensive utilisation and training of machine learning techniques on huge amount of text data using sentiment analysis. Model has the ability for producing text that are pertinent and logically conducted to the question asked by the user. The features of natural language processing are used by ChatGPT for suitable deployment as a chat mode or virtual assistant to provide data that facilitate the diversified issues and resolutions in the HR Management Domain (Sudirjo et al., 2023).

The variants of user generative artificial intelligence models that become the focal point in discussions of various domains about the potential advantages and disadvantages of society, environment, economy and management. Even though it has many advantages the use of these ChatGPT whether it leads to job creation or displacement is unclear. Mostly they are the replacement of human labour through practically irrelevant potentially trivial and generating new decisions and information. According to the perspective of ChatGPT CEO there may be an potential impact in AI technology family that has significant applications in research, business models, stakeholder,relationships and employment. The major consequences are largely uncertain and undiscovered. The introduction of potent and advanced text generative AI tools available in the market followed by the launch of ChatGPT as AI arms race that can create expansion of business applications, workers continue uncertainty several challenges such as security, ethical dilemmas, privacy issues, insensitivity of context misinformation, bias and well-being. These perspectives synthesise the use of generative AI in HRM processes outcomes relationships and practises (Budhwar et al., 2023).

1.1 Research Objectives

The main objectives of the research study are;

1. To analyse decision support system in human resource management with ChatGPT
2. To examine the decision support model for identifying the major components in HR management

2. LITERATURE REVIEW

Liao et al., (2024) examined the ChatGPT (Chatbot Generative Pre-Trained Transformer) can be considered as the promising one in decision support system that owes in the advanced level of sentiment analysis with interactive design and capabilities. Thus, ChatGPT focus primarily on learning the text semantics when compared with learning data of complex structures by conducting analysis of real-time data that necessitates intelligent DSS development by employing specialized machine learning techniques. Even though specific algorithms cannot be executed directly by ChatGPThelps in designing the DSS at text level. The study involves the drawbacks and benefits to employ ChatGPT considered as a tool of auxiliary design. The findings of the study indicate the collaboration of ChatGPT with human expertise has the potential for revolutionizing the development of effective and robust intelligent DSS.

Zhou & Cen., (2023) explore the digital platform of human resource management depending on ChatGPT that can offer advantages to provide advantages for personalized experiences of employees to enhance efficiency in decision making and to improve work productivity when compared with conventional platform of human resource management. The study methodology involves the structural model and working principle of ChatGPT elaborated with design introduction and approach for development for digital platform of human resource management with the use of ChatGPT. Additionally, every aspect of innovation in digital platform can be emphasized with personalized user experience and capabilities of text dialogue generation. The findings of the study indicates that digital platform of human resource management depending on ChatGPT by demonstrating significant advantages in holding of greater value and potential, work productivity, efficient decision-making process personalized experience of employees when compared with traditional platforms of human resource management. However, the successful application in the platform necessitates further refinement and exploration for addressing security concerns and data privacy to provide appropriate support and training for continuous innovation and improvement in performance and functionalities of platform. The study contributes with innovation and contribution of theory in human resource management.

3. METHODOLOGY AND SPECIFICATION OF MODEL

The proposed methodology includes the development of decision support system in HR management with ChatGPT. Initially, the key performance indicators should be identified for decision making in HRM practices. The methodology applied for decision making system establishes treasury among the associated components and the objectives of HR Management the frameworks is applied to management and performance measurement. HR management focus mainly on management and development of human resources and workforce within the company or an organisation. It aids for ensuring qualified organisation with highly competitive and motivated workforce for effective achievement of business goals. The major components of hr management using ChatGPT for decision making support are given below;

- Selection and recruitment of employees for the available vacancies in the organisation using ChatGPT who have appropriate qualifications, skill and experience. The applied resumes can be compared with the job requirements.

- The employee development after recruitment helps in development of skills and knowledge with the help of proper training using ChatGPT
- ChatGPT helps in performance management though feedback collection and employee performance. Their performance is tracked periodically
- ChatGPT helps in providing rewards and compensation for employees to motivate them
- Skills assessment and career guidance are provided ChatGPT based on the interest and skills of employees
- ChatGPT provides offboarding assistance to gather feedback and providing checklists

4. EMPIRICAL RESULTS

This section includes empirical results of the research study. The performance of the decision support model using chat GPT is evaluated using KNN and CNN that is impressive accuracy rate in CNN when compared with KNN algorithm. The novelty of the present study is the innovative approach of evaluating the developed decision support system for accurate and adoptable scoring model. The powerful chatbot ChatGPT helps in generation of human like text that helps in decision making process in hr management by analysing the text by comparing with the previously added resumes skills experience knowledge of the employees. The performance of the employees can be analysed for rewards and compensation to motivate themselves. It may help in productivity of the organisation. ChatGPT helps in providing proper training for performance management. Table 80.1 and Fig. 80.1 shows the accuracy of Decision support system using ChatGPT in various components of HRM.

Table 80.1 Decision support system using ChatGPT

Decision support system using ChatGPT	ChatGPT with KNN	ChatGPT with CNN
Selection and recruitment	92%	97%
Employee development	88%	92%
Performance management	92%	95%
Rewards and compensation	87%	98%
Skills assessment and career guidance	85%	95%
Offboarding assistance	90%	96%

Source: Author

According to our study the applications of ChatGPT in HR Management and strategic decision-making system several aspects are considered for maximisation of benefits by overcoming the potential challenges.

- Integration of ChatGPT with the existing business that allows easy access of relevant information and data to

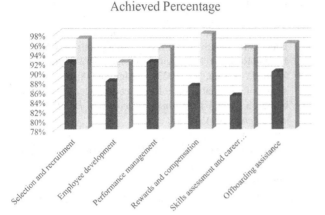

Fig. 80.1 Achieved percentage for Decision support system using ChatGPT

Source: Author

minimise operational disruptions and compatibility among the existing systems

• Employee development and training can be carried out using ChatGPT for providing analysis and accurate answers. AI based ChatGPT decision support system by providing relevant data by understanding the specific business context language

• The use of AI technologies has significant data exchange in operation and training process. Thus, the organisation must ensure the used data in the operations and management is protected and secured from unauthorised access

• ChatGPT has few ethical issues in biassed or false information sometimes. Thus the organisation must adopt few ethical guidelines in usage of charges to ensure that the decision support model used cannot causeway any social problems or harmful purposes

• ChatGPT helps in regular monitoring and supervision of interactions with employees and customers. The interactions can be monitored by the organisation or companies and to ensure that accurate responses provided are complied with company policies

• ChatGPT helps in supervised decision making for making important decisions in hr management with human responsibilities. ChatGPT is considered to be anvaluable assistant in the process of decision making. Even though, humans should evaluate the obtained results in the HRM processes such as Selection and recruitment, employee development, performance management, providing rewards and compensation for employees as well as offboarding assistance

• Employee development and training can be given using ChatGPT the decision support system helps in

providing proper training materials in the customised way according to the level of understanding and the individual needs of the employees

• ChatGPT helps in collecting feedback for the companies from the customers employees and the users about the experiences. Show that the organisation or companies can improve their service quality and performance by responding and listening to the feedbacks regularly.

5. CONCLUSION

Efficient and effective HR Management helps the organisation for optimal performance and to achieve their goals by involving every team member of that organisation. Even though HR Management faces huge number of challenges such as technological environment changes, business changes expectations and diverse need of the members as well as large team management. The main objective of the study to use ChatGPT as decision support system in HR management. The research study involves methods for interpreting and analysing ChatGPT as the effective decision support tool for its great potentiality transparency effectiveness and efficiency of HR processes. ChatGPT is an artificial intelligence model that provide virtual assistance in the form of text responses employee support process performance management employee development and providing assistance in recruitment process. The performance of the decision support model using ChatGPT is evaluated using KNN and CNN with impressive accuracy rate in CNN when compared with KNN algorithm. The novelty of the present study is the innovative approach of evaluating the developed decision support system for accurate and adoptable scoring model. The powerful chatbot ChatGPT helps in generation of human like text that helps in decision making process in hr management by analysing the text by comparing with the previously added resumes skills experience knowledge of the employees. The performance of the employees can be analysed for rewards and compensation to motivate themselves. It may help in productivity of the organisation. ChatGPT helps in providing proper training for performance management.

References

1. Gadzali, S. S., Gazalin, J., Sutrisno, S., Prasetya, Y. B., &Ausat, A. M. A. (2023). Human resource management strategy in organisational digital transformation. *JurnalMinfoPolgan*, *12*(1), 760–770.

2. Ausat, A. M. A., Azzaakiyyah, H. K., Permana, R. M., Riady, Y., &Suherlan, S. (2023). The role of ChatGPT in enabling MSMEs to compete in the digital age. *Innovative: Journal of Social Science Research*, *3*(2), 622–631.

3. Wahyoedi, S., Suherlan, S., Rijal, S., Azzaakiyyah, H. K., &Ausat, A. M. A. (2023). Implementation of Information Technology in Human Resource Management. *Al-Buhuts*, *19*(1), 300–318.

4. Sutrisno, S., Ausat, A. M. A., Permana, B., &Harahap, M. A. K. (2023). Do Information Technology and Human Resources Create Business Performance: A Review. *International Journal of Professional Business Review: Int. J. Prof. Bus. Rev.*, *8*(8), 14.

5. Sudirjo, F., Diawati, P., Riady, Y., Ausat, A. M. A., & Suherlan, S. (2023). The role of chatgpt in enhancing the information search and decision-making process of travellers. *JurnalMinfoPolgan*, *12*(1), 500–507.

6. Harahap, M. A. K., Junianto, P., Astutik, W. S., Risdwiyanto, A., & Ausat, A. M. A. (2023). Use of ChatGPT in Building Personalisation in Business Services. *JurnalMinfoPolgan*, *12*(1), 1212–1219.

7. Budhwar, P., Chowdhury, S., Wood, G., Aguinis, H., Bamber, G. J., Beltran, J. R., ... & Varma, A. (2023). Human resource management in the age of generative artificial intelligence: Perspectives and research directions on ChatGPT. Human Resource Management Journal, 33(3), 606–659.

8. Zhou, J., & Cen, W. (2023). Design and Application Research of a Digital Human Resource Management Platform based on ChatGPT. *Journal of Theory and Practice of Social Science*, *3*(7), 49–57.

9. Liao, Z., Wang, J., Shi, Z., Lu, L., & Tabata, H. (2024). Revolutionary potential of ChatGPT in constructing intelligent clinical decision support systems. *Annals of Biomedical Engineering*, *52*(2), 125–129.

Note: All the figures and tables in this chapter were made by the author.

*Recent Trends in Engineering and Science for Resource Optimization and
Sustainable Development – Prof. (Dr.) Dorota Jelonek et al. (eds)
© 2024 Taylor & Francis Group, London, ISBN 978-1-032-98030-0*

81

Role of ChatGPT in Selection and Recruitment Process: A Digital Transformation

Mrunalini U. Buradkar[1]

Assistant Professor, Electronics & Telecommunication Engineering,
St. Vincent Pallotti College of Engineering & Technology,
Nagpur, Maharashtra

S. R. Ganesh[2]

Assistant Professor, Department of Commerce,
Karnataka State Akkamahadevi Women University,
Vijayapura, Karnataka

Balaji M[3]

Associate Professor, Department of Post Graduate Studies,
MBA, Sheshadripuram College, Bengaluru, Karnataka

Pallavi Shetty[4]

Associate Professor, Department of MCA,
NMAM Institute of Technology, Nitte (Deemed to be University)

Melanie Lourens[5]

Deputy Dean Faculty of Management Sciences,
Durban University of Technology, South Africa

Shilpa Tandon[6]

Assistant Professor,
Management, Asian Business School, Noida

Abstract

The generative artificial intelligence (AI) models used by ChatGPT and its derivatives have gained significant traction in academic and media discourse over their possible advantages and disadvantages in relation to a range of areas of the economy, democracy, society, and environment. It is yet unclear if these technologies lead to the creation or displacement of jobs, or if they only change the nature of human employment by producing new, essentially meaningless, information and choices. The study's findings indicate that there is a lot of room for improvement in the efficacy, productivity, and transparency of HR procedures when using ChatGPT as a decision-support tool in HR management. ChatGPT and other AI models can function as text-based virtual assistants, supporting hiring, training, performance reviews, and employee assistance procedures. HR managers may make better decisions by using ChatGPT to collect insightful data about workers and the workplace.

Keywords

CHATGPT, Recruitment, Selection

[1]m.buradkar@gmail.com, [2]drganeshsr@kswu.ac.in, [3]balajim0405@gmail.com, [4]pallavirjsh@nitte.edu.in, [5]melaniel@dut.ac.za, [6]shilpaatandon@gmail.com

DOI: 10.1201/9781003596721-81

1. INTRODUCTION

The hiring and recruiting of employees and staff to engage in the workforce is referred to as recruitment. Four steps can be distinguished in the overall recruitment process. The first step is to verify the staffing needs. The employer applies, and the HR department evaluates the applications received by the department for the staff in need to identify the job title and the number of positions to be filled. The second step involves creating the recruiting strategy. During this phase, the HR division sets the candidate's minimum qualifications, including education, age, and skill set. It also authorises the base and budget salaries, validating the recruitment plan in its entirety. The third step involves choosing the applicants, confirming the methods for receiving the job offer, notifying the recipients, and carrying out the screening and interview procedures. The recruitment evaluation stage, which takes place in the fourth step, evaluates candidates in relation to the recruitment and determines the chosen individual. According to Baratelli and Colleoni (2022), organisations should prioritise streamlining their recruiting processes to lure top talent, since there is a direct correlation between these efforts and the potential value that prospective employees can provide to the organisation. AI as a technical tool can improve how work is done across a wide range of sectors.

AI has recently been used in the hiring process to aid in talent selection and enable a more efficient hiring procedure for the company. Hassan and Ibrahim (2019). Using AI solutions like ChatGPT is a fantastic way to streamline the hiring process. When implementing AI, companies must consider whether the improvements to the recruiting process will benefit both recruiters and job seekers and whether they can satisfy their intended needs. Herranen & Johansson, 2019. The human resources department will place limitations on the recruit's conditions throughout the stage of defining the recruitment plan. By counting the conditions of employees in the same position using the data, a more appropriate criterion can be determined. Throughout the hiring process, the job ad must clarify the role and list all requirements for the candidate. If the HR department can quickly produce content for the recruitment platform, it will be able to recruit successfully. Editing content for various job postings can be laborious when there are several openings, and the content posted on various job platforms may be modified accordingly. One of the key factors affecting recruiting efficiency is the time-consuming nature of the communication process between recruiters and candidates.

More innovation is made possible by the usage of ChatGPT in the hiring and personnel selection processes. By utilising this technology, businesses may incorporate artificial intelligence features into the initial phases of candidate evaluation, allowing them to review and assess thousands of applications swiftly and effectively. Furthermore, ChatGPT can be utilised during applicant interviews to help obtain more specific information about their backgrounds and experiences Dwivedi et al., (2023). This improves the effectiveness of the hiring process and produces richer data that helps decision-makers choose candidates who best fit the needs of the company. As a result, ChatGPT and other forms of technology have completely changed the way that hiring and selecting employees are done, opening new possibilities to improve efficiency and accuracy.

Numerous chances to improve different facets of human resource management arise from the usage of ChatGPT in the hiring and selection process for employees. First, Khan et al. (2023) found that ChatGPT is capable of efficiently screening and managing a high volume of applications and resumes that come into the organisation. ChatGPT's sophisticated text analysis capabilities enable it to swiftly find pertinent credentials and keywords based on business criteria, which minimises the time-consuming administrative workload. This makes it possible for HR departments to focus on candidates with greater potential and more wisely manage their resources.

Moreover, ChatGPT might prove to be a useful instrument for holding preliminary interviews with applicants. ChatGPT is a configurable chatbot that can interview candidates in a systematic and uniform manner to obtain comprehensive information about their qualifications, backgrounds, and goals. Limo et al., (2023). This method's primary benefit is that ChatGPT offers unbiased evaluations since it is unaffected by subjective opinions or emotional variables. Therefore, using ChatGPT to streamline the hiring and selection process for new employees not only increases productivity but also gives businesses more information into the calibre and potential of candidates.

2. LITERATURE REVIEW

A crucial step in human resource management, hiring and selecting employees is crucial to an organization's success. This procedure entails looking for, assessing, and choosing people whose potential and capabilities align with the organization's needs. The introduction of cutting-edge technology like ChatGPT is one of the major shifts that has completely changed the paradigm of employee recruitment and selection. The purpose of this study is to examine how ChatGPT has changed the hiring and selecting process for employee Raharjo, (2023). A qualitative literature review with an emphasis on a thorough comprehension of the subject from 2018 to 2023 is the methodology employed. This method's primary goal is to locate, evaluate, and compile pertinent scientific material that has been published in a variety of journals, conference proceedings, and other scholarly sources that may be accessed via Google Scholar. The study's findings demonstrate that implementing ChatGPT to change the hiring and selection procedures for new employees is a big step in the direction of improved efficacy and efficiency in human resource management. Nonetheless, we need to exercise caution while dealing with new effects and difficulties.

An innovative era in workforce management and employee engagement has begun with the integration of generative AI technologies such ChatGPT, into human resource management (HRM). This study explores the critical function that ChatGPT and related generative AI technologies fulfil in HRM, emphasising its importance in hiring, employee development, and internal communication Rane, (2023). These AI technologies use their NLP capabilities to expedite the hiring process, guaranteeing impartial candidate selection and raising the general effectiveness of HR departments.Additionally, ChatGPT supports customised learning experiences in training and development programmes, modifying content to match the needs of specific employees and promoting skill development and career advancement. However, there are certain difficulties with ChatGPT's broad use in HRM. To stop unfair behaviours and ensure that employees are treated fairly, ethical issues like algorithmic bias and data privacy must be thoroughly examined. Another major obstacle is the need for AI algorithms to be continuously monitored and improved to conform to changing organisational cultures and objectives.

The purpose of this study is to determine whether, in comparison to conventional HRM platforms, a digital HRM platform built on ChatGPT provides greater benefits in terms of individualised employee experiences, increased decision-making effectiveness, and increased productivity. First, the working principles and model structure of ChatGPT are explained. Next, the design and development process for the ChatGPT-based digital HRM platform is introduced. Additionally, the platform's novel features are highlighted, especially its capacity to generate discourse and provide a personalised user experience Zhou & Cen, (2023).According to the study's findings, the ChatGPT-based digital human resource management platform offers substantial benefits over traditional HRM platforms in terms of individualised employee experiences, decision-making effectiveness, and work productivity. However, for the platform to be successfully applied, more research and development is needed to solve data privacy and security concerns, offer suitable training, and support, and consistently innovate and improve platform perfor-

mance and features. The study advances the idea of HRM, provides more individualised and intelligent methods for managing human resources, and provides businesses with top-notch HR services.

Employers in the education sector, like universities, colleges, and training centres, are under pressure to find qualified candidates for a variety of roles, from teaching staff to administrative staff. To guarantee that educational institutions can uphold their quality standards and effectively serve the community's educational needs, an effective and efficient recruitment procedure is essential. The purpose of this study is to investigate how ChatGPT might be improved to improve hiring practices in the education industry. Employing a qualitative literature review methodology, the study makes use of Google Scholar data from 2016 to 2023. This method entails a thorough examination of numerous journal papers, articles, and related academic works that were released in this time frame Sulaeman et al., (2024). The results of the research suggest that implementing ChatGPT to optimise the personnel recruiting process in the education sector has the potential to significantly improve efficiency, effectiveness, and user experience in general. ChatGPT can be a useful tool for recruiting teams in choosing the best candidates for open positions by using AI to perform preliminary interviews, respond to applications quickly, and offer applicants with a personalised experience.

As technology advances, ChatsGPT is now recognised as a novel AI instrument. By understanding the language of the data, ChatGPT can produce text such that it may be written and even spoken to. Its cleverness has won the hearts and minds of a lot of users. HR professionals typically spend a lot of time posting vacancies, talking about the position, setting up interviews, and screening many interview candidates during the recruitment process Zhang, (2023). These procedures need a great deal of text creation and interaction; ChatGPT can assist with part of the task, but this can also lead to issues. This essay will examine ChatGPT's potential contribution to the HR hiring process as well as any potential drawbacks. Investigating the viability of utilising ChatGPT to support HR recruiting can aid in resolving certain current issues or facilitate more effective HR recruitment processes.

3. RESEARCH METHODOLOGY

There are several steps in the development of AI systems that can lead to bias developing. This study will explore the specifics of how bias may appear in AI, with an emphasis on algorithmic design, system implementation, and data collecting and preprocessing.

Bias can emerge in the data gathering stage when training data reflects past social injustices, creating biassed patterns. Data biases can originate from various sources, including human-generated labels, data sample methods, and data gathering procedures impacted by social biases or stereotypes. Although the construction of AI algorithms involves human decisions and subjective decision-making, these algorithms are not meant to function with human biases. When choosing features, creating optimisation goals, or selecting assessment criteria, algorithmic designers may introduce human biases into the process. Consequently, it becomes imperative to assess the design decisions rigorously and make sure that human biases are not unintentionally included into the algorithms.

Decisions made during algorithmic design can potentially be biassed in AI. The mathematical modelling and optimisation goals of AI algorithms might lead to biases. For example, if a model is trained to minimise prediction errors without taking fairness criteria into explicit consideration, it may unintentionally pick up on and reinforce discriminating trends seen in the data. Algorithmic biases can take many different shapes, like skewed predictions based on protected features or disparate error rates among distinct demographic groupings. Developing fairness-aware algorithms that include fairness measures and limitations into the optimisation process is necessary to combat algorithmic bias.

Furthermore, during the implementation stage, bias in AI can emerge. Biases in AI systems can be introduced or made worse by decisions made about system design, such as feature selection, performance measures, and decision thresholds. If some features have a disproportionate impact on the decision-making process or if fairness issues are not considered by performance measurements, biassed results may result. Furthermore, because decision criteria are subjectively interpreted and encoded, prejudices held by humans may unintentionally find their way into AI systems. Because implementing AI systems requires human agency and decision-making, biases may be introduced or reinforced. Human judgement affects things like how AI outputs are interpreted, how decision thresholds are defined, and how contextual considerations are considered. These arbitrary choices may unintentionally reinforce or magnify preexisting biases in the AI system. It is essential to have procedures in place to deal with potential biases throughout the system implementation stage and guarantee that fairness issues are considered when guiding human interventions. Careful system design, the integration of fairness-aware principles, and the maintenance of transparency in decision-making processes are all necessary to mitigate bias at the implementation level.

4. RESULTS

Table 81.1 Performance metrics of ChatGPT in recruitment process

Metric	Value
Accuracy	0.85
Precision	0.82
Recall	0.88
F1 Score	0.85

Table 81.2 Confusion matrix for ChatGPT in candidate screening

	Predicted Negative	Predicted positive
Actual negative	320	60
Actual positive	40	330

Table 81.3 Bias evaluation results

Bias Type	Detected Bias (%)
Gender Bias	2.3
Ethnicity Bias	1.8
Age Bias	0.5
Socioeconomic Bias	1.2

Table 81.4 Performance metrics on different data splits

Data Split	Accuracy	Precision	Recall	F1-Score
Train Set	0.90	0.92	0.88	0.90
Validation	0.85	0.88	0.82	0.85
Test Set	0.87	0.89	0.85	0.87

5. CONCLUSION

Like any new technology, AI might have unpredictably enormous influence on HRM, the field of research related to it, the economy, and society. While there is little doubt that it could be advantageous, its complete effects are unclear. It is true that it is unclear if true AI exists at all at this point, but even highly developed data compilation and problem-solving algorithms could bring both significant benefits and threats. The goal of HRMJ is to advance discussion and understanding regarding AI and its implications for HRM research and practice. The various theme presentations in this HRMJ perspectives editorial serve as a springboard for this exploration rather than offering a single, conclusive set of answers or conclusions. Its capacity to deliver prompt, customised responses simplify HRM procedures and promotes responsiveness and efficiency. Furthermore, by automating repetitive processes, these AI solutions free up HR experts to concentrate on more critical facets of their jobs. By guaranteeing smooth interactions that create enduring impressions, ChatGPT also helps to provide a favourable candidate experience during the recruitment process.

In HRM, generative AI technologies have a major impact on data processing and decision-making. ChatGPT helps HR professionals make well-informed decisions on employee performance, engagement, and organisational culture by analysing large amounts of data and identifying patterns. These insights make it easier to implement focused personnel acquisition, development, and retention strategies. But for ChatGPT integration in HRM to be effective, several issues need to be resolved. Concerns about data security and privacy are fundamentally ethical. HR departments handle confidential employee data; strong procedures and moral standards are necessary to establish confidence and protect this data. Furthermore, biases in AI systems are a problem. To guarantee fair procedures, HR professionals need to recognise and address biases. It is essential to have varied development teams, open communication, and ongoing monitoring.

Another worry is the problem of technical unemployment. ChatGPT automates work, yet concerns regarding employment displacement for humans are raised. It is critical to strike a balance between automation and human involvement. To meet the needs of the digital age, HR professionals should concentrate on reskilling and upskilling the workforce. It is crucial to foster an environment at work where technology is seen as an enhancement rather than a substitute. HRM has been completely transformed by ChatGPT and related generative AI technologies, which provide unmatched efficiency, engagement, and data-driven decision-making. A careful approach to resolving ethical issues, prejudices, and employment implications is necessary for their integration. HR practitioners need to take the lead in embracing these developments and promoting a peaceful coexistence of AI and human knowledge. This strategy guarantees a time when technology augments human capacities, resulting in a flexible, adaptable, and compassionate human resource management environment.

Although there are many opportunities, integrating ChatGPT and related AI models into HRM is not without its difficulties. HR professionals may embrace the power of generative AI while upholding human-centric values by proactively tackling concerns pertaining to ethics, bias, contextual understanding, trust, technology constraints, human roles, legal compliance, cybersecurity, and social effect. There are not many organisations using AI in recruitment or creating AI recruiting software because the field is still relatively young and has not seen much use of it. Since most businesses only employ AI to some level

in their hiring process, it is difficult to compile in-depth studies. Even while AI as a concept has been extensively studied, there are not enough businesses using AI specifically in their hiring practices, making it difficult to fully investigate AI's consequences and actual effectiveness. There could have been more interviewers for this study to be applicable.

References

1. Baratelli, G., & Colleoni, E. (2022). Does artificial intelligence (AI) enabled recruitment improve employer branding?. *International Journal of Business and Management*.
2. Ibrahim, W. M. R. W., & Hassan, R. (2019). Recruitment trends in the era of industry 4.0 using artificial intelligence: pro and cons. *Asian Journal of Research in Business and Management*, *1*(1), 16–21.
3. Johansson, J., & Herranen, S. (2019). The application of artificial intelligence (AI) in human resource management: Current state of AI and its impact on the traditional recruitment process.
4. Raharjo, I. B. (2023). ChatGPT's Role in Transforming Employee Recruitment and Selection Processes. *MALCOM: Indonesian Journal of Machine Learning and Computer Science*, *3*(2), 205–210.
5. Rane, N. (2023). Role and challenges of ChatGPT and similar generative artificial intelligence in human resource management. *Available at SSRN 4603230*.
6. Zhou, J., & Cen, W. (2023). Design and Application Research of a Digital Human Resource Management Platform based on ChatGPT. *Journal of Theory and Practice of Social Science*, *3*(7), 49–57.
7. Sulaeman, M. M., Nurcholidah, L., Handayati, R., & Wibowo, S. N. (2024). OPTIMIZING EMPLOYEE RECRUITMENT PROCESS IN THE EDUCATION BUSINESS THROUGH CHATGPT IMPLEMENTATION. *Technopreneurship and Educational Development Review (TENDER)*, *1*(1), 14–20.
8. Zhang, Y. (2023). The Impact of ChatGPT on HR Recruitment. *Journal of Education, Humanities and Social Sciences*, *19*, 40–44.
9. Dwivedi, Y. K., Kshetri, N., Hughes, L., Slade, E. L., Jeyaraj, A., Kar, A. K., ... & Wright, R. (2023). "So what if ChatGPT wrote it?" Multidisciplinary perspectives on opportunities, challenges and implications of generative conversational AI for research, practice and policy. *International Journal of Information Management*, *71*, 102642.
10. Khan, R. A., Jawaid, M., Khan, A. R., & Sajjad, M. (2023). ChatGPT-Reshaping medical education and clinical management. *Pakistan Journal of Medical Sciences*, *39*(2), 605.
11. Limo, F. A. F., Tiza, D. R. H., Roque, M. M., Herrera, E. E., Murillo, J. P. M., Huallpa, J. J., ... & Gonzáles, J. L. A. (2023). Personalized tutoring: ChatGPT as a virtual tutor for personalized learning experiences. *PrzestrzeńSpołeczna (Social Space)*, *23*(1), 293–312.

Note: All the tables in this chapter were made by the author.

82

Enhancing Cybersecurity in IOT-based Banking Systems Using Machine Learning

Balakrishnan S[1]

Assistant Professor, Department of Commerce,
Faculty of Science and Humanities,
SRM Institute of Science and Technology,
Ramapuram, Chennai

Netala Kavitha[2]

Assistant Professor, Department of Computer Science and
Engineering, Vignan's Institute of Information
Technology (A), Duvvada

Jnaneshwar Pai Maroor[3]

Assistant Professor, Department of Humanities,
NMAM Institute of Technology-Affiliated to NITTE
(Deemed to be University), Karnataka, India

Shweta Bajaj[4]

Assistant Professor, Department of Professional Communication,
Graphic Era Deemed to be University,
Dehradun, Uttarakhand

Ahmad Y. A. Bani Ahmad[5]

Department of Accounting and Finance, Faculty of Business,
Middle East University, Amman 11831, Jordan,
Applied Science Research Center,
Applied Science Private University,
Jordan

Patel Kalpeshkumar Laxmanbhai[6]

Research Scholar, Dept. of Computer Engineering,
Swarrnim Institute of Technology, Swarrnim Startup &
Innovation University, Gujarat

▬▬▬ Abstract

A technological revolution known as the Internet of Things (IoT) makes virtual data interchange possible for machine-to-machine and human-to-human communication. Low-cost sensors enabled by the Internet of entities enable us to locate,

[1]gurubalaji08@gmail.com, [2]netala.kavitha@gmail.com, [3]jnan_pai@nitte.edu.in, [4]Shwetabajaj.comm@geu.ac.in, [5]aahmad@meu.edu.jo,
[6]kalpeshlpatel@gmail.com

DOI: 10.1201/9781003596721-82

identify, and access the different entities and objects surrounding us. While there are numerous advantages to the IoT, there are also significant concerns, particularly about security and privacy. These issues require suitable solutions, and with the IoT, security and privacy are of utmost importance. This paper lists potential attacks against various kinds of networks along with defence strategies. This study educates vulnerability researchers and IoT network protection specialists how to prevent issues in real networks by simulating them and creating proactive solutions, which gives them useful information. The study concludes that IoT-based security solutions can improve the security of financial institutions like banks. Automation of security procedures, real-time monitoring, and enhanced efficiency are some benefits of IoT-based security systems. Careful planning is necessary for the adoption of these solutions, though, because banks must deal with several issues, including security breaches, a lack of standardisation, and high implementation costs. Consequently, it is advised that banks carefully evaluate their security requirements and apply IoT-based security solutions in a comprehensive manner.

■■■■■■■ **Keywords**

Cybersecurity, IoT, Banking systems

1. INTRODUCTION

Many of the devices we use in the modern era of computers are part of the IoT, which is connected to the Internet. The Internet, which is an insecure (open) communication medium, is how these gadgets share and send their data. Most of the time, this information is sensitive (social security numbers, banking, insurance, healthcare, and other finance-related data, for example) Butunet al., (2019). Malicious actors, like cybercriminals, are constantly looking for opportunities to play with things. They can execute attacks using techniques malware injection, and data modification, among others Lv et al., (2020). As an outcome, many researchers occasionally suggest different security procedures to lessen these threats.

The technique of teaching computer systems to learn from data and utilise algorithms to carry out activities without explicit programming is known as ML. As a subset of AI, DL is like ML Magaia et al., (2020). DL is founded upon an intricate collection of algorithms that are fashioned after the human brain. This makes it possible to process unstructured data, involving text, images. ML is the study of how a computer thinks and acts without the assistance of a human. But DL usually requires less continuous human interaction.

As a result, it performs superior analysis of photos, videos, and unstructured data than the conventional ML algorithms Parah et al., (2020).We can benefit from the combination of ML and cyber security in several ways. For instance, better performance from cyber security techniques, increased security for ML models, and efficient zero-day attack detection with reduced human involvement. Nevertheless, it could have several problems, including difficult security, which needs to be addressed properly.

The situation of "ML security," or cyber security in ML, is another name for it. Numerous phenomena are analysed

and predicted using machine learning models. Nevertheless, certain assaults, such as runtime interruption, privacy violation, membership inference, dataset poisoning, and model poisoning, might impact the performance of ML models (Sun et al., 2021). These attacks have the potential to cause ML models to anticipate related phenomena incorrectly. An adversary adds hostile examples (updated values) into a dataset in a "dataset poisoning attack," causing the machine learning method to generate incorrect predictions. The attacker's purpose in the "model poisoning attack" is to further contaminate the models by changing their parameters and interfering with their internal operations. The goal of a "privacy breach attack" is for the attacker to obtain important model information while also working to expose sensitive data. One type of privacy violation is the membership inference attack. Additionally, by assaulting the model's execution process, the attacker in a "runtime disruption attack" subverts the ML workflow and influences the precision of the prediction outputs. Therefore, to defend against these attacks, certain cyber security methods (like hashing algorithms, and signature generation and verification procedures) are required. The ML models are made secure by the implementation of these cyber security measures, and the predictions and outcomes we obtain are precise.

One of the most frequently targeted areas for cyberattacks, which can lead to data breaches, monetary losses, and reputational harm, is the financial sector. As a result, improving security in the financial industry is now imperative. Al-Fuqaha & Associates (2015). The IoT is one of the newest technical advancements that has the potential to totally change banking sector security.

These devices are perfect for using to improve security in the banking sector since they may be embedded with software, and other techniques that enable them to gather and transmit data. In the banking industry, IoT-based security solutions can offer predictive maintenance, automation of security procedures, and real-time monitoring and threat detection. Through the use of IoT-based security solutions, banks may increase customer satisfaction and confidence while better protecting their assets and client data. Blockchain technology, asset tracking, smart surveillance systems, biometric authentication, and other applications are examples of IoT-based security solutions in banking.

Banks that use fingerprint scanning or facial recognition technology to verify the identification of their customers can benefit from biometric authentication. IoT sensors can be used by smart surveillance systems to identify suspicious activity and provide alerts so that bank security staff can take preventative action. Asset tracking can assist banks in keeping an eye on and safeguarding their physical assets, while blockchain technology can help to maintain the integrity of transactions. Asghari et al., 2019). Nevertheless, there are inherent difficulties in putting IoT-based security into banking. The requirement to protect the privacy and security of data transferred between IoT devices is one of the primary obstacles. The expense of setting up and maintaining IoT-based security solutions, which can call for hefty infrastructure and training expenditures, is another difficulty.

According to Gubbi et al. (2013), IoT-based security has the power to completely change how security is applied in the banking industry. IoT-based security solutions can assist banks in better protecting their assets and client data, identifying and stopping fraudulent activity, and enhancing consumer trust and satisfaction in light of the financial sector's growing demand for increased security. To tackle the difficulties posed by IoT-based security solutions in the banking sector, proper design, execution, and upkeep are essential.

2. LITERATURE REVIEW

Cyberattacks have the potential to cause power outages, malfunctions with military equipment, and breaches of sensitive data; for example, if medical records fall into the wrong hands, they could be stolen. Because of the enormous financial value of the information it contains, the banking sector is especially vulnerable. The amount of digital footprints that banks have increases, increasing the attack surface available to hackers. The purpose of this study is to use the Banking Dataset to identify distributed denial-of-service (DDOS) attacks against financial institutions. Several categorization algorithms have been employed in this study to anticipate DDoS attacks. In order to improve the performance of generic models, we have made certain architectural changes Islam et al., (2022). We

have also utilised algorithms RF, K-Nearest Neighbours (KNN), and SVM. For the detection of (DDoS) assaults, the SVM shows an accuracy of 99.5%, while KNN and RF scored accuracy of 97.5% and 98.74%, correspondingly. After comparison, it was shown that the SVM outperforms KNN, RF, and other current ML and DL methods in terms of robustness.

ANNs are the source of DL, one of the key techniques for today's sophisticated cyber security systems and regulations. The advantages and disadvantages of utilising AI in cyber risk analytics to enhance organisational resilience and gain a deeper understanding of online threats. To intelligently highlight the variety of cyber security issues, one can employ multilayer perceptrons, CNN, generative adversarial networks, and deep reinforcement learning, as well as their ensembles and hybrid methods Ghillani (2022). The objective of the backpropagation method is to accurately maximise the network weights to convert the inputs into the desired outputs. Several optimisation technologies, involving Limited Memory BFGS (L-BFGS), and Stochastic Gradient Descent (SGD), are used during the training phase. Numerous cybersecurity issues may be solved with the help of these neural networks. Building reliable IoT systems, virus analysis, security threat analysis, and the identification of malicious botnet traffic are all done using MLP-based networks.

The IoT concept was generated to enhance people's lives by offering a wide variety of intelligent, networked gadgets and applications across multiple industries. But the biggest problems facing the devices in an IoT ecosystem are security risks. Although several state-of-the-art methods have been put out to safeguard IoT gadgets, further development is desired. ML has proven to be able to identify patterns when other approaches have failed. Saba et al., (2022). Using DL is one cutting-edge way to enhance IoT security. This creates a smooth choice for detection depending on anomalies. To use the power of the Internet of Things, this article describes a CNN-based technique for anomaly-based intrusion detection systems (IDS) that offers capabilities to effectively investigate whole traffic throughout the IoT. The suggested method demonstrates the capacity to identify anomalous traffic patterns and potential intrusions. Using the NID and BoT-IoT datasets, the model was trained and tested. Its accuracy was 99.51% and 92.85%, correspondingly.

The biometric features are examined within the framework of an IoT-based biometric cyber security case study that focuses on the banking industry's accomplishments and verification and safety concerns. In especially for the banking and finance industries, the need for highly secure identifying systems and person validates identification systems becomes crucially important because to the daily growth in security violations and fraudulent transactions. Biometric technology is viewed by many banking organisations as a very close solution to these security problems. The use of IoT-based biometric technology in banking security is still in its infancy, despite its popularity in industries like healthcare and numerous security activities. Mondol et al., (2022). Modern technologies bring a multitude of social, ethical, and systemic challenges because of the close interaction between biometric and human, physical, and behavioural aspects. A major financial business designed an IoT-based biometrically enabled banking safety system, and the case study's key success factors acted as guidelines. This prototype study shows that, rather than focusing on technical challenges associated with integrating biometrics into current information systems, it is crucial to develop a workable security plan that considers user privacy concerns using smart sensors, human endurance, institutional changes, and legitimate issues.

Concerns about computer network security and privacy are continually growing in today's environment. Computer security is now essential since information technologies are used in many aspects of daily life. Alongside the rise in Internet usage and the development of new technologies like the Internet of Things paradigm, there are fresh, contemporary attempts to get into computer networks and systems. Businesses are spending more money on research to improve the ability to identify these threats Da Costa et al., (2019). By contrasting the highest accuracy rates, institutions are using clever methods to test and validate. As a result, the focus of our research is on the extensive body of current literature on ML techniques used in intrusion detection and the Internet of Things for computer network security. With a focus on the Internet of Things and machine learning, the work intends to conduct a thorough and up to date analysis of pertinent works that address various intelligent techniques and their applied intrusion detection designs in computer networks. A study was conducted on over 95 works pertaining to security challenges in IoT contexts, covering various themes.

3. RESEARCH METHODOLOGY

Data Cleaning: To ensure data integrity and reliability, apply data cleaning procedures to eliminate errors, inconsistencies, and unnecessary information from the gathered datasets.

Feature Selection: To improve the precision of ML models in fraud detection, find and pick pertinent features like transaction patterns, user behaviour, and geography data.

Unsupervised Learning: For anomaly detection, use clustering techniques like K-Means and Isolation Forest to find abnormal patterns in unlabeled data that may indicate fraud. The method requires CatBoost because of its advantages and strengths. For example, the model can infer quickly thanks to the usage of symmetric trees. The use of mirrored trees in CatBoost's approach removes the requirement for individual tree validation. Moreover, the categorical preference of the model suggests that it handles these factors effectively. When processing categorical data, CatBoost frequently outperforms other algorithms in terms of complexity. The financial industry places a high value on fraud detection, so this business will benefit greatly from the algorithm's speedy learning rate. CatBoost also has a valuable function called parameter weighting, which can aid in real-time tuning. For these reasons, the CatBoost model was used in this investigation instead of the Linear Regression approach. It is simple to see how the differences in expected results could be useful in real-world situations.

These six criteria must be used to evaluate the quality of our data throughout the pre-processing stage. These standards can be used to evaluate our dataset's suitability for uses. Make sure the data is complete and free of any gaps or omissions before proceeding. The uniform representation of all attribute values in the dataset is the second prerequisite. Lastly, we must ensure that the values do not conflict and cause errors in our computations. Lastly, the dataset needs to be up to date and accurate. The fifth step is to look for any duplicates in the data collection. Lastly, we will ensure that no data is omitted or not properly cited.

4. RESULTS

Table 82.1 Performance comparison of machine learning models for anomaly detection

Model	Precision	Recall	F1 Score	Accuracy
Isolation Forest	0.92	0.87	0.88	0.91
One-Class SVM	0.89	0.82	0.85	0.88
Autoencoder	0.95	0.88	0.91	0.93

Table 82.2 Intrusion detection results

Model	Accuracy (%)	Precision (%)	Recall (%)	F1 Score (%)
Random Forest	94.1	92.5	96.8	94.5
Support Vector	93.7	91.2	94.5	92.8
Neural Network	96.5	94.8	97.2	96.0

Table 82.3 Performance of behaviour analysis models

Model	True Positive Rate	False Positive Rate	AUC Score
LSTM	0.87	0.05	0.91
GRU	0.90	0.08	0.88

Table 82.4 Predictive modeling results

Model	Accuracy	Precision	Recall	F1 Score
Random Forest	.94	.92	.89	.90
Gradient Boosting	.95	.94	.91	.92

Table 82.5 Overall performance metrics

Metric	Anomaly Detection	Behaviour Analysis	Predictive Modeling
Accuracy	-	-	0.945
Precision	0.93	-	0.93
Recall	0.85	-	0.90
F1 Score	0.88	-	0.91
AUC Score	-	0.895	-

5. CONCLUSION

The ever-changing and persistent nature of cybersecurity threats presents major problems to financial institutions globally in the quickly changing landscape of modern banking. This study examined the complex world of cybersecurity risks and the critical function that ML methods perform in fraud detection. Several important conclusions have been reached after thorough investigation, analysis, and evaluation of real-world issues. The results highlight the worrying level of complexity used by cybercriminals in their strategies, which range from sophisticated phishing techniques to sneaky malware attacks. The risks that financial organisations confront have been made even more severe by insider threats and flaws in mobile banking systems.

If these risks are not managed, they may result in significant monetary losses, a decline in consumer confidence, and serious harm to one's reputation. These algorithms are remarkably good at identifying complex patterns in large datasets, which makes fraud detection and prevention possible in real time. They are essential instruments in the contemporary financial environment because of their versatility and flexibility to change in response to new threats. The study emphasises how crucial preventative actions are in reducing cybersecurity threats. It is critical to make investments in employee training, multi-factor authentication that is reliable, ongoing monitoring, and cooperation with cybersecurity specialists and peers in the sector. Fortifying financial systems against emerging threats also requires regular ethical hacking exercises and the implementation of AI-driven security solutions. A well-informed clientele is essential in the fight against fraud.

Customers are better equipped to identify and report suspicious activity when they are informed about safe online behaviours and typical fraud tactics. Gaining and retaining the trust of customers is equally important. To maintain consumer trust in digital banking platforms, banks must show their dedication to cybersecurity through open communication and strong security measures. The banking sector's approach to cybersecurity needs to be alert, flexible, and creative as cyber threats continue to change. In conclusion, a proactive and multifaceted approach is required given the convergence of modern banking and cybersecurity. Banking institutions can confidently and resiliently traverse the complicated cybersecurity landscape by adopting cutting-edge technologies, encouraging collaboration, educating both customers and workers, and investing in continuous improvement. The foundation of a safe, reliable, and strong financial ecosystem in the digital era will be the future coexistence of human awareness and technical innovation.

References

1. Islam, U., Muhammad, A., Mansoor, R., Hossain, M.S., Ahmad, I., Eldin, E.T., Khan, J.A., Rehman, A.U. and Shafiq, M., 2022. Detection of distributed denial of service (DDoS) attacks in IOT based monitoring system of banking sector using machine learning models. *Sustainability*, *14*(14), p. 8374.

2. Ghillani, D., 2022. Deep learning and artificial intelligence framework to improve the cyber security. *Authorea Preprints*.

3. Saba, T., Rehman, A., Sadad, T., Kolivand, H. and Bahaj, S.A., 2022. Anomaly-based intrusion detection system for IoT networks through deep learning model. *Computers and Electrical Engineering*, *99*, p.107810.

4. Mondol, S.K., Tang, W. and Hasan, S.A., 2022, February. A Case Study of IoT-Based Biometric Cyber Security Systems Focused on the Banking Sector. In *International Conference on Expert Clouds and Applications* (pp. 249–261). Singapore: Springer Nature Singapore.

5. Da Costa, K.A., Papa, J.P., Lisboa, C.O., Munoz, R. and de Albuquerque, V.H.C., 2019. Internet of Things: A survey on machine learning-based intrusion detection approaches. *Computer Networks*, *151*, pp. 147–157.

6. Butun, I., Österberg, P. and Song, H., 2019. Security of the Internet of Things: Vulnerabilities, attacks, and countermeasures. *IEEE Communications Surveys & Tutorials*, *22*(1), pp. 616–644.

7. Lv, Z., Qiao, L., Li, J. and Song, H., 2020. Deep-learning-enabled security issues in the internet of things. *IEEE Internet of Things Journal*, *8*(12), pp. 9531–9538.

8. Magaia, N., Fonseca, R., Muhammad, K., Segundo, A.H.F.N., Neto, A.V.L. and de Albuquerque, V.H.C., 2020. Industrial internet-of-things security enhanced with deep learning approaches for smart cities. *IEEE Internet of Things Journal*, *8*(8), pp. 6393–6405.

9. Parah, S.A., Kaw, J.A., Bellavista, P., Loan, N.A., Bhat, G.M., Muhammad, K. and de Albuquerque, V.H.C., 2020. Efficient security and authentication for edge-based internet of medical things. *IEEE Internet of Things Journal*, *8*(21), pp. 15652–15662.

10. Sun, Y., Bashir, A.K., Tariq, U. and Xiao, F., 2021. Effective malware detection scheme based on classified behavior graph in IIoT. *Ad Hoc Networks*, *120*, p.102558.

11. Al-Fuqaha, A., Guizani, M., Mohammadi, M., Aledhari, M. and Ayyash, M., 2015. Internet of things: A survey on enabling technologies, protocols, and applications. *IEEE communications surveys & tutorials*, *17*(4), pp. 2347–2376.

12. Asghari, P., Rahmani, A.M. and Javadi, H.H.S., 2019. Internet of Things applications: A systematic review. *Computer Networks*, *148*, pp. 241–261.

13. Gubbi, J., Buyya, R., Marusic, S. and Palaniswami, M., 2013. Internet of Things (IoT): A vision, architectural elements, and future directions. *Future generation computer systems*, *29*(7), pp. 1645–1660.

Note: All the tables in this chapter were made by the author.

Recent Trends in Engineering and Science for Resource Optimization and
Sustainable Development – Prof. (Dr.) Dorota Jelonek et al. (eds)
© 2024 Taylor & Francis Group, London, ISBN 978-1-032-98030-0

83

The Impact of Various Leadership Styles on Employee Motivation in Contemporary Organizations

Arangarajan M[1]

Assistant Professor, Department of Mechanical Engineering,
Shri Angalamman College of Engineering and Technology,
Tiruchirapalli, Tamilnadu

Supriya Jain[2]

Assistant Professor, IBM, GLA University,
Mathura, Uttar Pradesh

Lalit Mohan Pant[3]

Assistant Professor, Department of Psychology,
Uttarakhand Open University Haldwani, Nainital

Ashish Tamta[4]

Assistant Professor, School of Tourism,
Hospitality and Hotel Management

Rekha Joshi

Assistant Professor, Department of Psychology,
MBPG College Haldwani Nainital, Uttarakhand

Ashwini V Rathi[5]

Assistant Professor, Department of Management Studies,
GH Raisoni University, Amravati,
Maharashtra, India

▬▬▬ Abstract

A company's leadership is a major factor in increasing its success. It goes without saying that an organization's leadership style affects how well it performs. According to the research's findings, workers who are given a sense of empowerment and who receive rewards and incentives for their efforts are more inspired and driven to complete their responsibilities inside an organisation. This study also suggests that organisations reconsider and update their leadership development initiatives to better meet the diverse demands of their workforce, depending on their motivation.

▬▬▬ Keywords

Employee motivation, Leadership, Organizations

[1]arangarajanm@gmail.com, [2]supriya.jain@gla.ac.in, [3]lmpant33@gmail.com, [4]ashishtamta@uou.ac.in, [5]ashwinirathi11@gmail.com

DOI: 10.1201/9781003596721-83

1. INTRODUCTION

A company's performance in achieving its goals is significantly influenced by the type of leadership it decides to employ. It is believed that a successful and long-lasting corporate enterprise is the product of astute leadership paired with an inspired staff Lofsten, (2016). People are the essential component of organisations since they give them their all and help them to succeed in their objectives Gibson, (2011). Every company in the world was started with the purpose of making a profit or giving social benefits to the society; sometimes, both objectives were achieved simultaneously. Consequently, the corporate objectives of all organisations are identical. Employers need workers to help them accomplish their goals and objectives, but employees require their leaders to inspire and motivate them.

Since employees are the cornerstone of the company, keeping them engaged and motivated is crucial Mullins, (2007). Since effective leadership is necessary to improve a company's performance, a corporation's success or failure is based on how well-led all organisational levels are. The beliefs, actions, and abilities required to influence others to accomplish organisational objectives are known as leadership traits Shirzad (2011). For this reason, one way to define leadership is as a tactic for getting people to cooperate to achieve a common goal.

In the modern workplace, there are many obstacles that leaders must overcome due to alterations in technology, globalisation, consumer demands, and the accessibility of information and options. Bolden, (2016). When taken as a whole, these have complicated not only the working environment but also its cultural element, requiring greater dedication and participation from the workforce. From a company standpoint, the purpose of a leader in performance management should ideally be to increase efficiency, customer happiness, and quality results.

Based on statistical data, a motivated workforce leads to improved organisational and employee performance as indicated by mission achievement, predetermined goals and objectives, sustainability, growth, and competitivenessNyberg et al., (2016). Developing a motivating plan necessitates considering components that propel workers to reach their maximum potential on an individual and group level, hence influencing the efficacy of employees Abbas & Saad, (2019).

There is no doubting that managers directly affect their employees' work ethics, efforts, contributions, and, eventually, output. Employees assume that their behaviour will immediately impact whether they get their desired results, especially whether their duties are completed Elliot et al.,

(2017). To reach good drivers of each employee and as a team, there are currently numerous techniques that specify how leaders should incorporate into organisational culture and personal beliefs. Whether it is inside or external to an organisation, motivation is a basic component that propels progress.

In summary, motivated people can complete any work, regardless of how small, or unachievable it may appear. Employee motivation varies in the workplace based on the organisational culture and leadership values upheld. However, the approach used may have a positive or negative influence on employees' commitment to their individual tasks, team goals, and organisational and personal objectives. Basically, the key to developing strategies for employee motivation is comprehending the variables that motivate individuals and teams. Research has shown that contented workers who are provided with a friendly work environment, flexible work hours, ample possibilities, and competitive benefits packages enhance their dedication and output by over 20%.

Many research on employee motivation have been done recently. These studies have looked at how it impacts both organisational and individual performance, what strategies leaders may use to enhance staff output, and how leaders can impact employee commitment and performance. Scholars have put forth a variety of strategies that leaders can incorporate into organisational culture to enhance performance on both an individual and team level. After a thorough review of the literature, it appears that leadership and employee dedication to tasks are closely related and have a substantial correlation. There is, however, a paucity of study on the influence of leadership style on employee motivators. According to this viewpoint, research is required to determine the implications held by leaders and the associated degree to which an employee's motivation and commitment are impacted by the leadership style that is implemented in the workplace. Crucially, research is required, and leaders in the sector must comprehend the influence that a leadership style has on employees' actions and behaviour, as well as how they view their own and the organization's objectives.

2. LITERATURE REVIEW

An overview of management styles and their significance in influencing workers' expectations of their superiors within the organisation is presented in this article. The purpose of the essay is to demonstrate the connection between the leadership style of the organisation and the expectations of workers for job completion. The study's foundation is a review of the literature and the findings of empirical research that involved 185 workers from ten various fields of

European businesses. When managers utilise a situational management approach, staff members anticipate complete autonomy in completing assignments. Employees expect guidelines for task performance when their superior exhibits an authoritarian style, but they do not want their work to be continuously under control. The correlations between the analysed data and the variables defining the employee's job, seniority, sex, education, and type of company were also incorporated in the analysis Drewniak et al., (2020). The findings show that a leader's personality, credentials, values, and management style have an impact on the organization's overall performance as well as the day-to-day operations and long-term success of its workforce. The character profile, abilities, and desired traits of modern leaders were ascertained in part by this investigation. The examination made it possible to spot patterns in the ways that modern leaders are changing in terms of their traits and communication style. As a result, the study adds to our knowledge of the uniqueness of leadership in the corporate context and provides a useful overview of a wide variety of topics pertaining to the traits of leaders.

According to a recent request for additional study. The purpose of this study is to calculate the effect of a classic, charismatic, rational-legal leadership style on employee engagement and workplace culture. Data was collected from 215 workers and 15 leaders using the survey approach Khaliq et al., (2021). Testing the underlying theory is the main goal of the current investigation. With the use of the basic random sample technique, 55% of the response rate was recorded. The results of this study show that leaders have a favourable and considerable impact on both employee motivation and workplace culture. The study's five variables were shown to have been significantly influential, and the findings have some real-world ramifications for leaders, employees, HRM departments, and organisations.

Organisational performance is correlated with profitability in terms of growth and profit over the long and short durations. Investors and businessmen want to make sure that their companies operate smoothly to increase profits, expand quickly, and get a competitive advantage in the market. Aside from material and immaterial resources, strong and dynamic leadership is essential for organisations to function efficiently. An organization's performance is significantly impacted by the leadership style of its head Adnan and Khan (2014). There is uncertainty because, while most academics in the past agreed that a leader's style and organisational effectiveness are related, some did not share this opinion. To clear up this misunderstanding, the researchers looked at the varied leadership philosophies and practices of Pakistani leaders in a range of organisations, as well as how these practices affected the effectiveness

of those organisations. There are three primary types of leadership styles: transactional, laissez-faire, and transformative. The third one has a detrimental effect on organisational performance, whereas the first two have a good but varying degree of impact. A questionnaire completed by leaders of industrial and service organisations was used to perform a quantitative study.

One crucial aspect of working at public universities is job satisfaction. Work satisfaction is regarded as an internal concept produced by different aspects of the workplace. Given the variety of factors that impact job happiness, educational institutions need to be mindful while putting the appropriate leadership philosophies into practice. Nevertheless, employees' aberrant reactions have increased because of the organization's disregard for this conduct (Al-maaitah et al., 2021). Therefore, the current study concentrated on the critical impact that leadership styles have in raising job satisfaction among public university staff members. The exact leadership style that is used in an organisation should receive a lot of attention. Similarly, leadership philosophies, as extrinsic variables, have a significant influence on public university workers' job happiness since they strengthen staff members' loyalty to the companies. Enhancing employee satisfaction and identifying an effective leadership style are key components of this programme, which aims to improve educational organisations.

The purpose of this essay is to examine the effects that various leadership philosophies have on workers' job happiness. The study was a cross-sectional investigation of workers in Ghana's mobile telecommunications industry. There were 400 complete and valid questionnaires collected. The primary statistical method used to assess the hypotheses was the multiple regression methodology. Except for idealised influence, which has no bearing on extrinsic pleasure, it is found that the three characteristics of a transformational leadership style—individualized consideration, inspiring motivation, and intellectual stimulation—have a positive impact and correlation with it. Except for active management by exception, which has a negligible link with extrinsic satisfaction, two aspects of transactional leadership style—contingent rewards and passive management by exception—show a favourable impact and correlation with it Brenyah and Tetteh, (2016). Additionally, except for individualised consideration, which showed no relationship at all, all three aspects of transformational leadership style—inspirational motivation, intellectual stimulation, and idealised influence—show a positive and significant relationship with intrinsic satisfaction. In contrast, the three dimensions of transactional leadership style show no relationship at all with job satisfaction. In general, job happiness is found to be significantly predicted by a leader's style.

3. RESEARCH METHODOLOGY

The purpose of the current study is to investigate how leadership styles, specifically within the context of the Los Angeles Department of Public Social Services (DPSS), influence employee motivation and work satisfaction. Managers and the board of supervisors will be able to recognise areas inside the organisation that require improvement owing to the current research. The findings will support the application of leadership strategies and/or trainings required to support managers in inspiring their staff, which may ultimately result in increased job satisfaction. These modifications have the potential to elevate the organisation to a new level by increasing efficacy and efficiency.

Secondary sources of data were used to collect and obtain the information and data. This process has been useful in assessing the data and pinpointing any gaps in the information. By suggesting it inside the organisation, the data that is gathered aids in identifying and estimating the consequences and the causes. Articles, periodicals, publications, and topic-related websites are the primary sources of the data.

3.1 Objectives

- To frame several machine learning and employment tactics that could be used to engage employees in HRM.
- To adapt the leadership styles that help in the employee's job performance and appraisal.
- To increase the organization's profit by balancing the work-life needs of people by ensuring both job happiness and customer satisfaction.
- To inspire staff members, highlight their abilities, and get them involved in their work. These concerns have a significant impact on how executives engage their workforce through the use of machine learning technologies and methodologies.

4. RESULTS

Table 83.1 Machine learning model evaluation results

Model	Accuracy	Precision	Recall	F1 Score
Linear Regression	.72	.74	.68	.71
Decision Tree	.81	.82	.79	.80
Random Forest	.85	.86	.83	.84
Support Vector	.79	.81	.76	.78

Table 83.2 Cross-validation results for RF model

Fold	Accuracy	Precision	Recall	F1 Score
1	.83	.84	.81	.82
2	.85	.86	.84	.85
3	.87	.88	.85	.86
4	.82	.83	.80	.81
5	.86	.87	.84	.85

Table 83.3 Regression model results

Leadership Style	R-squared	MAE	RMSE
Transformational	0.78	0.15	0.25
Transactional	0.62	0.20	0.30
Laissez-faire	0.45	0.25	0.35

Table 83.4 Clustering model results

Leadership Style	Silhouette Score
Transformational	0.72
Transactional	0.65
Laissez-faire	0.55

5. CONCLUSION

This study examined how different leadership ideologies affected employee motivation, placing a unique focus on dynamic leadership. Since leadership is a critical perception that influences and motivates both individuals and teams, a leader's efficacy and potential have a substantial impact on the motivation levels of a team. In addition, today's business climate demands dynamic leadership, and good leaders must also be adaptable and agile. Important leadership theories like transformational, transactional, authentic, and servant leadership have an impact on team motivation. Leadership style has a big impact on employee motivation. Employee requirements, conditions, the skills and understanding of leaders, and other factors all influence how motivated workers are.

Managers use a range of leadership philosophies, based on the characteristics of their followers and the objectives of the company. Each type of leadership has advantages and disadvantages of its own. Leadership and motivation are intimately intertwined. Effective leadership has a positive impact on worker motivation. Both job satisfaction and employee productivity and efficiency rise. It is feasible to conclude that leadership and team motivation are essential elements of the healthcare industry and can sustain worker satisfaction.

Increased employee loyalty and active participation can be a result of managers' leadership abilities, attitudes, and role modelling in insurance organisations. Support from management for staff development cultivates trust and opens lines of communication, both of which boost workers' sense of pride in their job inside the company. Employees are motivated to take initiative and make valuable contributions to the company by this sense of pride.

Two strategies to support employee empowerment are to give constructive criticism instead of creating a fear of failure and to inspire employees to make decisions on their own. Additionally, managers want to encourage cooperation, consideration, and camaraderie among employees. By placing greater emphasis on meeting employees' needs, you may further boost employee motivation and reduce costs associated with stress and turnover. To create a new system for encouraging work motivation across the country, the government needs to focus on the political environment and factors that support citizen engagement.

Furthermore, it is important to foster a humanitarian mindset in the workplace and in society at large in order to increase community cohesion, charitable giving, and individual generosity. For example, employees should be encouraged to work as a team to share duties with their colleagues or to help each other overcome obstacles at work. People will be inspired to strive harder for group objectives as a result, which will aid in the growth of organisations.

Whether an organisation is a joyful and healthy place to work depends largely on its culture. By clearly communicating and advocating the organization's vision to subordinates and gaining their acknowledgement, it is possible to influence their work habits and attitudes. Positive relationships between a leader and their followers encourage cooperation and communication, as well as inspire followers to fulfil the company's purposes and goals, all of which boost employee satisfaction.

References

1. Drewniak, R., Drewniak, Z. and Posadzinska, I., 2020. Leadership styles and employee expectations.
2. Khaliq, M., Usman, A. and Ahmed, A., 2021. Effect of leadership style on working culture and employees motivation. *The Journal of Educational Paradigms*, 3(1), pp. 166–170.
3. Khan, A.Z. and Adnan, N., 2014. Impact of leadership styles on organizational performance. *International Journal of Management Sciences*, 2(11), pp. 501–515.
4. Al-maaitah, D.A., Alsoud, M. and Al-maaitah, T.A., 2021. The role of leadership styles on staffs job satisfaction in public organizations. *The journal of contemporary issues in business and government*, 27(1), pp. 772–783.
5. Tetteh, E.N. and Brenyah, R.S., 2016. Organizational leadership styles and their impact on employees' job satisfaction: Evidence from the mobile telecommunications sector of Ghana. *Global Journal of Human Resource Management*, 4(4), pp. 12–24.
6. Löfsten, H., 2016. New technology-based firms and their survival: The importance of business networks, and entrepreneurial business behaviour and competition. *Local Economy*, 31(3), pp. 393–409.
7. Gibson, R., 2011. *Rethinking the future: rethinking business, principles, competition, control & complexity, leadership, markets and the world*. Hachette UK.
8. Mullins, L.J., 2007. *Management and organisational behaviour*. Pearson education.
9. Bolden, R., 2016. Leadership, management and organisational development. In *Gower handbook of leadership and management development* (pp. 117–132). Routledge.
10. Nyberg, A.J., Pieper, J.R. and Trevor, C.O., 2016. Pay-for-performance's effect on future employee performance: Integrating psychological and economic principles toward a contingency perspective. *Journal of Management*, 42(7), pp. 1753–1783.
11. Saad, G.B. and Abbas, M., 2019. The influence of transformational leadership on organizational and leadership effectiveness: An empirical case study of Pakistan. *Amazonia Investiga*, 8(21), pp. 117–129.
12. Elliot, A.J., Dweck, C.S. and Yeager, D.S. eds., 2017. *Handbook of competence and motivation: Theory and application*. Guilford Publications.

Note: All the tables in this chapter were made by the author.

Recent Trends in Engineering and Science for Resource Optimization and
Sustainable Development – Prof. (Dr.) Dorota Jelonek et al. (eds)
© 2024 Taylor & Francis Group, London, ISBN 978-1-032-98030-0

84

A Management Assessment on How Corporate Culture Affects Organizational Performance

Nukalapati Naresh Kumar[1]
Research Scholar, College of Law,
Koneru Lakshmaiah Education Foundation Green Fields,
Vaddeswaram, Guntur, Andhra Pradesh,

Petikam Sailaja[2]
Associate Professor, College of Law,
Koneru Lakshmaiah Education Foundation Green Fields,
Vaddeswaram, Guntur, Andhra Pradesh

Abstract

The corporate world currently is quite dynamic. However, an effective corporate culture and its distinct responsibilities with regard to performance blend many cultural structures, leading to enhanced production and satisfaction and ultimately boosting the financial performance of the company. In the last few years, artificial intelligence (AI) became significantly important for businesses looking to leverage corporate-related statistics and stay viable. While AI innovations have the capacity to enhance organizational performance, numerous businesses struggle to implement them because of inadequate organizational and AI capabilities. Hence this research investigates the corporate culture influence on organizational performance using machine learning algorithms. A variety of algorithms for selecting features and classification algorithm were developed, and the best edition was picked based on metrics such as accuracy, AUC score, F1 value etc. The Principal Component Analysis (PCA) approach was used to find the optimal number of attributes needed for an effective prediction. K-folds cross-validation was used to achieve a more balanced result by thoroughly evaluating all feasible pairs. Grid Search CV was also used to refine the selected hyper parameters for every method to attain the best possible results. Findings show a positive correlation between corporate culture and organizational performance using ML algorithms.

Keywords

Corporate culture, Artificial intelligence, Machine learning, and Organizational performance

1. INTRODUCTION

Numerous experts define culture as "the individuals and their distinctive personality and style of a company," "the manner we do activities around here," or "the emotive irrational characteristics of a company." Businesses are increasingly reconsidering what a workforce looks like. Employees are experiencing a blurring of work-life boundaries

[1]Nareshkumarb4u@gmail.com, [2]sailaja.petikam@gmail.com

DOI: 10.1201/9781003596721-84

(Akpa et al 2021). Facebook, for example, offers rewards to ensure employees never leave the workplace. Their "campuses" include amenities including ice cream stores, exercise facilities, arcades, barbershops, and restaurants. Contemporary organizations, such as Facebook, prioritize their staff' employment over their personal life. Employees go on corporate vehicles, consume three meals per day while working, and reside in company accommodation. Investments in corporate culture represent a significant departure from traditional cubicle layouts (Pathiranage et al 2020).

Corporate culture is a complex and dynamic phenomenon. Comprehending it is fraught with notions, ideas regarding information, and discussions that lead to different philosophies. The organization setting has been dynamic and complicated for both large and small organizations due to factors such as deregulation, technological advancements, competition, financial markets, and uncertain market conditions. Establishing a certain corporate culture is crucial for organizational performance (Cherian et al 2021).

Corporate culture is shaped by societal connections among people and categories, instead of being a singular notion. Culture has a substantial influence on employee behavior and corporate performance, including framework, compensation, and appraisal. Corporate management has prioritized economic factors over corporate culture when evaluating staff performance, despite concerns about culture and image. To increase corporate performance, it's important to implement metrics and methods that satisfy both clients and employees (Tedla 2016).

For years, society has seen technical advancements like the Industrial Age, the computer period, the web, and social networks. As technologies develops and data becomes more abundant, companies are repositioning themselves to capitalize on the possibilities of Artificial Intelligence (AI) innovations. Technological advancements have altered society's structure and interaction patterns (Pappas et al 2018).

Companies are held accountable for a variety of current societal concerns, which can have societal, ecological, and financial repercussions. Now that society is more conscious of the consequences of their use of various offerings, companies have been compelled to run in more environmentally friendly and open manner. Huge quantities of statistics have let firms understand that the statistics they control and how they employ it can provide a competitive advantage. Companies are making investments in Big Data-enabled technology, like AI. The ability to use data from diverse resources, share it with multiple stakeholders, and analyze it in various manners enables digital shifts and the building of environmentally friendly societies (Bley et al 2022).

According to an Appian study done in 2019, the most critical variables in obtaining impact from AI initiatives are changes to current IT and corporate cultures (Duan et al 2019). Previous research on technological adoption has identified corporate culture as a significant influencer of novel technological adoption. Corporate culture has a wide-ranging influence on the company and is seen as a crucial component in the failure of new technology efforts. Companies are integrating Big Data and AI technology to translate data into insight and employ that understanding to gain a competitive advantage. The most prevalent issues that companies confront are administrative and cultural, instead of those connected to technologies and data.

Hence this study aims to investigate the corporate culture affects organization performance using machine learning (ML) algorithm.

2. Literature Review

According to Stephen and Stephen (2016), an organization's corporate culture can impact staff performance, either beneficially or badly. A culture that values workers as integral to the organization's development leads to improved worker performance. Workers align their goals and ambitions with the companies to ensure its success.

Previous research has primarily focused on the technical aspects of embracing AI and big data analysis, rather than the cultural implications. Most empirical investigations presume a direct correlation between AI capabilities and performance. Nevertheless, there is a shortage of investigation that examines company culture as a key element (Mikalef, and Gupta 2021).

According to Shin (2022), while the Glassdoor data shows a correlation between employee perspective, corporate culture, and worker contentment, their effect on organizations' future performance varies by financial channel. Hales et al. (2018) proposed a stronger link between worker sentiment and organizational performance.

3. Research Methodology

Digital data is rising by the day as an outcome of networking sites, and data mining assessment can be used to extract important information from it. Natural language processing, or NLP, is an area of artificial intelligence that examines natural languages utilizing perception functions to collect valuable data. The investigation used data from the Kaggle database. Kaggle is an open data science platform that offers typical libraries for investigating fresh elements of AI.

If the data is not optimal attempts ought to be made to improve it, and various data analytics might be undertaken. When the data is considered sufficient for a help evaluation, it will be analyzed using machine learning computations. An appropriate classification algorithm is the core of any machine learning algorithm. This study chose four supervised algorithms because the research is being performed in a supervised environment: random forest (RF), support vector machine (SVM), extreme gradient boosting (XGB), and logistic regression (LR). Furthermore, each of the hyperparameters of these methods were adjusted using GridSearchCV to determine the best spectrum of data for the needs.

The test statistics was gathered from a secondary source and evaluated scientifically and inventively. The statistics is organized into three categories: Acquisition, Paidoff, and Collecting_paidoff. This data contains 11 employee details, such as his or her employee ID, the status of the advancement, the standard amount that the employee acquires, the duration of the time period for which the employee obtains the funding, and furthermore.This is testing data from Paidoff employee that paid their credit on schedule and settled within the stated term.

A semantic analyzer was created on a Core i7 Windows PC using Python 3.7, Keras, and the NLP tool PyTorch version 1.3. Adam, an optimal method, was chosen as the typical method for deep learning challenges. To avoid divergent behaviors, the optimal learning rate was chosen. The present research employed the binary cross-entropy technique to address the binary categorization problem. In the present investigation, proposed methodologies were evaluated for accuracy, recall, precision, AUC value, and F1-value to establish the organizational performance.

4. RESULT AND DISCUSSION

PCA has defined the performance metrics that will be analyzed after the train-test splitting, in which SMOTE is used just for the training batch while keeping the testing batch alone (Table 84.1 and Fig. 84.1). Furthermore, GridSearchCV was not employed in this instance.

Figure 84.1 indicates that the tree-oriented architecture with PCA outperforms any other arrangement used in this technique. The training-testing splitting helps to compre-

Table 84.1 PCA test batch

	RF	SVM	LR	XGB
Accuracy	0.86	0.93	0.95	0.88
Precision	0.95	0.99	0.99	0.97
AUC value	0.87	0.98	0.98	0.94
F1 value	0.93	0.981	0.97	0.98
Recall	0.91	0.984	0.96	0.89

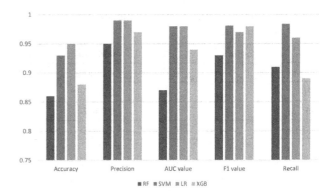

Fig. 84.1 PCA comparative models and training testing part

hend the architecture's generalization and evaluates the algorithm's adaption to new data. In this regard, tree-oriented models, followed by the SVM technique, surpass the LR technique.

In this case, accuracy will be assessed using 5-fold cross-validating, with GridSearchCV utilized for fine-tuning the hyperparameters (Table 84.2 and Fig. 84.2). This study shall focus primarily on accuracy generally.

Table 84.2 PCA cross-validation

	SVM	LR	XGB	RF
Accuracy	0.943	0.950	0.970	0.952
Precision	0.975	0.982	0.964	0.920
AUC value	0.970	0.975	0.968	0.911
F1 value	0.963	0.967	0.964	0.944
Recall	0.952	0.952	0.965	0.964

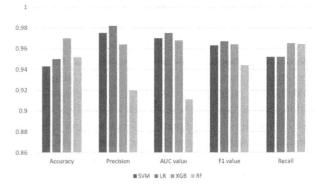

Fig. 84.2 Outcomes of PCA cross-validation

Figure 84.2 illustrates that the SVM method surpassed tree-oriented systems when the hyperparameters were changed employing GridSearchCV. The SVM technique's AUC and F1 values are both optimized. Table 84.3 and Fig. 84.3 show an examination of PCA variances across multiple techniques.

Table 84.3 PCA variance evaluations for several techniques

Algorithm	Variance
SVM	0.00106
LR	0.00655
RF	0.04074
XGB	0.01831

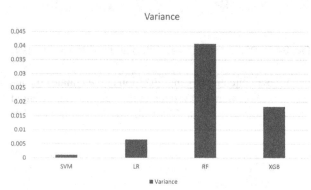

Fig. 84.3 PCA variance evaluations for several techniques

Table 84.3 displays the variance for all 5-folds created during the evaluation for every technique. The smaller the variance, the more precise and trustworthy the framework becomes. Furthermore, this study can see that the SVM technique outperforms the LR and XGB algorithms in both instances, with the RF approach following closely afterwards.

5. Conclusion

This paper efficiently showed how ML methods increased the estimated accuracy of corporate culture management assessments. As an outcome, the methods outperform traditional mathematical systems. They mitigated the negative impact of sample biases built into standard quantitative methods. ML methods, for instance, may evaluate massive volumes of data without being bound by hypotheses about data distribution. The approaches, which identify complex non-linear patterns in large datasets, have the potential to transform corporate culture management and enable precise organizational performance.

In the debate over which supervised learning technique to use, this study discovered that SVM or extreme gradient boosting techniques outperform alternate tree-based or linear methods if the investigation environment is equal to ours. Furthermore, in the debate over which dimensionality-reducing approach to use, this study methodology revealed that recursive character removal with 5-fold cross-validation might outperform methods based on PCA. In the future, this study hopes to execute more calculations

on real data collection that might be led by various means, as well as set up a GUI gateway dependent on this research which could produce interesting new information on the influence of corporate culture on organizational performance.

References

1. Akpa, V.O., Asikhia, O.U. and Nneji, N.E., 2021. Organizational culture and organizational performance: A review of literature. *International Journal of Advances in Engineering and Management*, *3*(1), pp. 361–372.
2. Pathiranage, Y.L., Jayatilake, L.V. and Abeysekera, R., 2020. A Literature Review on Organizational Culture towards Corporate Performance. *International Journal of Management, Accounting & Economics*, *7*(9).
3. Cherian, J., Gaikar, V., Paul, R. and Pech, R., 2021. Corporate culture and its impact on employees' attitude, performance, productivity, and behavior: An investigative analysis from selected organizations of the United Arab Emirates (UAE). *Journal of Open Innovation: Technology, Market, and Complexity*, *7*(1), p.45.
4. Tedla, T.B., 2016. *The impact of organizational culture on corporate performance*. Walden University.
5. Pappas, I.O., Mikalef, P., Giannakos, M.N., Krogstie, J. and Lekakos, G., 2018. Big data and business analytics ecosystems: paving the way towards digital transformation and sustainable societies. *Information systems and e-business management*, *16*(3), pp. 479–491.
6. Bley, K., Fredriksen, S.F.B., Skjærvik, M.E. and Pappas, I.O., 2022, September. The role of organizational culture on artificial intelligence capabilities and organizational performance. In *Conference on e-Business, e-Services and e-Society* (pp. 13–24). Cham: Springer International Publishing.
7. Duan, Y., Edwards, J.S. and Dwivedi, Y.K., 2019. Artificial intelligence for decision making in the era of Big Data–evolution, challenges and research agenda. *International journal of information management*, *48*, pp. 63–71.
8. Stephen, E.N. and Stephen, E.A., 2016. Organizational culture and its impact on employee performance and job satisfaction: A case study of Niger Delta University, Amassoma. *Higher Education of Social Science*, *11*(5), pp. 36–45.
9. Mikalef, P. and Gupta, M., 2021. Artificial intelligence capability: Conceptualization, measurement calibration, and empirical study on its impact on organizational creativity and firm performance. *Information & Management*, *58*(3), p.103434.
10. Shin, Y., 2022. The impact of organizational culture on employee communication satisfaction. *East Asian Journal of Business Economics (EAJBE)*, *10*(1), pp. 23–34.
11. Hales, J., Moon Jr, J.R. and Swenson, L.A., 2018. A new era of voluntary disclosure? Empirical evidence on how employee postings on social media relate to future corporate disclosures. *Accounting, Organizations and Society*, *68*, pp. 88–108.

Note: All the figures and tables in this chapter were made by the author.

*Recent Trends in Engineering and Science for Resource Optimization and
Sustainable Development – Prof. (Dr.) Dorota Jelonek et al. (eds)*
© 2024 Taylor & Francis Group, London, ISBN 978-1-032-98030-0

85

Strategic HRM: Matching Employees to Business Objectives

Sushmita Goswami[1]

Assistant Professor,
Institute of Business Management,
GLA University, Mathura, Uttar Pradesh, India

Shakti Awasthi[2]

Associate Professor,
Lala Lajpatrai Institute of management,
Mahalakshmi Mumbai – 34

Riza Bahtiar Sulistyan[3]

[3]Department of Management,
Institut Teknologi dan Sains Mandala, Indonesia

K. Vimala[4]

Assistant professor, Department of MBA,
Karpagam College of Engineering, Coimbatore

Swapnil Vichare[5]

Assistant Professor,
Department of Management,
Global Business School and Research Centre,
Dr. D Y Patil Vidyapeeth, Pune, Maharashtra

Ashwini V Rathi[6]

Assistant Professor,
Department of Management Studies,
GH Raisoni University, Amravati, Maharashtra, India

━━━━━ ■ **Abstract**

Human resources (HR) constitute a key basis of competitive edge in a market that is shifting quickly. Through the facilitation of business-specific competency growth, HR platforms may lead to a persistent competitive edge. The establishment of a connection between the strategy for HR and execution and the business's overarching strategic objectives is the focus of strategic human resource management (HRM). The majority of published data indicates a statistically significant correlation between increased HR strategy implementation and improved business success. Artificial intelligence speeds decision-making and corrects business processes and activities, facilitating the assessment of employees' job satisfaction

_____
[1]sushmita60485@gmail.com, [2]shakti.awasthi@yahoo.com, [3]rizabahtiars@gmail.com, [4]vimalark56@gmail.com, [5]swapnil.vichare@dpu.edu.in,
[6]ashwinirathi11@gmail.com

DOI: 10.1201/9781003596721-85

and improving management with a limited spectrum of jobs, types of work, and environments. Employees management is necessary to make sure the company recruits and keeps talented employees in regard to the increase in worldwide business activities and the number of businesses growing into exporting businesses. For many decades, businesses have battled to locate certified professionals to carry out the required training and responsibilities. Hence the objective of this study is to build acomputerized framework that measures employment satisfaction using an enhanced neural network approach. Artificial Neural Networks have been employed to supply the best attributes as data inputs to assess employee satisfaction rates. The improvement in precision, recall, and the F-value of the proposed work has recently been analysed in the sequence.

▬▬▬ Keywords

Human resources, Strategic human resource management, Business, Employees, and Employment satisfaction

1. INTRODUCTION

The economic landscape is shifting quickly, driven by factors including advancement, shifting investor and client requirements, and fiercer competition in the product marketplace. In order to thrive in this competitive landscape, companies must consistently enhance their operational efficiency through cost reduction, product and procedure innovation, and improvements in quality, productivity, and time to marketplace(Loon et al 2020).

Human resources, or the employees who comprise a business, are regarded as one of the most crucial resources for modern businesses. Since numerous additional drivers of competitive performance are no longer as potent as they once were, employees and the way they are handled are becoming increasingly crucial (Gerhart and Feng 2021). It is imperative to acknowledge that the foundation for competitive edge has shifted in order to formulate an alternative framework for analyzing matters related to strategy and human resource management (HRM) (Stone et al 2024). While traditional drivers of success like economies of size, secured marketplaces, and advances in product and procedure innovation continue to offer a business a competitive advantage, a business's human resources are more important to its long-term viability.

The level of satisfaction and pleasure employees have with their employment determines their level of employment satisfaction. The most common method for measuring employee satisfaction at job is through surveys. Survey-related factors that affect job retention could include pay, workload, management, opportunities, accessibility, teamwork, and facilities. These items are necessary for businesses that wish to keep their staff content (Davidescu et al 2020). Satisfaction does not make up for excellent work or performance. HR methods and approaches that emphasize ways to improve employment tenure demoralize excellent performers. The words "employee involvement" and "sat-

isfaction" seem to be synonymous, and both meanings are now commonly employed at the same time.

HRM has three main subdivisions: international HRM, strategic HRM (SHRM), and micro HRM. The subfunctions of HR regulations and practice are covered by micro HRM, which is divided into two primary classifications: handling people and smaller teams (e.g., hiring, selection, onboarding, development and training, performance administration, and compensation) and handling employment businesses and employee sound platforms (for instance, union-management interactions) (Cupcea and Bîrcă 2023).

HRM is becoming more important in the business order, in line with the realization that people are the business's most important asset. Ensuring that the business acquires and maintains the qualified, devoted, and highly driven personnel it requires is the goal of HRM. In order to improve and grow employees' innate capacities—their contributions, possibility, and employability—it is necessary to evaluate and meet their future requirements (Sypniewska et al 2023). Incentives for learning and ongoing growth are one way to do this. It entails carrying out hiring and choosing processes, developing managerial skills, and conducting training programs according to the demands of the business.

In the discipline of HRM, strategic human resource management, or SHRM, is a relatively recent development. HRM platforms' impact on business performance is a topic covered by SHRM, with a special emphasis on how to coordinate HR to obtain a competitive edge. Enterprises are realizing that effective HR strategies and procedures can boost all aspects of performance, including economic, excellence, and production (Boon et al 2018).

Combining various points of view, it is suggested that policies, values, culture, and practices are just a few of the numerous elements that make up SHRM. The aforementioned comments suggest that SHRM establishes connections, combines, and maintains coherence among different business tiers (Agustian et al 2023). Either overtly or covertly, it aims to better leverage HR in relation to the business's strategic requirements.

Intelligent systems are computerized replicas of natural intelligence procedures that possess the cognitive abilities of an electronic platform or a technology-driven device. Nevertheless, there aren't any apps accessible right now that can equal individual autonomy in larger areas or occupations requiring a significant quantity of daily knowledge, given the ongoing developments in computing speed and storage capability. HR units are currently embracing the digital age by utilizing cloud computing, artificial intelligence (AI), and data analytics, among other technologies, to streamline their operations (Vrontis et al 2022). These days, the majority of businesses employ HR bots or AI. AI has a significant potential impact on HRM. To achieve greater success in their recruitment efforts, businesses ought to leverage technological advances.

Hence this study aims to investigate the employment satisfactions using artificial neural network (ANN). Raw data from the categorization technique is fed into an ANN, which is used for predicting employment satisfaction.

2. LITERATURE REVIEW

An increasing amount of research demonstrating a favorable correlation between business performance and SHRM is fueling the increased interest in measuring (Zehir et al 2020). Over the past 20 years, the connection between HRM and business performance has been a highly discussed subject. The majority of main scientific study on this subject has come from America and, to some degree, from the UK. Scholars and businesses alike are working to demonstrate the beneficial effects of HRM on overall efficiency. The majority of published data indicates a statistically significant correlation between increased HR practice implementation and improved business success.

According to Kowalski and Loretto 2017, HRM has an impact on workers' well-being, which in turn affects business outcomes. The influence of HRM innovations on employee satisfaction and business efficiency is a topic of debate among scientists. According to investigators, HRM systems benefit both employers and workers (also known as the mutually beneficial viewpoint). Experts assert that the benefit of HRM systems is with corporations, not with employees, who are on the other end of the range.

In order to provide an empirical study on the conciliatory function of human capital in relation to the organizational architecture of human resources and sustained competitive edge of enterprises, Delery and Roumpi 2017 surveyed 779 businesses throughout China using an online survey.

3. RESEARCH METHODOLOGY

A crucial component of business statistical system layout for human resources is the creation of databases. It is essential to thoroughly review the specification and design papers, define the database design's material, and then choose the best database administration system in accordance with the material. Data about employees must be entered into the system. The employee id is typically set to a character kind with a range of 10 digits, per the system parameters in Table 85.1. When the individual starts working, they will be assigned this number.

Table 85.1   Employee details

Category	Employee Name	Identification Number	Gender	Phone number	Department	Feedback
Size	10	8	2	12	12	50
Type	Char	Int	Char	Char	Char	varChar
Properties	Yes	No	Yes	Yes	Yes	Yes

This investigation's main objective is to determine the proportion of satisfied employees across a range of sectors, such as technical, sales, IT, advertising, manufacturing, and accounting businesses. In order to obtain important information into a problem faced by human resources in each branch of any company, this study employed the HRA-Dataset available from the Kaggle repository for these kinds of job simulations of the recommended technique. With a single record per person, the collection contains biographical information about the staff of such a huge organization. The study finds that satisfaction rates are roughly 60% and effectiveness is about 70% on average. Each year, employees labor 200 hours on average, completing 3 to 4 activities.

3.1 Research Details

The results were found by simulating the operation in the Python environment. The coding division and spyder, an extremely powerful Python research setting, were created to help researchers carry out the activities of the suggested structure. The method of selecting rows will be genetic algorithm (GA). Each characteristic in the fitness functional is run through it to get the crossover, random mutation, and linear mutation level. The characteristic value changes from 0 to 1 if the characteristic meets the fitness operation. When the number of 1s exceeds the number of 0s, it indicates that most qualities are taken into account when determining the satisfaction degree. Regarding category-wise bifurcation, an employee's degree of satisfaction is regarded as a judgment category. Three categories are used to determine the satisfaction degree: low, moderate, and exceptional. The range of the satisfaction score is 0.09 to 0.90.

A significant amount of collected data is split up into distinct groups in order to measure the level of satisfaction with this activity depending on variables like pay, the most frequent assessment, the collection of tasks, etc. Massive amounts of data need to be initially classified and then optimized in order to extract the relevant data. For this purpose, GA is employed in this approach as an optimization, and in fact, the data classification is changed. Massive amounts of data need to be initially classified and then optimized in order to extract the relevant data. In this strategy, the resource class is changed and GA is utilized as an optimization for each of these goals.

4. Result and Discussion

This study aims to examine workplace employee satisfaction rates while accounting for the many previously discussed factors. The database's characteristic sets are included if the GA selects the complete row. For instance, if there are 500 rows in the fulfilled category with 8 characteristics in a database, 500*8 is the volume. If 80 rows are not chosen, 420*8 with group label 1 is supplied to the NN as feed. A training model is created using a feed-forward neural network. The suggested example has 3 class labels: unsatisfied, moderately satisfied, and satisfied. This framework will include the propagating values during neural training.

As shown in Table 85.2, the effectiveness of the previously described research is discussed in this part to enhance accuracy, recall, and the F-value over 1000 items.

Table 85.2 Outcomes of 1000 items precision

No. of Iteration	K-means (GA)	ANN (GA)
10	97.2	98.5
20	97.2	98.2
30	96	97.6
40	96.5	97.5
50	95	97.2

A comparison of the two methods is displayed in Fig. 85.1. One uses GA with K-means, while the other uses ANN for 1000 items of GA. The figure indicates that this method produces better outcomes than the alternative. Additionally, the level of precision decreases with an increase in the number of times. This could be the result of the tested sample being run repeatedly to provide an accurate level for evaluation.

The recall of 1000 items are displayed in Table 85.3. The evaluation of recall utilizing GA as a ML method to ANN

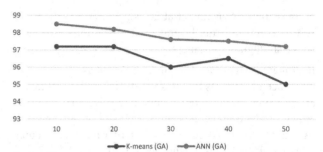

Fig. 85.1 Outcomes of 1000 items precision

Table 85.3 Outcomes of 1000 items recall

No. of Iteration	K-means (GA)	ANN (GA)
10	97	98.5
20	97.1	98.3
30	96	97.5
40	96.5	97.1
50	95	96

and K-means is shown in Fig. 85.2. The graph's blue and orange lines correspond to the recall scores for the K-means and ANN approaches to GA, accordingly. The figure shows that, with a larger recall score, ANN outperformed the K-means method in terms of performance.

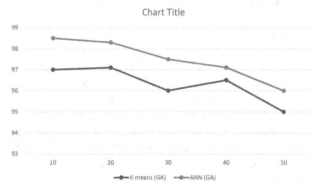

Fig. 85.2 Outcomes of 1000 items recall

The F-value for 1000 Items is displayed in Table 85.4. The F value comparing GA and ANN, and also K-means as a ML technique, is shown in Fig. 85.3. The F-value results for GA using an ANN and GA using a K-means tech-

Table 85.4 Outcomes of 1000 items F-value

No. of Iteration	K-means (GA)	ANN (GA)
10	97	98.4
20	97.3	98.1
30	96	97.6
40	96.6	97.2
50	95	96.4

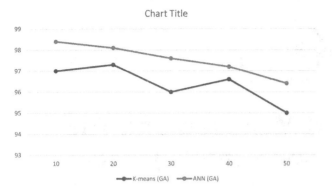

Fig. 85.3   Outcomes of 1000 items F-value

nique are shown in the graph by the blue and orange paths, correspondingly.

5. CONCLUSION

The field of HR has successfully adapted to the technological transition brought about by AI, but there is yet much work to be done. We must always look for ways to overcome any new development's challenges if we are to fully benefit from it. The HR field needs to take a similar approach. It is easy to arrive at that data is necessary for implementing AI to business operations depending on the many investigations included in this work. Thus, it is imperative that HR professionals take great care to ensure that accurate data is utilized. HRM strategies have integrated AI to enhance decision-making and preparation.

It is crucial to remember that resource production needs to be used as much as feasible in order to meet overall business' objectives. As a result, the employee's production determines and eventually achieves the business objectives. This study uses an ideal NN technique to develop a computerized framework for tracking employee engagement at workplace. Workers in HR will find the tool to be quite helpful in addressing concerns regarding general employment satisfaction. It provides an automatic data retrieval mechanism so that rapid decisions can be taken to raise the business's output or production. The proposed method, which saves time, has totally eliminated the human element of the study's base that was previously done directly to evaluate employees' performance. This study aims to develop a ML-based system that can automatically analyze employment performance. The proposed framework was effectively adapted for examining employee performance on the basis of precision, recall, and the f-value. The degree of satisfaction can be estimated employing three traits as input data and numerous neurons in the hidden layer. The outcomes indicate that combining GA and ANN improves forecast accuracy and performance.

References

1. Loon, M., Otaye-Ebede, L. and Stewart, J., 2020. Thriving in the new normal: The HR microfoundations of capabilities for business model innovation. An integrated literature review. *Journal of Management Studies, 57*(3), pp. 698–726.
2. Gerhart, B. and Feng, J., 2021. The resource-based view of the firm, human resources, and human capital: Progress and prospects. *Journal of management, 47*(7), pp. 1796–1819.
3. Stone, R.J., Cox, A., Gavin, M. and Carpini, J., 2024. *Human resource management*. John Wiley & Sons.
4. Davidescu, A.A., Apostu, S.A., Paul, A. and Casuneanu, I., 2020. Work flexibility, job satisfaction, and job performance among Romanian employees—Implications for sustainable human resource management. *Sustainability, 12*(15), p. 6086.
5. Cupcea, I. and Bîrcă, A., 2023. Theoretical-conceptual and comparative approaches to human resources management within organizations.
6. Sypniewska, B., Baran, M. and Kłos, M., 2023. Work engagement and employee satisfaction in the practice of sustainable human resource management–based on the study of Polish employees. *International Entrepreneurship and Management Journal, 19*(3), pp. 1069–1100.
7. Boon, C., Eckardt, R., Lepak, D.P. and Boselie, P., 2018. Integrating strategic human capital and strategic human resource management. *The International Journal of Human Resource Management, 29*(1), pp. 34–67.
8. Agustian, K., Pohan, A., Zen, A., Wiwin, W. and Malik, A.J., 2023. Human Resource Management Strategies in Achieving Competitive Advantage in Business Administration. *Journal of Contemporary Administration and Management (ADMAN), 1*(2), pp. 108–117.
9. Vrontis, D., Christofi, M., Pereira, V., Tarba, S., Makrides, A. and Trichina, E., 2022. Artificial intelligence, robotics, advanced technologies and human resource management: a systematic review. *The international journal of human resource management, 33*(6), pp. 1237–1266.
10. Zehir, C., Karaboğa, T. and Başar, D., 2020. The transformation of human resource management and its impact on overall business performance: big data analytics and AI technologies in strategic HRM. *Digital Business Strategies in Blockchain Ecosystems: Transformational Design and Future of Global Business*, pp. 265–279.
11. Kowalski, T.H. and Loretto, W., 2017. Well-being and HRM in the changing workplace. *The International Journal of Human Resource Management, 28*(16), pp. 2229–2255.
12. Delery, J.E. and Roumpi, D., 2017. Strategic human resource management, human capital and competitive advantage: is the field going in circles?. *Human Resource Management Journal, 27*(1), pp. 1–21.

Note: All the figures and tables in this chapter were made by the author.

Recent Trends in Engineering and Science for Resource Optimization and Sustainable Development – Prof. (Dr.) Dorota Jelonek et al. (eds)
© 2024 Taylor & Francis Group, London, ISBN 978-1-032-98030-0

86

Skilful Leadership and Management: The Importance of Emotional Intelligence

Ankita Saxena[1]

Assistant professor, Institute of Business Management, GLA University, Mathura, Uttar Pradesh

Sphurti Birajdar

Assistant Professor, Department of Management,
Lexicon Management Institute of Leadership and Excellence, Wagholi, Pune, Maharashtra

M. Vanisree[2]

Assistant Professor, Department of English, B V Raju Institute of Technology, Narsapur, Telanagana

P. B. Narendra Kiran[3]

Assistant Professor, Christ University, Bangalore

Prashant Kalshetti[4]

Associate Professor & Head, Department of BBA,
Dr. D Y Patil Vidyapeeth, Global Business School and Research Centre, Pune

Rajesh vemula[5]

Assistant Professor, Mittal school of business, Lovely Professional University

Abstract

Emotional intelligence (EI) has become more important in the study of organisational behaviour, particularly in relation to management and effective leadership. EI is the ability to identify, understand, and control one's own emotions as well as those of others. Those with high EI find it easier to navigate complex social interactions, build strong relationships, and resolve conflicts. EI is the ability to recognise, manage, and evaluate emotions. The ability to express one's emotions in a healthy way and to empathise with others is a sign of great emotional intelligence in a leader, and it will enhance both performance and workplace relationships. The study employed a range of machine learning (ML) methods, such as ANN, BRDT, Naive Bayes, and Random Forest, to predict EI based on behaviour credits. ML approaches have become more and more common. The results showed that the BRDT has the accuracy of 98.3 which is higher in all other machine learning models and gives better results. Seven behavioural attributes and seven additional individual attributes made up the prediction dataset.

Keywords

Emotional intelligence, Leadership, Accuracy, Machine learning

---

Corresponding authors: [1]ankitaaecsaxena@gmail.com, [2]vanisreemolugu@gmail.com, [3]narendra.kiran86@gmail.com, [4]prashantkalshetti@gmail.com, [5]raz5020@gmail.com

DOI: 10.1201/9781003596721-86

1. Introduction

Good leaders are still in high demand in the workplace because employers want someone who can steer their business and set a vibrant vision, and employees want someone who can set a good example or serve as a role model. This search is focused on people who control their emotions related to their work as well as those who execute job actions well. Those who have effectively blended strong emotional intelligence (EI) with excellent leadership are frequently the most sought-after leaders. According to Rahman, M., & Uddin, M. (2016), a large body of research on EI suggests that "we perform better when we are good at recognising and dealing with emotions in ourselves and others." As a result, emotional intelligence (EI) is getting more recognition in the leadership community as a crucial, essential quality of a successful leader. These affective competencies improve one's capacity for forming connections and have also been linked to more observable results.

The term "emotional intelligence" was coined by Peter Salovey and John Mayer to refer to a set of cognitive skills related to identifying, understanding, controlling, and strategically utilising emotions in a variety of contexts (Giao et al., 2020). Effectively detecting and interpreting emotions, using emotions to improve reasoning and problem-solving skills, realising the importance of emotions, and controlling emotions to influence behaviour are all parts of this process. Emotionally intelligent leaders are adept at perceiving and comprehending the feelings and thoughts of others around them. This talent allows them to react to their team's emotions in a thoughtful and productive way. The capacity for empathy fosters the growth of rapport and trust, which creates a favourable environment.

Furthermore, in addition to having a significant impact on leadership effectiveness, emotional intelligence also has a tremendous effect on team dynamics. The capacity of a leader to discern and regulate the emotions exhibited by their team members fosters a climate of psychological safety and trust. According to Karneli (2023), this fosters an environment that welcomes open communication, facilitates the exchange of ideas, and fosters a willingness to take risks. These components are necessary to promote innovation and creativity within the team. Moreover, those in leadership positions who possess a high aptitude for emotional intelligence are better equipped to manage and settle disputes as well as help to resolve conflicts.

The goal of this research is to increase employees' emotional intelligence by gathering more information on it and then using that information to provide recommendations for how to raise it. As a result, they are able to analyse themselves more thoroughly and develop. The chosen sample size reflected only a small portion of the total population.

2. Literature Review

2.1 Leadership and Emotional Intelligence

Leadership and emotional intelligence have a strong connection since EI shows how well a leader can train and support others in growing. This can be done in a setting where objectives are clear, responsibility, inspiration, respect for one another and faith are fostered, and each person's special role and contributions are acknowledged. Being a leader involves more than simply having authority and power; it also involves possessing spiritual values, emotional connections, and honesty (Oh, J., & Wang, J, 2020). EI is important for effective leadership, effective leadership styles and leader performance, as many studies have shown (Drigas, A., & Papoutsi, C, 2019, Drigas et al., 2021, Miao et al., 2018). Furthermore, a number of working variables, such as stress, job satisfaction, job performance, and intentions to leave the company, have been demonstrated to be positively impacted by EI (Papoutsi et al., 2019).

The development of emotions is the primary responsibility of leadership, which is a crucial component that upholds the dynamic interaction between emotional intelligence and leadership. More precisely, encouraging positive emotions in people they manage is one of a leader's main responsibilities (Goleman et al., 2013). This occurs when they foster cooperation and harmony, which generates a positive energy pool that brings out the best in people.

2.2 Developing Emotional Intelligence

The process of developing EI is a lifetime endeavour that can greatly enhance leadership abilities. By means of ongoing self-exploration and advancement, this route prepares leaders to handle the complexities of their roles with increased proficiency. Self awareness is the cornerstone of emotional intelligence, which comprises recognising and understanding one's own emotions (Goleman 2020). To make thoughtful decisions in their leadership responsibilities, leaders need to be able to comprehend their own emotions. Another crucial element of emotional intelligence is empathy, which calls for paying attention to stakeholders, coworkers, and team members in order to comprehend their needs, feelings, and points of view.

Developing emotional intelligence is a continuous process that improves one's capacity for leadership (Sheeba, M., & Rebekah, T. R, 2023). It entails empathy, self-awareness, and ongoing introspection, which eventually results in a leadership style that is genuine, empathetic, and flexible.

In today's ever changing environment, emotional intelligence has emerged as a critical characteristic that distinguishes outstanding leaders. It enables leaders to succeed by encouraging, motivating, and uplifting others under their direction, transcending organisational and industry borders. Increasing one's emotional intelligence adopting the ongoing practice of enhancing emotional intelligence produces remarkable results and fosters an environment where brilliance is valued. In general, excellent leaders and their organisations succeed because of their emotional intelligence.

3. METHODOLOGY

Replicating human intellect in machines is the aim of machine learning. Extraction of huge patterns from tiny amounts of data is its main objective. It is the study of algorithms used by computers that get better over time. Many disciplines, such as data mining, automaton and formal programming languages, recognising patterns, theory of graphs, artificial intelligence, neural networks, statistics, economics, behaviour in organisations, and development, are strongly linked to ML. Reinforcement learning, supervised learning, and unsupervised learning are the three primary methods in ML. One hybrid strategy that can be tailored to the problem a researcher is trying to address is semi-supervised learning. Every tactic has benefits and drawbacks, and some approaches are better than others at resolving specific types of problems. To create classification models that are utilised for prediction, classification algorithms are employed. Prior to being evaluated for prediction on the testing data that hasn't been seen, the model is trained using a set of data known as training data. The stages for creating a categorization model are listed below.

3.1 Data Collection

This work's brain is the analysis and interpretation of the data that has been gathered. The dataset utilised in this investigation was gathered from the following questions were used to gather information on emotional intelligence from respondents; the questions were constructed around a person's behavioural characteristics. People from a variety of multinational companies provide the data. Seven behavioural traits and seven individual characteristics found in the dataset are proven to be highly effective in determining emotional equilibrium. There were 600 samples with 16 characteristics in the dataset. There were 550 samples after duplicate data was removed. There were 550 samples with 14 characteristics after the superfluous attributes were removed. Only one can get at the conclusions with the associated problem revealed after data analysis. This conclusion, which was reached after analysis and interpretation, aids in identifying the problem's hidden answer and provides the best recommendations for solving issues.

Table 86.1 Description of characteristics

Characteristics	Estimate
Age	8
Gender	4
Qualification	16
Experience	14
Job designation	9
Industry sector	20
Results	12
Self esteem	6
Happiness	6
Empathy	6
Sociability	6
Self control	6
Relationship	6
Emotion perception	6

3.2 Preprocessing

Ascribes' basic information included age, gender, qualification, experience, designation, and IT sector. First, the unadulterated data were encoded. They were substituted by mathematical attributes. The entirety of the inquiry scores (from self-esteem to emotion perception) was included in the attribute called "Result". A copy of the columns was deleted. The entire dataset was then divided into preparation and testing sets. Fifteen percent was preserved for testing, while eighty-five percent was saved for preparation. Component scaling was utilised to place all of the data in the same reach and size.

3.3 Used Machine Learning Techniques

We look at several classifiers, such as keyword-based classifiers, BRDT, Navie Bayes, Random forest, and ANN.

- **Bayesian Rough Decision Tree (BRDT)**

    The Bayesian Rough Decision Tree (BRDT) is a hybrid of the Decision Tree and the Bayesian Rough set. Instead of calculating the entropy of the choice, the system determines the threshold value, which improves the decision tree. BRDT algorithms begin at the root node using a test scenario. An example of this would be to assign a specific feature to the next node based on the outcomes of the previous testing. The tested feature value is simultaneously sent to each node. This feature assignment and testing procedure is executed until the leaf node is reached. Sorting the feature values according to the leaf node class is the final step. BRDT uses information entropy as an index to calculate the uncertainty of the tested set.

- **Navie Bayes**

    The Naive Bayes classifier is a simple probabilistic classifier that uses strong norm objectivity to apply the Bayesian theorem. The naive Bayes classifier makes the assumption that a feature's existence in a class is unrelated to the existence of any other feature. It has been demonstrated that it works effectively in a number of intricate, challenging real-world Bayesian classification theories and is helpful for assessments that forecast future data patterns and help make wise decisions. Classification is one predictive machine learning technique that forecasts using historical data. Predictive models can determine a variable's class membership based on the known values of the other variables. By classification, data are placed into pre-defined groupings. It's called supervised learning as the classes are predetermined by one or more subject matter experts looking over the data.

- **Random Forest**

    Several trees' predictions are combined to create random forests. These trees are all trained separately. After training, the basic model is integrated during boosting using a complex weighting system. The trees are trained individually in this procedure, and the average is applied to the trees' predictions.

- **ANN**

    Artificial neural networks (ANN) are made up of units, also referred to as artificial neurons. These units are arranged in a series of layers to form the ANN of a system. Several dozen to millions of units can make up a layer, based on the amount of intricate neural networks are required to find the hidden patterns in the dataset. The typical architecture of ANN includes input, output, and hidden layers. The input layer collects data from the outside environment and feeds it into the neural network for analysis or learning. The input is subsequently transformed into meaningful data for the output layer by one or more hidden layers. Finally, the ANN response to the supplied input data is produced by the output layer.

4. EMPIRICAL RESULTS

4.1 Performance Evaluation

The model's accuracy affects how well it was trained and how it would behave in the actual world. Nevertheless, it won't provide any specifics of how it would be used to the issue. Accuracy only informs us of the effectiveness of the trained model.

$$Accuracy = \frac{TP + TN}{TP + FP + TN + FN} \qquad (1)$$

When assessing if there are more false positives than True positives, precision is important.

$$Precision = \frac{TP}{TP + FP} \qquad (2)$$

When there are more false negatives, recall is helpful. When false negatives occur more often than, our model's effectiveness suffers.

$$Recall = \frac{TP}{TP + FN} \qquad (3)$$

The F1-score is the weighted mean of recall and precision,

$$F1 - score = \frac{2 . precision . recall}{precision + recall} \qquad (4)$$

4.2 Results

All features have been chosen in experiments utilising a variety of machine learning approaches. To determine the specificity, sensitivity, and accuracy of the predicted model, all 16 attributes are chosen. Gaussian NB has been used in the Naive Bias approach's implementation. A combination of the Decision Tree and the Bayesian Rough set is the Bayesian Rough Decision Tree (BRDT). The Relu activation function, the 16 and 32 filters, the Adam Optimizer, the binary cross-entropy loss function, and 0.5 dropouts with 50 epochs are all used by ANN. Table 86.2 below lists the machine learning algorithm results that have been suggested.

Table 86.2 Comparison of machine learning algorithms

Approaches	Accuracy	Precision	Recall	F1 score
BRDT	98.3	98.4	98.2	98.3
RF	88.7	88.3	88.1	88.6
NB	90.8	90.2	90.5	90.7
ANN	93.8	93.3	93.5	93.7

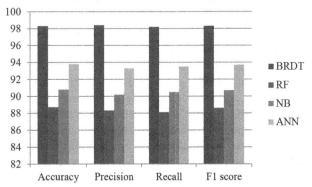

Fig. 86.1 Performance evaluation of different machine learning models

In Fig. 86.1, the BRDT based model produced the best accuracy; precision, recall, and F1 score result when compared to all other assessed model building procedures. These findings unequivocally imply that using a BRDT-based approach is preferable to using the other conventional ML classifier that was proposed in previous study for incorporating EI in the workplace.

4.3 Leadership Effectiveness and Emotional Intelligence

Effective leaders are able to develop a workforce that is not only engaged but also highly driven by using emotional intelligence with skill. This results in a revolutionary shift that is output- and creativity-driven and improves organisational outcomes. The notion of employee engagement is the cornerstone of this transformation. This crucial relationship between emotional intelligence and engagement highlights the enormous impact that emotionally intelligent leaders can have on their businesses, both individually and collectively.

Emotionally intelligent leaders are remarkably adept at handling crises with composure and fortitude. In the face of uncertainty, turmoil, or misfortune, they hold firm and steady. Their poise and fortitude serve as an inspiration, fostering trust within their groups, stakeholders, and the larger organisational ecosystem.

Because of this, emotional intelligence is essential to effective leadership and has the power to drastically alter organisations. Positive work environments with high levels of participation and team members motivated by a shared objective are fostered by emotionally savvy leaders. They are also adept at handling confrontations, resolving disputes through compromise, and preserving order in times of emergency. Emotional intelligence serves as a driving force for leaders to surpass conventional limits and foster an environment of trust, resiliency, and boundless potential.

5. CONCLUSION

A key component of effective leadership is emotional intelligence. Leaders that possess emotional intelligence are better at handling group dynamics, making informed judgements, and exerting overall leadership impact. Group dynamics and emotional intelligence are correlated. Developing emotional intelligence is a continuous process that improves one's capacity for leadership. It entails self-awareness, empathy, and ongoing introspection, which eventually results in a leadership style that is genuine, empathetic, and flexible.

This study tried to evaluate emotional intelligence leaders in the workplace using a variety of machine learning techniques. A range of performance evaluation metrics were used to analyse the effectiveness of the models that were put into practice. In contrast to NB, which included all of its feature characteristics after managing missing values, the increased result of the BRDT classifier was obtained while comparing the results with another recent study on similar subject. When all other evaluated model building procedures were taken into consideration, the BRDT-based approach in this research produced the highest accuracy outcome. These findings strongly show that EI in the workplace can be addressed with a BRDT-based approach instead of the other traditional ML classifier proposed in earlier studies.

References

1. Rahman, M., & Uddin, M. (2016). Role of emotional intelligence in managerial effectiveness: An empirical study. *Management Science Letters*, 6(3), 237–250.
2. Giao, H. N. K., Vuong, B. N., Huan, D. D., Tushar, H., & Quan, T. N. (2020). The effect of emotional intelligence on turnover intention and the moderating role of perceived organizational support: Evidence from the banking industry of Vietnam. *Sustainability*, 12(5), 1857.
3. Karneli, O. (2023). The Role of Adhocratic Leadership in Facing the Changing Business Environment. *Journal of Contemporary Administration and Management (ADMAN)*, 1(2), 77–83.
4. Oh, J., & Wang, J. (2020). Spiritual leadership: Current status and Agenda for future research and practice. *Journal of Management, Spirituality & Religion*, 17(3), 223–248.
5. Drigas, A., & Papoutsi, C. (2019). Emotional Intelligence as an Important Asset for HR in Organizations: Leaders and Employees. *International Journal of Advanced Corporate Learning*, 12(1).
6. Drigas, A., Papoutsi, C., & Skianis, C. (2023). Being an Emotionally Intelligent Leader through the Nine-Layer Model of Emotional Intelligence—The Supporting Role of New Technologies. *Sustainability*, 15(10), 8103.
7. Miao, C., Humphrey, R. H., & Qian, S. (2018). Emotional intelligence and authentic leadership: A meta-analysis. *Leadership & Organization Development Journal*, 39(5), 679–690.
8. Papoutsi, C., Drigas, A., & Skianis, C. (2019). Emotional intelligence as an important asset for HR in organizations: Attitudes and working variables. *International Journal of Advanced Corporate Learning*, 12(2), 21.
9. Goleman, D., Boyatzis, R. E., & McKee, A. (2013). *Primal leadership: Unleashing the power of emotional intelligence*. Harvard Business Press.
10. Goleman, D. (2020). *Emotional intelligence*. Bloomsbury Publishing.
11. Sheeba, M., & Rebekah, T. R. (2023). A STUDY ON THE ROLE OF EMOTIONAL INTELLIGENCE IN LEADERSHIP EFFECTIVENESS. *XIBA Business Review*, 6(1).

Note: All the figures and tables in this chapter were made by the author.

Recent Trends in Engineering and Science for Resource Optimization and Sustainable Development – Prof. (Dr.) Dorota Jelonek et al. (eds)
© 2024 Taylor & Francis Group, London, ISBN 978-1-032-98030-0

87

Handling Digital Advertising and Marketing in a Dynamic Setting

Nilesh Vitthal Limbore[1]

Assistant Professor, Sharadchandra Pawar Institute of
Management and Research, Someshwarnagar, Baramati, Pune

Rita Biswas

LLB, MBA (HR), PhD (Management), Senior Facilitator, Regenesys Business School

Anand Ashok Kopare[2]

Associate Professor, Management, Atlas SkillTech University Mumbai, Maharashtra

Ashwinkumar A. Santoki[3]

Assistant Professor, Department of Management,
Vanita Vishram Women's University, Surat, Gujarat

Sphurti Birajdar

Assistant Professor, Department of Management,
Lexicon Management Institute of Leadership and Excellence, Wagholi, Pune, Maharashtra

Nilesh Anute[4]

Associate Professor, ASM'S Institute of Business Management and Research, Pune

Abstract

Digital technologies and marketing are now widely used in a variety of global businesses. With the use of these resources, organizations can better create and communicate value to customers while also gaining real-time consumer data. The goal of this study is to have a deeper comprehension of the concepts and procedures related to ethical digital marketing by examining the scope of the field, obstacles to its use, and methods for enhancing local digital skills. Globally, the use of digital advertising has expanded quickly. Internet usage has a significant role in digital advertising. Digital advertisements draw in a large number of people. Therefore, through online advertising, the buying spectrum will be expanded. Key business players need to comprehend consumer behaviour and purchase preferences in order to attract more active customers and steady revenue streams. With the aid of data mining tools, information regarding purchase intentions and wants must be extracted in order to forecast the purchasing decisions of consumers. This research aims to investigate digital marketing data analytics through the application of machine learning (ML) and deep learning (DL) techniques.

Keywords

Digital advertising, Digital marketing, Machine learning, Deep learning

---

[1]nileshstat5@gmail.com, [2]anand.kopare@atlasuniversity.edu.in, [3]ashwinsantoki@gmail.com, [4]nileshanute@gmail.com

DOI: 10.1201/9781003596721-87

1. INTRODUCTION

Billionaires on a global scale now use the internet, social media, apps on their smartphones, and other digital communications tools on a daily basis. According to the most recent data (Statista, 2020a), there were 4.54 billion registered internet users globally as of the beginning of 2020, or 59% of the world's population. All throughout the world, using social networking sites has grown into an essential part of the lives of numerous individuals. In 2019, 2.95 billion individuals globally frequently utilized social media. By 2023, Statista (2020b) projects that number will increase to more than 3.43 billion. With the help of digital marketing and advertising, businesses can accomplish their marketing goals for comparatively less money (Ajina, 2019). Approximately more than 50 million firms have registered on Facebook, and 88% of businesses use Twitter for marketing. (Lister, 2017).

According to Grover et al. (2019), public services additionally, political campaigns have profited immensely from the extensive use of digital and advertising technology and applications. An increasing number of people are spending time online researching products and services, communicating with businesses, and sharing customer experiences with one another. Companies have included digital and social media platforms within their corporate advertising campaigns in response to this change in consumer behaviour (Stephen, 2016).

According to Shareef et al. (2019), businesses can reap substantial advantages by including social media marketing as a fundamental component of their overall company strategy. Lal et al. (2020) claim that social media provides companies with the chance to communicate with their customers, build brand awareness, influence consumer attitudes, ask for feedback, improve current products, and increase sales. Due to the decline of conventional means of communication and our society's dependence on brick-and-mortar operations, businesses now need to search for best practices when utilizing social media along with digital advertising methods in order to maintain and expand market share. Businesses face several challenges when developing their digital marketing campaigns and plans amid the new environment where consumers are more influential and aware of social and cultural norms. Negative electronic word-of-mouth about customer complaints can now reach millions of people instantly, potentially damaging the company in question. (Javornik et al., 2020).

Digital advertising refers to the promotion of items by marketers and rival businesses on websites like Facebook, Twitter, YouTube, and others. Although digital advertising is a type of mass communication that draws from tradi-

tional advertising strategies and modifies them to suit the needs of emerging media and technology. Through internet advertisements, people can simply access the things available online. Thus, a large number of businesses use online advertising these days.

The researcher attempted to investigate this sector of digital advertising from the perspective of the consumer and what they would expect from it. An increasing number of marketing firms are using online advertisements to promote their goods. Because of this, the researcher conducted this study from the perspective of consumers' attitudes both during and after making product decisions based on internet advertisements.

2. LITERATURE REVIEW

According to Susanne and Grabowska's (2015) research, "Online marketing strategies: the future is here," analysingthe behaviour of the consumer is essential for any marketing campaigns. Consumer behaviour is consistent whether they are shopping online or offline. To get clients' attention, a suitable approach should be taken. Without the internet, no business activity can occur. For the company's online presence, it is crucial to establish at least one online department. Establishing a distinct and well-defined brand that embodies the company's values and attracts both offline and online potential customers and devoted followers should be the primary objective of any business.

According to Kumar (2019), paper "Emerging trends in Digital marketing in India," customers are using the internet more and more to shop for the greatest deals on goods and services from vendors in India. Many digital marketing tactics, such as influencer marketing, content advertising, marketing for e-commerce, campaign marketing, social media marketing, search engine optimization (SEO), plus search engine marketing (SEM), and others, have contributed to a large increase in the pace at which customers shop. Social media usage has given digital marketers new ways to reach out to clients using online platforms. Online marketing offers a larger potential to obtain customer information than traditional marketing approaches, according to Niharika Satinder (2015) in their journal article "A study on internet marketing in India: Challenges and Opportunities." Online marketing will become more powerful in the coming years, transforming people's purchasing behaviors into world-class, efficient ones. Online buying got more convenient with the usage of credit cards.

Yurovsky, (2014) concluded that there are benefits and drawbacks to internet marketing in his study "Pros and Cons of Internet Marketing." The benefits include the following: building relationships, reaching a wider or international audience, empowering effect, removing geographical obstacles, target reaching, quick outcomes, cost effectiveness, measurable result, and availability 24 hours a day, 7 days a week. A few drawbacks of internet marketing include plagiarism, excessive use of copycats, a lack of credibility, unscrupulous perceptions, product nonconformity, excessive competition, harm from unfavorable reviews, reliance on technology, lack of acceptance by some, and loss of trust.

Many businesses today use analytics, including sentiment analysis, to more effectively comprehend and address customer feedback on their products and brand via internet marketing. For dynamic data analysis, commercial organizations will depend increasingly in the future on a variety of mining techniques and machine learning tools. Applications such as fraud detection, forecasting the stock market, managing client relationships, and summarization include analysis of sentiment, gender prediction, named entity recognition, keyword extraction, and other text mining methods. Pejic et al., (2019). In addition to being the third most popular website and the biggest platform for sharing videos online, Next to Google, YouTube is the 2nd-largest search engine globally. Billions of dollars are spent on marketing communications due to the widespread usage of YouTube. This demonstrates how social media platforms are crucial to online marketing. To attain better results, mechanisms that combine marketing tactics and machine learning tools might be implemented.

ML blends AI with statistics. In machine learning (ML), input data is learned from, and knowledge is generated as an example for intelligent decision-making on unknown test data, Miklosik et al., (2019). The microblogging age has

seen new developments in social data analysis, and amorphous data studies have been utilized to identify recurring patterns in behaviors. The primary analytical instruments are ML and AI. Many automated (machine) recognition, description, classification, and pattern grouping are areas of interest for engineering and scientific domains, including biology, psychology, healthcare, marketing, machine vision, artificial intelligence, and remote sensing. of patterns.

3. RESEARCH METHODOLOGY

E-commerce accounted for 30% of the fashion industry's 2019 production output. Sales from traditional ("brick-and-mortar") establishments are increasingly shifting to internet retailers as a result of the COVID pandemic. Fashion industry DL applications often involve algorithmic analysis and targeting for decision augmentation, including seasonality and trend tracking as well as consumer category segmentation (Chen et al., 2015). Numerous applications rely on text mining and picture classification, which enable autonomous object recognition in images. Here we will conclude the results on the basis of Traditional Machine learning and Deep learning.

3.1 Dataset for Image Classification

We made advantage of Fashion-MNIST, a freely accessible database of goods that Zalando's internal research team has selected (Xiao et al., 2017). A training collection of 62,000 samples and an evaluation set of 12,000 instances make up the data. Every example consists of a 29 x 29 grayscale image, with a total of 841 pixels, where each pixel corresponds to a single value between 0 and 260. Darker pixels are indicated by higher values. Every image is categorized into a single of the ten groups. In order to create a model that would categorize each clothing image into just one of the ten classes, we looked at a wide range of conventional ML and DL algorithms. Since CNNs have shown to be successful at classifying images, we decided to use them for this task. We employ a pretrained ResNet18 model and a simplified version of VGGNet10 for the sake of illustration (He et al., 2016). Because the photos in our dataset have a lesser resolution (29x29x1), we built the VGGNet variation of the network using three convolutional layers.

3.2 Text Mining

A few study fields have made text mining a central focus. Only thirty percent of the material on the internet is organized, with the remaining seventy percent being unstructured. Creating additional commercial value is the aim of content mining, which has historically provided high-quality information access. Text mining is the process of swiftly uncovering significant data patterns by analys-

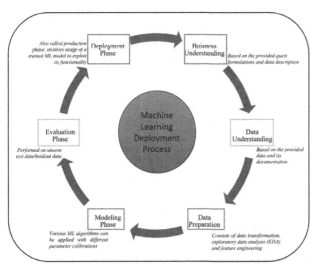

Fig. 87.1 Incorporating machine learning and deep learning to solve issues with digital marketing and advertisement

ing unstructured data. Words and sentences that can be misspelled are created by users on social networking sites like Twitter, Facebook, blogs, and WhatsApp. To enable the examination of the information using proper language and well-structured words, content mining is used. For instance, customer input is requested on each social media platform that promotes goods or services. Customers' writing styles vary, therefore they do not provide or discuss their criticism in an organized manner. The analyst gathers the feedback, examines it (a process known as response analysis), and organizes it into patterns that are useful for marketing purposes. Text mining methods are used in this work Zong et al., (2010).

3.3 Dataset for Textual Analysis

The Kaggle website provided us with access to the dataset, which we sampled to find only Positive (sentences classified as either positive or somewhat positive) and negative (sentences labeled as negative and moderately negative) are the two levels of polarity. Sorting a given sentence snippet's sentiment into either of these categories is the analysis task. 3364 processed negative sentences and 3704 processed positive words make up the final dataset sample. It is necessary to convert the text into an analysable form prior to doing ML. Word embedding offers a practical method for turning text words into numerical vectors.

4. RESULTS AND DISCUSSION

According to our findings, the DL approach (CNN) outperforms the classical ML in this picture classification challenge, performing at a higher accuracy level (94.63% vs. 85.52%) (see Table 87.1 for full results). We further present accuracy and recall scores, which quantify the proportion of pertinent outcomes and the overall proportion of relevant results that the algorithm properly classifies, as shown in Fig. 87.2. Additionally, we can raise the DL model's accuracy to 94.98% by utilizing pretrained ResNet18. A random estimate would yield an accuracy of only 10%, so take note that our accuracy rate is noticeably higher.

Table 87.1 Performance evaluation for image classification

Method	Model	Accuracy	Precision	Recall
Traditional ML	DT	80.19%	79.95%	79.95%
	SVC	82.35%	82.45%	82.03%
	RF	85.52%	85.93%	85.84%
	KNN	84.21%	84.29%	84.35%
DL	CNN	92.46%	92.18%	92.35%
	CNN dropout	94.63%	94.71%	94.31%
Proposed DL	**CNN+ResNet18**	**94.98%**	**95.00%**	**94.97%**

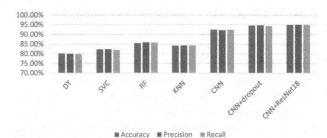

Fig. 87.2 Performance analysis for image classification

Decision-making processes in fashion companies can benefit from the use of such image recognition DL algorithms. Promotional methods may be created, for instance, by classifying and dividing different clothing categories (based on designs, hues, and forms) and connecting them to customer segments and age groups (Paolanti et al., 2019). Using information on trends and seasonality, managers can: (a) target more suitable client groups with their current products; (b) gain feedback on their present products from sales patterns; and (c) use this information to inform future product design decision.

We assessed the use of conventional ML and DL algorithms for textual sentiment of digital marketing in our analysis. We used pretrained BERT in conjunction with MLP for DL, drawing from a Devlin et al., (2019). In this work, we discovered that DL performs better than classical ML with greater accuracy (91.5 vs 80.32) (see Table 87.2 and Fig. 87.3).

Table 87.2 Performance evaluation for text mining

Method	Model	Accuracy	Precision	Recall
Traditional ML	SVM	80.32%	80.36%	80.08%
	RF	73.02%	74.34%	63.42%
	LR	77.06%	77.50%	76.81%
DL	MLP	81.47%	81.52%	81.35%
	BERT	**91.50%**	**90.14%**	**93.66%**

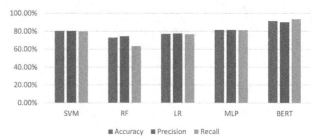

Fig. 87.3 Performance analysis for text mining

5. CONCLUSION

The two viewpoints on data/algorithmic and user-misbehaviour issues that are covered in this study are very closely related. Both have to deal with user-technology interactions, which can have both beneficial and harmful outcomes. In this work, we used publicly accessible datasets to conceptualize traditional machine learning (ML) and deep learning, to illustrate the fundamental ideas of deep learning, and to highlight the different parts of its algorithmic engine. It was shown that although classical machine learning has many benefits for businesses, managers must exercise great care, consideration, and understanding when implementing deep learning. There are some restrictions on our paper, but they also present chances for further research. First, we have provided a stylistic process for creating deep learning algorithms. Investigating organizational processes and structures that allow organizations to successfully incorporate DL into decision-making will be a primary goal of future empirical study. Further research must also address the ethical problems of prejudice and opacity, as well as strategies for resolving these problems, in light of the organizational difficulties that digital literacy faces.

References

1. Ajina, A. S. (2019). The perceived value of social media marketing: An empirical study of online word-of-mouth in Saudi Arabian context. Entrepreneurship and Sustainability Issues, 6(3), 1512–1527.

2. Chen, X., Tao, X., Zeng, X., Koehl, L., &Boulenguez-Phippen, J. (2015). Control and optimization of human perception on virtual garment products by learning from experimental data. *Knowledge-Based Systems*, *87*, 92–101.

3. Devlin, J., Chang, M. W., Lee, K., & Toutanova, K. (2019). BERT: Pre-training of deep bidirectional transformers for language understanding. NAACL HLT 2019–2019 Conference of the North American Chapter of the Association for Computational Linguistics. Human Language Technologies - Proceedings of the Conference, 1(Mlm), 4171–4186.

4. Grover, P., Kar, A. K., Dwivedi, Y. K., & Janssen, M. (2019). Polarization and acculturation in US Election 2016 outcomes–can twitter analytics predict changes in voting preferences. Technological Forecasting and Social Change, 145, 438–460.

5. He, K., Zhang, X., Ren, S., & Sun, J. (2016). Deep residual learning for image recognition. IEEE Conference on Computer Vision and Pattern Recognition, 770–778.

6. Javornik, A., Filieri, R., &Gumann, R. (2020). "Don't forget that others are watching, too!" the effect of conversational human voice and reply length on observers' perceptions of complaint handling in social media. Journal of Interactive Marketing, 50, 100–119.

7. Kumar, K. (2019). A Study of The Growth of Digital Marketing In Indian Scenario. *Pramana Research Journal*, *9*(1), 388–394.

8. Lal, B., Ismagilova, E., Dwivedi, Y. K., &Kwayu, S. (2020). Return on investment in social media marketing: Literature review and suggestions for future research. Digital and social media marketing (pp. 3–17). Cham: Springer.

9. Lister, M. (2017). The essential social media marketing statistics for 2017. Available at: http://www.wordstream.com/blog/ws/2017/01/05/social-media-marketing-statis tics.

10. Miklosik, A., Kuchta, M., Evans, N., & Zak, S. (2019). Towards the adoption of machine learning-based analytical tools in digital marketing. *Ieee Access*, *7*, 85705–85718.

11. Niharika Satinder (2015). A Study of Internet Marketing in India: Challenges and Opportunities. International Journal of Science Pramana Research Journal Volume 9,

12. Paolanti, M., Romeo, L., Martini, M., Mancini, A., Frontoni, E., & Zingaretti, P. (2019). Robotic retail surveying by deep learning visual and textual data. Robotics and Autonomous Systems, 118, 179–188.

13. Pejić Bach, M., Krstić, Ž., Seljan, S., &Turulja, L. (2019). Text mining for big data analysis in financial sector: A literature review. *Sustainability*, *11*(5), 1277.

14. Shareef, M. A., Kapoor, K. K., Mukerji, B., Dwivedi, R., & Dwivedi, Y. K. (2019a). Group behavior in social media: Antecedents of initial trust formation. Computers in Human Behavior.,

15. Statista. (2020a). Global digital population as of January 2020. Available at https://www. statista.com/statistics/617136/digital-population-worldwide/.

16. Statista. (2020b). Number of social network users worldwide from 2010 to 2023. Available at: https://www.statista.com/statistics/278414/number-of-worldwide-social-netwo rk-users/.

17. Stephen, A. T. (2016). The role of digital and social media marketing in consumer behavior. Current Opinion in Psychology, 10, 17–21.

18. Susanne Schwarzl and Monika Grabowska (2015). Online Marketing Strategies: the future is here. Journal of international studies, Volume: 8, No.2, pp. 187–196.

19. Xiao, H., Rasul, K., &Vollgraf, R. (2017). Fashion-MNIST: A novel image dataset for benchmarking machine learning algorithms

20. Yurovskiy, V. (2014). Pros and cons of internet marketing. *Research Paper*, 1–12.

21. Zhong, N., Li, Y., & Wu, S. T. (2010). Effective pattern discovery for text mining. *IEEE transactions on knowledge and data engineering*, *24*(1), 30–44.

Note: All the figures and tables in this chapter were made by the author.

*Recent Trends in Engineering and Science for Resource Optimization and
Sustainable Development – Prof. (Dr.) Dorota Jelonek et al. (eds)*
© 2024 Taylor & Francis Group, London, ISBN 978-1-032-98030-0

88

Impact of AI on Operations Management

K. Suresh Kumar[1]

Associate Professor, MBA Department,
Panimalar Engineering College, Varadarajapuram,
Poonamallee, Chennai

Vaishali Gupta[2]

Assistant Professor, GL Bajaj Institute of Technology &
Management, Greater Noida

Yugandhara Patil[3]

Assistant professor, MCA, International Institute of
Management Science, Chinchwad, Pune

Gaurav Jindal[4]

Associate Professor, Department of Master of
Computer Applications, G L Bajaj Institute of Technology and
Management, Gr. Noida

N. Rao Cheepurupalli

Department of Mineral Processing and metallurgical engineering,
Faculty of Mines, Aksum University, Ethiopia

Mahesh Manohar Bhanushali[5]

Assistant Professor, Management Studies,
University of Mumbai, VPM's Dr. V. N. Bedekar Institute of
Management Studies, Thane

Abstract

Operational management's use of AI signifies a dramatic transformation in company, altering decision-making processes, efficiency, and competitive dynamics across industries. AI reduces errors and increases productivity by automating manual tasks. Examples of this include supply chain optimisation algorithms and AI chatbots for customer support. This study compares various machine learning (ML) strategies for supply chain demand prediction, one of the most popular artificial intelligence (AI) approaches. In the current study, support vector machines (SVMs) and artificial neural networks (ANNs) are used in conjunction with more conventional time series prediction methods, such as exponential smoothing and moving average, to predict the supply chain long-term demand. The largest Indian automaker's component supplier's data set is then

Corresponding author: [1]pecmba19@gmail.com, [2]Vaishali.gargsohan@gmail.com, [3]yugandharapatil52@gmail.com, [4]gauravjindal05@gmail.com, [5]maheshbhanu87@gmail.com

DOI: 10.1201/9781003596721-88

used to implement this research. The comparison reveals that the forecasts generated by ML algorithms are substantially more accurate and closer to the real data than those generated by conventional methods for predicting.

◼◼◼◼◼ **Keywords**

Operation management, AI, ML, Prediction, Supply chain, Demand

1. INTRODUCTION

In commercial circles, the phrase "artificial intelligence" (AI) is becoming often employed. It's a technique that works with machine intelligence, particularly in computer systems. AI is a vast field of science that straddles the boundaries of arts and sciences, computer science, mathematical information, statistics, operations management and philosophy. The goal of artificial intelligence (AI) is to create nonbiological systems—like computers and machines—that are capable of carrying out tasks that normally call for human intelligence. On the other hand, ML is a branch of AI that concentrates on statistical learning methods.

Operational management's use of AI signifies a dramatic transformation in company, altering decision-making processes, efficiency, and competitive dynamics across industries. Algorithms for supply chain optimisation and AI chatbots for customer service are examples of how AI automates manual tasks, increasing efficiency but also potentially increasing error rates. In addition to improving data-driven insights, this automation frees up resources for strategic endeavours. Artificial intelligence's capacity to examine vast volumes of data leads to a deep understanding of consumer behaviour, market dynamics, and internal operational patterns, which improves strategic decision making. Thanks to AI's predictive analysis, businesses can quickly adjust to changes in the market by foreseeing and mitigating future issues (Kamble et al., 2018).

The ideal approach to solving this large data-related challenge is to use AI approaches also referred to approaches that employ large datasets to automatically identify and extract trends across parameters are known as ML approaches (Biggio & Roli 2018). Machine learning algorithms are capable of producing new insights, pointing researchers in the correct path, and uncovering patterns in data that had not been noticed before. The use of ML techniques can be advantageous for a number of industries, particularly operations, manufacturing, healthcare, and housing (Mansouri et al., 2021).

Moreover, machine learning is widely employed in the administration of many supply chain aspects and domains. Recently, there has been an increase in research interest in ML methods and their potential applications in supply chain management. Owing to the limitations of traditional methods for analysing large volumes of data, scientists are now using ML strategies, which are extremely effective in analysing and interpreting large amounts of data.

The goal of the research is to predict a time-series including pattern and periodic tendencies for an Indian auto parts supplier. Particular ML techniques like SVM and ANN are contrasted with the Mean Absolute Percentage Error(MAPE) index and more conventional time-series prediction methods like moving averages and exponential smoothing with and without patterns. The capacity of these techniques to simulate trends and seasonal variations found in suppliers' data led to their selection.

1.1 Objective of Study

1. To demonstrate the use of ANNs and SVMs in wholesaler sales prediction.
2. To illustrate the differences in the accuracy of wholesaler sales forecasts between different time-series prediction techniques.

2. LITERATURE REVIEW

2.1 Management of Operations (OM)

Three modules make up operation management (OM): "in the door," "out of door," and any management actions that don't fall into one of these three categories. "In the doors," the initial module, handles the administrative tasks necessary to obtain the necessary inputs. The primary responsibilities in this module are sourcing, purchasing, logistics, and supplier selection. "Out of door," the second module, handles the administrative tasks necessary to deliver products and services to clients. The distributor, retailer, and consumer are the three entities that are the focus of this module (Santiváñez and Melachrinoudis 2020).

According to Morikawa (2017), businesses in both the manufacturing and non-manufacturing sectors anticipate positive effects from AI. As a result, Proposition 1a of this study looked into the possible use of AI systems for Proposition 1b's quality function deployment and product inspection. All procedures, including inventory control, logistics, reverse logistics, and outsourcing, are included in the supply chain (Subramanian and Ramanathan 2012; Quiroz and Wamba 2019). Downstream and upstream supply chains are the two segments that make up a supply chain. In the upstream supply chain, selecting a supplier is a frequent decision-making process (Kar, 2015).

2.2 Artificial Intelligence

Artificial Intelligence (AI) is the capacity of a computer to precisely learn from outside inputs and use that knowledge to carry out specific tasks and goals (Haenlein and Kaplan 2019). The learning methods that the framework can employ are semi-supervised, supervised, or unguided (Kar 2016). As per Kumar et al. (2019), artificial intelligence (AI) is a technology that provides an abundance of information and possibilities that may be sorted down to personalised targeting.

Because AI can automate monotonous tasks like scheduling, data entry, and order processing, resources may be allocated to more proactive and value-adding projects. In addition to improving operational efficiency, this shift creates an inventive and adaptable organisational culture—both crucial for being competitive in the fast-paced manufacturing sector (Kinkel et al., 2022).

The evaluation of AI to supply chain operations represents a substantial advancement in the management and optimisation of supply chains by businesses. AI is particularly useful for supply chain optimisation, as it increases accuracy and efficiency in key areas. One such field is demand forecasting, where artificial intelligence algorithms can analyse enormous volumes of data to better correctly estimate future product demand and assist businesses in avoiding excess production or overstocks (Helo and Hao 2022).

3. METHODOLOGY

Forecasting a time-series with pattern and seasonal shifts for an Indian auto parts supplier is the aim of this project. For long-term demand predicted, we employ two ML methods: are SVM and ANN. We predict the same data using conventional time series prediction techniques, such as moving average and exponential smoothing, as a benchmark for comparing ML algorithms.

3.1 Supper Vector Machine (SVM)

Unlike neural networks and linear regression, SVMs, a more recent family of universal function correlates, are not predicated on the idea of empirical risk reduction. They are depending on the structural risk reduction concept of theory statistical learning. Structural risk reduction attempts to minimise the true error on an unseen, chose at random test

instances, whereas MLR and NN minimise the error for the scenarios that are currently visible. SVM use a higher-dimensional projection of the data to maximise interclass margins and minimise regression error margins. Because the margins are soft, it is possible to find a solution even in cases when the training set contains contradicting samples. The Radial Basis Function (RBF) kernel is one of the kernels that can be utilised to enable higher dimensional space from mapping of non-linear and adjusting the number of errors in relation to the model complexity using a complexity parameter. The method translates into the minimising of the subsequent function:

$$F(f) = \frac{P}{N}\sum_{j=1}^{N}\left|Z_j - f(x_j)\right|_j + \frac{1}{2}|f|^2 \qquad (1)$$

The fact that this function prevents over-fitting by assigning zero loss to mistakes smaller than e is a crucial point. Put another way, this function tube fits with a data radius rather than a precise value. This is comparable to a fuzzy function description. This loss function's second noteworthy feature is that it minimises a least modulus rather than least squares. As we'll see later, the e option also has a significant impact by giving the data a sparse form. Under very broad circumstances, the objective functions minimize can be expressed as follows:

$$F(x) = \sum_{j=1}^{N} p_j Q(x, x_j) \qquad (2)$$

Where is the quadratic problem's solution, denoted by p_j. The referred to as kernel function, (x,xj), is the same as the X. This often utilised instrument facilitates nonlinear mapping and yields the generalised inner product. A variety of options, including Gaussian, sigmoid, polynomial, and splines, are available for the kernel function.

3.2 Artificial Neural Network (ANN)

ANN is a type of generalised nonlinear nonparametric algorithms that are motivated by studies on the nervous system and brain. The majority of predicting systems are drawn to ANNs due to their proven ability to be universal approaches. Furthermore, for modelling unknown functions, ANNs are less expensive than linear subspace techniques like polynomial and trigonometric series. It is widely recognised that a feed-forward network may arbitrarily well imitate any constant functioning, with the output module's transfer function being easily recognised and the middle-layer cells' logistic activities given an adequate number of middle-layer units. As a result, this study employed a three-layer feedforward network.

The parameters that in the multivariate regression framework represent the regressors are connected to the output that represents the regress by using a middle layer. The network model can be expressed as follows:

$$R_t = f(X_t, \beta, \gamma) + \varepsilon_t \qquad (3)$$

3.3 Moving Average

This technique forecasts the upcoming period using the average of n prior periods. Finding the ideal value for n is the issue.

3.4 Exponential Smoothing

In order to mitigate transient fluctuations in the data, these algorithms employ a weighted average of past values. The weights decrease rapidly over time. The forecasting formula is as follows:

$$F_{t+1} = F_t + \beta(B_t - F_t) \qquad (4)$$

Where:

F_t = Forecasted demand at time t+1

B_t = Real demand at time t;

4. EMPIRICAL RESULTS

Here, SVM's suitable kernel function that can forecast with the least amount of error is found. To this end, four different kernel functions are studied: polynomial, sigmoid, RBF, and linear. The MAPE index is used to express the inaccuracies in these forecasts.

The findings displayed in Table 88.1 indicate that the linear kernel function type is the best choice for this type of data. Figure 88.1 displays the results of various functions of kernel together with the time it took to solve them.

Table 88.1 Various kernel functions' output

Kernel Functions	MAPE Value
RBF	173.937
Linear	173.392
Polynomial	188.384
Sigmoid	193.762

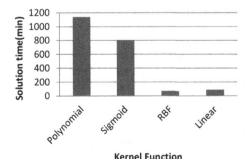

Fig. 88.1 Kernel functions solution time

We can utilise Bayesian regularisation in the training procedure to find out how many parameters the network is using efficiently, irrespective of the overall network parameters number. However, after experimenting with a number of different middle-layer unit and layer counts, we discovered that the last quantity of units of middle-layer and layers are set to three and three, respectively.

Table 88.2 Displays the outputs of each case's ANN MAPE index

Number of units in all layers	MAPE	
	3- layer	4- layer
2	167.54	172.19
3	164.71	169.76
4	165.29	165.29
5	167.92	170.58
6	166.47	170.58

We initially examined a values range for n and then MAPE index were computed to determine the ideal value of n. As a result, we determined that 200 was the ideal value for n, and MAPE = 169.842. Table 88.3 displays the outcome of utilising a moving average.

Table 88.3 Outcome of applying the moving average method

N	MAPE
2	180.357
4	178.871
10	173.896
30	171.957
60	169.956
80	170.186
100	168.843
150	169.989
200	169.842
300	172.021
500	173.438
1000	174.134

The demand was projected using exponential smoothing on the data set, and it was discovered that the ideal value for n is 0.01 0. Using this number, the forecasting error was calculated, and the best combination's MAPE index was found to be 1 67.797(Table 88.4).

The suggested approaches are evaluated for stability using sets of raw data, and the score of MAPE is recalculated for the outcomes. Table 88.5 displays the least amount of MAPE that was produced by each procedure. The research's suggested artificial neural network (ANN) can be used to

Table 88.4 Outcome of applying the exponential smoothing approach

B	MAPE
.001	169.961
.005	169.963
.010	170.156
.050	169.685
0.100	172.216
0.150	174.632
0.300	175.549
0.450	175.978
0.600	176.531
0.750	176.864
0.900	178.461

Table 88.5 Using the MAPE index for comparison and model validity

Techniques of prediction	Training Data	Testing Data
ANN	172.253	165.425
SVM	180.735	169.642
Moving Average	182.632	169.686
Exponential Smoothing	183.874	173.174

estimate demand in a supply chain more effectively than earlier traditional approaches, according to the findings.

5. CONCLUSION

In summary, artificial intelligence (AI) has had a profound and transformative effect on business operations across a wide range of industries. Businesses have been able to improve customer experience, optimise operations, and make data-driven decisions that have improved their overall performance and competitiveness thanks to AI technologies and applications.

In order to forecast demand in the supply chain one of the operation management part, this research employs a few ML techniques, which are a subset of AI, including ANN and SVM. The procedure consisted of two steps. The first step involved training an ANN with three layers and three middle units employing sensitivity analysis. Four distinct functions of kernel were then used to determine the optimal function of kernel and parameter arrangement for the SVM technique. Two conventional forecasting techniques were then employed to make the forecast and evaluating predicting errors for all methods using the MAPE index. The outcomes demonstrated that artificial neural networks are more accurate forecasters than previous techniques. The optimal set of parameters from each method is used in the next step for simulation validity and evaluation. The effectiveness of the suggested models is assessed using raw data in the second step. The outcomes demonstrated once

more how accurately artificial neural networks can forecast when compared to SVM and other conventional forecasting techniques.

References

1. Kamble, S. S., Gunasekaran, A., & Gawankar, S. A. (2018). Sustainable Industry 4.0 framework: A systematic literature review identifying the current trends and future perspectives. *Process safety and environmental protection*, *117*, 408–425.
2. Biggio, B., & Roli, F. (2018, October). Wild patterns: Ten years after the rise of adversarial machine learning. In *Proceedings of the 2018 ACM SIGSAC Conference on Computer and Communications Security* (pp. 2154–2156).
3. Mansouri Musolu, F., Sadeghi Darvazeh, S., & Raeesi Vanani, I. (2021). Deep learning and its applications in medical imaging. *Internet of Things for Healthcare Technologies*, 137–153.
4. Santiváñez, J. A., & Melachrinoudis, E. (2020). Reliable maximin–maxisum locations for maximum service availability on tree networks vulnerable to disruptions. *Annals of Operations Research*, *286*(1), 669–701.
5. Morikawa, M. (2017). FIRMS'EXPECTATIONS ABOUT THE IMPACT OF AI AND ROBOTICS: EVIDENCE FROM A SURVEY. Economic Inquiry, 55(2), 1054–1063.
6. Queiroz, M. M., & Wamba, S. F. (2019). Blockchain adoption challenges in supply chain: An empirical investigation of the main drivers in India and the USA. *International Journal of Information Management*, *46*, 70–82.
7. Subramanian, N., & Ramanathan, R. (2012). A review of applications of Analytic Hierarchy Process in operations management. *International Journal of Production Economics*, *138*(2), 215–241.
8. Kar, A. K. (2015). A hybrid group decision support system for supplier selection using analytic hierarchy process, fuzzy set theory and neural network. *Journal of Computational Science*, *6*, 23–33.
9. Haenlein, M., & Kaplan, A. (2019). A brief history of artificial intelligence: On the past, present, and future of artificial intelligence. *California management review*, *61*(4), 5–14.
10. Kar, A. K. (2016). Bio inspired computing–a review of algorithms and scope of applications. *Expert Systems with Applications*, *59*, 20–32.
11. Kumar, V., Rajan, B., Venkatesan, R., & Lecinski, J. (2019). Understanding the role of artificial intelligence in personalized engagement marketing. *California Management Review*, *61*(4), 135–155.
12. Kinkel, S., Baumgartner, M., & Cherubini, E. (2022). Prerequisites for the adoption of AI technologies in manufacturing–Evidence from a worldwide sample of manufacturing companies. *Technovation*, *110*, 102375.
13. Helo, P., & Hao, Y. (2022). Artificial intelligence in operations management and supply chain management: An exploratory case study. *Production Planning & Control*, *33*(16), 1573–1590.

Note: All the figures and tables in this chapter were made by the author.

Recent Trends in Engineering and Science for Resource Optimization and
Sustainable Development – Prof. (Dr.) Dorota Jelonek et al. (eds)
© 2024 Taylor & Francis Group, London, ISBN 978-1-032-98030-0

Navigating the Digital Workforce through Integrating Machine Learning and AI for HR Success in Stock Market Finance

Ahmad Y. A. Bani Ahmad[1]

Department of Accounting and Finance, Faculty of Business, Middle East University,
Amman 11831, Jordan,Applied Science Research Center, Applied Science Private University, Jordan

Bandham Saidulu

Assistant Professor, Mallareddy Engineering College for Women, Hyderabad

Avinash Hanmant Ghadage[2]

Associate Professor, D. Y. Patil Vidyapeeth's, Global Business School and Research Centre, Tathawade, Pune

Ajim Shaikh

Professor, Yashwantrao Mohite Institute of Management, Bharati Vidyapeeth (Deemed to be University) Pune, Maharashtra

Melanie Lourens[2]

Deputy Dean Faculty of Management Sciences, Durban University of Technology, South Africa

Jigneshkumar B. Patel[3]

Assistant Professor, Kalol Institute of Management, KIRC Campus, Kalol, Dist. Gandhinagar, Gujarat

▬▬▬ Abstract

Human resource is an important department of each of the organisation and these modern technologies are improving the functioning of the department. The human resource refers to the department in a company which helps to manage the employees and the process related to managing the workforce. The key technologies which the HR are using nowadays to integrate their department are Artificial intelligence and Machine learning. AI is the advanced technology which learns about the data and patterns and solves complex problems. On the other hand, ML is a process or algorithm which tracks the human data and imitates and forms similar results. This research navigates the implications of the digital workforce through integration of Machine learning and AI for the success of the HR in the stock market finance. The major findings of the research include the theories of stock market and HRM, the use of AL and ML in human resources, the parts of the stock market that are being impacted by AI and ML and lastly the advantages and difficulties with AI and ML Integration in HR. The report has come up with the findings that it is extremely important for the companies to invest in these advanced technologies like AI and ML if they are looking forward to success in the age of digitisation and stock finance. Companies which use AI and machine learning can easily improve their talent acquisition process, increase the staff engagement, and education and training initiatives. This will result in a more flexible, competent, and driven workforce.

▬▬▬ Keywords

Machine learning, Human resource, Artificial intelligence, Stock market, Finance

[1]aahmad@meu.edu.jo, [1]navinghadage@gmail.com, [2]melaniel@dut.ac.za, [3]jigneshpatel2411@gmail.com

DOI: 10.1201/9781003596721-89

1. INTRODUCTION

Human resources (HR) are evolving in the age of digital media from a managerial position to an essential resource in promoting company success. HR's function is becoming ever more focused on incorporating technology to improve performance, efficiency, and worker happiness. A key component of this change is the inclusion of machine learning (ML) and artificial intelligence (AI) into the HR process. HR has historically been in charge of handling employee-related duties like recruitment, instruction, and progress reviews. Nonetheless, with an emphasis on managing talent, company development, and staff planning, HR is taking on a greater significance in businesses in the digital era (Okatta, Ajayi and Olawale, 2024). To remain successful in the digital world, firms must include AI and ML in their HR operations. These innovations have the power to completely transform HR procedures by customizing customer service, automating repetitive operations, and evaluating data to deliver knowledge. HR can improve its efficacy, productiveness, and influence on overall corporate success by using AI and ML.

Since the beginning of the 1990s, research on company governance has gained popularity, with research studying how labour influences company behaviour and strategic choice-making. Some writers have diagnosed a shift in the management structures of continent-wide Europe and Japan regarding American or British models, citing shifts in stock exchange legislation and financial sector availability as the causes of managerial greater attention to minority investors and stock market values. This highlights the issue of what effects this tendency will have on HRM.

1.1 Aim and Objectives

Aim: The study focuses on navigating the Digital Workforce through Integrating Machine Learning and AI for HR Success in Stock Market Finance.

Objectives:

- To explore Some theories regarding the stock market and HRM.
- Recognizing AI and ML in Human Resources.
- To discuss the parts of the stock market that are being impacted by AI and ML.
- To focus on advantages and difficulties with AI and ML Integration in HR.

2. LITERATURE REVIEW

2.1 Some Theories Regarding the Stock Market and HRM

The research on the "Different kinds of Capitalist" emphasizes how the stock market and HRM processes are relat-

ed. It is believed that a flexible labour market with little job security works well with an economic structure that encourages rapid funding redirection. On the other hand, fewer liquid stock markets encourage a certain amount of job security inside businesses, which is advantageous for the growth of certain human resources. HRM procedures of mentioned and non-listed businesses should not vary significantly in a nation that has a purely "insider" type of management (Mallick, 2021). But if the share market's impact grows, more noticeable changes in HR operations between the two sets of businesses might be seen. The present shifts in business governance have enhanced the economic viability standards placed on public corporations, which is their greatest immediate impact (Asfahani, 2024). Every business operating in a capitalist economy must reach a certain level of success to pay for its capital expenses. Investment strategies aim to provide their investors with the most revenue (within a certain degree of risk), which has implications for the firms they acquire shares. Financial institutions thus support novel approaches to company governance that aim to maximize "shareholder return."

Accounting research may be used to understand the many options available for achieving the highest level of monetary effectiveness. It is possible to differentiate between two tactics: the offensive, high roadway, which recognizes that some immediate expenditures might raise total revenue and, therefore, both net outcome and revenue over a greater period, and the protective, low roadway, which involves minimizing labour expenses. The influence of profit limits on compensation levels and training costs is still up for debate (Sakka, El Maknouzi and Sadok, 2022).

2.2 Recognizing AI and ML in Human Resources

In today's lightning-fast corporate climate, artificial intelligence (AI) and computerization are now critical tools for recruiting, hiring, and managing staff. Screening chatbots and computerized social networking scanning tools are examples of common artificial intelligence algorithms which offer substandard or poor indications of an applicant's chances of succeeding in the firm. FirstJob developed Mya, an AI recruitment employee that communicates with candidates to confirm they fulfil job criteria, respond to inquiries, and update them on the progress of their resumes. Via the candidate's characteristics, social media harvesting techniques gather an infinite amount of information that is then used to forecast future involvement rates and other particular actions.

Uses of in-between artificial intelligence that gather data straight from the candidate include testing, gaming, and exercises. Pymetrics was utilized to create an efficient AI scanning procedure for Unilever, that demands candidates to play sports based on neurobiology for within 20 minutes. These tests assess an applicant's recall, focus, risk tolerance, and capacity to interpret environmental information. While they don't concentrate on particular job indicators, these moderate AI apps often provide recruiters with a reasonable indicator of whether an applicant is a suitable match for the role. Customized algorithms are used by powerful artificial intelligence systems to connect applicants who best possess these attributes to specific position-related criteria. A sophisticated artificial intelligence (AI) human resource management business, The Hiring Company, has created a video interviewing software that evaluates each candidate's responses, voice, posture, feelings, and vocabulary. Additionally, Affectiva has developed software for recognizing emotions, which aids in assessing someone's mental agility and sincerity. These sophisticated AI algorithms can recognize the look of hatred that a candidate gets while talking about their former employer.

2.3 Parts of the Stock Market are Being Impacted by AI and ML

By improving managing finances, buy-side movement, and operating efficiency, artificial intelligence and ML are revolutionizing the stock market. Larger financial professionals and financial institutions are the main users of it, although it may also be employed in trading to increase the difficulty of traditional algorithmic trades. AI and ML algorithms develop into computer-programmed algebra by learning from the information supplied, which allows them to recognize and carry out transactions without the need for human interaction (Pandey, Balusamy and Chilamkurti, 2023). AI can improve the running of huge orders with little Influence on the marketplace, and it may additionally boost stability control in increasingly digitized marketplaces like the FX and stock markets. To increase efficiency and expedite operation, dealers may also use AI and ML for HR to request flow control and handle risks. However, the widespread use of computational models by financial experts may encourage herding conduct and one-sided sectors, increasing the danger to fluidity and rigidity, especially in stressful situations. During regular periods, AI and ML algorithm trading may boost availability; yet, it can also cause integration, rapid failures, and more market instability. It is hard to reduce such hazards because of the complexities and challenges of understanding and imitating the decision-making process of AI and ML models and algorithms.

3. METHODOLOGY

This research follows an empirical process where data has been collected from the previously published journals and articles. The secondary data collection has been helpful in gathering important information about the topic. Mathematical expressions and equations have been performed through own understanding and analysis. This empirical research collected data from sites like PubMed, ResearchGate, IEEE and Google Scholar.

4. ANALYSIS

4.1 How Linear Regression, K-Nearest Neighbours and AI can help the HR in Employment

- **Linear Regression**

Linear regression is an important tool that helps in predicting the financial information and also predict the trends of the employees. This modelling techniques help in regression modelling in order to stimulate a linear relationship between various dependent and independent variables (Lawal, Yassin and Zakari, 2020, December). The linear regression model is helpful as it helps in creating a best-fit line which explains about the relationship between the independent factors and the dependent variable.

In this technique, a straight line is created which is represented here,

$$O = S_x + K$$

where O can be referred to as the, S_x can be considered as the slope, and K represents the constant.

The line is drawn using the above mathematical expression and is also ensured that this line is useful in crossing the highest number of possible data sets in the points (Sonkavde et al., 2023). When the values of the data are organized as a chart on a graph, a straight line is created and this line is fitted between the points in a manner so that the difference between each point or the square of the numbers is the smallest. For each of the factors assigned to x, the hypothesis lines derive and forecast the value of the y. This forecasting method is thus known as regression. Further, in order to evaluate the results and also to check how well the model fits the line, various parameters such as "RMSE, MAE, MSE, and R-squared" are used (Dospinescu and Dospinescu, 2019).

In the case of HR, they can use this model to identify how various variables can affect employee performance. This will help to get a better all-round approach of the performance.

- **K-Nearest Data**

KNN can be referred to as a classification and regression technique and it is often denoted as the lazy

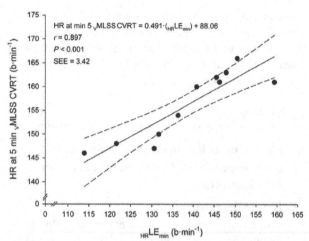

Fig. 89.1 Linear regression as a measuring employee trends

learner as it does not require a huge amount of time for learning. KNN is one of the easiest ML algorithms and the only value which is required to be calculated is the value of K and the Euclidean distance (Tanuwijaya and Hansun, 2019). The slow learning aspect of this ML makes it much quicker to perform than other algorithms. However, for analysing the big data, it cannot be used as it skips the learning step The Euclidean distance calculation is given below,

$$D(h_i, p_r) = \sqrt{\sum_{l=1}^{n} (P_r - h_i)^2}$$

Here, P_r stands for the predicted value and hi represents the data value.

- **AI or artificial intelligence**

  AI has the advantage of learning about the vast amount of data and it can identify the patterns while adapting to the changing conditions. AI can also be extremely helpful in predicting the market more accurately and effectively. AI is a great tool which is not only re-shaping the landscape of stock market prediction, but also, it is helping the HR to predict the trends of the employees and thus helping in better recruitment. AI uses methods like "data-driven insights, algorithmic trading, sentiment analysis, predictive analytics, and risk management". These processes can instantly help in improving the decision-making process in a stock firm and improve investment outcomes.

4.2 Advantages of Machine Learning and AI Integration in HR

Incorporating AI and ML into HR may greatly improve HR management and promote corporate performance. Here are few of the advantages which the HR can enjoy by using AI and ML in the business process.

- **Removing Bias**

  Artificial intelligence is a great tool which can be used in the HR department as it helps to ensure data-based decisions making (Khair et al., 2020). These are extremely crucial processes in the functions of the HR department such as recruitment, performance management and also leadership planning. An HR has the role of ensuring that the background of the candidates do not affect the employment process and the AI system actually helps to remove any kind of biases and helps to establish this aspect.

- **Employee Satisfaction**

  With the help of artificial intelligence technology, the HR of a company can ensure the maximum employee satisfaction by giving them a personalized experience. The queries of the employees can be solved easily with the help of an AI-powered chatbot and this also decreases the workload of the HR. Also, it makes the functioning of the HR department smooth as the employees do not really have to wait for an appointment with the HR to get a reply for their queries. Along with this, with the AI system, the employees can also get a personalized onboarding as well as training suggestions.

- **Better and improved efficiency**

  The companies which are choosing to use AI in HR can easily increase their organisational efficiency and improve the functionalities of the human resource department (Niehueser and Boak, 2020). The greatest advantage of using AI is that it automates most of the repetitive tasks. For example, the HR can easily conduct the attendance and leave management process with AI and it eliminates all kind of paperwork. HR can obtain the correct data as the human involvement is reduced and also the chances of the errors.

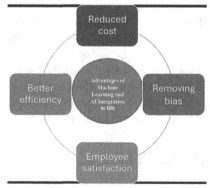

Fig. 89.2 The advantages of machine learning and AI integration in HR

- **Reduced Costs**

  The AI and ML systems helps in ensuring reduced time and workforce investment in manual record-keeping or documentation. This helps to reduce the overall costs of the system.

4.3 Difficulties with AI and ML Integration in HR

Although these programs are interesting, there are certain concerns to be aware of, since they all remain very new. The main one is that " Instruction information" is necessary for AI to function. Stated differently, the algorithms include historical data. In the event that the present leadership processes are too centralized, harsh, prejudiced, or unfair, that can end up setting up everything that find problematic. Visible and "tuneable" AI is necessary so that humans can check the algorithms to be sure things are operating correctly. In the same way that early cars weren't always reliable, our beginning programs will need "adjusting knobs" and "bumpers" as we figure out methods to improve their accuracy.

Bias may be established by the structures. Let's say that the company has extremely limited African American experts and has not previously recruited any women for technical positions inevitably the AI hiring algorithm would draw the conclusion that black and female technicians had lower chances of rising to executive positions. It will need some time to properly eliminate this kind of prejudice from the software.

Both data disclosure and unintentional usage are risks. Think about a typical use of statistics where we attempt to forecast the chance of an outstanding worker quitting the organization. Supervisors may behave incorrectly if we inform them that an individual "carries a significant possibility of quitting"; they may choose to overlook the employee or handle them improperly. Properly applying psychological economics is a skill that we must acquire. AI is currently not a stand-alone making choices system, but rather a "instrument" for development and recommendation.

Fig. 89.3 Difficulties with AI and ML integration in HR

Entelo's leaders of AI spoke on the necessity to develop "interpretive" and "open" artificially intelligent machines. Stated differently, the framework need to provide an explanation for every choice it chooses so that we, as human beings, may determine if the standards it relied upon remain relevant. This constitutes a single of the most crucial requirements for new instruments, yet sadly, the majority of artificial intelligence on the market today are total black boxes.

5. CONCLUSION

As much as AI stays on to change the world of human resource management. Visibility and these intellectual tech developments must be balanced by HR departments. To reduce the risk of unintentionally introducing prejudice into their projects, HR executives and professionals require to possess a thorough grasp of the decision-making process. Encouraging workers to believe modern technology will depend on this openness.

As the study probably figured out right now, there are quite a lot of benefits to using artificial intelligence (AI) regarding hiring and HR requirements. However, nothing in this world is flawless, especially when it comes to a technological advancement that hasn't yet reached its full capabilities. A lot of the resources and amenities may not have all the features that people are used to in the guide non-AI globally including the option to provide comments to candidates after an online screening or the capacity to analyze CVs according to certain standards. Sensitivity and human connection, or the chance to actually choose candidates and connect with them throughout the hiring procedure, are two important things that artificial intelligence (AI) lacks. These things are now unachievable when depending only on AI technologies to do all of the job for the sake of the company. People experience feelings, while AI receives statistics, and this difference won't be altering anytime soon.

References

1. Asfahani, A.M., 2024. Fusing talent horizons: the transformative role of data integration in modern talent management. Discover Sustainability, 5(1), pp. 1–14. https://link.springer.com/article/10.1007/s43621-024-00212-7
2. Chen, Z., 2023. Collaboration among recruiters and artificial intelligence: removing human prejudices in employment. Cognition, Technology & Work, 25(1), pp.135-149. https://link.springer.com/content/pdf/10.1007/s10111-022-00716-0.pdf
3. Dospinescu, N. and Dospinescu, O., 2019. A PROFITABILITY REGRESSION MODEL IN FINANCIAL COMMUNICATION OF ROMANIAN STOCK EXCHANGE'

S COMPANIES. Ecoforum Journal, 8(1). http://www.eco-forumjournal.ro/index.php/eco/article/download/884/557

4. Khair, M.A., Mahadasa, R., Tuli, F.A. and Ande, J.R.P.K., 2020. Beyond Human Judgment: Exploring the Impact of Artificial Intelligence on HR Decision-Making Efficiency and Fairness. Global Disclosure of Economics and Business, 9(2), pp. 163–176. http://i-proclaim.my/journals/index.php/gdeb/article/download/730/660

5. Lawal, Z.K., Yassin, H. and Zakari, R.Y., 2020, December. Stock market prediction using supervised machine learning techniques: An overview. In 2020 IEEE Asia-Pacific Conference on Computer Science and Data Engineering (CSDE) (pp. 1–6). IEEE. https://www.researchgate.net/profile/Zaharaddeen-Lawal/publication/351169516_Stock_Market_Prediction_using_Supervised_Machine_Learning_Techniques_An_Overview/links/627e637637329433d9adf157/Stock-Market-Prediction-using-Supervised-Machine-Learning-Techniques-An-Overview.pdf

6. Mallick, A., 2021. Application of Machine Learning (ML) in Human Resource Management. New Business Models in the Course of Global Crises in South Asia: Lessons from COVID-19 and Beyond, pp. 209-220. https://link.springer.com/chapter/10.1007/978-3-030-79926-7_12

7. Niehueser, W. and Boak, G., 2020. Introducing artificial intelligence into a human resources function. Industrial and commercial training, 52(2), pp. 121–130. https://ray.yorksj.ac.uk/id/eprint/4512/1/PDF_Proof%20(1).PDF

8. Okatta, C.G., Ajayi, F.A. and Olawale, O., 2024. Navigating the future: integrating ai and machine learning in hr practices for a digital workforce. Computer Science & IT Research Journal, 5(4), pp. 1008–1030. https://scholar.googleusercontent.com/scholar?q=cache:CWwQGb8HbsAJ:scholar.google.com/+Navigating+the+Digital+Workforce+through+Integrating+Machine+Learning+and+AI+-for+HR+Success+in+Stock+Market+Finance&hl=en&as_sdt=0,5&as_ylo=2020

9. Pandey, A., Balusamy, B. and Chilamkurti, N. eds., 2023. Disruptive artificial intelligence and sustainable human resource management: Impacts and innovations-The future of HR. CRC Press. https://books.google.co.in/books?hl=en&lr=&id=bh_dEAAAQBAJ&oi=fnd&pg=PT14&dq=Navigating+the+Digital+Workforce+through+Integrating+Machine+Learning+and+AI+-for+HR+Success+in+Stock+Market+Finance&ots=E-R4u-AddlR&sig=njopePSO-xfBEzlYrezTV7hQeGQ&redir_esc=y#v=onepage&q&f=false

10. Sakka, F., El Maknouzi, M.E.H. and Sadok, H., 2022. Human resource management in the era of artificial intelligence: future HR work practices, anticipated skill set, financial and legal implications. Academy of Strategic Management Journal, 21, pp. 1–14. https://scholar.googleusercontent.com/scholar?q=cache:egQwI1NyIKkJ:scholar.google.com/+Navigating+the+Digital+Workforce+through+Integrating+Machine+Learning+and+AI+for+HR+Success+in+Stock+Market+Finance&hl=en&as_sdt=0,5&as_ylo=2020

11. Sonkavde, G., Dharrao, D.S., Bongale, A.M., Deokate, S.T., Doreswamy, D. and Bhat, S.K., 2023. Forecasting stock market prices using machine learning and deep learning models: A systematic review, performance analysis and discussion of implications. International Journal of Financial Studies, 11(3), p. 94. https://www.mdpi.com/2227-7072/11/3/94/pdf?version=1690521261

12. Stahl, B.C., Andreou, A., Brey, P., Hatzakis, T., Kirichenko, A., Macnish, K., Shaelou, S.L., Patel, A., Ryan, M. and Wright, D., 2021. Artificial intelligence for human flourishing–Beyond principles for machine learning. Journal of Business Research, 124, pp. 374–388. https://www.sciencedirect.com/science/article/pii/S0148296320307839/pdfft?md5=003aca7081d3419d7201d16c98aaee30&pid=1-s2.0-S0148296320307839-main.pdf

13. Tanuwijaya, J. and Hansun, S., 2019. LQ45 stock index prediction using k-nearest neighbors regression. International Journal of Recent Technology and Engineering, 8(3), pp. 2388–2391. https://www.researchgate.net/profile/Julius-Tanuwijaya/publication/336715759_LQ45_Stock_Index_Prediction_using_k-Nearest_Neighbors_Regression/links/5dae7e5ca6fdccc99d929d4a/LQ45-Stock-Index-Prediction-using-k-Nearest-Neighbors-Regression.pdf

14. Zhang, Y., Xu, S., Zhang, L. and Yang, M., 2021. Big data and human resource management research: An integrative review and new directions for future research. Journal of Business Research, 133, pp. 34–50. https://nscpolteksby.ac.id/ebook/files/Ebook/Journal%20International/Business%20Administration/Journal%20of%20Business%20Research%20-%20Volume%20133%2C%20September%202021%2C%20Pages%2034-50.pd

Note: All the figures in this chapter were made by the author.

Recent Trends in Engineering and Science for Resource Optimization and
Sustainable Development – Prof. (Dr.) Dorota Jelonek et al. (eds)
© 2024 Taylor & Francis Group, London, ISBN 978-1-032-98030-0

Internet of Things and Machine Learning Integration for Safe Data Sharing in Finance and HR

Anantha Murthy[1]

Assistant Professor, Department of MCA, NMAM,
Institute of Technology, NITTE (Deemed to be University), India

A. Vinay Bhushan[2]

Associate Professor, Business Analytics, Kirloskar Institute of Management, Yantrapur, Harihar

Uttara Bhattacharya[3]

Assistant Professor, Dr. D Y Patil Institute of Management and Research, Pune

Ahmad Y. A. Bani Ahmad[4]

Department of Accounting and Finance, Faculty of Business, Middle East University,
Amman 11831, Jordan, Applied Science Research Center, Applied Science Private University, Jordan.

Melanie Lourens[5]

Deputy Dean Faculty of Management Sciences, Durban University of Technology, South Africa

Jigneshkumar B. Patel[6]

Assistant Professor, Kalol Institute of Management, KIRC Campus, Kalol, Dist. Gandhinagar, Gujarat

▬▬▬ Abstract

Technological innovations in communication and information are evolving quickly thanks to advances in the Internet of Things (IoT), big data, and cloud-based computing. This is speeding up their incorporation into a variety of industries, including smart materials, electricity, and healthcare. Machine learning, or ML, is a technology built on data to improve the efficiency, immediate, changing, and massive nature of data. It covers techniques such as the GMM algorithm and Deep learning. The link between ML and the IoT, the importance of combining these advances, and the use of huge amounts of data in the sector are all examined in this paper. This paper aims to find out how integration of the Internet of Things and Machine Learning can help in Safe Data Sharing in Finance and HR. The objectives of the paper focus on the benefits of IoT and ML in financial services and the HR services and also the discussion of the limitations of those technologies. The paper finds that with IoT and ML, the HR and finance department can automate their services, get real time insights and thus provide better support to the stakeholders. Also, these technologies help in easy fraud detection and mitigation of the potential risks. However, there are also some potential challenges of these systems like low computational capacity of the technologies, high energy use, risks of confidentiality and safety and various others. Thus, it is estimated that ML devices and the IoT together can increase the social order's intellect and provide limitless growth opportunities.

[1]anantham2004@gmail.com, [2]avbvinay@gmail.com, [3]Uttara.Bhattacharya@gmail.com, [4]aahmad@meu.edu.jo, [5]melaniel@dut.ac.za, [6]jigneshpatel2411@gmail.com

DOI: 10.1201/9781003596721-90

■■■■■■ **Keywords**

Internet of things, Machine learning, Data sharing, Finance, Human resources

1. INTRODUCTION

The fields of managing human resources (HRM) are being revolutionized by the Internet of Things (IoT) along with machine learning (ML). These innovations provide previously unseen possibilities for improving decisions, expediting HR procedures, and changing worker engagement. The combination of IoT and machine learning is an important shift in the way HR activities are thought out and carried out, not merely a technical improvement. Due to its substantial impact on transforming both the job and everyday life, the phrase "Internet of Things" (IoT) is starting to gain popularity among businesses and consumers. IoT is altering not only how people live in the house but also how HR is managed and how businesses operate. This research investigates the effects of the Internet of Things on HRM, like an overview of research on adaptable technology for businesses, electronic HR, HR procedures, IoT information and data evaluation, and data utility. Artificial intelligence (AI) and robotics are being progressively integrated into financial institutions, such as banks, to improve the user experience. Digital developments and the Internet of Things enable banks to improve customer satisfaction, boost productivity, and simplify processes. To pinpoint possible regions for development, banks, for instance, may use detection devices and tags with RFID to track consumer behaviour precisely, watch client behaviour, and gather and evaluate immediate information. Additionally, they are using AI-powered technology to forecast wants and foresee client demands.

As a result, IoT and ML aren't just simply breakthroughs in technology; they also signal an evolution in the way that HR tasks are thought out and carried out. Through exploring the changing function of these advances, companies may get an accurate understanding of the opportunities and upcoming developments influencing HRM in a future where artificial intelligence and machine learning rule the day.

1.1 Aim and Objectives

Aim: The study aims to explore the Internet of Things and Machine Learning Integration for Safe Data Sharing in Finance and HR

Objectives:

- To explore the integration of IoT and ML in financial services and the HR sector.
- To explore the benefits of IoT and ML in financial services and the HR services
- To focus on discussing and limitations of those technologies.

2. LITERATURE REVIEW

2.1 IoT in Human Resource

By offering businesses greater effectiveness, safety, and accuracy, the IoT is revolutionizing HR management and other sectors. Consequently, studies have been conducted on the connection between IoT and HRM, especially as it relates to HRM. While it's remaining in its infancy, connecting IoT devices with HRM is going to grow in popularity over time. Automated settings have the potential to bring about changes in work settings, and research has looked at how applying information technology to recruiting and creativity assets might boost the success of a business. Modern human resource management (HRM) includes a wide range of topics, including hiring, integrating new hires, organizing and preparing tasks, job growth, learning and growth, assistance, and more. The usage of digital technology to handle all HRM duties is growing, and HRM procedures are being applied more and more to keep HRM evolving (Pathak, and Solanki, 2021). Cloud computing via IoT provides versatility, affordability, and adaptation. Depending on how their company grows and uses the cloud, businesses may scale up or decrease their technological needs. Smaller businesses can battle with bigger ones about managing employees, hiring, and upkeep techniques thanks to online determining technological value for money, which removes the need for costly software updates or equipment. When cloud-based computing and IoT operate together, HR managers may profit from several things. For example, sharing a lot of data safely about people and the processes they belong to can create a more adaptable workplace and improve making decisions. Finally, the way that businesses operate and handle their workforces is about to face a radical change because of the convergence of IoT and HRM.

2.2 IoT in Financial Services

With the additional benefit and safety of wireless banking, users may access their financial institutions instantly from any smartphone or tablet. For safety and identity reasons in smartphone banking and getting entry to prohibited locations or systems, biometric indicators like fingerprinting and recognition of face are adopted. With encrypting limiting unwanted viewing of private data and lowering the chance of fraud, electronic wallets are becoming growing in popularity as a safe way to make transactions online (Unal, et al. 2021). Organizations can monitor client interactions and evaluate user habits by storing recordings of all user activity in a central database. The use of technological advances or online resources for illegal purposes, such as data or financial theft, fraud, or illicit activity, is known as cyber criminality. More precise details about consumers' purchasing patterns, financial objectives, and budgetary restrictions are provided by IoT-generated facts, which enables banks to provide customized services.

Fig. 90.1 IoT in financial services

2.3 ML in Human Resource

Prediction analytics, also acquiring talent, satisfaction with work, managing performance, and education and growth are just a few areas where ML has had a big influence on Human Resources. Workers in HR may make more informed judgments by using this tool, which can evaluate enormous volumes of human resources data to find possible applicants and forecast their likelihood of getting picked for a position. Using application data, job postings, and applications as input, machine learning techniques simplify hiring procedures. AI bots and virtual assistants may help HR departments tremendously, releasing assets for important tasks, thanks to advances in the processing of natural language (NLP) methods (Khalil, et al. 2021). Examples of these apps include Siri or Alexa. By analysing information from assessments of success and worker polls, machine learning may also assist in identifying and resolving problems related to low staff engagement.

2.4 ML in Financial Services

Machine learning (ML) technologies are being used by financial analysts more and more for creative scientific investigations since they enable the application of new datasets and investigate novel problem areas, especially those related to events forecasting. In finance, machine learning (ML) is a scientific and theoretical breakthrough that expands on the concept of deductive reasoning and opens up new study areas where information may be used to forecast activities. Financial companies and hedge fund executives utilize automated trading, which uses machine learning (ML) systems to evaluate massive amounts of data at once and execute thousands of deals per day. This approach does not depend on feelings or goals, enabling quick trading judgments. Feelings evaluation in finance is used primarily to examine how potential earnings are impacted by the mood of the stock market (Li, 2021). Sentiment evaluation should be used in conjunction with other economic and tactical evaluations, since doing it solo may be dangerous. Sentiment analysis-based influence on markets presents moral questions, necessitates complying with regulations, and calls for responsible data usage.

3. METHODOLOGY

A secondary data collection and analysis process has been used for this empirical research. The research provides valuable insights about how ML and IoT services are helping in ML and financial services. Most of the data collected for this research are from journals and articles published in PubMed, ResearchGate, Google Scholar and many more. The most recent data has been used for developing the analysis of the report.

4. ANALYSIS

4.1 Benefits of IoT in Human Resources

- **Employee Wellbeing**

  The employee wellbeing and employee health are some of the top concerns for employers. All the employees want to make sure that their workers can deliver the best quality services without compromising their health. A sick employee can face health issues which leads to decreased productivity in the company and leads to low generation of revenue. This is where the IoT devices come for a greater help for the HR teams. There are a wide range of IoT devices which help in easy tracking of the health of the employees with just a connected device.

- **Better Recruitments**

  Recruitment is another important aspect of the human resource department of any company. With IoT technology, the HR department can easily track the responses of the employees during the interview and recruitment. Thus, with advanced technology, the HR department can improve the efficiency of their recruitment process.

- **Enhanced Employee Productivity**

  IoT is one of the most advanced technologies which has a feature of eye-tracking (Gunawardena, Ginige and Javadi, 2022). This helps to detect eye movement using some kind of sensor. Thus, the human resource department can use the technology to monitor the motions of the employees, track their total working hours and also know about their potential distractions.

4.2 Benefits of IoT in Financial Services

- **Fraud detection**

  Banks and financial institutions face the challenges of hacking and other cybercrimes the most. IoT devices along with machine learning (ML) applications can be of greater help in finance as it helps to learn about anomalies. These systems can easily collect data from various machines and also web applications which include information of payment portals, servers, and teller machines. Thus, this helps to easily detect the fraud at once and thus helps the companies to take quick decisions.

- **Auditing**

  Accounting and auditing are the processes which help to track the irregularities in the financial statements and budgets and thus uncover the fraud. Most of the traditional process of accounting and auditing methods includes a ton of paperwork. By using the IoT technologies, the companies can simplify their auditing process as all the transactions can be easily tracked in real-time and thus can be sent to the accounting department. This not only ensures that the auditing process is error free but also streamlines the accounting process and orifices efficient results.

- **IoT payments**

  Using the IoT devices, the financial companies can streamline the payment process. IoT payments do not include any human intervention and it is the payment which is made through a device. With IoT payments, the consumers can pay their invoices from various devices like contactless cards, smartphones, smartwatches and many more (Agrawal, 2021). Thus enabling this IoT payment process, the companies can improve the consumer experience while making the payment process smoother.

- **Improves efficiency**

  IoT is a process which helps in easy tracking of employee and business performance in real-time (Alti and Almuhirat, 2021). There are a wide range of IoT devices like "wearables track teams' productive hours" which can alert the leaders of something is missing. It also provides opportunities to check whether the de-

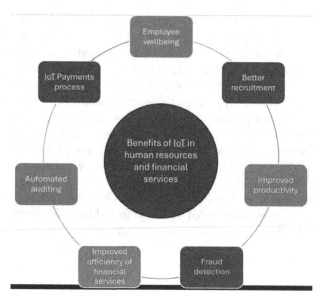

Fig. 90.2 Benefits of IoT in human resources and financial services

vices such as ATMs and also the customer kiosks are working efficiently.

4.3 Benefits of ML in Human Resources

Machine Learning (ML) comes with a wide range of advantages in HR processes as it helps in time saving and also reduces the potential decision-making risk which is pivotal in maintaining a balanced workforce. In the process of talent acquisition, AI and ML are of greater help in tracking the relevant skills and traits of the applicants by analysing a wide range of CVs. Also, the machine language easily streamlines the overall communication with candidates. ML systems like chatbots can be helpful in scheduling and arranging interviews for the applicants (Laiq and Dieste, 2020, August). Also, machine learning can improve the routine tasks of a company like sending emails, generating interview questions for various positions, and also monitoring process evolution. This application process of machine learning in HR also helps in easy and faster documentation of interviews, audio files, videos and many more. Overall, in this era of personalized services and interactions, the ML devices are of greater help as it meets the expectations of the HR and gives a chance of real time access to the resources. Also, ML can provide customised training opportunities for the employees and thus foster an effective communication in the organisation (Sofia et al., 2023).

4.4 Benefits of ML in Financial Services

The most important aspect of Machine Learning is its feature of processing and also analysing the raw data for creating valuable insights (Subasi, 2020). In the financial institutions, daily, a large amount of data gets generated which includes information of the market, customer, and transactional data. Machine learning algorithms can be helpful in processing this data and thus help to generate better insights which are not easy to track or identify manually. Thus, with the ML devices, the companies can better manage the risks and also make more informed investment decisions, and improve customer service. Another important benefit of ML in financial services is the ability of automation. Machine learning helps to easily automate a wide range of financial processes and helps in easy "fraud detection, anomaly detection, and tax optimization" (Ali et al., 2022). It also reduces the chances of error while also saving time. Machine learning tools help in improving the decision making as these systems can track the historical data and the patterns. This system comes to help in the process of trading, as ML helps to develop various trading strategies and thus execute trades automatically. Lastly, machine learning helps to improve the overall customer service. ML leveraged robot-advisors, can provide advice regarding personalized investment to the customers and thus help them to better manage their finances. This improves the overall consumer satisfaction and decreases the challenges and workloads of the financial advisors.

4.5 Challenges of Integration of IoT and ML in Financial and HR Services

Several issues arise when ML and IoT innovations are integrated, such as restricted computational capacity, energy use, confidentiality and safety, immediate analysis, scaling, device interoperability, and data collecting and administration. Effective installation requires proper data gathering and administration, particularly for devices located in far, away, or challenging-to-reach areas. Another major barrier is the accuracy of data, as variables like climate and humidity may cause IoT gadgets to provide data that is incorrect, lacking, or untrustworthy. Major challenges include immediate processing and delay, particularly for sensors located

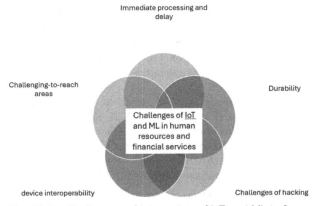

Fig. 90.3 Challenges of integration of IoT and ML in financial and HR services

in distant or difficult-to-reach areas with insufficient internet access. IoT gadgets may be hacked or targeted because of the private information that they gather and transmit. For this reason, security and confidentiality are essential. Durability is another major challenge because when additional IoT gadgets are connected to systems, the system has to handle the increasing processing and data throughout.

5. DISCUSSION

Enhancing speed and reliability for intelligent Internet of Things (IoT) applications has become the focus of ML research. To mitigate error predictions and promote computer-type interfaces, Eldeeb et al. presented a supporting vector machine (SVM) inspired FUG allocating approach. Using controlled, uncontrolled, and combined machine learning techniques, Bhatti et al. created the if Ensemble outlier identification tool for Wi-Fi indoor positioning (Mishra, and Tyagi, 2022). The safety of IioT gadgets may be enhanced by using machine learning along with big data analysis, which are crucial instruments for evaluating and protecting IoT items. In their discussion of machine learning's application to standard IioT guidelines, Zolanvari et al. addressed the potential of cyber-vulnerability.

The Alves et al.-proposed ML4IoT is intended to coordinate machine learning processes on enormous amounts of data collection. Along with adaptability, robustness, and efficiency, the platform allows the creation of several ML frameworks, each having its process. The structure saves and maintains IoT information in holders, while the ML4IoT data administration module uses developed applications to simplify the performance of ML operations (Ali, et al. 2021). Conclusively, ML within the context of the Internet of Things has promise for enhancing the security of processes, finances, and efficiency via the use of machine learning methods to examine massive volumes of current information. Several elements of our lives are improved by the Industry Internet of Things (IoT), which links different equipment, gadgets, and applications. To guarantee privacy and confidentiality, Internet of Things solutions need for additional planning and monitoring. Security risks like loss of assistance, violent takeover, file type challenging material monitoring, scanning, and improper configuration are predicted using ML techniques. Having a rate of acquisition of 0.01 and an estimation time of 34.51 milliseconds, Latif et al.'s ultralight randomized neural network (RaNN)-based forecasting system was able to forecast these assaults with a success rate of 99.20 %. In contrast to cutting-edge ML techniques for IoT safety, the suggested strategy improves identifying threats' effectiveness by a median of 5.65%. Use, program, system, and physical layering make up the four levels of the Internet of Things framework (Jamil, et al. 2021). We need extremely adaptable technologies that address confidentiality, safety, and authorized, and societal problems. The Internet of Things solutions create vast amounts of data, making traditional data analysis techniques ineffective.

6. Conclusion

Identifying fraud and risk prevention have been greatly enhanced by the combination of M and IoT in the financial and HR sectors. Financial companies may keep on top of scammers by using current information, identifying trends, and creating customized models. To preserve confidence in the industry, confidentiality and safety of data concerns are essential. A greater secure economic environment may be created by the careful application of IoT and ML, which can enhance risk control, identify fraud, and comply with the law.

References

1. Agrawal, S., 2021. Integrating Digital Wallets: Advancements in Contactless Payment Technologies. International Journal of Intelligent Automation and Computing, 4(8), pp. 1–14. https://research.tensorgate.org/index.php/IJIAC/article/download/111/105

2. Ali, A., Abd Razak, S., Othman, S.H., Eisa, T.A.E., Al-Dhaqm, A., Nasser, M., Elhassan, T., Elshafie, H. and Saif, A., 2022. Financial fraud detection based on machine learning: a systematic literature review. Applied Sciences, 12(19), p. 9637. https://www.mdpi.com/2076-3417/12/19/9637/pdf?version=1664239132

3. Ali, M., Karimipour, H. and Tariq, M., 2021. Integration of blockchain and federated learning for Internet of Things: Recent advances and future challenges. Computers & Security, 108, p.102355. https://www.sciencedirect.com/science/article/abs/pii/S0167404821001796

4. Alti, A. and Almuhirat, A., 2021. An Advanced IoT-Based Tool for Effective Employee Performance Evaluation in the Banking Sector. Ingénierie des Systèmes d Inf., 26(1), pp. 103–108. https://www.researchgate.net/profile/Alti-Adel/publication/350553163_An_Advanced_IoT-Based_Tool_for_Effective_Employee_Performance_Evaluation_in_the_Banking_Sector/links/6065fbe892851c91b1985c55/An-Advanced-IoT-Based-Tool-for-Effective-Employee-Performance-Evaluation-in-the-Banking-Sector.pdf

5. Gunawardena, N., Ginige, J.A. and Javadi, B., 2022. Eye-tracking technologies in mobile devices Using edge computing: a systematic review. ACM Computing Surveys, 55(8), pp. 1–33. https://scholar.archive.org/work/rocjkid3l-fayrlsuunnk4iibsq/access/wayback/https://dl.acm.org/doi/pdf/10.1145/3546938

6. Jamil, F., Kahng, H.K., Kim, S. and Kim, D.H., 2021. Towards secure fitness framework based on IoT-enabled blockchain network integrated with machine learning algorithms. Sensors, 21(5), p. 1640. https://www.mdpi.com/1424-8220/21/5/1640

7. Khalil, R.A., Saeed, N., Masood, M., Fard, Y.M., Alouini, M.S. and Al-Naffouri, T.Y., 2021. Deep learning in the industrial internet of things: Potentials, challenges, and emerging applications. IEEE Internet of Things Journal, 8(14), pp. 11016–11040. https://ieeexplore.ieee.org/abstract/document/9321458/

8. Laiq, M. and Dieste, O., 2020, August. Chatbot-based interview simulator: a feasible approach to train novice requirements engineers. In 2020 10th International workshop on requirements engineering education and training (REET) (pp. 1–8). IEEE. http://oa.upm.es/63783/1/TFM_%20MUHAMMAD_LAIQ.pdf

9. Li, H., 2021. Optimization of the enterprise human resource management information system based on the internet of things. Complexity, 2021, pp. 1–12. https://scholar.googleusercontent.com/scholar?q=cache:LYiEFy9t6goJ:scholar.google.com/+Internet+of+Things+and+Machine+Learning+Integration+for+Safe+Data+Sharing+in+Finance+and+HR&hl=en&as_sdt=0,5&as_ylo=2020

10. Mishra, S. and Tyagi, A.K., 2022. The role of machine learning techniques in internet of things-based cloud applications. Artificial intelligence-based internet of things systems, pp. 105–135. https://link.springer.com/chapter/10.1007/978-3-030-87059-1_4

11. Pathak, S. and Solanki, V.K., 2021. Impact of internet of things and artificial intelligence on human resource development. Further advances in internet of things in biomedical and cyber physical systems, pp. 239–267. https://link.springer.com/chapter/10.1007/978-3-030-57835-0_19

12. Sofia, M., Fraboni, F., De Angelis, M., Puzzo, G., Giusino, D. and Pietrantoni, L., 2023. The impact of artificial intelligence on workers' skills: Upskilling and reskilling in organisations. Informing Science: The International Journal of an Emerging Transdiscipline, 26, pp. 39–68. https://cris.unibo.it/bitstream/11585/917132/1/InfoSciV26p039-068Morandini8895.pdf

13. Subasi, A., 2020. Practical machine learning for data analysis using python. Academic Press. https://library.kre.dp.ua/Books/2-4%20kurs/%D0%9C%D0%BE%D0%B2%D0%B8%20%D0%BF%D1%80%D0%BE%D0%B3%D1%80%D0%B0%D0%BC%D1%83%D0%B2%D0%B0%D0%BD%D0%BD%D1%8F/Python/Python%20%D1%80%D0%B5%D0%BA%D0%BE%D0%BC%D0%B5%D0%BD%D0%B4/practical-machine-learning-analysis-python.pdf

14. Unal, D., Hammoudeh, M., Khan, M.A., Abuarqoub, A., Epiphaniou, G. and Hamila, R., 2021. Integration of federated machine learning and blockchain for the provision of secure big data analytics for Internet of Things. Computers & Security, 109, p. 102393. https://www.sciencedirect.com/science/article/abs/pii/S0167404821002170

Note: All the figures in this chapter were made by the author.

Recent Trends in Engineering and Science for Resource Optimization and Sustainable Development – Prof. (Dr.) Dorota Jelonek et al. (eds)
© 2024 Taylor & Francis Group, London, ISBN 978-1-032-98030-0

91

Improving Cybersecurity in Business Settings: The Challenges and Solutions of Internet of Things

Lakshmana Rao
Associate Professor, School of Law, Gitam (Deemed To Be University), Visakhapatnam

Keerthiraj
Assistant Professor of Political Science, School of Liberal Arts, Alliance University, Bengaluru, Karnataka, India

Madhulika Mishra[1]
Assistant Professor, GLA University, Mathura

Apoorva Misra
Assistant Professor, School of Law, Alliance University, Bengaluru, India

Amandeep Singh Arora[2]
Associate Professor, CS/IT, Don Bosco Institute of Technology

Vikas Rao Vadi[3]
Professor, CS/IT, Don Bosco Institute of Technology

▬▬ Abstract

Cybersecurity which is mostly about IoT applications in the corporate world comes with a number of challenges that are topmost among [the] priority of stakeholders. The development of more IoT devices in business operations has increased the attacked field greatly, thus posing different cyber threats, compared to traditional security methods. These problems range from compromise of the security of IoT Devices, weak authentication, intrusion of privacy, and complexity of multi-domain IoT ecosystem management. This research, as the paper reflects, analyses issues related to cybersecurity in IoT-powered enterprise environments and comes up with solutions. One of the strategies to implement in the case of the gateway is to adopt multiple-element authentication, regular updates of the IoT firmware, as well as thorough risk assessment. Also, the paper underlines paying no less attention to cybersecurity training and awareness of crews to eliminate mistakes and insider threats. Agencies can shift the situation from this demanding environment to rather a safer posture by enforcing these proposed solutions, which will eventually aid in protecting sensitive information from cyber-attacks. On the other hand, these types of groups are required to bear their substantial SAT-IV axis, safeguard their IoT networks, and adapt their cybersecurity measures to the ever-changing threats in the dynamic digital environment.

▬▬ Keywords

Cybersecurity, Internet of things (IoT), Business settings, Challenges, Solutions

[1]madhulika.bhadra@gmail.com, [2]amandeep.dbit@gmail.com, [3]vikasraovadi@gmail.com

DOI: 10.1201/9781003596721-91

1. INTRODUCTION

IoT brought the revolution with groups, they know no bounds of boundaries and the data passes seamlessly from device to device and system. In a nutshell, the Internet of Things or IoT is the world of electronic devices that are combined with software and connectivity, which monitor, gather, and share weather information. IoT integration was used by businesses to yield tremendous boosts of productivity and efficiency which era enterprise environments. Companies will be able to use IoT solutions to automate their approaches, monitor their inventory, and make sensible decisions based on the data provided by analytics which will lead to a streamlined process and better decision-making. For example, the IoT-enabled machines at packaging plants may be used to control the production line to prevent downtime, and hence better operational performance. The advent of the Internet of Things (IoT) integration at the enterprise level will not only bring in a lot of new opportunities but also mean a bigger security challenge. It is natural that this is the number one issue that concerns Internet of Things users: the raised attack floor that is due to the cross-connectivity of IoT devices. Each connected device is the door for cybercriminals to take all the advantages of the situation and breach the data, disrupt the machine's work, and create economic losses (Kuzminykhet al., 2021). IoT devices majority of the time are not considered with strong security measures that leave them exposed to a variety of vulnerabilities which include weak authentication mechanisms, unencrypted data transmission, and firmware that is not secure. Thus, the cyber world gives criminals with opportunities such as unauthorized access, malware infection, and any other kind of cyber security attack. The IoT's systems heterogeneity, with dedicated individual devices, protocols, and suppliers happens to add yet another layer of difficulty to the current cybersecurity attempts. This causes the system to run smoothly, meaning that it doesn't fall into disorder, and also ensuring the use of strict methods along with regular monitoring requires careful planning, strong protocols, and ongoing monitoring, respectively. Amongst these obstacles, companies need to give an emphasis on the importance of cybersecurity when formulating their IoT strategies. This involves the employment of access control systems that are firmly justifiable, encryption of facts both at rest and in transit, as well as updating device firmware to patch away vulnerabilities, and ultimately creating a responsible cybersecurity culture among employees (Raimundo and Rosário., 2022). Hereby the absence of appropriate cybersecurity arrangements can lead to problems to such an extent that the whole enterprise can become endangered and that can negatively influence the capacity of Internet of Things integration.

2. LITERATURE REVIEW

One of the main cybersecurity implications concerning IoT in groups which the literature on the subject provides is the challenge of evaluating and managing complexities and risks that arise as groups increasingly employ IoT. IoT ecologies' security evokes a lot of discussions among researchers about which of the emerging technologies have the capabilities of balancing the high risks the IoT ecosystems come with. Among the issues pointed out in the literature and essential for further research is the vulnerability of IoT gadgets that is inherent for such an environment. Not every IOT device has a backup function, to which encryption protocols and authentication mechanisms are partially contributing, thus becoming foolish to the cyber-attacks. For example, the Mirai botnet which happened in 2016 was able to exploit vulnerable IoT appliances including cameras and routers to create one of the most significant botnets of the time that sent thousands of connections to important network services. Such a grand-scale invasion brought to the fore the urgency of higher-level security measures in IoT devices for the purpose of guarding against such a big-to-be bridge. The complexity of the IoT lexicon exacerbates the cybersecurity difficulties for enterprises. Create your own shortcut (Abiodun et al., 2021). It will help you memorize the sentence. That's it. Managing the! many multitudes of gadgets, protocols, and providers that are inside an audibly connected community need good cybersecurity approaches. Research by Gartner makes a case for developing an all-around IoT security framework which includes tools for authentication, encryption, access managing, and tracking for attacks of unauthorized access and data breaches.

As a case study, the researchers also showed special cyber-attacks that were designed for the IoT in the industrial arena. In 2008, the now-famous Stuxnet computer virus employed communal engineering systems architecture, along with Internet of Things devices, to destroy target Iranian nuclear applications. This incident lent credence to the fact that the cyber-attacks now have to a large extent become infrastructural-based (in terms of critical infrastructure) and are conducted via vulnerable IoT devices thus causing the issue of IoT security in business deployments. Vendors have a critical responsibility of security enhancement, which they should demonstrate by integration of secure measures at every stage from manufacturing to software development. This involves embedded functions such as stable boot policies that guard the integrity of the device at some point in time during the startup, software updates that are reliable and that allow the fixing of the vulnerabilities and improvement of resilience, and components that support security hardware such as Trust Platforms

Modules (TPMs) that provide cryptographic keys' control and secure data storing (Djenna et al., 2021). Through the application of security design patterns right at the device layer, manufacturers can preemptively and effectively rebuff various threats to capacity, shrink attack floors, and elevate the overall cybersecurity of IoT devices which in turn establishing a more solid and trustworthy IoT ecosystem is a possible link. DividingIoT networks into separate segments to isolate vital belongings from much less secure gadgets, lowering the effect of potential breaches and limiting lateral motion by attackers. Implementing AI-pushed behavioral analytics to hit upon peculiar patterns and anomalies in IoT tool conduct, permitting early detection of capability protection threats. Exploring the usage of blockchain for stable IoT device authentication, facts integrity, and decentralized protection protocols to enhance acceptance as true with and transparency in IoT ecosystems. Adhering to regulatory standards and tips, along with GDPR and NIST cybersecurity framework, to ensure IoT deployments comply with felony and protection necessities, defensive sensitive data, and privacy rights.

The literature assessment underscores the complicated nature of IoT-associated cybersecurity demanding situations for organizations and emphasizes the need for proactive techniques, collaboration among stakeholders, and nonstop innovation to address these dangers effectively. Future studies should pay attention to emerging threats, evolving technologies, and best practices for securing IoT environments in dynamic business settings.

3. DATA AND VARIABLES

For this study on enhancing cybersecurity in IoT for organizations, records become gathered from diverse sources consisting of enterprise reports, educational studies, and cybersecurity incident databases. The primary attention was on analyzing cybersecurity demanding situations, solutions, and their effect on exclusive commercial enterprise sectors.

The variables taken into consideration within the analysis are classified as follows:

This includes more than a few IoT devices typically utilized in enterprise settings inclusive of sensors, actuators, smart cameras, industrial manipulation systems (ICS), and wearable devices. The analysis consists of extraordinary cybersecurity protocols and measures hired in IoT environments, which include encryption requirements (e.g., AES, TLS), authentication methods (e.g., PKI, OAuth), and intrusion detection/prevention structures (IDS/IPS) (Jhanjhi et al., 2021).

Table 91.1 The variables considered

Variable	Description
Types of IoT Devices	Sensors, Actuators, Smart Cameras, ICS, Wearable Devices
Cybersecurity Protocols	AES, TLS, PKI, OAuth, IDS/IPS
Business Sectors	Healthcare, Manufacturing, Transportation, Retail, Smart Cities

This well-known method (i.e. roundtable discussions) was applied among particular types of devices creating a conducive environment for a complete analysis of cybersecurity issues and solutions across the different protocols and also in the different enterprise sectors. The resulting analysis from this approach is imperative for improving cybersecurity approaches at multiple work settings.

4. METHODOLOGY AND MODEL SPECIFICATION

This research methodology considered a blend of both qualitative and quantitative techniques to be used to enhance the cybersecurity situation of the internet for the commercial enterprise.

4.1 Data Collection

Questionnaires were performed by the IT professionals, cyber security specialists, and officially appointed representatives towards the universities to get into the requirements, the present methods and the measures adopted in IoT environments. In the analysis of the real-international case research on cyber-attacks on internet-of-things devices in enterprise, we can understand more about the impact and the effects of the breach (HaddadPajouh et al. 2021). Visual data from organizational reports, as well as statistical information from cybersecurity incident databases have been statistic-analyzed to make oneself aware of the trends, correlations, and patterns that are related to the cybersecurity of the IoT.

4.2 Model Specification

In this study, we apply a model known as the Cybersecurity Risk Assessment Model (CRAM) to evaluate cyber threats in the IoT industry and prioritize risks for enterprise systems. CRAM includes an inner fault probability, asset risk vulnerability, seriousness coefficient, cost and effect values to give it a risk score. CRAM's formula for the conveyance of both information and entertainment is

Risk Score = Vulnerability Severity × Threat Likelihood)/ Asset Value × Impact Magnitude

For the purpose of computing, there shall be a computation made using the given data that will come up with an average number of cyber-attacks per quarter yearly.

For Year 1:

Average incidents per quarter = (50+55+65+70)/4 = 240/4
= 60 incidents

For Year 2:

Average incidents for each quarter = (60+70+80+90)/4
= 60/4 = 60 incidents a
quarter.

Next, we can calculate the percentage increase in cybersecurity incidents from Year 1 to Year 2:

Percentage of Increase = (Year 2 Inmate Incidents – Year 1 Inmate Incidents)/100 × Year 1 Inmate Incidents.

A percentage difference is calculated by a formula
(75 – 60)/60 × 100 = 15/60 × 100 = 25%.

This means that there is a huge difference as the number of security incidents increased by 25% compared to Year 1 during Year 2, which sheds light on the dominant challenges of cybersecurity issues in IoT ecosystem growth over a period. Such calculations can produce important enabling information for purposes of looking at the direction and scale of cyber-attacks.

Threat Likelihood means the probability of a cybersecurity hazard to happen, Vulnerability Dependent measures the magnitude influence of being exploited, Asset Value signifies the importance of target to the business, and Impact Magnitude shows the gravity of the effect on business processes on the chance of a cyber-attack (Nayyar et al., 2020). Frameworks including the NIST Cybersecurity Framework and ISO/IEC 27001 have been hired to manual evaluation of cybersecurity controls, change control processes, and compliance with industry standards in IoT deployments within commercial enterprise environments.

Table 91.2 Cybersecurity incidents per quarter over a span of two years

Quarter	Year 1	Year 2
Q1	50	60
Q2	55	70
Q3	65	80
Q4	70	90

By combining both quantitative analysis and mounted frameworks/models, this technique guarantees a comprehensive assessment of cybersecurity-demanding situations and allows informed decision-making for enforcing powerful answers in IoT-enabled commercial enterprise settings.

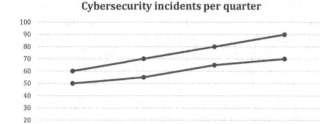

Fig. 91.1 Cybersecurity incidents per quarter over a span of two years graph

5. EMPIRICAL RESULT

The empirical findings of this studies shed light on the popular cybersecurity troubles in IoT for businesses and the effectiveness of various cybersecurity solutions applied to mitigate those risks.

5.1 Prevalence of Cybersecurity Issues

The research found out that a huge portion of businesses utilizing IoT technologies face cybersecurity challenges. The maximum not unusual problems encompass

Many IoT gadgets that lack strong security features, making them liable to exploitation by way of cybercriminals. IoT appliances, exposed to attackers, are discovered riddled with weak authentication mechanisms and default credentials, opening the doors to compromise and unauthorized access. Numerous vulnerabilities, such as insufficient encryption and privacy protocols, may result in data leaks (Rekha et al., 2023). An example of the effects: The IoT equipment handling along with the protocols and manufacturers' issues leads to a remainder of a stable environment landscape.

Businesses have applied numerous cybersecurity answers to address IoT-associated risks, with varying tiers of effectiveness:

With the intrusion created by mere manufacturers and hence introducing security features continuously throughout the manufacturing stage the level of resilience is known to increase which in turn hampers the attacks. Adopting rigid encryption apps along with smart card readers gives rise to info protection thereby giving way to access management (Serror et al., 2020). A consistent process of updating the firmware for IoT devices strengths the software steadfastness and ensures an all-around high level of security. The parts of IoT known in different locations are exposed to other domains and do not compromise each other as the result of cyber breaches. The AI-based behavioral analytics companies find and draw attention to anomalous styles and as such permit rapid chance detection and response capabilities. Cybersecurity standards and rules will help to identify who is before committing criminal offences, and as such the security features will be even stronger.

5.2 Conceptual Framework

The empirical evidence states that those at more advanced stages of the conceptual framework's theories show a greater ability to resist and deal with the cybersecurity issues resulting from the Internet of Things. Utilizing the complete policy features integrated with a robust architecture concept like policy configuration, encryption, authentication, firmware updates, network segmentation, behavioral analysis, and regulations compliance is paramount in moderating the risks and fulfilling the standard successfully (Corallo et al., 2022). To sum up, the next phase of research must be geared toward analyzing the long-term impact of such cybersecurity measures and the changes of the threat landscapes in these IoT-operated business environments.

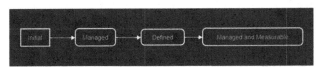

Fig. 91.2   Conceptual framework

6. CONCLUSION

This study has illuminated the vital cybersecurity demanding situations faced by way of agencies in IoT environments and assessed the effectiveness of various mitigation techniques. The occurrence of vulnerabilities in IoT gadgets, inadequate authentication mechanisms, statistics privacy issues, and the complexity of handling IoT ecosystems necessitate robust cybersecurity measures. The findings underscore the importance of a proactive technique to cybersecurity, incorporating secure with the aid of design principles, encryption, authentication, firmware updates, community segmentation, behavioral analytics, and regulatory compliance. Businesses ought to prioritize cybersecurity consciousness and schooling for personnel to mitigate human mistakes and insider threats. Moving ahead, destiny studies have to be aware of several regions (Carr and Lesniewska., 2020). Firstly, continuous assessment and improvement of cybersecurity frameworks and requirements tailor-made in particular for IoT environments are essential. Additionally, exploring rising technologies including blockchain for stronger IoT protection deserves interest. Collaborative efforts between enterprises, academia, and policymakers are crucial for addressing evolving cyber threats in IoT-enabled business settings efficaciously. Improving cybersecurity in IoT environments is an ongoing endeavor that calls for comprehensive strategies, collaboration, and adaptability to guard businesses against rising cyber threats and ensure the ongoing boom and innovation of IoT technology.

References

1. Kuzminykh, I., Ghita, B. and Such, J.M., 2021, August. The challenges with Internet of Things security for business. In International Conference on Next Generation Wired/Wireless Networking (pp. 46–58). Cham: Springer International Publishing.
2. Raimundo, R.J. and Rosário, A.T., 2022. Cybersecurity in the internet of things in industrial management. Applied Sciences, 12(3), p. 1598.
3. Abiodun, O.I., Abiodun, E.O., Alawida, M., Alkhawaldeh, R.S. and Arshad, H., 2021. A review on the security of the internet of things: challenges and solutions. Wireless Personal Communications, 119, pp. 2603–2637.
4. Djenna, A., Harous, S. and Saidouni, D.E., 2021. Internet of things meet internet of threats: New concern cyber security issues of critical cyber infrastructure. Applied Sciences, 11(10), p. 4580.'
5. Jhanjhi, N.Z., Humayun, M. and Almuayqil, S.N., 2021. Cyber security and privacy issues in industrial internet of things. Computer Systems Science & Engineering, 37(3).
6. HaddadPajouh, H., Dehghantanha, A., Parizi, R.M., Aledhari, M. and Karimipour, H., 2021. A survey on internet of things security: Requirements, challenges, and solutions. Internet of Things, 14, p. 100129.
7. Nayyar, A.N.A.N.D., Rameshwar, R.U.D.R.A. and Solanki, A.R.U.N., 2020. Internet of Things (IoT) and the digital business environment: a standpoint inclusive cyber space, cyber crimes, and cybersecurity. In The evolution of business in the cyber age (pp. 111–152). Apple Academic Press.
8. Rekha, S., Thirupathi, L., Renikunta, S. and Gangula, R., 2023. Study of security issues and solutions in Internet of Things (IoT). Materials Today: Proceedings, 80, pp. 3554–3559.
9. Serror, M., Hack, S., Henze, M., Schuba, M. and Wehrle, K., 2020. Challenges and opportunities in securing the industrial internet of things. IEEE Transactions on Industrial Informatics, 17(5), pp. 2985–2996.
10. Corallo, A., Lazoi, M., Lezzi, M. and Luperto, A., 2022. Cybersecurity awareness in the context of the Industrial Internet of Things: A systematic literature review. Computers in Industry, 137, p. 103614.
11. Carr, M. and Lesniewska, F., 2020. Internet of Things, cybersecurity and governing wicked problems: learning from climate change governance. International Relations, 34(3), pp. 391–412.
12. Algarni, M., Alkhelaiwi, M. and Karrar, A., 2021. Internet of things security: A review of enabled application challenges and solutions. International Journal of Advanced Computer Science and Applications, 12(3).

Note: All the figures and tables in this chapter were made by the author.

*Recent Trends in Engineering and Science for Resource Optimization and
Sustainable Development – Prof. (Dr.) Dorota Jelonek et al. (eds)*
© 2024 Taylor & Francis Group, London, ISBN 978-1-032-98030-0

92

The Journey of English Language: Exploring the Manipulation of the Language in Connection to ICT

P. Sasikumar

Assistant Professor (English), Department of Social Sciences, SRM College of Agricultural Sciences,
SRM Institute of Science and Technology, Baburayanpettai, Chengalpattu – 603201, Tamilnadu, India

P. Mohanraj*

Assistant Professor (English), Department of Agricultural Extension and Communication,
SRM College of Agricultural Sciences, SRM Institute of Science and Technology, Baburayanpettai,
Chengalpattu – 603201, Tamilnadu, India

S. Suganya Karpagam

Department of English, Emerald Heights College for Women, Ooty – 643006, Tamilnadu, India

S. S. Soundarya

Assistant Professor, Department of English, RVS College of Arts and Science,
Sulur, Coimbatore 641402, Tamilnadu, India

K. Nagarathinam

Associate Professor & Head, Department of English, RVS College of Arts and Science,
Sulur, Coimbatore 641402, Tamilnadu, India

T. Manickam

Assistant Professor, Department of English, RVS College of Arts and Science,
Sulur, Coimbatore 641402, Tamilnadu, India

Abstract

Many people currently speak English. The term 'global language' has frequently been used to describe a language that is a standard means of communication in the contemporary era. It is the language most commonly taught as a second language worldwide. The English language has experienced numerous transformations since its inception and remains in a state of constant evolution. In contemporary times, there is a growing imperative to prioritize the development of bilingual individuals who can proficiently utilize English for effective communication. It is due to the recognition that English transcends its status as a language solely associated with literature. It has become the preferred language of communication in a wide range of industries and sectors. The English language's presence in India represents individuals' aspirations for excellence in education and enhanced engagement in domestic and global contexts. The observable consequence of the prevalence of English is that it is currently being sought after by individuals at the earliest levels of education. Given that India is a linguistically diverse country, English is commonly called the lingua franca in inter-state mobility and communication. This paper provides an overview of historical and contemporary developments in English Language Teaching, explicitly focusing on the impact of Information and Communication Technology on the enhancement of English teaching methodologies in India.

*Corresponding author: mohanrajcandy@gmail.com

DOI: 10.1201/9781003596721-92

■■■■■■■ **Keywords**

English, Technology, Media, Methods, Online teaching, Pandemic

1. INTRODUCTION

Language is a beautiful system. English is an associate language, but it is considered the most significant language in India. English became widely known in our country due to its historical connection with the British. It started when the British arrived in India after establishing the East India Company in 1600. It gained popularity due to Lord Macaulay's educational policies [1]. English has had quite a journey. It started as a disliked tool of control, but now it is widely spoken and even considered a fancy language by some. It would not be wrong to say that it is regarded as a first language for certain groups of people in Indian society [2]. Just like how language constantly evolves, the field of English Language Teaching (ELT) has also undergone many changes. During the late years of the 20th century, something exciting happened in Indian classrooms [3]. English started to become a global language, which brought about a tremendous change in the learning environment for students. While the previous environment lacked opportunities to learn English, everyone was excited. Hindi is widely spoken in India and is a prominent language of oral and written communication [4]. However, English communicates with the international community and inter-state and intra-state interactions within India. Indians perceive English as a symbol representing superior education, an enriched culture, and heightened intellectual capabilities[5,6]. Figure 92.1 illustrates the basic block diagram of Education 4.0.

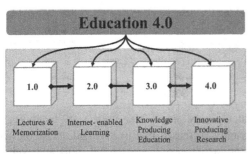

Fig. 92.1 Evolution of education

Many individuals from India who possess proficiency in English frequently incorporate elements of Indian languages into their spoken discourse[7]. It is common among individuals of Indian descent to transition abruptly into speaking fluent English during their conversations. English also functions as a means of communication among individuals

in India who speak diverse languages [8]. The significance of the English language is pronounced in various sectors, such as legal, financial, educational, and business domains, within the Indian context.

The process of liberalization in the Indian economy has presented a multitude of justifications for acquiring proficiency in the language. In the past, individuals who pursued a specialization in English typically headed toward careers in education or public administration [9]. However, contemporary times have witnessed the emergence of a diverse array of professional opportunities in this field. Currently, there are call centers that necessitate trainers to impart communication skills to their personnel, global corporations that employ marketing staff who must undergo instruction in spoken English, and medical transcription centers that demand proficient translators and reporters. Individuals aspiring to immigrate to Western countries often rely on the expertise of professionals to successfully navigate and excel in standardized assessments like the International English Language Testing System (IELTS) [10]. Consequently, the prospects for English Language Teaching (ELT) in India are abundant and boundless.

The initial observation of the change occurred within social, political, and economic domains. Abruptly, the English language no longer served as a symbol of social prestige for the elite people. In the past, English was predominantly utilized in everyday interactions by individuals from the upper classes and selected smaller cohorts [11]. The middle class exclusively allocated it for official functions or social events at which they aimed to make a lasting impact. The lower socioeconomic strata perceived English language proficiency as unattainable, and due to the lack of initiatives by government schools in India to impart spoken English skills, this particular demographic still needs to have exposure to the language. However, circa 1995, a significant shift in the paradigm occurred. Multinational firms have emerged due to economic liberalization, resulting in substantial changes. These include creating diverse job opportunities that require proficiency in the English language, an expansion of English-language television channels, a growing number of English publications, and an increasing appeal for international lifestyles [12].

In the socio-linguistically diverse context of India, English frequently serves as a medium of communication, connecting individuals from various linguistic backgrounds. English is the official primary language for many educated people in India [13]. Therefore, within the current context, the English language is crucial in facilitating the convergence of individuals from diverse regional languages in India, fostering closer interactions within social, educational, and administrative networks. As mentioned earlier, the tool is a linguistic mechanism to enhance national organizational coherence [14]. English is widely recognized as the principal medium of instruction in higher education. English is taught as a secondary language at all levels of education throughout all states in India. English is now considered a vital component of the curriculum at all levels of instruction, from elementary school to University. Despite this, the status of English in various boards, universities, and other institutions is influenced by multiple social, political, and cultural issues. There are a few states that place English as the third language on their list of official languages, giving precedence to the state's regional language as well as the national language. The majority of states, however, place English as the second language [15].

The extensive use of English in India's governmental operations indicates its notable status. According to the Indian Constitution, English is the official language for all orders, rules and regulations, bye-laws, and other related matters. English-medium education in India holds significant prestige due to the reputation of Anglophone school learning and the prevalent use of English in most Indian universities. The English language is employed to disseminate information across various academic disciplines, including but not limited to medicine, engineering, technology, and the sciences.

2. E-LEARNING TECHNOLOGIES

The establishment and growth of English as a worldwide language have brought about significant changes in approaches to teaching English in a contemporary, technology-driven global society. Due to its growing user base, English has become an international language, serving as a medium for education, commerce, and various facets of human life. The emergence and development of this widely spoken language can be attributed to the historical influence of colonialism, which laid the foundation for its growth, and the subsequent assertive post-colonial strategies that further fostered its expansion [16]. Nevertheless, due to the various factors linked to globalization in the 20th century, the English language emerged as a dominant force on a global scale. The evolving nature of English in education is transitioning it from a subject of study to a medium of instruction. It is anticipated that this trajectory will persist in the foreseeable future. The integration of computers into the education sector began in the early 1990s.

2.1 ICT Components

A few years ago, the advent of smartphone technologies increased the usage of Information and Communication Technology (ICT). ICT refers to telecommunication technologies for worldwide information access. The components of ICT are hardware, software, data, internet access, cloud computing and communication technologies. ICT enables modern computing for communicating in real-time scenarios—the convergence of multiple technologies such as audio, video and telecommunication networks with computer networks. Figure 92.2 depicts the components of ICT.

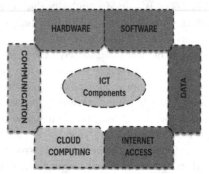

Fig. 92.2 Components of ICT

The framework for access models to ICT in Fig. 92.3 consists of three components for ICT accessing. Devices such as mobile phones, computers and laptops. Due to its inequality in resources, it is based on the price. Rich people have high access to technology when compared to poor people. The second component is conduits, referring to telephone lines and internet connections.

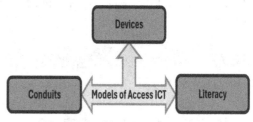

Fig. 92.3 Models of access ICT

In contemporary times, the impact of technology has significantly increased due to its rapid evolution and the proliferation of various educational tools and web-based applications. The worldwide society is envisioning the fourth industrial revolution, 'Industry 4.0.' Data Analysis, Artificial Intelligence, and the Internet of Things are the future of humanity [17]. Digital tools play a crucial role in the established domain of Computer-Assisted Language Learning (CALL) and have become more integral to English Language Teaching (ELT). The approach to computer-assisted English language teaching has become more varied due to practical experiences in the field.

The use of social media platforms makes it possible for members of online communities and networks to share fresh ideas, thoughts, interests, opinions, and suggestions with one another, in addition to up-to-date information, a variety of forms of material such as papers, images, and videos, and the interchange of information that is no longer current. As a direct result, social media platform users can instantly share content with users in different parts of the world. Globally, more than two-thirds of adult learners use social media. WhatsApp, Facebook, Messenger, YouTube, Twitter, Instagram, WeChat, QZone, Sino Weibo, Likee, LinkedIn, Skype, QQ, and other emerging social media platforms are popular today [18]—Fig. 92.4 illustrates various technologies available to learn ELT.

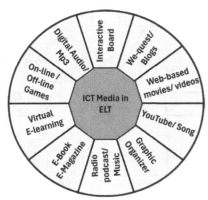

Fig. 92.4 Ways to Learn ICT

Social media usability challenges existing paradigms. It undermines hierarchical English Language (EL) teaching and learning techniques. Social media allows equitable access and communication for all learners. Figure 92.5 illustrates the importance of multimedia learning and education 4.0 for professional English learning. Thus, this technology can improve English language skill development through improved communication, collaboration, and distance learning. Eight of ten adult students search social media, web links, and apps for terms, things, and content. Actively seek and obtain current information and enthusiastically connect with fellow learners. Google Survey Monkey and Facebook Polls can help adult learners collect accurate statistics on their stakeholders' ESL efforts. Adult learners can create Google Blogger, Google Classroom, Google Meet, G Suite for Education, blogs, and e-portfolios. Adult learners can write, edit, and extend on these platforms. It allows adjustments and additions [19]. ESL assignments and quizzes can be found on Flickr and Pinterest. Slide-Share also makes presentations. Social media is a dynamic online platform with continually changing usability, tools, techniques, and strategies. Adult users' natural intelligence

Fig. 92.5 Importance of social media in education

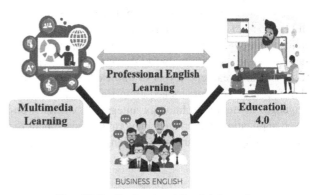

Fig. 92.6 Professional English Learning

improves social media features' efficacy and ease. The efficacy of this pedagogical approach, which integrates electronic and digital technologies, has been demonstrated to be highly effective in facilitating remote learning within the context of the digital world.

The COVID-19 pandemic has increased digital use in education, affecting English language instruction and the overall system. The role of technology in language learning has experienced significant transformation during the recent COVID-19 pandemic. The perspective on learning has undergone considerable transformation and revolution due to the closure of educational institutions during the lockdown period. Technology integration into the education system has rapidly increased, resulting in its omnipresence and indispensability.

The transition occurred suddenly, propelling our society to rely more on technology. E-learning and other online teaching methods have replaced traditional classroom instruction and become the prevailing standard. The emergence of web-based virtual classrooms has unexpectedly become crucial, revealing the significant potential for incorporating technology into English language teaching and learning. The influence of technological advancements on conventional higher education was relatively constrained and contingent upon the instructor's preferences before the advent of the COVID-19 pandemic. The concept of e-learning encompasses the utilization of technology. It contains various forms of CALL, extending from Massive Open Online Courses (MOOCs) to Virtual Learning Environments (VLEs) and Learning Management Systems (LMSs) [20].

During the COVID-19 pandemic, E-learning methods showcased the effectiveness of technology-enhanced English Language Teaching (ELT) in delivering consistent guidance and facilitating successful knowledge acquisition. Figure 92.7 depicts the application of various technologies during the COVID-19 pandemic. As a result of the implementation of online teaching, course assignments can be structured and tailored to accommodate individualized and specialized needs. The instructor has the autonomy to create electronic content catering to the diverse needs of various learner cohorts, tailoring the learning materials to align with the learners' levels of knowledge. It is particularly relevant for learners acquiring English as a second language. Utilizing technology integrated within web-based applications has created a favorable environment for implementing diverse language teaching methodologies.

Fig. 92.7 ICT applications during COVID-19

Technology integration into English instruction effectively stimulates students' interest and enthusiasm due to their status as digital natives. The advent of Internet facilities has presented many opportunities for language acquisition, as it provides convenient access to a wide array of instructional resources and study materials. Figure 92.8 illustrates the benefits of English as a Second Language (ESL). ESL helps students to be productive with the use of fun games and stimulating activities. Students can be awarded for active participation in these game-based activities.

Fig. 92.8 Benefits of english as a second language

Activity-based learning breaks the obstacles induced by traditional learning techniques within the class. It provides an opportunity to review the ICT methods students practice for language improvement. ESL benefits both students

and teachers with a good rapport. Figure 92.9 illustrates top 10 games for kids to improve ESL. As a result of the language's birth and spread over the world, teaching English in a modern, technologically-driven global culture has undergone significant modifications. As a result of its growing user base, English is now widely utilized worldwide for business, education, and other facets of daily life. This widely spoken language owes its roots to the historical effect of colonialism, which established the conditions for its spread, and to the harsh post-colonial practices that followed, which pushed it even more. Nonetheless, the English language became a dominant force on a worldwide scale in the twentieth century as a result of a variety of globalization-related factors. English is transitioning from a subject of study to a medium of instruction in education. This trend is predicted to continue in the near future.

Fig. 92.9 Top 10 ESL games for kids

Our culture's reliance on technology underwent a rapid transformation as a result of this. The standard classroom setting has been displaced by e-learning and other forms of online instruction. Surprisingly, online virtual classrooms have become more common, showcasing the enormous potential of technology in English language instruction. Activity-based learning removes the limits that are imposed on students by traditional methods of instruction. This opens up a discussion about how pupils might use ICT tools to hone their linguistic abilities. Educator-student interactions improve greatly in ESL classrooms. There is a need for trainers to teach communication skills at call centers, for marketing specialists who are bilingual in written and spoken English at multinational enterprises, and for skilled reporters and translators at medical transcription companies. Using online education, course assignments may be tailored to each student's specific strengths and weaknesses. The instructor is free to develop digital materials to meet the varying demands of their classes, adapting course materials to their students' skill sets. For individuals learning English as a second language, it is essential reading.

Social media sites provide a forum for members of virtual communities to share and debate their latest ideas, thoughts, interests, opinions, and recommendations. They can also share outdated information, as well as a variety of objects such as papers, images, and videos. Consequently, users all around the world can instantaneously exchange content on social media sites. More than two-thirds of all adult students around the world make use of social media. Using social media forces us to rethink the way we do things. It challenges the traditional order of English language (EL) education. The features, functions, and accessibility of social media are always evolving along with the network itself. The experience and maturity of their users elevates the utility and efficiency of social media. This kind of teaching, which makes use of both digital and electronic tools, has proven particularly useful for aiding distance education in the modern digital era. After schools were closed for the duration of the lockdown, there was a radical shift and revolution in how people viewed education. The fast rise in the usage of technology in the classroom has made it ubiquitous and indispensable. Students are excited and interested in learning English when technology is successfully integrated into the classroom. Because it makes a large range of study materials and instructional tools publicly accessible, the Internet has opened up several paths for language acquisition.

3. CONCLUSION

This research delves into the significance of ELLs' use of social media and how Web 2.0 practices and technology tools are evolving to meet their needs as adults. Social media ESLL is helpful for adult learners. Mobile ESL education is made possible by devices like tablets and personal computers, apps, and broadband internet. The COVID-19 outbreak is making improvements in school systems all throughout the world. The dominance of online language learning over more conventional methods of teaching a second language has unleashed previously unexplored forms of imagination. Recent developments around the world have highlighted the value and use of e-learning platforms. The use of CALL is inevitable, and the role of the teacher has changed as a result of the proliferation of online communication. Technology employed by teachers of English as a foreign language needs to be updated to satisfy the pedagogical demands of the present and the future. The course in English as a Second Language looked at India's beginnings, development, and the role of English from the time of British rule till independence. This debate implies that India's most significant contribution to the world is the English language. It has worked out well for Indians. The English language opens up new horizons and opportunities for learning for people all across the world. This language

not only improves our intelligibility, but also provides us with numerous career prospects. Learning English is a must if you want to make it in the modern world..

References

1. Akçayır, Gökçe. "Why do faculty members use or not use social networking sites for education?." Computers in human behavior 71 (2017): 378–385

2. Alghamdi, Abdulelah A. "Impact of the COVID-19 pandemic on the social and educational aspects of Saudi university students' lives." PLoS One 16.4 (2021): e0250026.

3. Al-Jarrah, Tamer Mohammad, et al. "The role of social media in the development of English language writing skill at school level." International Journal of Academic Research in Progressive Education and Development 8.1 (2019): 87–99.

4. Asselin, Marline, and Ray Doiron. "Towards a transformative pedagogy for school libraries 2.0." School Libraries Worldwide (2008): 1–18.

5. Boyd, Danah M., and Nicole B. Ellison. "Social network sites: Definition, history, and scholarship." Journal of computer-mediated Communication 13.1 (2007): 210–230.

6. Nawi, Nur Syafiqa Mohd, and Nurul Asma'Amani Muhmad Nor. "The Challenges in the Teaching of English Literature." Journey: Journal of English Language and Pedagogy 6.1 (2023): 130–147.

7. Freeman, Yvonne, and David Freeman. "Four keys for school success for elementary English learners." International handbook of English language teaching (2007): 349–364.

8. Garcia, Eugene E., Kerry Lawton, and Eduardo H. Diniz De Figueiredo. "The education of English language learners in Arizona: A history of underachievement." Teachers College Record 114.9 (2012): 1–18.

9. Carjuzaa, Jioanna, and William G. Ruff. "American Indian English language learners: Misunderstood and under-served." Cogent Education 3.1 (2016): 1229897.

10. Fithriani, Rahmah. "Discrimination behind nest and nnest dichotomy in ELT professionalism." KnE Social Sciences (2018): 741–755.

11. Pearson, William S. "Critical perspectives on the IELTS test." ELT Journal 73.2 (2019): 197–206.

12. Bordoloi, Ritimoni, Prasenjit Das, and Kandarpa Das. "Perception towards online/blended learning at the time of Covid-19 pandemic: an academic analytics in the Indian context." Asian Association of Open Universities Journal 16.1 (2021): 41–60.

13. Thompson, Christopher G., and Sam von Gillern. "Video-game based instruction for vocabulary acquisition with English language learners: A Bayesian meta-analysis." Educational Research Review 30 (2020): 100332.

14. Sah, Pramod K. "English medium instruction in South Asia's multilingual schools: unpacking the dynamics of ideological orientations, policy/practices, and democratic questions." International Journal of Bilingual Education and Bilingualism 25.2 (2022): 742–755.

15. Aithal, P. S., and Shubhrajyotsna Aithal. "Analysis of higher education in Indian National Education Policy Proposal 2019 and its implementation challenges." International Journal of Applied Engineering and Management Letters (IJAEML) 3.2 (2019): 1–35.

16. Sah, Pramod K., and Jeevan Karki. "Elite appropriation of English as a medium of instruction policy and epistemic inequalities in Himalayan schools." Journal of Multilingual and Multicultural Development 44.1 (2023): 20–34.

17. Srivani, V., et al. "Impact of Education 4.0 among engineering students for learning English language." PLoS One 17.2 (2022): e0261717.

18. Goswami, Mohit, and Yash Daultani. "Make-in-India and Industry 4.0: Technology readiness of select firms, barriers and socio-technical implications." The TQM Journal 34.6 (2022): 1485–1505.

19. Kundu, Arnab, and Tripti Bej. "Experiencing e-assessment during COVID-19: an analysis of Indian students' perception." Higher Education Evaluation and Development 15.2 (2021): 114–134.

20. Joshi, Amit, Muddu Vinay, and Preeti Bhaskar. "Impact of coronavirus pandemic on the Indian education sector: perspectives of teachers on online teaching and assessments." Interactive technology and smart education 18.2 (2021): 205–226.

Note: All the figures in this chapter were made by the author.

Recent Trends in Engineering and Science for Resource Optimization and Sustainable Development – Prof. (Dr.) Dorota Jelonek et al. (eds)
© 2024 Taylor & Francis Group, London, ISBN 978-1-032-98030-0

93

The Impact of Merger and Acquisition on Customers Service of PNB Bank During COVID-19 Pandemic

M. Shahid Ahmed*, Badhusha M H N

Department of Commerce, Jamal Mohamed College (Autonomous)
(Affiliated to Bharathidasan University),
Tiruchirappalli

Abstract

Customers' perceptions on employees' mergers and acquisitions and COVID-19 pandemic attitudes are examined in this study. The main goal is to examine how customer support and financial institutions have affected the banking business. The report uses methodological methodologies to examine the most significant M&As between 10 Indian retail and saving banks between 2022 and 2023. The study relies on primary data from a 36-question questionnaire to get pertinent data from respondents. Respondents were randomly sampled. Employees' perceptions of how M&A affects customer service growth suggest that bank M&A negatively impacts costs, branch location and proximity, and financial sector routines and positively impacts goods and services after M&A, covid-19 pandemic. The analysis examines factors affecting Punjab National Bank's merger and acquisition, hence it should not be generalized. Post-Covid-19 epidemic employee attitudes and customer service and long-term banking partnerships are examined in this article.

Keywords

Pandemic, Analysis of customers, Employees' attitude, Merger & Acquisition

1. INTRODUCTION

The merger will lead to high incidence of rationalization of branches, Mr. Rao said (THE HINDU OCT 5 2020). "If there are two branches within 500 meters, we will club them together and then use the license to open a new branch in the South or in the West.". A month after the country's biggest bank amalgamation exercise, integration seems to have taken place only on paper as the nationwide lockdown to contain the spread of Covid-19 has proved a big hur-

dle. On April 1, Punjab National Bank (PNB) took over Oriental Bank of Commerce (OBC) and United Bank of India (UBI) to become the country's second-largest lender after State Bank of India (SBI) in relation to business. The monetary policy of 2015 was another route for mergers and acquisitions in the banking sector, as it directed them to almost four times raise their paid-up cash. Commercial banks were required to raise their paid-up capital to Rs 8 billion, from the existing Rs 2 billion to Rs. 2.5 billion and from Rs. 640 million to national Rs. 800 million, from Rs.

*Corresponding author: shahidprofessional99@gmail.com

DOI: 10.1201/9781003596721-93

200 million to Rs. Banks today stressed more financially than human factors for merger. The analysis reveals the influence of M&A and Covid-19 on consumers since banks and banks (BFIs) are service providers, and their performance depends primarily on customer satisfaction. The introduction, update or implementation of post-M&A technologies has to be carefully designed to ensure minimum service interruption (Fashola, 2014). M&A is the most favored long-term organizational turnaround and reinforcement approach in the current globalised environment. The key explanation behind M&A is to build a synergy of one plus one, and is more than two. The banking sector plays a key role in the nation's economic and social development. Globalizations, globalization of the market and technical growth have significantly altered the banking environment, but consumer loyalty still has to be taken into account. In the increasingly evolving climate, however the banking sector relies on the mechanism of expansion, organizational transformation and strengthening to stay competitive and viable.

Chen and Vashishtah (2015) the merger and acquisition flows in the US banking sector were analysed in order to research the impact of a banking business system on the disclosure of lenders. In a differentiated methodology to examine shifts in the timescale of bank mergers, analysts have noticed a substantial rise in the declaration of borrowers as their leading banks partake in mergers and acquisitions. In comparison, while the rise in transparency is greater for creditors, and raises public debt dependence, it is not confined to and also expresses itself for borrowers who have no public debt rise or a reduction in the supply of credit.

Joash and Njangiru (2015) Argue that merger and takeover boost the organization's flexibility and productivity of activities, thus improving consumer loyalty. In addition, mergers and acquisitions increase the company's financial base to give the business access to more equity and credit lines to ensure liquidity during the year. This increases the reliability and efficiency of the company's activities to increase consumer loyalty.Ojha and Walsh (2016) they claimed that in recent years, mergers have helped most financial companies grow their resources and become more successful. However the key implications after integration have been customer rights security and the banking industries are properly safeguarded in crucial financial roles.

2. STATEMENT OF THE PROBLEM

The new corona virus pandemic that has harmed our economy had many speculating that the government might delay consolidation. According to a PTI study, the PNB merger will make it larger than BoB. By March 2019, the combined PNB market share of deposits rises to 8% from 5.2 percent after a $2 billion fraud and money laundering case. Sitharaman said the merger will help the banks expand their operations and market penetration by increasing their capital bases.

2.1 Research Methodology

This study examined PNB bank customers service effect characteristics and randomly sampled 65 Coimbatore clients. A main data-questionnaire survey of 65 selected brand consumers proved reliable using Cronbach's Alpha. Later, SPSS version 25 is utilized to pilot the major study with 65 consumers' questionnaire data. Rader-Diagram, Friedman's test for k-related samples, Chi-square, and Multiple Regression were utilized for analysis. This study uses secondary data from reference books, journals, research papers, and the internet.

2.2 Test of Reliability

In the study, the cronbach's alpha reliability of co-efficient was applied based on the primary data collected all over the survey the reliability is 8.61.

Table 93.1 Reliability statistics

Cronbach's Alpha	N of Items
.861	17

3. FACTORS IMPACTED BY PNB BANK CUSTOMERS SERVICES

Factors affected by PNB Bank customer's services during Merger, Acquisition and covid-19 Pandemic in Coimbatore district have been recognized. PNB Bank customer's service is measures by eighteen variables. Based on the conformity given by the selected respondents, factor analysis with main component method using vari-max rotation was applied to group the variables in to factors. The result of the KMO measures of samples adequacy and Bartlett's test of sphericity indicates that application of factors analysis is appropriate for the data. The KMO measures of sampling adequacy was 0.935 and it was significant (p<.001). Seventeen variables are reduced into fewer factors by analyzing correlation between variable (PNB customer service). In this case, seventeen variables are reduced toseven factors.

3.1 Demographic Breakdown of PNB Bank Customers

In this study, 73.8% of the respondents were male and 26.2% respondents were female customers, Monthly income Below Rs 15,000 is (13.8%), 15,000 to 30,000 is (30.8%), 30,000 to 45,000 is (24.6%) above 45,000

Table 93.2 Communalities

Factors	Communalities	Extraction
Factors 1 Customer Identity	1.1 New Account Number	.862
	1.2 Email Address/Physical Address	.801
	1.3 Mobile	.860
Factors 2 Re-submission of customers information for auto-Credits/Debits	2.1 Your Account Number to employer	.877
	2.2 IFSC Code for transactions	.661
	2.3 Auto Credit	.811
	2.4 Auto Debit	.883
Factors 3 Local branches and ATMs Positive Impacts	3.1 ATM services will be larger	.763
	3.2 ATM Charges low due to merger	.743
Factors 4 Borrowers: Deposit, lending rates to be decided by merged entity	4.1 Personal Loan Rate Reset	.643
	4.2 Home Loan Rate Reset	.650
Factors 5 New Credit/Debit cards influenced me during PNB merger	5. New Credit/Debit cards affected me during merger	.704
Factors 6 Increase Paperwork for Fixed Deposit	6. FD Maturity withdrawal high process	.713
Factors 7 PNB Banking Services were affected during covid-19	Transaction is Ease during covid-19 pandemic	.732
	Row Without any fear I used ATM services	.674
	I faced problems, merger especially during covid-19 period	.779
	PNB cares customers during Merger, Acquisition and Pandemic period	.720
	Extraction Method: Principal Component Analysis.	

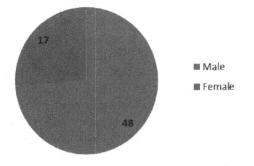

Fig. 93.1 Rader diagram showing the frequency of Gender

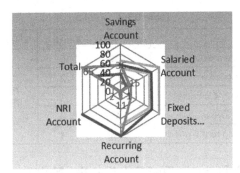

Fig. 93.2 Rader diagram showing frequency of type of bank accounts

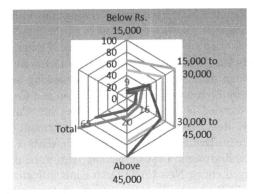

Fig. 93.3 Rader diagram showing frequency of Monthly Income

(30.8%) and total 65 respondents. PNB bank customers have different type of accounts Savings account (55.4%), salaried account (23.1%), fixed deposits account(1.5%), current account (16.9), NRI Accounts(3.1%), and total 65 respondents.

3.2 Friedman's Test for K-related Samples

Frequency of PNB Bank often visit is divided in to five parts Just once aweek, At least once amonth, Once in 2-4 months details, Once in ayear and According to theneed are examined with Friedman's test for k-related samples the test result and discussions were presented below. Selected customers in Coimbatore have ranked, to test the impact of various factors influencing PNB Bank often visited frequency; Friedman's test for k-related samples was applied to study the association with reasons for PNB Bank often visited frequency is impacted.

Null Hypothesis H01:

The entirefactors are having equal impact on PNB Banking services of Coimbatore customers.

The results of the Friedman's test showing that the null hypothesis is rejected at 1% level. Not all the factors have equal impact on Customers services of PNB banks.

Table 93.3 Ranks

	Mean Rank	Ranking	Chi-Square
Just once a week	2.08	5	
At least once a month	3.51	2	
Once in 2- 4 months	3.55	1	73.405 (P<.001)
Once in a year	3.46	3	
According to the need	2.41	4	

3.3 Chi-square Test Relating the Major Account Traction with PNB Customer is Demographic

To testing the Major accounting transaction with PNB Bank customer's services experiences of the respondents the scales used are "Strongly Agree, Agree, Disagree, and Strongly Disagree." The PNB Bank Customerswere asked to spot out theirservice experiences on banking services offered by PNB during merger Acquisition and Covid-19 is are secreted in to seven factors. They are Customer Identity, Re-submission of customers information for auto-Credits/Debits, Local branches and ATMs Positive Impacts, Borrowers: Deposit, lending rates to be decided by merged entity, New Credit/Debit cards influenced me during PNB merger, Increase Paperwork for Fixed Deposit and Banking services during pandemic period. The hypothesis structures to spot out the impact on customers service during Merger, Acquisition and Covid-19 Pandemic situation "Thereisnosignificantimpactbetween the Major Account transaction with Gender, Monthly Income, Qualification, Occupation and type of account of PNB customers". Chi-square test was utilized to recognize thecrash between factors. The consolidated results of the chi-square test are shown in Table 93.2.

The results of the study between "Major Account transactions Impacted with PNB Banking Services to its customers" are obtainable in the Table 93.2. "H_0: There is no significant impact between Major account transaction with demographic variables of PNB bank customers" except gender and Type of account, the other PNB bank services were impacted. Onthediffering, the calculated value of Monthly income, Qualification, and occupation variables is more than the table value at 5% level of significance. Hence, the null hypothesis is rejected and concluded that the customer perception on TVS products was significantly influenced during Merger, Acquisition, and Covid-19 Pandemic period of the Coimbatore district.

3.4 Multiple Linear Regresstion Analysis with Linear

By considering Merger, Acquisition and Covid-19 impact of PNB Bank Customers service, Independent variables

Table 93.4 Consolidated results of chi-square test

Sl. No	Major Account transaction Impacted with	Table Value	d.f	Calcu-lated Value	Level of Signifi-cance	Result
1	Gender	3.841	1	2.671	5	Not Significant
2	Monthly Income	9.488	4	1.449	5	Significant
3	Qualification	5.991	2	7.532	5	Significant
4	Occupation	11.070	5	12.793	5	Significant
5	Type of Account	9.488	4	8.297	5	Not Significant

Customer Identity, Re-submission of customers information for auto-Credits/Debits, Local branches and ATMs Positive Impacts, Borrowers: Deposit, lending rates to be decided by merged entity, New Credit/Debit cards influenced me during PNB merger, Increase Paperwork for Fixed Deposit and Banking services during pandemic period as predictor variable and customers service impacts during merger and pandemic as Dependent variable. Multiple regression analysis is conducted to examine the impactduring Merger, Acquisition and covid-19 pandemic period customer service provided by PNB Bank branches in Coimbatore.

H0: There is no significant impact between independent variables with dependent variable

Table 93.5 ANOVA[b]

Model	Sum of Squares	df	Mean Square	F Calculat-ed Value	F Tabu-lated	Sig.	R	R²
1 Regres-sion	4690.583	9	521.176	829.535	4.95	.000		.000[a]
Residual	34.555	55	.628					
Total	4725.138	64					.9	.1
a. Dependent Variable: Total Impacts DV								
b. Predictors: (Constant), Customer Identity, Re-sub-mission of customers information for auto-Credits/Debits, Local branches and ATMs Positive Impacts, Borrowers: Deposit, lending rates to be decided by merged entity, New Credit/Debit cards								

In the above Table the 6, calculated (829.535) is greater than F tabulated (4.95). Therefore: The null hypothesis is rejected, with significant value=.000<0.005. There is positive impact between the independent variables Customer Identity, Re-submission of customers information for auto-Credits/Debits, Local branches and ATMs Positive Impacts, Borrowers: Deposit, lending rates to be decided by merged entity, New Credit/Debit cards influenced me during PNB merger, Increase Paperwork for Fixed Deposit and Banking services during pandemic period as predictor variable and dependent variable PNB customer service during Merger, Acquisition and Covid-19 R value =(0.9), Which refers to coefficient of correlation of the independent variable are highly impacted with dependent variables.

Table 93.6 Major summary

Model Summary				
Model	R	R Square	Adjusted R Square	Std. Error of the Estimate
1	.996ª	.993	.991	.79264

Source: Primary data

The above table shows the model summary of customers services offered by PNB bank branchesduring Merger, Acquisition and Covid-19 services (predictor) and it explains the 99.6 % of PNB customers impacted (R^2=0.993).

4. DISCUSSION AND CONCLUSION

Merger and consolidation are necessary in order to integrate financial services, reinforce the banking system, modernize services, compete with global banks and strengthen the field of employment. Although the phase of merger and acquisition is not uncomfortable, PNB banks and financial institutions have been aggressively adopting the strategy of merger and acquisition. The result of the KMO measures of samples adequacy and Bartlett's test of sphericity indicates that 0.935 and it was significant (p<.001) application of factors analysis is appropriate for the data. Since the entire business depends on customer positive feedback and satisfaction, good customer service is essential in this kind of business. The outcome of the Friedman's test results that the null hypothesis is rejected at 1% level. Not all the factors have equal impact on Customers services of PNB banks. (chi-squ-73.405 (P<.001)) The level of satisfaction can also vary depending on other options the customer may have and other products against which the customer can compare the organization's products. (Nippatlapalli, 2013). The null hypothesis is rejected and states that customer service provided by PNB products significantly influenced during Merger, Acquisition, and Covid-19 Pandemic period of Coimbatore district in regard to the demographic factors. The result suggests that merger and acquisition in PNB banks and financial institutions are more benefited to credit clients than depositors and so the calculated values of Gender (2.671) Monthly Income (1.449) Qualification (7.532) Occupation (12.793) and type of account (8.297) @ 5% level of significance Which is similar to (Prompitak, 2009). Prompitak, 2009 concludes merged

banks can obtain efficiency gains through mergers and can pass these benefits to their customers in the form of lower lending rates and interest margins. As revealed by the data, the ANNOVA table shows that F, calculated (829.535) is greater than F tabulated (4.95). The strengthened capital base has increased technological improvements linked to ATM, extended network of branches and favorable interest rates. Fusion and acquisition appears to be advantageous for developing countries like India as fusion enriches the workspace of regional and technical growth, which eventually benefits the extension of the customer's network. The regulatory authorities involved should also be cautious to support clients by economies of scale in services sectors.

References

1. Acton, D. D., C. Fagan and K. Mamano. 2012. Planning for a not-for-profit combination: Insight into federal, state, and funding considerations. *The CPA Journal* (May): 64–67.
2. Bansi, N. and G. Tuff. 2012. Managing your innovation portfolio. *Harvard Business Review* (May): 66–74.
3. Cai, Y., Y. Kim, J. C. Park and H. D. White. 2016. Common auditors in M&A transactions. *Journal of Accounting and Economics* (February): 77–99.
4. Fashola, O. I. (2014). Customer reactions to bank M&A: Evidence from the Nigerian banking industry. European Journal of Business and Management, 6(25), 43–65.
5. Chen, Q., & Vashishtha, R. (2015), The effect of bank merger on corporate information disclosure. Proceeding of Colorado Summer Accounting Conference (2014) and AAA Annual Meeting (2014), University of British Columbia.
6. Joash, G. O., & Njangiru, M. J. (2015). The effect of mergers and acquisitions on financial performance of banks: A survey of commercial banks in Kenya. International Journal of Innovative Research and Development, 4(8), 101–113.
7. Ojha, S., & Walsh, J. (2016).Merger policy and its impact on Nepalese banks. International Review of Management and Development Studies, 1(2), 117–134.
8. Raj, Amruth. (2013). A Study On Customer Satisfaction of Commercial Banks: Case Study on State Bank of India. IOSR Journal of Business and Management. 15. 60–86. 10.9790/487X-1516086.
9. Asogwa, Cosmas & Uzuagu, Anthonia & Okereke, Godwin & Omeje, Hyginus & Ige, Samson & Azubuike, Roseline & Chukwuma, Joseph. (2018). Impact of the Emerging Financial Holding Company Model on Small Business Borrowers' Financial Welfare: Contemporary Evidence from Nigeria Based on the Monti-Klein Approach: The Emerging Financial Holding Company Model. African Development Review. 30. 56–70. 10.1111/1467-8268.12312.

Note: All the figures and tables in this chapter were made by the author.

*Recent Trends in Engineering and Science for Resource Optimization and
Sustainable Development – Prof. (Dr.) Dorota Jelonek et al. (eds)*
© 2024 Taylor & Francis Group, London, ISBN 978-1-032-98030-0

94

Risk Management Strategies for Biotech Startups: A Comprehensive Framework for Early-Stage Projects

Varun Choudhary[1]

Department of Project Management,
Harrisburg University of Science and Technology, USA

Kaushal Patel[2]

Department of Supply Chain Management,
Northeastern University, USA

Moazam Niaz[3]

Department of Supply Chain Management,
Wichita State University, USA

Mrunal Panwala[4]

Northeastern University,
Boston Campus Industrial Engineering

Anirudh Mehta[5]

Rutgers University, New Jersey
Department of Chemical and Biochemical Engineering

Kartikeya Choudhary[6]

Rajasthan Technical University, Kota

■■■■■■ **Abstract**

The paper on "Risk Management Strategies for Biotech Startups" explores the complex difficulties encountered by nascent biotechnology companies, due to their substantial expenses in experimental research and development, financial limitations, and ever-changing market circumstances. The research used thematic analysis to examine a comprehensive risk management framework. This analysis is conducted utilizing secondary data obtained from literature and research sources. The identified themes include the Risk Profile Concept, which explores the genetic intricacies and dangers faced by biotech companies, the Impact of Risk on startups in relation to regulatory obstacles and economic instability, and the Risk Reduction Schedule, which outlines systematic methods to minimize environmental, procedural, and industrial risks. The study delves into Financial Risk Mitigation, with a focus on strategic financial management to achieve sustainable growth. It also examines Operational Impact and Risk Control, which deals with the complexities of operations and

[1]choudhary.neu@gmail.com, [2]kaushalpatel14497@gmail.com, [3]moazam.memon@gmail.com, [4]mrunal2531@gmail.com, [5]anirudhmehta158@gmail.com, [6]Kartikeyachoudhary@gmail.com

DOI: 10.1201/9781003596721-94

challenges related to compliance. Furthermore, it discusses Strategic Risk Assessment, which highlights the importance of aligning strategic goals with risk management. The methodology entails doing a qualitative thematic analysis of the current literature, which enables a comprehensive comprehension of risk management in the biotechnology industry. The findings support the development of robust startups that are capable of navigating the complex biotechnology sector and positioning themselves for global competitiveness.

▬▬▬ Keywords

Biotech startups, Risk management, Risk profile concept, Financial risk mitigation, Risk reduction schedule, Strategic risk assessment, Genetic challenges, Early-stage projects, Sustainability, Procedural risks, Environmental risks, Industrial risks, Global competitiveness, Financial management

1. INTRODUCTION

1.1 Background

The industry of biotechnology is challenging because of its high Experimental research and development (R&D) expenses and earnings-generating. However, risks are raised by rising expenses, limited funds, and shifting customer perceptions. According to a 2001 PWC survey, just 28% of the highest-ranking executives have risk prevention measures in place, and only 40% of them are dedicated to risk control. Mistakes of the management team and low financial flow pose major dangers to the biotechnological sector, especially during its early stages stage. Regulatory and assurance issues, worldwide monetary instability, growing markets, major deals, rising costs, and the effects of a growing population are among the top 10 hazards. Biotechnology firms need solid corporate leadership as well as funds to create a workable approach to risk control to reduce danger. The entrepreneurial climate, shipping of products, attention to products, intellectual property (IP) laws, and expertise in the price of assets are critical success elements in the biotechnology business. Having a thorough understanding of these variables may aid in developing efficient risk management plans that reduce companies' and sectors' hazards (Corea, et al. 2021).

A start-up firm that focuses on technology has to carefully choose when to introduce an item to the industry. This involves going through many stages of manufacturing, such as investigation, licensing, permission, creative growth and marketing. In the biotechnology sector, funding is essential since it normally takes approximately fifteen years for an item to reach the market. Funding requirements and having a strong product portfolio are essential elements of its launch phase.

1.2 Aim and Objectives

Aim

The main aim of the study is to describe the risk management strategies for early-stage biotechnology startups.

Objectives

1. To analyse the Risk profile concept for the biotechnology sector.
2. To describe The Impact of risk in biotechnology startups.
3. To know how Risk can be reduced through the risk reduction schedule.

2. LITERATURE REVIEW

2.1 Risk Profile Concept for the Biotechnology Sector

Due to the absence of an organizational strategy to reduce danger in the beginning stages of the firm lifecycle, the biotechnology market is very sensitive to risk. Protecting intellectual property (IP), demonstrating the sustainability of innovation, and managing finances are three significant hazards at the start of the life of a business cycle. While market uncertainty is more noticeable as a result of low-cost rivalry and governmental regulations, the whole risk picture reduces as the firm expands. Since there is more competition in lower-cost regions and there are cultural impacts, the total risk assessment marginally rises when the firm expands into fresh markets. As a sector with a 90% breakdown rate, recognizing and handling these obstacles is critical (Santi, and Neves, 2020).

Businesses that survive acknowledge the dangers and adjust to focusing on fresh dangerous goals, such as too much regulation, low-priced rivalries, and availability of skilled workers. The accessibility of critical talents, inexpensive opposition, excessive regulation, and the cost of raw materials were among the worries raised by CEOs of new economies. Lower on the priority list and maybe unworthy of concern are major outside factors like terrorist activity, climate change, and worldwide epidemic dangers. To create a risk control firm portfolio, it is essential to analyse these sector-specific risk elements against the background of the economic situation.

2.2 Forming the Plan for Risk Mitigation

It is essential to know how the marketplace's threat profile affects businesses at the corporate level. An overview of a few of the most important things to think about while creating a strategy for reducing risks. If a firm is developing innovation just for spin-offs or if its item actually will be comparable in the marketplace, that company's risk acceptance and ability to survive will rely on those factors. Offering innovation in auction helps a business operate more profitably and reduces a number of the risks related to the massive costs of developing new products (Schlapp, 2020). However, this advantage must be stood against the impact of losing money on the marketing of goods and the reduced chances to build a manufacturing pipeline. The business has to explain the word "impact" in the context of its goals and core values. Risk assessments identify outside hazards that might influence the sustainability of the corporation's structure and deal with insider hazards that could influence the strategy's performance, allowing for the measurement of influence over time. They also indicate how an event might lead a business's success to differ from Its strategic goals. The mechanics of relationships within the corporate setting are equally crucial. Growing knowledge of how culture affects business relationships has been seen in modern business leadership during the last ten years. Businesses are realizing in today's globalized business landscape how important culture is to their achievement and how It sets an atmosphere for business interactions.

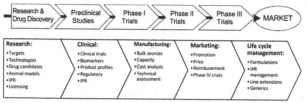

Fig. 94.1 Value chain of biotech business

The primary obstacle to gaining entry to new sectors, according to a 2007 PWC CEO poll, is cultural boundaries. To address this, experts said that the deal's closeness would help to get across the barrier (Shimasaki, 2020). An organization may profit from unpredictability see possibilities and hazards more effectively, and make decisions with confidence and effectiveness when they have an effective risk-handling plan in place.

2.3 The Impact of Risk in the Biotechnology Start-Ups

Organizations are subject to risk at every level, which includes economic, marketplace, and commercial risks. Within the biotechnology sector, the risk is further ex-plained as being subjected to unpredictable conditions or possible departure from planned or projected results. In the field of biotechnology, high levels of failure are approved as the rule. The risk level is a key factor when investing in the biotechnology industry. Significant earnings were historically based on the significant risk associated with biotechnology, although this is changing as the sector reacts to shareholder spending, rising market rivalry, and demands to cut costs (Bieske, et al. 2023). These changes raise the need for organizations to identify risks in the context of the planning process, build a risk overview, and design a plan for risk control that fits the business to fulfil its objectives.

Risk Related to Investing

When selecting whether to make investments in biotechnology, one must consider advertisement, technological, and human risk. Ralph Kristofferson, an advisor at Morgenthaler Ventures, calculates that around 40% of risks are related to innovation and that just 10% of total risk levels are related to the marketplace. Investor trust in the possibility of the company strategy is influenced by the CEO and CFO's reputation.

Risk Related to Return on Investment

As opposed to other sectors if risk is focused only on return on investment, or ROI, in marketplaces, biotech risk is dependent on personnel, technological advances, and the marketplace. To mitigate the significant chance of technical breakdown, the majority of venture capitalists (VCs) see biotechnology as exciting funding with a minimum 30% return on investments. In Canada, the biotechnology industry is seen as "risky" and receives funding appropriately. Aside from technological failure, industrial risks might include selling too much, doubtful investigation, a lack of skills or expertise inside the company, and a shortage of preparation for common threats like fire or shutdown of business.

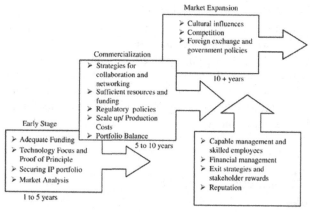

Fig. 94.2 Risk management in biotech start-ups

Risk Related to Emergency Plans

Few biotechnology firms have emergency-ready plans, even though 63 per cent of 100 biotechnology managers surveyed said they considered a risk control strategy essential to their company's performance and 50% said it was essential to their start-up capital investors. Businesses with strong handling risks programs attract more capital than their competitors because buyers feel secure knowing that their money is protected against unexpected emergencies. Several biotechnology businesses have created risk control and protection plans because they notice this, but as the outcomes of the survey show, there is still additional work to be done in this field by others.

Risk Related to Executive Management's Opinion

Executive management's opinion on the main survey results was used to create the risk score. The risk analysis identifies the important factors, but it cannot be believed they are the only ones that top executives should take into account and talk about. By comparing such several research pools, one may get a broad overview of risk and take into account risk factors that are not taken into account by the current approach. The biotechnology sector must be aware of the many risk signs and how deciding key success elements can help businesses manage risk and make wise choices that will enable them to grow and succeed on both a global and local level.

2.4 Risk Reduction through the Risk Reduction Schedule

In the biotechnology sector, risk control is a complex procedure that takes into account several variables, including commercial image environmental, moral, societal, and faith-related issues, as well as challenges. A new drug's production may take over a decade and will expense anything from $800 million to $1.7 billion at launch. A business must create a risk management strategy as an element of its marketing strategy and balance those risks against strict business governance guidelines (Bruneo, et al. 2023).

$$\tilde{d} = \arg\max\left\{I_3 - S(\tilde{d}) \middle| I_3 - S(\tilde{d}) \geq c * I_1\right\}$$

Establishing an atmosphere of efficient methods for managing both physical and intangible property with a spotlight on keeping and improving the brand's image must be the goal of a strategy for risk management. A company-wide corporate risk management plan—a systematic procedure that adjusts staff, technologies, plans, and skills to assess and handle uncertainties—is necessary for the biotechnology industry.

Reducing Risk on Ecology and Procedures

Risk can be broadly classified into two categories: Ecological hazards, which are related to outside influences like the marketplace and financial conditions, instability in society and politics, new regulations, and technological advancements; and procedural risks, which are related to inside influences like company activities, financing, and chances for growth.

Reduction of Risks in Industrial Sectors

Diverse industrial sectors—such as those related to wellness, crops and agriculture, the natural world, chemicals, or forestry—have various risk histories and ranges of risk. For any organization to get accurate information to make decisions, every risk factor must be ranked according to significance. In the pharmaceutical business, for instance, medicines are subject to strict rules, require a great deal of testing, and provide consumers with a choice. Agriculture-related items, on the other hand, are subject to simpler regulations.

Classical Risk Reducing

Classical risk consists of recalls from goods and responsibility, which are often uncontrolled and permanent. If resolving moral issues is necessary to get the desired outcome, there is a higher risk involved. Greater hazards include, for instance, the direct negative impact on people's lives and the moral problems raised by human alteration of genes.

Additional Risk Reduction

For biotechnology companies, additional risks include terrorist activity, political and social insecurity, and difficulties in the worldwide economy. For biotechnology companies, reducing these risk factors is essential, particularly in terms of obtaining private capital investment, finishing clinical trials properly, and producing drugs (Zemlyak, et al. 2021).

After the risk portfolio is created, it requires ongoing attention and clear protocols that promptly deal with and reduce any risks. Additionally, it has to be flexible sufficiently to be effective in the face of a shifting economic environment, since outside forces like new regulations may still cause a firm to struggle. The plan for risk reduction has to take into account not just the effects of starting a company, but also the way those effects may be used to gain an advantage in the international marketplace.

3. METHODOLOGY

The "Risk Management Strategies for Biotech Startups" study used theme analysis to collect and analyse secondary data from literature and research. The qualitative method of thematic analysis was used to identify risk management themes and patterns in biotechnology companies' dynamic context. The technique began with a thorough assessment of academic journals, research papers, and industry reports (Massaro *et al.,* 2021). This phase examined biotechnology risk management knowledge, theories, and frameworks. The thematic analysis was based on selected literature.

The literature was extensively analysed for risk management strategies, issues, and frameworks. This involves collecting biotech startup risk profiles, impact assessments, and risk reduction timetables. The retrieved data was thematically coded to identify and categorise risk management themes and patterns (Rana *et al.,* 2022). Codes were allocated to concepts and ideas to organise analysis. Then, patterns in the coded themes were identified and evaluated to reveal biotechnology risk management insights. This stage helped identify study similarities, discrepancies, and subtleties. The investigation helped create a conceptual framework for biotech startup risk management techniques. This framework organised theme area relationships.

Validating the thematic analysis by cross-referencing findings from numerous sources and verifying theme interpretation uniformity improved reliability. The insights gained strength via this approach. Thus, thematic analysis revealed a comprehensive understanding of risk management in biotech startups via methodical study and interpretation of secondary data topics. This methodical notion gave the study depth and context, providing biotechnology industry insights for application as well as discussion.

4. ANALYSIS

4.1 Risk Profile Concept in Biotechnology

The risk profile idea in biotechnology involves an all-encompassing test of the hereditary difficulties fought by corporations working in this place rapidly changing industry. The risk profile includes an all-encompassing analysis of the determinants that enhance the complicated and high-risk character of the area. Biotechnology start-ups face significant costs for exploratory tests with research and development (R&D), in addition to limited financial resources, together with uniformly changeful client ideas (Silva et al., 2021).

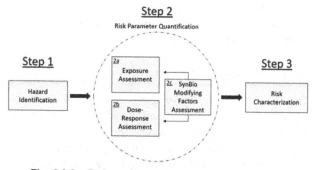

Fig. 94.3 Risk profile concept in biotech start-ups

These troubles create more complex apiece difficult structure of the biotechnology subdivision in allure first stages. The risk profile contains supervisory and security concerns, global business-related excitability, arising markets, significant undertakings, climbing expenses, and the significant belongings of an extending global culture. Recognising these parts is essential for biotechnology associations to have a robust risk administration plan.

Within this foundation, the plan of risk profile emphasises the important meaning of powerful corporate guidance, smart monetary planning, and bureaucratic rules of a persuasive risk control design to manage the hereditary instabilities in biotechnology manufacturing (Mirvis and Googins, 2022). By getting a deep understanding of these components, biotech firms grant permission to constitute effective risk administration plans that boost association resilience and support tenable progress.

4.2 Impact of Risk on Biotechnology Start-ups

The power of risk on biotechnology start-ups spans an expansive range of troubles that have a solid impact on the management and profit of these beginning firms. Regulatory impediments, general economic excitability, and the changeable traits of cultivating markets are a few of the primary risk determinants that considerably impact the beginning stages of biotech undertakings (Smith *et al.,* 2021).

The financial results are meaningful, as growing expenses and lacking money present weighty risks to the fiscal movements of these firms. Errors produced by apiece management groups can bring about an order of negative results, intensifying the risks faced by biotechnology companies (Dell'Acqua et al., 2023). In addition, manufacturing faces challenges formal by elaborate intellectual property requirements, evolving consumer stances, and the essentiality to supply pioneering products immediately.

Moreover, the results of an immediately increasing worldwide population infuriate the troubles, needing a deliberate approach to efficiently control these demographic changes. Biotech companies must efficiently handle these complex risks, making their competency to conform to the changing environment detracting (Smith and Miller, 2023). Gaining an inclusive understanding of the extensive influence of risk is important for evolving robust trade planning's that weaken hazards and position these trades for lasting success in a uniformly changeful biotechnology environment.

4.3 Risk Reduction Schedule in Biotech Start-ups

The risk reduction arrangement in biotech parties is an organized and proactive plan planned to lessen the miscellaneous risks inherent in the business. This involves a set of projected conduct and schedules aimed at accurately directing environmental, procedural, along industrialized hazards, guaranteeing a more logical progress through the critical aspects of a start-up's growth (Dvorak et al., 2022).

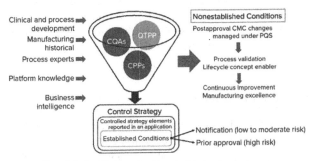

Fig. 94.4   Risk reduction in biotech start-ups

The approach orderly evaluates and mitigates environmental hazards to guide market dynamics, monetary positions, sociological and governmental impacts, supervisory changes, and technological bettering. Internal organisational movements, finance determinations, and growth prospects are carefully governed to check procedural risks and boost functional elasticity.

The prioritisation of industrial risks in various fields of biotechnology, including healthcare, farming, and forestry, is contingent on their importance and potential influence on the start-up (Dwivedi et al., 2021). The risk decline schedule requires the prioritisation of miscellaneous criteria, and permissive parties to capably control resources together with making well-informed determinations.

An enthused approach is essential in biotech manufacturing, where the complex and behind-growth processes demand an itemized grasp of possible challenges. Biotech associations can reinforce their strength to overcome risks and guarantee sustained progress and viability in a remarkably competitive and cultivating field by executing a risk reduction agenda.

4.4 Financial Risk Mitigation

In the context of the strategic management of biotechnology firms, the minimization of financial risk is of utmost importance since it acknowledges the specific challenges that are associated with the various stages of development. In the impulsiveness of the biotechnology manufacturing, which is characterized by expensive experimental research and development in addition limited financial resources, it is very necessary to place a high priority on effective financial risk reduction to achieve sustainable development (De Vrieze et al., 2020).

Around it many important components are involved in the solutions for mitigating financial jeopardy. Management

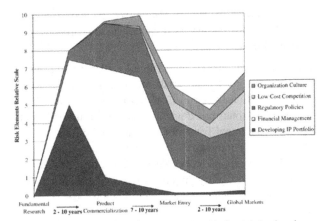

Fig. 94.5   Financial risk management in biotechnology firms

in a company that is both effective besides efficient is essential for ensuring that available finances are distributed and managed suitably. In the direction of outstanding an equilibrium amid the requirement for invention and the restraints compulsory by financial resources, it is robust to involve in careful financial preparation and demeanour an inclusive valuation of risk (Kayode-Ajala, 2023). Through gaining an understanding of the intricate financial components of the biotechnology industry, new businesses can improve their capacity to endure risks, successfully overcome financial restraints, and develop a firm foundation for long-term accomplishment. In an industry where the complexity of finances has a significant impact, biotechnology businesses need to reduce financial risk, strategic leadership, and the development of a robust product portfolio as their top priorities to secure their long-term viability.

4.5 Operational Impact and Risk Control

Biotech start-ups must cautiously analyse the functional effects and implement persuasive risk control measures to favourably manoeuvre the intricate type of business. Emerging firms engaged in biotechnology face solid operational risks on account of the blend of their vital character, absolute supervisory necessities, and intellectual property questions (Wang *et al.,* 2023).

Operational impact refers to the results of risks on constant movements, such as test exertions, licensing agreements, and market launches. Regulatory barrier's ability to interfere with operational workflows, need adept risk administration methods to guarantee smooth compliance and commodity progress.

Risk control explanations in the operational rule circumscribe management of strong processes to handle supervisory hurdles in addition to intellectual property concerns (Snyder, 2023). It is of extreme importance to authorize exact codes for emergencies and guarantee that they are by manufacturing standards. Moreover, the expression of possibility plans empowers trades to capably manage unexpected hurdles.

The desire emphasises the critical need for adopting a proactive posture towards risk administration in ordinary operations. Biotech startups can nurture flexible and flexible organisational breeding by comprehending and dealing with functional barriers (Bocken and Geradts, 2020). This approach guarantees the uninterrupted functioning of lively movements and protects against disruptions that concede the possibility impede progress in the indistinctly controlled and inventing biotech production.

4.6 Strategic Risk Assessment

Strategic risk appraisal is an essential facet for biotech companies, concentrating on the study of risks within the more expansive foundation of their company plan. This idea accepts that the prosperity of biotech actions is carefully tied to their volume to handle clever obstacles in a promptly changeful atmosphere.

Strategic risk assessment, in this place framework, entails a proactive test of internal and extrinsic ingredients that could conceivably deter the attainment of complete aims (Bode *et al.*, 2022). It includes a comprehensive understanding of by what method various risks, to a degree law making obstacles and advertise action, might impact the overall affluence of the association.

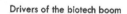

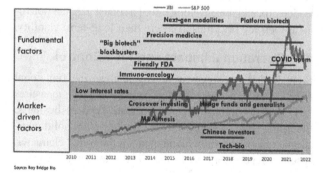

Fig. 94.6 Strategic risks in biotechnology

Biotech start-ups encounter the strategic crisis of efficiently managing two together novelty and supervisory compliance together while guiding along the display dynamics (Faber, 2023). The idea explores how these producers can actively synchronise their clever aims with risk administration, recognising attainable impediments and adjusting their trade plans therefore.

$$I_2 = p^n * (R^{max} - r') + p * (r' - R_{d0})$$

$$r' = \frac{p^n - p}{d_0 * \tilde{d}}$$

Biotech companies can reinforce their accountable processes and achieve tenable development by integrating a calculated risk amount. This approach enables bureaucracy to take convenience, efficiently address questions, and optimise their accomplishment. The subject emphasises the significance of bearing the skill to anticipate and regulate, permissive start-up's to successfully guide along the complex landscape of the biotechnology business accompanying elasticity and strategic adaptability.

5. CONCLUSION

A sustainable business must have effective risk management, which should be an ongoing, collaborative process of development. Businesses that are aware of their risk factors are better able to control risks, which boosts profits and reduces operating expenses. The leadership team may better focus on risk at different periods of development with the use of risk portfolios along with growth frameworks. The strategy process which involves risk reduction strategies enables management to adapt and be ready for shifting risk categories. The methodology promotes communication among interested parties, including businesses, the governing body, and the sector, to be responsible for dealing with and decreasing potential hazards. By putting management education programs into place, businesses may better connect their objectives with sector risk categories and solve complicated challenges. Canada will have an edge in competition in the worldwide knowledge economy of the twenty-first century if it can identify dangerous sectors. To maintain its position as a leading candidate in the worldwide biotechnology industry, Canada has to reduce creativity risk, engage in foreign direct investment, enhance Development and research output, and close the innovation-to-commercialization gap.

ACKNOWLEDGMENT

I would want to extend my heartfelt appreciation to all the scholars and industry professionals who have made substantial contributions to our comprehension and progress in the field. Consistent with this, I would want to convey my appreciation to the organizations who have generously granted me access to invaluable data. Their commitment is crucial for the implementation of this undertaking. I would like to express my gratitude to the entire scientific community for fostering collaboration and facilitating the exchange of knowledge in this significant field of study.

References

1. Bieske, L., Zinner, M., Dahlhausen, F. and Trübel, H., 2023. Trends, challenges, and success factors in pharmaceutical portfolio management: cognitive biases in decision-making

and their mitigating measures. Drug Discovery Today, p. 103734. https://www.sciencedirect.com/science/article/abs/pii/S1359644623002507

2. Bocken, N.M. and Geradts, T.H., 2020. Barriers and drivers to sustainable business model innovation: Organization design and dynamic capabilities. *Long range planning*, *53*(4), p.101950. https://www.sciencedirect.com/science/article/pii/S0024630119301062

3. Bode, C., Macdonald, J.R. and Merath, M., 2022. Supply disruptions and protection motivation: Why some managers act proactively (and others don't). *Journal of Business Logistics*, *43*(1), pp. 92–115. https://onlinelibrary.wiley.com/doi/abs/10.1111/jbl.12293

4. Bruneo, H., Giacomini, E., Iannotta, G., Murthy, A. and Patris, J., 2023. Risk and return in the biotech industry. International Journal of Productivity and Performance Management. https://www.emerald.com/insight/content/doi/10.1108/IJPPM-04-2023-0179/full/html

5. Corea, F., Bertinetti, G. and Cervellati, E.M., 2021. Hacking the venture industry: An Early-stage Startups Investment framework for data-driven investors. Machine Learning with Applications, 5, p.100062. https://www.sciencedirect.com/science/article/pii/S2666827021000311

6. De Vrieze, J., De Mulder, T., Matassa, S., Zhou, J., Angenent, L.T., Boon, N. and Verstraete, W., 2020. Stochasticity in microbiology: managing unpredictability to reach the Sustainable Development Goals. *Microbial biotechnology*, *13*(4), pp. 829–843. https://ami-journals.onlinelibrary.wiley.com/doi/abs/10.1111/1751-7915.13575

7. Dell'Acqua, F., McFowland, E., Mollick, E.R., Lifshitz-Assaf, H., Kellogg, K., Rajendran, S., Krayer, L., Candelon, F. and Lakhani, K.R., 2023. Navigating the jagged technological frontier: Field experimental evidence of the effects of AI on knowledge worker productivity and quality. *Harvard Business School Technology & Operations Mgt. Unit Working Paper*, (24-013). https://papers.ssrn.com/sol3/papers.cfm?abstract_id=4573321

8. Dvorak, Z., Rehak, D., David, A. and Cekerevac, Z., 2020. Qualitative approach to environmental risk assessment in transport. *International Journal of Environmental Research and Public Health*, *17*(15), p. 5494. https://www.mdpi.com/1660-4601/17/15/5494

9. Dwivedi, Y.K., Hughes, L., Ismagilova, E., Aarts, G., Coombs, C., Crick, T., Duan, Y., Dwivedi, R., Edwards, J., Eirug, A. and Galanos, V., 2021. Artificial Intelligence (AI): Multidisciplinary perspectives on emerging challenges, opportunities, and agenda for research, practice and policy. *International Journal of Information Management*, *57*, p.101994. https://www.sciencedirect.com/science/article/pii/S026840121930917X

10. Faber, P.W., 2023. Value generation of digital solutions for small to medium sized biotech companies through commercialization. https://research-api.cbs.dk/ws/portalfiles/portal/92147943/1620652_Master_s_Thesis_Final.pdf

11. Kayode-Ajala, O., 2023. Establishing cyber resilience in developing countries: an exploratory investigation into institutional, legal, financial, and social challenges. *International Journal of Sustainable Infrastructure for Cities and Societies*, *8*(9), pp. 1–10. https://vectoral.org/index.php/IJSICS/article/view/27

12. Massaro, M., Secinaro, S., Dal Mas, F., Brescia, V. and Calandra, D., 2021. Industry 4.0 and circular economy: An exploratory analysis of academic and practitioners' perspectives. *Business Strategy and the Environment*, *30*(2), pp. 1213–1231. https://onlinelibrary.wiley.com/doi/abs/10.1002/bse.2680

13. Mirvis, P. and Googins, B., 2022. *Sustainability to social change: Lead your company from managing risks to creating social value*. Kogan Page Publishers. https://books.google.com/books?hl=en&lr=&id=lyJhEAAAQBAJ&oi=fnd&pg=PP1&dq

14. Rana, I.A., Khaled, S., Jamshed, A. and Nawaz, A., 2022. Social protection in disaster risk reduction and climate change adaptation: A bibliometric and thematic review. *Journal of Integrative Environmental Sciences*, *19*(1), pp. 65–83. https://www.tandfonline.com/doi/abs/10.1080/1943815X.2022.2108458

15. Santi, C. and Neves, G.T., 2020. Startups and prominent business. In New and Future Developments in Microbial Biotechnology and Bioengineering (pp. 209–223). Elsevier. https://www.sciencedirect.com/science/article/abs/pii/B978044464301800010X

16. Schlapp, J., Löchel, H. and Varntanian, S., Strategic pharma/biotech R&D portfolio management: Analysis of internal and external project planning. https://scholar.googleusercontent.com/scholar?q=cache:2-M1vxt4BXkJ:scholar.google.com/+Risk+Management+Strategies+for+Biotech+Startups:+A+Comprehensive+Framework+for+Early-Stage+Projects&hl=en&as_sdt=0,5&as_ylo=2020

17. Shimasaki, C. ed., 2020. Biotechnology Entrepreneurship: Leading, Managing and Commercializing Innovative Technologies. Academic Press. https://books.google.co.in/books?hl=en&lr=&id=K5fgDwAAQBAJ&oi=fnd&pg=PP1&dq=+Risk+Management+Strategies+for+Biotech+Startups:+A+Comprehensive+Framework+for+Early-Stage+Projects&ots=pXnCbXCzq6&sig=K1S-5GatLi-yYtXJPPpTwpGZ4gq8&redir_esc=y#v=onepage&q&f=false

18. Silva, D.S., Ghezzi, A., Aguiar, R.B.D., Cortimiglia, M.N. and ten Caten, C.S., 2021. Lean startup for opportunity exploitation: adoption constraints and strategies in technology new ventures. *International Journal of Entrepreneurial Behavior & Research*, *27*(4), pp. 944–969. https://www.emerald.com/insight/content/doi/10.1108/IJEBR-01-2020-0030/full/html

19. Smith, M. and Miller, S., 2023. Technology, institutions and regulation: towards a normative theory. *AI & SOCIETY*, pp. 1–11. https://link.springer.com/article/10.1007/s00146-023-01803-0

20. Smith, V., Wesseler, J.H. and Zilberman, D., 2021. New plant breeding technologies: An assessment of the political economy of the regulatory environment and implications for sustainability. *Sustainability*, *13*(7), p. 3687. https://www.mdpi.com/2071-1050/13/7/3687

21. Snyder, D.V., 2023. Contracts as an Instrument of International Management and Governance. *Contracts for Responsible and Sustainable Supply Chains: Model Contract Clauses, Legal Analysis, and Practical Perspectives, Susan A. Maslow & David V. Snyder, eds. (2023).* https://papers.ssrn.com/sol3/papers.cfm?abstract_id=4530887

22. Wang, Z., Sun, H., Ding, C., Xin, L., Xia, X. and Gong, Y., 2023. Do Technology Alliance Network Characteristics Promote Ambidextrous Green Innovation? A Perspective from Internal and External Pressures of Firms in China. *Sustainability*, *15*(4), p. 3658. https://www.mdpi.com/2071-1050/15/4/3658

23. Zemlyak, S.V., Sivakova, S.Y. and Nozdreva, I.E., 2021. Risk Assessment Model of Government-Back Venture Project Funding: The Case of Russia. J. Legal Ethical & Regul. Isses, 24, p. 1. https://heinonline.org/HOL/LandingPage?handle=hein.journals/jnlolletl24&div=368&id=&page=

Note: All the figures in this chapter were made by the author.

Recent Trends in Engineering and Science for Resource Optimization and Sustainable Development – Prof. (Dr.) Dorota Jelonek et al. (eds)
© 2024 Taylor & Francis Group, London, ISBN 978-1-032-98030-0

95

The Use of E-Learning in Organizational Learning Processes

Dorota Jelonek[1]
Faculty of Management,
Czestochowa University of Technology, Poland

Sanjar Mirzaliev[2]
Department and Doctoral School,
Tashkent State University of Economics

Edyta Kowalska[3]
Bielanski Hospital, Poland

▬▬▬ Abstract

The concept of learning organizations is one of the most recent management methods, aimed at continuous learning and achieving an advantage in a turbulent environment.

The simplified systematic literature review conducted showed that research on organisational learning is most often concerned with knowledge management and minimises the importance of organisational learning. The aim of this article is to verify the following hypothesis: E-learning has a significant and positive impact on the organizational learning process.

An S-O-D research model was developed. Survey data were collected using questionnaire surveys (n=241). The results were analysed using the structural equation model (SEM) method.

The findings support the results of research verifying the effectiveness of e-learning training and they enrich the knowledge of factors moderating the relationship between e-learning and organizational learning.

▬▬▬ Keywords

Organizational learning processes, E-learning, SEM method

1. INTRODUCTION

The concept of learning organizations is one of the most recent management methods, aimed at continuous learning and achieving an advantage in a turbulent environment.

Such organizations are proficient in performing activities related to creating, acquiring, and processing knowledge and modifying their behavior based on the knowledge and experience gained.

[1]dorota.jelonek@pcz.pl, [2]s.mirzaliev@tsue.uz, [3]kowalska-edyta@outlook.com

DOI: 10.1201/9781003596721-95

In the post-pandemic economy, e-learning is becoming a common form of learning in organizations and universities. The use of this method can significantly improve the management of the entire learning process and strengthen competitive advantages in the market. Achieving these goals is possible if e-learning is seen as one of the directions of development (Jelonek, Dunay, Illes, 2017). E-learning encompasses a wide range of tools and activities that support development processes (Machalska, 2019) as well as remote work (Nowacka, Jelonek 2022). Digital learning tools are evolving and improving, and their implementation should be treated as an innovation deployment, ensuring a conducive climate for innovation (Bylok et al. 2019). Delloite's principal analyst David Mallon analyzed the future of work and, in particular, the problem of the development of e-learning in organizations. He proposed mapping the company to identify areas that can be upgraded or planned for future development.

To date, studies exploring the relationships between organizational development and e-learning mostly have confirmed the positive impact of e-learning on enterprise development. At the same time, the analysis of research on this impact shows that e-learning, along with technologies and strategies such as augmented reality, artificial intelligence, gamification, and microlearning will play a key role in the organizational environment so that they can remain viable and competitive and meet the expectations of their customers (Capytech, 2021).

The aim of this article is to verify the following hypothesis: E-learning has a significant and positive impact on the organizational learning process.

An S-O-D research model was developed. Survey data were collected using questionnaire surveys (n=241). The results were analysed using the structural equation model (SEM) method.

The article is organised as follows: introduction and identification of the research gap in the literature using a simplified systematic literature review method. The research model, description of the research sample, results and conclusions are then presented.

2. IDENTIFYING THE RESEARCH GAP OF E-LEARNING IN THE ORGANIZATIONAL LEARNING PROCESS

The topic of organizational learning has been gaining popularity since the early 1990s. Figure 95.1 summarizes the number of publications from the Web of Science (WoS) database on organizational learning over the years.

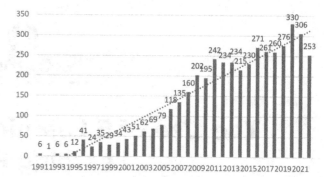

Fig. 95.1 Number of publications in the WoS database devoted to organizational learning from 1991 to 2022

Source: Authors' own elaboration

The first papers in WoS describing organizational learning are from 1991. In subsequent years, the number of publications fluctuated, but there was a noticeable upward trend from 1991 to 2019. A decline in the number of publications was observed in 2021 and 2022. A possible reason for this is the increased interest of researchers in the impact of the COVID-19 coronavirus pandemic on specific areas of organizations' operations.

At the same time, it is important to note the growth of researchers' interest in the subject of e-learning, particularly evident in the early 21st century. (Fig. 95.2).

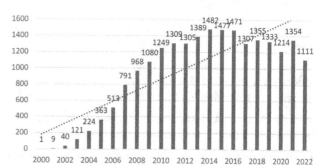

Fig. 95.2 Number of publications in the WoS database devoted to e-learning from 2000 to 2022

Source: Authors' own elaboration

The number of publications in the WoS database increased uninterruptedly from 2000 (1 publication) to 2011 (1309 publications). It then began to fluctuate, with values ranging from 1111 (2022) to 1482 (2014). Despite the decline in interest in this topic observed in 2022, there was an upward trend in the period studied. The order of magnitude of the number of publications is also important: starting in 2009, the WoS database has been enriched every year with more than 1,000 literature items. This demonstrates the high topicality of the problem studied.

A systematic literature review was conducted to identify the research gap in the use of e-learning in organizational learning. The review was conducted on 24 January 2023 using the resources of WoS and Scopus databases. Four keywords were selected: organizational learning, organizational learning, e-learning and elearning, and then the search formula was determined: ("organisational learning" OR "organizational learning") AND ("e-learning" OR "elearning"). The list of publications has been narrowed to the years 2010-2022. The steps of the systematic literature review are shown in Table 95.1.

Table 95.1 Number of publications after using individual search criteria

Criteria	WoS	Scopus
Keywords: "e-learning" OR "elearning"	21507	97540
Keywords: "organisational learning" OR "organizational learning"	4437	6869
Keywords: ("organisational learning" OR "organizational learning") AND ("e-learning" OR "elearning")	109	121
Pulication year: 2010-2022	79	84
Sum (WoS+Scopus)	163	
Removing duplicate publications	143	
Analysis of titles and abstracts	139	

Source: Authors' own elaboration

Based on the criteria, 143 publications were retrieved, which in turn were analyzed by title and abstract. Based on the analysis, it was noted that a large part of the selected items refers to knowledge transfer (Dhaouadi 2017; Masuda et al. 2016), knowledge management (Talukder 2022; Alsharhan et al. 2021; Soltani et al. 2020; Manai and Yamada 2017, Pinto et al. 2017; Zhang et al. 2015) and satisfaction of those using e-learning within the organization (AlMulhem 2020; Lin et al. 2019; Al-Qahtani et al. 2013). Applications used by organizations for e-learning have also been described (Ismailova et al. 2021; Zhang et al. 2015; Smaliukiene and Bekesiene 2011). Furthermore, researchers have referred to the problems of organizational innovation (Stan 2018; Quamari et al. 2019; Weitze 2015) and described a specific model for implementing e-learning (Vallerand 2017). They also focused on the effects of e-learning within organizations (Liu et al. 2012; Hattinger et al. 2018; Li and Tsai 2020). Aparicio et al. (2014) conducted a bibliometric study on e-learning concept trends from 2008 until 2013. The researchers highlighted 22 concepts related to e-learning, including computer support for collaborative learning in organizations. The increase in interest in e-learning was driven by the COVID-19 pandemic and the need for distance learning (Hadavi and Wakefield 2021), remote working, and communication.

Soltani et al. (2020) examined a sample of 290 members of tax administration staff and confirmed the positive impact of the factors such as organizational learning, knowledge management, and e-learning systems on organizational intelligence. The relationship of e-learning with organizational learning has not been studied but it seems to be an interesting research problem. Therefore, the present study attempted to fill this research gap.

3. RESEARCH MODEL

Based on an extensive review of research in this area, it was noted that studies on the variable of learning to date have concerned, among other things, knowledge management or technological solutions, minimizing the importance of the organizational learning process.

Given the results of research on organizational learning and e-learning and looking for variables moderating this relationship, the following hypothesis was formulated:

H: E-learning has a significant and positive impact on the organizational learning process.

A research model based on the above relationships between organizational learning and e-learning in enterprises of different sizes from the fast-moving consumer goods (FMCG) sector and operating in a nationwide market is presented in Fig. 95.3.

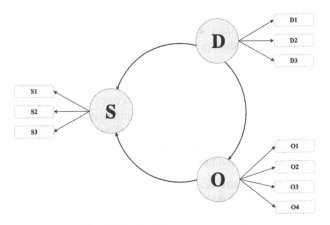

Fig. 95.3 S-O-D research model

Source: Authors' own elaboration

Figure 95.3 shows a schematic diagram of the model of the dependence of the variables S (reinforcing the elements of learning organization development) and O (organizational learning) on variable D (e-learning).

4. Description of the Research Sample and Research Methodology

The achievement of the set goal and the verification of the accepted hypothesis was achieved based on the primary research, which in its scope covered various aspects related to organizational learning. For the purpose of this paper, only a selected part of the obtained data was used to determine whether there is a relationship between organizational learning and e-learning in enterprises..

A questionnaire survey and interview technique were used to conduct quantitative research. A survey questionnaire was prepared for the purpose of the study and distributed to 300 companies. The survey used computer-assisted telephone interview (CATI) and computer-assisted web interview (CAWI) methods. The selection of the research sample was purposive and based on a preliminary interview. The respondents were employees at various levels of companies manufacturing fast-moving consumer goods. The sector is growing despite high uncertainty about the economic situation, which is holding businesses back from spending.

During the research conducted, the target group was organizations that fit the characteristics of a learning organization. A total of 217 fully completed questionnaires were received. The author's survey questionnaire consisted of 49 statements, which were rated by the respondent on a five-point Likert scale. The statements corresponded to variables defined in the organizational learning process. The research tool was supplemented with several items of the respondent data section that focused on the business sector, voivodeship, position held, and job seniority.

The survey covered organizations of which 9.2% were small businesses, 15.7% were medium-sized enterprises, and 75.1% were large companies. The size of the enterprise was determined based on the employment criterion. The threshold values were taken from the European Commission's recommendation for defining small, medium-sized, and large entities.

Another characteristic of the surveyed companies was job seniority at the enterprise the employees surveyed worked for. Job seniority was grouped and transformed by recoding, which made it possible to reduce the number of values adopted by the variable. The purpose of range-based recoding was to allow for the comparison of a group of people with job seniority within ranges of up to 5 years (45.4% of respondents), 6 to 20 years (22.6%), and more than 20 years (23%). In the entities surveyed, the job seniority of employees ranged from 1 year to 35 years. The average number of years the employees worked was 14.

The last variable studied in the respondent data was the group of positions held. Each referred to the hierarchy of positions within the organization. Managers of various levels made up the largest group of respondents (54.4%).

In light of the research results and the learning and knowledge-sharing process, an important element is the training method. The pervasiveness of modern technology is driving the needs of learners, leading to changes in forms of teaching and learning. E-learning is a modern form of learning and has many applications in teaching, training, retraining, or evaluation. The COVID-19 pandemic has revolutionized most industries around the world, fundamentally changing the way people work and learn. The role of e-learning has increased significantly. As a result, respondents were asked whether e-learning had been used in their organizations before the outbreak of the pandemic and whether the organizations increased its use as a training method. Of the 217 organizations surveyed, as many as 185 (85.3%) had used e-learning before the pandemic as a method of employee learning. The detailed composition of the research sample with the division into small, medium-sized, and large entities is presented in Table 95.2.

Table 95.2 Distribution of responses to the statement: Organization used e-learning before the outbreak of the COVID-19 pandemic

	Small	Medium-sized	Large	Total
The organization used e-learning before the outbreak of the COVID-19 pandemic	4	18	163	185
The organization did not use e-learning before the outbreak of the COVID-19 pandemic	16	16	0	32

Source: Authors' own elaboration

The distribution of responses from the respondent group varied widely. According to more than half of the respondents (73.73%), e-learning training is effective. It was rated highest among middle-level managers with job seniority between 5 and 20 years (16.13%). In contrast, it was rated lowest among the group of lowest-level employees with job seniority of more than 20 years (7.83%).

The remaining part of the paper presents a statistical analysis of the results obtained, which was carried out using the structural equation model (SEM) method.

5. Results

Based on the structural equation modeling, survey data obtained from the survey questionnaire were standardized.

Observed variables were recorded as deviations from mean values. Latent variables were also presented as deviations from the mean values, thus the mean values of these variables are equal to zero. Furthermore, the model makes other structural equation modeling assumptions that are consistent with standard SEM model assumptions.

In the model in question, the exogenous observed variables (D1;D2;D3) are indicators of the exogenous variable D. Similarly, the endogenous observed variables are indicators of the variable O, that is (O1; O2; O3; O4) and variable S (S1; S2; S3). Figure 95.1. shows a schematic diagram of the model of the dependence of the variables S (reinforcing the elements of learning organization development) and O (organizational learning) on variable D (e-learning). SEM model: S-O-D consists of three measurement sub-models. The first sub-model determines relationships between observed variables S1; S2; and S3, and a latent variable D which corresponds to the profile of information technologies. The second concerns latent variable O, corresponding to organizational learning, and related observed variables O1; O2; O3; O4. Explanations of individual variables are presented in Table 95.3. Table 95.4 shows the results for selected model parameters.

Table 95.3 Names of SOD model variables

Variable name	Symbol
supporting employees' professional development and career paths	S1
effective motivating of employees	S2
developing the organizational culture that encourages knowledge sharing	S3
organization's strategy	O1
leadership	O2
organizational culture	O3
unlimited access to knowledge resources	O4
content distribution	D1
training content	D2
distance learning services	D3

Source: Authors' own elaboration

The table above shows values of parameters for individual paths of relationships between specific variables presented in the form of arrows in the figure that shows the model. The symbolic designation of the path of relationships is placed in the first column of the table (in accordance with the adopted notation discussed earlier). Those exceeding the set threshold values are printed in red.

Table 95.4 also contains a standard error for evaluation of the parameter, t statistic, and probability level, which for all

Table 95.4 Evaluation of selected parameters in the model

Model path	Model evaluation			
	Parameter evaluation	Standard error	t-statistic	Probability level
(D)->[D1]	1,950	0,213	9,149	0,000
(D)->[D2]	1,539	0,123	12,510	0,000
(D)->[D3]	2,117	0,158	13,391	0,000
(O)->[O1]	1	-	-	-
(O)->[O2]	2,457	0,175	14,047	0,000
(O)->[O3]	1,080	0,095	11,318	0,000
(O)->[O4]	0,947	0,116	8,175	0,000
(S)->[S1]	1	-	-	-
(S)->[S2]	0,425	0,037	11,573	0,000
(S)->[S3]	0,651	0,046	14,007	0,000
(D)->(O)	0,675	0,105	6,436	0,000
(D)->(S)	0,563	0,147	3,838	0,000
(O)->(S)	1,752	0,165	10,623	0,000

Source: Authors' own elaboration

the parameters is lower than the value of significance level adopted for calculations (0.05). For parameters concerning observed variables, this probability is negligibly small as it is for the parameter concerning the relationship between D and O and D-S.

Tables 95.5 and 95.6 present the basic statistics related to parameter evaluation and selected indices of model fit to the data.

Table 95.5 Indices of the noncentrality-based model

Index	Noncentrality-based indices		
	Lower 90% confidence interval	Estimated point	Upper 90% confidence interval
Population noncentrality parameter	1.005	1.245	1.520
Steiger-Lind RMSEA index	0.077	0.097	0.118
McDonald's index of noncentrality	0.468	0.537	0.605
Population Gamma index	0.867	0.901	0.933
Adjusted population Gamma index	0.599	0.657	0.712

Source: Authors' own elaboration

In the case of the S-O-D model, root mean square error of approximation (RMSEA) and population Gamma index do not reach values that exceed the limits of goodness of fit.

Other indices are slightly lower than what is considered a limit: 0.95. The value of the chi-squared statistic given in the table below and the corresponding probability level (p<0.0001) indicate a poor fit of the model to the data. Testing the hypothesis of a perfect fit of the model to the data using the chi-squared statistic is a method that is very sensitive to the statistical quality of the data. In research practice, there are a large number of outliers from the required probability distributions, and it is often difficult to meet all the strict assumptions concerning the measurement data. Many authors believe that being guided only by this measure of estimation quality can lead to the rejection of a valid model. It is assumed that if the value of the chi-squared statistic divided by the number of degrees of freedom is less than 5 (4.93 in this case) and the other indices discussed before take correspondingly good values, the model fit can be considered sufficient.

Table 95.6 Selected statistics for the S-O-D model

Statistic	Summary of basic statistics
	Value
Divergence function	1.380
Max residual cosine	0.000
Max. gradient modulus	0.000
ICSF criterion	0.000
ICS criterion	0.000
ML chi-squared	158.005
Degrees of freedom	32.000
p	<0.0001
Standardized residual RMS	0.106

Source: Authors' own elaboration

Based on the results of SEM modeling, the hypothesis that there is no causal relationship between e-learning and the variable corresponding to organizational learning should be rejected. The coefficient that corresponds to this correlation is 0.563.

The parameter determining the relationship between O and S, i.e. the effect of organizational learning on reinforcing the elements of learning organization development, is statistically significant (1.75, at p<0.0001).

6. CONCLUSION

The problems discussed by the authors are part of the discussion on the increasing importance of e-learning as a result of the strategic focus on management by the development of the learning organization. The concepts addressed may prove useful to researchers as well as business practitioners considering investments in e-learning.

Furthermore, the findings of the present study support the results of research verifying the effectiveness of e-learning training and they enrich the knowledge of factors moderating the relationship between e-learning and organizational learning since the study focused on three moderators that, to date, have not been studied. The results of the survey prove that e-learning in FMCG enterprises has a significant and positive impact on organizational learning.

Among the suggestions for further research in this area, the moderating variables of the sector and business scope seem to be of particular interest.

When considering the results of the present study, it is important to keep in mind that the study has several limitations. The results obtained may have been influenced by the sample structure. Furthermore, the results of the survey may have also been affected by respondents' subjective evaluations.

References

1. Al Mulhem, A. (2020). "Investigating the effects of quality factors and organizational factors on university students' satisfaction of e-learning system quality,. Cogent Education, 7(1), 1787004.
2. Al-Qahtani, M., Al-Qahtani, M., & Al-Misehal, H. (2013, April). "Learner satisfaction of e-learning in workplace: Case of oil company in Middle East", In 2013 10th International conference on information technology: New generations (pp. 294–298). IEEE.
3. Alsharhan, A., Salloum, S., & Shaalan, K. (2021). "The impact of eLearning as a knowledge management tool in organizational performance", Advances in Science, Technology and Engineering Systems Journal, 6(1), 928–936.
4. Aparicio, M., Bacao, F., & Oliveira, T. (2014). "Trends in the e-learning ecosystem: A Bibliometric study", In AMCIS 2014 Proceedings, Twentieth Americas Conference on Information Systems. AIS.
5. Bylok, F., Jelonek, D., Tomski, P., & Wysłocka, E. (2019). Role of Social Capital in Creating Innovative Climate in Enterprises: The Case of Poland. In Proceedings of the ICICKM 2019 16th International Conference on Intellectual Capital Knowledge Management & Organisational Learning, Sydney, Australia, pp. 72–79.
6. Capytech. 2021. "7 E-learning trends for 2022,"
7. Dhaouadi, N. (2017). "Knowledge Transfer Through E-learning: Case of Tunisian Post", In Digital Economy. Emerging Technologies and Business Innovation: Second International Conference, ICDEc 2017, Sidi Bou Said, Tunisia, May 4–6, 2017, Proceedings 2 (pp. 85–94). Springer International Publishing.

8. Hadavi, Y., & Wakefield, R. L. (2021). "Fostering E-Learning Satisfaction during COVID: Do Interactivity and Engagement Help?", In AMCIS 2021 Proceedings, Twenty-Seventh Americas Conference on Information Systems, Montreal.

9. Hattinger, M., Eriksson, K., Hegli, P., & Henriksen, N. (2018). "Management Strategies For Knowledge Transformation: A Study Of Learning Effects In Industry-Union-University Collaborative E-Learning Initiatives", In ICERI2018 Proceedings (pp. 10049–10057). IATED.

10. https://www.capytech.com/index.php/2021/12/30/7-e-learning-trends-for-2022/ (access: 02.09.2023).

11. Ismailova, R., Medeni, T. D., Medeni, I. T., Muhametjanova, G., & Soylu, D. (2021). "Organizational learning management system application via micro PC hardware: A case study in Kyrgyzstan", International Journal of Virtual and Personal Learning Environments (IJVPLE), 11(1), 54–63.

12. Jelonek, D., Dunay, A., Illes, B. (2017). "Academic e-learning management with e-learning scorecard," Polish Journal of Management Studies (2017 Vol. 16: 2), pp. 122–132.

13. Li, D. C., & Tsai, C. Y. (2020). „Antecedents of employees' goal orientation and the effects of goal orientation on e-learning outcomes: the roles of intra-organizational environment", Sustainability, 12(11), 4759.

14. Lin, C. Y., Huang, C. K., & Zhang, H. (2019). "Enhancing employee job satisfaction via E-learning: the mediating role of an organizational learning culture", International Journal of Human–Computer Interaction, 35(7), 584–595.

15. Liu, Y. C., Huang, Y. A., & Lin, C. (2012). "Organizational factors' effects on the success of e-learning systems and organizational benefits: An empirical study in Taiwan", The International Review of Research in Open and Distributed Learning, 13(4), 130–151.

16. Machalska, M. 2019. "Digital learning Od e-learningu do dzielenia się wiedzą", Warszawa: Wydawnictwo Wolters Kluwer.

17. Manai, O., & Yamada, H. (2017, March). "How can we accelerate dissemination of knowledge and learning? developing an online knowledge management platform for networked improvement communities", In Proceedings of the Seventh International Learning Analytics & Knowledge Conference (pp. 548–549).

18. Masuda, A., Morimoto, C., Matsuodani, T., & Tsuda, K. (2016, April). "A Case Study of Team Learning Measurements from Groupware Utilization", In Proceedings of the 8th International Conference on Computer Supported Education (pp. 193–198).

19. Nowacka, A., Jelonek, D., (2022), The impact of the multi-variant remote work model on knowledge management in enterprises. Applied tools, Proceedings of the 17th Conference on Computer Science and Intelligence Systems, FedCSIS 2022, pp. 827–835

20. Qamari, I. N., Dewayani, J., & Ferdinand, A. T. (2019). "Strategic Human Resources Roles and Knowledge Sharing: How do enhancing Organizational Innovation?", Calitatea, 20(168), 86–92.

21. Soltani, Z., Zareie, B., Rajabiun, L., & Agha Mohseni Fashami, A. (2020). „The effect of knowledge management, e-learning systems and organizational learning on organizational intelligence", Kybernetes, 49(10), 2455–2474.

22. Stan, O. M. (2018). "Quality Time and Online Training Innovation for the Public-Sector Managers", In Transylvanian International Conference in Public Administration, Cluj-Napoca: Accent (pp. 443–461).

23. Talukder, B. (2022). "Digital Skill Transformation and Knowledge Management Challenge in a Global IT Service Firm: An Empirical Study", Journal of Information & Knowledge Management, 2250090.

24. Vallerand, J., Lapalme, J., & Moïse, A. (2017). "Analysing enterprise architecture maturity models: a learning perspective", Enterprise Information Systems, 11(6), 859–883.

25. Weitze, C. L. (2015). "Pedagogical innovation in teacher teams: An organisational learning design model for continuous competence development", In ECEL 2015: The 14th European Conference on E-Learning (pp. 629–638). Academic Conferences and Publishing International.

26. Zhang, X., Gao, Y., Yan, X., de Pablos, P. O., Sun, Y., & Cao, X. (2015). "From e-learning to social-learning: Mapping development of studies on social media-supported knowledge management", Computers in Human Behavior, 51, 803–811.

Note: All the figures and tables in this chapter were made by the author.

Recent Trends in Engineering and Science for Resource Optimization and Sustainable Development – Prof. (Dr.) Dorota Jelonek et al. (eds)
© 2024 Taylor & Francis Group, London, ISBN 978-1-032-98030-0

96

Evolution of ERP Systems in the Perspective of Process Maturity

Damian Dziembek[1],
Tomasz Turek[2]
Czestochowa University of Technology

■ Abstract

Enterprise Resource Planning systems play significant role in the optimization and improvement of business processes e.g. production, logistics or human resources ones. As it was mentioned in the paper defining the concept of ERP system process maturity it is crucial to determine the efficiency and efficacy of ERP systems in redefinition and implementation of processes in accordance with customer's needs and requirements. The levels of maturity distinguished in the paper embrace such main categories as: ERP supporting processes, process-oriented ERP, process ERP systems and advanced process ERP systems of which appropriate characteristics were provided together with its clarification.

■ Keywords

Process maturity, Enterprise resource planning systems, Business processes improvement

1. INTRODUCTION

The requirements of the contemporary environment, the pursuit of improvements in business processes as well as rational and effective resource management determine the application of various IT systems in enterprises (Uwizeyemungu, Raymond 2010). An important role in supporting enterprise management is currently played by integrated ERP (Enterprise Resource Planning) IT systems. ERP IT systems allow for support for and integration of nearly all areas of the enterprise activity, facilitate the collection, processing and use of an increasing amount of information resources as well as support managers and specialists in monitoring and analyzing business processes as well as making rapid and accurate decisions.

One of the directions of the development of ERP systems is to take into account the process approach in its organization and the principles of operation. In the opinion of the authors, growing concentration in ERP systems on the process paradigm and the creation of a number of tools supporting process management, cooperating or fully integrated with ERP systems – allow for introducing the concept of process maturity of ERP systems as well as identifying the levels of process maturity of ERP systems. The objective of the article is to define the concept of process maturity of ERP systems and to present, in model terms, the levels of process maturity. Moreover, the outlined levels of process maturity will be related to selected ERP systems available in the market. The following research methods have been utilized in the study: literature analysis, observation of phenomena, deduction and induction.

[1]damian.dziembek@pcz.pl, [2]tomasz.turek@pcz.pl

DOI: 10.1201/9781003596721-96

2. The Concept and Essence of Process Maturity of ERP Systems

Increasing customer requirements, stronger and stronger market competition, pressure to increase the efficiency of increasingly complex business processes and striving to increase the flexibility and rapidity of the actions taken, determine the implementation of various IT solutions (including integrated ERP IT systems) (Klaus, Rosemann, Gable 2000). ERP systems are implemented and utilized in various types of enterprises, diversified in terms of industry, size, form of ownership or scope of operation (Motwani, Subramanian, Gopalakrishna 2005). Concisely, ERP systems can be defined as an enterprise information system designed to integrate and optimize business processes and transactions in a corporation (Addo-Tenkorang, Helo: 2011) or as a cross-functional enterprise backbone that integrates and automates many internal business processes and information systems within the manufacturing, logistics, distribution, accounting, finance, and human resource functions of a company (O'Brien, Marakas: 2010). The ERP system can be also defined as a set of integrated business applications or modules to carry out most common business functions, including inventory control, general ledger accounting, accounts payable, accounts receivable, material requirements planning, order management, and human resources. ERP modules are integrated, primarily through a common set of definitions and a common database, and the modules have been designed to reflect a particular way of doing business, that is, a particular set of business processes. ERP systems play an important role in supporting enterprise management since they enable an organization to integrate all the primary business processes in order to enhance efficiency and maintain a competitive position (Hurbean, Fotache 2014).

The issue of the process approach in enterprise management has gained practical importance in recent years as a response to dynamic changes in the environment. The scale, complexity, dynamics and diversity of business processes taking place in contemporary enterprises require the application of appropriate tools, e.g. ERP IT systems (Palade, Møller 2023). By means of ERP systems, it is possible to increase the efficiency of both main and auxiliary processes in the company. Basically, ERP systems enable proper organization and operation of a set of processes in the enterprise through their planning, mapping and implementation, analysis and control of their course and improvement.

In the opinion of the authors, the support for the implementation of business processes by ERP systems for many years allows for defining the concept of process maturity of ERP systems. In the subject literature, the concept of process maturity is most frequently related to the organization and defined as Business Process Maturity Model (BPMM). The BPMM model was developed by Object Management Group (OMG), originating from the model of Capability Maturity Model the component of which it was initially. The objective of BPMM is to provide an opportunity to make comprehensive assessment of the process maturity of any organization and to define the direction of operations which allow organizations for operation and management based on business processes (Business Oriented Organization). The Gartner international consulting company has also developed its Business Process Maturity Model (BPMM). Another model in the area of process management is Process and Enterprise Maturity Model (PEMM), developed by Michael Hammer (Power: 2007).

Process maturity of ERP systems can be defined as the ability of this class systems to support different stages of process management, i.e. definition, design, implementation, measurement, control and their improvement. In order to define process maturity of ERP systems, it is necessary to assess the advancement of methods, techniques and technologies used in them in the field of process management through the prism of appropriately selected criteria. Moreover, it is essential to specify how ERP systems support the management of the company's processes, including, among others, determining how efficiently and effectively ERP systems are able to redefine and implement processes due to variable customer requirements or changes in process objectives, established at the strategic level. An important potential determining process maturity of ERP systems is equally an unambiguous and clear way of defining processes (e.g. notation system) and analyzing resources used for their various implementation options (e.g. measurement of time, labor, costs).

3. Levels of Process Maturity of ERP Systems

Due to the role and significance of the process viewpoint in ERP systems, it is possible to suggest the following stages of evolution of the process approach in this class of integrated IT systems supporting management:

3.1 ERP Systems Supporting Processes (Partially or Comprehensively)

These are integrated management IT systems in which it is possible to map processes in the form of related documents and data. This group of ERP systems most often operates on the common dedicated database, which implies that the data and documents once introduced can be visible and used by other users of the system (depending on allocated allowances). The implemented data and generated documents reflecting the course of the business process cannot be freely modified and deleted in the ERP system. This is not only out of necessity to maintain the continuity of the business process but also due to risk of the loss of consistency of the database. In this group of ERP systems, processes can take place independently of the modelled (or unmodelled) - in the organization - process maps and models (no close link between the process model and its mapping in the ERP system). Process maps and models are usually created in other external applications not integrated with the ERP system (no data exchange model supporting communication between applications). Moreover, in this group of ERP systems, there is no possibility to graphically map the process and conduct its thorough analysis However, at this level, it is possible to precisely define the processes (graphically or descriptively) functioning in the enterprise.

This group may also include the systems containing standardized process maps (functioning in the model organizations of the industry), which facilitate the mapping of the circulation of data and documents in the enterprise and relationships between individual business areas. This allows for gaining knowledge of the proven course of processes in other and often competitive businesses and, as a result, avoiding errors at the stage of the analysis and subsequently optimizing own processes based on the practices used by industry leaders (Kharunnisa, Noerlina, Meiryani 2023).

The division into ERP systems supporting processes partially or comprehensively results from the complexity of ERP systems and the amount of supported areas. In practice, one may come across ERP systems with a large number of modules, dedicated mainly to large organizations which may almost completely support the area of operations of the company (e.g. IFS) or ERP systems possessing a number of modules but not making it possible to map all and often very complex processes. A specific case is ERP systems dedicated to medium (ENOVA) or smaller enterprises (CDN OPTIMA), containing a smaller number of modules adjusted and consciously narrowed by producers to support the most important areas of activity.

3.2 ERP Systems Oriented to Processes (Cooperating with Single or a Larger Number of External Modelers)

It is a group of ERP systems which extends the functionality of the previously distinguished group of systems to the possibility of downloading modelled processes from external applications (e.g. BPMS or Workflow). ERP systems oriented to processes have no possibility of the analysis and optimization of business processes. All the operations in terms of graphic design, modelling and simulation are carried out in external applications (e.g. Aris Toolset for SAP, TM Workflow Comarch ERP XL).

Within the framework of this group of systems it is possible to distinguish ERP systems which cooperate exclusively with one specific modeler (and the process description format) and ERP systems, which enable collaboration with more than one modeler (and simultaneously a larger number of formats describing processes) operating as an external application. The number of cooperating and integrated external modelers is therefore the criterion for the division of ERP systems oriented to processes and is the characteristic enabling the assessment of its development potential.

At this stage of development of ERP systems, business processes, due to the functionality and capabilities of modelers, may be fully professionally presented and recorded, allowing for understanding and undertaking further improvement actions in the enterprise. External modelers enable the measurement, control, simulation and optimization of processes, eventually affecting the improvement in processes implemented in ERP systems.

3.3 Process ERP Systems (Containing One or More Notations and Presentations of Processes)

It is a group of ERP systems which have been closely integrated with the process modelling tools. Process modelling takes place directly in the ERP system in this case or in one of its modules. It should be pinpointed that the modelled process simultaneously determines the mode and form of its implementation in the ERP system. Therefore, it is not possible that the maps and processes created are not used in such an ERP system. Moreover, the change in the process is automatically mapped in its course and implementation in the ERP system. Process ERP systems also depart from a rigid division into modules and functions. The most important characteristics of strictly-process systems are the following: one application instead of modules, visible only the functions resulting from the role of the user in the system, possibility of starting processes and not functions, personalized (also mobile available) desktop, visible tasks to perform, possibility of modelling new processes adjusted to the needs of the enterprise, they include sets of indicators for the process analysis, they include graphical process analysis tools, possibility of the process time analysis, possibility of modification of the already existing processes, they have tools for continuous process improvement and work in transparent and ergonomic interface.

The summary of differences between process ERP systems and process support / process-oriented systems is included in Table 96.1.

Table 96.1 Differences between process ERP systems and process support/process-oriented systems

ERP systems	
Process support/ process-oriented ERP systems	**Process ERP systems**
Modules	Processes
Activating functions	Activating processes
Modelling processes using additional tools	Direct modelling of processes in the ERP system
Analysis of processes in the ERP system – obstructed	Full and direct possibility of the process analysis
Indirect process improvement in the ERP system	Possibility of direct process improvement in the ERP system
Process improvement in the ERP system often based on intuition	Process improvement based on quantitative indicators
ERP system users may have a problem with the identification of their role in the organization and process	Users know their role in the organization and process

Source: Own study

Process ERP systems allow managers to respond dynamically to any changes occurring within the enterprise or its environment. Full insight into the organization and the course of processes allows for ordering and improving processes which, in turn, should improve efficiency and effectiveness of the company. A rapid change in business processes in the ERP system is particularly important in the case of dynamic development or the need to restructure the enterprise. The analysis and ongoing control of processes in this group of ERP systems makes it possible to identify the reasons for failure, too long implementation or stopping the process. The authorized users of the process ERP system (e.g. managers) gain an opportunity to currently introduce necessary changes eliminating the problems emerging in the course of the process implementation. The analysis of processes in the ERP system is the basis for the automation of operations and, as a result, allows for optimization of processes implemented in the enterprise in cooperation with partners and customers.

Process ERP systems usually possess the built-in group of predefined processes, characteristic of individual types of industries. On the basis of the initially defined processes, the enterprise may use the knowledge of the supplier and

introduce some of their improvements or build its own set of processes. Such systems have significant capacities to adapt them to the needs of users (e.g. they can see their tasks to perform within the process or learn the overall course of the process and other people involved in its implementation), having transparent and user-friendly interface. The properly selected and implemented process ERP system ought to provide all the benefits associated with the application of the process approach in the company. The effect of the use of the process ERP system may be in particular: improved quality of customer service, improvements in rapidity and efficiency of business operations, increased competitiveness of the company in the dynamically developing market and reducing costs of implementation of organizational changes. The constraint of process ERP systems can be costly and time-consuming workload for preparing the process map, designing all processes reflecting the specificity of the enterprise and their optimization. Another problem may be the complicated operation of the process ERP system (too many options available) and the need to constantly improve the processes.

Generally, among various notations of business processes, one may distinguish BPMN, EPC, ERM, CMMN, UML, DFD. The criterion for the division of process ERP systems is the possibilities of the modelers implemented in them, cooperating with:

- one notation of process presentation (e.g. BPMN),
- more than one notation of process presentation (e.g. BPMN and EPC).

The larger the number of notations in which the course of processes can be mapped in the modeler integrated with the ERP system the higher the level of its development. At present, process ERP systems are usually integrated with one modeler of processes supporting BPMN notation (e.g. Merit). Progress in information technologies and further development of ERP systems as well as an increase in the popularity of different methods for process description will definitely affect the possibility of support for various notations in modelers available in ERP systems.

At this level of development, business processes may be fully designed and implemented in ERP systems (no need to use external modelers). Moreover, the measurement, analysis and control of the course of processes is possible directly in ERP systems, which will allow for their improvement as well as improvements in the performance of the enterprise.

3.4 Advanced process ERP systems

It is a group of ERP systems which, in addition to the possibilities of process ERP systems, has significantly developed tools for modelling processes that enable carrying out advanced simulations and using many different methods of optimization and continuous process improvement. Thus, it is possible to analyze weaknesses in order to identify the reasons for their occurrence, which allows for preventing the emergence of defects of the process and preventing their occurrence in other processes.

The greater the possibilities of simulation of operations, actions and various resources the higher the level of development of the ERP system. Moreover, the more the methods and techniques for optimization and improvement in processes the greater the development potential of the ERP system. Advanced process ERP systems extend the possibilities of process ERP systems to the analysis, control and improvement of not only internal (own) processes of the company, but they also extend their possibilities to design, monitoring and optimization of the processes of cooperators (the enterprise without borders).

At this stage of development, ERP systems provide the implementation of the full and systematized process management cycle (planning, design, modelling, simulation, optimization, implementation and improvement) on the basis of various and advanced methods, techniques and concepts. The enterprise, through the opportunities generated by such ERP systems, is fully focused on the improvement in processes. Systematic changes are made throughout the whole system of processes, aimed at optimization of results generated by individual processes and the organization as a whole. ERP systems, through the implementation of universality of process management, enable the combination of the activities of employees with the results of the company's processes.

Moreover, ERP systems, at this stage, also allow for:
- appropriate management of knowledge of processes (collection, codification, transfer, use and development of knowledge) as well as the development of appropriate organizational culture and values shared;
- the use of advanced and cooperating ITC (e.g. cloud computing, Big Data, IoT, mobile technologies, software agents and artificial intelligence) enabling rapid access and processing of information resources from different sources in order to collect information on the state, disabilities, opportunities to improve business processes (Ali, Nguyen, Gupta 2023), (Cheng 2018) (Jelonek 2017).

The ERP system, through the use of modern ICT, will support intelligent networks combining not only internal and external processes but also integrating different machines, systems, products, customers and suppliers.

Thus, it will be possible to deepen automation further, continuously optimize products and processes, collect and process a vast amount of data in real time, operate machinery and equipment and also to quickly adapt to changes in the market situation. Moreover, advanced process ERP systems will possess self-adapting capabilities/adapting their processes to changing market requirements (Jelonek, Stępniak, Ziora 2019).

4. QUANTIFICATION OF PROCESS MATURITY OF SELECTED ERP SYSTEMS

Practice indicates that there are at least a few large suppliers of ERP systems in the global market. The examples can be the solutions of SAP, Oracle, Microsoft, IFS, etc. These systems are usually adjusted to the specificity of the market in which they are offered, and they have a few language versions. In addition to such solutions, there are ERP systems provided by national producers of business software. These solutions most frequently have the limited amount of language versions (most often one). Instead, they are much better adapted to legal provisions, tax regulations and specific management conditions in the country in which they have been produced. The ERP systems of local suppliers are therefore easier to implement in small and medium organizations which implement internal and external processes within the framework of a single language and cultural background.

In this study, an attempt will be made to quantify selected ERP systems and classify them into the previously indicated levels of process maturity (levels 1 to 4). The selection of systems is subjective and results from business, teaching and training experience of the authors. The selection of classification systems is significant but the research methodology consisting in the identification of the main features of ERP systems which determine their qualification (or not) to the specific level of maturity is equally important. In the research, two systems were taken into account, offered by the international suppliers of ERP systems as well as two local systems. "International" systems are SAP ERP and Microsoft Dynamics solutions. "Local" solutions are the systems by Polish manufacturers: Comarch ERP XL and Macrologic Merit by Asseco. The main features of ERP systems in terms of their quantification and process maturity are presented in Table 96.3.

As indicated in Table 96.2, all the analyzed ERP systems show the features characteristic of the first level of maturity. Business processes are mapped in data processed in the systems and generated documents. All the systems use the common database to support processes and record economic event.

Table 96.2 The quantification of the process maturity of selected ERP systems

Characteristics of the ERP system determining the level of maturity	SAP ERP	Microsoft Dynamics	Comarch ERP XL	Macrologic Merit
I. ERP supporting processes				
- the possibility to map processes in documents and data	X	X	X	X
- common database	X	X	X	X
- processes independent of process maps	X	X	X	
II. Process-oriented ERP				
- cooperates with the modeler	X	X		X
- modeler as an external application	X	X		
- lack of the possibility to analyze processes	X	X	X	
- graphic design of processes outside the system	X	X	X	
- impossible or difficult operations improving processes	X	X	X	
III. Process ERP system				
- close integration of the system and processes				X
- modeler as a module (component) of the ERP system		X		X
- process model determines the mode of process implementation				X
- predefined processes in the ERP system		X		X
- possible measurement, analysis, improvement in processes in ERP				X
IV. Advanced process ERP system				
- possible process simulation and optimization				
- optimization of internal and external processes	X	X		
- knowledge of processes, process organizational culture				
- process integration of machinery, equipment, people	X	X		

Source: Own study

On the other hand, there are not so many similarities at the second level of Table 96.2. Strong process orientation is indicated by the SAP ERP and Microsoft Dynamics systems. Both manufacturers of business software have ensured the possibility of cooperation of their solutions with external modelers both in the area of graphic modelling and other possibilities of process formalization. Due to the lack of close integration of the process modeler with the ERP system, the process models do not determine the course of the process in the system. If they are formalized and recorded in the company's records, they become a type of recommendation for the course of events and tasks. Most of the features of the process-oriented system are also indicated by the Comarch ERP XL system. A typically process ERP system – the third level – has been found the Macrologic Merit system. This system has a set of more than 300 predefined business processes supporting all functional areas, divided into the areas: logistics, production, finance, human resources and management. Process models can be activated in the form as they are, or they can be modified using the built-in process modeler. Such a philosophy of operation of the Merit system has consequences. Firstly, there is close integration of models and processes. This means that the process model enforces a specific sequence of events and tasks in accordance with the formal record. Secondly – it allows the analysis of processes, their modernization and continuous improvement.

The properties of the process system are also indicated by the Microsoft Dynamics system. However, the philosophy of operation is different in this case. The system has the tool of Business Process Flows, the task of which is only to analyze the course of the process. The process modification is limited. Ready-to-use processes cannot be modified and improved to such an advanced degree as in the case of the Merit system. Table 2 indicates that the fourth level, concerning process-advanced ERP systems, is a direction of development of business software rather than the actually operating systems. The systems of the class available currently do not allow the simulation based on optimization of business processes. If such operations are conducted, it is rather in external tools and systems and only then the results are adapted to ERP systems. Also, contemporary ERP systems do not have extensive knowledge of processes or organizational culture in the area of processes.

Larger suppliers of business software (SAP, Microsoft) make sure there is a possibility of integration and optimization of external processes. This is due to the fact that their solutions are very often used in distributed organizations and network enterprises. Due to the multi-module nature and wide functionality, these systems allow for the integration of not only people but also machinery and equipment (e.g. using IoT models).

5. Conclusions

Everyone strives for process perfection, being aware of the significance of the process approach in the contemporary economy. The paper is mainly of a conceptual character. It is an important and difficult task to determine the future of ERP systems, which constitute the basis for supporting processes in enterprises. The development of ERP systems can be perceived through the prism of many criteria, a special role is played by the increasing possibilities of supporting business processes. The authors' proposal consists in distinguishing the stages of ERP systems development along with indicating the characteristic features of each stage such as ERP systems supporting processes partially or comprehensively, ERP systems oriented to processes, process ERP systems and advanced process ERP systems. The reference of the distinguished stages of ERP systems development to selected ERP systems available on the market is presented. The presented development stages can become the basis for comparison and evaluation of ERP systems.

References

1. Addo-Tenkorang R, Helo P.: Enterprise Resource Planning (ERP): A Review Literature Report, Proceedings of the World Congress on Engineering and Computer Science 2011 Vol II, WCECS 2011, October 19–21, 2011, San Francisco, USA (2011)

2. Ali I., Nguyen N.D.K., Gupta S. (2023), A multi-disciplinary review of enablers and barriers to Cloud ERP implementation and innovation outcomes, Journal of Enterprise Information Management, 36 (5), pp. 1209–1239

3. Cheng Y.-M. (2018), What drives cloud ERP continuance? An integrated view, Journal of Enterprise Information Management, 31 (5), pp. 724–750

4. Hurbean L., Fotache D.(2014), ERP III: The Promise of a New Generation, Proceedings of the IE 2014 International Conference

5. Jelonek, D., (2017), Big Data Analytics in the Management of Business, MATEC Web of Conferences, 125, 04021

6. Jelonek, D., Stępniak, C., Ziora, L., (2019), The meaning of big data in the support of managerial decisions in contemporary organizations: Review of selected research, Advances in Intelligent Systems and Computing, 886, pp. 361–368

7. Kharunnisa A., Noerlina, Meiryani (2023), ERP Implementation and Foreign Ownership in High-Tech Industry to Achieve Sustainable Development Goals, 5th International Conference of Biospheric Harmony Advanced Research, ICOBAR 2023, Vol. 426

8. Klaus, H., M. Rosemann, and G. G. Gable (2000), "What is ERP?" Information Systems Frontiers, 2(2), 141–162

9. Motwani J., Subramanian R., Gopalakrishna P. (2005), Critical factors for successful ERP implementation: Exploratory findings from four case studies, Computers in Industry 56

10. O'Brien, J.A. and Marakas, G.M.: Management Information Systems: Managing Information Technology in the Business Enterprise. 10th Edition, McGraw Hill, New York (2010)

11. Palade D., Møller C. (2023), Guiding Digital Transformation in SMEs, Management and Production Engineering Review, 14 (1), pp. 105–117

12. Power B. (2007), Michael Hammer's Process and Enterprise Maturity Model

13. Uwizeyemungu, S., Raymond L. (2010), Linking the Effects of ERP to Organizational Performance: Development and Initial Validation of an Evaluation Method. Information Systems Management 27

Note: All the tables in this chapter were made by the author.

Recent Trends in Engineering and Science for Resource Optimization and Sustainable Development – Prof. (Dr.) Dorota Jelonek et al. (eds)
© 2024 Taylor & Francis Group, London, ISBN 978-1-032-98030-0

97

Criminogenically Safe Environment in Regions of Georgia: Statistics and Reality

Nino Abesadze*

Faculty of Economics and Business,
I. Javakhishvili Tbilisi State University,
Tbilisi, Georgia

Otar Abesadze

Faculty of Business and Technology,
Georgian National University (SEU),
Tbilisi, Georgia

Rusudan Kinkladze

Faculty of Business Technologis,
Georgian Technical University,
Tbilisi, Georgia

Natalia Robitashvili

Faculty of Economics and Business,
Batumi Shota Rustaveli State University,
Batumi, Georgia

━━━━━ **Abstract:**

The criminogenic situation is an important catalyst for the development of any country. The degree of democratization of the country, the social background of the population, and the scale of political and economic development largely depend on its level.

The **main goal** of the research was to reveal the main tendencies of crime in the regions.

Methodology. Both general methods (analysis, synthesis, induction, deduction) and specific statistical analysis methods were used in the research process.

Results. In 2013-2022, the number of crimes registered throughout Georgia is characterized by an increasing trend; The most common crimes in Georgia are: theft, violence, threats and illegal manufacture, production, purchase, storage, transportation, transfer or storage of narcotic drugs. Among the regions, the highest crime rate is recorded in Tbilisi and Adjara, and the lowest in Guria and Mtskheta-Mtianeti. The main part of criminals belongs to the age group of 25-34 years. In 2022, the number of male criminals in all regions will significantly exceed the number of female criminals. On average, every year in all regions last In ten years, crime rates are increasing. At present, from the point of view of crime, the situation is serious in Tbilisi, as well as in Adjara and Samegrelo-Zemo Svaneti. The correlation between the number

*Corresponding author: nino.abesadze@tsu.ge

DOI: 10.1201/9781003596721-97

of crimes and the level of unemployment, as well as between the number of crimes and the average monthly income is a strong connection. The number of victims is increasing in most regions of Georgia.

■■■■■ **Keywords**

Crime, Analysis, Trend, Crime Situation, Region

1. INTRODUCTION

The criminogenic situation is an important catalyst for the development of any country. The degree of democratization of the country, the social background of the inhabitants, the scale of political and economic development, etc., depend a lot on its level. In fact one of the most important conditions for the democratic development of the country is independent justice. Statistical indicators related to the court's activity are an important indicator in the evaluation of the development trends of the independence of the courts data. The methodology of calculating indicators in legal statistics is specific and requires a different approach depending on the peculiarities of the field. (Abesadze & Mchedlishvili, 2016)

At the modern stage of society's development, in the wake of the development of democratic processes, the aggravation of the criminogenic situation and the problem of gender equality in Georgia are more and more intensively in the center of public attention. It is a pity, but it is a fact that there are increased facts of crime detection, which has a negative effect on the qualitative indicators of the society in the aggravated social background.. (Abesadze O. , 2023)

It is a fact that the quality of the country's democracy, along with other important indicators, is determined by the quality of public information about the criminogenic situation in the country. Analyzing the geographic structure of crime is a constant area of interest for law enforcement agencies. Public interest in his research is growing every year, because it affects not only the political situation of the country, but also the economy and the image of the country in general. (Abesadze & Mazanishvili, Statistical trends of criminality according to the regions of Georgia, 2023)

Quantitative analysis of delinquency indicators requires an appropriate information base. Statistical databases are important for determining and analyzing the causes of crime, to identify cause-and-effect relationships based on this.. (Abesadze, Lortqipanidze, & Abesadze, Correlation-regression analysis of factors affecting the number of criminal offenses, 2018)

For a comparative analysis of the level of crime by region, it is necessary to study crime in a regional perspective, because it provides an opportunity to conduct a comparative analysis, as a result of which the most "safe" regions in the country can be identified. **Therefore, the main goal** of the research was to reveal the main tendencies of crime in the regions.

2. METHODOLOGY

Both general methods (analysis, synthesis, induction, deduction) and specific statistical analysis methods were used in the research process. After the first two stages of statistical research: observation and grouping, averages, relative indicators, correlation-regression analysis, dynamic rows, etc. were used in the analysis. methods. Absolute crime data provide an opportunity to reveal a trend in dynamics, but to characterize the intensity of change, it is necessary to calculate relative indicators, because the intensity of crime is a complex quantitative and qualitative parameter of the criminological situation in the country, region, city and settlement, which indicates the manifestations of crime and, accordingly, the level of public danger. The crime rate is one of the basic indicators of criminological statistical analysis, which is per 100,000 people, although in relatively small cities or regions, the crime rate may be calculated per 10,000 or 1,000 people. (Marshava & Mindorashvili, 2008) Taking into account the specifics of the population of the regions of Georgia, to ensure comparability, the crime rates are calculated per 10,000 people. Relative quantities form precisely interrelated statistics system of indicators (Gloveli, 2014)

3. RESULTS

In dynamics, the picture of criminality in the country has been changing interestingly in the last 10 years. For example, if in 2013, the largest number of investigation cases was in Tbilisi (24,723), then in Imereti (4,325) and Adjara (4,158). , then in Adjara (125) and Shida Kartli (108). And the lowest rate was recorded in Racha (68) . (National Statistics Office of Georgia, 2023) (Prosecutor's

Office of Georgia) In 2014-2015, based on the crime rate, the highest number of cases under investigation was recorded in Tbilisi (168 and 150), Samegrelo Zemo-Svaneti (154 and 132) and Adjara (121 and 127), and the lowest in Guria (70) and Imereti (82).

In 2016-2017, the highest number of investigated cases per 10,000 inhabitants was recorded in Tbilisi (160, 170), Adjara (131, 127) and Samegrelo-Zemo Svaneti (109), and the lowest in Guria (79 and 72). (Prosecutor's Office of Georgia) In 2018-2019, the highest number of investigated cases per 10,000 inhabitants was recorded in Adjara (288, 276, respectively), Tbilisi (243, 275), in 2018 in Kvemo Kartli (158), in 2019 in Racha-Lechkhumi in Kvemo Svaneti (169), and the smallest Mtskheta-Mtianeti (99) and Kakheti (93).

In 2020-2022, the highest number of cases under investigation per 10,000 inhabitants was recorded in Adjara (according to the years, 256, 259, 226) and Tbilisi. (211, 202,186) and the smallest in Mtskheta-Mtianeti (respectively 83.99). But in 2022, the lowest rate was detected in Kakheti (114). In 2022, criminality increased in Samegrelo Zemo-Svaneti and Imereti, for which the crime rate was 143 and 15, respectively. (Prosecutor's Office of Georgia, 2023) (https://police.ge/ge/useful-information/statistics)

Therefore, it can be concluded that the highest rate of initiation of investigation is recorded mainly in Tbilisi and Adjara, and the lowest in Guria and Mtskheta-Mtianeti.

In general, on the basis of the average annual growth rates of crime calculated for the ten-year period for the study period, it was determined that the number of cases registered in 2013-2022 is increasing in all regions, except for Kakheti and Tbilisi.

In Kakheti and Tbilisi, the number of cases under investigation decreases annually by 1.6% and 1.2%, respectively. The highest growth rate was recorded in Adjara (6.6%). It should be noted here that in 2022, compared to 2021, the sharpest increase of 15% of the cases started in investigation was recorded in Shida Kartli, and the sharpest decrease of 17% in Kakheti. (See Fig. 97.2)

It is interesting how the number of criminals changes. What is the change in the number of persons who have been prosecuted for the case that has been investigated.

The majority of people who have started persecution per 10,000 people is traditionally the highest in Tbilisi, Adjara and Samegrelo-Zemo Svaneti. And the lowest was recorded in Racha-Lechkhumi and Kvemo Svaneti regions and Mtskheta-Mtianeti.

But in 2019-2020, the majority of persons who started persecution per 10,000 people was the highest in Samegrelo Zemo Svaneti (respectively 64, 52), Tbilisi (63.8, 50) and Adjara (59.7, 45.8), while the lowest rate was recorded in Kakheti and Mtskheta-Mtianeti region.

In the end, if we analyze the number of persons who started persecution, it turns out that the majority of people who

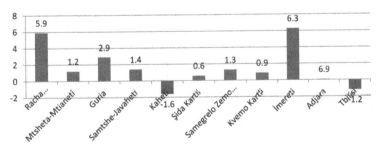

Fig. 97.1 The average annual growth rate of cases under investigation in the regions of Georgia (%) (National Statistics Office of Georgia, 2023)

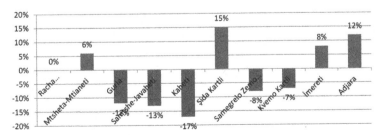

Fig. 97.2 Change in the number of cases under investigation in 2022 compared to 2021 (%) (National Statistics Office of Georgia, 2023) (Prosecutor's Office of Georgia)

started persecution each year come from Tbilisi, Adjara and Imereti. However, the crime rate, calculated for every 10,000 people of the population of each region, showed a completely different result. It turned out that in 2013-2022, Imereti was not among the first three regions with a high rate of crimes. The criminogenic situation was steadily more difficult until 2021 in the following three regions: Tbilisi, Adjara and Samegrelo Zemo-Svaneti. In 2021, the number of persons who started persecution per 10,000 people was the highest in Adjara (65), Racha-Lechkhumi (62.5) and Tbilisi (57.5). In 2022, the same high rate was recorded in Shida Kartli (62.8). The regional analysis also showed that during the last 10 years, the criminological level was steadily low in the Mtskheta-Mtianeti region. Accordingly, we can say that Mtskheta-Mtianeti is one of the safest regions from a legal point of view. (Abesadze & Mazanishvili, Statistical trends of criminality according to the regions of Georgia, 2023)

It is also interesting to see the age structure of persons starting prosecution in 2022 according to the regions of Georgia.

Data analysis showed that, in Tbilisi, Imereti, Kvemo Kartli and Mtskheta Mtianeti in 2022, the main part of criminals belonged to the age category of 25-34 years. In Adjara, Samegrelo-Zemo Svaneti, Kakheti, Shida Kartli, Guria and Samtskhe-Javakheti, criminals in the age category of 35-44 years were predominantly identified. As for Racha-Lechkhumi, the majority of criminals here are 45-54 years old. As expected, most of the criminals are men. The percentage of men varies between 91-92% in all regions (Prosecutor's Office of Georgia) It is also worth noting that the average annual growth rate of criminals for all regions varies between 0.1-6.4%, which means that the number of persons prosecuted increases every year.

Since several regions have been identified with the most severe situation in terms of criminogenicity, it will be interesting to analyze the most common types of crime in these regions.

The most common crimes in the Adjara region are theft, violence, threats, family violence, and transporting, sending or taking narcotics. In Tbilisi, theft dominates among crimes and has the highest rate compared to other crimes. Cases of domestic violence and threats are constantly increasing. If in 2017, 135 people were prosecuted for threats, in 2020 this number has already increased to 1137. Accordingly, the average annual increase in threats amounted to 20.8%.

The third region where the severity of the criminological situation was highlighted is Samegrelo Zemo-Svaneti region. The structure of widespread crimes is significantly different from Tbilisi and Ajara regions. Drug-related crimes dominate here. Crimes such as the illegal purchase, possession, transportation, delivery, distribution, and/or non-prescription use of cannabis or marijuana experienced a significant increase in 2019 compared to 2018. The number of persons prosecuted for this crime has increased by 525%, that is, the number has doubled. Since 2020, crime has started to decrease. The rate of theft is also at a high level, compared to other crimes, this rate is constantly on the first place. In Samegrelo-Zemo Svaneti region, the facts of family violence were also revealed, which was increasing from 2013 to 2022, although it decreased in 2022 (68.8%). The average annual growth rate of domestic violence in Samegrelo-Zemo Svaneti region is 12.7%. (Abesadze & Mazanishvili, Statistical trends of criminality according to the regions of Georgia, 2023)

Since several regions have been identified with the highest number of persons who have been prosecuted per 10,000 of the population, it will be interesting to see if this is related to the increase in the level of unemployment in the mentioned regions or the decrease in the average monthly income per capita. The existence of these connections can be easily determined by extracting the correlation coefficient. Using the SPSS program, we calculated the correlation for the mentioned three regions and got the following result: (Abesadze & Mazanishvili, Statistical trends of criminality according to the regions of Georgia, 2023)

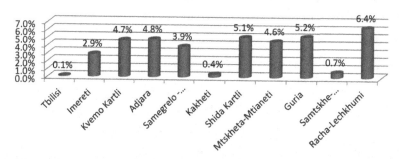

Fig. 97.3 Average annual growth rates of persecuted persons in 2013-2022. (National Statistics Office of Georgia, 2023) (Prosecutor's Office of Georgia)

In the Adjara region, the Pearson coefficient of correlation between the number of persons who have started persecution and the level of unemployment is equal to 0.182, and the correlation between the number of people who have started persecution and the average monthly income per capita is 0.078. In both cases, there is a very weak connection, and we can say that the increase in the number of persons who have started persecution in the Adjara region is not related to the change in either the unemployment level or the average monthly income.

In the Samegrelo-Zemo Svaneti region, the Pearson coefficient of correlation between the number of persons who have started persecution and the level of unemployment is equal to 0.694. The correlation coefficient is quite strong, which means that the number of people who have started persecution is greatly influenced by the level of unemployment. If unemployment goes up, so will the number of people being prosecuted. And the correlation between the number of persons who started persecution and the average monthly income per capita is 0.485, which indicates a medium-level relationship, that is, to some extent, the increase in average monthly income per capita may increase the number of people who started persecution in Samegrelo-Zemo Svaneti region.

As for the number of affected persons, there is an increasing trend in most regions of Georgia.

The number of affected persons in Tbilisi is increasing by 6.5% on average annually, the most persons were affected in 2021, and the least in 2013. In 2022, compared to 2021, the number of affected persons decreased by 19%. (National Statistics Office of Georgia, 2023) (Prosecutor's Office of Georgia, 2023)

The average annual growth rate of affected persons in the Adjara region was 8.2%. The most individuals were recorded in 2021, and the least in 2017. In 2022, compared to the previous year, the affected persons decreased by 7.1%.

The average annual growth rate of affected persons in the Imereti region was 2.4%. The most individuals were recorded in 2021, and the least in 2014. In 2022, compared to the previous year, the affected persons decreased by 17.7%.

The number of affected persons in Kvemo Kartli region increases by 2.9% on average annually, the most persons were affected in 2022, and the least in 2015. In 2022, compared to 2021, the number of affected persons increased by 3.8%.

The average annual growth rate of affected persons in the Kakheti region was 4.7%. The most individuals were recorded in 2022, and the least in 2017. In 2022, compared

to the previous year, the number of affected persons increased by 17.4%.

The average annual growth rate of affected persons in the Shida Kartli region was 2.8%. The most people were affected in 2021, and the least in 2016. In 2022, compared to the previous year, the number of affected persons decreased by 19.2%.

The average annual growth rate of affected persons in Samegrelo-Zemo Svaneti region was 2.7%. The most people were affected in 2021, and the least in 2019. In 2022, compared to the previous year, the number of affected persons decreased by 0.9%.

In the Samtskhe-Javakheti region, the average annual growth rate of affected persons was 3.2%. The most people were affected in 2021, and the least in 2016. In 2022, compared to the previous year, the number of affected persons decreased by 12.5%.

In the Mtskheta-Mtianeti region, the average annual growth rate of affected persons was 8.1%. The most people were affected in 2022, and the least in 2020. In 2022, compared to the previous year, the number of affected persons increased by 17.5%.

The average annual growth rate of affected persons in the Guria region was 7.3%. The most people were affected in 2016, and the least in 2014. In 2022, compared to the previous year, the number of affected persons decreased by 4.6%.

In Racha-Lechkhumi and Kvemo Svaneti region, the average annual growth rate of affected persons was 3.2%. The most people were affected in 2018, and the least in 2014. In 2022, compared to the previous year, the number of affected persons decreased by 12.1%.

The highest average annual increase in the number of affected persons was recorded in Adjara and Mtskheta-Mtianeti, and the smallest in Imereti and Samegrelo-Zemo Svaneti. It should also be noted that no average annual decrease in the number of affected persons was observed in any region.

4. CONCLUSION

In 2013-2022, the number of crimes registered throughout Georgia is characterized by an increasing trend; 5 main crimes, which are the most relevant in Georgia, have been identified, namely, theft, violence, domestic violence, threats, and the illegal manufacture, production, purchase, storage, transportation, transfer or possession of narcotic drugs or their analogues; In addition, the types of crimes change according to the age groups of the defendants, for example, if the rate of theft and violence dominates the crimes among persons under the age of 54, from the age of 55, such crimes as fraud, threats, production of forged documents come to the first place; The statistical analysis of the criminal situation in the regional perspective revealed that the highest rate of initiation of investigation is recorded mainly in Tbilisi and Adjara, and the lowest in Guria and Mtskheta-Mtianeti. In 2013-2022, the number of registered cases increases in all regions, except Kakheti and Tbilisi. In Kakheti, the number of cases under investigation decreases annually by 1.6%, and in Tbilisi by 1.2%. The highest increase was observed in Adjara (6.6%). In 2022, compared to 2021, the sharpest increase of 15% of the cases started in investigation was recorded in Shida Kartli, and the sharpest decrease of 17% in Kakheti; In the case of persons who started persecution, it was found that the majority of people who started persecution each year come from Tbilisi, Adjara and Imereti. However, according to the crime rate, it was found that in 2013-2022, Imereti was not among the first three regions with a high rate of crimes in any year; The criminogenic situation until 2021 was steadily more difficult in Tbilisi, Adjara and Samegrelo Zemo-Svaneti. In 2022, the same high rate was recorded in Shida Kartli (62.8); The regional analysis also showed that the level of criminology was consistently low in the Mtskheta-Mtianeti region; According to the latest data, the main part of criminals belongs to the age group of 25-34 years. In Adjara, Samegrelo-Zemo Svaneti, Kakheti, Shida Kartli, Guria and Samtskhe-Javakheti, the age category of criminals was 35-44. As for the majority of criminals belong to the age category of 45-54 years; In 2022, the number of male offenders in all regions dramatically exceeds the number of female offenders; The average annual growth rate of delinquency showed increasing rates for all regions.

From the criminogenic point of view, the situation is difficult in Tbilisi, as well as in Adjara and Samegrelo-Zemo Svaneti, the most common crime is theft. The relationship between the number of persecuted persons and the level of unemployment, the number of the persecuted population and the average monthly income per capita, the correlation coefficient calculated to determine Kavsiri confirmed that in Tbilisi, there is a strong relationship between the persecuted persons and the average monthly income, and it was also found in the Samegrelo-Zemo Svaneti region. that the number of persons who have started persecution is closely related to both the level of unemployment and the level of income of the population.

The number of affected persons is increasing in most regions of Georgia.

References

1. Abesadze, N., & Mazanishvili, T. (2023). Statistical trends of criminality according to the regions of Georgia. *Challenges of Globalization in Economy and Business" collection of papers* (pp. 1–7). Tbilisi: TSU.

2. Abesadze, N., & Mchedlishvili, L. (2016). The main directions of improvement of criminal law statistics methodology in Georgia. *Challenges of globalization in economy and business* (pp. 21–26). Tbilisi: TSU.

3. Abesadze, N., Lortqipanidze, E., & Abesadze, O. (2018, February 14). Correlation-regression analysis of factors affecting the number of criminal offenses. 36-41. Tbilisi, Georgia. Retrieved from "Statistics teaching and statistical research in Georgia.".

4. Abesadze, O. (2023). Statistical characteristics of criminal offenders in Georgia at the modern stage. *The world economy in the post-pandemic period:* (pp. 18–25). Tbilisi: TSU.

5. Abesadze, N., & Mazanishvili, T. (2023). Globalization and modern statistical trends of the criminogenic situation in Georgia. Tbilisi

6. Gloveli, E. (2014). Statistical analysis of criminal offenses and Forecasting in Georgia. Tbilisi, Georgia. *https://police.ge/ge/useful-information/statistics*. (n.d.). Retrieved from https://police.ge/ge/useful-information/statistics.

7. Marshava, K., & Mindorashvili, M. (2008). *Legal statistics.* Tbilisi: TSU.

8. *Prosecutor's Office of Georgia.* (n.d.). Retrieved from Prosecutor's Office of Georgia.

9. *National Statistics Office of Georgia.* (2023). Retrieved from National Statistics Office of Georgia.

10. *Prosecutor's Office of Georgia.* (2023). Retrieved from Prosecutor's Office of Georgia.

Note: All the figures in this chapter were made by the author.

Recent Trends in Engineering and Science for Resource Optimization and
Sustainable Development – Prof. (Dr.) Dorota Jelonek et al. (eds)
© 2024 Taylor & Francis Group, London, ISBN 978-1-032-98030-0

98

The Dilemma of Brand Vs. Generic Medication: Factors that Impact the Purchase Attitude of Consumers in Egypt

Rania Mohy Nafea[1]

Professor,
School of Human Resources and Global Business,
Seneca Polytechnic

Ahmed Phargaly

Maastricht School of Management (MsM)

Walid El Leithy[2]

Associate Professor, University of Hertfordshire

■■■■■■ **Abstract**

This deductive paper studies the "Purchase Attitude" of consumers in Egypt towards Brand versus Generic medication by combining the Elaboration Likelihood Model and the Tripartite model of Attitude. Although researchers have previously examined this topic in the field, they have solely focused on policymakers and company managers, and hence the role of consumers as decision-makers is not much considered. Therefore, the current study intends to bridge that gap. The significance of the relationship between five independent variables was tested. Only physician prescription and disease risk were noted as good predictors of change in Attitude and justified 36.7% of the variance.

■■■■■■ **Keywords**

Purchase attitude, Brand versus generic medication, Tripartite model of attitude, Egypt

1. INTRODUCTION

Attitude is a little thing that makes a tremendous difference," stated Winston Churchill. Healthcare spending, especially on medication, has been a priority for governments in various countries for some time (Borger et al., 2006) to minimize medication budgets. Some countries have public insurance, whereas others pay for private insurance. Since global inflation began, buyers have cut spending on all things, including medication (Borger et al., 2006).

As a result, people are switching to generic drugs (Ess et al., 2003). This plan was welcomed by policymakers worldwide as a relief from the high cost (Ess et al., 2003). The World Health Organization, which routinely updates the *essential drug list* (a comprehensive list of safe and

Corresponding author: [1]ranianafie@gmail.com, [2]walid.elleithy@gmail.com

DOI: 10.1201/9781003596721-98

essential medications needed in a healthcare system), includes numerous generic drugs to promote their use at affordable costs and help health systems save (WHO, 2008). While Generics helped policymakers expand coverage to more people, other stakeholders saw it differently. Physicians questioned the quality of generics (De Run and Felix, 2006), but pharmacists praised their efficacy compared to patented medications. Similarly, due to their lower cost, well-off consumers think Generics are inferior to Brand drugs (Burton et al., 1998).

Hence, understanding the factors that may influence Egyptian consumers' purchasing attitudes is crucial, but this is a gray area that has not been studied. The few research works on this topic targeted policymakers and company managers. Previous papers ignored consumers as decision-makers. This study extends the variables addressed in the Elaboration Likelihood Model and the Tripartite model of Attitude and adds a new construct to be tested in one theoretical framework, which the researchers believe has never been tested collectively. Thus, this study adopts a deductive approach to examine the elements that affect Egyptian consumers' medicine purchase attitudes (Brand Vs. Generic).

1.1 Research Objective

The study examines the association between the elements of the theoretical framework and consumer attitudes regarding Generics. Understanding this attitude helps stakeholders. The government can examine the budgetary impact and take steps according to the purchase direction to reduce healthcare spending. Pharma corporations, on the other hand, will use physicians' information to assess customers' attitudes regarding pharmaceuticals kinds and refocus their promotional efforts on the most influential stakeholders (physicians and pharmacists).

2. LITERATURE REVIEW

The WHO, the international public health advocate, defines Generics as a type of medicine having three essential qualities to distinguish it from others (WHO, 2008). It must be interchangeable with the original brand, created without brand corporate approval, and marketed after the patent expires (WHO, 2008). The WHO defines interchangeability as two or more pharmacological products in the same therapeutic class with different active ingredients but the same therapeutic effect (WHO, 2008).

The FDA states that generic pharmaceuticals are identical to brand-name drugs in dosage form, safety, strength, mode of administration, quality, performance, and intended use (FDA, 2021, 2018). It should be substitutable

and therapeutically similar to the brand name medicine, employing the same active components and having the same clinical impact (FDA, 2021, 2018). The FDA accepts that inactive ingredients differ from the branded product, but they must meet FDA criteria (FDA, 2021).

In 2015, Cristancho et al. examined how nations define and classify Generic pharmaceuticals. 21 countries, including Egypt, with 4 billion people, were sampled (Alfonso-Cristancho et al., 2015). A total of 13 nations have a Generic drug definition, but the remainder either didn't or used the WHO definition (Alfonso-Cristancho et al., 2015). A total of 16 of 21 nations have special registration requirements for Generic drugs, usually linked to WHO inter-changeability or EMA bioequivalence (Alfonso-Cristancho et al., 2015). Generic registration requires patent expiration for the original brand in all countries (Alfonso-Cristancho et al., 2015).

Q1. What is the relationship between medicine price and the Purchase Attitude toward medicine types?

Lichtenstein et al. (1993) defined price as having two roles: a negative role that affects buying attitude as consumers sacrifice the economic cost of higher prices and a positive role if buyers associate price with product quality; hence higher prices suggest higher quality. Burton et al. (1998) found that reduced prices positively affect buying attitudes and increase when personal income decreases. Price-conscious buyers will always choose lower pricing above other factors. Value consciousness, which balances quality and price, is the second construct. Value-conscious consumers want products that offer more for their money. The last component is the price-quality relationship, which assumes that some consumers may use product price as an indicator of quality. Hence low-priced products are low-quality.

Q2. What is the relationship between physician prescriptions and the Purchase Attitude toward medicine types?

Physicians influence customers' medicine preferences (De Run and Felix, 2006). Prescription medications play a bigger role than OTC (over-the-counter) ones (Ladha, 2007). Although Egypt does not adequately regulate the administration of drugs based on legitimate prescription (Taher et al., 2012), doctors' recommendations are the key influencer on consumers in many other nations (Tolken, 2011). Ladha (2007) emphasizes the importance of medical prescription and purchase attitude. She claims that the presence of a prescription is the key factor determining a consumer's attitude toward a drug and that the attitude altered between OTC and prescription pharmaceuticals proves that prescriptions matter.

Q3. What is the relationship between the level of disease risk and the Purchase Attitude toward medicine types?

According to Podulka et al. (1989), risk perception affects medicine choice. The risk level is the possibility that the Generic medicine consumed is not effective enough. OTC medications are for less significant health conditions and can be chosen by the patient, while prescription pills are riskier and require a doctor's consultation. 100 Chicago customers were surveyed. The target audience was asked about their comfort in taking Generic pharmaceuticals for a range of sickness medications. Headache aspirin was the least disease-risky, and chronic heart medication was the most. Consumers were less comfortable taking Generic prescribed drugs than Generic OTC medicines. However, consumers were more likely to take Generic pain medications and less likely to take thyroid or heart medications. The most crucial portion is that as disease severity grew, Generic acceptability decreased as follows: 63% for pain medicine, 60% for antibiotics, 45% for antianxiety, 40% for diabetes, and 34% for heart medicine.

Q4. What is the relationship between insurance coverage and the Purchase Attitude toward medicine types?

Insurance firms absorb and cover people's risks (Borger et al., 2006), acting as risk managers. The key stakeholder in the pharma cycle is the insurance third party, which can affect pharmaceutical buying attitudes through insurance coverage policies (Visser and Mirabile, 2004). Pauly (1968) linked insurance to consumption attitudes. "Moral Hazard" suggests that medical insurance coverage may increase medical care use. The author explained that this attitude is more of a rational economic action than misbehavior. Individual medical care costs are divided among all insured persons; therefore, he/she cannot limit the use of the service. Thus, the insured pays a premium that is usually less than the service's worth, which increases utilization. Shrank et al. (2009) were able to confirm medical insurance coverage as an important determinant for the Attitude of purchase toward Generics. The researcher assumes that the increase in usage postulated by the Moral hazard theory may not be limited to extra services rendered but may extend to a kind of preference between Brands (usually more expensive) and Generics. To verify this assumption, the construct will be tested.

Q5. What is the relationship between customer demographics and (Age, gender, education, income) the Purchase Attitude toward medicine types?

An exploratory study in America examined how elderly and younger people view Generic drug risk (Bearden et al., 1979). The survey examined 1979 prescription medications. The authors defined risk as uncertainty (probability of loss) and consumer importance. Performance, financial, social, physical, psychological, and convenience risks were examined. Two hundred people were asked about Generic medicine hazards, and 105 gave reliable answers. 56% were senior (55+), and 44% were youth. The senior group focused on finances, whereas the young group focused on social, psychological, and physical elements. Although the elderly requires financial alleviation, they prefer brand medications due to the perceived risk. The authors concluded that age group affects Generic purchase attitudes, especially from a risk perspective.

2.1 Research Hypotheses

H1: The Price of medicine has a significant relationship with the Purchase Attitude toward medicine types.

H2: Physician prescriptions have a significant relationship with the Purchase Attitude toward medicine types.

H3: The level of disease risk has a significant relationship with the Purchase Attitude toward medicine types.

H4: The existence of insurance coverage has a significant relationship with the Purchase Attitude toward medicine types.

H5: Customer demographics (age, gender, education, income) have a significant relationship with the Purchase Attitude toward medicine types.

3. METHODOLOGY

3.1 Research Design

In order to answer the research questions and test the developed hypothesis, the researchers used a mixed approach whereby interviews were performed with pharmacists and pharmaceutical company managers, and sales reps to validate the problem.

3.2 Quantitative Approach

The researchers relied on the quantitative method to determine the correlations between the independent variables and Purchase Attitude. Since attitude is a broad measure, the study cannot simply use qualitative methods. To verify the hypothesis and establish the relationships, the quantitative method was chosen.

3.3 Data Collection

A survey design was used, and a questionnaire of 36 closed-ended items was adopted from previous research and tested for validity and reliability. A total of 900 surveys were launched through Qualtrics during the first half of 2021 in the Greater Cairo area through Linkedin and WhatsApp groups. Out of all the distributed surveys, only

172 completed responses proceeded for data analysis after excluding the improper and missing responses. Finally, the results were discussed with stakeholders of the Egyptian pharma industry. This triangulation method was used to explain the data and understand its relevance to different stakeholders.

3.4 Statistical Analysis

For statistical analysis, SPSS software was used. The researchers started by assessing the reliability of questions and the consistency of responses among the different groups. This was followed by using descriptive analysis. The inferential analysis started with correlation testing to measure the strength of dependency of the dependent variable on each of the independent variables, after which regression analysis using the stepwise method to estimate the change in the dependent variable based on the change in predictors was employed, and finally, ANOVA analysis was run to assess the fit of the model and conclude the percentage of variance explained out of the proposed model.

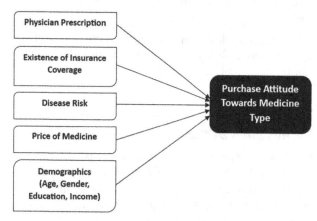

Fig. 98.1   Conceptual framework adapted

4. Results and Discussion

4.1 Correlation Test

A bivariate Correlation test was run on the primary data using SPSS, resulting in 5 out of 8 variables having a significant relationship with the dependent variable. The p-value for disease risk, physician prescription, income, insurance coverage, and age are below 0.05, which indicates a significant relationship with Purchase Attitude. On the other hand, the correlation coefficient determines the significance of the relationship. Disease risk seems to be the most significant, with a 0.531 Pearson coefficient, followed by physician prescription. The correlation shows that all variables have a positive relationship except for

insurance coverage which indicates a negative relationship with Attitude. Less significant relationships include income level, insurance coverage, and age, as shown in Table 98.1.

Table 98.1   Correlation results

Score of Attitude			
	Pearson Correlation	Sig. (2-tailed)	N
Score of Disease Risk	.531**	.000	172
Score of Physician Prescription	.424**	.000	172
Personal Income Level	.305**	.000	172
Medical Insurance Coverage	-.264**	.000	172
Age	.254**	.001	172
Score of Price	-.141	.065	172
Educational Level	.130	.089	172
Gender	-.071	.358	172

** Correlation is significant at the 0.01 level (2-tailed).

The researchers noticed the existence of a significant relationship between a couple of the independent variables, which may predict collinearity, rejecting some of the variables when running the regression analysis. Table 98.2 shows that the 4 demographic independent variables (age, gender, education, and income) have a significant relationship with each other, with the highest of them being between age and income level (0.792).

4.2 Regression Analysis

Based on the results of the correlation and having identified 5 independent variables out of 8 to have a significant relationship with the dependent variable, it is important to study the regression analysis to calculate the percentage that the independent variables can justify the variance. The researchers used Linear Regression Stepwise, allocating Purchase Attitude as the dependent variable and the 5 variables resulting from the correlation analysis as having a significant relationship as the independent variables.

Regression testing resulted in 2 models, one inferring the disease risk only as a predictor of the variance and the second adding the physician prescription along with the disease risk to be good predictors of the variance. Looking at the ANOVA results for the 2 models, both resulted in p-values less than 0.05, which indicates that both models are accepted. Last is the Coefficients, which resulted as well in a p-value less than 0.05, indicating the acceptability of both models.

Therefore, the researcher selected the second model, which includes both disease risk and physician prescription as predictors of the variance in Purchase Attitude. The R-value

Table 98.2 Correlation results of the demographic independent variables

		Age	Gender	Educational Level	Personal Income Level
Age	Pearson Correlation	1	-.286**	.366**	.792**
	Sig. (2-tailed)		.000	.000	.000
	N	172	172	172	172
Gender	Pearson Correlation	-.286**	1	-.267**	-.368**
	Sig. (2-tailed)	.000		.000	.000
	N	172	172	172	172
Educational Level	Pearson Correlation	.366**	-.267**	1	.444**
	Sig. (2-tailed)	.000	.000		.000
	N	172	172	172	172
Personal Income Level	Pearson Correlation	.792**	-.368**	.444**	1
	Sig. (2-tailed)	.000	.000	.000	
	N	172	172	172	172

**. Correlation is significant at the 0.01 level (2-tailed)

Table 98.3 Linear regression step-wise results

Model	R	R Square	Adjusted R^2	Std. Error of the Estimate
1	.531	.282	.278	5.584
2	.606	.367	.360	5.257

a. Predictors: (Constant), Score of disease risk
b. Predictors: (Constant), Score of disease risk, Score of Physician prescription

Table 98.4 ANOVA results – dependent variable: score of attitudes

Mo	del	Sum of Squares	df	Mean Square	F	Sig.
1	Regression	2084.681	1	2084.681	66.869	.000ᵃ
	Residual	5299.848	170	31.176		
	Total	7384.529	171			
2	Regression	2713.746	2	1356.873	49.095	.000ᵇ
	Residual	4670.783	169	27.638		
	Total	7384.529	171			

a. Predictors: (Constant), Score of Disease risk
b. Predictors: (Constant), Score of Disease risk, Score of Physician prescription

of 0.606 indicates a good degree of correlation between both independent and dependent variables. R^2 value infers the applicability of the model and the ability to justify 36.7% of the total variance. Nevertheless, the adjusted R^2 showed 0.36, which is less than 0.7%, indicating a low redundancy in the predictors selected for the model.

5. CONCLUSION

From the above quantitative data, the researchers can conclude that the Purchase Attitude of consumers is a complex topic, especially in a market like Egypt where several factors are at play. First is the high population count, causing a strain on government resources and budget allocation to healthcare spending of 4.74% in 2019. Second, the disparity in income distribution in the country, not only between the different social classes but also between the different governorates, including rural and urban areas. Third, healthcare facilities, including the distribution of medication, remain an issue in Egypt, whereby Brands are not offered everywhere due to their price, forcing consumers to only purchase Generics regardless of the disease risk. Finally, since Egypt is an

Table 98.5 Beta coefficients for the predictors of the dependent variable

Model		Unstandardized Coefficients		Standardized Coefficients	t	Sig.
		B	Std. Error	Beta		
1	Constant	16.860	1.492		11.301	.000
	Score of Disease risk	.748	.091	.531	8.177	.000
2	Constant	11.784	1.762		6.687	.000
	Score of Disease risk	.633	.089	.450	7.081	.000
	Score of Physician prescription	.868	.182	.303	4.771	.000

emerging economy, most of the population is not under the government of private insurance coverage, forcing them to go for the cheaper, more affordable option of Generics. Even though the results indicate that 5 independent variables were found to be significant, the regression formula below highlights that disease risk and physician prescriptions explain only 36.7% of the variance. Accordingly, the researchers recommend a more comprehensive discussion with stakeholders in the future through qualitative analysis to explain the remaining 63.3%. Unfortunately, this needs to be conducted with the WHO or the Egyptian Ministry of Health to ensure a proper geographic outreach and inclusion of all income and age groups. Moreover, it is imperative to include the uneducated and illiterate consumers, something which this research is lacking given the use of Qualtrics to collect the data.

$$\text{Purchase Attitude} = 11.784 + 0.633 \text{ (Disease risk)} + 0.868 \text{ (Physician prescription)}$$

5.1 Funding

This research did not receive any specific grant from funding agencies in the public, commercial, or not-for-profit sectors.

References

1. Alfonso-Cristancho, R., Andia, T., Barbosa, T., Watanabe, J.H., 2015. Definition and classification of generic drugs across the world. Appl. Health Econ. Health Policy 13, 5–11. https://doi.org/10.1007/s40258-014-0146-1

2. Bearden, W.O., Mason, J.B., Smith, E.M., 1979. Perceived risk and elderly perceptions of generic drug prescribing. Gerontologist 19, 191–195. https://doi.org/10.1093/geront/19.2.191

3. Borger, C., Smith, S., Truffer, C., Keehan, S., Sisko, A., Poisal, J., Clemens, M.K., 2006. Health spending projections through 2015: Changes on the horizon. Health Aff. 25, 61–73. https://doi.org/10.1377/hlthaff.25.w61

4. Burton, S., Lichtenstein, D.R., Netemeyer, R.G., Garretson, J.A., 1998. A scale for measuring attitude toward private label products and an examination of its psychological and behavioral correlates. J. Acad. Mark. Sci. 26, 293–306. https://doi.org/10.1177/0092070398264003

5. De Run, E.C., Felix, M.-K.N., 2006. Patented and generic pharmaceutical drugs: perception and prescription. Int. J. Bus. Soc. 7, 55–78.

6. Ess, S.M., Schneeweiss, S., Szucs, T.D., 2003. European healthcare policies for controlling drug expenditure. Pharmacoeconomics 21, 89–103. https://doi.org/10.2165/00019053-200321020-00002

7. FDA, 2021. Generic Drug Facts [WWW Document]. Food Drug Adm. URL https://www.fda.gov/drugs/generic-drugs/generic-drug-facts (accessed 8.2.23).

8. FDA, 2018. What We Do [WWW Document]. Food Drug Adm. URL https://www.fda.gov/about-fda/what-we-do (accessed 8.2.23).

9. Ladha, Z., 2007. Marketing strategy: Are consumers really influenced by brands when purchasing pharmaceutical products? J. Med. Mark. 7, 146–151. https://doi.org/10.1057/palgrave.jmm.5050072

10. Lichtenstein, D.R., Ridgway, N.M., Netemeyer, R.G., 1993. Price perceptions and consumer shopping behavior: A field study. J. Mark. Res. 30, 234–245. https://doi.org/10.1177/002224379303000208

11. Pauly, M. V, 1968. The Economics of Moral Hazard: Comment. Am. Econ. Rev. 58, 531–537.

12. Podulka, M., Krautkramer, K., Amerson, D., Phillips, B., Dolinsky, D., 1989. Consumers' attitudes toward generic drugs. J. Pharm. Mark. Manage. 4, 93–104. https://doi.org/10.3109/J058v04n01_06

13. Shrank, W.H., Cox, E.R., Fischer, M.A., Mehta, J., Choudhry, N.K., 2009. Patients' perceptions of generic medications. Health Aff. 28, 546–556. https://doi.org/10.1377/hlthaff.28.2.546

14. Taher, A., Stuart, E.W., Hegazy, I., 2012. The pharmacist's role in the Egyptian pharmaceutical market. Int. J. Pharm. Healthc. Mark. 6, 140–155. https://doi.org/10.1108/17506121211243068

15. Tolken, R., 2011. An explorative study of consumers' attitudes towards generic medications. University of Pretoria, Pretoria.

16. Visser, P.S., Mirabile, R.R., 2004. Attitudes in the social context: The impact of social network composition on individual-level attitude strength. J. Pers. Soc. Psychol. 87, 779–795. https://doi.org/10.1037/0022-3514.87.6.779

17. WHO, 2008. Measuring medicine prices, availability, affordability and price components. World Health Organization Press, Geneve.

Note: All the figures and tables in this chapter were made by the author.

Printed in the United States
by Baker & Taylor Publisher Services